Depth Map and 3D Imaging Applications:

Algorithms and Technologies

Aamir Saeed Malik
Universiti Teknologi Petronas, Malaysia

Tae-Sun Choi
Gwangju Institute of Science and Technology, Korea

Humaira Nisar
Universiti Tunku Abdul Rahman, Perak, Malaysia

Information Science
REFERENCE

Managing Director: Lindsay Johnston
Book Production Manager: Sean Woznicki
Development Manager: Joel Gamon
Development Editor: Michael Killian
Acquisitions Editor: Erika Carter
Typesetters: Mackenzie Snader
Print Coordinator: Jamie Snavely
Cover Design: Nick Newcomer

Published in the United States of America by
Information Science Reference (an imprint of IGI Global)
701 E. Chocolate Avenue
Hershey PA 17033
Tel: 717-533-8845
Fax: 717-533-8661
E-mail: cust@igi-global.com
Web site: http://www.igi-global.com

Library of Congress Cataloging-in-Publication Data

Depth map and 3D imaging applications: algorithms and technologies / Aamir Saeed Malik, Tae Sun Choi, and Humaira Nisar, editors.
p. cm.
Summary: "This book present various 3D algorithms developed in the recent years to investigate the application of 3D methods in various domains, including 3D imaging algorithms, 3D shape recovery, stereoscopic vision and autostereoscopic vision, 3D vision for robotic applications, and 3D imaging applications"-- Provided by publisher.
Includes bibliographical references and index.
ISBN 978-1-61350-326-3 (hardcover) -- ISBN 978-1-61350-327-0 (ebook) -- ISBN 978-1-61350-328-7 (print & perpetual access) 1. Algorithms. 2. Three-dimensional imaging. I. Malik, Aamir Saeed, 1969- II. Choi, Tae Sun, 1952- III. Nisar, Humaira, 1970- IV. Title: Depth map and three-D imaging applications.
QA9.58.D47 2012
621.36'7015181--dc23

2011031955

British Cataloguing in Publication Data
A Cataloguing in Publication record for this book is available from the British Library.

Editorial Advisory Board

List of Reviewers

Table of Contents

Section 2
Shape From X: Algorithms & Techniques

Section 3
Stereoscopy & Autostereoscopy

Section 4
Robotic Vision

Section 5
3D Imaging Applications

Foreword

Imaging is as old as human intelligence. Indeed, anthropologists identify the point of departure between animal and human at the point where the creature felt the need to create an image. The creation of images in prehistoric times was a means of teaching hunting techniques, recording important events, and communicating (Figure1). It is from those elementary images that hieroglyphs evolved and eventually alphabets. Imaging has always been part of human culture. Its decorative nature was perhaps less important than its role in recording significant events, mainly for impressing the masses for the importance and glory of its rich and powerful patrons. In the last 200 years or so, technology-based imaging started to co-exist in parallel with manual imaging, restricting the role of the latter mainly to art. Technology based imaging is nowadays very much a major part of our everyday life, through its medical applications, routine surveillance, or entertainment. However, imaging has always been haunted by the need to depict a 3D world on a 2D medium. This has been a problem that pertains to paintings throughout the millennia: from the ancient Egyptians, who were painting full eyes even when seen sideways, to Pi-

Figure 1.

casso and the cubists, who tried to capture all 3D aspects of the depicted object on a 2D canvas, imaging in 3D has been the holy grail of imaging. Modern technology has at last matured enough to allow us to record the 3D world as such, with an enormous range of applications: from medicine and cave technology for oil exploration, to entertainment and the 3D television. This book is dedicated exactly to these modern technologies, which fascinate and excite. Enjoy it!

Maria Petrou
Informatics and Telematics Institute, CERTH, Greece & Imperial College London, UK

Maria Petrou *studied Physics at the Aristotle University of Thessaloniki, Greece, Applied Mathematics in Cambridge, UK, and obtained her PhD and DSc degrees both from Cambridge University in Astronomy and Engineering, respectively. She is the Director of the Informatics and Telematics Institute of CERTH, Thessaloniki, Greece, and the Chair of Signal Processing at Imperial College London, UK. She has co-authored two books, "Image Processing, the fundamentals" and "Image Processing dealing with texture", in 1999 (second edition 2010) and 2006, respectively, and co-edited the book "Next generation artificial vision systems, reverse engineering the human visual system." She has published more than 350 scientific articles on astronomy, computer vision, image processing and pattern recognition. She is a Fellow of the Royal Academy of Engineering.*

Preface

This book has three editors, and all of us are involved in image processing and computer vision research. We have contributed to the 3D imaging research, especially in the field of passive optical 3D shape recovery methods. Over the last decade, significant progress had been made in 3D imaging research. As a result, 3D imaging methods and techniques are being employed for various applications. The objective of this book is to present various 3D algorithms developed in the recent years and to investigate the application of 3D methods in various domains.

This book is divided into five sections. Section 1 presents various 3D imaging algorithms that are developed in recent years. It covers quite a variety of research fields including 3D mapping, holography, and 3D shape compression. Six chapters are included in Section 1. Section 2 deals with 3D shape recovery methods that fall in the optical passive as well as active domains. The topics covered in this section include shape from focus, shape from heating, and shape from fluorescence. Section 2 includes 5 chapters.

Section 3 is dedicated to stereoscopic vision and autostereoscopic vision. The dedication of a whole section to stereoscopic and autostereoscopic vision emphasizes the importance of these two technologies. Seven chapters are included in this section. Section 4 discusses 3D vision for robotic applications. The topics included in this section are 3D scene analysis for intelligent robotics and usage of stereo vision for various applications including fire detection and suppression in buildings. This section has three chapters.

Finally, Section 5 includes a variety of 3D imaging applications. The applications included in this section are 3D DMB player, 3D scanner, 3D mapping, morphological study of meteorite impact rocks, 3D tracking, 3D human body posture estimation, 3D face recognition, and 3D thumbnails for 3D videos. A total of nine chapters are included on several of the above mentioned applications in this section.

There are 31 chapters in this book. Chapter 1 is not included in any of the sections as it provides an introduction to 3D imaging. Chapter 1 briefly discusses the classification for 3D imaging. It provides an overview of the 3D consumer imaging products that are available commercially. It also discusses the future of 3D consumer electronics.

SECTION 1

Chapter 2 to Chapter 7 are included in this section. Chapter 2 discusses multi-view stereo reconstruction as well as shape from silhouette method. Multiple images are used with multiple views for 3D reconstruction. This chapter can be included in both Section 2 and Section 3 since Section 2 deals with methods like shape from silhouette while Section 3 covers stereovision. However, we decided to put it as the first chapter of section I because it presents an algorithm dealing with 3D shape reconstruction and also because we want to emphasize the importance of these two topics at the very beginning of this book.

Chapter 3 deals with the iterative reconstruction method that can be used in various medical imaging methods like X-ray, Computed Tomography, Positron Emission Tomography, Single Photon Emission Computed Tomography, Dose-calculation in Radiotherapy, and 3D-display Volume-rendering. This chapter is included in the book to emphasize on the importance of 3D transmissive methods that have greatly influenced our present day life style by improving the healthcare services.

Chapter 4 provides methods for generating 3D maps of the environment surrounding us. These maps are especially useful for robot navigation. This chapter especially discusses 3D map registration in detail.

Chapter 5 emphasizes the importance of compression for data storage and transmission for large chunks of 3D data. It describes a 3D image compression method that could reduce the data storage and transmission requirements.

Chapter 6 addresses holographic images. The future of true 3D lies in the holographic imaging technology. The holographic images are marred with noise and low quality. Hence, restoration and enhancement are very important for holographic images. This chapter summarizes related issues and provides solution for the restoration and enhancement of the holographic images.

Chapter 7 is the last chapter in section I. This chapter deals with an active optical 3D shape recovery method. For active fringe patterns projection, off-the-shelf projector is used in order to reduce the cost of the system.

SECTION 2

Chapter 8 to Chapter 12 are included in Section 2. Chapter 8 gives a very good introduction of the 3D shape recovery approaches. It includes the geometric approaches, photometric methods, and the real aperture techniques. Details are provided for various methods and techniques falling under one of the three approaches.

Chapter 9 discusses the focus measures in detail. A total of eleven focus measures are discussed, and they are categorized under four major classes. A very detailed comparison is provided for the eleven focus measures. The performance comparison is provided with respect to several types of noise, varying illumination and various types of textures.

Chapter 10 uses S-Transform for developing a focus measure method. High frequency components in the S-transform domain are targeted by the developed focus measure. The focus measure is used as a shape from focus technique to recover the 3D shape.

Chapter 11 uses genetic programming for developing a focus measure. An optimal composite depth function is developed, which utilizes multiple focus measures to get the optimized depth map for 3D shape recovery.

Chapter 12 provides two methods for recovering 3D shape of the transparent objects. Using normal optical methods, the 3D shape of transparent objects cannot be recovered accurately and precisely. This chapter discusses shape from heating and shape from fluorescence techniques to recover the 3D shape. These are new methods and have been introduced recently.

SECTION 3

Chapter 13 to chapter 19 are included in Section 3. Chapter 13 to Chapter 17 are related to stereoscopic vision, while the last two chapters in this section are on autostereoscopic vision. Although these two topics can be placed under Section 2, they have been placed in a separate section because of their importance in terms of consumer electronics.

Chapter 13 discusses a stereoscopic algorithm which treats the stereovision as modular approach. Hence, the stereovision algorithm can be divided into various stages and each of the stage can be implemented individually.

Chapter 14 and Chapter 15 discuss applications of the stereovision. Offroad intelligent vehicle navigation using stereovision in the agricultural environment is dealt in chapter 14 while chapter 15 discusses visually induced motion sickness (VIMS) that is associated with stereoscopic movies.

Chapter 16 provides details of viewpoint interpolation methods that are used for synthesizing the in-between views from few views that are captured by few fixed cameras. Chapter 17 presents a reversible watermarking based algorithm to deal with the high costs of memory, transmission bandwidth and computational complexity for 3D images.

Chapter 18 and Chapter 19 deal with autostereoscopic vision. Stereoscopic displays require 3D glasses to view in 3D while the autostereoscopic displays do not require any 3D glasses. Chapter 18 introduces the basic concepts of autostereoscopic displays and discusses several of its technologies. Chapter 19 addresses the very important issue of bandwidth for high resolution multi-view autostereoscopic data.

SECTION 4

Chapter 20 to Chapter 22 are included in section IV. This is the shortest section in this book. Although, all the three chapters in this section could easily be included in Section 3 but we decided to allocate a separate section to emphasize the topic of robotic vision.

Chapter 20 is an invited chapter. It deals with intelligent robotics by capturing and analysing a scene in 3D. Real time processing is important for robotic applications and hence this chapter discusses limitations for the analysis of 3D data in real time. This chapter provides very good description of various technologies that address the limitation issues for real time processing.

Chapter 21 and Chapter 22 use the stereovision for robotic applications. Chapter 21 discusses the autonomous operation of robots in real working environments while chapter 22 deals with the specific application of fire detection and suppression in the buildings.

SECTION 5

Chapter 23 to Chapter 31 are included in this section. Nine chapters deal with nine different 3D applications. It is the last section of the book. However, some of the applications dealing with stereovision, robotics and compression are also discussed in earlier sections. We placed them in those sections because we think that they are more relevant to the topics in those sections.

Chapter 23 discusses a 3D DMB player. DMB stands for digital multimedia broadcasting, and it is used for terrestrial-DMB (T-DMB) systems. The chapter also introduces an approximation method to create auto-stereoscopic images in the 3D DMB player. Hence, this chapter is also related to section III where autostereoscopic vision is discussed.

Chapter 24 presents a detailed overview of the 3D scanning technologies. Comparison of several 3D scanning methods is provided based on accuracy, speed, and the applicability of the scanning technology.

Chapter 25 deals with 3D mapping in outdoor environments, while chapter 26 presents 3D scanning method to study morphology of a meteorite rock. For 3D mapping, examples are taken from pavement runway inspection and urban mapping. For 3D scanning, meteorite rock is selected from the Karikkoselkä impact crater (Finland).

Chapter 27 discusses 3D tracking for mixed reality. 3D tracking is one of the active research areas in 3D imaging. This chapter addresses 3D tracking in mixed reality scenario. Mixed reality deals with virtual objects in real scenes. It is a very important topic with applications in medical, teaching, and gaming professions. Multi-sensor fusion methods for mixed reality with 3D camera tracking are discussed in this chapter.

Chapter 28 uses stereovision for the reconstruction of 3D human body posture that is further utilized in human activity recognition. Human activity recognition is of vital importance for visual surveillance applications. Hence, interest in human activity recognition research has increased manifolds in the recent years.

Chapter 29 deals with 3D face recognition, while chapter 30 discusses 3D face expression recognition. In Chapter 29, a method for 3D face recognition is presented based on adaptive non-uniform meshes. In chapter 30, a feature extraction method is discussed that does not require any neutral face for the test object.

Chapter 31 is the last chapter of this section, as well as the last chapter of the book. Chapter 31 introduces a thumbnail format for 3D videos with depth. A framework is presented in the chapter that generates 3D thumbnails from layered depth video (LDV) and video plus depth (V+D).

FINAL WORDS

The work on this book started in November 2009 and it has taken about one and a half years to complete it. All the chapters in this book went through multiple reviews by the professionals in the field of 3D imaging and 3D vision. All the chapters had been revised based on the comments of multiple reviewers by the respective authors of the chapters. Contributors for the book chapters come from all over the world, i.e., Japan, Republic of Korea, China, Australia, Malaysia, Taiwan, Singapore, India, Tunisia, Turkey, Greece, France, Spain, Belgium, Romania, Netherlands, Italy, and United States. This indicates that this book covers a topic of vital importance for our time, and it seems that it will remain so at least for this decade.

3D imaging is a vast field and it is not possible to cover everything in one book. 3D research is ever expanding and the 3D research work will go on with the advent of new applications. This book presents state of the art research in selected topics. We hope that the topics presented in this book attract the attention of researchers in various research domains who may be able to find solutions to their problems in 3D imaging research. We further hope that this book can serve as a motivation for students as well as researchers who may pursue and contribute to the 3D imaging research.

Aamir Saeed Malik, Tae-Sun Choi, Humaira Nisar

Acknowledgment

The editors would like to thank all members of the Editorial Advisory Board. Their contributions and suggestions have made a positive impact on this book. Specifically, due recognition goes to Fabrice Meriaudeau of University of Bourgogne, Naeem Azeemi of COMSATS Institute of Information Technology, Kishore Pochiraju of Stevens Institute of Technology, Martin Reczko of Synaptic Ltd., Iftikhar Ahmad of Nokia, Nidal Kamel of Universiti Teknologi Petronas, Umer Zeeshan Ijaz of University of Cambridge, and Asifullah Khan of Pakistan Institute of Engineering and Applied Sciences.

The editors would also like to acknowledge all the reviewers for providing their professional support to this book through their valuable and constructive reviews. Each chapter in the book went through multiple reviews, and the editors appreciate the time and the technical support provided by the reviewers in this regard. The reviewers include Abdul Majid of Pakistan Institute of Engineering and Applied Sciences, Andreas F. Koschan of University of Tennessee, Antonios Gasteratos of Democritus University of Thrace, Aurelian OvidiusTrufasu of Politehnica University of Bucharest, Fakhreddine Ababsa of University of Evry Val d'Essonne, Hiroki Takada of University of Fukui, Ibrahima Faye of Universiti Teknologi Petronas, Mannan Saeed of Gwangju Institute of Science and Technology, Mercedes Farjas of Universidad Politécnica de Madrid, Muzaffar Dajalov of Yeungnam University, Song Zhang of Iowa State University, and Tae-Seong Kim of Kyung Hee University.

The editors acknowledge support of Department of Electrical and Electronic Engineering at Universiti Teknologi Petronas, Bio Imaging Research Center at Gwangju Institute of Science and Technology, Department of Electronic Engineering, Faculty of Engineering and Green Technology at Universiti Tunku Abdul Rahman, Perak, Malaysia, and Center for Intelligent Signal and Imaging Research at Universiti Teknologi Petronas.

Finally the editors express their appreciation for IGI Global who gave us the opportunity for editing this book. We would like to acknowledge IGI Global and its entire staff for providing professional support during all the phases of book development. Specifically, we would like to mention Michael Killian (Assistant Development Editor), who provided us assistance during all the phases in the preparation of this book.

Aamir Saeed Malik, Tae-Sun Choi, Humaira Nisar

Chapter 1
Introduction to 3D Imaging

Aamir Saeed Malik
Universiti Teknologi Petronas, Malaysia

Humaira Nisar
Universiti Tunku Abdul Rahman, Malaysia

ABSTRACT

With the advent of 3D consumer products in the electronics market, 3D imaging is all set to take off. Last decade had seen a lot of research activity with respect to 3D imaging. It will not be wrong to say that this decade will be the decade of 3D imaging. This chapter briefly introduces 3D imaging with respect to various 3D consumer products and 3D standardization activity. It also discusses the challenges and the future of 3D imaging.

INTRODUCTION

3D imaging is not a new research area. Researchers are working with 3D data for the last few decades. Even 3D movies were introduced using the cardboard colored glasses. However, the consumers did not accept the results of that 3D research because of low quality visualization of 3D data. The researchers were limited by the hardware resources like processing speed and memory issues. But with the advent of multicore machines, specialized graphics processors and large memory modules, 3D imaging research is picking up the pace. The result is the advent of various 3D consumer products.

3D imaging methods can be broadly divided into three categories, namely, contact, reflective and transmissive methods. The contact methods, as the name implies, recover the 3D shape of the object by having physical contact with the object. These methods are generally quite slow as they scan every pixel physically and they might modify or damage the object. Hence, they cannot be used for valuable objects like jewellery, historical artifacts etc. However, they provide very accurate and precise results. An example is the CMM (coordinate measuring machine) which is a contact 3D scanner (Bosch 1995). Such scanners are common in manufacturing and they are very

DOI: 10.4018/978-1-61350-326-3.ch001

precise. Another application of contact scanners is in the animation industry where they are used to digitize clay models.

On the other hand, reflective and transmissive methods do not come in physical contact with the object. The transmissive methods are very popular in the medical arena and include methods like CT (Computed Tomography) scanning, MRI (Magnetic Resonance Imaging) scanning and PET (Positron Emission Tomography) scanning (Cabeza, 2006). CT scanners are now installed in almost all the major hospitals in every country and they use X-rays for scanning. MRI and PET are more expensive then CT and are not as frequently used as CT scanners, especially in the third world countries. However, because of its usefulness MRI has become quite popular and is now available at major hospitals in third world countries. These technologies have revolutionized the medical profession and they help in accurate diagnosis of the diseases at an early stage. Apart from the medical profession, these 3D scanning technologies are used for non-destructive testing and 3D reconstruction for metals, minerals, polymers etc.

The reflective methods are based either on the optical or the non-optical sources. For non-optical based methods, radar, sonar and ultrasound are good examples which are now widely accepted and mature technologies. They are used by rescue services, medical professionals, environmentalists, defense personnel etc. They have wide range of applications and their cost varies from few hundred to hundred of thousands of dollars.

The optical based reflective methods are the ones that have direct effect on the everyday consumer. These methods are the basis for commercialization of consumer products including 3D TV, 3D monitors, 3D cameras, 3D printers, 3D disc players, 3D computers, 3D games, 3D mobile phones etc. The optical based reflective methods can be active or passive. Active methods use projected lights, projected texture and patterns for acquiring 3D depth data. Passive methods utilize depth cues like focus, defocus, texture, motion, stereo, shading etc to acquire 3D depth data. Passive methods are also used in conjunction with active methods for better accuracy and precision.

3D TELEVISION

We start with the introduction of 3D TV because it is the motivation for most of the other 3D consumer technologies. The first version of the TV was black-and-white TV. Although, there were multiple gray levels associated with it but the name associated with it was black-and-white TV. The first major transition was from black-and-white TV to color TV. It was a big revolution when that transition occurred. The earlier color TVs were analog. Then, digital color TVs were introduced followed by transition from standard resolution to high definition (HD) resolution of the images.

However, the era of 2D HDTV appears to be short because we are now witnessing the advent of 3D HDTV (Wikipedia HDTV). These, 3D HDTV are based on the stereoscopic technology and hence are known as stereoscopic 3D TV or S3D TV. Since, they also support high definition resolution; hence, they can be called S3D HDTV. All the major TV manufacturers have introduced S3D HDTV in the consumer market. They include various models from leading manufacturers like Sony, Panasonic, Mitsubishi, Samsung, LG, Philips, Sharp, Hitachi, Toshiba and JVC.

S3D HDTV can be switched between the 2D and 3D imaging modes hence maintaining the downward compatibility with 2D images and videos. Additionally, they provide software that can artificially shift the 2D images and videos to produce the stereo effect and hence the TV programs can be watched in 3D. However, the quality still needs to be improved. At this moment, the best 3D perception is achieved by the images and videos that are produced in 3D. As mentioned above, these products are based on stereovision.

Hence, they require the usage of 3D glasses for watching in 3D.

3D MONITORS AND PHOTO FRAMES

In addition to S3D HDTV, 3D monitors are also available based on the same stereoscopic technology (Lipton 2002, Mcallistor 2002). Hence, they are available with 3D glasses. The 3D glasses are discussed in detail in the next section. 3D photo frames are now also being sold in the electronics market. However, they are based on stereoscopic vision with 3D glasses as well as on autostereoscopic vision technology which does not require glasses. At this moment in time, autostereoscopic displays are only available in small sizes and they are restricted because of the viewing angle in large sizes.

3D GLASSES

S3D HDTV relies on stereovision. In stereovision, separate images are presented to each of our eye, i.e., left and right eye. The images of the same scene are shifted similar to what our left and right eye see. As a result, the brain combines the two separate shifted images of the same scene and creates the illusion of the third dimension. The images are presented at a very high refresh rate and hence the two separate images are visualized by our eyes almost at the same time. Our brain cannot tell the difference of the time delay between the two images and they appear to be received by our eyes at the same time. The concept is similar to video where static images are presented one after the other at a very high rate and hence our brain visualizes them as continuous.

For separate images to be presented to our left and right eye, special glasses are required. These glasses had come to be known as 3D glasses. In early days, cardboard glasses were used. These cardboard glasses had different color for each of the lens with one being magenta or red and the other being blue or green. On the 3D display system, two images were shown on the screen with one is red color and the other in blue color. The lens with the red color filter absorbed red color and allowed blue image to pass through while the lens with the blue filter allowed the red image to enter the eye. Hence, one eye looked at the red colored image while the other eye watched the blue colored image. The brain received two images and hence 3D image created. However, two separate images were based on two separate colors. Therefore, true color movie is not possible with this technique. So, the image quality of early 3D movies was quite low.

Current 3D Glasses Technology

The current 3D glasses can be categorized into two classes: active shutter glasses and polarized glasses. Samsung, Panasonic, Sony and LG use the active shutter glasses. High refresh rate is used so that two images can be projected on the TV alternately; one image for the right eye and one for the left eye. Generally, the refresh rate is 120 hertz for one image and 240 hertz for both the images. The shutters on the 3D glasses open and close corresponding to the projection of images on the TV. There is a sensor between the lenses on the 3D glasses that connect with the TV in order to control the shutter on each of the lens. The brain received two images at very high refresh rate and hence it combines them to achieve the 3D effect. By looking away from the TV, one may see the opening and closing of the lenses and hence it might cause irritation for some viewers. The active shutter glasses are expensive compared to polarized glasses.

JVC uses polarized glasses to separate the images for the right eye and the left eye. The famous movie, Avatar, was shown in US with the polarized glasses. These glasses are very cheap compared to the active shutter glasses. Two images of the scene, each with a different polarization, are

Figure 1. 3D Blu-Ray disc player

projected on the screen. Since, the 3D polarized glasses have lenses with different polarization, hence, only one image is allowed in each eye. The brain receives two images and creates the 3D image out of them.

3D DISC PLAYERS

In the last decade, Sony won the standards war for the new disc player with blu-ray disc player being accepted as the industry standard. All the manufacturers accepted the standard with Blu-Ray Disc Association as the governing body for the Sony based HD technology. Recently, the Blu-Ray Disc Association has embraced the 3D (Figure 1). As a result, Sony, Samsung and other leading manufacturers have already released 3D blu-ray disc players. Additionally, Sony is also offering Sony Play station 3 upgrade to 3D, via a firmware download.

3D GAMES

Games have already moved to the 3D arena. Sony is selling Play Station with 3D gaming capability. However, to play 3D games, 3D TV with 3D glasses are required. The first four Play Station 3 3D games are Wipeout HD, Motor Storm Pacific Rift, Pain, and Super Stardust HD. Microsoft Xbox has similar plans.

Nintendo has introduced the new handheld model replacing the existing DS model. The new handheld Nintendo has 3D screen. This screen is not based on stereoscopic vision technology. Rather, it's based on autostereoscopic vision. Autostereoscopic displays do not require glasses. At this moment in time, the autostereoscopic technology is limited to small sized displays. Hence, Nintendo is taking advantage of this technology by introducing handheld gaming consoles based on autostereoscopic vision (Heater 2010) (Figure 2).

3D CAMERAS

The camera manufacturers have already launched various 3D camera models. One of the first 3D cameras was launched by Fuji in 2009. That camera was a 10 Mega Pixel camera with two CCD sensors. In September 2010, Sony launched two different 3D camera models. They were Cyber-shot DSC TX9 (a 12 Mega Pixel camera) and WX5. Both of the cameras provided 3D sweep panorama in addition to 2D sweep panorama. The images acquired by the 3D cameras can be seen on 3D TV, 3D computer and 3D photo frames.

3D COMPUTERS

3D computers are nothing more than the combination of 3D TV technology and 3D disc players. Similar to 3D TVs, the current 3D display technology is based on stereovision. Hence, 3D glasses are required. Again, some manufacturers

Figure 2. Sony Play Station 3

Figure 3. 3D computer

provide 3D computers with active shutter glasses while the others provide the polarized glasses. 3D blue-ray disc player is standard with most of the 3D computers. One of the earliest 3D computers is from Acer and Asus (Figure 3). Acer provided their first laptop with 15.6 inch widescreen 3D display in December 2009. Acer 3D laptop used a transparent polarizing filter overlaid on the screen and hence it required corresponding 3D polarized glasses. Asus provided the 3D laptops with software Roxio CinePlayer BD which had the ability to convert 2D titles to 3D. LG is also entering the market of 3D laptops. In 2011, about 1.1 million 3D laptops are expected to sell. This number is expected to increase to about 14 million by 2015.

3D PRINTERS

Normal 2D printers are part of our everyday life. They are based on various technologies like laser, inkjet etc and provide printouts in grayscale or color depending on the printer model. Some of the big names in printer technology are HP, Brother and Epson. The concept of 3D printer is to produce an object in 3D. Soon there will be huge data available in 3D within very short span of time as the 3D cameras will proliferate the market. Hence, the demand for producing 3D objects will increase. 3D printers are currently available but they are very expensive with the cheapest model in thousands of dollars. However, with the increase in 3D data and the demand for 3D printing, it is not far that 3D printers will become cheaper. HP has already taken a step in this direction by buying a 3D printer company with the aim of mass producing 3D printers in near future.

3D MOBILE PHONES

Mobile phones have changed the culture of the world today. It is a strong mini-computer in hand with the ability to take pictures, make videos, record sound and upload them instantaneously on the web. They are playing great role in human rights protection, cultural revolutions, political upheaval, news, tourism and almost every other thing in our daily lives. As mentioned earlier, autostereoscopic displays work well in small sizes and they do not require glasses. Hence, 3D mobile phones are based on autostereoscopic displays. 3D cameras are already available and it is just matter of time that they become part of the 3D mobile phones. Sky is the limit of our imagination for a 3D device that can capture as well as display in 3D, transmit in 3D, record in 3D and can serve as a 3D gaming platform.

In 2009, Hitachi launched a mobile phone with stereoscopic display. However, it is the autostereoscopic technology that will lead the way for 3D mobile phones. In April 2010, Sharp introduced 3D autostereoscopic display technology that does not require glasses. However, the image shown through that display was as bright as it would be on standard LCD screen. Sharp used parallax barrier technology to produce 3D effect. Later in chapter 18, the autostereoscopic technology is discussed in detail. Sharp announced mass production of these small autostereoscopic displays for mobile devices. At the time of the announcement, the device measured 3.4 inches (8.6 cm) with a resolution of 480 by 854 pixels, brightness (500 cd/m^2) and the contrast ratio (1000:1).

AUTOSTEREOSCOPIC 3D TV

Autostereoscopic 3D TV is also known as A3D TV (Dodgson 2005). A3D TV is multi-view displays which do not require any glasses. It has large 3D viewing zone, hence, multiple users can view in 3D at the same time. Currently, A3D TV is based on two types of technologies, namely, lenticular lenses and the parallel barrier. In case of lenticular lenses, tiny cylindrical plastic lenses transparent sheets are pasted on the LCD screen. The tiny cylindrical plastic lenses project two images, one for each of our eye, hence producing 3D effect. Since, these sheets are pasted on LCD screen, so the A3D TV based on this technology can only project in 3D and 2D display is not possible with this technology.

The other technology is called parallel barrier technology. Sharp and LG are the front runners pursuing this technology. Fine gratings of liquid crystal with slits corresponding to certain columns of pixels are used in front of the screen. These slits result in separate images for the right and left eye when voltage is applied to the parallax barrier. The parallax barrier can also be switched off, hence allowing A3D TV to be used in 2D mode. Chapter 18 discusses in detail the autostereoscopic displays.

3D PRODUCTION

3D TVs are of no use without the 3D production of movies, dramas, documentaries, news, sports and other TV programs. Conversion of 2D to 3D with software does not provide good 3D visualization results. Many production companies are investing in 3D production. ESPN is currently using cameras with two sets of lenses for their live 3-D broadcasts. In 2007, Hellmuth aired live the NBA sports tournament in US in 3D HD and it is leading the 3D HD production. Professional tools are now available from Sonic for encoding videos and formatting titles in blue-ray 3D format.

Various movies were released in last few years in 3D. They include the release of Monsters vs. Aliens by DreamWorks Animation in September 2009, Disney/Pixar's "Up" and 20th Century Fox's "Ice Age: Dawn of the Dinosaurs" etc. In 2009, US$1 billion was generated at box offices worldwide before the release of Avatar in late 2009. Avatar alone generated about $2.7 billion at box offices worldwide (Wikipedia-Disney) After that, the production in 3D is becoming more of a routine production. Hence, the quality of 3D production is bound to increase with the passage of time.

3D STANDARDS

There are various companies and organizations that are competing for the 3D standards. Some of them include:

- Standard for 3-D mastering and distribution (Society of Motion Pictures and Television Engineers, SMPTE)
    - http://www.smpte.org/home/
- MPEG's Multiview Video Coding (Moving Pictures Experts Group – MPEG)
    - http://mpeg.chiariglione.org/
    - http://www.mpegif.org/
- 3D Consortium (Japan)
    - http://www.3dc.gr.jp/english/
- 3D Working Group for 3D home entertainment (Digital Entertainment Group)
    - The members of the 3D Working Group for 3D home entertainment include Microsoft, Panasonic, Samsung Electronics, Sony, 20th Century Fox

  Home Entertainment, Walt Disney Studios Home Entertainment and Warner Home Entertainment Group
  - http://www.degonline.org/
- The Wireless HD Consortium
  - They provide Wireless HD standard for in-room cable-replacement technology
  - The original throughput standard is based on 4Gbps for high-definition video up to 1080p
  - In the 1.1 spec, throughput is increased to more than 15Gbps for streaming 3D video formats mentioned in the HDMI 1.4a specification
  - http://www.wirelesshd.org/
- The 3D@Home Consortium
  - This is for the advancement of 3D technology into the home
  - http://www.3dathome.org/
- The Blue-ray Disc Association
  - In December 2009, it announced the agreement that allows for full 1080p viewing of 3-D movies on TVs
  - To create the 3D effect, two images in full resolution will be delivered by the Blue-ray disc players.

3D TV: MARKET FORECAST

According to a survey by In-Stat in September 2009, 67% said that they are willing to pay more for a 3D version of a Blue-ray disc then a 2-D version. In another survey by a research firm GigaOM in September 2009, there will be 46 million 3D TV units sold worldwide by 2013. In December 2009, another research firm, Display Search, forecasted the 3D TV market to grow to US$15.8 billion by 2015. It is expected that Sony will be selling about 40% to 50% 3D TVs out of its all TV units by end of 2012. LG is expected to be selling close to 4 million 3D TVs in 2012. These forecast figures show that there is no turning back now and all the leading manufacturers are investing heavily in 3D technology.

CONCLUSION AND FUTURE DIRECTIONS

The 3D imaging products have already started appearing in the consumer market since 2009. With the wide availability of 3D cameras and 3D mobile phones, 3D data will soon proliferate the web. The 3D movies and other 3D content are already changing our viewing culture. In near future, the shift will be from stereoscopic displays with 3D glasses to autostereoscopic displays without the glasses. The gaming culture is also shifting to 3D gaming. Within next five years till 2015, 3D imaging will become part of our everyday life from cameras to mobile phones to computers to TV to games. Hence, intelligent algorithms and techniques will be required for processing of 3D data. Additionally, bandwidth requirements will increase for transmission. Good compression methods will be required as we move to multi-view imaging displays. The ultimate goal for imaging displays is to generate 3D views like we, ourselves, see in 3D. That will be accomplished by research in holography. However, that is something to be discussed in the next decade. This decade is for the stereoscopic displays, autostereoscopic displays and for all the technology that is associated with them.

ACKNOWLEDGMENT

This work is supported by the E-Science grant funded by the Ministry of Science, Technology and Innovation (MOSTI), Government of Malaysia (No: 01-02-02-SF0064).

REFERENCES

Bosch, J. A. (Ed.). (1995). *Coordinate measuring machines and systems*. New York, NY: M. Dekker.

Cabeza, R., & Kingstone, K. (Eds.). (2006). *Handbook of functional neuroimaging of cognition*. MIT Press.

Dodgson, N. A. (2005). Autostereoscopic 3D displays. *IEEE Computer*, *38*(8), 31–36. doi:. doi:10.1109/MC.2005.252

Heater, B. (2010, March 23). Nintendo says next-gen DS will add a 3D display. *PC Magazine*. Retrieved from http://www.pcmag.com/ article2/ 0,2817, 2361691,00.asp

Lipton, L., & Feldman, M. (2002). A new autostereoscopic display technology: The SynthaGram. *Proceedings of SPIE Photonics West 2002: Electronic Imaging,* San Jose, California.

McAllister, D. F. (2002). Stereo & 3D display technologies, display technology. In Hornak, J. P. (Ed.), *Encyclopedia of imaging science and technology* (pp. 1327–1344). New York, NY: Wiley & Sons.

Wikipedia. (n.d.). *Disney*. Retrieved from http:// en. wikipedia.org/ wiki/ Disney_ Digital_ 3-D

Wikipedia. (n.d.). *High definition television*. Retrieved from http://en.wikipedia.org/ wiki/ High_ definition_ television

ADDITIONAL READING

Inition website. http://www.inition.co.uk

3DHOME. http://www.3dathome.org

3DTV TECHNOLOGY. http://www. 3dtvtechnology. org.uk/ polarization

KEY TERMS AND DEFINITIONS

Stereoscopic: It refers to 3D using two images just like our eyes. It requires 3D glasses to view in 3D.

Autostereoscopic: It refers to 3D displays that do not require 3D glasses to view in 3D.

Section 1
3D Imaging Methods

Chapter 2
Multi-View Stereo Reconstruction Technique

Peng Song
Nanyang Technological University, Singapore

Xiaojun Wu
Harbin Institute of Technology Shenzhen, China

ABSTRACT

3D modeling of complex objects is an important task of computer graphics and poses substantial difficulties to traditional synthetic modeling approaches. The multi-view stereo reconstruction technique, which tries to automatically acquire object models from multiple photographs, provides an attractive alternative. The whole reconstruction process of the multi-view stereo technique is introduced in this chapter, from camera calibration and image acquisition to various reconstruction algorithms. The shape from silhouette technique is also introduced since it provides a close shape approximation for many multi-view stereo algorithms. Various multi-view algorithms have been proposed, which can be mainly classified into four classes: 3D volumetric, surface evolution, feature extraction and expansion, and depth map based approaches. This chapter explains the underlying theory and pipeline of each class in detail and analyzes their major properties. Two published benchmarks that are used to qualitatively evaluate multi-view stereo algorithms are presented, along with the benchmark criteria and evaluation results.

INTRODUCTION

High quality 3D models have large and wide applications in computer graphics, virtual reality, robotics, and medical imaging, etc. Although many of the 3D models can be created by a graphic designer using specialized tools (e.g., 3D Max Studio, Maya, Rihno), the entire process to obtain a good quality model is time consuming and tedious. Moreover, the result is usually only an approximation or simplification. At this place, 3D modeling technique provides an alternative and has already demonstrated their potential in several application fields.

DOI: 10.4018/978-1-61350-326-3.ch002

In general, 3D modeling technique can be classified into two different groups: active and passive methods. The active methods try to acquire precise 3D data by laser range scanners or coded structured light projecting systems which project special light patterns onto the surface of a real object to measure the depth to the surface by a simple triangulation technique. Although such 3D data acquisition systems can be very precise, most of them are very expensive and require special skills. Compared to active scanners, passive methods work in an ordinary environment with simple devices and flexibilities, and provide feasible and comfortable means to extract 3D information from a set of calibrated pictures. According to the information contained in images which is used to extract 3D shape information, passive methods can be categorized into four classes: shape from silhouette, shape from stereo, shape from shading (Zhang, 1999), and shape from texture (Forsyth, 2001; Lobay, 2006). This chapter will mainly focus on shape from stereo technique that tries to reconstruct object models from multiple calibrated images by stereo matching. Shape from silhouette technique is also introduced since it outputs a good shape estimate which is required by many shape from stereo algorithms.

In order to generate 3D model of a real object, digital cameras are used to capture multi-view images of the object which are obtained by changing the viewing directions to the object. Once the camera has been calibrated, a number of images are acquired at different viewpoints in order to capture the complete geometry of the target object. In many cases, the acquired images need to be processed before surface reconstruction. Finally, these calibrated images are provided as input to various multi-view stereo algorithms which seek to reconstruct a complete model from multiple images using information contained in the object texture. The major advantage of this technique is that it can output high quality surface models and offer high flexibility of the required experimental setup.

This chapter is structured as follows. Next section gives a brief introduction to camera calibration followed by the section that discusses several issues about how the original pictures should be taken and processed. Then, shape from silhouette concept and approaches are explained in detail, along with a discussion of its applications. After that, a section mainly focuses on the classification of shape from stereo approaches and introduces the pipeline, theory and characteristics of each class. Final section presents two published benchmarks for evaluating various multi-view stereo algorithms.

CAMERA CALIBRATION

Camera calibration is the process of finding the true parameters of the camera that produced a given photograph or video. Camera calibration is the crucial step in obtaining an accurate model of a target object. The calibration approaches can be categorized into two groups: full-calibration and self-calibration. Full-calibration approaches (Yemeza, 2004; Park, 2005) assume that a calibration pattern with precisely known geometry is presented in all input images, and computes the camera parameters consistent with a set of correspondences between the features defining the chart and their observed image projections. While the self-calibration approaches (Hernandez, 2004; Eisert, 2000; Fitzgibbon, 1998) are proposed to reduce the necessary prior knowledge about the scene camera geometry only to a few internal and external constraints. In these approaches, the intrinsic camera parameters are often supposed to be known a priori. However, since they require complex optimization techniques which are slow and difficult to converge, their accuracy is not comparable to that of the fully-calibrated systems. In practice, many applications such as 3D digitization of cultural heritage prefer to fully-calibrated systems since maximum accuracy is a very crucial requirement while self-calibration approaches are preferred when no Euclidean information is available such as reconstruction of a large scale outdoor building.

IMAGE ACQUISITION AND PROCESSING

There are many important issues about how the original pictures should be taken and processed, which eventually determine the final model quality. In this section only three issues that are closely related to multi-view stereo reconstruction technique are discussed: uniform illumination, silhouette extraction, and image rectification.

One of the most obvious problems during image acquisition is that of highlights. Highlights depend on the relative position of object, lights and camera which means that they change position along the object surface from one image to the other. This can be problematic in recovering the diffuse texture of the original object. Highlights should be avoided in the original images by using a diffuse and uniform lighting. Moreover, multi-view stereo matching will also be influenced by uniform illumination. In order to make sure the uniform lighting condition for each image, the target object should be illuminated by multiple light sources at different positions.

To facilitate silhouette segmentation, it is better to use a monochrome background in the setup of image acquisition. This facilitates the identification of the object silhouette using standard background subtraction method which needs two consecutive acquisitions for the same scene, with and without the object, keeping the camera and the background unchanged. However, standard background subtraction method may in some cases fail when the background color happens to be the same with the object color which will cause erroneous holes inside the silhouettes. However, if the transition between the background and the object is sharp, the correct silhouette can still be found. Some manual processing is needed to fix the erroneous holes. In practice, it is better to select a background color with high contrast to the object color which will make image segmentation simple.

In practice, multi-view stereo algorithms always rectify image pairs to facilitate stereo matching. Stereo rectification determines a transformation of each image plane such that pairs of conjugate epipolar lines become parallel to the horizontal image axes. Using projection matrices of the reference and primary images, we can rectify stereo images by using the rectification technique proposed by (Fusiello, 2000). The important advantage of rectification is that computing stereo correspondences is simpler, because search is done along the horizontal lines of the rectified images.

SHAPE FROM SILHOUETTE

Shape from silhouette approaches try to create a 3D representation of an object by its silhouettes within several images from different viewpoints. The 3D representation named visual hull (Laurentini, 1994) is constructed by intersection of the visual cones formed by back-projecting the silhouettes in the corresponding images. The visual hull can be very close to the real object when much shape information can be inferred from the silhouettes (see Figure 1 left). Since concave surface regions can never be distinguished using silhouette information alone, the visual hull is just an approximation of the actual object's shape, especially if there are only a limited number of cameras. The visual hull of a toy dinosaur demonstrated in Figure 1 right shows that a concave region on the dinosaur body cannot be correctly recovered (illustrated by the red square).

3D Bounding Box Estimation

Many visual hull computation approaches need the target object's 3D bounding box, e.g. volumetric approach takes it as a root node when building visual hull octree structure, deformable model

Figure 1. The visual hull of a toy alien model (left) and a toy dinosaur model (right)

approach needs a 3D bounding volume to construct an initial surface.

The 3D bounding box can be estimated only from a set of silhouettes and the projection matrices. In practice, an accurate 3D Bounding Box can improve the precision of the final model. We can calculate the 3D bounding box only from a set of silhouettes and the projection matrices. This can be done by considering the 2D bounding boxes of each silhouette. The bounding box of the object can be computed by an optimization method for each of the 6 variables defining the bounding box, which are the maximum and minimum of x, y, z (Song, 2009). On the other hand, the 3D bounding box can also be estimated using an empirical method. When the image capture system has been constructed, the origin of the world coordinate is defined. If we know the approximate position of the origin, the center of bounding box can be estimated. The size of the bounding box is simple to estimate since we can just make it large enough to contain the object. Then this estimated initial bounding box can be applied to compute the visual hull mesh. In practice, the resulting visual hull mesh also has a bounding box which is very close to the object's real bounding box.

Visual Hull Computation

The main problem for visual hull computation is the difficulty in designing a robust and efficient algorithm for the intersection of the visual cones formed by back-projecting the silhouettes. Various algorithms have been proposed to solve this problem, such as volumetric (Song, 2009), polyhedral (Matusik, 2000; Shlyakhter, 2001), marching intersection (Tarini, 2002), and deformable model approaches (Xu, 2010). This section gives a brief introduction to volumetric approach.

In the volumetric approach, the 3D space is divided into elementary cubic elements (i.e., voxels) and projection tests are performed to label each voxel as being inside, outside or on the boundary of the visual hull. This is done by checking the contents of its projections on all the available binary silhouette images. The output of volumetric methods is either an octree (Szeliski, 1993; Potmesil, 1987), whose leaf nodes cover the entire space or a regular 3D voxel grid (Cheung, 2000). Coupled with the marching cubes algorithm (Lorensen, 1987), a surface can be extracted. Since these techniques make use of a voxel grid structure as an intermediate representation, the vertex positions of the resulting mesh are thus limited to the voxel grid. The most important part for volumetric approach is projection test, which is a process to check the projection of a voxel on all the available binary silhouette images. The test result classifies the voxel as being inside, outside or on the boundary of the visual hull. Specifically, if the projection of the voxel is in all the silhouettes, the corresponding voxel is inside the visual hull surface; if the projection is completely out of at least one silhouette, its type is out; else, the voxel is on the visual hull surface.

Discussion

The visual hull is an approximation of the real object shape and the level of satisfaction obviously depends on the kind of object and on the number and position of the acquired views. However, it still has many applications in the field of shape analysis, robotic and stereo vision etc. Firstly, it offers a rather complete description of a target object and can be directly fed to some 3D applications as a showcase. Moreover, the generated visual hull model can be sensibly improved from the appearance point of view by means of color textures obtained by the original images. Secondly, the visual hull is an upper bound of a real object which is big advantage for obstacle avoidance in the field of robotic or visibility analysis in navigation. Finally, it provides good initial model for many reconstruction algorithms, e.g. the snake-based multi-view stereo reconstruction algorithm uses it as an initial surface since it can capture the target object's topology in most case.

MULTI-VIEW STEREO RECONSTRUCTION

Multi-view stereo technique seeks to reconstruct a complete 3D object model from a collection of calibrated images using information contained in the object texture. In essence, the depth map of each image is estimated by matching multiple neighboring images using photo-consistency measures which operate by comparing pixels in one image to pixels in other images to see how well they correlate. The position of corresponding 3D point is then computed by a triangulation method. In practice, the image sequence captured for surface reconstruction contains many images, from one dozen to more than one hundred and the camera viewpoints may be arranged arbitrarily. Therefore, a visibility model is needed to determine which images should be selected for stereo matching. Multi-view stereo reconstruction algorithms can be mainly categorized into four classes according to the taxonomy of (Seitz, 2006): 3D volumetric, surface evolution, feature extraction and expansion, and depth map based approaches. We introduce the pipeline of each class first and then take one typical algorithm of each class to explain the implementation details. Finally, the characteristics of each class is summarized some of which are validated by the evaluation results on the Middlebury benchmark.

3D Volumetric Approach

3D volumetric approaches (Treuille, 2004) first compute a cost function on a 3D volume, and then extract a surface from this volume. Based on the theoretical link between maximum flow problems in discrete graphs and minimal surfaces in an arbitrary Riemannian metric established by (Boykov, 2003), many approaches (Snow, 2000; Kolmogorov, 2002; Vogiatzis, 2005; Tran, 2006; Vogiatzis, 2007) use graph cut to extract an optimal surface from a volumetric Markov Random Field (MRF). Typically, graph cut based approaches first define a photo consistency based surface cost function on a volume where the real surface is embedded and then discretize it with a weighted graph. Finally, the optimal surface under this discretized function is obtained as the minimum cut solution of the weighted graph.

In the graph cut based approach proposed in (Vogiatzis, 2005), they first build a base surface S_{base} as the visual hull and the parallel inner boundary surface S_{in} which define a volume C enclosed by S_{base} and S_{in} The photo-consistency measure $\rho(x)$ used to determine the degree of consistency of a point x with the images is the NCC value between patches centered on x. And the base surface S_{base} is employed for obtaining visibility information by assuming that each voxel has the same visibility as the nearest point on S_{base} The cost function associated with the photo-consistency of a candidate surface S is the integral of $\rho(x)$ on the surface,

$$E_{surf}[S] = \iint_s \rho(x)dA \quad (1)$$

If the base surface S_{base} is not far from the real surface, then voxels that lie on the real surface would have smallest ρ values. Therefore, surface reconstruction can be formulated as an energy minimization problem which tries to find the minimal surface $S_{\min}$ in the volume C. The minimal surface under this function is obtained by computing the minimum cut solution of the graph. In order to obtain a discrete solution, 3D space is quantized into voxels of size $h \times h \times h$. The graph nodes consist of all voxels whose centers are in C. Each voxel is a node in the graph, G, with a 6-neighbor system for edges. The weight for the edge between voxel (node) v_i and v_j is defined as,

$$w(v_i, v_j) = \frac{4\pi h^2}{3} \rho(\frac{x_i + x_j}{2}) \quad (2)$$

where h is the voxel size. The voxels that are part of S_{in} and S_{base} are connected with the source and sink respectively with edges of infinite weight. With the graph G constructed this way, the graph cut algorithm is then applied to find $S_{\min}$ in polynomial time.

Since the graph cut algorithm usually prefers shorter cuts, protrusive parts of the object surface is easy to cut off. In this case, a shape prior that favors objects that fill the space of the visual hull more can be applied. The main problem for graph cut based approach is that for high resolutions of the voxel grid, the image footprints used for consistency determination become very small which often results in noisy reconstructions in textureless regions.

Surface Evolution Approach

Surface evolution approaches (Hernandez, 2004; Zaharescu, 2007; Kolev, 2009) work by iteratively evolving a surface to minimize a cost function, in which the surface can be represented by voxels, level sets, and surface meshes. Space carving (Matsumoto, 1997; Fromherz, 1995) is a technique that starts from a volume containing the scene and greedily carves out non-photoconsistent voxels from that volume until all remaining visible voxels are consistent. Since it uses a discrete representation of the surface but does not enforce any smoothness constraint on the surface, the reconstructed results are often quite noisy. Level set techniques (Malladi, 1995) start from a large initial volume and shrink inward to minimize a set of partial differential equations defined on a volume. These techniques have an intrinsic capability to freely change the surface topology while the drawbacks are the computation time and the difficulty to control the topology. Topology changes have to be detected and taken care of during the mesh evolution which can be an error prone process. Snake techniques formulate the surface reconstruction as a global energy minimization problem. The total energy term E is composed of an internal energy E_{int} to obtain a final well-shaped surface, and an external energy E_{ext} to make the final surface confirm the shape information extracted from the images. This energy minimization problem can be transformed to a surface iteration problem in which an initial surface mesh is driven by both the internal force and external force that iteratively deform to find a minimum cost surface.

Since the snake approach of (Hernandez, 2004) wants to exploit silhouettes and texture for surface reconstruction, the external energy is composed of the silhouette related energy E_{sil} and the texture related energy E_{tex}. The minimization problem is posed as finding the surface S of R^3 that minimizes the energy $E(S)$ defined as follows:

$$E(S) = E_{ext}(S) + E_{\text{int}}(S) = E_{tex}(S) + E_{sil}(S) + E_{\text{int}}(S) \tag{3}$$

And this energy minimization problem can be transformed to a surface iteration problem as follows:

$$S^{k+1} = S^k + \Delta t(F_{tex}(S^k) + F_{sil}(S^k) + F_{\text{int}}(S^k)) \tag{4}$$

To completely define the deformation framework, this approach needs an initial surface S_0 that will evolve under the different energies until convergence. Since snake deformable models maintain the topology of the mesh during its evolution, the initial surface must capture the topology of the object surface. The visual hull is a quite good choice in this case. The texture force F_{tex} contributes to recovering the 3D object shape by exploiting the texture of the object to maximize the image coherence of all the cameras that see the same part of the object which is constructed by computing a Gradient Vector Flow (GVF) filled (Xu, 1998) in a volume merged from the estimated depth maps. The silhouette force F_{sil} is defined as a force that makes the snake match the original silhouettes of the sequence which can be decomposed into two different components: a component that measures the silhouette fitting, and a component that measures how strongly the silhouette force should be applied. The internal force F_{int} contains both the Laplacian and biharmonic operators that try to smooth the surface during surface evolution process. The deformable model evolution process at the kth iteration can then be written as the evolution of all the vertices of the mesh v_i.

$$v_i^{k+1} = v_i^k + \Delta t(F_{tex}(v_i^k) + \beta F_{sil}(v_i^k) + \gamma F_{\text{int}}(v_i^k)) \tag{5}$$

where Δt is the time step and β and γ are the weights of the silhouette force and the regularization term, relative to the texture force. The time step Δt has to be chosen as a compromise between the stability of the process and the convergence time. Equation 5 is iterated until convergence of all the vertices of the mesh is achieved.

Snake deformable offers a well-known framework to optimize a surface under several kinds of constraints extracted from images such as texture, silhouette, and shading constraints. However, its biggest drawback is that it cannot change the topology of the surface during the evolution. Moreover, since the snake approach is evolved based on surface mesh, they have to deal with artifacts like self intersections or folded-over polygons. The resolution of the polygon mesh has to be adjusted by tedious decimation, subdivision and remeshing algorithms that keep the mesh consistent. Finally, large distances between the initial and the true surface (e.g. in deep concavities) often lead to slow convergence of the deformation process.

Depth Map Based Approach

Generally, depth map based approaches (Goesele, 2006; Bradley, 2008; Campbell, 2008; Liu, 2009; Song, 2010; Li, 2010) involve two separate stages. First, a depth map is computed for each viewpoint using binocular stereo. Second, the depth maps are merged to produce a 3D model. In these methods, the estimation of the depth maps is crucial to the quality of the final reconstructed 3D model. Since the estimated depth maps always contain lots of outliers due to miscorrelation, an outlier rejection process is always required before final surface reconstruction.

Song et al. (Song, 2010) proposed a depth map based approach to reconstruct a complete surface model using both texture and silhouette information contained in images (see Figure 2 for illustration). Firstly, depth maps are estimated from multi-view stereo efficiently by an expansion-based method. The outliers of the estimated depth maps are rejected by a two-step approach. Firstly, the visual hull of a target object is incorporated as a constraint to reject 3D points out of the visual hull. Then, a voting octree is built from the estimated point cloud and a threshold is selected to eliminate miscorrelations. To downsample the 3D point cloud, for each node at the maximum depth of the voting octree, the point with largest confidence value is extracted in the corresponding voxel to construct a new point cloud on the object surface with few outliers and smaller scale. The surface normal of each point in the point cloud is estimated from the positions of the neighbors and the viewing direction of each 3D point is employed to select the orientation of estimated surface normal. The resulted oriented point cloud is called point cloud from stereo (PCST). In order to restore the textureless and occluded surfaces, another oriented point cloud called point cloud from silhouette (PCSL) is computed by carving the visual hull octree structure using the PCST. Finally, Poisson surface reconstruction approach (Kazhdan, 2006) is applied to convert the oriented point cloud both from stereo and silhouette (PC-STSL) into a complete and accurate triangulated mesh model.

The computation time of depth map based methods are dominant by the depth map estimation step which can vary from few minutes to several hours for the same input dataset. Since these approaches use an intermediate model represented by 3D points, they are able to recover accurate details on well textured region while result in noisy reconstructions in textureless regions.

Feature Extraction and Expansion Approach

The idea behind this class (Habbecke, 2007; Goesele, 2007; Jancosek, 2009; Furukawa, 2010) is that a successfully matched depth sample of a given pixel provides a good initial estimate

Figure 2. Overall approach of (Song, 2010). From left to right: one input image, visual hull, PCST, PCSL, PCSTSL, the reconstructed model.

for depth and normal for the neighboring pixel locations. Typically, these algorithms use a set of surface elements in the form of patch with either uniform shape (e.g. circular or rectangular) or non-uniform shape known as patch model. A patch is usually defined by a center point, a normal vector, and a patch size to approximate the unknown surface of a target object or scene. The reconstruction algorithm always consists of two alternating phases. The first phase computes a patch model by matching a set of feature points to generate seed patches and expanding the shape information from these seed patches. Note that a filtering process can be done simultaneously with the expansion process or as a post process for the patch model. The second phase converts the patch model into a triangulated model.

Recent work by Furukawa and Ponce (Furukawa, 2010) proposes a flexible patch-based algorithm for calibrated multi-view stereo. The algorithm starts by computing a dense set of small rectangular oriented patches covering the surfaces visible in the images by a match, expand and filter procedure: (1) matching: features found by Harris and difference-of-Gaussians operators are first matched across multiple pictures to generate a sparse set of patches associated with salient image regions, (2) expansion: spread the initial matches to nearby pixels and obtain a dense set of patches, (3) filtering: visibility and a weak form of regularization constraints are then used to eliminate incorrect matches. Then the algorithm converts the resulting patch model into an initial mesh model by PSR approach or iterative snapping: (1) PSR approach directly converts a set of oriented points into a triangulated mesh model, (2) iterative snapping approach computes a visual hull model and iteratively deforms it towards reconstructed patches. Note that the iterative snapping algorithm is applicable only to object datasets with silhouette information. Finally, an optional final refinement algorithm is applied to refine the initial mesh to achieve even higher accuracy via an energy minimization approach (Furukawa, 2008). Since this algorithm takes into account surface orientation properly in computing photometric consistency, which is important when structures do not have salient textures, or images are sparse and perspective distortion effects are not negligible, it outputs accurate object and scene models with fine surface detail despite low-texture regions or large concavities.

Since this class of approach takes advantage of the already recovered 3D information, the patch model reconstruction step is quite efficient. And they do not require any initialization in the form of a visual hull model, a bounding box, or valid depth ranges. Finally, these approaches are easy to find correct depth in low-textured regions due to its expansion strategy and patch model representation, i.e., use large patches in homogeneous area while small patches for well textured region.

Discussion

We have introduced the pipeline, theory and characteristics of each class for multi-view stereo algorithm. With the development of this area, some approaches take the advantages of several existing methods and modify each existing method in an essential way to make them more robust and accurate. For example, Vu et al. (Vu 2009) proposed a multi-view stereo pipeline to deal with large scenes while still producing highly detailed reconstructions. They first extract a visibility consistent mesh close to the final reconstruction using a minimum s-t cut from a dense point cloud merged from estimated depth maps. Then a deformable surface mesh is iteratively evolved to refine the initial mesh to recover even smaller details. In fact, this approach combines the characteristic of depth map based, 3D volumetric, and surface evolution classes. However, since the accuracy of the final mesh basically depends on the estimated depth maps, this approach is classified as depth map based class in this chapter.

Shape from stereo is based on the assumption that the pixel intensity of a 3D point does not differ significantly when projected onto different camera views. However, this assumption does not hold in most practical cases due to shading, inhomogeneous lighting, highlights and occlusion. Therefore, it is difficult to obtain robust and reliable shape by using only stereo information. This method relies substantially on the object's texture. When a target object lacks texture, structured light can be used to generate this information.

BENCHMARK

Multi-view 3D modeling datasets can mainly be classified into two categories. The first category is object datasets in which a single object is photographed from viewpoints all around it and usually fully visible in acquired images. The uniqueness of datasets of this category is that it is relatively straightforward to extract the apparent contours of the object and thus compute its visual hull. The other category is scene datasets in which target objects may be partially occluded and/or embedded in clutter, and the range of viewpoints may be severely limited. The characteristic of datasets of this category is that it is hard to extract the apparent contours of the object to compute its bounding volume. Typical examples are outdoor scenes such as buildings or walls. Two benchmarks have been published to evaluate various multi-view stereo algorithms quantitatively: the Middlebury benchmark for object datasets and the large scale outdoor benchmark for scene datasets.

Middlebury Benchmark

The Middlebury benchmark (Seitz, 2006) datasets consist of two objects, *temple* and *dino*. The *temple* object (see Figure 3 left) is a 159.6 mm tall, plaster reproduction of an ancient temple which is quite diffuse and contains lots of geometric structure and texture. While the *dino* object (see Figure 3 right) is a 87.1mm tall plaster dinosaur model which has a white, Lambertian surface without obvious texture. The images of the datasets were captured by using the Stanford spherical gantry and a CCD camera with a resolution of 640×480 pixels attached to the tip of the gantry arm. From the resulting images, three datasets were created for each object, corresponding to a full hemisphere, a single ring around the object, and a sparsely sampled ring. A more detailed description of the *temple* and *dino* datasets can be found in (Seitz, 2009). In order to evaluate the submitted models, an accurate surface

Figure 3. The Middlebury benchmark: temple (left) and dino (right) objects

model acquired from laser scanner is taken as the ground truth model with 0.25mm resolution for each object.

The reconstruction results for the Middlebury benchmark datasets are evaluated on the accuracy and completeness of the final result with respect to the ground truth model, as well as processing time. The accuracy is measured by distance *d* such that a given percentage, say *X%*, of the reconstruction is within *d* from the ground truth model and the completeness is measured by percentage *Y%* of the ground truth model that is within a given distance *D* from the reconstruction. The default value is *X*=90 and *D*=1.25. In order to compare computation speed fairly, the reported processing time will be normalized according to the processor type and frequency. We present the results of quantitative evaluation of current state-of-the-art multi-view stereo reconstruction algorithms on this benchmark datasets shown in Table 1. Please note that only the published approaches are considered for the accuracy ranking, ignoring the evaluation results of unpublished papers. Since Furukawa and Ponce evaluate the submissions of the same approach twice for two different publications (Furukawa, 2007; Furukawa, 2010), only the result of (Furukawa, 2010) is included for accuracy ranking. The algorithms listed in Table 1 are grouped using the classification method presented in previous section in order to validate the characteristic of each class.

Table 1 shows that the accuracy and completeness rankings among the algorithms are relatively stable. Since most of the algorithms in this benchmark generate complete object models, the completeness numbers were not very discriminative. We mark the top three most accurate algorithms for each data set in Table 1 using red, green, and blue color respectively. First of all, we can find that the evaluation results of the depth map based approaches on the *temple* object is very good for the reason that this class is adapt in reconstructing well textured object with many slight details. While the property that depth map based approach cannot handle textureless region quite well has also been demonstrated by the Figure 4 (see the region marked by the red square). Secondly, the approach of (Furukawa, 2010) outperforms all other submitted for all the three datasets of the *dino* object since the feature extraction and expansion approaches can recover correct shape information for low-textured objects.

Large Scale Outdoor Benchmark

This benchmark data (Strecha, 2008) contains outdoor scenes and can be downloaded from (Strecha, 2010). Multi-view images of the scenes are captured with a Canon D60 digital camera

Table 1. Quantitative evaluation results of current state-of-the-art multi-view stereo algorithms

		Temple			Dino		
		Full	**Ring**	**SparseR**	**Full**	**Ring**	**SparseR**
3D Volumetric Approach	Vogiatzis 2005	1.07, 90.7%	0.76, 96.2%	2.77, 79.4%	*0.42*, 99.0%	0.49, 96.7%	1.18, 90.8%
	Tran 2006		1.12, 92.3%	1.53, 85.4%		1.12, 92.0%	1.26, 89.3%
	Vogiatzis 2007	0.5, 98.4%	0.64, 99.2%	0.69, 96.9%			
Surface Evolution Approach	Hernandez 2004	**0.36**, 99.7%	0.52, 99.5%	0.75, 95.3%	0.49, 99.6%	0.45, 97.9%	0.6, 98.5%
	Zaharesu 2007		0.55, 99.2%	0.78, 95.8%		0.42, 98.6%	***0.45***, 99.2%
	Kolev 2009		0.72, 97.8%	1.04, 91.8%		0.43, 99.4%	0.53, 98.3%
Depth Map-based Approach	Goesele 2006	***0.42***, 98.0%	0.61, 86.2%	0.87, 56.6%	0.56, 80.0%	0.46, 57.8%	0.56, 26.0%
	Bradley 2008		0.57, 98.1%	**0.48**, 93.7%		***0.39***, 97.6%	*038*, 94.7%
	Campbell 2008	*0.41*, 99.9%	***0.48***, 99.4%	0.53, 98.6%			
	Liu 2009			0.65, 96.9%			0.51, 98.7
	Vu 2009		**0.45**, 99.8%			0.53, 99.7%	
	Li 2010		0.64, 98.2%			0.43, 99.7%	
	Song 2010		0.61, 98.3%			*0.38*, 99.4%	0.54, 95.5%
Feature Extraction And Expansion	Habbecke 2007	0.66, 98.0%			***0.43***, 99.7%		
	Goesele 2007	0.42, 98.2%			0.46, 96.7%		
	Jancosek 2009	0.65, 85.8%	0.7, 78.9%	***0.59***, 74.9%	0.91, 73.8%	0.71, 76.6%	0.66, 74.9%
	Furukawa 2010	0.49, 99.6%	*0.47*, 99.6%	0.63, 99.3%	**0.33**, 99.8%	**0.28**, 99.8%	**0.37**, 99.2%

Figure 4. The dino models reconstructed by depth map based approaches. From left to right, (Goesele, 2006), (Vu, 2009), (Li, 2010), and (Song, 2010).

with a resolution of 3072 × 2028 square pixels. Figure 5 shows two datasets of this benchmark. The ground truth which is used to evaluate the quality of image based results is acquired by a laser scanner, outlier rejection, normal estimation and Poisson based surface reconstruction process.

Evaluation of the multi-view stereo reconstructions is quantified through relative error histograms counting the percentage of the scene recovered within a range of 1 to 10 times an estimated noise variance σ which is the standard deviation of depth estimates of the laser range

Figure 5. Large scale outdoor benchmark, Fountain-P11 (left) and Herz-Jesu (right) datasets

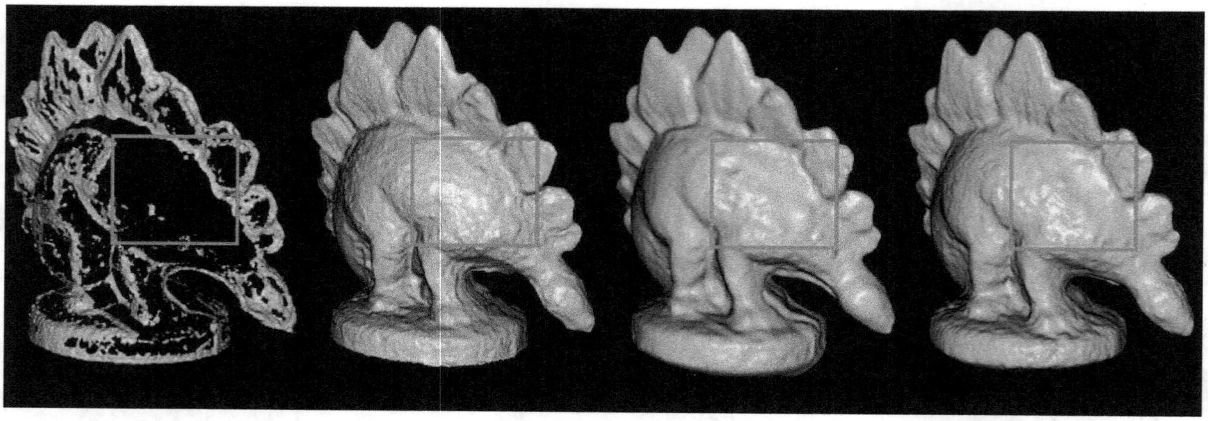

Table 2. Completeness measures for the Fountain dataset

	σ	2σ	3σ	4σ	5σ	6σ	7σ	8σ	9σ	10σ
Zaharescu 2007	14.6	38.8	55.5	65.1	70.4	73.7	75.9	77.3	78.3	79.0
Furukawa 2007	14.8	41.1	58.0	66.9	71.7	74.6	76.5	77.8	78.8	79.6
Vu 2009	18.0	47.7	67.9	78.7	84.2	87.2	88.8	89.8	90.4	90.9
Jancosek 2009	7.9	24.6	42.0	56.5	66.6	72.1	75.0	76.7	77.8	78.6

scanner used in the experiments. Table 2 present the results of quantitative evaluation of current state-of-the-art multi-view stereo reconstruction algorithms on the fountain dataset of this benchmark. Each entry in the table shows the percentage of the laser-scanned model that is within σ distance from the corresponding reconstruction. Since the feature extraction and expansion approaches do not require any initialization in the form of a visual hull model or a bounding box, they are very appropriate for scene datasets reconstruction. Another finding is that (Vu, 2009) achieves the best performance for this dataset since this approach combines advantages of several existing approaches.

FUTURE RESEARCH DIRECTIONS

Further development of multi-view stereo technique could move in many directions. A few of them are indicated as follows: firstly, research will focus on recovering 3D models with even higher accuracy to know the maximum accuracy that can be achieved by this technique; secondly, this technique will be more and more broadly employed for outdoor 3D model acquisition, which is a great challenge; finally, most shape from stereo algorithms assume that an object or a scene is lambertian under constant illumination, which is certainly not true for most surfaces in practice. Therefore, it is important to know whether this technique can recover a high quality 3D model of an object with arbitrary surface reflectance properties under real lighting conditions. Due to the accumulation of solid research results and many years' experience, it is firmly believed that multi-view stereo technique will be greatly advanced in the future.

CONCLUSION

This chapter gives a brief introduction to the multi-view stereo technique, ranging from camera calibration, image acquisition to various reconstruction algorithms. Several hundreds of reconstruction algorithms have been designed and applied for various applications and can be mainly categorized into four classes. The underlying theory and pipeline of each class are explained in detail and the properties of each class are also analyzed and validated by the evaluation results on the published benchmarks. Although we are still far away from the dream to recover a 3D model of an arbitrary object from multi-view automatically, multi-view stereo technique provides us a powerful alternative to acquire complex 3D models from real world. This technique has become more powerful in recent years, which has been confirmed by evaluation results on the introduced benchmarks.

REFERENCES

Boykov, Y., & Kolmogorov, V. (2003). *Computing geodesics and minimal surfaces via graph cuts*. In International Conference on Computer Vision *2003*.

Bradley, D., Boubekeur, T., & Heidrich, W. (2008). *Accurate multi-view reconstruction using robust binocular stereo and surface meshing*. IEEE Computer Society Conference on Computer Vision and Pattern Recognition.

Campbell, D. F., Vogiatzis, G., Hernández, C., & Cipolla, R. (2008). Using multiple hypotheses to improve depth-maps for multi-view stereo. In *Proceedings 10th European Conference on Computer Vision, LNCS 5302*, (pp. 766-779).

Cheung, K. M., Kanade, T., Bouguet, J., & Holler, M. (2000). A real time system for robust 3D voxel reconstruction of human motions. *IEEE Computer Society Conference on Computer Vision and Pattern Recognition, 2,* 714-720.

Eisert, P., Steinbach, E., & Girod, B. (2000). Automatic reconstruction of stationary 3-D objects from multiple uncalibrated camera views. *IEEE Transactions on Circuits and Systems for Video Technology*, *10*(2), 261–277. doi:10.1109/76.825726

Fitzgibbon, A. W., Cross, G., & Zisserman, A. (1998). Automatic 3D model construction for turn-table sequences. *Lecture Notes in Computer Science*, *1506*, 155–170. doi:10.1007/3-540-49437-5_11

Forsyth, D. A. (2001). Shape from texture and integrability. *International Conference on Computer Vision*, (pp. 447-452).

Fromherz, T., & Bichsel, M. (1995). *Shape from multiple cues: Integrating local brightness information*. International Conference for Young Computer Scientists.

Furukawa, Y., & Ponce, J. (2006). *3D photography dataset*. Retrieved from http://www.cs.washington.edu/ homes/ furukawa/ research/ mview/ index.html

Furukawa, Y., & Ponce, J. (2007). Accurate, dense, and robust multi-view stereopsis. *IEEE Computer Society Conference on Computer Vision and Pattern Recognition*, (pp. 1-8).

Furukawa, Y., & Ponce, J. (2008). Carved visual hulls for image-based modeling. *International Journal of Computer Vision*, *81*(1), 53–67. doi:10.1007/s11263-008-0134-8

Furukawa, Y., & Ponce, J. (2010). Accurate, dense, and robust multi-view stereopsis. *IEEE Transactions on Pattern Analysis and Machine Intelligence*, *32*(8), 1362–1376. doi:10.1109/TPAMI.2009.161

Fusiello, A., Trucco, E., & Verri, A. (2000). A compact algorithm for rectification of stereo pairs. *Machine Vision and Applications*, *12*(1), 16–22. doi:10.1007/s001380050120

Goesele, M., Curless, B., & Seitz, S. M. (2006). Multi-view stereo revisited. *IEEE Computer Society Conference on Computer Vision and Pattern Recognition*, (pp. 2402-2409).

Goesele, M., Snavely, N., Curless, B., Hoppe, H., & Seitz, S. M. (2007). Multi-view stereo for community photo collections. *International Conference on Computer Vision*, (pp. 1-8).

Habbecke, M., & Kobbelt, L. (2007). A surface-growing approach to multi-view stereo reconstruction. *IEEE Computer Society Conference on Computer Vision and Pattern Recognition*, (pp. 1-8).

Hernandez Esteban, C., & Schmitt, F. (2004). Silhouette and stereo fusion for 3d object modeling. *Computer Vision and Image Understanding*, *96*(3), 367–392. doi:10.1016/j.cviu.2004.03.016

Jancosek, M., Shekhovtsov, A., & Pajdla, T. (2009). Scalable multi-view stereo. *International Conference on 3D Digital Imaging and Modeling*, (pp. 1526-1533).

Kazhdan, M., Bolithp, M., & Hoppe, H. (2006). Poisson surface reconstruction. *Eurographics Symposium on Geometry Processing*, (pp. 61-70).

Kolev, K., Klodt, M., Brox, T., & Cremers, D. (2009). Continuous global optimization in multivew 3D reconstruction. *International Journal of Computer Vision*, *84*(1), 80–96. doi:10.1007/s11263-009-0233-1

Kolmogorov, V., & Zabih, R. (2002). Multi-camera scene reconstruction via graph-cut. *European Conference on Computer Vision, 3*, (pp. 82-96).

Lander, P. (1998). *A multi-camera method for 3D digitization of dynamic, real-world events.* PhD dissertation, Carnegie Mellon University, Pittsburgh, PA.

Laurentini, A. (1994). The visual hull concept for silhouette based image understanding. *IEEE Transactions on Pattern Analysis and Machine Intelligence*, *16*(2), 150–162. doi:10.1109/34.273735

Li, J., Li, E., Chen, Y., Xu, L., & Zhang, Y. (2010). Bundled depth-map merging for multi-view stereo. *IEEE Computer Society Conference on Computer Vision and Pattern Recognition*, (pp. 2769-2776).

Liu, Y., Cao, X., Dai, Q., & Xu, W. (2009). Continuous depth estimation for multi-view stereo. *IEEE Computer Society Conference on Computer Vision and Pattern Recognition*, (pp. 2121-2128).

Lobay, A., & Forsyth, D. A. (2006). Shape from texture without boundaries. *International Journal of Computer Vision*, *67*(1), 71–91. doi:10.1007/s11263-006-4068-8

Lorensen, W. E., & Cline, H. E. (1987). Marching cubes: A high resolution 3D surface construction algorithm. *International Conference on Computer Graphics and Interactive Techniques, 21*, (pp. 163-169).

Malladi, R., Sethian, J. A., & Vemuri, B. C. (1995). Shape modeling with front propagation: A level set approach. *IEEE Transactions on Pattern Analysis and Machine Intelligence*, *17*(2), 158–175. doi:10.1109/34.368173

Matsumoto, Y., Terasaki, H., Sugimoto, K., & Arakawa, T. (1997). A portable three-dimensional digitizer. *International Conference on Recent Advances in 3D Imaging and Modeling*, (pp. 197-205).

Matusik, W., Buehler, C., & McMillan, L. (2001). Polyhedral visual hulls for real-time rendering. *12th Eurographics Workshop on Rendering*, (pp. 115-125).

Matusik, W., Buehler, C., Raskar, R., Gortler, S., & McMillan, L. (2000). Image-based visual hulls. *International Conference on Computer Graphics and Interactive Techniques*, (pp. 369-374).

Narayanan, P., Rander, P., & Kanade, T. (1998). Constructing virtual worlds using dense stereo. *International Conference on Computer Vision*, (pp. 3-10).

Park, S.-Y., & Subbarao, M. (2005). A multiview 3D modeling system based on stereo vision techniques. *Machine Vision and Applications, 16*, 148–156. doi:10.1007/s00138-004-0165-2

Potmesil, M. (1987). Generating Octree models of 3D objects from their silhouettes in a sequence of images. *Computer Vision Graphics and Image Processing, 40*(1), 1–29. doi:10.1016/0734-189X(87)90053-3

Seitz, S. M., Curless, B., Diebel, J., Scharstein, D., & Szeliski, R. (2006). A comparison and evaluation of multi-view stereo reconstrucion algorithms. *IEEE Computer Society Conference on Computer Vision and Pattern Recognition, 1,* (pp. 519-526).

Seitz, S. M., Curless, B., Diebel, J., Scharstein, D., & Szeliski, R. (2009). *Dino and Temple datasets*. Retrieved from http://vision.middlebury.edu/mview/

Shlyakhter, I., Rozenoer, M., Dorsey, J., & Teller, S. (2001). Reconstructing 3D tree models from intrumented photographs. *IEEE Computer Graphics and Applications, 21*(3), 53–61. doi:10.1109/38.920627

Snow, D., Viola, P., & Zabih, R. (2000). Exact voxel occupancy with graph cuts. *IEEE Computer Society Conference on Computer Vision and Pattern Recognition*, (pp. 345-353).

Song, P., Wu, X., & Wang, M. Y. (2009). A robust and accurate method for visual hull computation. *IEEE International Conference on Information and Automation*, (pp. 784-789).

Song, P., Wu, X., & Wang, M. Y. (2010). Volumetric stereo and silhouette fusion for image-based modeling. *The Visual Computer Journal, 26*(12), 1435–1450. doi:10.1007/s00371-010-0429-y

Strecha, C., von Hansen, W., Van Gool, L., Fua, P., & Thoennessen, U. (2008). On benchmarking camera calibration and multi-view stereo for high resolution imagery. *IEEE Computer Society Conference on Computer Vision and Pattern Recognition*, (pp. 2838-2845).

Strecha, C., von Hansen, W., Van Gool, L., Fua, P., & Thoennessen, U. (2010). *Outdoor dataset.* Retrieved from http://cvlab.epfl.ch/ ~strecha/ multiview/ denseMVS.html

Szeliski, R. (1993). Rapid octree construction from image sequences. *Computer Vision, Graphics, and Image Processing . Image Understanding, 58*(1), 23–32. doi:10.1006/ciun.1993.1029

Tarini, M., Callieri, M., Montani, C., Rocchini, C., Olsson, K., & Persson, T. (2002). *Marching intersections: An efficient approach to shape-from-silhouette* (pp. 283–290). Vision, Modeling, and Visualization.

Tran, S., & Davis, L. (2006). 3D surface reconstruction using graph-cuts with surface constraints. *European Conference on Computer Vision*, (pp. 219-231).

Treuille, A., Hertzmann, A., & Seitz, S. (2004). Example-based stereo with general BRDFs. *European Conference on Computer Vision, 2,* (pp. 457-469).

Vogiatzis, G., Hernández, C., Torr, P. H. S., & Cipolla, R. (2007). Multi-view stereo via volumetric graph-cuts and occlusion robust photo-consistency. *IEEE Transactions on Pattern Analysis and Machine Intelligence, 29*(12), 2241–2246. doi:10.1109/TPAMI.2007.70712

Vogiatzis, G., Torr, P. H. S., & Cipolla, R. (2005). Multi-view stereo via volumetric graph-cuts. *IEEE Computer Society Conference on Computer Vision and Pattern Recognition, 2,* (pp. 391-398).

Vu, H., Keriven, R., Labatut, P., & Pons, J.-P. (2009). Towards high-resolution large-scale multi-view stereo. *IEEE Computer Society Conference on Computer Vision and Pattern Recognition*, (pp. 1430-1437).

Xu, C., & Prince, J. L. (1998). Snakes, shapes, and gradient vector flow. *IEEE Transactions on Image Processing*, 359–369.

Yemeza, Y., & Schmitt, F. (2004). 3D reconstruction of real objects with high resolution shape and texture. *Image and Vision Computing, 22*, 1137–1153. doi:10.1016/j.imavis.2004.06.001

Zaharescu, A., Boyer, E., & Horaud, R. (2007). TransforMesh: A topology-adaptive mesh-based approach to surface evolution. *Proceedings of the 8th Asian Conference on Computer Vision, 2*, (pp. 166-175).

Zhang, R., Tsai, P.-S., Cryer, J. E., & Shah, M. (1999). Shape from shading: A survey. *IEEE Transactions on Pattern Analysis and Machine Intelligence, 21*(8), 690–706. doi:10.1109/34.784284

ADDITIONAL READING

Bouguet, J.-Y. (2010). Camera calibration toolbox for matlab. http://www.vision. caltech.edu/ bouguetj/ calib_doc

Forsyth, D. A. (2001). Shape from texture and integrability. *International Conference on Computer Vision*, 447-452.

Horn, B. K. P. (1970). *Shape from Shading: A Method for Obtaining the Shape of a Smooth Opaque Object from One View*. PhD dissertation, Massachusetts Institute of Technology, Boston.

Laurentini, A. (1994). The Visual Hull Concept for Silhouette Based Image Understanding. *IEEE Transactions on Pattern Analysis and Machine Intelligence, 16*(2), 150–162. doi:10.1109/34.273735

Niem, W. (1999). Automatic reconstruction of 3D objects using a mobile camera. *Image and Vision Computing, 17*, 125–134. doi:10.1016/S0262-8856(98)00116-4

Seitz, S. M., Curless, B., Diebel, J., Scharstein, D., & Szeliski, R. (2006). A comparison and evaluation of multi-view stereo reconstrucion algorithms. *IEEE Computer Society Conference on Computer Vision and Pattern Recognition*, 1, 519-526.

Strecha, C., von Hansen, W., Van Gool, L., Fua, P., & Thoennessen, U. (2008). On Benchmarking Camera Calibration and Multi-View Stereo for High Resolution Imagery. *IEEE Computer Society Conference on Computer Vision and Pattern Recognition*, 2838-2845.

KEY TERMS AND DEFINITIONS

Benchmark: Something whose quality or quantity is known and which can therefore be used as a standard with which other things can be compared.

Camera Calibration: The process of finding the intrinsic and extrinsic parameters of the camera that took photographs.

Image Processing: A technique in which the data from an image are digitized and various mathematical operations are applied to the data in order to create an enhanced image that is more useful or pleasing to a human observer, or to perform some of the interpretation and recognition tasks usually performed by humans.

Multi-View Stereo Reconstruction: A shape reconstruction technique that tries to extract the 3D shape of a scene from two or more images taken at known camera positions by stereo matching different images.

Shape from Silhouette: A shape reconstruction technique by intersection of the visual cones formed by back projecting the silhouettes in the corresponding images.

Visual Hull: An approximate shape representation of an object created by shape from silhouette 3D reconstruction technique.

Chapter 3
Forward Projection for Use with Iterative Reconstruction

Raja Guedouar
Higher School of Health Sciences and Technics of Monastir, Tunisia

Boubaker Zarrad
Higher School of Health Sciences and Technics of Monastir, Tunisia

ABSTRACT

Modelling the forward projection or reprojection, that is defined as the operation that transforms a 3D volume into series of 2D set of line integrals, is of interest in several medical imaging applications as iterative tomographic reconstruction (X-ray, Computed Tomography [CT], Positron Emission Tomography [PET], Single Photon Emission Computed Tomography [SPECT]), dose-calculation in radiotherapy and 3D-display volume-rendering. As forward projection is becoming widely used, iterative reconstruction algorithms and their characteristics may affect the reconstruction quality; its accuracy and performance needs more attention. The aim of this chapter is to show the importance of the modelling of the forward projection in the accuracy of medical tomographic data (CT, SPECT and PET) reconstructed with iterative algorithms. Therefore, we first present a brief overview on the iterative algorithms used in tomographic reconstruction in medical imaging. Second, we focus on the projection operators. Concepts and implementation of the most popular projection operators are discussed in detail. Performance of the computer implementations is shown using the well-known Shepp_Logan phantom. In order to avoid possibly confounding perspective effects implied by fan or cone-beam, this study is performed in parallel acquisition geometry.

DOI: 10.4018/978-1-61350-326-3.ch003

INTRODUCTION

Tomographic reconstruction is the technique underlying nearly all of the key diagnostic imaging modalities, including X-ray, CT, PET, SPECT, some acquisition methods for Magnetic Resonance Imaging (MRI), and newly emerging techniques such as electrical impedance tomography (EIT) and optical tomography. During the last decades, various algorithms have been developed for both 2D and 3D tomographic reconstruction such as the analytical and the iterative methods. The analytical algorithms, the most used, have advantage to be fast, but they are not able to model the characteristics of the data acquisition process. While the iterative algorithms are able to precisely model the physical and statistical characteristics of the data acquisition process, independent of the dimensionality of the image. The ability to perform accurate iterative reconstruction relies fundamentally on the modelling of the forward projection. Some examples where modelling the forward projections have been found worthwhile to explore include: Redundant data, better noise models, incomplete data, resolution recovery, beam hardening correction and metal artifact reduction.

In general, more detailed models result in higher image quality but also in higher computational load, which can become especially cumbersome in 3D problems. Some of numerical methods for implementing forward and backprojections reduce total processing time by simplifying the process used in determining the actual value to be backprojected or reprojected but they result in varying degrees of approximation errors. These simplifications and approximations limit the absolute accuracy of the reconstruction, contribute to image reconstruction errors and may negate the advantages of an iterative reconstruction. Conversely, more accurate interpolation techniques tend to impose added requirements of the reconstruction algorithms, and thus longer processing times. As the projection is becoming widely used with iterative reconstruction algorithms and their characteristics may affect the reconstruction quality, its accuracy and performance needs more attention for better understanding. In this context, this chapter aims to detail the implementing of forward projection using the most models that are frequently used in medical tomography reconstruction with focus on errors generated by the geometrical models. It is structured as follows. After this introduction, a brief overview on the iterative algorithms used in tomographic reconstruction in medical imaging is first presented. Second, the concepts and implementation of the most popular projection operators will be detailed. Their performances have been shown using the well-known Shepp_Logan phantom (Shepp & logan, 1994).

ITERATIVE RECONSTRUCTION TECHNIQUES

During the last decades, various algorithms have been proposed for both 2D and 3D tomographic reconstruction such as the analytical and the iterative methods. The analytical algorithms, the most used, have advantage to be fast, but they are not able to model the characteristics of the data acquisition process. Iterative tomographic reconstruction which is the process of recovering 3D image data from a set of integrals of that data over 2D subspaces, provide an attractive solution for tomographic imaging modalities over analytic techniques and they have been successfully used in medical imaging (Ziegler, 2008; Suetens, 2002), including computed tomography (CT), single photon emission computed tomography (SPECT), positron emission tomography (PET), tomosynthesis and projection mode 2D magnetic resonance imaging (MRI). The iterative methods aim to minimize or maximize a cost function between reconstructed slices $\boldsymbol{T}$ and measured projection $\boldsymbol{P}$ and have the advantage to incorporate imaging geometry and physics effects into

Figure 1. Representation of reconstruction steps at n^{th} iteration with iteration process. The notions used are: $\hat{T}^{(n)}$ is the reconstructed image, P and $\hat{P}^{(n)}$ are the measured and the calculated projection data, R and R^t are the forward and back projection matrix, ε_p^n projection error that measured the discrepancy between P and $\hat{P}^{(n)}$, and ε_T^n is its backprojected image.

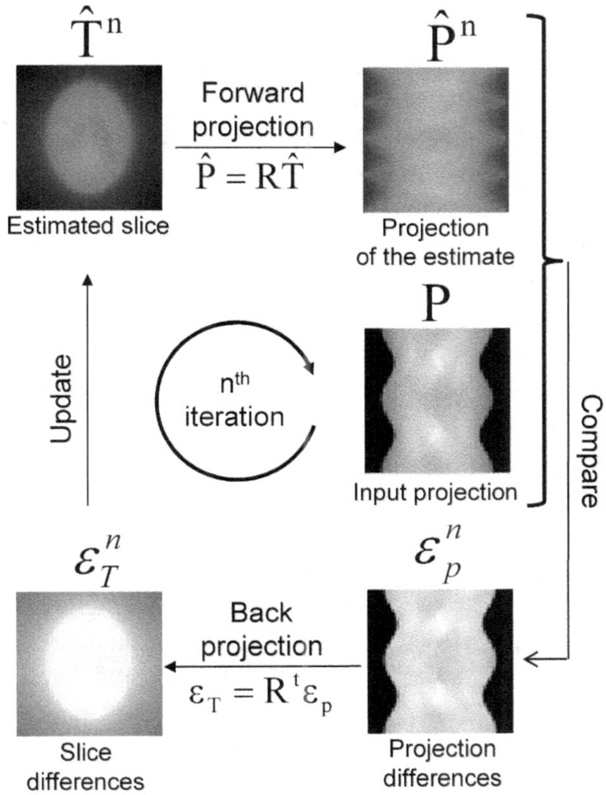

the forward projection operator ***R*** that results in quantitatively improved reconstruction images. All iterative methods begin with initial guess for solution and successively improve it until solution is as accurate as desired. In theory, infinite number of iterations might be required to converge to exact solution. In practice, iteration terminates when some measure of error is as small as desired. Figure 1 illustrates steps of implementation of an iterative reconstruction algorithm where both forward projection matrix and back projection matrix (the reverse model of forward projection) are needed to achieve one iteration.

A large variety of iterative techniques are proposed and applied in medical tomographic reconstruction that differ from each other in the way the correction terms are derived and how the update to the new estimate is calculated. Iterative algorithms can be classified mainly into two classes (Vandenberghe et al., 2001): conventional algebraic reconstruction techniques and iterative statistical methods.

Conventional Algebraic Reconstruction Techniques

Conventional algebraic reconstruction techniques aim to minimize weighted square norms (Jiang and Wang, 2003). The oldest of this family is due to S. Kaczmarz works and is known as the Algebraic

Reconstruction Technique (ART) (Gordon, 1973; Gordon & Herman, 1970; Herman, 2009) which has a simple intuitive basis. Each projected density is thrown back across the reconstruction space in which the densities are iteratively modified to bring each reconstructed projection into agreement with the measured projection. Assuming that the pattern being reconstructed is enclosed in a square space of n x n array of small pixels contain grayness or density number, which is uniform within the pixel but different from other pixels. A "ray" is a region of the square space which lies between two parallel lines. The weighted ray sum is the total grayness of the reconstruction figure within the ray. The projection at a given angle is then the sum of non-overlapping, equally wide rays covering the figure. The ART algorithm consists of altering the grayness of each pixel intersected by the ray in such a way as to make the ray sum agree with the corresponding element of the measured projection.

Other versions of these algorithms are Simultaneous Algebraic Reconstruction Technique (SART) (Gilbert, 1972), Simultaneous Iterative Reconstruction Technique (SIRT) (Gilbert, 1972) and the Iterative Least-Squares Technique (ILST) (Goitein, 1972). Conceptually, these techniques differ in the procedure of updating $\hat{T}_i^{(n)}$. Beginning with initial guess $T^{(0)}$, ART solves one measurement at a time by updating all corresponding (image) pixels or voxels using the following equation:

$$\hat{T}_i^{(n+1)} = \hat{T}_i^{(n)} + a'_{ik} \cdot \left(\left(P_k - \sum_{i=1}^{N} a_{ik}\hat{T}_i^{(n)}\right) \Big/ \sum_{k=1}^{M} a_{ik}^{2}\right) \tag{1}$$

where $\hat{T}_i^{(n)}$ is the value of reconstructed image at the pixel i for the nth iteration, N is the total number of reconstructed image pixels, P_k is the measured projection data at k^{th} bin, M is the total number of projection image bins, a_{ik} and a'_{ik} are the forward and back projector weighting coefficients, respectively, that map the ith pixel to k^{th} bin. While, the Additive Simultaneous Iterative Reconstruction Technique (ASIRT) (P. Gilbert, 1972) is a version of SART, computes next iteration by solving for each component of T as below:

$$\hat{T}_i^{(n+1)} = \hat{T}_i^{(n)} + \left(1 \Big/ \sum_{k=1}^{M} a'_{ik}\right) \cdot \sum_{k=1}^{M}\left(a'_{ik}\left(\left(P_k - \sum_{i=1}^{N} a_{ik}\hat{T}_i^{(n)}\right) \Big/ \sum_{i=1}^{N} a_{ik}\right)\right) \tag{2}$$

Statistical Image Reconstruction Methods

Statistical image reconstruction methods reconstruct images by iteratively maximizing a likelihood function (Nuyts, 2001; Green, 1990; Herbert & Leahy, 1989). They take the noise on the measurement data into account. Therefore they use a statistical modelling of the measurement process. The best-known example is the Maximum Likelihood (ML-EM) algorithm (Shepp and Vardi, 1982; Vardi et al., 1985) that takes the Poisson nature of the data into account according to the following formula:

$$\hat{T}_i^{(n+1)} = \frac{\hat{T}_i^{(n)}}{\sum_{k=1}^{M} a_{ik}} \cdot \sum_{k=1}^{k} a_{ik} \left(\frac{P_k}{\sum_{i=1}^{N} a'_{ik}\hat{T}_i^{(n)}} \right) \tag{3}$$

Other examples of these techniques are the Maximum A Posteriori (MAP) method mostly used to guarantee good noise reduction and edge preservation (Alenius & Ruotsalainen, 1997; Alenius et al. 1998; Herbert & Leahy, 1989), the convex method (Lange & Fessler, 1995) and the ordered subsets convex (OSC) method (Kamphuis & Beekman, 1998). These methods are known to produce images with better signal to noise ratio at the cost of increased computation time, and many recent developments toward faster methods make

Algorithm 1.

0: Algorithm "Iterative reconstruction"

1: Initiation: make a guess on the data to be reconstructed (usually assuming that all pixels have the same value) ($\hat{T}$), set the iteration index it = 0,

2: Forward projection: Estimate projection data based on the current guess at n^{th} iteration $\hat{P}_k^{(n)} = \sum_{i=1}^{N} a_{ik} \hat{T}_i^{(n)}$ where a_{ik} is the coefficient of the forward projection operator that maps the ith data pixel to k^{th} projection bin;

3: Comparison: calculate the discrepancy ε_p (error) between acquired projections and reprojected ones;

4: Backprojection: backproject the discrepency ε_p over the image space: $\varepsilon_{T_i}^{(n)} = \sum_{k=1}^{M} a'_{ik} \varepsilon_{pk}^{(n)}$ where a'_{ik} is the coefficient of the backprojection operator that maps the kth projection bin to i^{th} data pixel;

5: Modification: update the current data by incorporating weighted backprojection in a specific way according to the defined algorithm (addition or multiplication);

6: Evaluation: evaluate the reconstruction error between $\hat{T}_i^{(n-1)}$ and $\hat{T}_i^{(n)}$, if the error is not sufficiently small, set it = it+1, and repeat steps 2 to 5 ;

7: end "Iterative reconstruction"

these methods promising as using the simplified cost function (T. Köhler, 2003; Thibault, 2007).

A third class of iterative algorithms called the Iterative Filtered Backprojection (IFBP) methods can be considered (Xu et al., 1993). These methods are based on iterative algebraic application of Filtered Backprojection (FBP) methods. For IFBP, the step of backprojection in equation (1) is replaced by a filter followed by a backprojector, the same operation normally performed in FBP. Instead of converging to the least squares solution, IFBP converges to a weighted least-squares solution with the reconstruction filter being the weighting function. (Xu et al., 1993; Lalush & Tsui, 1993). Since the FBP method is used in each iteration, certain artifacts are very rapidly suppressed. Therefore, for the purpose of suppressing such artifacts, IFBP methods are usually much faster than other iterative methods.

Below follows a brief algorithmic description of iterative reconstruction implementation (Algorithm 1):

The advantage of the iterative algorithms are that they are able to precisely model the physical and statistical characteristics of the data acquisition process, independent of the dimensionality of the image, and can easily accommodate any data acquisition geometry. Their major disadvantages are that the processing is time consuming and the computational burden is high since one projection and one backprojection operation (the reverse of forward projection) have to be performed at each iteration. Moreover, the accuracy of iterative reconstructed images is dependant highly on the choice and the implementation of these operations that require a model for the imaging system at hand. A variety of efficient forward and backprojection algorithms are currently available in clinical, in industrial, and research-oriented applications of tomography, and they differ in accuracy and computational speed. Some of numerical methods for implementing forward and backprojections reduce total processing time by simplifying the process of interpolation used in determining the actual value to be backprojected or reprojected but they result in varying degrees of approximation errors. These simplifications and approximations limit the absolute accuracy of the reconstruction, contribute to image reconstruction errors and may negate the advantages of an iterative scheme. Conversely, more accurate interpolation techniques tend to impose added requirements of the reconstruction algorithms, and thus longer processing times. Therefore, the selection of an appropriate projection model for a specific application requires the knowledge of the method's accuracy and computational complexity. In literature, there are many papers considering the reconstruction algorithms and reporting empirical comparisons of various approaches, but the algorithm implementation and the effects of the forward projection matrix are not often described in detail. Only few papers have recently described them (De Man and Basu, 2004). However, less attention has been made to characterize errors generated by geometrical projection modelling. In the following, the concepts and the algorithm implementation of forward projection operator are discussed in detail.

Forward Projection Operators

A forward projection or reprojection, that is defined as the operation that transforms a 3D volume into a series of 2D set of line integrals, is of interest in several medical imaging applications like (Boag, 2000) as iterative tomographic reconstruction (CT, SPECTand PET) (Lewitt, 2003; Ollinger, 1990; Ziegler, 2008; Zeng, 1994), dose-calculation in radiotherapy (Bortfeld, 1994) and 3D-display volume-rendering (Chidlow, 2003). It is also useful in industrial and research-oriented applications of tomography. The projector is a system matrix of weighting coefficients a_{ik} that maps the image pixels T_i to projection bin P_k and models the imaging process as:

$$P_k = \sum_{i=1}^{N} a_{ik} T_i \tag{4}$$

It is the key element in calculating projection data from a discretized image. The main issue is how to evaluate the contribution of a given pixel from an imaged data in a projection bin from the obtained projection. The accurate calculation of projection matrix is probably the most important step in iterative tomography reconstruction algorithms, in which repeated applications of the forward and reverse model are used to solve for the image that best fits the measurements according to an appropriate objective function. It defines (1) how the continuous function to be estimated is represented by a finite set of parameters; and (2) how projection data are calculated from this continuous function. These require the modelling of the projection matrix including the geometry of the reconstruction problem and a number of other physical parameters. A variety of efficient models have been proposed to simulate the tomography projection process. Some of the methods can be described as procedures for forward projection but they are not all based on explicit models. Another family of methods is based on basis function of intensity coefficient distribution (Herman, 1976; Lewitt, 2003). Generally, forward projection models are varying on the choice of the image basis function that models the voxel shapes and the integration function that is related to the acquisition geometry.

The choice of basis function affects the result of an iterative method. A good basis function should (1) be able to accurately represent a constant function; (2) allow for cost-effective implementation of forward projection and backprojection operations; and (3) contain a minimal amount of aliasing artifacts. A lot of basic functions have been investigated include the following: square

Figure 2. Representation of common geometrical integration functions (a) linear integration (b) strip integration

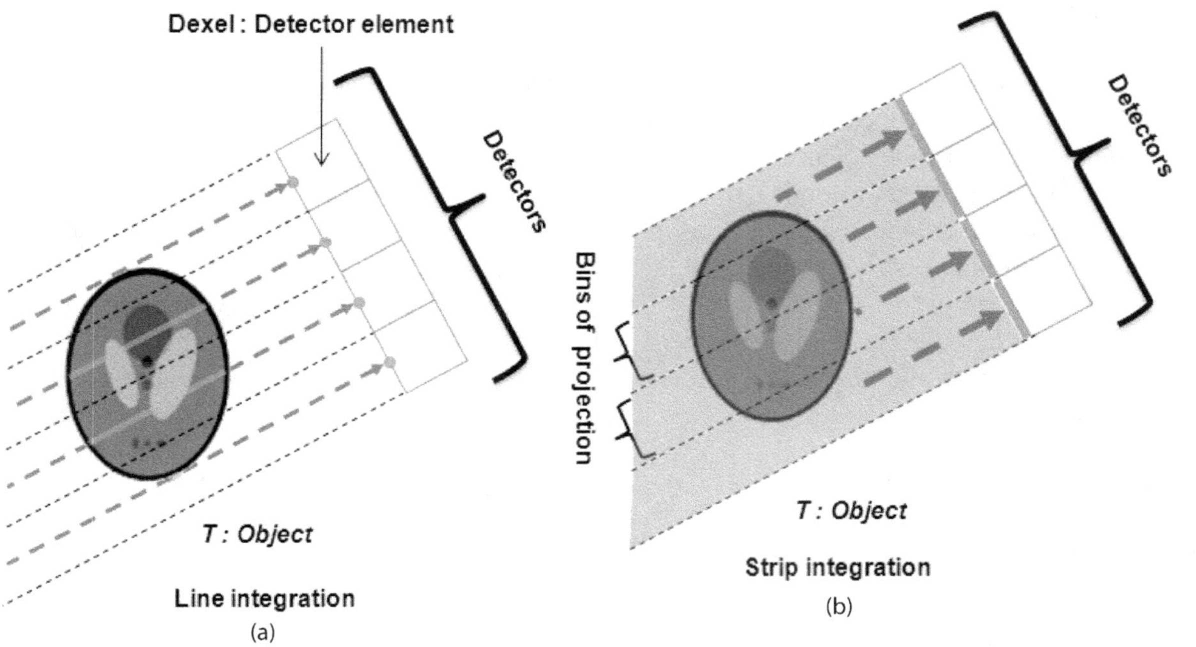

basis function (Peters, 1981; Schwinger et al., 1986; Thibault, 2007), Fourier series, circular harmonics, wavelets, “natural pixels”, B-splines, Dirac impulses, Gaussian functions and organ-based basis functions. Other related representations include polygons, polar grids, logarithmic polar grids, tetrahedral meshes and rotationally symmetric basis functions (Lewitt, 1992; Matej, 1996). Kaiser-Bessel functions called ‘blobs’ that consider non zero values only in a circular disk around the origin, and smoothly decreases from a positive value at the origin to zero at the edge of the disk, have been a particularly popular choice of rotationally symmetric basis. Although in the context of SPECT imaging, blobs were not found to be advantageous (Yendiki, 2004), more favorable results have been reported in CT and PET (Lewitt, 1992; Matej, 1996; Ziegler, 2008). Naturally, the fineness of the grid can affect edge artifacts and aliasing (De Man, 2000; Zbijewski, 2006). More recent papers, have enhanced and augmented these basic approaches, and reader is referred to (De Man & Basu, 2004) for a more complete list of references.

The most common type of geometrical integration function is a Dirac line (Figure 2 (a)) transforming the volume integral into a line integral along the line corresponding to measurement (Cormack, 1964). With this configuration, aliasing may occur during projection in particularly with high voxel density (Hsieh et al., 1998). Therefore, using other types of integration functions that consists either of several Dirac lines or a strip (Figure 2 (b)), can be used for suppressing aliasing in the projection generation process.

Regarding all these possibilities, a variety of forward projection models can be defined as a combination between a selected basis functions and an appropriate geometrical integration function. Some models consider that the intensity within each pixel is uniformly distributed whereas others assume that it is concentrated at pixel center. Moreover, the projection may be performed either as line integrals or over finite width paths (strip projection bin). This is where the various models differ from each other in mostly three ways:

- Either they assume that the voxels are solid blocks or that the voxels are infinitesimal thin spikes (or sample points).
- Either they trace rays emerging from the bins, or a few sub-bins within a pixel, or they trace beams, usually bounded by the bin boundaries.
- Either they trace the rays (or beams) across the volume from the bins or they project the voxels onto the projection plane.

FORWARD PROJECTION ALGORITHMS: CONCEPT AND IMPLEMENTATION

In this section, we propose to detail the implementation of the forward projection matrix using models that are frequently used in medical tomographic reconstruction today with an iterative scheme. In order to avoid possibly confounding perspective effects implied by fan or cone-beam, all algorithms are performed in parallel acquisition geometry. Although we only show the implementation for the 2D case, for 3D rendering, the drawings would extend into 3D (which turns every linear interpolation into a bilinear interpolation).

To simplify illustrations, all implementations will be shown in the case of 2D functions assuming that each image element (pixel or voxel) value will be distributed into two adjacent projection bins using the notation defined in Figure 3. To be noted that, the pixel value can be distributed in maximum into three adjacent projection bins. For all projection methods, the backprojection is defined as the transpose operation and the weight factors a_{ik} remain the same, but the detector values are weighted and assigned to the image pixels as:

$$T_i = \sum_{k=1}^{M} a_{ki} P_k \tag{5}$$

where M is the total number of bins.

Since computational time is not important because of the availability of fast processer, therefore, we neglect it in our study. We pre-computed system matrix of projection and store it in random access memory. These techniques have better computational efficiency and an even greater advantage in 2D over techniques that calculate coefficients in real time and at the same time do reconstruction task. The major drawback of these techniques is in 3D where they require huge memory storage.

Two approaches for implementing system matrix with iterative algorithms are proposed and used: one approach is to pre-compute it beforehand and store in random access memory and the other one is to calculate its coefficients on the fly at the same time as of the reconstruction task. The computed method on fly is well suited for hardware implementation because no coefficient is stored after being calculated and used. However, this approach increases computational time since an additional step is added for each projection bins and at each iteration. If the system matrix needs to be computed only once as is often the case in PET, computation time is not an issue and the weighting coefficients can be calculated on the fly (real time) and at the same time do reconstruction task. The pre-computing techniques have better computational efficiency and an even greater advantage in 2D over techniques that calculate coefficients in real time. The major drawback of these techniques is in 3D because they require huge memory storage. However, some geometric acquisitions present interesting symmetric properties and the effective number of weighting coefficient need to be pre-calculated is reduced according to the symmetry degree which can considerably decrease the size of the useful memory. For example, if a parallel tomographic acquisition is done over 360 degree, the weighting functions are symmetric about 45° (number of views multiples of 4 for the 360°) and thus only need to be pre-calculated for one-eighth of the total number of projection angles between 0°

Figure 3. Representation of notations to be used to define the relation between the image pixels and projection bins: T_i: pixel i value to be reprojected, P_k, P_{k+1}: projection bins (detector elements) k and k+1 of the projected image P (sinogram), θ: projection angle, $a_{i,k}$: weighting factor of the contribution of pixel i to projection bin k computed using a given projection model.

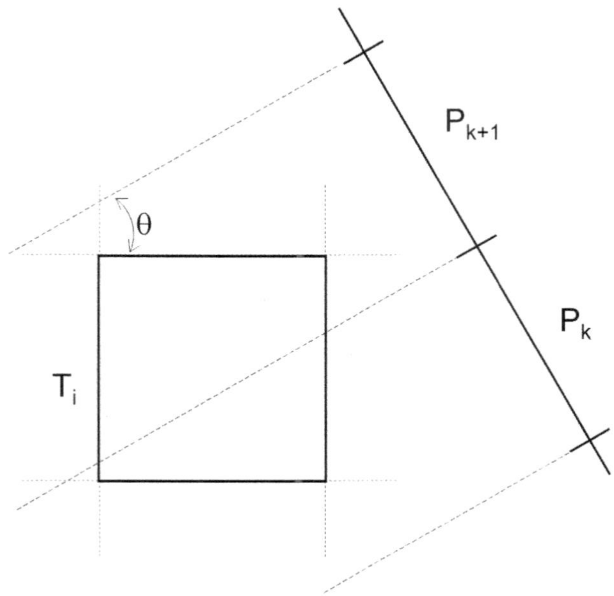

and 45° (Schwinger et al., 1986). The following steps are performed to implement the projection operation if the coefficients of the system matrix are calculated on fly (Algorithm 2).

In case of pre-calculated system matrix, for all projection angles, the corresponding weighting coefficients and their projection bins are pre-calculated and stored for each pixel, and the program looks up these pre-calculated values. To generate the backprojection, the same pre-calculated couples (weighting coefficient and the corresponding projection bin) are used to back project bin values into all pixel slices for all projection angles. For implementation, the following steps are performed (Algorithm 3).

In the following, we present the concept and the implementation of the system matrix of the most popular forward projection modelling in a unified framework under the pre-calculation approach in order to perform the projection step in iterative shame.

Ray-Driven Methods

(Herman, 1980; Siddon, 1985; Zhuang et al, 1994; Zeng and Gullberg, 1993) are perhaps the most intuitive approach to approximating the line integrals. They consist of tracing one or more equispaced ray paths through each projection bin. The total length of intersection between the ray paths and each pixel is used as weighting factor either in 2D or in 3D. The projection value for projection line k can be written as a summation:

$$P_k = \sum_{i=1}^{N} l_{i,k} * T_i$$

where $l_{i,k}$ represents an effective intersection length of projection line k with pixel i. This is illustrated in Figure 4 where each bin is divided into 2 sub-bins.

Algorithm 2.

```
Algorithm "Projection operation using weighting coefficients calulated on fly"
   for all projection angles θ do
       for all image slices do
           for all pixel slices i do
               determine the bin k in which the considered pixel i is contributed
               calculate the corresponding weighting coefficient a_ik
               update P_k=P_k+a_ik T_i
           end for
       end for
   end for
end "Projection operation on using weighting coefficients calulated on fly"
```

Algorithm 3.

```
Algorithm "Projection operation with pre-calulated weighting coefficients"
   for all projection angles θ do
       for all image slices do
           upload the matrix of weighting coefficient a_ik and the corresponding bins k from the storage table.
           for all pixel slices i do
               update P_k=P_k+a_ik T_i.
           end for
       end for
   end for
end "Projection operation with pre-calulated weighting coefficients"
```

Ray-driven methods are generally well-suited for projection, but tend to introduce artifacts (Moiré patterns) in the backprojection (De Man and Basu 2002). Their accuracy can be improved by increasing the number of ray-paths that are traced per projection bin (sub-bins) (Zhuang et al.).

Below follows a brief algorithmic description of ray-driven forward projector implementation where each bin is divided into M sub-bins (Algorithm 4).

It should be noticed that if no sub-division is considered (conventional ray-driven method), M should be replaced by 1 in this algorithm.

Ray-Driven with Linear Interpolation (Joseph's Method) (Joseph, 1983)

The coefficients are computed in 2D as the row intersection length combined with the linear interpolation between the two nearest voxels within that row, and in 3D as the slab intersection length combined with bilinear interpolation between the

Figure 4. Representation of the ray-driven method (2 sub-bins). The two corresponding weighting coefficients are: $a_{i,k} = L_3/2$ *and* $a_{i,k+1} = (L_1+L_2)/2$.

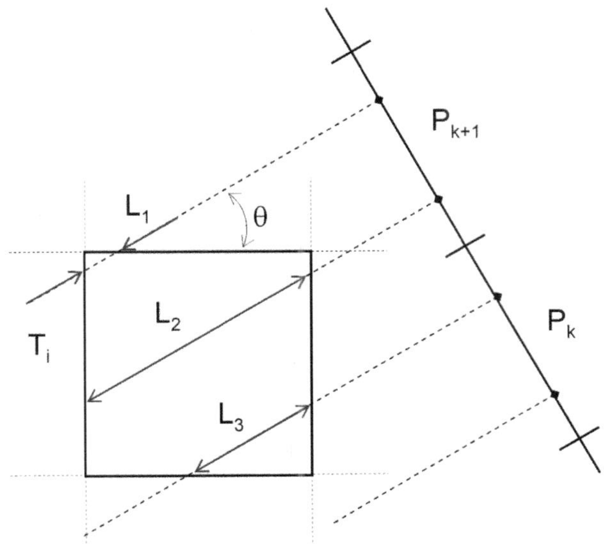

Algorithm 4.

```
0: Algorithm "Ray driven projection operator with M sub-bins"
    1: for all projection angles θ do
        2: for all image slices do
            3: for all pixel slices i do
                4: for all intersecting rays (bins) k do
                    5: for all sub-rays m do
                        6: calculate the length of intersection between the considered ray and the contributing pixel i: L_m,
                    7: end for
```

8: calculate weighting coefficient $a_{ik} = \left(\sum_{m=1}^{M} L_m\right) \Big/ M$

```
                    9: save the couple (k, a_ik) in an access memory file
                10: end for
            11: end for
        13: end for
    12: end for
11: end "Ray driven projection operator with M sub-bins"
```

Figure 5. Representation of the interpolated ray-driven method (Joseph's linear interpolation method). The two corresponding weighting coefficients corresponding to the contribution of voxels Ti to projection line k and k+1 are computed as: $a_{i,k} = l_\theta * d_1/d$ *and* $a_{i,k+1} = l_\theta * d_2/d$ *where* $d = d_1 + d_2$

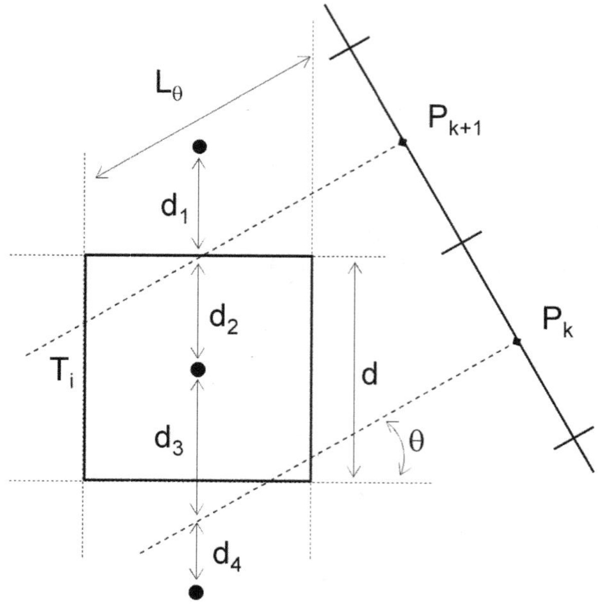

four nearest voxels within that slab. This is shown schematically in Figure 5.

To implement Joseph's method, the same steps as ray-driven algorithm are performed with some modifications (Algorithm 5).

A more general projection with a trilinear interpolation is also used in3D where the projection line is divided into a number of segments with fixed step size (Wang 1999). At each step, the contribution to projection line i is computed as the product of the step-size and a voxel value is obtained by trilinear interpolation of eight neighboring voxels. If the used step is equal to the column or raw width, this method is equivalent to Joseph's method.

Pixel-Driven Methods (Herman, 1980; Peters, 1981; Zhuang et al, 1994)

The pixel-driven method owes its name to the fact that the index of the main loop is the image pixel index. For each image pixel, the center of the pixel is projected onto the detector array along the projection direction, and a value is obtained from, or accumulated in, the detector by (typically linear) interpolation. The projection accuracy can be increased by dividing each pixel into sub-pixels and forward projection is accomplished simply by determining the projection bin within which the centre of each sub-pixel is located. This is shown schematically in Figure 6 where each pixel is divided into 2×2 sub-pixels.

Below follows a brief algorithmic description of pixel-driven forward projector implementation in case of M sub-pixels (Algorithm 6).

If no sub-division is considered (conventional pixel-driven), previous algorithm is performed with M=1. However, simple pixel-driven projection is rarely used because it introduces high-frequency artifacts (Zeng and Gullberg 1993, De Man and Basu 2002). To improve accuracy, linear interpolation is performed between dis-

Algorithm 5.

```
0: Algorithm "Ray driven projection operator with linear interpolation"
    1: for all projection angles θ do
        2: for all image slices do
            3: calculate the length of intersection between the considered ray and the contributing column i:
                if θ ∈ [-45°; 45°] or θ ∈ [135°; 225°], then Lθ=pixel width/ cos θ
                else Lθ,=pixel width/ sin θ
            4: for all pixel slices i do
                5: for all intersecting rays k do
                    6: calculate the distances from projection bin k to the center pixel i following the contributing column direction: di,k,
                    7: calculate weighting coefficient aik=(Lθx dm)/ pixel width,
                    8: save the couple (k, aik) in an access memory file.
                9: end for
            10: end for
        11: end for
    12: end for
13: end "Ray driven projection operator with linear interpolation"
```

Figure 6. Representation of the pixel-driven method (xXx sub-pixels). The two corresponding weighting coefficients corresponding to the contribution of voxels T_i to projection line k and k+1 are computed as $a_{i,k}$ = ¼ and $a_{i,k+1}$ = 3/4.

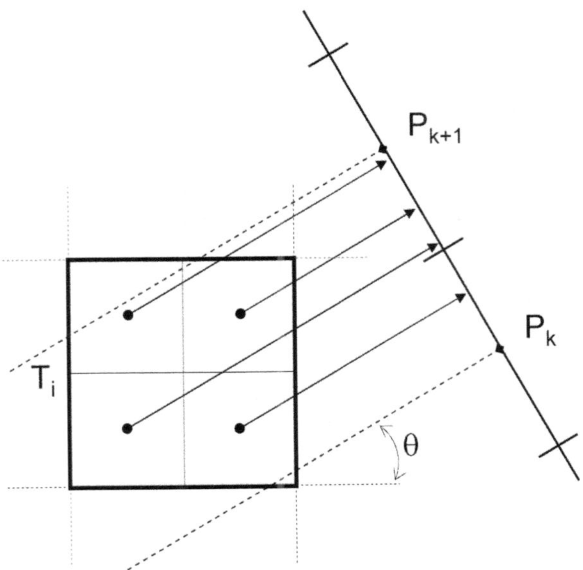

Algorithm 6.

```
0: Algorithm "Pixel driven projection operator with M sub-pixels"
     1: for all projection angles θ do
          2: for all images slices do
               3: for all pixel slices do
                    4: for all sub-pixels m do
                         5: determine bin k in which the con-
sidered sub-pixel center i is projected.
                         6: increment a_ik= a_ik +1
                    7: end for
                         8: calculate the total number
of the contributed pixel: a_ik= a_ik /M,
                         9: Save the couple (k, a_ik) in
an access memory file
               10: end for
          11: end for
     12: end for
13: end "Pixel driven projection operator with M sub-pixels"
```

tances from each pixel center to the centers of the two nearest projection bins.

Bilinear Interpolation Projection

It is the standard method for computing projections. Projections are computed by interpolation based upon the distances from each center pixel to the centers of two nearest projection bins. As with pixel-driven methods, projection accuracy can be increased by dividing each pixel into sub-pixels and applying bilinear interpolation projection to the sub-pixels. In terms of basis function, the basic functions obtained from bilinear interpolation are pyramid shaped, each with a support extending over a square region with size of four pixels. This is shown schematically in Figure 7 where each pixel is divided into 4 sub-pixels.

Projection Based Upon Square Voxel

It considers the intensity within pixel distributed uniformly in a square areas (in 2 D) or in cubic volume (in 3D) and the pixel contribution to projection bin is proportional to the intersection area between the square and the strip bin (Peters, 1981; Schwinger et al., 1986, Thibault, 2007). This is shown schematically in Figure 8.

Below follows a brief algorithmic description of square forward projector implementation (Algorithm 7).

Below follows a brief algorithmic description of square forward projector implementation (Algorithm 8).

Projection Based Upon Overlapping Circles (Disks) (Shepp, 1982; Zhuang, 1994) or Spheres (Balls) (Reyes 2007)

These consider the intensity within pixel distributed uniformly in a circular area in 2 D (or in sphere area (in 3D)) rather than square area and the pixel contribution to projection bin is proportional to the intersection area between the disk and the strip bin. The potential advantage of

Figure 7. Representation of pixel-driven method with bilinear interpolation projection. In this case, the interpolation is done between the values of the four nearest sub-bin pixels. The two weighting coefficients corresponding to the contribution of voxels Ti to projection line k and k+1 is computed as: $a_{i,k} = 1-(d_1+d_2+d_3+d_4)/(4d)$ *and* $a_{i,k+1} = (d_1+d_2+d_3+d_4)/(4d)$.

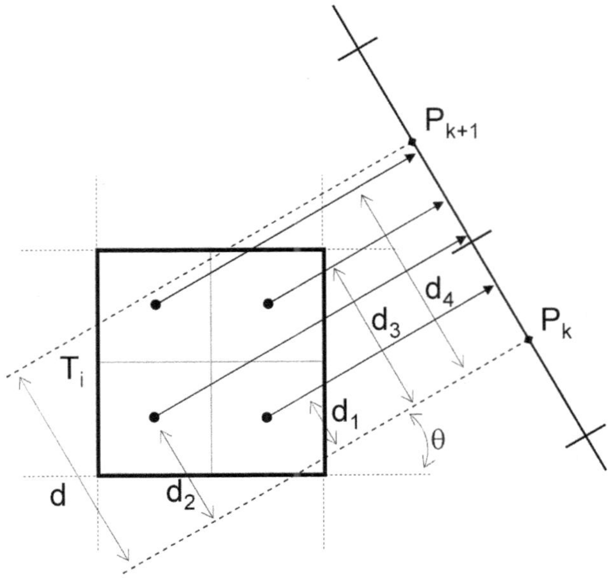

Figure 8. Representation of projection based upon square pixels. The two weighting coefficients corresponding to the contribution of voxels Ti to projection line k and k+1 is computed as: $a_{i,k} = s_2/(s_1+s_2)$ *and* $a_{i,k+1} = s_1/(s_1+s_2)$.

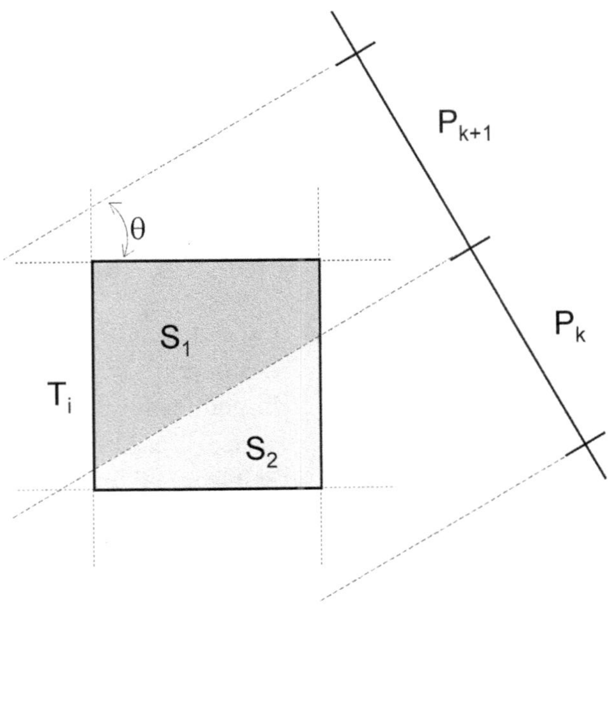

Algorithm 7.

```
0: Algorithm "Square projection operator"
    1: for all projection angles θ do
        2: for all images slices do
            3: for all pixel slices do
                4: calculate the intersection area between the pixel and the strip projection bin (trapezoid surface)
                5: calculate the weighting coefficients: a_ik= a_ik /square area
            6: end for
            7: save the couple (k, a_ik) in an access memory file
        8: end for
    9: end for
10: end "Square projection operator"
```

Algorithm 8.

```
0: Algorithm "Square projection operator"
    1: for all projection angles θ do
        2: for all images slices do
            3: for all pixel slices do
                4: calculate the intersection area between the pixel and the strip projection bin (trapezoid surface)
                5: calculate the weighting coefficients: a_ik= a_ik /square area
            6: end for
            7: save the couple (k, a_ik) in an access memory file
        8: end for
    9: end for
10: end "Square projection operator"
```

this approach is that the forward projection of a circle is independent of projection angle, while the forward projection of a square pixel is angle dependant. The use of circular pixels produces a reasonably fast projection algorithm. One only needs to identify the location of the forward projection of the center of the circular pixel. Then, the analytical computation of the portion of the circular pixel lying within each projection bin is straightforward. This method is equivalent to bi-nonlinear interpolation. To improve accuracy, each disk can be divided into xXx sub-disks. This is illustrated in Figure 9 (a) and (b) where each disk is divided into 2×2 sub-disks.

The two corresponding weighting coefficients are: $a_{i,k} = s_2/(s_1+s_2)$ and $a_{i,k+1}= s_1/(s_1+s_2)$ in case (a)

Figure 9. Representation of (a) projection based upon disks (b) projection based upon 2×2 sub-disks

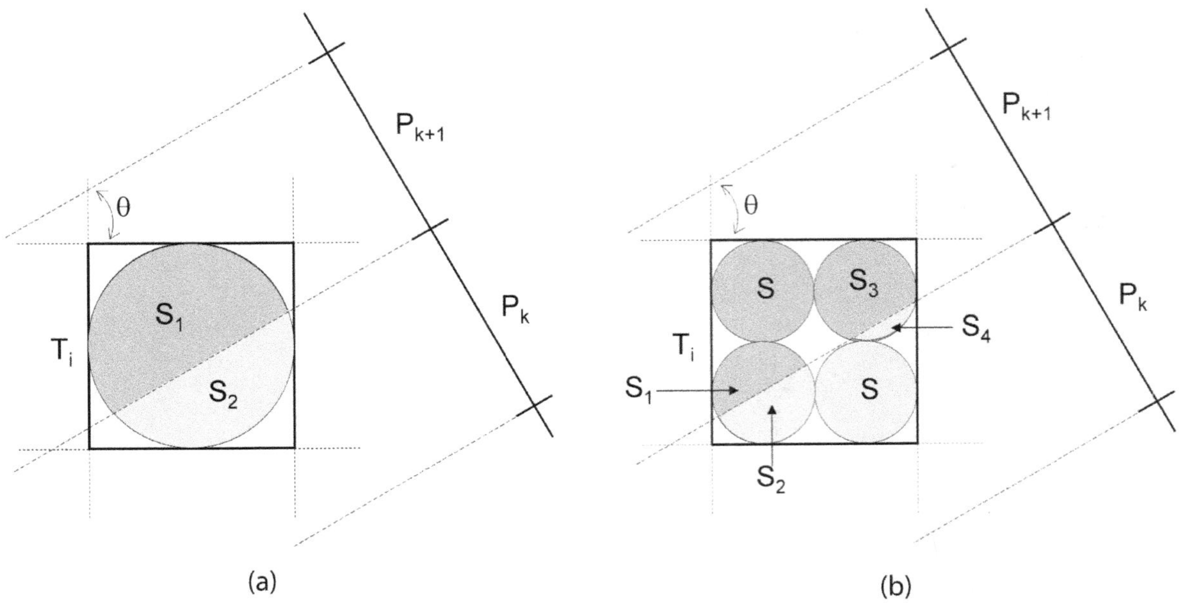

and $a_{i,k} = (s+s_2+s_4)/(4s)$ and $a_{i,k+1} = (s+s_1+s_3)/(4s)$ in case (b).

Below follows a brief algorithmic description of disk forward projector implementation in case of *M* sub-disks (Algorithm 9).

Distance-Driven Projector Method

It was recently proposed in (De Man 2002, De Man 2004). It works by mapping pixel and detector boundaries to a common axis and the coefficients are computed as the row or slab intersection length combined with the overlap coefficient (the length of overlap). In 3D, the overlap area is computed as the product of the overlap lengths in x-y and in z respectively. This method resembles to the Joseph interpolation but instead of a triangular interpolation function, their interpolation employs two convolved rectangle functions with different widths. These widths were chosen to be the image sampling distance and the spatial dependent ray distance respectively. Thus, in terms of basis and irradiation functions, the first rectangle would correspond to the basis function and the second to the integration function. This method is illustrated in Figure 10 and below follows a brief algorithmic description of distance-driven forward projector implementation.

Below follows a brief algorithmic description of distance-driven forward projector implementation (Algorithm 10).

As the concept and the implementing of forward projection algorithms frequently used in medical tomographic vary from one method to other, their performance will also vary. In fact, all these methods result in varying degrees of approximation errors and cause artifacts and/or introduce noise in the projection and reconstructed slices that limit their absolute accuracy. The pixel-driven approach is well suited for hardware implementation, but pixel-driven projection is rarely used in reconstruction, because it introduces high-frequency artifacts. These high-frequency artifacts can be prevented by using a more accurate model (disks, square, bilinear interpolation), but this further increases the arithmetic complexity. Ray-driven methods are generally well-suited for projection, but tend to introduce artifacts in the

Algorithm 9.

```
Algorithm "Disk projection operator with M sub-disks"
            for all projection angles θ do
                 for all images slices do
                      for all pixel slices do
                           for all sub-disks m do
                                calculate the contribution a_{i,k} is proportional to the intersection area between the pixel and the strip projection bin: S_m
                                increment a_ik = a_ik + S_m
                           end for
                           calculate the weighting coefficients: a_ik = a_ik / disk areas
                           save the couple (k, a_ik) in an access memory file
                      end for
                 end for
end "Disk projection operator"
```

backprojection (De Man & Basu, 2002, 2004). Furthermore, ray-driven methods generally have highly non-sequential memory access patterns. The distance-driven method avoids the artifact characteristics of ray-driven backprojection and pixel-driven projection (De Man & Basu, 2002, 2003, 2004). Recent studies have shown that a better image quality can be obtained by using more appropriate basis functions, but at the expense of a longer reconstruction time.

Figure 10. Representation of distance-driven method. The two corresponding weighting coefficients are: $a_{i,k} = d_2/d$ and $a_{i,k+1} = d_1/d$.

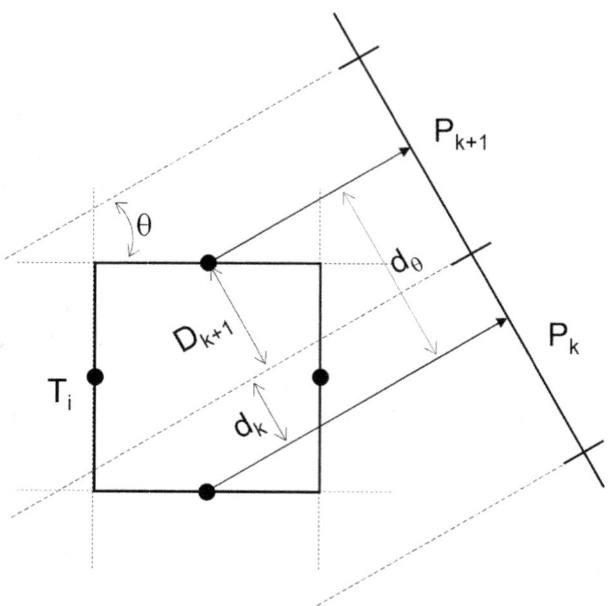

Algorithm 10.

```
0: Algorithm "Distance driven projection operator"
    1: for all projection angles θ do
        2: for all images slices do
            3: calculate the distance d between two opposite middle points of pixel width orthogonally to ray direction:
               if θ ∈ [-45°; 45°] or θ ∈ [135°; 225°], then d= pixel width* cos θ
               else d= pixel width*sin θ
            4: for all pixel slices do
                5: determine the bin k in which the border of pixel i (middle point of pixel width) is projected.
                6: calculate the length of overlap between pixel and projection bin boundaries: d_k
                7: calculate the contribution a_{i,k}=d_k/d
                8: save the couple (i, k, a_{ik}) in an access memory file
            9: end for
        10: end for
    11: end for
13: end "Distance driven projection operator"
```

EFFECT OF MODELLING OF FORWARD PROJECTOR ERRORS ON ITERATIVE RECONSTRUCTION METHODS

As the projection is becoming widely used with iterative reconstruction algorithms and their characteristics may affect the reconstruction quality, its accuracy and performance needs more attention and they need to be better understood. Regardless of all approaches of forward projector modelling; errors are inevitable in the forward projection matrix. Thus, it is important to understand the effect of modelling errors on iterative reconstruction methods. In this context, we compare the forward projection algorithms implemented in a unified framework as described above on a projection task and on iterative reconstruction. Reprojection and reconstructed slices with MLEM and ASIRT techniques are shown using a standard slice (2D) of 3D Shepp-Logan phantom which is considered as a standard test for different reconstruction methods. All algorithms and data are simulated using a user-friendly interface on PC for visualization and reconstruction of tomographic data (Guedouar et al., 2011).

Figure 11 shows the projection absolute error images between the reprojected sinogram and the standard reference of the Shepp-Logan phantom. Figure 12 and 13 show results of tomographic reconstruction of Shepp-Logan phantom using MLEM and ASIRT. The same backprojector is associated with investigated forward projection to form the pair of reconstruction unless with distance-driven method in order to have faithful comparison regardless of the approaches of matched and mismatched reconstruction pairs (Zeng et al., 2000; Guedouar & Zarrad, 2010a; Guedouar & Zarrad, 2010b). In the case of distance-driven method, its reverse model is used

Figure 11. Projection absolute error images between the reprojected sinogram and the standard reference of the Shepp-Logan phantom. forward projector models used are: (a) Interpolated ray-driven; (b) simple ray-driven; (c) ray-driven using 3 sub-bins; (d) interpolated pixel-driven; (e) bilinear interpolation; (f) projection based on simple disk; (g) projection based on 4 sub-disks; (h) distance-driven; (k) projection based on square pixels. Images presented in this figure were thresholded to the interval (0,3) to improve the displayed internal errors. Images show that error mainly concentrates on the edge part in the sinogram. Operator that reduces noise inside sinogram increases edge errors and vice versa.

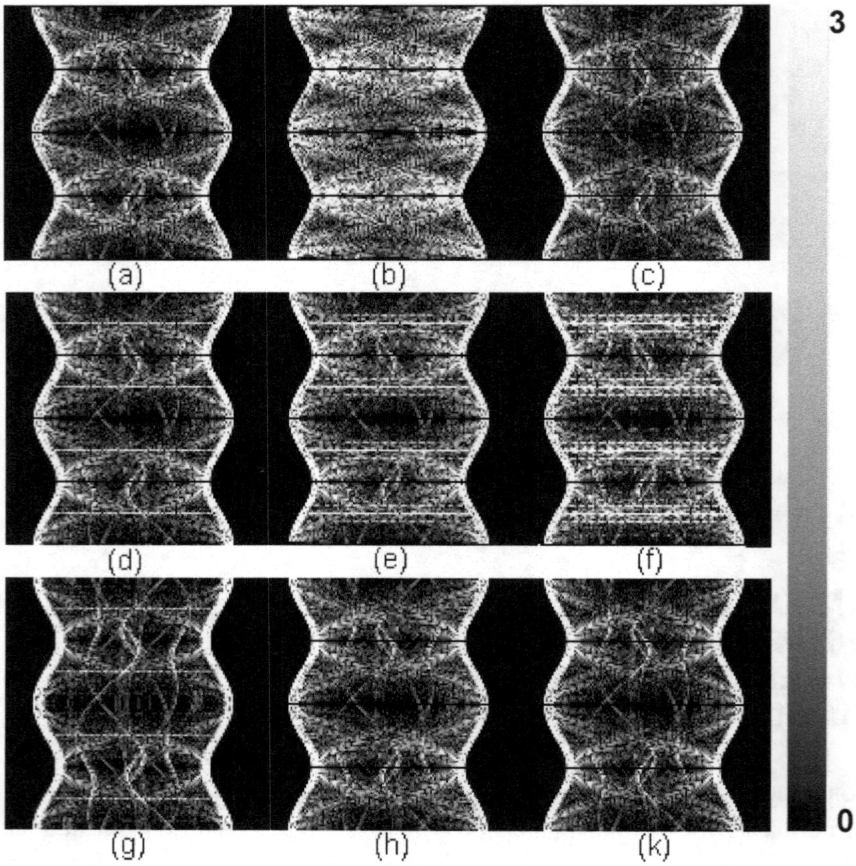

as in (De Man & Basu, 2004). Backprojection accuracy was not evaluated.

Visual inspection of the projections generated by the most effective methods (not shown here) looked similar and comparable to the analytic projections except with the simple pixel-driven model which increases artifacts. However, projection absolute error images (Figure 11) show that all methods result in varying degrees of errors and cause artifacts and/or introduce noise in the reprojected sinogram that limit the absolute accuracy of projection process. Also, errors introduced to projections are mainly concentrated on the edge part and no models can reduce both of errors in internal and edge regions of the projection. It can be noted that the model which provides the least internal errors, increases edge errors as the bilinear interpolation models, whereas the operator which provides the least edge errors, increases those in internal region as ray-driven

Figure 12. Results of tomographic reconstruction of Shepp-Logan phantom using MLEM. The top image corresponds to the standard exact calculated noiseless slice. Reconstructed slices correspond to the convergence iteration (optimal solutions). MLEM reconstruction is performed using the forward projectors: (a) Interpolated ray-driven; (b) simple ray-driven; (c) ray-driven using 3 sub-bins; (d) interpolated pixel-driven; (e) bilinear interpolation; (f) projection based on simple disk; (g) projection based on 4 sub-disks; (h) distance-driven; (k) projection based on square pixels. Interpolated pixel-driven is used as backprojector except with distance-driven where its reverse model is used. Images presented in this figure were thresholded to the interval (4.8,5) to improve the displayed contrast and aliasing.

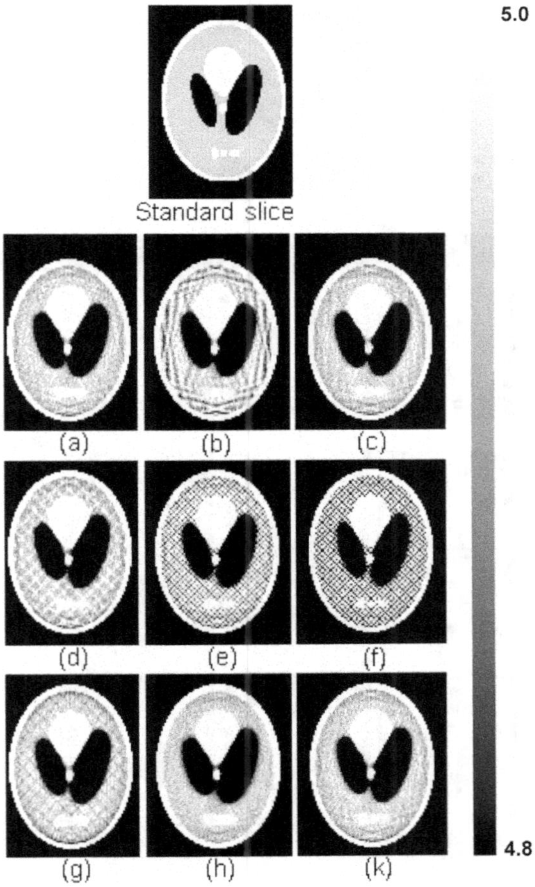

with no subdivision. The projection approaches such as oversampling (models with sub-division), interpolation (linear and bilinear) and realistic basis function (square and disks) can perform better reprojection than the conventional methods and decrease the difference between the reprojected and exact projections but they show the same dependency in the region to be reprojected with high or low transition. Smoothing projectors that use interpolation between pixels or detector elements (bins) are efficient to reduce noise but they increase high frequency noise near edge (i.e. appearance of interpolation artifacts) which unfortunately may degrade image resolution.

Therefore, no current projection approach can give the least errors in all regions. The interpolated ray-driven method seems to provide the best trade of between internal and edge errors.

Figure 13. Results of tomographic reconstruction of Shepp-Logan phantom using ASIRT. The top image corresponds to the standard exact calculated noiseless slice. Reconstructed slices correspond to the convergence iteration (optimal solutions). ASIRT reconstruction is performed using the forward projectors: (a) Interpolated ray-driven; (b) simple ray-driven; (c) ray-driven using 3 sub-bins; (d) interpolated pixel-driven; (e) bilinear interpolation; (f) projection based on simple disk; (g) projection based on 4 sub-disks; (h) distance-driven; (k) projection based on square pixels. Interpolated pixel-driven is used as backprojector except with distance-driven where its reverse model is used. Images presented in this figure were thresholded to the interval (4.8,5) to improve the displayed contrast and aliasing.

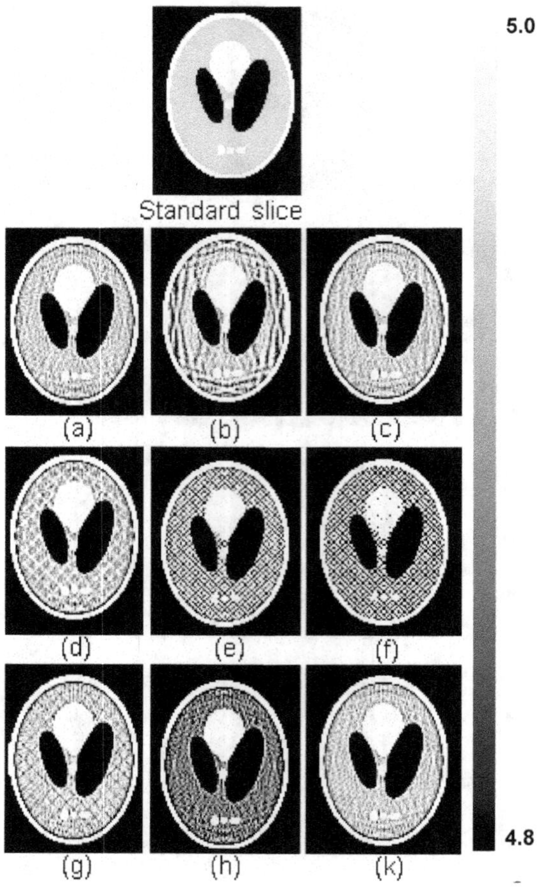

Experiments from Figures 12 and 13, show the evidence propagation of forward projection modelling errors from projection into reconstruction. Artifacts and noise are greater for the iterative algorithms, which is due to the fact that small projection errors might accumulate through the iterative process. It is clear that the forward projector affect the severity of the edge artifacts which means that the appearance and severity of these artifacts is highly dependent on the details of the implementation used to compute (Snyder et al 1987). The forward projection models that increase the noise and the aliasing in the internal region of reconstructed images decrease the edge errors. Smoothing projectors decrease noise but introduce important errors in the edge region which effects spatial resolution.

ACCURACY OF MODELLING OF FORWARD PROJECTOR: ISSUE AND SOLUTION

A wealth of publications exists that discuss and compare performance of the forward projection by using a defined criterion or by assessing a trade-off between numerical accuracy, visual differences in reconstructions, and computational speed. Siddon (Siddon, 1985), Joseph (Joseph, 1983), Herman (Herman, 1980) and Lewitt (Lewitt & Matej, 2003) have described interpolation and integration mechanisms that are frequently used with the forward projection in CT. Zhuang et al. (Zhuang et al., 1994) evaluate projectors of iterative reconstruction and proposed a simple modification that can be applied to any projector to increase the numerical accuracy of the method. The accuracy of any ray-driven projector can be improved by increasing the number of ray-paths traced within each image pixel into a number of smaller sub-pixels and applying the pixel-driven projection method to the sub-pixels and applying the pixel-driven projection method to the sub-pixels. Yu and Huang (Yu & Huang, 1993) analyzed loss of resolution due to reprojection technique, comparing a square-pixel area weighted convolution and a Gaussian pixel method using a nearest-neighbor forward-projection model with sub-binning. Bella (Bella et al., 1995) evaluated different implementations of the method of shears for image rotation. He examined use of various interpolation methods for method of shears, including nearest neighbor interpolation, up sampled nearest-neighbor interpolation (four sub-bins), linear interpolation, and cubic interpolation; standard bilinear and bicubic interpolation were used for reference standards. Wallis proposed in (Wallis & Miller, 1997) an optimal rotator for iterative reconstruction. Recently, De Man and Basu (De Man & Basu, 2004) presented a 3D distance-driven method for projection and backprojection and compared its performance in terms of artifact generation, loss of resolution and computational burden with the two most used methods (interpolated pixel driven and ray driven). They have shown that it eliminates the artifacts seen in ray-driven backprojection and pixel-driven projection.

All these publications have shown that, the choice in the calculation method for the coefficient matrix is critical and may affect significantly the final reconstructed images with iterative techniques. These methods result in varying degrees of approximation errors and cause artifacts and/or introduce noise in the reprojected sinogram and reconstructed slices that limit their absolute accuracy. Recent studies have shown that a better image quality can be obtained by using more appropriate basis functions, but at the expense of a longer reconstruction time. To overcome this difficulty, implementation using graphics hardware is proposed. However, a major disadvantage of using graphics hardware in the reconstruction process is the lack of precision of the hardware. The tradeoff between noise and spatial resolution in reconstructed images can be considered as the most important criteria to make good choice of forward projection modelling.

It has been shown that if the projection data is corrupted by noise, the reconstructed images will in turn be corrupted by noise. The artifacts in the images resulting from this noise can produce corruption especially at the boundaries of objects in the images (edge artifacts). In particular, images reconstructed with MLEM seem to be seriously affected by edge artifacts that appear as severe over and undershoot in the regions of sharp intensity transitions. As the true pixel value in the reconstructed images is influenced by these artifacts, their quantitative analysis is difficult. This significantly limits the clinical usefulness of the images, both for diagnostic and therapeutic purposes, since an accurate knowledge about locations of object boundaries is crucial in applications such as computer-assisted surgery, and radiotherapy. Also, post processing such as noise reduction, binarization, or segmentation of image information is significantly complicated by the presence of such artifacts. It has been shown that the appearance and severity of these artifacts is highly dependent on the details of the implementation used for computation (Figure 12).

The removal of edge and aliasing artifacts from the reconstructions is one of the issues of crucial importance if statistical reconstruction is to be utilized. These artifacts can be prevented by using more sophisticated weighting schemes but at the expense of a longer reconstruction time. Low pass filters are used but unfortunately, filtering methods may also substantially increase high frequency noise which again degrades image resolution. Some regularization methods within the iterative techniques are proposed but they lead to added computational burden. Hence, it is obvious to reduce the artifact by modifying the forward implementation that they disappear in their environment. If this is achieved, the artifacts in tomographic slice will disappear. An earlier study (Guedouar & Zarrad, 2010b) compares the performance of the most used forward projection regarding the compromise between noise and spatial resolution. Unlike the existing work that often focuses on a specific type of modelling error, such as geometric response, attenuation or scatter, this study evaluated the forward projection errors generated by the different geometrical models in two regions of interest (regions with high and low transition) via numerical sense (its RMSE). Error propagation from the forward projection matrix into reconstructed images with iterative techniques was shown. Based on this comparison study, a new projection method was proposed in order to preserve edges without increasing noise. Preliminary results show that this method can promise more accuracy in term of RMSE and aliasing reduction of the reconstructed images. Combining acceleration schemes and availability of faster computers will decrease the execution time to an acceptable level and this method can be easily extended to 3D data sets.

CONCLUSION

The choice in the calculation method for the forward projection matrix is critical and affects significantly the final reconstructed images with iterative techniques. All projection methods result in varying degrees of errors and cause artifacts and/or introduce noise in the reprojected sinogram that limit the absolute accuracy of projection process. Errors introduced to projections are mainly concentrated on the edge part and no models can reduce both of errors in internal and edge regions of the projection. The appearance and severity of these artifacts is highly dependent on the details of the implementation used to compute the forward projector modelling. Therefore, the tradeoff between noise and spatial resolution in iterative reconstruction can be reduced by using an appropriate forward projection modelling according to the goal of the slice to be reconstructed regardless of the execution time needed for the reconstruction.

REFERENCES

Alenius, S., & Ruotsalainen, U. (1997). Bayesian image reconstruction for emission tomography based on median root prior. *European Journal of Nuclear Medicine*, *24*, 258–265. doi:10.1007/BF01728761

Alenius, S., Ruotsalainen, U., & Astola, J. (1998). Using local median as the location of the prior distribution in iterative emission tomography image reconstruction. *IEEE Transactions on Nuclear Science*, *45*, 3097–3104. doi:10.1109/23.737670

Bella, E. V. R. D., Barclay, A. B., Eisner, R. L., & Schafer, R. W. (1995). Comparison of rotation-based methods for iterative reconstruction algorithms. *IEEE Nuclear Science Symp. Medical Imaging Conf.*, 2, (pp. 1146–50).

Boag, A., Bresler, Y., & Michielssen, E. (2000). A multilevel domain decomposition algorithm for fast O(N2 log N) reprojection of tomographic images. *IEEE Transactions on Image Processing*, *9*(9), 1573–1582. doi:10.1109/83.862638

Bortfeld, T., Bürkelbach, J., Boesecke, R., & Schlegel, W. (1990). Methods of image reconstruction from projections applied to conformation radiotherapy. *Physics in Medicine and Biology*, *25*(10), 1423–1434. doi:10.1088/0031-9155/35/10/007

Chidlow, K., & Möller, T. (2003). Rapid emission tomography reconstruction. *Volume Graphics*, *45*, 15–26.

Cormack, A. M. (1964). Representation of a function by its line integrals, with some radiological applications. *Journal of Applied Physics*, *35*(10), 2908–2913. doi:10.1063/1.1713127

De Man, B., & Basu, S. (2004). Distance-driven projection and backprojection: Extension to three dimensions and analysis. *Physics in Medicine and Biology*, *49*, 2463–2475. doi:10.1088/0031-9155/49/11/024

Geman, S., & McClure, D. E. (1987). Statistical methods for tomographic image reconstruction. *Bull Int Stat Inst*, *52*(4), 5–21.

Gilbert, P. (1972). Iterative methods for the three-dimensionl reconstruction of an object from projections. *Journal of Theoretical Biology*, *36*, 105–117. doi:10.1016/0022-5193(72)90180-4

Girodias, K. A., Barrett, H. H., & Shoemaker, R. L. (1991). Parallel simulated annealing for emission tomography. *Physics in Medicine and Biology*, *36*(7), 921–938. doi:10.1088/0031-9155/36/7/002

Goitein, M. (1972). Three-dimensional density reconstruction from a series of two-dimensional projections. *Nuclear Instruments and Methods*, *101*, 509–518. doi:10.1016/0029-554X(72)90039-0

Gordon, R., & Herman, G. T. (1973). Reconstruction of pictures from their projections. *Communications of the ACM*, *14*, 759–768. doi:10.1145/362919.362925

Guedouar, R., Ben Salah, R., & Zarrad, B. (2011). User-friendly Interface on PC for Visualization and Reconstructing of Tomographic. *Annual Meeting of the Society for Imaging Informatics in Medicine (SIIM)*. Washington, DC, USA.

Guedouar, R., & Zarrad, B. (2010a). A comparative study between matched and mis-matched projection/backprojection pairs used with ASIRT reconstruction method. *Nuclear Instrumentations and Methods A*, *619*, 225–229. doi:10.1016/j.nima.2010.02.077

Guedouar, R., & Zarrad, B. (2010b). A new reprojection method based on a comparison of popular reprojection models. *Nuclear Instrumentations and Methods A*, *619*, 270–275. doi:10.1016/j.nima.2010.02.074

Hansen, E. W. (1981). Theory of circular harmonic image reconstruction. *Journal of the Optical Society of America*, *71*(3), 304–308. doi:10.1364/JOSA.71.000304

Herman, G. T. (2009). *Fundamentals of computerized tomography: Image reconstruction from projection* (2nd ed.). Springer.

Herman, G. T., & Lent, A. (1976). Iterative reconstruction algorithms. *Computers in Biology and Medicine*, *6*(4), 273–294. doi:10.1016/0010-4825(76)90066-4

Hsieh, Y.-L., Zeng, G. L., & Gullberg, G. T. (1998). Projection space image reconstruction using strip functions to calculate pixels more natural for modelling the geometric response of the SPECT collimator. *IEEE Transactions on Medical Imaging*, *17*(1), 24–44. doi:10.1109/42.668692

Hudson, H. M., & Larkin, R. S. (1994). Accelerated image reconstruction using ordered subsets of projection data. *IEEE Transactions on Medical Imaging, 13*(4), 601–609. doi:10.1109/42.363108

Jiang, M., & Wang, G. (2003). Convergence studies on iterative algorithms for image reconstruction. *IEEE Transactions on Medical Imaging, 22*(5), 569–579. doi:10.1109/TMI.2003.812253

Joseph, P. M. (1983). An improved algorithm for reprojecting rays through pixel images. *IEEE Transactions on Medical Imaging, 1*(3), 192–196. doi:10.1109/TMI.1982.4307572

Kamphuis, C., & Beekman, F. J. (1998). Accelerated iterative transmission CT reconstruction using an ordered subsets convex algorithm. *IEEE Transactions on Medical Imaging, 17*(6), 1101–1105. doi:10.1109/42.746730

Köhler, T., Proksa, R., & Nielsen, T. (2003). SNR-weighted ART applied to transmission tomography. *IEEE Nuclear Science Symposium Conference Record, 4*, 2739–2742.

Lalush, D. S., & Tsui, B. M. W. (1993). Improving the convergence of iterative filtered backprojection algorithms. *Medical Physics, 21*, 1283–1286. doi:10.1118/1.597210

Lewitt, R. (1992). Alternatives to voxels for image representation in iterative reconstruction algorithms. *Physics in Medicine and Biology, 37*, 705–716. doi:10.1088/0031-9155/37/3/015

Lewitt, R. M. (1990). Multidimensional digital image representations using generalized Kaiser-Bessel window functions. *Journal of the Optical Society of America. A, Optics and Image Science, 7*(10), 1834–1846. doi:10.1364/JOSAA.7.001834

Lewitt, R. M., & Matej, S. (2003). Overview of methods for image reconstruction from projections in emission computed tomography. *Proceedings of the IEEE, 91*(10), 1588–1611. doi:10.1109/JPROC.2003.817882

Matej, S., & Lewitt, R. M. (1996). Practical considerations for 3-D image reconstruction using spherically symmetric volume elements. *IEEE Transactions on Medical Imaging, 15*(1), 68–7. doi:10.1109/42.481442

Ollinger, J. M. (1990). Iterative reconstruction-reprojection and the expectation maximization algorithm. *IEEE Trans. Med. Imag., 9*(94).

Peters, T. M. (1981). Algorithms for fast back- and re-projection in computed tomography. *IEEE Transactions on Nuclear Science, 28*(4), 3641–3647. doi:10.1109/TNS.1981.4331812

Reyes, M., Malandain, G. P., Koulibaly, M., Gonzalez-Ballester, M. A., & Darcourt, J. (2007). Model based respiratory motion compensation for emission tomography image reconstruction. *Physics in Medicine and Biology, 52*(12), 3579–3600. doi:10.1088/0031-9155/52/12/016

Schwinger, R. B., Cool, S. L., & King, M. A. (1986). Area weighted convolution interpolation for data re-projection in single photon emission computed tomography. *Medical Physics, 13*(3), 350–353. doi:10.1118/1.595959

Shepp, L. A., & Logan, B. F. (1974). The Fourier reconstruction of a head section. *IEEE Transactions on Nuclear Science, 21*, 21–43.

Shepp, L. A., & Vardi, Y. (1982). Maximum likelihood reconstruction for emission tomography. *IEEE Transactions on Medical Imaging, 1*(2), 113–122. doi:10.1109/TMI.1982.4307558

Siddon, R. (1985). Fast calculation of the exact radiological path length for a three-dimensional CT array. *Medical Physics, 12*, 252–255. doi:10.1118/1.595715

Thibault, J.-B., Sauer, K., Bouman, C., & Hsieh, J. (2007). A three-dimensional statistical approach to improved image quality for multi-slice helical CT. *Medical Physics, 34*(11), 4526–4544. doi:10.1118/1.2789499

Wallis, J. W., & Miller, T. R. (1997). An optimal rotator for iterative reconstruction. *IEEE Transactions on Medical Imaging, 16*, 118–123. doi:10.1109/42.552061

Wang, G., Vannier, M. W., & Cheng, P.-C. (1999). Iterative X-ray cone-beam tomography for metal artifact reduction and local region reconstruction. *Microscopy and Microanalysis*, *5*(1), 58–65. doi:10.1017/S1431927699000057

Xu, F., Mueller, K., Jones, M., Keszthelyi, B., Sedat, J., & Agard, D. (2008). On the efficiency of iterative ordered subset reconstruction algorithms for acceleration on GPUs. MICCAI (Workshop on High-Performance Medical Image Computing & Computer Aided Intervention), New York.

Yu, D., & Huang, S. (1993). Study of reprojection methods in terms of their resolution loss and sampling errors. *IEEE Transactions on Nuclear Science*, *40*, 1174–1178. doi:10.1109/23.256732

Zbijewski, W., & Beekman, F. J. (2006). Comparison of methods for suppressing edge and aliasing artifacts in iterative x-ray CT reconstruction. *Physics in Medicine and Biology*, *51*(7), 1877–1890. doi:10.1088/0031-9155/51/7/017

Zeng, G., & Gullberg, G. (1993). A ray-driven backprojector for backprojection filtering and filtered backprojection algorithms. *IEEE Nuclear Science Symp. Medical Imaging Conf.* San Francisco, (pp. 1199–1201).

Zeng, G., & Gullberg, G. (2000). Unmatched projector/backprojector pairs in an iternative reconstruction algorithm. *IEEE Transactions on Medical Imaging*, *19*(5), 548–555. doi:10.1109/42.870265

Zeng, G. L., Hsieh, Y. L., & Gullberg, G. T. (1994). A rotating and warping projector/backprojector for fan-beam and cone-beam iterative algorithm. *IEEE Transactions on Nuclear Science*, *41*, 2807–2811. doi:10.1109/23.340651

Zhuang, W., Gopal, S. S., & Hebert, T. J. (1994). Numerical evaluation of methods for computing tomographic projections. *IEEE Transactions on Nuclear Science*, *41*(4), 1660–1665. doi:10.1109/23.322963

ADDITIONAL READING

Bender, R., & Herman, G. T. (1970). Algebraic reconstruction techniques (ART) for three-dimensional electron microscopy and x-ray photography. *Journal of Theoretical Biology*, (29): 471–481.

Danielsson, P.-E., Magnusson, M., & Sunneg˚ardh, J. (2005). Basis and window functions in CT. *8th International Meeting on Fully Three-Dimensional Image Reconstruction in Radiology and Nuclear Medicine*. Salt Lake City, Utah, USA.

De Man, B., & Basu, S. (2002). Distance-driven Projection and Backprojection. *IEEE Nuclear Science Symposium and Medical Imaging Conference*, Norfolk, Virginia.

De Man, B., & Basu, S. Thibault, J-B. Hsieh, J., Fessler, J.A., Bouman, C., & Sauer, K.(2005). A study of different minimization approaches for iterative reconstruction in X-ray CT. *IEEE Nuclear Science Symposium and Medical Imaging Conference,* Puerto Rico.

De Man, B., & Fessler, J. A. (2010). Statistical iterative reconstruction for X-ray computed tomography . In Jiang, M., Censor, Y., & Wang, G. (Eds.), *Biomedical Mathematics: Promising Directions in Imaging, Therapy Planning and Inverse Problems* (pp. 113–140). Madison, WI: Medical Physics Publishing.

Defrise, M., & Gullberg, G. T. (2006). Image reconstruction. *Physics in Medicine and Biology*, *51*, 139–154. doi:10.1088/0031-9155/51/13/R09

Fessler, J. A. (2000). *Statistical image reconstruction methods for transmission tomography. In, M. Sonka and J. Michael Fitzpatrick, editors, Handbook of Medical Imaging, Volume 2. Medical Image Processing and Analysis (pp.* 1–70). SPIE, Bellingham.

Hanson, K. M., & Wecksung, G. W. (1985). Local basis-function approach to computed tomography. *Applied Optics*, *24*(23), 4028–4039. doi:10.1364/AO.24.004028

Herman, G. T. (1980). *Image Reconstruction from Projections, the Fundamentals of Computerized Tomography*. New York: Academic Press.

Herman, G. T., & Meyer, L. B. (1993). Algebraic reconstruction techniques can be made computationally efficient. *IEEE Transactions on Medical Imaging*, *12*(3), 600–609. doi:10.1109/42.241889

Hsieh, J. (2003). *Computed Tomography: Principles, Design, Artifacts, and Recent Advances* (1st ed.). Bellingham, Washington: SPIE.

Jinyi, Q., & Leahy, R. M. (2006). Iterative reconstruction techniques in emission computed tomography . *Physics in Medicine and Biology*, *51*, 541–578. doi:10.1088/0031-9155/51/15/R01

Jinyi, Q. A., & Huesman, R. H. (2004). Propagation of errors from the sensitivity image in list mode reconstruction. *IEEE Transactions on Medical Imaging*, *23*, 1094–1099. doi:10.1109/TMI.2004.829333

Jinyi, Q. A., & Huesman, R. H. (2005). Effect of errors in the system matrix on maximum a posteriori image reconstruction. *Physics in Medicine and Biology*, *50*, 3297–3312. doi:10.1088/0031-9155/50/14/007

Lange, K., & Fessler, J. A. (1995). Globally convergent algorithms for maximum a posteriori transmission tomography. *IEEE Transactions on Image Processing*, *4*(10), 1430–1438. doi:10.1109/83.465107

Leahy, R., & Qi, J. (2000). Statistical approaches in quantitative positron emission tomography. *Statistics and Computing*, *10*, 147–165. doi:10.1023/A:1008946426658

Lewitt, R. M. (1990). Multidimensional digital image representations using generalized Kaiser-Bessel window functions. *Journal of the Optical Society of America. A, Optics and Image Science*, *7*(10), 1834–1846. doi:10.1364/JOSAA.7.001834

Lewitt, R. M. (1992). Alternatives to voxels for image representation in iterative reconstruction algorithms. *Physics in Medicine and Biology*, *37*(3), 705–716. doi:10.1088/0031-9155/37/3/015

Long, Y., Fessler, J. A., & Balter, J. M. (2010). 3D forward and back-projection for X-ray CT using separable footprints. *IEEE Transactions on Medical Imaging*, *29*(11), 1839–1850. doi:10.1109/TMI.2010.2050898

Matej, S., Fessler, J. A., & Kazantsev, I. G. (2004). Iterative tomographic image reconstruction using Fourier-based forward and back- projectors. *IEEE Transactions on Medical Imaging*, *23*(4), 401–412. doi:10.1109/TMI.2004.824233

Matej, S., & Lewitt, R. M. (1995). Efficient 3D grids for image reconstruction using spherically symmetric volume elements. *IEEE Transactions on Nuclear Science*, *42*(4), 1361–1370. doi:10.1109/23.467854

Mitchell, J. R., Dickof, P., & Law, A. G. (1990). A comparison of line integral algorithms. *Computers in Physics*, *17*, 172.

Natterer, F. (1980). Efficient implementation of optimal algorithms in computerized tomography. *Mathematical Methods in the Applied Sciences*, *2*, 545–555. doi:10.1002/mma.1670020415

Nuyts, J., Michel, C., & Dupont, P. (2001). Maximum-likelihood expectation-maximization reconstruction of sinograms with arbitrary noise distribution using NEC-transformations. *IEEE Transactions on Medical Imaging*, *20*, 365–375. doi:10.1109/42.925290

Snyder, D. L., Miller, M. I., Thomas, J. L., & Politte, D. G. (1987). Noise and edge artifacts in maximum-likelihood reconstructions for emission tomography. *IEEE Transactions on Medical Imaging*, *6*(3), 228–238. doi:10.1109/TMI.1987.4307831

Suetens, P. (2002). *Fundamentals of medical imaging*. Cambridge University Press.

Vandenberghe, S., Asselera, Y. D., Van de Wallea, R., Kauppinenb, T., Koolea, M., & Bouwensa, L. (2001). Iterative reconstruction algorithms in nuclear medicine. *Computerized Medical Imaging and Graphics*, (25): 105–111. doi:10.1016/S0895-6111(00)00060-4

Vardi, Y., Shepp, L. A., & Kaufman, L. (1985). A statistical model for positron emission tomography. *Journal of the American Statistical Association*, (80): 8–20. doi:10.2307/2288030

Xu, X., Liow, J. S., & Strother, S. C. (1993). Iterative algebraic reconstruction algorithms for emission computed tomography: a unified framework and its application to positron emission tomography. *Medical Physics*, (20): 1675–1684. doi:10.1118/1.596954

Yendiki, A., & Fessler, J. A. (2004). A comparison of rotation- and blob-based system models for 3D SPECT with depth-dependent detector response. *Physics in Medicine and Biology*, *49*(11), 2157–2168. doi:10.1088/0031-9155/49/11/003

Zbijewski, W., & Beekman, F. J. (2006). Comparison of methods for suppressing edge and aliasing artifacts in iterative x-ray CT reconstruction. *Physics in Medicine and Biology*, *51*(7), 1877–1890. doi:10.1088/0031-9155/51/7/017

Ziegler, A., Köhler, T., Nielsen, T., & Proksa, R. (2005). Iterative cone-beam CT image reconstruction with spherically symmetric basis functions. *8th International Meeting on Fully Three-Dimensional Image Reconstruction in Radiology and Nuclear Medicine*. Salt Lake City, Utah, USA

Chapter 4
Algorithms for 3D Map Segment Registration

Hao Men
Stevens Institute of Technology, USA

Kishore Pochiraju
Stevens Institute of Technology, USA

ABSTRACT

Many applications require dimensionally accurate and detailed maps of the environment. Mobile mapping devices with laser ranging devices can generate highly detailed and dimensionally accurate coordinate data in the form of point clouds. Point clouds represent scenes with numerous discrete coordinate samples obtained about a relative reference frame defined by the location and orientation of the sensor. Color information from the environment obtained from cameras can be mapped to the coordinates to generate color point clouds. Point clouds obtained from a single static vantage point are generally incomplete because neither coordinate nor color information exists in occluded areas. Changing the vantage point implies movement of the coordinate frame and the need for sensor position and orientation information. Merging multiple point cloud segments generated from different vantage points using features of the scene enables construction of 3D maps of large areas and filling in gaps left from occlusions. Map registration algorithms identify areas with common features in overlapping point clouds and determine optimal coordinate transformations that can register or merge one point cloud into another point cloud's coordinate system. Algorithms can also match the attributes other than coordinates, such as optical reflection intensity and color properties, for more efficient common point identification. The extra attributes help resolve ambiguities, reduce the time, and increase precision for point cloud registration. This chapter describes a comprehensive parametric study on the performance of a specialized Iterative Closest Point (ICP) algorithm that uses color information. This Hue-assisted ICP algorithm, a variant developed by the authors, registers point clouds in a 4D (x, y, z, hue) space. A mobile robot with integrated 3D sensor generated color point cloud used for verification and performance measurement of various map registration techniques. The chapter also identifies various algorithms required to accomplish complete map generation using mobile robots.

DOI: 10.4018/978-1-61350-326-3.ch004

INTRODUCTION

Complete and dimensionally accurate maps of the environments are of interest to many domains including surveying, search and rescue, security, defense and construction. Laser based scanning devices (Light Detection And Ranging-LIDAR) are generally used to generate point clouds that describe spatial information in the form of numerous discrete point coordinate measurements. Point data are acquired by measuring time of flight of scattered light or phase shift between incident and reflected light to find the distance between the object surface and the scanning device (Blais, 2004). The speed of scanning discrete points can be enhanced by pulse and phase based measurement technologies (Blais, 2004). Precise rotation mechanisms with high-resolution encoders spin a 2D LIDAR device to generate a 3D point cloud. Point cloud scanners have been mounted on airplanes (Browell et. al. 1990) and ground vehicles (Gebre, et al. 2009) to create large area terrain maps. When vision sensors are integrated with the laser ranging systems, point clouds can also contain the color information of the scene. Optical imagery from the camera is associated with point coordinates to produce color point clouds (Andresson, 2007).

A 3D point cloud obtained from a single vantage point is seldom adequate to construct a complete map. Generation of a complete map of an environment requires merging or registration of map segments taken from various vantage points. The registration enables construction of large-scale global 3D maps (Thrun, 2003). Registering the map segments is trivial if precise position and orientation of the sensor are accurately known about a global reference frame. Position sensors such as inertial measurement units or those relying on global positioning systems are prone to errors and can be highly inaccurate under certain conditions. The map registration process determines the rigid body translation and rotation of the sensor as its output (Thrun, 1993, 2003). The map registration quality varies depending upon the sensor resolution and the extent of overlap between the map segments. Different techniques exist for merging 3D maps by exploiting geometric features and measuring surfaces. The most popular registration algorithm for point cloud registration is the iterative closest point (ICP) algorithm (Thrun, 2003). In ICP, the corresponding closest points in different point clouds are associated and optimal rigid transformation required to minimize a mean-square error of separation between the associated points (Bsel, 1992) is iteratively found. The color attributes of the sampled point can be utilized in ICP progress to increase computational speed and provide higher accuracy. Anderson (2007) filtered the point set data based on hue before conducting traditional ICP. Houng et al., (2009) processed images to extract corresponding visual features that are used in registration process.

In this chapter, we examine the algorithms required for a mobile robot to generate a dimensionally accurate and complete map of an area without prior information about the area. We focus particularly on the techniques for registration of map segments taken from various vantage points. The chapter also describes a mobile robotic system with a color point cloud scanner and various algorithms required for accomplishing the mission of generating a complete and dimensionally accurate map of an area.

MOBILE MAPPING WITH COLOR POINT CLOUD SCANNERS

Color point clouds are created by synchronizing range sensors such as the LIDAR with video/still cameras. LIDAR devices discretely measure the distance between a light source and a reflection target at a high frequency. By changing the path of the light through mirrors and actuators, a point cloud of a 3D space is produced. A calibrated vision sensor maps the color information to the sampled points. Installing such a scanning sensor

Figure 1. 3D scanning devices built with 2D commercial scanners

on a mobile platform extends its range and enables mapping of large areas.

3D Color Scanner

The 3D color scanner used in this effort consists of a 2D LIDAR and two 1.3 megapixel high-frame rate video cameras installed on the LIDAR scanning plane. The LIDAR and the cameras move such that the scan plane is rotated about an axis within the plane, thus generating 3D color point clouds. Figure 1 shows that the LIDAR consists of a rotating mirror which is driven about Y axis (degree of freedom: θ) and the scan plane is rotated about Z axis (φ: degree of freedom). The rotations are controlled by servomotors installed on the axes. The cameras are calibrated to be on the LIDAR scan plane and a forty-pixel wide image stripe is extracted from the cameras. The color information is then matched, in real-time, to the points ranged by the LIDAR. The relative distance between cameras and LIDAR is pre-configured and images are pre-aligned. The 2D LIDAR generates scans at a frequency of 38 Hz and the cameras provide imagery at 60 frames per second. Time synchronization establishes that the pixel color is mapped to each ranged point. Use of two cameras reduced occlusions due to the offset between the LIDAR mirror and the camera lens. All areas visible to the LIDAR are visible to one of the two cameras. The 2D range measurement along with the scanner rotation position (φ) is used to generate the coordinate in a spherical coordinate system, which is transformed to Cartesian system as necessary. Figure 1 also shows a picture of a compact version of the system.

The 3D color scanner is mounted on a mobile vehicle for mapping large areas. This mobile mapping system generates color point cloud data. Figure 2 shows the mobile system with the scanner installed on top of the vehicle. The vehicle has no global positioning devices other than wheel encoders. Cameras and short-range infrared sensors enable observation of terrain conditions, collision avoidance and allow a remote operator to drive the vehicle. Map data and video feeds are transmitted using an on-board wireless communication system. This mapping system performs scans only when it is stopped. The vehicle can localize itself from the map observations and moves directly from one vantage point to the next and acquires additional map information. This system can generate color point cloud maps with 0.25° angular resolution in the vertical scanning direction with a coverage angle of 100°. In the rotation (φ) direction, the resolution is at 0.1° with coverage angle 300°. The map segment from one

Figure 2. Mobility platform used for 3D color map construction in large area (Gebre et al., 2009)

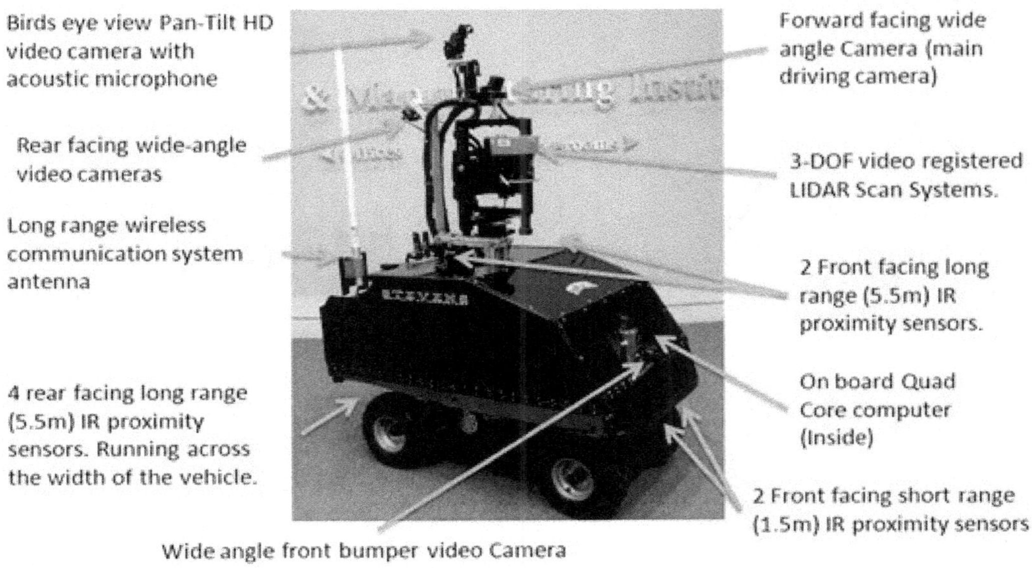

vantage point covers a maximum radius of 80 meters.

The data elements produced by the scanner are shown in Figure 3. Figure 3(a) shows the camera image taken from the vantage point depicting scene visible to the scanner. The 3D color point cloud generated at that vantage point is shown in Figure 3(b). In this figure, the coordinate (x,y,z) and the color (r,g,b) for all the pixels are known. The point density (spatial resolution of the point cloud) varies on the left and right sides of the color scan scene depending upon the distance between the scanned point and the scanner. The closest area to the scanner has the highest density of points. The scanner also records the optical reflection intensity of laser beam. The intensity information is combined with range measurement data and shown in Figure 3(c). The object surface material, color and distance towards scanner cause variations in intensity data. Similarities between intensity point cloud and color point cloud can be observed between Figure 3(b) and (c) on edges, doors, and windows.

Algorithms for Complete Mapping

An autonomous robot with the color point cloud scanner can reduce the surveying and map building cost and time. However, several methodologies for robust self-localization, map completeness evaluation, map based navigation and 3D map registration must evolve before a high degree of autonomy can be achieved.

A mapping robot deployed at initial start position must go through the four phases of the mapping processes as shown in Figure 4. The robot must be able to localize itself so it can navigate the scene. This can be accomplished by 2D SLAM (Simultaneous Localization and Mapping) techniques or other methods. Methodologies for establishing the map completeness and detection of occluded areas are necessary. Determination of the optimal vantage point for filling in the occluded areas and exploring unmapped areas is also a critical step. As the navigation is based on imprecise mapping and localization information, the map segment registration based on 3D color point clouds is the last but crucial step in building the complete map of a given area. In this subsec-

Figure 3. High dimensional point cloud map segment taken from a single vantage point. (a) Image of an urban building.(b) Color point cloud map. (c) Laser reflection intensity map.

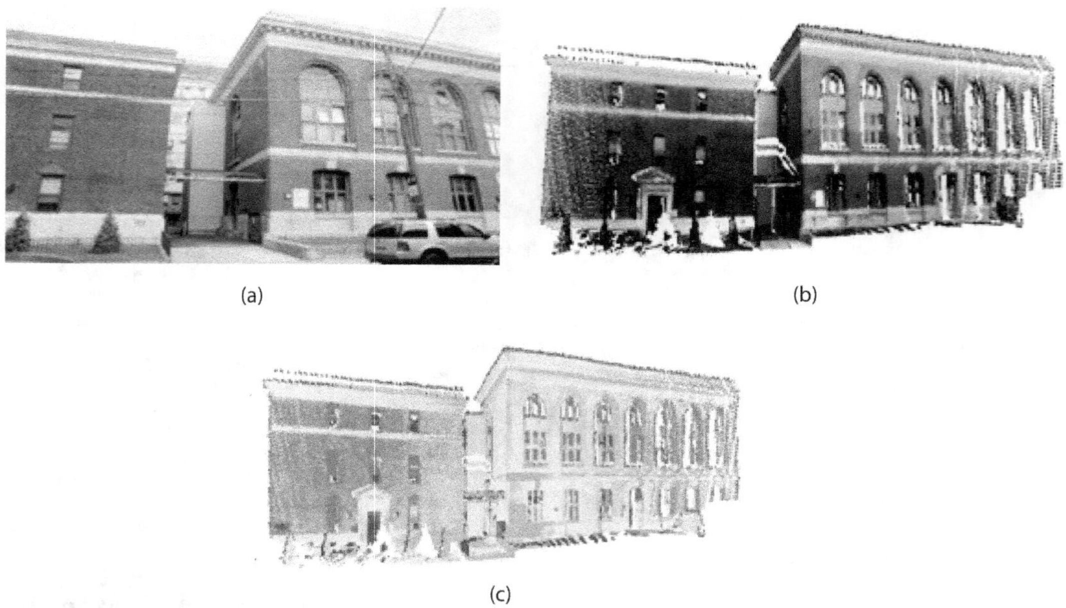

tion, we discuss the algorithms that address each of these tasks.

Robot Self-Localization

The self-localization problem requires mobile mapping robot to determine its location in an unknown environment. Localization is critical because robot cannot effectively navigate to the next waypoint without the location information. Map registrations require location and pose estimates. Usually robot is equipped with multiple position and orientation sensors like GPS, Inertial Measurement Unit (IMU), odometer, and wheel encoders to measure real-time pose and position. Multiple position and location sensors return robot position information with certain level of error due to reasons like sensor precision, GPS signal noise and errors, sensor drift for IMU and inaccurate measurements from other sensors.

The main challenge for robot localization is to escape location sensor noise, drift errors, and constantly provide accurate location and position reference for the robot. Probabilistic self-localization techniques based on maximum-likelihood estimation have been applied to address this problem. These techniques assume that the noise of position sensor follows certain probabilistic distribution, which can be described mathematically. They also assume that two subsequent map results are highly comparable to each other and several landmarks can be quickly identified. Therefore, accurate relative position and location can be solved by comparing current map with a previous map in short time intervals, and probabilistically maximizing similarity between two maps (Olson, 2000). Map could be generated by different sensors like stereo cameras, sonar or laser range finders. Landmarks extracted from maps are commonly applied in the self-localization process to reduce computation cost. Whyte and Bailey (2006) utilize the relative localization results between two neighbor vantage points to merge the two maps.

A two-step process, termed as Simultaneous Localization and Mapping (SLAM), typically localizes the robot. The robot position is established

Figure 4. Map completeness orientated robotic mapping process

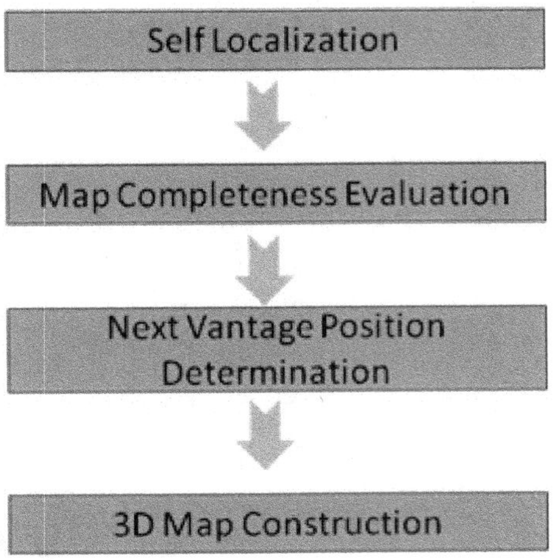

from multiple but imprecise sensor measurements and comparison of landmarks in the scene. The position sensor data is improved using sensor fusion techniques by Spletzer (2003). Location information is estimated based on previous location, driving command information and current sensor measurements. In SLAM, probabilistic methods are applied to reduce sensor noise effects. Extended Kalman filter and particle filters and noise models improve the location estimates (Montemerlo, et al. 2003). The SLAM solution has been expanded into 3D space with a six degree of freedom (6DOF) SLAM which applies sensor measurement and robot kinematics models (Nücher, 2005). Landmark extraction and map comparison entail the major computation effort during the SLAM progress. Real-time SLAM has been demonstrated with stereovision sensors (Davison, 2003).

The SLAM technique simultaneously considers the localization and mapping mission (Thrun, et al., 2000). The SLAM problem can be described by a joint posterior:

$$P(x_t, m \mid z_{0:t}, u_{0:t}, x_0) \quad (1)$$

Where, $\boldsymbol{x}_t$ is the state vector representing the robot location and orientation, $\boldsymbol{m}_i$ is the vector representing the ith landmark location, $\boldsymbol{z}_{it}$ is the robot mapping measurement about i^{th} landmark at time t, and $\boldsymbol{u}_t$ is the control vector applied at t-1 time to drive robot to state $\boldsymbol{x}_t$ at time t.

The SLAM problem requires that equation (1) be solved for the time, t, and the latest robot state vector $\boldsymbol{x}_t$ be computed. Solving the joint posterior from, 0-t requires an observation model and a motion model based on Bayes Theorem (Whyte and Bailey, 2006). The observation model determines the probabilistic distribution of observation $\boldsymbol{z}_t$ with known vehicle state and landmarks location as:

$$P(z_t \mid x_t, m) \quad (2)$$

The robot motion model describes probability on state transition of robot state vector, $\boldsymbol{x}_t$ with known previous state $\boldsymbol{x}_{t-1}$ and control input $\boldsymbol{u}_t$

$$P(x_t \mid x_{t-1}, u_t) \quad (3)$$

The transition of state vector is assumed as a Markov process, implying that the next robot state $\boldsymbol{x}_t$ can only be determined on previous state $\boldsymbol{x}_{t-1}$ and latest control input $\boldsymbol{u}_t$ and not the history of states. The state of robot is independent of both observations and landmarks. Equation (1) can be recursively solved in a Prediction (time update) and Correction (Measurement update) form.

Prediction is shown in Box 1.

Correction:

$$P(x_t, m \mid z_{0:k}, u_{0:t}, x_0) = \frac{P(z_t \mid x_t, m) P(x_t, m \mid z_{0:t-1}, u_{0:t}, x_0)}{P(z_t \mid z_{0:t-1}, u_{0:t})} \quad (5)$$

Equation (4) and equation (5) recursively solve latest robot state joint posterior. Robot state can be predicted from the motion model $P(x_t \mid x_{t-1}, u_t)$

Box 1.

$$P(x_t, m \mid z_{0:t-1}, u_{0:t}, x_0) = \int P(x_t \mid x_{t-1}, u_t) P(x_{t-1}, m \mid z_{0:t-1}, u_{0:t-1}, x_0) dx_{t-1} \quad (4)$$

and control input at time *t*. The observation model $P(z_t \mid x_t, m)$ is applied to correct state prediction with observation and mapping at time *t*.

In order to find solutions to the SLAM problem, proper practical descriptions about motion and observation model in equation (2) and equation (3) should be provided with reliability and efficiency. Extended Gaussian Filter (EKF) is applied to represent these models on state-space model with additive Gaussian noise (Welch and Bishop, 1995). The EKF based SLAM simplifies motion model as:

$$x_t = f(x_{t-1}, u_t) + w_t \quad (6)$$

$f(\mathbf{x}_{t-1}, \boldsymbol{u}_t)$ is the robot kinematics model and $\mathbf{w}_t$ is the additive uncorrelated Gaussian disturbances with zero mean and covariance $\boldsymbol{Q}_t$. The observation model can be described as:

$$z_t = h(x_t, m) + v_t \quad (7)$$

In which, $h(\mathbf{x}_t, \boldsymbol{m})$ is the observation geometry description and $\boldsymbol{v}_t$ is the additive uncorrelated Gaussian disturbance with zero mean and covariance $\boldsymbol{R}_t$. Eqs. (6) and (7) can be applied to the SLAM prediction and correction. In EKF-SLAM process, the mean and covariance of both motion model and observed motion should be updated at every time *t*. Other probabilistic methods such as Particle Filter (PF) (Montemerlo et al., 2003) and Graph Filter (GF) are used to solve the SLAM problem. A typical SLAM method is implemented on 2D space, however, SLAM in 3D space with 6 Degree of Freedom (6DOF) on robot kinematics have been implemented by expanding landmarks state, motion model and observation model into 3D space (Nücher, 2005).

Map Completeness Evaluation

The map completeness problem can be addressed with several methodologies including grid occupancy, obstacle recognition and object view completion detection. The completeness of map is calculated by occupancy grid map (Thrun, 2003), which entails projecting the acquired map on an occupancy grid and calculating the occupancy level. Possible mapping area is determined based on the contour of the objects and separating the map into areas that can be potentially mapped or impossible to map (Oh et al., 2004). Terrains are extracted from current incomplete map for possible paths for navigation. The map evaluation also returns possible explorative area that is accessible to the mobile robot but not mapped. If map completeness is the most important factor for the mission, algorithms that evaluate latest exploration status after every scan may require assessment of the complete map and not just the current map segment. There are many techniques to evaluate the completeness of mapping, namely, grid based occupancy map (Thrun, 2003), network/graph, cell based map (Zelinsky, 1994) and template based completeness evaluation (Oh et al, 2004).

The occupancy grid map is one of the most commonly used methods to determine map completeness. Area of interest is gridded and acquired maps from different vantage position are transferred into or projected onto the grid. Grid is marked as occupied when data exists on this grid, every grid should be represented with certain level of occupancy, which is computed by density of point cloud map on this grid. Map can be assumed as complete all the mapped objects form self-closed contours or closed contours with the boundaries of the mapped area.

A major challenge in map completeness evaluation is deciding whether an area can be mapped. For example, when mapping robot is performing indoor exploration, space behind wall of the hallway may not be accessible. Contours extracted from latest global map may be used to determine possible navigation paths. Possible mapping area exists for contours with gaps. Ascertaining that the gaps in map contours are indeed traversable paths requires discerning traversable pathways in the map.

Map-Based Navigation

Determination of the next vantage point may depend upon several criteria: best view, coverage of unmapped areas, areas of overlap with current map, localization, accessibility and traveling costs. Two steps are required for determination of the next vantage position. The first step is the generation of candidate positions and second step is the selection of optimal vantage point from the list. The candidate vantage positions can be created based on frontier exploration algorithm (Basilico & Amigoni, 2009) considering obstacles, position and terrain conditions. The vantage position is selected between candidate positions that have the best view coverage and shortest traveling cost. Next vantage point should be decided based on the best view to fill occluded regions and cover as much new area as possible. Frontier based exploration algorithm provides vantage point candidates for the best view point, these candidate points are evaluated to determine best vantage point for next mapping.

Computing vantage position for mapping based on previous vantage positions and incomplete map is known as the Next Best View (NBV) problem (Yamauchi, 1997; Basilico & Amigoni, 2009). NBV algorithms navigate robot to acquire maximum uncovered area. A certain level overlapped area ensures that the robot has enough landmarks to navigate between the current and the next best view vantage point. Frontier based algorithm can be applied to provide candidate positions for the next best view point. Based on the regions on the boundary between mapped and unmapped space, the frontier can be extracted. Considering the range for mapping sensor constraints, next mapping position on the frontier can then be generated. Current frontier should be evaluated in occupancy grid map so that the frontier grid positions that cover more unoccupied can be selected to accelerate the coverage of the area. These candidate points can be evaluated based on the criteria for the exploration and time and power requirements for reaching the vantage point.

The map data acquired from various vantage points must be registered into global map space using various registration algorithms. Although this section describes the various algorithms required for complete map generation, the focus of this chapter remains on the registration aspect of the mapping exploration.

ALGORITHMS FOR REGISTERING MAP SEGMENTS

Three-dimensional point cloud segments acquired from different locations have to be combined together as complete large-scale map. Position and orientation information required for registration can be provided directly by mobile platform sensors such as GPS and IMU (Thrun 1993). In most cases, position information acquired from sensor is reasonably accurate. However, the orientation information is costly and relatively imprecise because orientation sensor measurement can be affected by external disturbances like magnetic field variations and sensor integration drift with time. Position and orientation information can also be provided by indirect techniques based on both

Figure 5. Map segments generated from two vantage points (Top) and registered map (Bottom)

rough position sensor measurement and common geometric feature identification. Figure 5 shows two maps generated from separate vantage points. The left map on the top row shows map generated with robot facing towards one side of the building, the right map shows the map generated from the second vantage point. The bottom figure shows the map data from the first vantage point registered into the coordinate system of the second location. Registering the two segments produces the complete map of the façade of the building.

Comparing with the SLAM algorithm, map registration techniques focus on generating accurate map details rather than localization of the robot in a global coordinate system (Arun, 1987; Bsel, 1992; Lorusso, 1995; Rusinkiewucz, 2001). Discrete range points received from color point cloud sensor contain detailed spatial information about the environment. Different techniques exist for merging such point clouds together by exploiting geometric features and measuring surfaces. Map registration techniques such as Iterative Closest Point (ICP) algorithm proposed by Bsel (1992) has been applied to stitch two neighbor 3D point cloud maps together into one map based on their common coverage area. Upon convergence, ICP algorithm terminates at a minimum. Several algorithms are in existence for calculating the minimum average distance between two point clouds. Singular Value Decomposition (SVD) method by Arun (1987), eigen-system methods that exploit the orthonormal properties of the rotation matrices, and unit and dual quaternion techniques were adopted in ICP process. Quaternion based algorithms have been used in ICP for map fusion by Bsel (1992), SVD based algorithms are widely used in ICP and 6DOF SLAM (Arun 1987, Nucher, 2005, Joung et al., 2009) as they are robust to reach local minimum and easy to implement. Several variants of ICP are reported by Rusinkiewucz (2001) to increase the speed and precision. Corresponding points sampling, matching, weighting and rejecting are some methods used to accelerate the ICP algorithm. In the ICP algorithm, associating corresponding points in two point cloud data sets is the most critical step. Nearest neighbor search in 2D or 3D space is commonly used for associating the corresponding points. Parallel ICP algorithms have been developed by Robertson (2002) to accelerate computation speed. Point to plane registration method (Lorusso, 1995, Rusinkiewucz, 2001, Salvi et al., 2007) accelerates the ICP iteration and convergence.

Other techniques include the point signature method by Chua (1997), which uses signature points to describe curvature of point cloud and matches corresponding signature points during the registration process. Spin image based methods compute 2D spin image to represent surface characterization and solve the registration problem by finding best correspondence between two different scan spin images (Johnson 1997). Other methods like principle component analysis (Chung and Lee, 1998) and algebraic surface model (Tarel et al., 1998) are based on the point cloud surface geometrical features. The normal vector distribution can be translated into an orientation histogram in an Extended Gaussian Image (EGI) (Makadia & Daniilidis, 2006). Rigid motion required to register two point clouds is solved from the cross covariance function (Chibunichev & Vilizhev, 2008) of the two EGI images. Rigid motion could also be solved in Fourier domain by computing Discrete Fourier Transform on Rotation Group on SO(3) (SOFT) (Joistekecm and Ricjnirem, 2008).

Registration of color point clouds has been considered (Ferbabdez, et al., 2007; Druon, 2007; Newman et al., 2006; Anderson, 2006, 2007). By applying proper calibration on the hybrid sensor system (Joung et al., 2009; Newman, Cole, Ho, 2006), range measurement and visual information can be integrated together to construct a visually accurate representation of the scene. Color mapped 3D data was used in map registration by weighted red, green, blue data. The corresponding point search during the ICP is conducted on both the coordinate and color data (Johnson, Kang, 1997). Hue filters were also used to constrain the closest point search in every ICP iteration (Druson, 2007). Color data can be used to estimate initial alignment of pair wise scans using Scale Invariant Feature Transform (SIFT) techniques. Color attributes transferred in YIQ color model can also be weighted to construct new variant together with range information for ICP fine registration. Depth-interpolated Image Feature (DIFT) algorithm solves corresponding points between two images and registers color point clouds based on extracted correspondences (Anderson, Lilienthal, 2010).

In this chapter, we introduce hue assisted ICP algorithm for registration of color point clouds. The criteria for association are defined on a 4D space rather than 3D geometric space. The fourth dimension selected is the hue, representing the intrinsic color values of the pixel. While achieving the effect of a hue-based filter, hue-association reduces the nearest neighbor search burden considerably (Men & Pochiraju, 2010). The remaining sections of the paper describe the approach and the performance of the algorithm under several hue distributions in the scene.

HUE-ASSISTED ITERATIVE CLOSEST POINT (H-ICP) ALGORITHM

The primary hypothesis of this algorithm is that the hue value can be applied to increase the accuracy of point association and accelerate the registration process. The major time and computation cost during ICP is finding the correct points pairs. Closest spatial distance is typically applied in 3D ICP method. The distance value in 3D space can be expanded into 4D space by adding weighted hue value as the 4th dimension. By integrating hue value into the closest point search, accuracy of point association can then be improved.

Figure 6. Rubik's cube camera images take from two vantage points

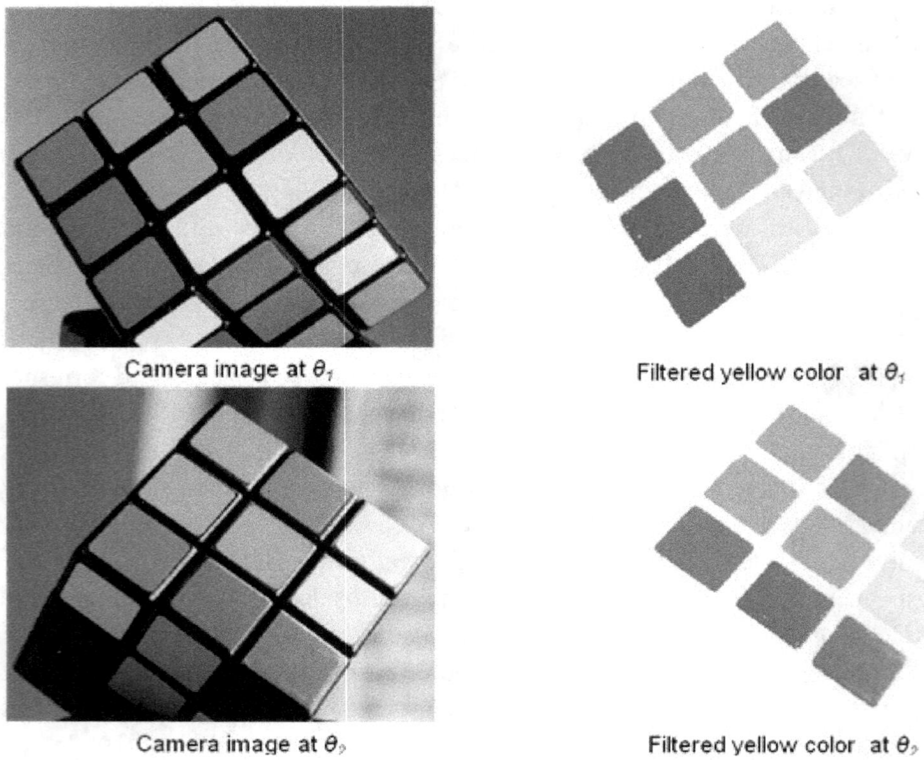

Hue Invariance with Vantage Point

Hue value remains consistent about the same point between images taken from two vantage points, while the color values represented in red, green and blue quantities usually differ because of variation in light conditions. In order to apply color to improve the association process, lighting effect should be removed. Color raw data are transformed into representation of separate chroma, lightness and brightness value. Figure 6 shows two camera images of different angles of a color palette on a Rubik's cube, four colors are used on the same surface. Figure 6 also shows the color pixels with the background and black frame removed. Histograms showing the red, green and blue value in RGB space for all the pixels are shown in Figure 7. In the RGB histogram, R, G, and B distributions of the image vary considerably with the vantage point. When the RGB color space is transformed into HSL space and histograms of hue, lightness and saturation are plotted in Figure 8, the hue values remain relatively invariant with the position of the camera. Therefore, hue value of the pixel, taken from the Hue-Saturation-Lightness (HSL) model, is used as the fourth dimension in the point association process. In Figure 9, the hue rendered point cloud of color point cloud in Figure 3(b) is shown. Hue values are normalized between 0 and 1. The hue distribution is typically similar to the color distribution in Figure 3(b).

Construction of a Weighted 4D Search Space

Both hue and range value have to be combined together in the H-ICP variant as $\{x_o, y_o, z_o, h_w\}$ for point association. x_o, y_o, z_o are the original coordinate values with distance units and h_w is the weighted hue value. Hue values are normalized

Figure 7. RGB distribution varies with camera positions

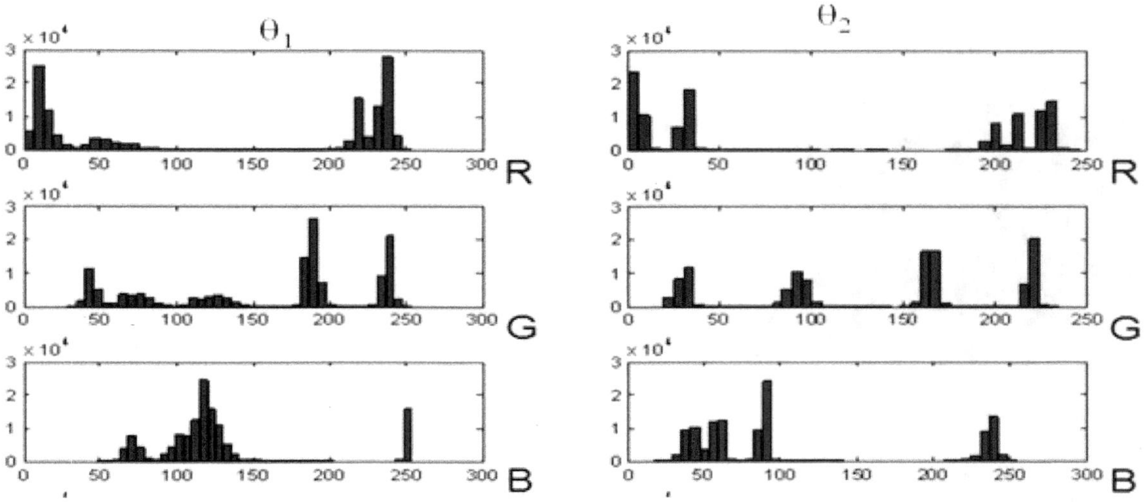

Figure 8. HSL distribution: hue remains invariant

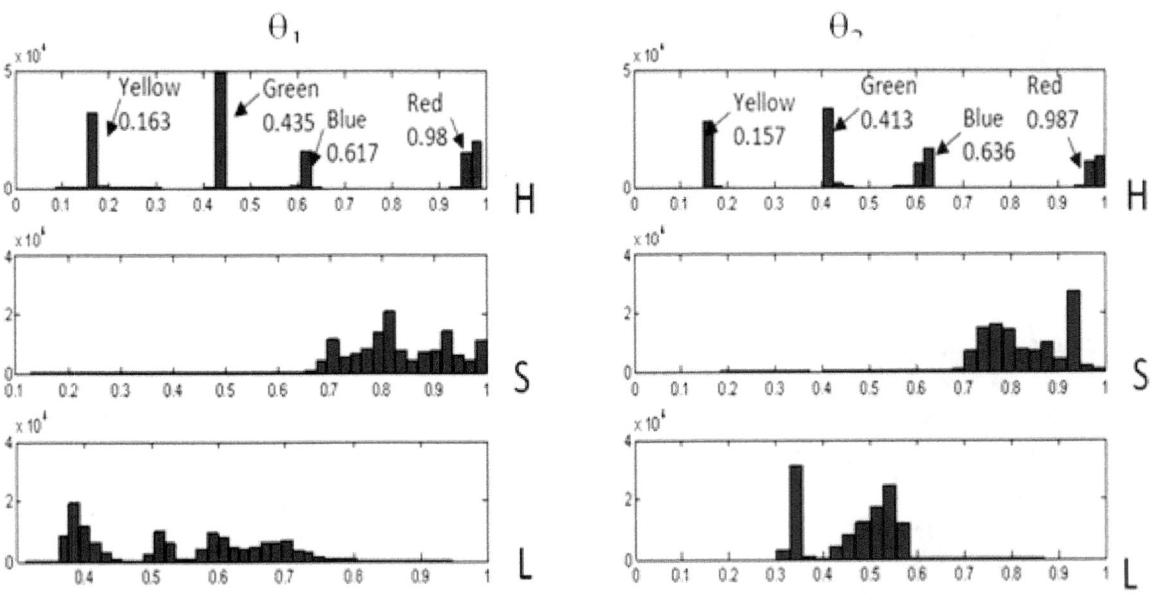

to a 0-1 range and must be weighted during the closest point search in the four-dimensional space. In order to normalize the coordinates, we find the bounding box for each point cloud segment and the coordinate space is rescaled to a 0-1 range. The normalized variant for point association is $\{x, y. z, h_w\}$, where $x=x_o/r_x$, $y=y_o/r_y$, $z=z_o/r_z$. r_x, r_y, r_z are the dimensions of the bounding box in x, y, z directions.

The weight value for the hue dimension should be properly selected for point association. Since both range and hue value are normalized from 0 to 1. Weight for hue represents its influence in the nearest neighbor search process. Low weight

Figure 9. Hue rendered point cloud of the scene shown in Figure 3

Figure 10. Point association based on nearest distance (dotted) and nearest distance and hue (solid)

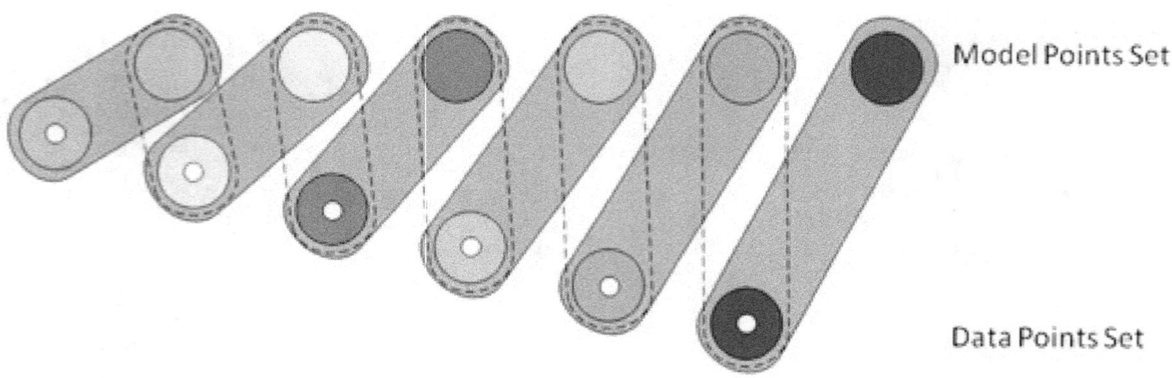

biases the point association towards the range data and a high weight towards the hue values. Small weight values for the hue correspond to the traditional 3D-ICP. Hue weight should be selected between 10% and 35% for accurate point association. Error in H-ICP will be evaluated by the average mean square root distance of normalized associated point pairs.

k-d Tree Based Point Cloud Association

In 3D ICP algorithm, corresponding points are searched according to the closest distance rule. This may cause incorrect matching during single iteration loop as Figure 10. Dashed line circle illustrates range based nearest point association results, in which all points in data set look for nearest neighbor in 3D space. It takes more than one iteration to pair correct nearest neighbor points for given data points set. Grey circle denotes the H-ICP nearest point search that also uses the correct hue property in finding the best neighbor in the model. Depending on the correct color information, corresponding point can be locked with less iteration.

The ICP computation speed and precision are highly dependent on association process. Use of a k-d tree for closest point search and association or the Nearest Neighbor Search (NNS) problem

Figure 11. k-d tree construction and nearest neighbor search in 2D space. (a) k-d tree construction in 2D space. (b) 2D space nearest neighbor search in k-d tree.

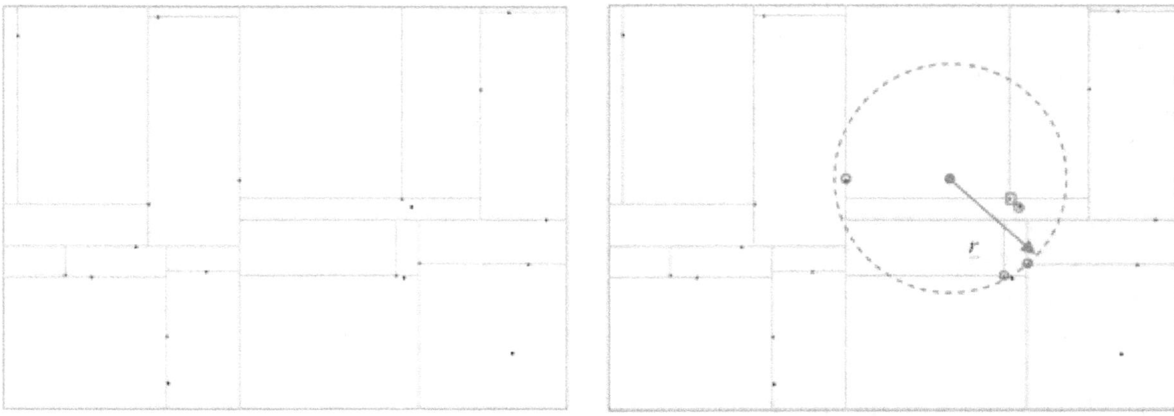

(a) (b)

increases the speed and efficiency of the search. The k-d tree is a spatial partitioning data structure that stores and organizes data in a *k* dimensional space. The k-d tree is a generalized type of binary tree, with every leaf node is a k-dimensional data point that splits the hyperspace into two subspaces. Splitting is done sequentially from the first dimension to the kth dimension. A typical k-d tree in 2D space is shown in figure 11(a). Each point in the 2D space divides the space sequentially into a left-right spaces (about x-axis) or into a top-bottom spaces (about y-axis).

Nearest neighbor search can be done very efficiently on k-d trees. For a given point with known coordinates in the data point cloud and a search radius, the algorithm recursively moves down the tree and follows the same procedure as insertion. Search stops at a leaf node of the tree and the points in the model tree within the search radius are identified. The nearest point is obtained using distance computation. Figure 11(b) shows the nearest neighbor (red square) for the search point at the center of the circle. The nearest point is then regarded as the point associated with the search point.

In 3D closest point search, the distance between 2 points between 2 point clouds is:

$$r_{ij} = \sqrt{(m_{ix} - d_{jx})^2 + (m_{iy} - d_{jy})^2 + (m_{iz} - d_{jz})^2} \quad (8)$$

in which, $\boldsymbol{d}_i\{d_{ix}, d_{iy}, d_{iz}\}$ and $\boldsymbol{m}_j\{m_{jx}, m_{jy}, m_{jz}\}$ are point spatial coordinates in data and model point cloud respectively.

In 4D space, the 4th dimension for each point should be weighed hue value d_{hw} or $m_{hw.}$ The spatial value of points should be normalized by 3D search radius r_{ij} as mentioned in section 4.1. In order to accomplish closest point search in 4D space, the distance between two normalized points $\boldsymbol{d}_i\{d_{ix}, d_{iy}, d_{iz,}, h_{ihw}\}$ and $\boldsymbol{m}_j\{m_{jx}, m_{jy}, m_{jz}, m_{jhw}\}$ should be:

$$r_{ij}^{'} = \sqrt{(m_{ix} - d_{jx})^2 + (m_{iy} - d_{jy})^2 + (m_{iz} - d_{jz})^2 + (m_{ihw} - d_{jhw})^2} \quad (9)$$

or

$$r_{ij}^{'} = \sqrt{r_{ij}^2 + \Delta h_{ijw}^{\ 2}} \quad (10)$$

In the ICP process, search radius effects the computation time and final result. A constant search radius is applied for all iteration loops. Once the search radius is large, too many points will be included as candidates and increases the computational burden. Candidate points cloud will be missed if search radius is too small. The search radius is determined by the density of point cloud. In 4D k-d tree search, the search radius comprises of two parts -- a distance part and weighted hue part as seen in equation (9). The search range for 3D distance is selected such that it ensures about 50 candidate points within search radius. As hue value is not transformed at iteration, hue search is analogous to filtering. If the weight for hue is high, k-d tree search will bias toward hue dimension. Therefore, appropriate hue weighting ensures that spatial search dominates over hue filtering.

The ICP algorithm iteratively converges at minimum error, which is described by mean square root of the spatial distance between paired points. At each iteration, a rigid transformation matrix is computed so that the distance error metric between the associated points is minimized. Data point cloud is transformed into the model space using the computed transformation matrix. This iteration continues until error metric converges.

Use of hue as a fourth dimension is significant in those instances where the coordinate based matching results in a non-unique registration. For example, if the points in the model and the data point clouds belong to a plane, traditional coordinate based ICP results in non-unique association of points. In such cases, using the hue value may result in unique registration of the points. The color assisted ICP algorithm in this paper can be described as follows.

1. Estimate the initial transformation matrix ***R*** and ***T***;
2. Construct k-d tree of model point cloud $\boldsymbol{M\{m_1,m_2,m_3,\dots m_M\}}$, hue value has been weighted as the 4th dimension;
3. ***While*** merging error *ε>preset error*

Use ***R*** and ***T*** to transfer data point cloud $\boldsymbol{D\{d_1,d_2,\dots d_N\}}$.

$$\vec{D} = R\vec{D} + T$$

4. ***For*** *i*=1 ***to*** length of data point cloud

Search closest point for point $\boldsymbol{d_i}\{d_{ix}, d_{iy}, d_{iz}, d_{ih}\}$ in model *k*-d tree

If closest point $\boldsymbol{m_j}$ exists in search range *r*

Pair $\mathbf{d_i}$ and $\mathbf{m_j}$ as $\{\mathbf{d_k}, \mathbf{m_k}\}$;

k++;

End If

End For

Acquire paired point cloud $\boldsymbol{D_p}$ and $\boldsymbol{M_p}$, contain *N* Points, calculate normalized mean square root distance *εas* error,

$$\varepsilon = \frac{1}{N}\sum_{i=1}^{N}\sqrt{(d_{ix}-m_{ix})^2+(d_{iy}-m_{iy})^2+(d_{iz}-m_{iz})^2}$$

6. Construct orthonormality matrix ***H*** (Equation14) and solve rigid rotation ***R*** and translation ***T*** (Equation15, 16) for next iteration;

End While

Solving Rigid Transformation

ICP algorithm is an iteration process to calculate rigid transformation matrix based on associated point clouds. $M_i = \{m_{ix}, m_{iy}, m_{iz}\}$ represent the coordinates of the ith point in the model point cloud and $d_j = \{d_{jx}, d_{jy}, d_{jz}\}$ is the jth point in data point cloud. Rigid transformation (*R*) that minimizes the error measure *E(R,T)* shown in Equation (11) is determined.

$$E(R,T) = \frac{1}{N}\sum_{i=1}^{N}\left\|m_i - (Rd_i + T)\right\| \qquad (11)$$

A centroid for the associated points is calculated as the first step (Equation12) and associated points are translated into centroid relative coordinates (Equation13). Orthonomal matrix of associated points can then be constructed as shown in Equation14. Rotation ***R*** and translation ***T*** are decoupled based on the gravity center of associated points. Using Singular Value Decomposition (SVD) methods, ***R*** can be determined as shown in Equation15. Translation ***T*** is computed using Equation16.

$$\overline{m} = \frac{1}{N}\sum_{i=1}^{N} m_i, \overline{d} = \frac{1}{N}\sum_{i=1}^{N} d_i \tag{12}$$

In which, $\overline{m} = \{\overline{m}_x, \overline{m}_y, \overline{m}_z\}$ and $\overline{d} = \{\overline{d}_x, \overline{d}_y, \overline{d}_z\}$ are the center points of associated points in model and data point clouds. N is the amount of point pairs. The coordinated of associated point in center point relative space should be

$$m_i' = m_i - \overline{m}, d_i' = d_i - \overline{d} \tag{13}$$

In which, $m_i' = \{m_{ix}', m_{iy}', m_{iz}'\}$ and $d_i' = \{d_{ix}', d_{iy}', d_{iz}'\}$ are the ith associated point with center relative coordinates. The orthonormality matrix H can be constructed based on m' $\{m_i'$, *i=1... N*} and d' $\{d_i'$, *i=1... N*}.

$$H = \begin{bmatrix} S_{xx} & S_{xy} & S_{xz} \\ S_{yx} & S_{yy} & S_{yz} \\ S_{zx} & S_{zy} & S_{zz} \end{bmatrix}$$

Where

$$\begin{aligned} S_{xx} &= \sum_{i=1}^{N} m'_{ix} d'_{ix} \\ S_{yy} &= \sum_{i=1}^{N} m'_{iy} d'_{iy} \\ S_{zz} &= \sum_{i=1}^{N} m'_{iz} d'_{iz} \\ S_{xy} &= \sum_{i=1}^{N} m'_{ix} d'_{iy} \end{aligned} \tag{14}$$

Singular value decomposition is performed on constructed ***H*** matrix for optimal rotation **R**

$$H = U \Lambda V^T \tag{15}$$

where optimal rotation $R = VU^T$.

The translation ***T*** can be calculated as

$$T = \overline{m}^T - R\overline{d}^T \tag{16}$$

Convergence Criteria

Convergence and stopping criterion for the H-ICP iteration are discussed in this sub-section. An association stability criteria is introduced as the one of the convergence criteria. Association stability, denoted as *S*, is defined as the number of points which changed their paired point in any iteration. If a point comes into association or changes its nearest neighbor, S is incremented. Large value of *S* signifies that point association has not stabilized. H-ICP iteration is terminated when *S* vanishes and the distance error converges.

A pair wised color point H-ICP registration was accomplished based on above criteria. Model point cloud contained 122,409 points with color attributes. Data point cloud is extracted from model point cloud with a known rotation (θ_z=5°). The H-ICP registration process is compared with 3D ICP, error as shown in Figure 12(a). The associated point number reaches maximum after

Table 1. Hue and RGB values for several common colors

Color	R	G	B	Hue
Gray	128	128	128	0
Yellow	255	255	0	60
Green	0	255	0	120
Cyan	0	255	255	180
Blue	0	0	255	240
Magenta	255	0	255	300
Red	255	0	0	360

the 5th iteration (Figure 12(b)), but error has not converged. From Figure 12(c) the association stability (S) reaches 0 after 15th and 26th iteration for H-ICP and 3D ICP respectively. Error and rigid transformation are shown in Figure 12(a) and Figure 13. The known transformation (θ_z=5°) is recovered by the H-ICP and ICP algorithms.

MAP REGISTRATION WITH ICP AND H-ICP

The hue distribution or the color of the model is generally independent of the geometry. If the entire body is painted with a color of a single hue, H-ICP is as effective as the traditional ICP. In this section, we describe the performance of the algorithm under various hue distribution scenarios. The Stanford bunny point cloud is considered as the benchmark data set. In HSL color space, hue value varies from 0- 360. The color correspondence between RGB and hue is given in Table 1.

Environments with Fixed Hue Distributions

For the first experiment, we textured the Stanford bunny point cloud model as shown in Figure 14(a). In this model, the hue varies from 0 to 360 with from bottom to top at Z direction in seven segments. Figure 14(b) also shows the initial registration of the model and data point clouds used for this simulation.

The H-ICP registration progress is shown in Figure 15(a) and Figure 15(b). Figure 15(a) shows the mean square error during the ICP process and Figure15 (b) shows the number of points associated during iteration loops. Both data and model point cloud after registration is shown in Figure 14(b). The hue-assisted ICP registers the point and data clouds faster than the traditional coordinate based ICP.

Continuously Varied Hue Along One Dimension

In the second simulation, a continuous hue distribution is assigned to the bunny model. The hue value is varied from 0 to 360, smoothly, along the z (vertical) direction. The resultant model and data clouds are shown in Figure 16 (a), (b). Saturation and lightness value have been set as constant at every point inside dataset. Hue value can be calculated by equation (17).

$$h = 360\frac{z_j - z_{\min}}{z_{\max} - z_{\min}} \qquad (17)$$

h is the hue value at range point i, z_i is the coordinate distance for i^{th} point at z direction, z_{max} and z_{min} are maximum and minimum coordinate of the point cloud at z direction.

Figure 12. Building color point cloud registration comparison between H-ICP and 3D ICP algorithm. (a) Comparison of error convergence. (b) Association number convergence. (c) Association stability convergence.

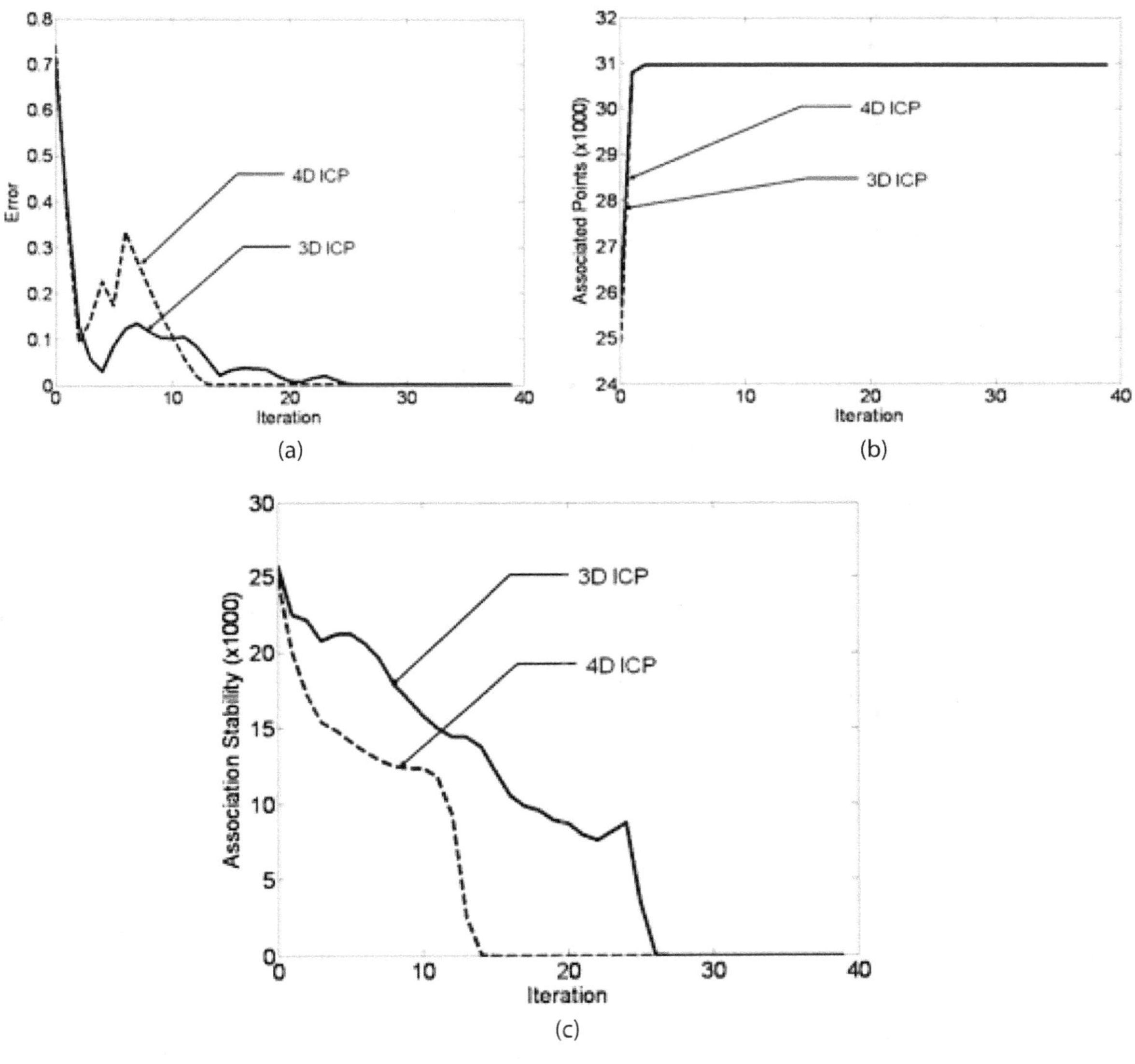

Continuous hue distribution on point cloud data is registered together (Figure 16 (c)) and the results are shown in Figure 17. A comparison of model performance on discrete and continuous distribution of hue on the same model shows the expected acceleration in performance due to uniform distribution of hue on the model.

Randomized Hue on the Model

In this case, the model considered has a continuously distributed hue but with a randomized and noisy pattern. In this case, there is no geometric pattern for the color on the object. The color point clouds are rendered in Figure 18 (a, b). The merged cloud point cloud after registration is shown in Figure 18(c). Figure 19 shows the

Figure 13. Convergence of translation and rotation estimates during registration

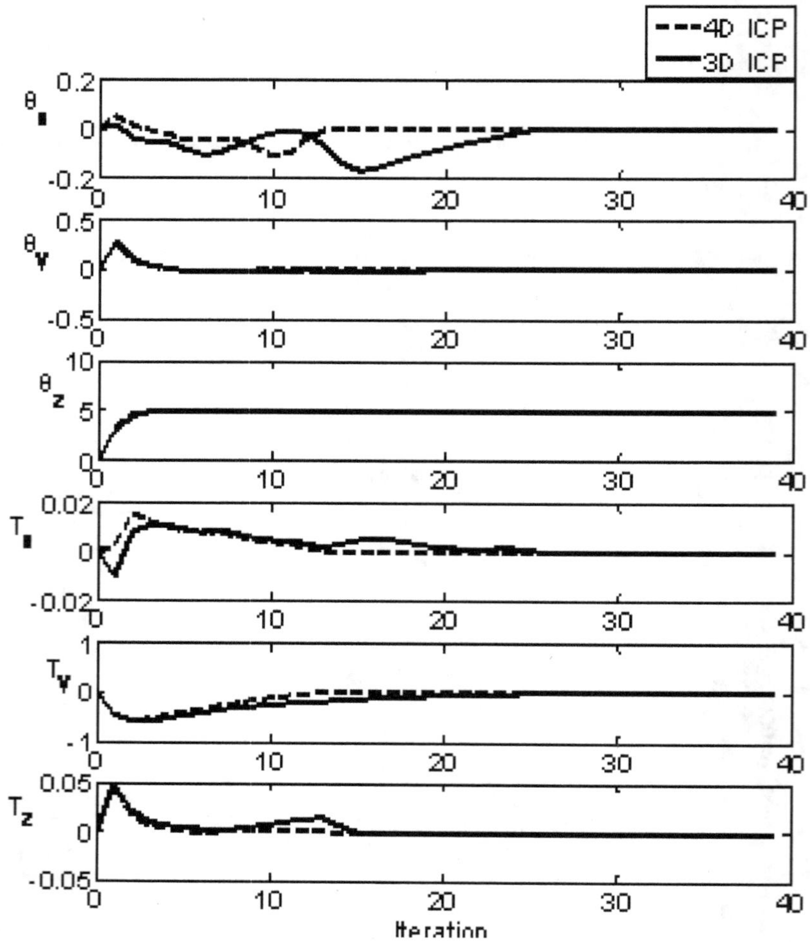

error minimization iteration and comparison with the seven-segment hue distribution model. In this case, the hue confuses the nearest neighbor search. The registration accuracy is also not as good as a patterned hue case.

Effect of Imaging Noise

In the previous simulation, the imaging sensor is assumed perfect. The hue on a point is assumed to be recorded by the imaging sensor perfectly in both model and data clouds. Some noise in the color measurement can be expected when the point clouds are generated from two vantage points (Gebre et al., 2009). Considering this situation, we colorized the bunny model but with 50% noise in the sensor. The points in the model and data clouds differ in color by as much as 50%. The resulting point clouds are shown in Figure 20(a, b). The merged color point cloud is shown as Figure 20(c).

Hue assisted color ICP matching result in camera noise color point cloud is compared with 3D ICP matching performance. From Figure 21, noise in hue decreases the matching accuracy and reduces the iteration efficiency. Two groups of cloud point clouds are selected to evaluate the performance of H- ICP algorithm compared with typical 3D ICP. A known transformation point cloud data pair was generated by transforming

Figure 14. Registration of point clouds with uniformly distributed hues (a) Stanford Bunny point clouds with hue distributed as seven distinct stripes (b) Registered color point cloud

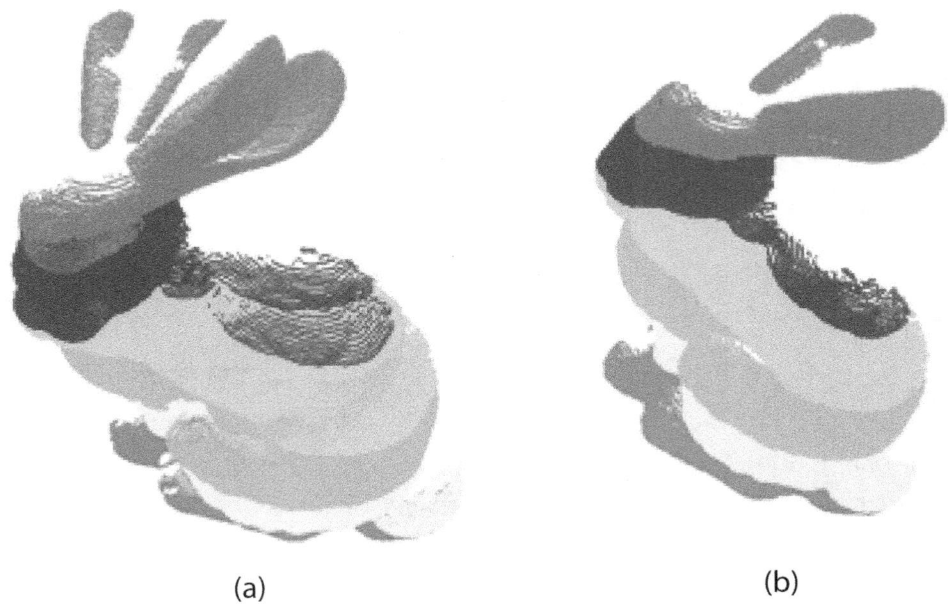

Figure 15. Registration comparisons between H-ICP and 3D ICP algorithm (a) Mean square error comparison. (b) Associated point number comparison.

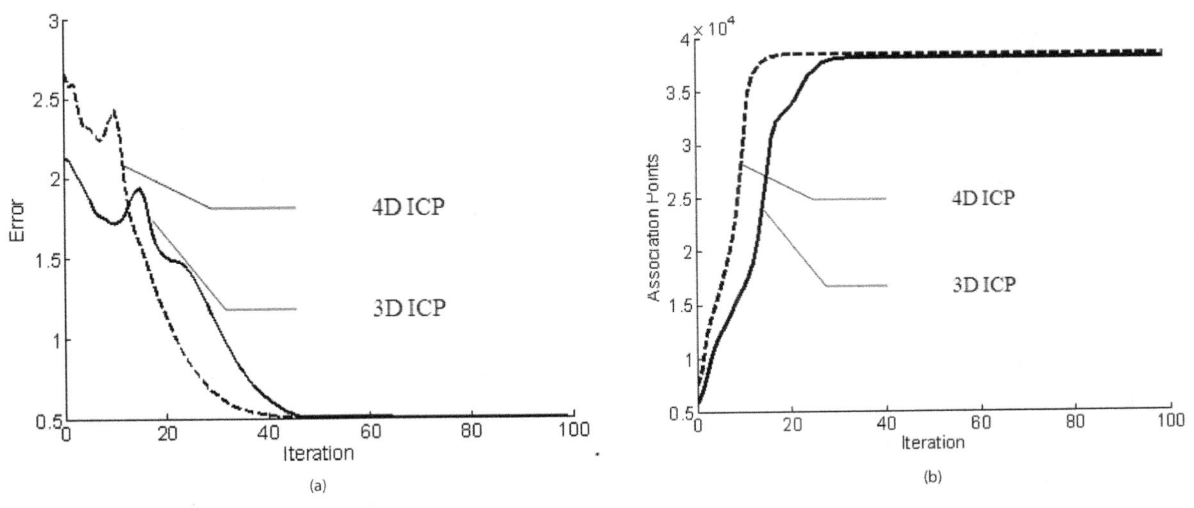

model point cloud at 6DOF to compare the convergence speed and registration accuracy as the rigid transformation is already known. Outdoor large scale area pair wised registration includes 8 pair wised data registration.

Registration with Six DOF Rotation

In this experiment, registration speed between 3D ICP and H-ICP are compared using data and model point clouds with known (and exact) registration

Figure 16. Bunny model with continuous hue variation in one axis (a) Data point cloud. (b) Model point cloud. (c) Merged View.

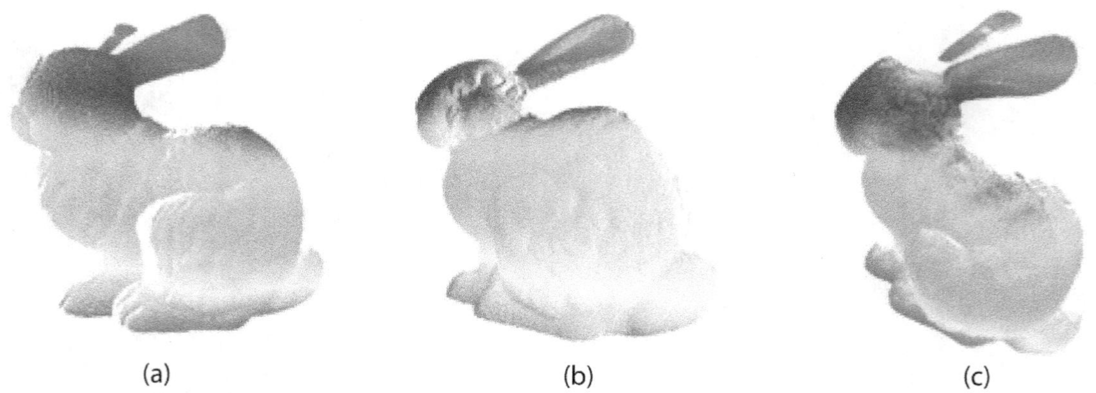

Figure 17. Registration comparisons between 7 segment hue model and continuous hue model. (a) Mean square error comparison (b) Associated point number comparison.

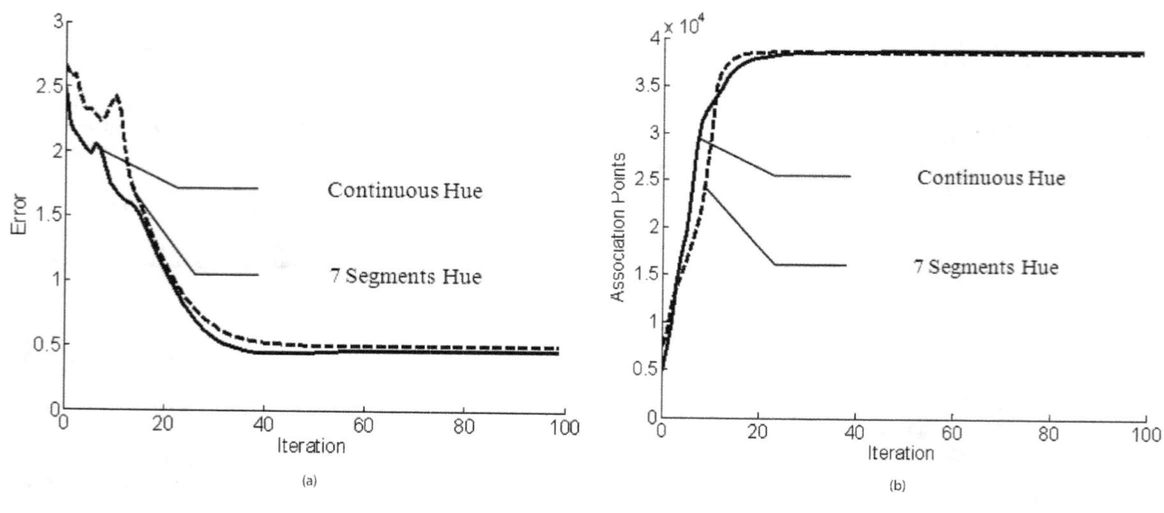

transformation. Both H-ICP algorithm and 3D ICP algorithm have been applied on a building data set (Gebre et al., 2009). The data point cloud is taken from a view position that is 10° off in Y and Z axis from the model point cloud. Translation between the point clouds is known to be 2.46, 2.612 and 0.347 along the X, Y, and Z respectively. Same parameters for registrations are selected to be the same as in the previous 1-DOF registration. Error comparison and associated point number comparison are shown in Figure 22(a) and (b). Association stability is shown in Figure 22(c). The evolution of rigid transformation during ICP is shown in Figure 23. The H-ICP completes registration after 102th iteration and the traditional 3D ICP after the 164th iteration, which demonstrates the effectiveness of H-ICP for registering complex and realistic point clouds. The merged color point cloud about building is shown in Figure 24.

Figure 18. Bunny point cloud with randomized hue distribution (a) Data point cloud (b) Model point cloud (c) Merged View

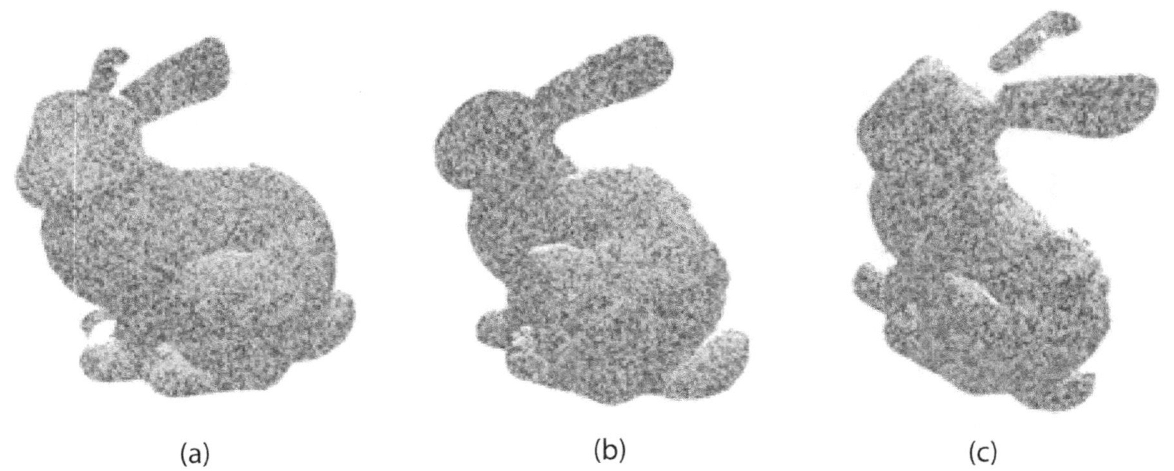

Figure 19. Comparison between discrete and random hue distribution case (a) Mean square error comparison (b) Associated point number comparison

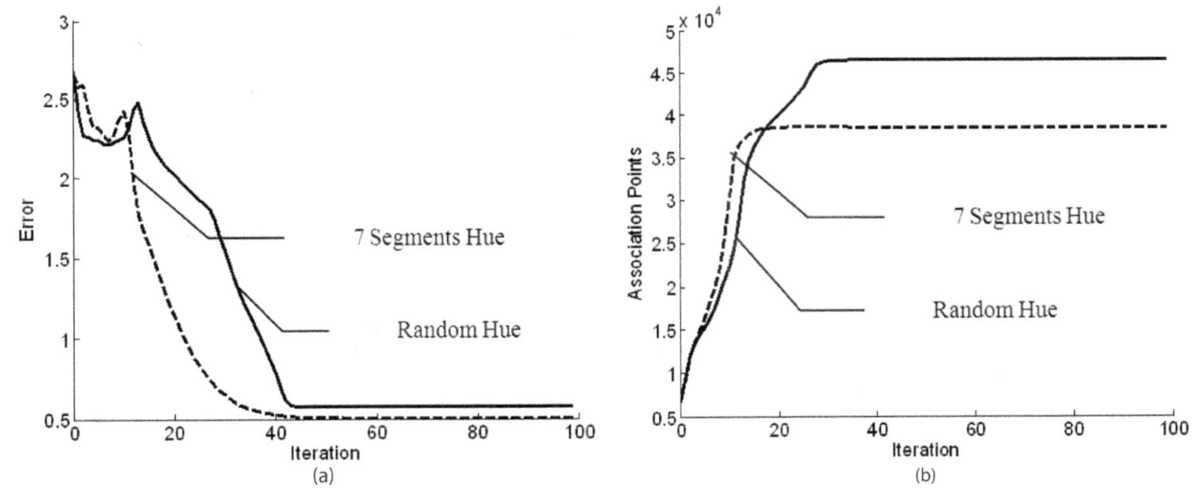

Sequential Registration of Multiple Point Clouds

3D ICP and H ICP algorithms have been applied on several outdoor map segments. Color point clouds taken from eight different vantage points have been registered together to construct a large scale color point cloud map. Figure 25 shows the top view of outdoor mapping area in aerial image. This scene includes trees, road, electrical poles and buildings. Figure 26 shows the registered map and the vantage points from which map segments are obtained. Pair-wise registration is applied to construct a single map about the reference coordinate of the first map segment. 3D search radius in k-d tree was set as 1.5 and the 3D range data was normalized based on this radius. Hue value was normalized to a 0-1 range, hue search radius

Figure 20. Hue mapped with noise. (a) Data point cloud. (b) Model point cloud. (c) Merged View.

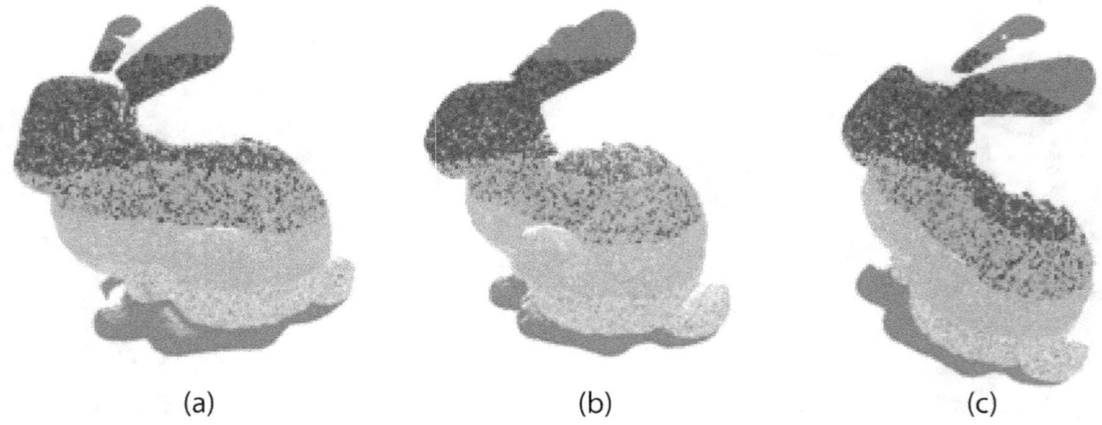

Figure 21. Comparison between H- ICP in and 3D ICP for noisy hue case. (a) Mean square error comparison. (b) Associated point number comparison.

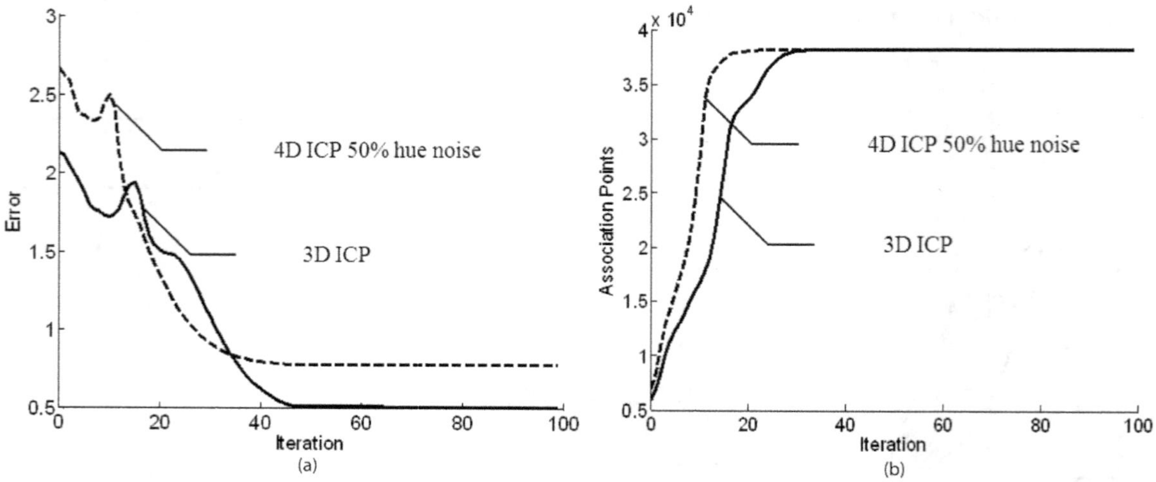

Table 2. Sequential registration of multiple point cloud maps

Position	3D ICP Iterations	H- ICP Iterations	3D ICP Error	H- ICP Error
2	45	35	0.842	0.856
3	54	44	0.929	0.961
4	77	54	0.039	0.290
5	49	43	0.104	0.319
6	66	59	0.165	0.179
7	73	69	0.129	0.128
8	99	95	0.068	0.070

Figure 22. Registration comparisons between 3D ICP and H-ICP algorithm. (a) Mean square error comparison. (b) Associated point number comparison. (c) Stability Comparison.

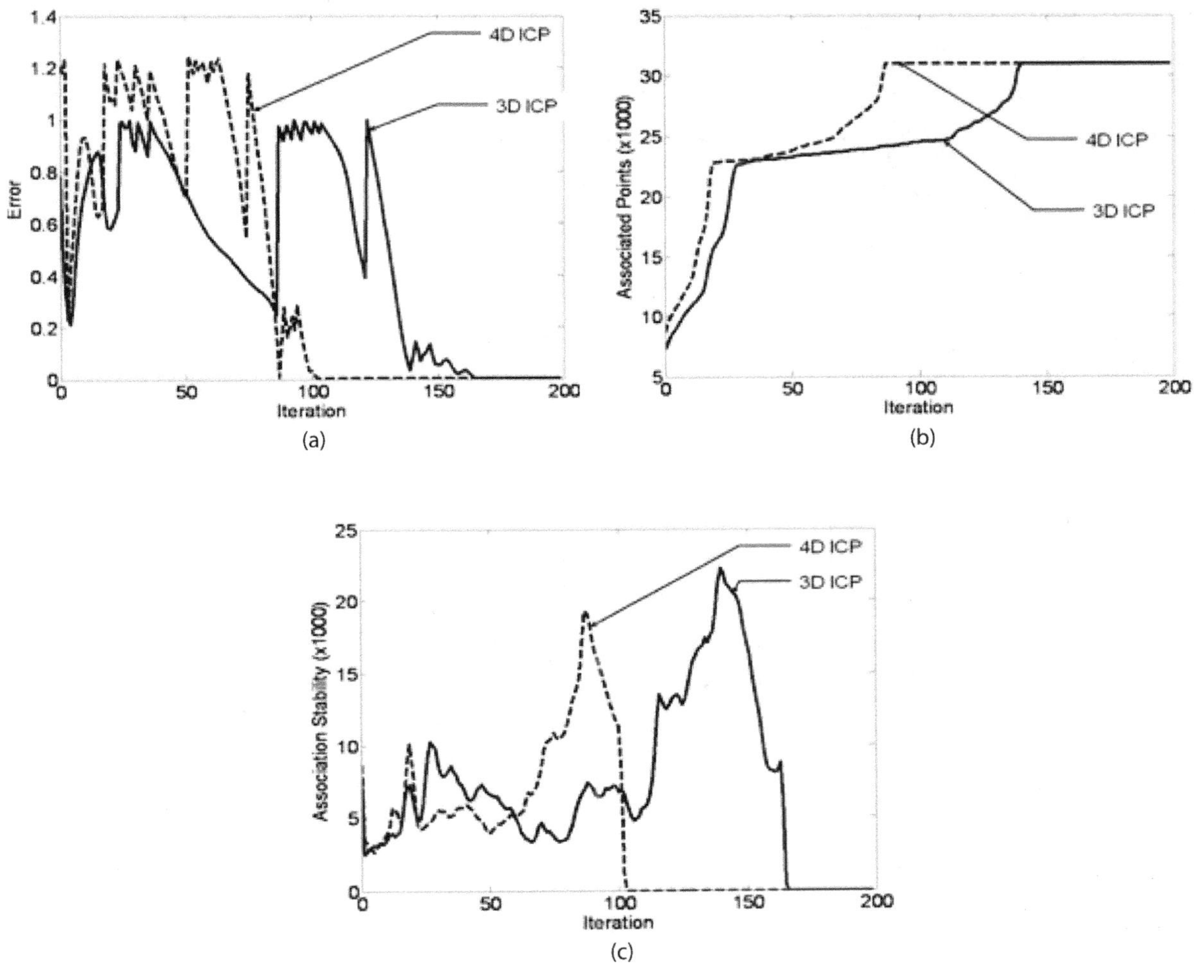

was set to be 0.15, and hue weight was set to 5.0. The final error and the number of iterations required to register the point clouds is shown in Table 2. H-ICP requires less number of iterations than 3D ICP.

This experiment proves that faster registration will be conducted by adding color value into registration progress. Position 3 and 4 acquired point clouds have been registered together and shown in Figure 27, Figure 27(a) describes two different point clouds with two different colors; point cloud at position 4 (black) has been registered into position 3 point cloud (blue). Combined point clouds with color are shown in Figure 27(b).

FUTURE RESEARCH DIRECTIONS

Point clouds are inefficient representations of geometry. Some of the future research directions can include:

a. Efficient generation of higher order geometric representations --- lines, surfaces and solids from the point cloud data;

Figure 23. Convergence of translation and rotation estimates during registration

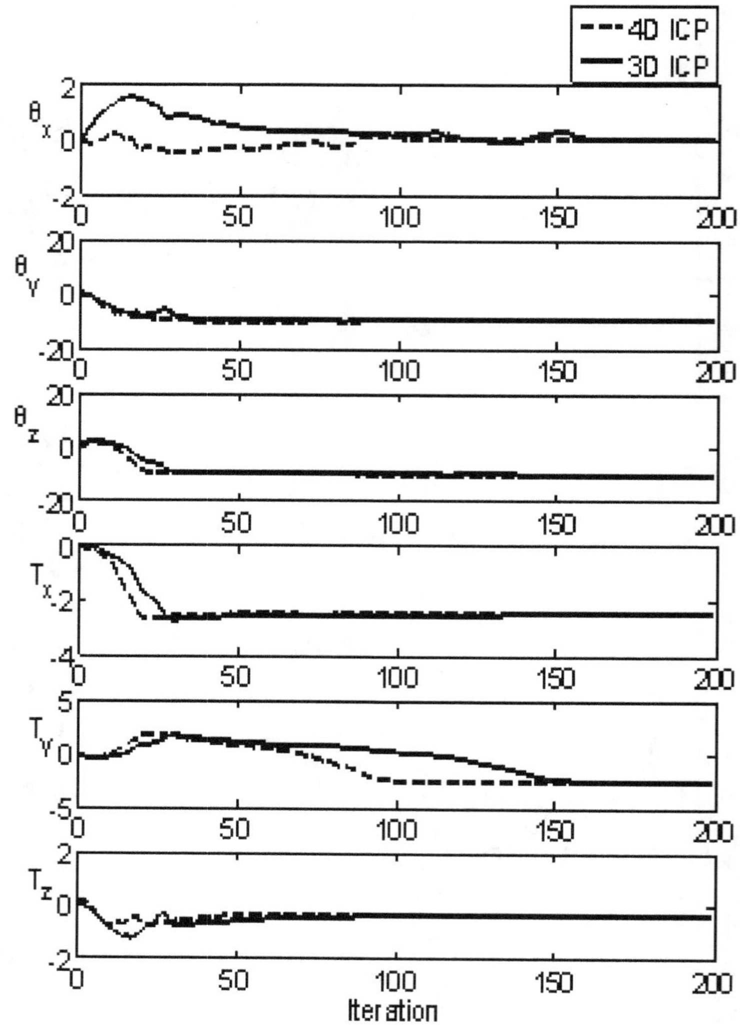

b. Map completeness measures that predict the geometry missing in the occluded areas based on a knowledge-base; and
c. Extra sensing modalities such as infrared or thermal imaging, acoustic/ultrasonic and radio frequency imaging to help determination of materials in the scene.

Architecture, surveying and engineering fields have considerable needs for automatic or semi-automatic conversion of 3D point clouds into higher order line, surface and solid models that are compatible with commercial CAD software. This enables bringing the point cloud data into existing business processes like generation of drawings for code compliance, additions and modifications to existing built areas and remodeling interior spaces.

CONCLUSION

This chapter describes an algorithm to introduce color attribute into point cloud registration process and fundamental algorithms for autonomous ro-

Figure 24. Registered data and model point clouds

Figure 25. Aerial image of outdoor mapping area and vantage positions

Figure 26. Top view of eight sequentially registered color point cloud maps

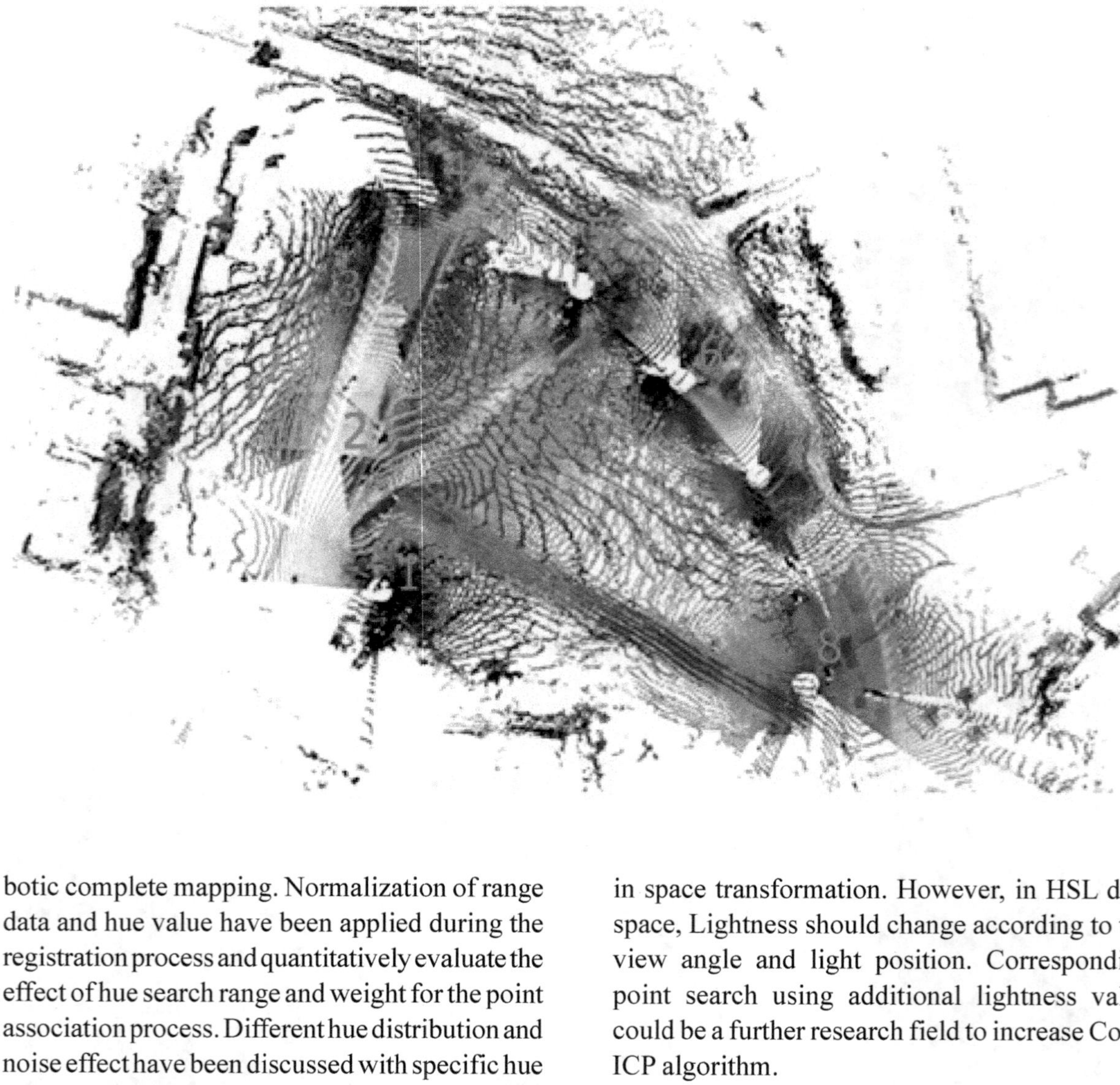

botic complete mapping. Normalization of range data and hue value have been applied during the registration process and quantitatively evaluate the effect of hue search range and weight for the point association process. Different hue distribution and noise effect have been discussed with specific hue rendered color point clouds. A building data set and large-scale outdoor point cloud has been registered using image data assisted algorithm. Use of the hue value to assist the point association and error minimization is shown to be effective during the ICP iteration schemes. Higher dimensional point association based on weighted hue and range data leads more accurate point matching result, conduct earlier convergence of ICP progress, and reduce computation time. When rigid transformation is been application in every iteration loop during the ICP period, hue value does not change in space transformation. However, in HSL data space, Lightness should change according to the view angle and light position. Corresponding point search using additional lightness value could be a further research field to increase Color ICP algorithm.

ACKNOWLEDGMENT

The authors acknowledge the partial support from US/ARMY ARDEC for the development of ROAMS robot. We acknowledge the help of Mr. Biruk Gebre for generating some of the scan point cloud maps used in this chapter.

Figure 27. Map registered from scans taken from two vantage points. (a) Registered position 4(black) point cloud into position 3 (blue) point cloud. (b) Color point cloud after registration.

REFERENCES

Andreason, H., & Lilienthal, A. J. (2010). 6D scan registration using depth-interpolated local image features. *Robotics andAutonomous Systems, 59*, 157-165.

Andresson, H. (2007). *Vision aided 3D laser scanner based registration*. Paper presented at European Conference on Mobile Robots: ECMR.

Arun, K. S., Huang, T. S., & Blostein, S. D. (1987). Least square fitting of two 3D-Point sets. *IEEE Transactions on Pattern Analysis and Machine Intelligence*, *9*(5), 698–700. doi:10.1109/TPAMI.1987.4767965

Basilico, N., & Amigoni, F. (2009). *Exploration strategies based on multi-criteria decision making for an autonomous mobile robot.* Presented at European Conference on Mobile Robots, Mlini/Dubrovnik, Croatia.

Blais, F. (2004). Review of 20 years of range sensor development. *Journal of Electronic Imaging, 13*, 231–243. doi:10.1117/1.1631921

Browell, E. V., Butler, C. F., & Ismail, S. (1990). Airborne lidar observations in the wintertime arctic stratosphere: Polar stratospheric clouds. *Geophysical Research Letters, 17*, 385–388. doi:10.1029/GL017i004p00385

Bsel, P. J. (1992). A method for registration of 3D shapes. *IEEE Transactions on Pattern Analysis and Machine Intelligence, 14*, 239–256.

Chibunichev, A. G., & Velizhev, A. B. (2008). *Automatic matching of terrestrial scan data using orientation histograms*. Presented at the International Archives of the Photogrammetry, Remote Sensing and Spatial Information Sciences, Beijing, China.

Chua, C. J. R. (1997). Point signatures: A new representation for 3D object recognition. *International Journal of Computer Vision, 25*, 63–85. doi:10.1023/A:1007981719186

Chung, D., & Lee, Y. D. S. (1998). Registration of multiple range views using the reverse calibration technique. *Pattern Recognition, 31*(4), 457–464. doi:10.1016/S0031-3203(97)00063-0

Davison, A. J. (2003). *Real-time simultaneous localization and mapping with a single camera*. Presented at the 9th IEEE Computer Society Conference on Computer Vision and Pattern Recognition, Madison, Wisconsin.

Druon, S. (2007). *Color constrained ICP for registration of large unconstructed 3D/Color data sets*. Presented at IEEE International Conference on Robotics and Automation, Roma, Italy.

Fernandez, J. C., & Singhania, A. (2007). *An overview of lidar point cloud processing software*. GEM Center Report, University of Florida.

Gebre, B., Men, H., & Pochiraju, K. (2009). *Remotely operated and autonomous mapping system (ROAMS)*. Paper presented at the 2nd Annual IEEE International Conference on Technologies for Practical Robot Applications, Woburn, MA.

Johnson, A. E., & Kang, S. B. (1997). *Registration and integration of textured 3D data*. Presented at International Conference on Recent Advances in 3D Didital Imaging and Modeling, Ottawa, Canada.

Johnson, A. (1997). *Spin-images: A representation for 3d surface matching*. Doctoral dissertation, Carnegie Mellon University, USA.

Joung, J. H., An, K. H., Kang, J. W., et al. (2009). *3D environment reconstruction using modified color ICP algorithm by fusion of a camera and a 3D laser range finger*. Presented at the 2009 IEEE International Conferrnce on Intelligent Robos and Systems, St. Louis, USA.

Kostelec, P. J., & Rockmore, D. N. (2008). FFT on the rotation group. *Journal of Fourier Analysis and Application, 14*, 145–179. doi:10.1007/s00041-008-9013-5

Lorusso, A. (1995). *A comparison of four algorithms for estimating 3D rigid transformations*. Presented at British Machine Vision Conference.

Makadia, A., Iv, E. P., & Daniilidis, K. (2006). *Fully automatic registration of 3D point clouds*. Presented at the 2006 Computer Society Conference on Computer Vision and Pattern recognition, New York, NY.

Men, H., & Pochiraju, K. (2010). *Hue assisted registration of 3D point clouds*. Presented at ASME 2010 Interaional Design Engineering Technical Conference, Montreal, Canada.

Montemerlo, M., Thrun, S., et al. (2003). *FastSLAM 2.0: An improved particle filtering algorithm for simultaneous localization and mapping that provably converges*. Presented at the International Joint Conference on Artificial Intelligence, Acapulco, Mexico.

Newman, P., Cole, D., & Ho, K. (2006). Outdoor SLAM using visual appearance and laser ranging. Presented at IEEE International Conference on Robotics and Automation, Orlando, Florida.

Nüchter, A. (2005). *6D SLAM with approximate data association*. Presented at IEEE International Conference on Robotics and Automation, Barcelona, Spain.

Nuchter, A. (2007). *Catched k-d tree search for ICP algorithms*. Presented at 6th International Conference on 3-D Digital Imaging and Modeling, Montreal, Canada.

Oh, J. S., & Choi, Y. H. (2004). Complete coverage navigation of cleaning robots using triangular-cell-based map. *IEEE Transactions on Industrial Electronics*, *51*, 718–727. doi:10.1109/TIE.2004.825197

Olson, C. F. (2007). Probabilistic self-localization for mobile robot. *IEEE Transactions on Robotics and Automation*, *16*, 55–67. doi:10.1109/70.833191

Robertson, C. (2002). Parallel evolutionary registration of range data. *Computer Vision and Image Understanding*, *87*, 39–50. doi:10.1006/cviu.2002.0981

Rusinkiewucz, S. (2001). *Efficient variants of the ICP algorithm*. Presented at 3rd International Conference on 3-D Digital Imaging and Modeling, Quebec City, Canada.

Salvi, J., Matabosch, C., & Fofi, D. (2007). A review of recent range image registration methods with accuracy evaluation. *Image and Vision Computing*, *25*, 578–596. doi:10.1016/j.imavis.2006.05.012

Spletzer, J. R. (2003). *Sensor fusion techniques for cooperative localization in robot teams*. Doctoral dissertation, University of Pennsylvania.

Tarel, J., Civi, H., & Cooper, D. (1998). *Pose estimation of free-form 3D objects without point matching using algebraic surface models*. Presented at IEEE Workshop on Model-based 3D.

Thrun, S. (1993). *Exploration and model building in mobile robot domains*. Paper presented at IEEE International Conference on Neural Networks, San Francisco, USA.

Thrun, S. (2003). Robotic mapping: A survey. In Lakemeyer, G. (Ed.), *Exploring artificial intelligence in the new millennium* (pp. 1–35). Morgan Kaufmann Publishers Inc.

Thrun, S., Fox, D., Burgard, W., & Dellaert, F. (2000). Robust Monte Carlo localization for mobile robots. *Artificial Intelligence*, *128*(1-2), 99–141. doi:10.1016/S0004-3702(01)00069-8

Welch, G., & Bishop, G. (1995). *An introduction to the Kalman filter. Technical Report TR 95-041*. University of North Carolina, Department of Computer Science.

Whyte, H. D., & Bailey, T. (2006). simultaneous localization and mapping (slam): part i- the essential algorithms, *IEEE Robotics and Automation Magazine*.

Yamauchi, B. (1997). *A frontier based approach for autonomous exploration*. Presented at IEEE International Conference on Autorotation and Robotics, Albuquerque, NM.

Zelinsky, A. (1994). Planning paths of complete coverage of an unstructured environment by a mobile robot. *The International Journal of Robotics Research*, *13*(4), 315. doi:10.1177/027836499401300403

ADDITIONAL READING

Blais, F. (2004). Review of 20 Years of Range Sensor Development. *Journal of Electronic Imaging*, *13*, 231–243. doi:10.1117/1.1631921

Druon, S. (2007), *Color Constrained ICP for Registration of Large Unconstructed 3D/Color Data Sets*, Presented at IEEE International Conference on Robotics and Automation, Roma, Italy.

Gebre, B., Men, H., & Pochiraju, K. (2009), *Remotely Operated and Autonomous Mapping System(ROAMS)*, Paper presented at the 2nd Annual IEEE International Conference on Technologies for Practical Robot Applications, Woburn, MA, Fernandez, J.C., Singhania, A. et Al (2007), An Overview of Lidar Point Cloud Processing Software, *GEM Center Report,* University of Florida.

Lorusso, A. (1995), *A Comparison of Four Algorithms for Estimating 3D Rigid Transformations*, Presented at British Machine Vision Conference.

Makadia, A., Iv, E. P., & Daniilidis, K. (2006), *Fully Automatic Registration of 3D Point Clouds*, Presented at the 2006 Computer Society Conference on Computer Vision and Pattern recognition, New York, NY.

Men, H., & Pochiraju, K. (2010), *Hue Assisted Registration of 3D Point Clouds*, presented at ASME 2010 Interaional Design Engineering Technical Conference, Montreal, Canada.

Montemerlo, M., Thrun, S., et al. (2003), *FastSLAM 2.0, An Improved Particle Filtering Algorithm for Simultaneous Localization and Mapping that Provably Converges*, Presented at the International Joint Conference on Artificial Intelligence, Acapulco, Mexico.

Nüchter, A. (2005), *6D SLAM with Approximate Data Association*. Presented at IEEE International Conference on Robotics and Automation, Barcelona, Spain.

Nuchter, A. (2007), *Catched k-d tree Search for ICP Algorithms*, Presented at 6th International Conference on 3-D Digital Imaging and Modeling, Montreal, Canada.

Rusinkiewucz, S. (2001), *Efficient Variants of the ICP Algorithm*, Presented at 3rd International Conference on 3-D Digital Imaging and Modeling, Quebec City, Canada.

Salvi, J., Matabosch, C., & Fofi, D. (2007). A Review of Recent Range Image Registration Methods with Accuracy Evaluation. *Image and Vision Computing, 25*, 578–596. doi:10.1016/j.imavis.2006.05.012

Thrun, S. (2003). *Robotic Mapping: A Survey, Exploring artificial intelligence in the new millennium* (pp. 1–35). Morgan Kaufmann Publishers Inc.

KEY TERMS AND DEFINITIONS

3D Map: 3D map represents certain level of spatial information about the geometry features in specific area. The map is dimensionally accurate and may have a relative or absolute reference.

3D Mapping: 3D mapping is the process of applying measurement devices to construct 3D map about specified environment.

Color Point Cloud: Discrete points group with both dimensiaonl accurate measurement and texture property, normally generated by both ranging device and color camera.

Map Registration: A process to accurately stitch pair or multiple point clouds together into single point cloud.

Point Cloud: Discrete points group with accurate 3D coordinates describing object surface dimensional measuremnts, usually contructed by laser ranging devices.

Chapter 5
3D Shape Compression Using Holoimage

Nikolaus Karpinsky
Iowa State University, USA

Song Zhang
Iowa State University, USA

ABSTRACT

As 3D becomes more ubiquitous with the advent of 3D scanning and display technology, methods of compressing and transmitting 3D data need to be explored. One method of doing such is depth mapping, in which 3D depth data is compressed into a 2D image, and then 2D image processing techniques may be leveraged. This chapter presents a technique of depth mapping 3D scenes into 2D images, entitled Holoimage. In this technique, digital fringe projection, a special kind of structured light technique from optical metrology, is used to encode and decode 3D scenes pixel-by-pixel. Due to the pixel-by-pixel 3D data processing nature, this technique can be used on parallel hardware to realize real-time speed for high definition 3D video encoding and decoding.

INTRODUCTION

Advancements in 3D imaging and computational technology have made acquisition and display of 3D data simple. Techniques such as structured light, stereovision, and light detection and ranging (LIDAR) have led the path in 3D data acquisition (Gorthi & Rastogi, 2010). Stereoscopic displays have made the display of 3D data a reality. However, as these fields and techniques evolve, a growing problem is being confronted; how can 3D data be efficiently stored and transmitted?

DOI: 10.4018/978-1-61350-326-3.ch005

Storage and transmission of 3D data has become a large problem due to the file sizes associated with 3D geometry. Standard 3D file storage techniques do not lend themselves nicely to high detailed, high frame rate scenes. Instead, traditional 3D file storage techniques aim to store models and then animated models based on constraints of a few points, typically skeletal points

(Forstmann et al., 2007). This does not hold true for 3D scenes captured from 3D scanners, as they consist of a large array of 3D coordinates, all of which are animated; this animation is inherently unstructured and unconstrained making typical 3D file storage difficult. Introducing 3D models into 3D scenes only further exacerbates the problem, as both modalities need to be accounted for. How then can such scenes with both types of 3D data be encoded in a unifying way, which provides not only an efficient storage and transmission medium, but also a quick encoding and decoding of high definition (HD) data?

One solution to this problem is to use depth mapping to encode 3D scenes consisting of unstructured scanned data and structured models into 2D images, and then rely on 2D image compression and transmission techniques. The benefit of doing this is that decades of research and development in 2D image processing can be leveraged, utilizing existing compression and transmission techniques along with existing infrastructure. Existing video services such as Youtube and Vimeo can be used with slight modifications; only the video renderer needs to be modified to decode and display 3D scenes rather than 2D images.

Holoimage (Gu et al, 2006) is a technique developed to accomplish the task of depth mapping an entire HD 3D scene. Utilizing techniques developed in optical metrology, Holoimage creates a virtual fringe projection (a special kind of structured light) system which can depth map an entire 3D scene point-by-point into 2D images. The benefits of such a technique include: (1) using existing research in the field of optical metrology; (2) leveraging existing research in the field of image processing; and (3) achieving point-by-point computation though the whole process. Employing parallel hardware such as that of a graphics processing unit (GPU), HD 3D scenes can be encoded and decoded in real-time. Thus Holoimage meets the requirements of encoding and decoding a 3D scene with little speed hindrance, lending itself nicely to 3D video and other high-speed, high-resolution 3D applications (Zhang & Huang, 2006a; Zhang & Yau, 2006; Zhang & Yau, 2007). This chapter will delve into the details of the Holoimage technique, will show compression results, and will discuss the advantages and shortcomings.

BACKGROUND

Related Work

To compress point cloud data, two different classes of encoders have been developed: progressive coders, and single-rate coders. Progressive coders encoded point clouds with a coarse representation and then progressive refinements. This allows for the coarse representation to be displayed almost immediately, and then gradual streaming of refinements to occur when they become available. These schemes typically involve building a tree of vertices in memory, such as a kd or spanning tree, followed by entropy encoding using predictive heuristics, and finalized with run level or Huffman encoding. This allows for high levels of lossless compression such as 27:1, but with slow encoding times for dense point clouds. Schnabel and Klein developed such a technique that uses Octrees, with coarse representations using approximately 2 bits per pixel, and then refinements using up to 6 bits per pixel (Schnabel & Klein, 2006).

The other class of encoders, single-rate coders, requires the entire file before decoding can commence. This class of encoders typically consists of a simple decoder, which can quickly decode and display a file. Fast decoding makes these techniques viable for real-time applications such as 3D video, but compression rates typically are not as high as seen in progressive encoders. Chai et al. developed such an encoder, which encodes a depth map for a scene along with a triangular-mesh (Chai, Sethuraman, Sawhney & Hatrack, 2004). They were able to achieve compression ratios ranging from approximately 2:1 all the way

Table 1. 3D File formats compared to uncompressed 2D image format all with 640 x 480 points. Note the closest format is still over 5 times as large as its 2D counterpart. Also 3D formats contain only vertices and connectivity if required; no point normals or texture coordinates are stored. DAE – Digital Asset Exchange; OBJ – Wavefront Object file; STL – Stereolithography file format.

	Bitmap image	PLY	DAE	OBJ	STL
File size:	1.2MB	6.5MB	10.6MB	12.8MB	17MB
Ratio:	1: 1	1: 5.42	1: 8.83	1: 10.67	1: 14.17

to approximately 27:1. When rendering a large number of triangles, their implementation fell far below real-time, achieving only 8.8 frames per second for approximately 125k triangles. The coder presented in this chapter is classified as a single-rate coder.

Depth Mapping

Unlike 2D images, 3D geometry conveys much more information, albeit at the price of increased data size. An example of this extra information is facial identification; 2D facial identification systems rely on the texture of a 2D image, whereas 3D facial identification systems utilize the structure of 3D geometry (Abate et al 2007). Thus, if lighting conditions change, or the subject's texture changes such as applying makeup to circumvent the system, 2D facial identification systems will fail. 3D facial identification systems relying on facial structure on the other hand will not, as the subject would have to change their physical facial structure in order to fool the 3D facial identification system. The cost of all this added information is large in comparison with 2D though in terms of file size.

To illustrate this point, consider a simple 2D color image with three color channels and 8-bit color depth resulting in 24 bits per pixel or 3 bytes per pixel. Now consider a simple 3D point cloud with a 4 byte floating point number for each component of a 3D coordinate (x, y, z) resulting in 12 bytes per point. Already the 3D geometry is on an order of 4 times as large. Now include connectivity information which is standard in most 3D data formats, along with point normals, texture coordinates, and associated texture data; the resulting 3D geometry is on the order of 10-20 times larger than a 2D image with the same number of points. This illustrates the point that where even one of the smallest 3D formats, polygon file format (PLY) is still over 5 times larger than an uncompressed 2D image. Now employ 2D image compression such as portable network graphics (PNG) compression and the result is staggering (Table 1).

To overcome this problem for large static models, computer graphics has employed what is known as depth mapping for some time. The idea behind depth mapping is to encode 3D geometry into 2D images, which can then later be decoded back to 3D, known as image based rendering (Krishnamurthy et al, 2001). Typically, the model being depth mapped is aligned with a plane such as the XY plane, and then the Z component is encoded in a 2D image known as a depth map. The result of the process is a 2D image, which assumes that it is XY axis aligned, the points are uniformly spaced, and each pixel encodes the depth at the point or the Z value. Performing this operation allows for the use of decades of existing research in 2D image processing to be leveraged as the 3D geometry has been encoded into a 2D image. Thus, storage and transmission of the geometry are simplified. Typically, large static models in a 3D world are terrain models, and depth mappings are often employed to quickly generate these models at photorealistic levels. Due to the nature of 3D scanners and their use of

2D images to generate 3D data, a natural connection with depth mapping is apparent. In a typical 3D scanning system, the 3D coordinates are recovered from 2D images captured by cameras. Therefore, there should be a way to convert these recovered 3D scenes back into 2D images.

Another technique that is often employed with depth mapping is level of detail (LOD) triangle meshes (Lindstrom et al, 1996). As a camera moves farther from its subject, the perspective makes the subject appear to get smaller and smaller. As the subject gets smaller from the camera's viewpoint, less camera pixels can capture the subject, thus the subjects level of detail can be decreased without affecting the detail at the camera's viewpoint. LOD triangle meshes take advantage of this principle and reduce the number of vertices in the overall mesh to display an appropriate level of detail, while displaying as few vertices as possible for speed. Since depth maps are applied per vertex, the fewer vertices the faster the decoding process. Thus LOD meshes work nicely with depth mapping, decoding and displaying only the needed level of detail in a mesh.

HOLOIMAGE TECHNIQUE

Principle

The principle behind Holoimaging is borrowed from optical metrology and is known as fringe projection (Gu et al, 2006). Figure 1 shows a basic fringe projection system, which consists of a projector and a camera. A projector projects a structured pattern or structured light onto an object, and a camera captures the resulting scene. As the structured pattern from the projector lands on the objects in the scene, the 3D geometry distorts the pattern, which is what the camera captures. Assuming that the geometric relationship between the projector pixels and the camera pixels are known, the 3D geometry can be reconstructed from the distortion between each image. Thus 3D geometry is transformed into a single 2D image, and then the 2D image can be used to reconstruct the 3D geometry.

In the Holoimaging system setup, it differs slightly from a real 3D fringe projection system in that the camera and projector are both virtual orthogonal devices instead of perspective ones. In a real system the pinhole camera model, a perspective projection, is used which complicates the technique of encoding and decoding. The camera and projector lens distortion usually brings 3D shape measurement errors. Thus, using an orthogonal ideal projection simplifies the process further. Another difference is that in real 3D fringe projection system, the light usually cannot pass through an opaque object, but in a virtual fringe projection (Holoimage) system, the fringe patterns can pass through any object to generate fringe patterns for 3D shape recovery. Moreover, since the position of the virtual camera and projector can be precisely configured, the geometric relationship between the two can be precisely defined resulting in no need to calibrate the camera and projector. This is usually a very complicated process with a real 3D fringe projection system (Zhang & Huang, 2006b). With the Holoimaging setup, 3D shape reconstruction is significantly simplified and is highly precise, resulting in a quick and efficient way to depth map an HD 3D scene. In our previous work, we have demonstrated that the Holoimaging system can precisely recover a 3D scene (Zhang & Yau, 2008), and can be used to recover arbitrary 3D shapes (Karpinsky & Zhang, 2010a), albeit via different phase-shifting techniques.

Three-Step Phase-Shifting Algorithm

The structural pattern, or fringe pattern, that is used in the Holoimaging system is a sinusoidaly-varying pattern, which is typical of a fringe projection system (Zhang & Huang, 2006a). Phase shifting is usually used to achieve pixel-by-pixel spatial resolution during 3D shape recovery.

Figure 1. Virtual fringe projection setup, otherwise known as a Holoimaging setup

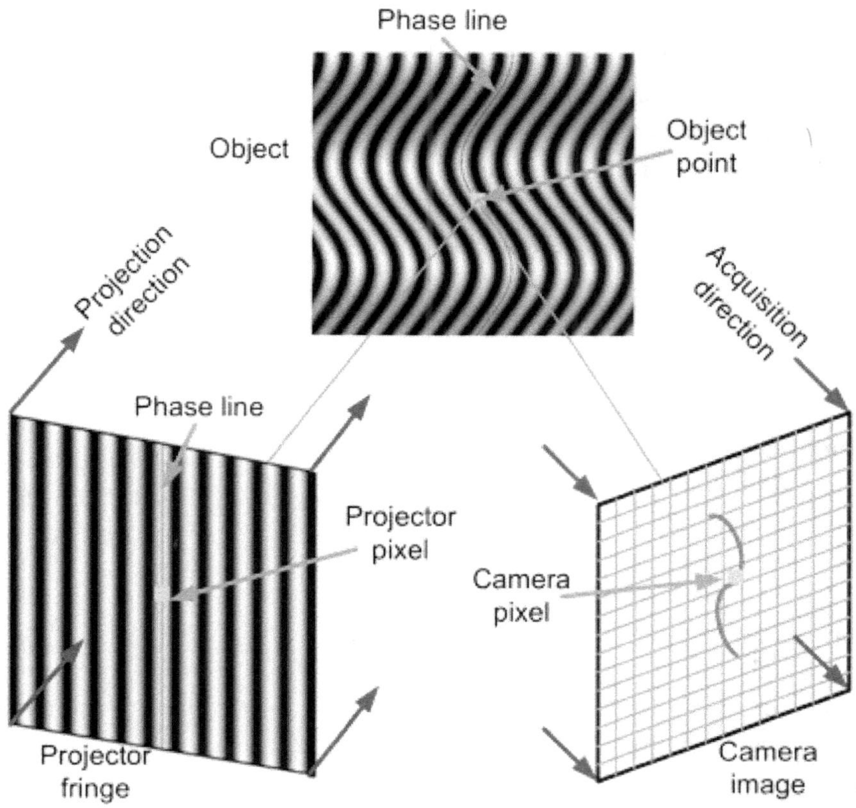

Phase-shifting algorithms are extensively used in optical metrology because of their measurement accuracy and speed. Over the years, a number of phase shifting algorithms have been developed including three-step, four-step, least square algorithms etc (Schreiber & Bruning, 2007). All these algorithms differ in the number of fringe images required and the amount of phase shift, but they all share the same properties: (1) high measurement speed, because it only requires a minimal amount of images to recover one 3D shape; (2) high spatial resolution, because the phase can be obtained pixel by pixel, thus the measurement can be performed pixel by pixel; (3) less sensitivity to surface reflectivity variations, since the calculation of the phase will automatically cancel out the DC components.

In a real world 3D shape measurement system using a fringe projection technique, a three-step phase-shifting algorithm is typically used in high-speed applications as it requires the least number of fringe patterns for 3D shape recovery. The fringe images of a three-step phase-shifting algorithm with equal phase shift can be described as

$$I_1(x,y) = I'(x,y) + I''(x,y)\cos[\phi(x,y) - 2\pi / 3], \tag{1}$$

$$I_2(x,y) = I'(x,y) + I''(x,y)\cos[\phi(x,y)], \tag{2}$$

$$I_3(x,y) = I'(x,y) + I''(x,y)\cos[\phi(x,y) + 2\pi / 3]. \tag{3}$$

Here, $I'(x,y)$ is the average intensity, $I''(x,y)$ the intensity modulation, and $\phi(x,y)$ the phase to be solved for. $I'(x,y)$ stands for the background light, the surface reflectivity, and the projected average light. $I''(x,y)$ indicates the fringe quality. The phase can be obtained by simultaneously solving Equation (1)-(3):

$$\phi(x,y) = \tan^{-1}\left\{\frac{\sqrt{3}\left[I_1(x,y) - I_3(x,y)\right]}{2I_2(x,y) - I_1(x,y) - I_3(x,y)}\right\}. \tag{4}$$

The phase value provided from the arctangent function only ranges from $-\pi$ to $+\pi$, which will result in 2π phase discontinuities. To obtain a continuous phase map, a phase unwrapping algorithm is usually needed (Ghialia & Pritt, 1998). The phase unwrapping step is essentially to detect the 2π phase jumps and remove them by adding or subtracting multiples of 2π. In other words, the unwrapped phase can be written as

$$\Phi(x,y) = \phi(x,y) + 2\pi \times k(x,y). \tag{5}$$

Here $\Phi(x,y)$ is the unwrapped phase, and *k(x,y)* is integer, which might differ for different pixels. The phase unwrapping step is essentially to find correct *k(x,y)* for each point. Once the continuous phase map is obtained, 3D information can be recovered if the system is calibrated (Zhang & Huang, 2006b).

As can be seen in Equation (4), the phase is calculated pixel-by-pixel, thus the 3D information can be obtained pixel-by-pixel, which is advantageous over most other 3D imaging techniques. Therefore, this technique allows for pixel-level spatial resolution. Since only three images are required, it is possible to achieve high-speed (Zhang, 2010a).

Holoencoding: Coordinate-to-Phase Conversion

Because of the background lighting and random noise effect, three phase-shifted fringe patterns are typically required in order to perform 3D shape measurement in a real world system. In contrast, within a virtual Holoimaging system, all environmental variables can be precisely controlled; therefore only two phase-shifted fringe patterns are needed in order to solve for the phase $\phi(x,y)$. These two fringe patterns can be modeled and encoded into two primary color channels of the projector. Since the background light can be precisely controlled, the fringe images can be ideally sinusoidal and described in the following two equations:

$$I_r = \frac{255}{2} \times [1 + \sin\Phi(x,y)], \tag{6}$$

$$I_g = \frac{255}{2} \times [1 + \cos\Phi(x,y)]. \tag{7}$$

From these two equations, the wrapped phase $\phi(x,y)$ may be obtained point-by-point by

$$\phi(x,y) = \arctan\left(\frac{I_r - \frac{255}{2}}{I_g - \frac{255}{2}}\right). \tag{8}$$

Similarly, this yields a phase value for each pixel that ranges from [-π, +π), which can later be used to reconstruct the 3D geometry. The unwrapped phase $\Phi(x,y)$ can be obtained by adopting a phase unwrapping algorithm to find *k(x, y)*. However, since there are three primary color channels and the blue channel is not yet utilized in the Holoimaging system, we can encode *k(x, y)* into the third channel by projecting it along with the fringe patterns. In practice, the third color channel is encoded as

Figure 2. Diagram of Holoimage fringe image. (A) Red color channel (I_r) given in Equation (6); (B) Green color channel (I_g) given in Equation (7); (C) Blue color channel (I_b) given in Equation (9); (D) Holoimage with all three color channels combined. Note it is rendered in grayscale but is a RGB color image in the actual Holoimage system.

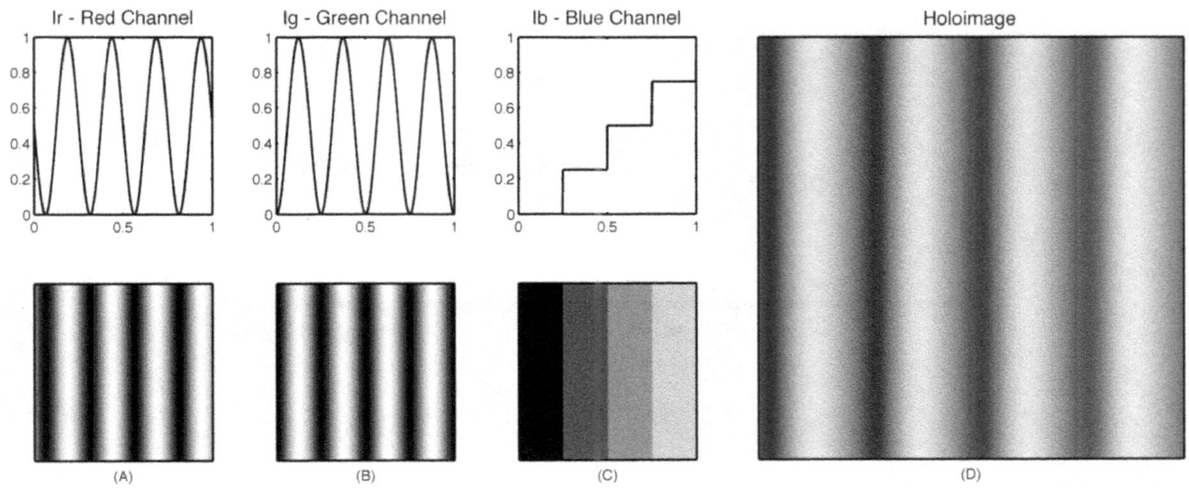

$$I_b = k(x,y) \times stepHeight = \left\lfloor \frac{\Phi(x,y)}{2\pi} \right\rfloor \times stepHeight. \quad (9)$$

Embedding these functions in the red, green and blue color channels, a gradient image to be projected is created; this image is seen in Figure 2. Due to the exactness of the virtual system these jumps can be mathematically determined, and a third function can be used which simply specifies the number of periods of the function or the multiple of 2π to add at each point (Karpinsky & Zhang, 2010b). This function is given above as I_b. By referring to this image, the continuous phase map can be obtained by

$$\Phi(x,y) = 2\pi I_b(x,y) \,/\, (stepHeight) + \phi(x,y). \quad (10)$$

Given that there are only three images, these functions can be encoded into the three primary color channels (red, green, blue, or RGB) of a 2D image and projected in the virtual system at once, achieving depth mapping of 3D geometry into a 2D image. Because the 3D information can be encoded into a single color image, it drastically reduces the size of storing 3D geometry data. In addition, because the phase at each point can be solved for point-by-point without referring to any neighboring point, the decoding can be achieved in parallel. With a highly parallel computation device, such as GPU, the decoding step can be realized in real-time.

Holodecoding: Phase-to-Coordinate Conversion

Decoding a Holoimage is achieved through a very simple triangulation. To explain the concept in the context of a digital fringe projection system, Figure 3 is given which decodes a single depth value z using a reference plane (a flat surface with $z = 0$). In other words, the depth z value is relative to the flat plane. The ultimate goal is to be able to calculate the z value for each point in a point-by-point manner from the computed phase value in Equation (10).

Figure 3. Schematic diagram for phase to coordinate conversion. In order to decode the depth value from the Holoimage the projection angle and fringe pitch used during encoding must be known.

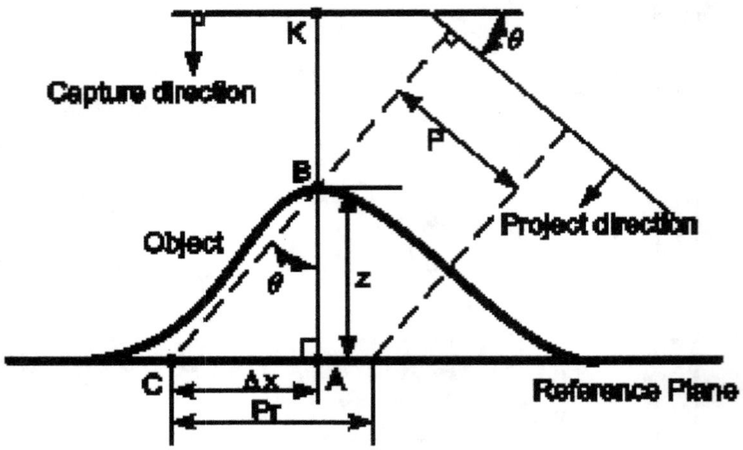

To begin, from Figure 2, we can use basic trigonometry to find z in terms of Δx_{C-A} and $\tan\theta$, where θ is the angle between the capture plane and the projection plane.

$$z = \frac{\Delta i_{C-A}}{\tan\theta}. \tag{11}$$

To simplify the 3D rendering, the graphics pipeline is usually set up in a way that the rendered scene gets visualized within a unit cube, thus the size of a pixel is $\frac{1}{W}$, where W is the total number of pixels horizontally in the unit cube. If the origin of the coordinate system for the unit cube is aligned with the origin of the image then x can be found by simple scaling, that is

$$x = \frac{i}{W}, \tag{12}$$

where i is the index of the pixel being decoded in the Holoimage. Therefore, the distance between C and A in the unit cube is actually:

$$\Delta x_{C-A} = \frac{i_{C-A}}{W}. \tag{13}$$

At this point Equations (11) - (13) can be combined yielding the following:

$$z = \frac{\Delta i_{C-A}}{W\tan\theta}. \tag{14}$$

This gives z in terms of the change of index from point C to point A, along with the number of pixels horizontally, and the angle between the projection and capture planes. Since there is no easy way to find i for point C and point A given a point, we will have to look further to see if the phase value can be leveraged.

For an arbitrary pixel K in the Holoimaging system, the point A on the reference plan would have a phase value of Φ_A^r. From the camera perspective or the Holoimage perspective, point B would be in the place of point A and the phase value would be Φ_B or just Φ. From the projector's perspective, point B and point C (on the reference plane) have the same value, i.e. $\Phi = \Phi_B = \Phi_C^r$. Since the fringe stripes are uniformly distributed on the reference plane we have the following equation.

$$\Delta\Phi = \Phi_C^r - \Phi_C^r = \Phi_B - \Phi_A^r. \quad (15)$$

The phase of a point on the reference plane can be defined as a function of the index i and the fringe pitch (number of pixels per period of fringe) on the reference plane.

$$\Phi^r = \frac{2\pi i}{P_r}. \quad (16)$$

Here P_r is the fringe pitch on the reference plane. Again using more trigonometry, we can define the fringe pitch on the reference plane in terms of the fringe pitch of the projector.

$$P_r = \frac{P}{\cos\theta}. \quad (17)$$

Here, P is the fringe pitch that the projector actually projects. In other words, P here is the computer generated fringe pattern pitch number. Combining Equation (16) and Equation (17), we obtain the phase of a point on the reference plane in terms of the fringe pitch P and the angle between the capture plane and projection plane θ.

$$\Phi^r = \frac{2\pi i\cos\theta}{P}. \quad (18)$$

Furthermore, Equation (15) and Equation (18) can be combined to obtain

$$\Delta\Phi = \frac{2\pi i_C\cos\theta}{P} - \frac{2\pi i_A\cos\theta}{P} = \Phi - \frac{2\pi i_A\cos\theta}{P}, \quad (19)$$

or in another means as,

$$\Delta\Phi = \frac{2\pi\cos\theta\Delta i_{C-A}}{P} = \Phi - \frac{2\pi i_A\cos\theta}{P}. \quad (20)$$

Rearranging the first part of Equation (20) yields

$$i_{C-A} = \frac{\Delta\Phi P}{2\pi\cos\theta}. \quad (21)$$

From here we can go back to where we left off with Equation (14) and substitute in Equation (21).

$$z = \frac{\Delta\Phi P}{2\pi W\cos\theta\tan\theta}, \quad (22)$$

or

$$z = \frac{\Delta\Phi P}{2\pi W\sin\theta}. \quad (23)$$

Substituting in $\Delta\Phi$ from Equation (20) we obtain:

$$z = \frac{P\left(\Phi - \frac{2\pi i_A\cos\theta}{P}\right)}{2\pi W\sin\theta}. \quad (24)$$

Now we relate the depth information z with the projected fringe patterns, the Holoimage pixel index, and the setup of the Holoimaging system, that is

$$z = \frac{P\Phi - 2\pi i_A\cos\theta}{2\pi W\sin\theta}. \quad (25)$$

This yields a value z in terms of P the fringe pitch; i_A the index of the pixel being decoded in the Holoimage; θ the angle between the capture plane and the projection plane; Φ the phase at the current pixel being decode in the Holoimage; and W the number of pixels horizontally. Because the system is an orthogonal system and the rendering is performed within a unit cube, the x and y coordinates can be calculated by scaling the j and i as,

$$x = \frac{j}{W}, \tag{26}$$

$$y = \frac{i}{W}. \tag{27}$$

All of these terms are specific to the point at which the Holoimage is being decoded, thus making the decoding a point-by-point function; given parallel hardware, a Holoimage scene can be decoded in parallel giving a large speed boost.

Holoimage Example

To verify the performance of the proposed approach, we first tested a simulated pyramid object with a known shape and dimension (unit can be any since it is normalized into a unit cube) as shown in Figure 4. This object is then sent to the Holoimging system to generate the Holoimage, as shown in Figure 4(A). In this example, the Holoimaging system was configured as the follows: stair step height of 32 grayscale values, projection angle of $\theta = 30^\circ$, fringe pitch of P = 16 pixels, display window size of 512 X 512 pixels, and the rendering is performed within a unit cube. During the rest of this chapter, all experiments are performed under the same Holoimaging system setup.

From the red and green channels, the wrapped phase map can be calculated using Equation (8), which is shown in Figure 4(B). Figure 4(C) shows the blue channel stair image that is then applied to unwrap the phase point by point using Equation (9). The unwrapped phase map is shown in Figure 4(D). Once the phase map is known and the configuration of the system is pre-defined, the *(x, y, z)* coordinates for each pixel can be calculated from the unwrapped phase map point by point using Equations (25)-(27). Figure 4(E) shows the recovered 3D shape. To verify the accuracy of the Holoimaging system, the difference map between the original data and the reconstructed one was obtained, as shown in Figure 4(F). The error is approximately 0.05%, which can be negligible in comparison with the quantization error: representing the depth map with 8-bit grayscale images in one channel generates error of 0.39%, or $\frac{1}{2^8} \times 100\% = 0.39\%$.

To further demonstrate the accuracy of the Holoimaging system, an actual scanned 3D object is then tested for the proposed technique. Figure 5 shows the experimental result. The original shape is shown in Figure 5(A). It can be seen that the 3D shape is a typical statue face with very detailed 3D structures. Due to the nature of the Holoimage technique, all the details can be recovered. Figure 5(D) shows the Holoimage generated for the object. From the Holoimage, the wrapped phase, unwrapped phase map, and the 3D shape can be obtained. Figure 5(C) shows the recovered 3D shape. If the original shape and the recovered one are rendered in the same window, the results are shown in Figure 5(D) in shaded mode. It clearly demonstrates that the recovered 3D shape and the original one are almost perfectly aligned, that is, the recovered 3D shape and the original 3D shape do not have significant difference. The difference map is further calculated and plotted in Figure 5(E). The error was found to be 0.004%. The error is again negligible in comparison with quantization error.

Both simulation and the real data shows that the Holoimging system can be used to accurately recover the original 3D geometry with a single color image. Because 3D geometry can be represented with a single color imaging, it poses potential for 3D shape compression, which will be detailed next.

Figure 4. 3D recovery using the single color fringe image. (A) Fringe image; (B) Phase map using red and green channels of the color fringe image; (C) Stair image (blue channel); (D) Unwrapped phase map; (E) Recovered 3D shape; (F) Difference map between the recovered 3D shape and the original one (RMS error 0.05%).

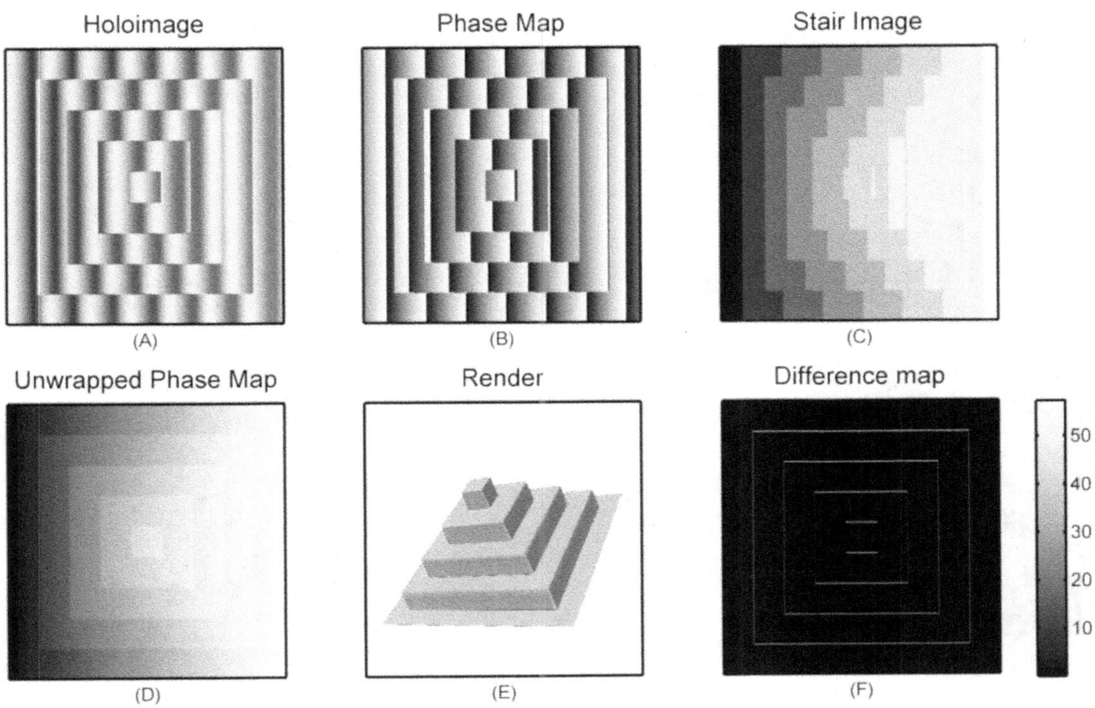

Table 2. Comparison of PLY model format to Holoimage encoded in various image formats. Note that even an uncompressed bitmap format still yields a compression ratio of over 6:1.

	PLY	BMP	TIFF	PNG	JPEG+PNG
File Size (bytes):	4,838,400	786,488	395,194	141,344	81,090
Ratio:	1: 1	6.15: 1	12.24: 1	34.23: 1	59.67: 1

Holoimage Compression

Once a scene is Holoencoded the resulting 3D color image can be compressed using standard 2D image compression techniques. An example face shown in Figure 5 that is 512×512 pixels has been Holoencoded and then compressed using different techniques. As an example, we used Bitmap, PNG, tagged image file format (TIFF), and differing compression levels of joint photograph experts group (JPEG) + PNG compression to store the Holoimage. Table 2 compares different compression techniques in comparison with storing the geometry in the PLY format, which is a typical highly compressed 3D format. Note that the even storing the geometry in uncompressed BMP format still yields a compression ratio of approximately 6:1.

One caveat of the technique is that the image is encoded 3D geometry, thus lossy image

Figure 5. 3D recovery using the color fringe image for scanned data. (a) 3D scanned original data; (b) Color fringe image; (c) 3D reconstructed shape; (d) Overlap original 3D shape (light gray) and the recovered 3D shape (dark gray) in shaded mode; (e) Difference map (RMS error 0.004%).

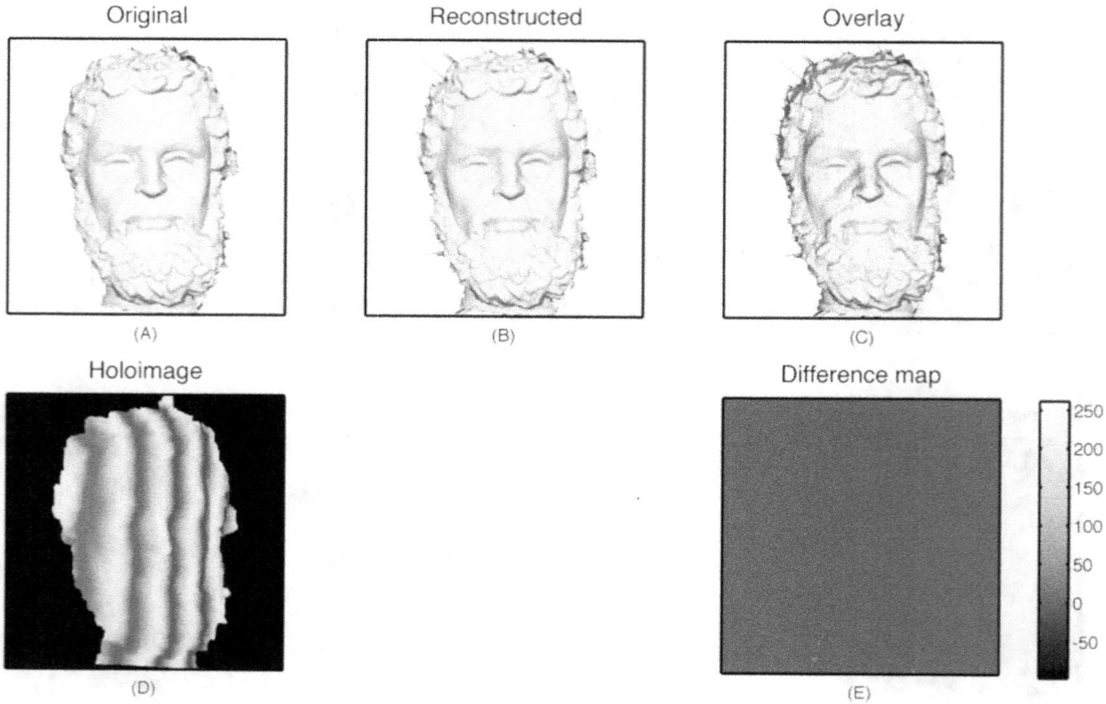

compression can create artifacts in the reconstructed geometry much like it does in the actual 2D image. An example of such compression is JPEG encoding; high compression rates can be achieved, but at the cost of blocking artifacts. Blocking artifacts occur due to the fact that the JPEG compression standard performs its cosine transform on 8 X 8 blocks of pixels in the image. The edges of these blocks can have sharp discontinuities at high levels of compression. These discontinuities from the blocking artifacts lead to what is known as spiking noise, which is shown in Figure 6. Currently JPEG encoding cannot be directly implemented with Holoimages as the blocking artifacts cause significant problems with the blue color channel which is used to unwrap the phase and is intolerant of noise. To alleviate this problem, the red and green color channels can be encoded with JPEG encoding and the blue color channel can be encoded with a lossless format such as PNG. This modified 2D image format allows for the high compression rates seen in JPEG and other lossy image formats, without introducing errors in the blue color channel.

As can be expected, the higher the compression rate on lossy formats, the more apparent the blocking artifacts become, resulting in more spikes and ripple noise. Because these spikes only appear along the edges of the stair, which usually shifts one pixel left or right, they can be removed through filtering (e.g., median filtering) on the phase map before triangulation (Karpinsky & Zhang, 2010b). The third row of Figure 6 shows a median filtered mesh, where most of the spikes have been removed. Median filtering leads to the loss of point-by-point processing for the decoding procedure, but one

Figure 6. Comparison of reconstructed geometry under varying levels of JPEG compression. The first row shows the Holoimage used to compress the geometry; row two shows the reconstructed geometry before median filtering; row three shows the reconstructed geometry after median filter. Note that median filtering has removed most of the spiking noise, but some ripples have formed on the model such as on the forehead. Column (A) shows the uncompressed Holoimage and the associated results; Column (B) shows the 90% compressed Holoimage and the associated results; Column (C) shows the 70% compressed Holoimage and the associated results; and Column (D) shows the 50%, compressed Holoimage and the associated results.

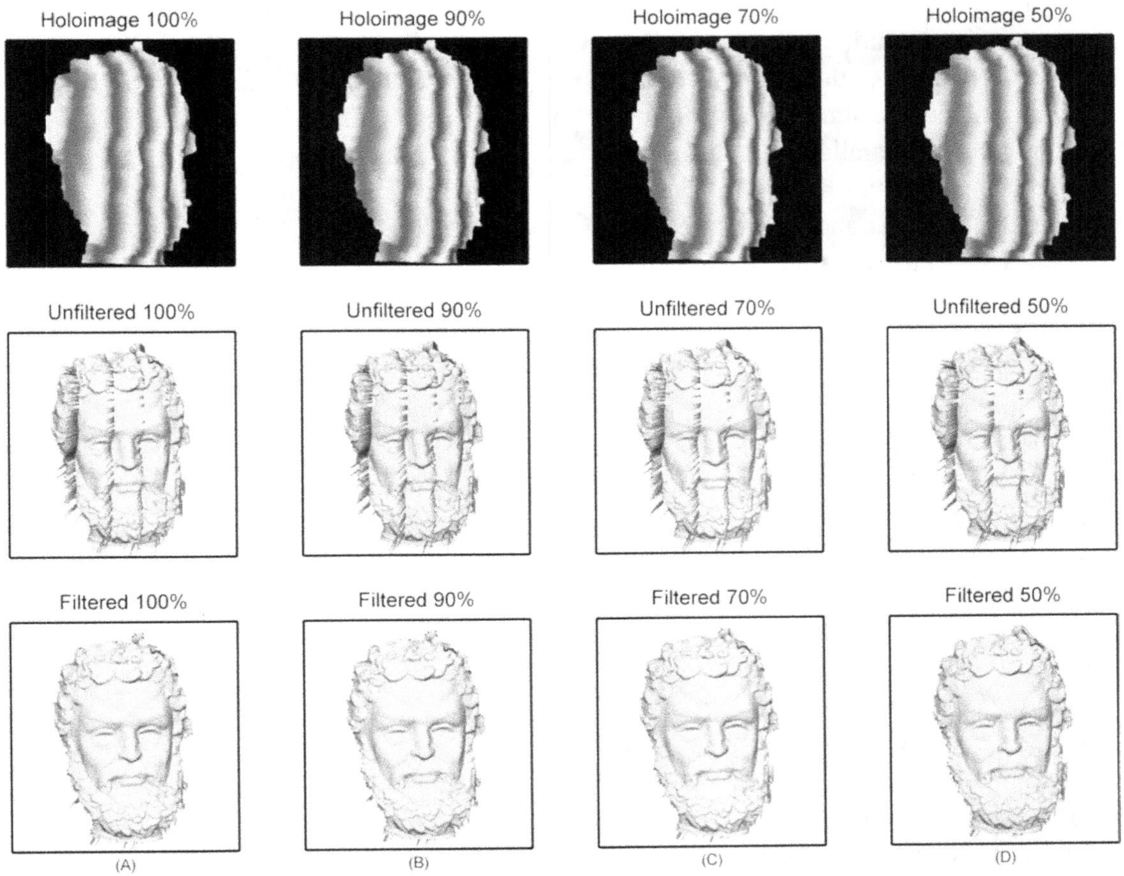

should note that the spiky noise is only one pixel in width, thus if filtering is implemented correctly it can still be implemented in parallel. We have demonstrated that compression ratios of 36:1 can be achieved without significant loss in the quality of the data (Karpinsky & Zhang, 2010b). Because the compression ratio is very high and the 3D scenes will be converted into 2D images, this compression technique would easily allow for streaming of high frame rates of compressed Holoimages.

FUTURE RESEARCH DIRECTIONS

One known problem in Holoimaging is subpixel shift in which during the encoding process the sinusoidal fringe gets quantized into an RGB pixel. This results in some error, which at times can create what is known as spiking error (as illustrated in Figure 6). For 3D HD scenes, this error is unacceptable and must be filtered out. A 3×3 median filtering on the phase map can be performed to help eliminate small amounts of this noise, but at the cost of losing the point-by-point decoding of Holoimaging. Parallel processing can still be maintained if implemented correctly, but this is not an optimal solution. If instead 2D image filtering can be applied to the Holoimage before the decoding, this noise could be removed without losing point-by-point parallelism of decoding.

This leads into a major direction of research in terms of Holoimages and depth mapping in general, which is filtering of the depth map images. Since Holoimages are encoding information, standard 2D image filtering has differing effects, which can sometimes be adverse. How then can 2D image filtering be applied to the Holoimages to retain point-by-point parallel decoding and achieve the removal of spiking noise? If this question can be answered it has the potential to solve subpixel shift along with enabling Holoimages to be saved in a highly compressed lossy image format.

Another source of noise in Holoimaging occurs when compressing the Holoimages. If the Holoimages are stored in a lossy image format, compression artifacts are introduced, resulting in erroneous decoded spiking noise. An example of this is with JPEG compression, which introduces blocking artifacts as it divides the image into 8×8 blocks before applying the discrete cosine transform and quantizing. With high compression levels the blocking artifacts are apparent to the human eye. With small amounts of blocking artifacts, Holoimages can be affected adversely. Again filtering can be used to reduce this such as median or Gaussian filtering, but this is typically applied to the decoded phase map and not the actual Holoimage. Also there might be some directions that could encode the 3D geometry in another way so that the blue channel will not contain sharp edges. By circumventing the problems with information loss in JPEG compression, the problem of spiky noise might be completely eliminated. This, of course, requires the investigation of the exact manner of JPEG compression.

If research is done to develop methods to overcome these shortcomings then Holoimaging has the potential to compress 3D scenes into 2D images and then store the resulting depth maps in lossy image formats. Once in this format, the images could be encoded in into video via the wide variety of 2D video codecs. This could allow for wide spread adoption of 3D video without the need to create or adapt new storage and transmission techniques.

CONCLUSION

Holoimaging yields an effective way to encode, transmit, and decode 3D scenes. Encoding relies on techniques borrowed from optical metrology, namely fringe projection. The fundamental behind the technique is to project a specially designed sinusoidal structured pattern onto objects in a 3D scene and then capture how the objects distort the pattern. Being a virtual system, all environmental variables can be controlled giving a precise known relationship between camera and projector, removing the calibration step seen in real-world fringe projection systems. Also, being a virtual system only two fringe images need to be used which can be embedded in the red and green color channel of an image, along with a stair image in the blue color. The stair image allows for point-by-point decoding lending the technique to be easily implemented in parallel architectures.

Once compressed into a 2D depth map, existing image compression techniques can be applied to compress and transmit the depth map. One of these compression techniques, JPEG encoding, has been explored which allows for high compression but at the loss of data. In order to save Holoimages in this format, the JPEG had to be augmented with a lossless compression technique for the blue color channel. This was due to the fact that the blue color channel was very intolerant of noise. As compression rates increased, blocking artifacts due to the JPEG compression in the Holoimage led to spiking noise, but this could be easily removed though median filtering of the phase map. Filtering on the phase map causes Holoimaging to lose its point-by-point nature, but if implemented correctly can still be performed on parallel architecture. One major area to explore in Holoimaging would be filtering of the actual Holoimage, which would be more efficient in terms of filtering 2D data vs. 3D data and keeping the pipeline point-by-point. Another area would be different methods to encode the Holoimage so that the spiky problem will be fundamentally and completely eliminated.

REFERENCES

Abate, A., Nappi, M., Riccio, D., & Sabatino, G. (2007). 2D and 3D face recognition: A survey. *Pattern Recognition Letters*, *28*(14), 1885–1906. doi:10.1016/j.patrec.2006.12.018

Chai, B., Sethuraman, S., Sawhney, H., & Hatrack, P. (2004). Depth map compression for real-time view-based rendering. *Pattern Recognition Letters*, *25*(7), 755–766. doi:10.1016/j.patrec.2004.01.002

Forstmann, S., Ohya, J., Krohn-Grimberghe, A., & McDougall, R. (2007). Deformation styles for spline-based skeletal animation. *SCA '07: Proceedings of the 2007 ACM SIGGRAPH/Eurographics Symposium on Computer Animation* (pp. 141-150). San Diego, California.

Ghiglia, D. C., & Pritt, M. D. (1998). *Two-dimensional phase unwrapping: Theory, algorithms, and software*. New York, NY: John Willey & Sons.

Gorthi, S., & Rastogi, P. (2010). Fringe projection techniques: Whither we are? *Optics and Lasers in Engineering*, *48*(2), 133–140. doi:10.1016/j.optlaseng.2009.09.001

Gu, X., Zhang, S., Huang, P., Zhang, L., Yau, S., & Martin, R. (2006). Holoimages. In *Proceedings of the 2006 ACM Symposium on Solid and Physical Modeling*. Cardiff, Wales, United Kingdom.

Huang, P., Zhang, C., & Chiang, F. (2003). High-speed 3-D shape measurement based on digital fringe projection. *Optical Engineering (Redondo Beach, Calif.)*, *42*(1), 163–168. doi:10.1117/1.1525272

Huang, P., & Zhang, S. (2006). Fast three-step phase-shifting algorithm. *Applied Optics*, *45*(21), 5086–5091. doi:10.1364/AO.45.005086

Karpinsky, N., & Zhang, S. (2010). Composite method for discontinuous 3-D surface measurement: simulations. In P. Rastogi & E. Hack (Eds.), *International Conference on Advanced Phase Measurement Methods in Optics and Imaging* (pp. 438-442). Monte Verita (Ascona), Switzerland.

Karpinsky, N., & Zhang, S. (2010b). Composite method for 3-D shape compression. *Optical Engineering (Redondo Beach, Calif.)*, *49*(6), 063604. doi:10.1117/1.3456632

Krishnamurthy, R., Chai, B., Tao, H., & Sethuraman, S. (2001). Compression and transmission of depth maps for image-based rendering. *IEEE International Conference on Image Processing* (pp. 828-831).

Li, Y., Jin, K., Jin, H., & Wang, H. (2010). High resolution, high speed 3D measurement based on absolute phase measurement. In P. Rastogi & E. Hack (Ed.), *International Conference on Advanced Phase Measurement in Optics and Imaging* (pp. 389-394). Monte Verita (Ascona), Switzerland.

Lindstrom, P., Koller, D., Ribarsky, W., Hodges, L. F., Faust, N., & Turner, G. A. (1996). Real-time, continuous level of detail rendering of height fields. *SIGGRAPH '96: Proceedings of the 23rd Annual Conference on Computer Graphics and Interactive Techniques* (pp. 109-118). New Orleans, LA.

Schnabel, R., & Klein, R. (2006). Octree-based point-cloud compression. In M. Botsch & B. Chen (Eds.), *Eurographics Symposium on Point-Based Graphics*. Vienna, Austria.

Schreiber, H., & Bruning, J. H. (2007). Phase shifting interferometry. In Malacara, D. (Ed.), *Optical shop testing* (pp. 547–666). New York, NY: John Willey & Sons. doi:10.1002/9780470135976.ch14

Zhang, S. (2010a). Recent progresses on real-time 3-D shape measurement using digital fringe projection techniques. *Optics and Lasers in Engineering*, *48*(2), 149–158. doi:10.1016/j.optlaseng.2009.03.008

Zhang, S. (2010b). High-resolution, high-speed 3D dynamically deformable shape measurement using digital fringe projection techniques. In M. K. Sharma (Ed.), *Advances in measurement systems* (pp. 29-50). Vukovar, Croatia: In-tech.

Zhang, S., & Huang, P. S. (2006a). High-resolution, real-time three-dimensional shape measurement. *Optical Engineering (Redondo Beach, Calif.)*, *45*(12), 123601. doi:10.1117/1.2402128

Zhang, S., & Huang, P. S. (2006b). Novel method for structured light system calibration. *Optical Engineering (Redondo Beach, Calif.)*, *45*(8), 083601. doi:10.1117/1.2336196

Zhang, S., & Yau, S.-T. (2006). High-resolution, real-time 3-D absolute coordinate measurement based on a phase-shifting method. *Optics Express*, *14*(11), 2644–2649. doi:10.1364/OE.14.002644

Zhang, S., & Yau, S.-T. (2007). High-speed three-dimensional shape measurement system using a modified two-plus-one phase-shifting algorithm. *Optical Engineering (Redondo Beach, Calif.)*, *46*(11), 113603. doi:10.1117/1.2802546

Zhang, S., & Yau, S.-T. (2008a). Three-dimensional data merging using Holoimage. *Optical Engineering (Redondo Beach, Calif.)*, *47*(3), 033608. doi:10.1117/1.2898902

ADDITIONAL READING

Chyou, J., Chen, S., & Chen, Y. (2004). Two-dimensional phase unwrapping with a multichannel Least-mean-square algorithm. *Applied Optics*, *43*(30), 5655–5661. doi:10.1364/AO.43.005655

Curless, B., & Levoy, M. (1996). A volumetric method for building complex models from range images. *SIGGRAPH '96: Proceedings of the 23rd Annual Conference on Computer Graphics and interactive Techniques*. (pp. 303-312). New Orleans, LA.

Fehn, C. (2003). A 3D-TV approach using depth-image-based rendering. In Hamza, M. H. (Ed.), *Visualization, Imaging, and Image Processing (396-084)*. Benalmadena, Spain.

Fleishman, S., Cohen-Or, D., Alexa, M., & Silva, C. (2003). Progressive point set surfaces. *ACM Transactions on Graphics*, *22*(4), 997–1011. doi:10.1145/944020.944023

Gumhold, S., Kami, Z., Isenburg, M., & Seidel, H. (2005). Predictive point-cloud compression. *SIGGRAPH '05: ACM SIGGRAPH 2005 Sketches*. Los Angeles, California.

Huang, P. S., Zhang, C., & Chiang, F.-P. (2002). High-speed 3-d shape measurement based on digital fringe projection. *Optical Engineering (Redondo Beach, Calif.)*, *42*(1), 163–168. doi:10.1117/1.1525272

Karpinsky, N., Lei, S., & Zhang, S. (2009). High-resolution, real-time fringe pattern profilometry. In C. Quan, K. Qian, A. Asundi, and F. Chau (Ed.), *Forth International Conference on Experimental Mechanics* (pp. 75220E). Singapore, Singapore.

Kauff, P., Atzpadin, N., Fehn, C., Müller, M., Schreer, O., Smolic, A., & Tanger, R. (2007). Depth map creation and image-based rendering for advanced 3DTV services providing interoperability and scalability. *Image Communication, 22*(2), 217–234.

Luebke, D., Watson, B., Cohen, J. D., Reddy, M., & Varshney, A. (2002). *Level of detail for 3D graphics*. San Francisco, CA: Morgan Kaufman Publishers.

Pauly, M., Gross, M., & Kobbelt, L. P. (2002). Efficient simplification of point-sampled surfaces. *VIS '02: Proceedings of the Conference on Visualization* (pp. 163-170). Boston, Massachusetts.

Pauly, M., Keiser, R., Kobbelt, L. P., & Gross, M. (2003). Shape modeling with point-sampled geometry. *ACM Transactions on Graphics* [San Diego, California.]. *Proceedings of SIGGRAPH, 2003*, 641–650. doi:10.1145/882262.882319

Rusinkiewicz, S., Hall-Holt, O., & Levoy, M. (2002). Real-time 3D model acquisition. *ACM Transactions on Graphics, 21*(3), 438–446. doi:10.1145/566570.566600

Rusinkiewicz, S., & Levoy, M. (2000). QSplat: a multiresolution point rendering system for large meshes. *SIGGRAPH '00: Proceedings of the 27th Annual Conference on Computer Graphics and Interactive Techniques* (pp. 343-352). New York, New York.

Salvi, J., Fernandez, S., Pribanic, T., & Llado, X. (2010). A state of the art in structured light patterns for surface profilometry. *Pattern Recognition, 43*(8), 2666–2680. doi:10.1016/j.patcog.2010.03.004

Shade, J., Gortler, S. J., He, L., & Szeliski, R. (1998). Layered depth images. *SIGGRAPH '98: Proceedings of the 25th Annual Conference on Computer Graphics and Interactive Techniques* (pp. 231-242). Orlando, Florida.

Yang, Z., Cui, Y., Anwar, Z., Bocchino, R., Kiyanclar, N., & Nahrstedt, K. (2006). *Real-time 3D video compression for tele-immersive environments*. San Jose, California: Proceedings of Multimedia Computing and Networking.

Zhang, C., & Chen, T. (2004). A survey on image-based rendering-representation, sampling and compression. *Signal Processing Image Communication, 19*(1), 1–28. doi:10.1016/j.image.2003.07.001

Zhang, L., Snavely, N., Curless, B., & Seitz, S. (2004). Spacetime faces: high resolution capture for modeling and animation. *SIGGRAPH '04: ACM SIGGRAPH 2004 Papers*, 548-558. Los Angeles, California.

Zhang, S. (2010). Recent progresses on real-time 3-D shape measurement using digital fringe projection techniques. *Optics and Lasers in Engineering, 48*(2), 149–158. doi:10.1016/j.optlaseng.2009.03.008

Zhang, S., & Huang, P. (2004). High-resolution, real-time 3-D shape acquisition. *CVPRW '04: Proceedings of the 2004 Conference on Computer Vision and Pattern Recognition Workshop* (pp. 28). Washington, DC.

Zhang, S., & Huang, P. (2006). High-resolution, real-time three-dimensional shape measurement. *Optical Engineering (Redondo Beach, Calif.), 45*(12). doi:10.1117/1.2402128

Zhang, S., & Huang, P. (2007). Phase error compensation for a 3-D shape measurement system based on the phase-shifting method. *Optical Engineering (Redondo Beach, Calif.), 46*(6). doi:10.1117/1.2746814

Zhang, S., Li, X., & Yau, S. (2007). Multilevel quality-guided phase unwrapping algorithm for real-time three-dimensional shape reconstruction. *Applied Optics*, *46*(1), 50–57. doi:10.1364/AO.46.000050

Zhang, S., & Yau, S. (2007). Generic nonsinusoidal phase error correction for three-dimensional shape measurement using a digital video projector. *Applied Optics*, *46*(1), 36–43. doi:10.1364/AO.46.000036

Zhang, S., & Yau, S. (2008). Three-dimensional shape measurement using a structured light system with dual cameras. *Optical Engineering (Redondo Beach, Calif.)*, *47*(1). doi:10.1117/1.2835686

KEY TERMS AND DEFINITIONS

Fringe Projection: Projecting sinusoidaly varying fringe patterns (structured light) onto an object.

Holoimaging: The technique of virtually applying fringe projection to encode 3D geometry into 2D images and then be able to decode back into 3D a scene via a phase shifting technique.

Phase Shifting: Process of taking multiple fringe images and shifting the sinusoidal fringe patterns spatially in the phase domain.

Phase Unwrapping: Finding and removing 2π discontinuities resulting from the arctangent function used in phase wrapping.

Phase Wrapping: Process to retrieve the phase from fringe pattern(s). This is typically done by adopting an arctangent function, which yields a phase map containing 2π discontinuities.

Spiky Noise: Noise in the mesh that results from problems such as subpixel shift or blocking artifacts in 2D compression. This noise can typically be removed by median filtering on the decoded mesh.

Structured Light: Light that is structured into a pattern, which can be used to encode a scene.

Chapter 6
Restoration and Enhancement of Digitally Reconstructed Holographic Images

Rajeev Srivastava
Banaras Hindu University, India

ABSTRACT

Holograms can be reconstructed optically or digitally with the use of computers and other related devices. During the reconstruction phase of a hologram by optical or digital methods, some errors may also be introduced that may degrade the quality of obtained hologram, and may lead to a misinterpretation of the holographic image data, which may not be useful for particular application. The basic common errors are zero-order diffraction and speckle noise. These errors have more undesirable effects in digital than in optical holography because the systems of recording and visualization used in the digital holography are extremely sensitive to them or inclusively increase them. The zero-order diffraction can be removed by using high pass filters with low cut-off frequencies and by subtracting the average intensity of all pixels of the hologram image from the original hologram image. Further, the speckle noise introduced during the formation of digital holographic images, which is multiplicative in nature, reduces the image quality, which may not be suitable for specific applications. As the range of applications get broader, demands toward better image quality increases. Hence, the suppression of noise, higher resolution of the reconstructed images, precise parameter adjustment, and faster, more robust algorithms are the essential issues. In this chapter, the various methods available in literature for enhancement and speckle reduction of digital holographic images have been discussed, and a comparative study of results has been presented.

DOI: 10.4018/978-1-61350-326-3.ch006

INTRODUCTION

The basic concept (Srivastava, 2010) of holography was first introduced by Gabor in 1948 to improve the resolving power of electron microscope for coherent imaging. After the advent of laser technologies, the Goodman in 1967 had conceived the basic idea of digital holography. Since one of the main drawbacks of electron microscopes is that the higher the spatial resolution, the lower is depth of focus which imposes restrictions in imaging for a specific application. This problem of microscopy can be resolved by holography. Holography is the science of producing holograms which is an advanced form of photography that allows an image to be recorded in three dimensions (3D) and the technique of holography can also be used to optically store, retrieve, and process information. Holography is related to measuring the wave field followed by reconstruction of the wave field, i.e. both the amplitude and the phase of the light wave scattered by the object. Due to advancements in digital optics, CCD and CMOS cameras and computers, it became possible not only to record the digital holograms but also to reconstruct them. Further, with the advancement and use of digital image processing and optical information processing methods for further processing of digital holographic images, nowadays it is possible to generate realistic digital holograms with no defects that may be used in different areas of applications. Holography which was originally invented to solve problems in electron-microscopy, now in its new form of digital holography, can be used to solve problems of optical microscopy. Holography is capable of recording 3-D information and optical reconstruction is then possible with visual 3-D observation. Since there are no wet chemical processing and other time consuming procedures, digital holography can be done in almost real time through numerical reconstruction which offers great flexibility on controlling some parameters, such as focusing, image size and resolution.

When image of an object is observed through a microscope or the object's diffraction pattern, the information about the phase of the emanated wave is lost. However, if one records the interference pattern of light coming from an object called the object beam with a reference beam which has the same wavelength as the object beam and of a known phase distribution such as a plane wave or a spherical wave then it is possible to reconstruct both the phase and amplitude of the object beam (See Figure 1) (Yaroslavskii & Merzlyakov, 1989). This reconstruction of the object beam can be done optically by taking the hologram, which is the recorded fringe pattern obtained from interfering the object and reference beam, and shining the reference beam at it and the hologram in turn diffracts the light so that an image of the object is visible. As an example, we can consider the recording of the hologram of a spherically scattered wave like the light scattered from a Rayleigh scatterer where the spherical wave coming from the object interferes with a plane wave and as result a pattern of concentric rings are formed which resembles a Fresnel zone plate and like a Fresnel zone plate, the fringes focus a plane wave illuminating it to a point. The holograms can be reconstructed optically or digitally with the use of computers and other related devices. Figure 2 shows the steps involved in digital reconstruction and image processing of holograms. The various components of the setup contain following components.

Hologram Sensor which captures original hologram in analog form; Analog-to-digital converter which converts the analog form of recorded hologram in digital form for further processing with computers; pre-processing of digital hologram which involves the preparation of hologram data in some specific desired format etc; image reconstruction is associated with Digital reconstruction of holograms by applying various steps such as use of transformations (DFT, Fresnel's) etc; followed by image processing step which is responsible for producing realistic

Figure 1. Digital holographic microscopy principles

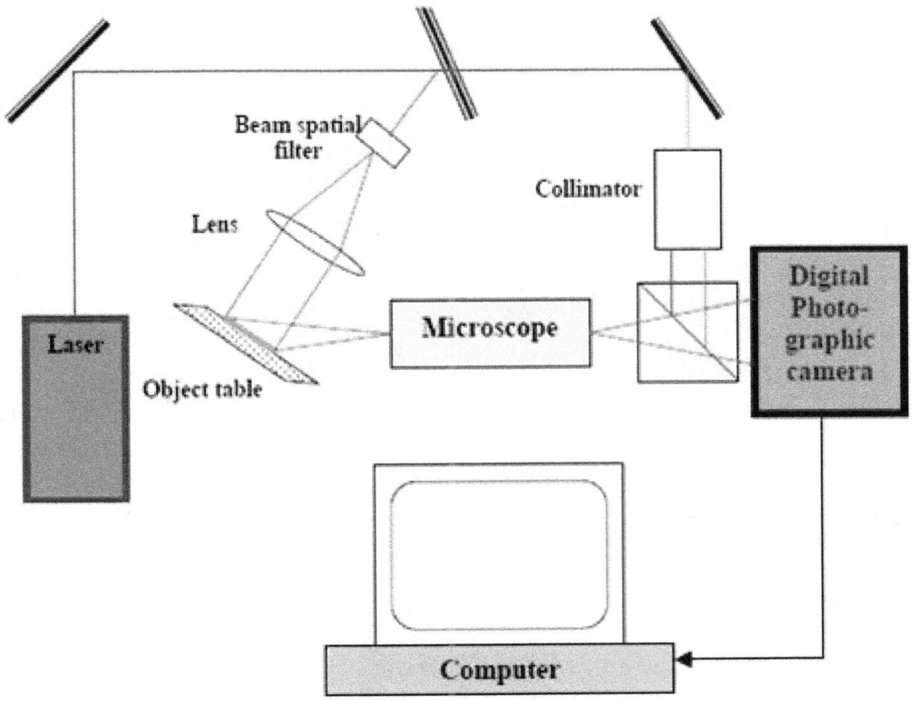

Figure 2. Steps in digital reconstruction of hologram

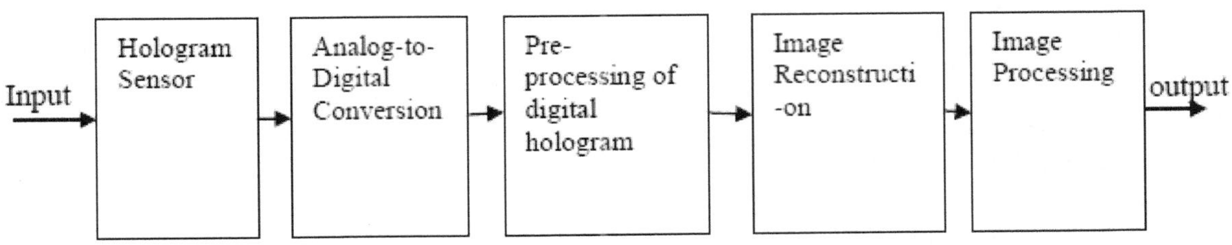

holograms by applying various image contrast enhancement techniques, speckle reduction techniques and zero-order diffraction removal techniques etc.

Difference Between Digital Holography and Digital Photography

The basic difference between ordinary digital photography and digital holography is that in digital photography only intensity, i.e. amplitude distribution of light coming from an object being imaged is recorded on a particular plane because the camera lens can be focussed only in a particular plane and the details of the field nearer and farther than the focussed plane are discarded whereas in digital holography both the amplitude and the phase distribution of light coming from object being imaged can be recorded in any plane between the object and the observer, producing the complete realistic field of view as originally observed.

Advantages of Digital Holography

The advantages of digital holographic microscopy include (Yaroslavskii, & Merzlyakov, 1989): real time image reconstruction for visual analysis, flexibility in retrieving arbitrary focal plane and focal plane data fusion, availability of digital image processing technology for sensor data calibration and processing reconstructed images, and direct availability of data for numerical analysis.

Image Processing Problems in Digital Holography

Some of the image processing problems involved in digital holography include: digital representation of holograms and optical transforms, hologram sensor signal correction, fast reconstruction algorithms, image contrast enhancement, speckle reduction, encryption, zero-order diffraction removal, phase unwrapping etc.

Application Areas

The various areas of applications of digital holography include: Copyright protection, security systems, holographic interferometry, microscopic examination of certain kinds of biological specimen, stereoscopic holography, high capacity system for image storage and re-examination, applications using short-coherence length light such as light-in-flight measurements & short coherence tomography, particle distribution measurement, endoscopic digital holography, optical reconstruction of digital holograms, comparative digital holography, encrypting of information with digital holography, synthetic apertures and many more.

In this chapter, some of the standard techniques for speckle reduction and enhancement of digitally reconstructed holographic images are explained, implemented and their performance comparisons are presented. Next section presents basic concepts of digital holography and speckle formation followed by the section discussing various techniques for speckle reduction and enhancement of digital holographic images. Then, results and performance comparison of various techniques for enhancement and speckle reduction of digital holographic images are presented. Finally, conclusion and future directions of research are discussed.

DIGITAL HOLOGRAPHY: GENERAL PRINCIPLES AND FORMATION OF SPECKLE NOISE

The digital holography framework (García-Sucerquia, Herrera, & Velasquez, 2005) for hologram recording and reconstruction has three planes namely object plane, hologram plane and real Image plane separated by a distance *d* and involves two diffraction processes one from the object plane to hologram plane and another from the hologram plane to the image plane. The object located at the object plane $z=0$ is coherently illuminated and the optical field scattered by it interferes with the plane reference wave in such a way that the interference pattern is recorded in a CCD camera located at a distance $z=d$ in hologram plane where only the intensity impinging upon the CCD is recorded. The optical field at the image plane located at a distance *d* from the hologram plane is calculated by means of calculating the diffraction process of the plane reference wave when it illuminates the transmittance represented by intensity incident upon the CCD.

Holographic Recording and Reconstruction Process

During the recording process of digital hologram, to produce a recording of the phase of the light wave at each point in an image, holography uses a reference beam which is combined with the light from the object known as object beam. Optical interference between the reference beam and the object beam, due to the superposition of the light waves, produces a series of intensity fringes that can be recorded on standard photographic film. These fringes form a type of diffraction grating on the film, which is called the interference pattern or hologram. These recorded fringes not only directly represent their respective corresponding points in the space of a scene but also an individual section of even a very small size on a hologram's surface contains enough information to reconstruct the entire original scene as viewed through that point's perspective. This is possible because during holographic recording, each point on the hologram's surface is affected by light waves reflected from all points in the scene, rather than from just one point. In holographic reconstruction process, once the film is processed, if illuminated once again with the reference beam, diffraction from the fringe pattern on the film reconstructs the original object beam in both intensity and phase as both the phase and intensity are reproduced, the image appears three dimensional (3D) and the viewer can move his viewpoint and see the image rotate exactly as the original object would. The holography typically uses a laser in production because of the need for interference between the reference and object beams. The light from the laser is split into two beams, one forming the reference beam, and one illuminating the object to form the object beam. A laser is used because the coherence of the beams allows interference to take place. Before the invention of laser, early holograms were made by other coherent light sources such as mercury-arc lamps. In simple holograms, the coherence length of the beam determines the maximum depth the image can have. A laser will typically have a coherence length of several meters, sufficient for a deep hologram.

The mathematical models (Yaroslavskii, & Merzlyakov, 1989) of recording and reconstruction of holograms assume that monochromatic coherent radiation that is described by its complex amplitude as a function of spatial coordinates is used for hologram recording and reconstruction and object characteristics defining its ability to reflect or transmit incident radiation are described by radiation reflection or transmission factors which are also functions of spatial coordinates. If $I(x, y, z)$ is complex amplitude of the object illumination radiation at point (x, y, z); $O(x, y, z)$ is object reflection or transmission factor then complex amplitude $a(x, y, z)$ of the radiation reflected or transmitted by the object at this point is defined as:

$$a(x, y, z) = I(x, y, z)O(x, y, z) \tag{1}$$

If $\alpha(x_h, y_h) = A_{obj}\exp(i\varphi_{obj})$ and $R(x_h, y_h) = A_{ref}\exp(i\varphi_{ref})$ denote complex amplitudes of the object and reference beams, respectively, at point (x_h, y_h) of the hologram plane, then intensity recorded by the recording medium at this point is a squared module of their sum which reads

$$I_h = |\alpha + R|^2 = \alpha R^* + \alpha^* R + \alpha\alpha^* + RR^*. \tag{2}$$

where * denotes complex conjugate. This intensity is a hologram signal, or a hologram. The first term in the sum in the right hand side of equation (2) is proportional to the object's beam complex amplitude which is called the mathematical hologram.

Alternatively, equation (2) can be written as:

$$\begin{aligned} I_h &= |A_{obj}\exp(i\varphi_{obj}) + A_{ref}\exp(i\varphi_{ref})|^2 \\ &= A_{obj}A_{ref}\exp(i(\varphi_{obj} - \varphi_{ref})) + A_{obj}A_{ref}\exp(-i(\varphi_{obj} - \varphi_{ref})) + A_{obj}^2 + A_{ref}^2 \end{aligned} \tag{3}$$

Hologram reconstruction consists in applying to the mathematical hologram a transform that implements wave back propagation from the hologram plane to object. For this, one has either to eliminate, before the reconstruction, other three

Figure 3. Schematic diagram for digital hologram recording and reconstruction

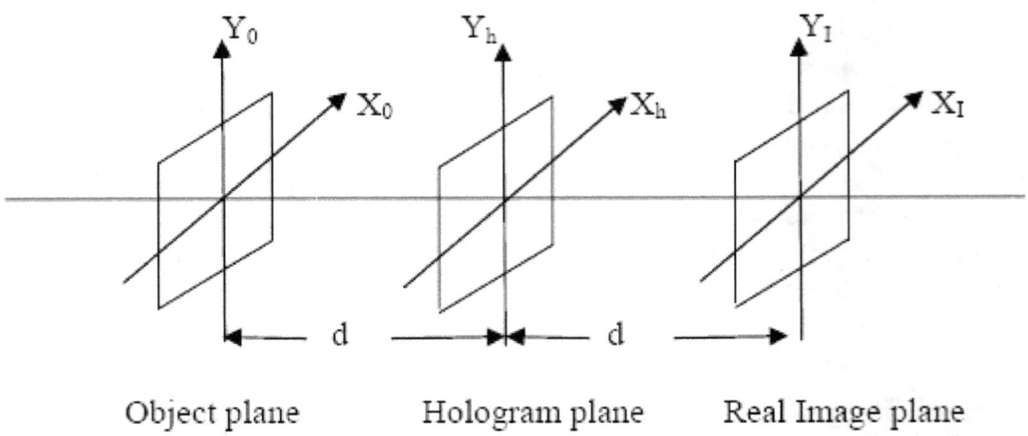

terms or to apply the reconstruction transform to the entire hologram and then separate the contribution of other terms in the reconstruction resulted from that of the mathematical hologram term.

The mathematical model for reconstructing hologram reads

$$I = I_h \cdot A_{ref} \exp(i\varphi_{ref})$$
$$= A_{ref}^2 \cdot A_{obj} \exp(i\varphi_{obj}) + (A_{obj}^2 + A_{ref}^2) \cdot A_{ref} \exp(i\varphi_{ref})$$
$$+ A_{ref}^2 \exp(i2\varphi_{ref}) \cdot A_{obj} \exp(-i\varphi_{obj} \quad (4)$$

Wave back propagation transformations used to reconstruct mathematical holograms are linear transformations and they are mathematically modelled as integral transformations which are also known as diffraction transform. In digital holography, this diffraction process can be described by the Kirchhoff–Fresnel diffraction integral and for the given setup in Figure 3, this integral can be described by Fresnel–Fraunhofer approximation. With a finite size of hologram the image I_h reconstructed from it is characterized by readings of the optical field E(x,y,z) which are linked to the hologram readings I(k,l) and can be described by a discrete Fresnel transformation (García-Sucerquia, Herrera, & Velasquez, 2005) which reads

$$E(m,n,z) = \frac{iE_0}{\lambda z} \exp(-\frac{i\pi}{\lambda z(\frac{m^2}{N_x^2 \Delta x^2} + \frac{n^2}{N_y^2 \Delta y^2})})$$
$$\times \sum_{k=0}^{N_x - 1} \sum_{l=o}^{N_y - 1} I(k,l) \exp(-\frac{i\pi}{\lambda z(k^2 \Delta x^2 + l^2 \Delta y^2)}) \exp(i2\pi)(\frac{km}{N_x} + \frac{\ln}{N_y}) \quad (5)$$

Where $\Delta x_h \times \Delta y_h$ is the resolution of the rectangular CCD having $N_x \times N_y$ pixels which registers the hologram; $m = 0,1,\ldots N_x - 1$ and $n = 0,1,\ldots N_y - 1$ and image pixel dimensions $\Delta x_i \times \Delta y_i$ are related to the pixel CCD dimensions $\Delta x_h \times \Delta y_h$ by $\Delta x_i = \frac{\lambda z}{N_x \Delta x_h}$ and $\Delta y_i = \frac{\lambda z}{N_y \Delta y_h}$.

Further it can be observed that equation (3) is the discrete Fourier transform (DFT) of $I(k,l) \exp(-\frac{i\pi}{\lambda z(k^2 \Delta x^2 + l^2 \Delta y^2)})$.

From equation (3), the intensity and phase of the optical field can be obtained by

$$I(m,n,z) = \left|E(m,n,z)\right|^2 = \mathrm{Re}[E(m,n,z)]^2 + \mathrm{Im}[E(m,n,z)]^2 \tag{6}$$

And

$$\varphi(m,n,z) = \arctan\left(\mathrm{Im}[E(m,n,z)] \Big/ \mathrm{Re}[E(m,n,z)]\right). \tag{7}$$

where *Re* and *Im* denotes real and imaginary part of the optical field and due to this the digital holography allows us to compute the intensity and phase of a reconstructed digital hologram for a particular distance *z* from the hologram plane.

In paper (Xiao-ou, 2008), it has been shown by the authors that the diffraction wave forming the real image I_{real} is given by

$$I_{real}(x,y) = \iint_{\Sigma} \left|R(x,y)\right|^2 \alpha^*(x,y)\exp(i\frac{2\pi}{\lambda})d'(x_i,y_i,z_i,x,y) \tag{8}$$

where α (x_h, y_h) and $R(x_h, y_h)$ denote complex amplitudes of the object and reference beams, respectively, at point (x_h, y_h) of the hologram plane and $d'(x_i,y_i,z_i,x,y)$ optical path length between hologram plane and image plane, see Figure 3, with the Fresnel approximation expression reads

$$d'(x_i,y_i,z_i,x,y) \approx \frac{x^2+y^2}{2d} + \frac{x_i^2+y_i^2}{2d} - \frac{xx_i+yy_i}{d}. \tag{9}$$

(x_i,y_i) is a point on image plane. According to equation (8), after the de-convolution operation, the distribution of light intensity can be expressed by (Xiao-ou, 2008)

$$\begin{aligned} I &= I_{ireal} I_{ireal}^* \\ &= \left|R\right|^2 \left|R\right|^2 \sum \left|a(x_o,y_o,z_o)\right|^2 \left|O(x_o,y_o,z_o)\right|^2 \\ &= cI_0 I_m \end{aligned} \tag{10}$$

where $I(x, y, z)$ is complex amplitude of the object illumination radiation at point (x, y, z); $O(x, y, z)$ is object reflection or transmission factor ; $a(x, y, z)$ complex amplitude of the radiation reflected or transmitted by the object at this point defined by the equation(1) and $R(x_h, y_h)$ denote complex amplitude of the reference beam at point (x_h, y_h) of the hologram plane;(x_o, y_o, z_o) is a point on object plane; I_0 is the intensity of the ideal object light and I_m is the intensity of the speckle noise. Therefore, from equation (10), it may be seen that speckle noise in digital holography is multiplicative in nature.

Based on analysis as above and as presented in paper (Cai, 2008), it can be concluded that speckle noise in reconstruction of digital holography is mainly due to interference illumination. Further, the formation of speckle noise in digital holography can be categorized in three parts:

a. The speckle noise forms on the surface of the recording object due to its optical roughness when illuminated by the coherent light and it is multiplicative in nature as shown in equation (10).
b. A speckle hologram creates with the interference of the object beam and the reference light in hologram plane.
c. In the reconstruction process, the speckle noise is modulated in the various diffraction orders. Since the reconstructed image of the hologram is the convolution result of the original object light and Fourier transformation of the hologram aperture function, and the small size of hologram aperture diffraction aggravates the speckle noise in the reconstructed image (Cai, 2008).

REDUCTION OF SPECKLE NOISE FROM DIGITAL HOLOGRAPHIC IMAGES

The speckle noise aggravated by the small size of hologram aperture diffraction can be reduced by setting an appropriate aperture function matching the recording parameter and aperture size of the hologram and de-convolve the reconstructed image with it (Cai, 2008).

Approaches for Speckle Reduction From Digitally Reconstructed Holographic Images

For further elimination of speckle noise from digitally reconstructed holographic images, there are two basic approaches.

Homomorphic Filtering

The first approach, converts the multiplicative speckle noise to additive one by using homomorphic filtering approach explained as follows:

i. Apply the logarithmic transform on equation (10) to convert the multiplicative noise into additive one. Suppose C is a constant in equation (10) and it is one, then equation (10) after logarithmic transform reads

$$\begin{aligned} \log I &= \log I_0 + \log I_m \\ \Rightarrow v(x,y) &= I(x,y) + \eta_s(x,y) \end{aligned} \tag{11}$$

where $v(x,y) = \log I$ is the observed hologram image in log domain, $I(x,y) = \log I_0(x,y)$ is the noiseless hologram image in log domain that is to be recovered and $\eta_s(x,y) = \log I_m(x,y)$ the amount the of the speckle noise which is now an additive noise and is to be minimized.

ii. In this step, an additive noise removal filter such as Wiener filter, median filter, PDE based diffusion filters etc is applied to remove or minimize the additive noise $\eta_s(x,y)$.

iii. Finally, the restored holographic image, $I_{restored}$ can be obtained by taking the exponentiation of output obtained in step ii.

$$I_{restored} = \exp(I(x,y)). \tag{12}$$

The most common filters that can be used for removal of additive noise is median filter, Wiener filter, Wavelet based filters etc. In recent years, partial differential equation (PDE) based filters have been developed that reduces the additive noise. Some of these PDE based filters are based on 2D diffusion or heat equation and its extensions.

Specialized Speckle Reduction Filters

The second approach uses specialized speckle reduction filter to directly reduce the speckle noise. Some examples of these types of filters successively applied for multiplicative speckle noise in other digital imaging modalities such as in ultrasound imaging, synthetic aperture radar imaging includes Lee Filter, Lee-Sigma Filter, Frost Filter, Kuan Filter, Speckle reducing anisotropic diffusion (SRAD) filter etc. These filters can also be used to reduce speckle noise from digitally reconstructed holographic images. The brief descriptions of various filters are as follows:

Mean Filter

The Mean Filter is a simple one and does not remove the speckles but averages it into the data and it is the least satisfactory method of speckle noise reduction as it results in loss of detail and resolution. However, it can be used for applications where resolution is not the first concern.

Median Filter

The Median filter is also a simple one and removes pulse or spike noises. Pulse functions of less than one-half of the moving kernel width are suppressed or eliminated but step functions or ramp functions are retained.

Wiener Filter (Jain, 2006)

The Wiener filter is the MSE-optimal stationary linear filter for images degraded by additive noise and blurring. The calculation of the Wiener filter requires the assumption that the signal and noise processes are second-order stationary (in the random process sense). Wiener filters are often applied in the frequency domain. Given a degraded image *x*(*n*,*m*), one takes the Discrete Fourier Transform (DFT) to obtain *X*(*u*,*v*). The original image spectrum is estimated by taking the product of *X*(*u*,*v*) with the Wiener filter *G*(*u*,*v*).

Lee-Sigma and Lee Filters (Lee, 1981; Lee, 1983)

The Lee-Sigma and Lee filters utilize the statistical distribution of the DN values within the moving kernel to estimate the value of the pixel of interest. These two filters assume a Gaussian distribution for the noise in the image data. The Lee filter is based on the assumption that the mean and variance of the pixel of interest is equal to the local mean and variance of all pixels within the user-selected moving kernel. The scheme for computing digital number output (DN_{out}) is as follows:

$$k = \frac{\operatorname{var}(x)}{(mean)^2 \sigma^2 + \operatorname{var}(x)} \tag{13}$$

where

$$\operatorname{var}(x) = \left[\frac{\sigma_w + \mu_w^2}{\sigma^2 + 1} \right] - \mu_w^2 \tag{14}$$

μ_w and σ_w are the mean and variances of pixels within chosen window. The Sigma filter is based on the probability of a Gaussian distribution. It is assumed that 95.5% of random samples are within a two standard deviation range. This noise suppression filter replaces the pixel of interest with the average of all DN values within the moving kernel that fall within the designated range.

Frost Filter (Frost et al., 1982)

The Frost filter replaces the pixel of interest with a weighted sum of the values within the n×n moving kernel. The weighting factors decrease with distance from the pixel of interest. The weighting factors increase for the central pixels as variance within the kernel increases. This filter assumes multiplicative noise and stationary noise statistics and follows the following formula:

$$DN = \sum_{nxn} k\alpha e^{-\alpha|t|} \tag{15}$$

Where

$$\alpha = \left(\frac{4}{n\bar{\sigma}^2}\right)\left(\frac{\sigma^2}{\bar{I}^2}\right) \tag{16}$$

Where DN is the digital number defined as above, *k* = normalization constant, *I* = local mean, σ =local variance, $\bar{\sigma}$ = image coefficient of variation value, $|t| = |X-X_0| + |Y-Y_0|$, and n = moving kernel size.

Kuan Filter (Kuan, 1987)

Kuan filter first transforms the multiplicative noise model into a signal-dependent additive noise model. Then the minimum mean square error criterion is applied to the model. The resulting filter has the same form as the Lee filter but with a different weighting function. Because Kuan filter made no approximation to the original model, it can be considered to be superior to the Lee filter.

The resulting grey-level value R for the smoothed pixel is:

$$R = I_c * W + I_m * (1 - W) \quad (17)$$

where:

$$W = (1 - C_u^2 / C_i^2) / (1 + C_u^2)$$

$$C_u = \sqrt{\frac{1}{NumberofLooks}}$$

$$C_i = \frac{S}{I_m}$$

I_c = center pixel in filter window, I_m = mean value of intensity within window, and S = standard deviation of intensity within window.

The Kuan filter is used primarily to filter speckled radar data. It is designed to smooth out noise while retaining edges or shape features in the image. Different filter sizes will greatly affect the quality of processed images. If the filter is too small, the noise filtering algorithm is not effective. If the filter is too large, subtle details of the image will be lost in the filtering process. A 7×7 filter usually gives the best results. The *NumberofLooks* parameter is used to estimate noise variance and it effectively controls the amount of smoothing applied to the image by the filter. Theoretically, the correct value for *NumberofLooks* should be the effective number of looks of the radar image. It should be close to the actual number of looks, but may be different if the image has undergone re-sampling. The user may experimentally adjust the *NumberofLooks* value so as to control the effect of the filter. A smaller *NumberofLooks* value leads to more smoothing; a larger *NumberofLooks* value preserves more image features.

Speckle Reducing Anisotropic Diffusion (SRAD) Filter

In this chapter (Yu & Acton, 2002), the authors provides the derivation of speckle reducing anisotropic diffusion (SRAD), a diffusion method tailored to ultrasonic and radar imaging applications. SRAD is the edge-sensitive diffusion for speckled images, in the same way that conventional anisotropic diffusion is the edge-sensitive diffusion for images corrupted with additive noise. At first authors had shown that the Lee and Frost filters can be cast as partial differential equations, and then SRAD filter is derived by allowing edge-sensitive anisotropic diffusion within this context. SRAD exploits the *instantaneous* coefficient of variation, same as the Lee and Frost filters utilize the coefficient of variation in adaptive filtering. The *instantaneous* coefficient of variation is a function of the local gradient magnitude and Laplacian operators.

Speckle Reduction Using Wavelet Transform

In paper (Sharma, Sheoran, Jaffery, & Moinuddin, 2008), authors have introduced a method for improvement of signal-to-noise ratio in digital holography using wavelet transform. The basic problem in optical and digital holography is the presence of speckle noise in the reconstruction process, which reduces the signal-to-noise ratio (SNR). The presence of speckle noise is serious drawback in optical and digital holography since it substantially reduces the SNR in the reconstructed image. This issue has been addressed in this chapter.

Other methods for speckle reduction from digital holographic images include:

In paper (Monroy, & Garcia-Sucerquia, 2009), authors have introduced a method for incrementing lateral resolution in digital holography by speckle noise removal. Experimental features such as wavelength, camera specifications and reconstruction distance determine the theoretical limit for lateral resolution in digital holography. However, the actual experimental resolution limit is about 50% below such theoretical limit due to the high-contrast speckle noise presented in the reconstructed holograms. In this chapter, the proposed method is based on extended work presented in paper (Garcia-Sucerquia et al., 2006). By this approach of reducing the contrast of the speckle noise, it is experimentally shown that an improvement of the order of 50% can be reached when 100 reconstructed images are superimposed.

PDE-Based Filters

In recent years, several PDE based methods have been developed for removal of additive noise from images (Perona, & Malik, 1990; Gilboa et al., 2004; You, & Kaveh, 2004) which can be used by homomorphic filters to reduce speckle noise. The basic idea behind PDE based noise removal are based on energy minimization techniques discussed as follows:

In PDE based noise removal techniques (Romeny, 1994; Caselles et al., 1998), suppose *I* is a 2D scalar noisy image that we want to restore and the noise can be considered as high frequency variations σ with low amplitude, added to the pixels of the regular image.

$$I_{noisy} = I_{regular} + \sigma \tag{18}$$

To regularize I_{noisy} a common idea is to minimize its variations estimated by gradient norm of image:

$$\| \nabla I \| = \sqrt{(I_x^{\ 2} + I_Y^{\ 2})} \tag{19}$$

Then the corresponding variational problem is the minimization of energy functional

$$\min_{I:\Omega \to R} E(I) = \int_{\Omega} \| \nabla I \|^2 d\Omega \tag{20}$$

The necessary condition for minimizing the energy functional *E(I)* described by equation (20) can be obtained using Euler-Lagrange minimization that results in following heat equation

$$\frac{\partial I}{\partial t} = c\nabla^2 I = c(I_{xx} + I_{YY}) \tag{21a}$$

With initial condition as the observed noisy image given as:

$$I_{(t\text{-}0)} = I_{noisy} \tag{21b}$$

where $\nabla^2 I$ is Laplacian of image *I* and *c* is the diffusion constant and I(x, y, t) = I(x, y) .This equation describes the isotropic diffusion process. The basic disadvantage of the isotropic diffusion is that in addition to noise removal it may also blur the edges and fine structures present in the image after certain iterations.

Perona and Malik (1990) proposed a nonlinear diffusion method to avoid blurring and localization problem of linear diffusion filtering which is termed as anisotropic diffusion. Anisotropic diffusion is the opposite of isotropic, i.e. to designate a regularization process that does not smooth the image with the same weight in all the spatial directions.This achieves both noise removal and edge enhancement through the use of a non-uniform diffusion which acts as unstable inverse diffusion near edges and as linear heat equation like diffusion in homogeneous regions without edges. In paper (Perona & Malik, 1990), authors have used the anisotropic diffusion process to avoid blurring and localization problem of linear diffusion filtering to remove additive noise from images. In anisotropic diffusion based filter, the basic idea is that heat equation (21) for linear diffusion can be written in divergence form:

$$\frac{\partial I}{\partial t} = \nabla^2 I = div(gradI) = \vec{\nabla}.\vec{\nabla} I \tag{22}$$

The introduction of a conductivity coefficient c in the above diffusion equation makes it possible to make the diffusion adaptive to local image structure [PM]:

$$\frac{\partial I}{\partial t} = \vec{\nabla}.c\vec{\nabla}I = c\nabla^2 I + \nabla c.\nabla I \quad (23)$$

The two possible choices for diffusion coefficient c are:

$$c_1 = \exp\left(-\frac{\|\nabla I\|}{k^2}\right) and c_2 = \frac{1}{1+\frac{\|\nabla I\|^2}{k^2}}; \quad (24)$$

where k>0.

Both expressions are equal up to first order approximation and *k* is a fixed gradient threshold that differentiates homogeneous area and regions of contours and edges. The value of conductivity coefficient ranges in between 20-50.

In anisotropic diffusion based model (Perona, & Malik, 1990), if real time factor *t* is replaced by complex time factor *it*and the diffusion coefficient $c(\|\nabla I_t^n\|)$ by *c*(Im(*I*)) then it leads to following complex diffusion equation (Gilboa *et al.*, 2004) originally proposed for image enhancement and additive noise removal from digital images.

$$\frac{\partial I}{\partial t} = div\left(c(\mathrm{Im}(I))\nabla I\right) \quad (25)$$

There are two variants of complex diffusion based filter. First one is linear complex diffusion based filter, and the second one is nonlinear complex diffusion based filter. In *linear* complex diffusion based filter for image enhancement and de-noising, the authors (Gilboa *et al.*,2004) proposed to replace the diffusion coefficient term in equation (25) with a complex diffusion coefficient $c = \exp(i\theta)$, and for *nonlinear* complex diffusion, the diffusion coefficient is defined as follows (Gilboa *et al.*, 2004):

$$c(\mathrm{Im}(I)) = \frac{e^{i\theta}}{1+\left(\frac{\mathrm{Im}(I)}{k\theta}\right)^2} \quad (26)$$

Here k is the edge threshold parameter. The value of k ranges from 1 to 1.5 for digital images. The value of k is fine tuned according to the application in hand. For experimentation purposes value of θ is chosen to be $\frac{\pi}{30}$.

In a recent work (Srivastava, Gupta, & Parathasarthy, 2010), authors have proposed a partial differential equation (PDE)-based homomorphic diffusion filter to reduce speckle noise from digitally reconstructed holographic images. For digital implementations, the proposed scheme was discretized using finite differences scheme. Further, the performance of the proposed PDE based technique is compared with other speckle reduction techniques such as homomorphic anisotropic diffusion filter based on extended concept of(Perona, & Malik, 1990), homomorphic Weiner filter, Lee filter, Frost filter, Kuan filter, speckle reducing anisotropic diffusion (SRAD) filter and hybrid filter in the context of digital holography. For the comparison of various speckle reduction techniques, the performance is evaluated quantitatively in terms of all possible parameters that justify the applicability of a scheme for a specific application. The chosen parameters are mean-square-error (MSE), normalized mean square error (NMSE), peak-signal-to-noise ratio (PSNR), speckle index, average signal-to-noise ratio (SNR), effective number of looks (ENL), correlation parameter (CP), mean structure similarity index map (MSSIM) and execution time in seconds. For experimentation and computer simulation, MATLAB 7.0 has been used and the performance is evaluated and tested for various sample holographic images for varying amount of speckle variance. The results obtained justify the applicability of proposed schemes.

RESULTS AND COMPARISONS

In this section, the results of various filters are presented for speckle reduction from digital holographic images and a comparative study has been shown. The performance is measured in terms of speckle index (SI) and Average signal-to-noise ratio (SNR) defined as follows:

Speckle Index (SI)

Since speckle noise is multiplicative in nature, average contrast of an image may be treated as a measure of speckle removal. Speckle index (SI) is defined as,

$$SI = \frac{\sqrt{\mathrm{var}(I)}}{E(I)}. \qquad (27)$$

and its discrete version for an image reads,

$$SI = \frac{1}{mn}\sum_{i=1}^{m}\sum_{j=1}^{n}\frac{\sigma(i,j)}{\mu(i,j)}$$

where $m \times n$ is the size of the image, μ is the mean and σ is the standard deviation. The speckle index can be regarded as an average reciprocal signal-to noise ratio (SNR) with the signal being the mean value and noise being the standard deviation.

Average SNR=1/SI. (28)

Figure 4 shows the visual results of various speckle reduction filters for the sample digital holographic image holographic.jpg. Table 1 shows the performance comparison of various filters in terms of Avg. SNR and speckle index for the same image. From Figure 4 and Table 1, it can be seen that homomorphic complex diffusion based method is outperforming all methods in consideration and it may be considered as an optimal choice for speckle reduction from digital holographic images.

FUTURE RESEARCH DIRECTIONS

Some of the open problems related to digital holographic images include:

1. Devising techniques for speckle reduction and zero- order diffraction from holographic images using partial differential equation (PDE) based approaches in variational framework.
2. Devising wavelet based techniques for enhancement, restoration and speckle reduction.
3. Devising techniques for encrypting of information with digital holography using wavelet based and PDE based approaches.
4. Use of hybrid techniques such as one from Fourier's optics based and another one involving wavelet or PDE-based approaches and many more.

CONCLUSION

In this chapter the basic concepts of digital holography, difference between digital holography and photography, advantages of digital holography and its applications have been discussed in brief in introduction part. The general principles of holographic recording and reconstruction and principles of formation of speckle noise in digital holographic images have been discussed in section followed by introduction. The various techniques for speckle reduction available in literature are

Figure 4. (a) Original speckled holographic image, holographic.jpg; and Filtered image using(b) Homomorphic complex diffusion based method (c) Homomorphic anisotropic diffusion method (d) Homomorphic Wiener Filter (e) Lee Filter (f) Frost Filter (g) Kuan Filter (h) SRAD filter.

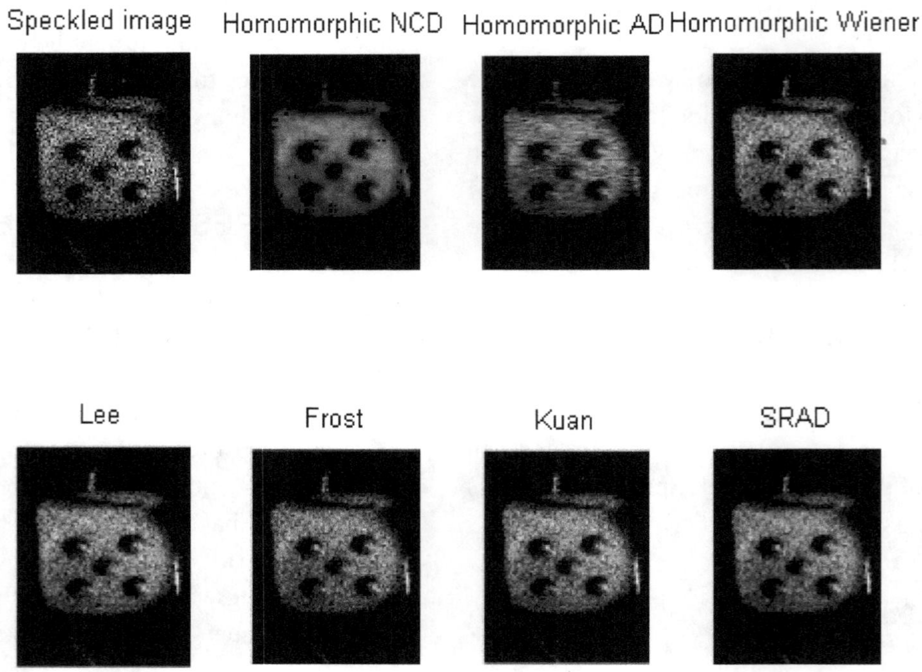

Table 1. Comparison of performances of various speckle reduction filters for the sample digital holographic image, holographic.jpg, SNR of original speckled image= 225.5583, Speckle Index of original speckled image

Speckle Reduction Filters	SNR of restored image	Speckle Index of Restored Image
Homomorphic complex diffusion based method	256.8098	0.00380
Homomorphic anisotropic diffusion method	249.3765	0.00401
Homomorphic Wiener Filter	256.410	0.00390
Lee Filter	238.0952	0.00420
Frost Filter	243.9024	0.00410
Kuan Filter	246.9135	0.00405
SRAD filter	255.1020	0.00392

also discussed. The various important techniques discussed in this chapter for speckle reduction include homomorphic filtering approach, and various specialized filters such as speckle reducing anisotropic diffusion based filter, PDE based methods. Further, the implementation and performance analysis of various speckle reduction techniques are presented. The homomorphic complex diffusion based speckle reduction method performs better in comparison to other methods in consideration.

REFERENCES

Cai, X. O. (2008). Reduction of speckle noise in the reconstructed image of digital holography. *Optik (Stuttgart)*, *121*(4), 394–399. doi:10.1016/j.ijleo.2008.07.026

Cai, X. O., & Wang, H. (2008). The influence of hologram aperture on speckle noise in the reconstructed image of digital holography and its reduction. *Optics Communications*, *281*(2), 232–237. doi:10.1016/j.optcom.2007.09.030

Frost, V. S., & Stiles, J. A. (1982). A model for radar images and its application to adaptive digital filtering of multiplicative noise. *IEEE Transactions on Pattern Analysis and Machine Intelligence*, *4*(2), 157–166. doi:10.1109/TPAMI.1982.4767223

García-Sucerquia, J., Herrera, J., & Velasquez, D. (2005). Reduction of speckle noise in digital holography by using digital image processing . *Optik (Stuttgart)*, *116*, 44–48. doi:10.1016/j.ijleo.2004.12.004

Gilboa, G., Sochen, N., & Zeevi, Y. Y. (2004). Image enhancement and denoising by complex diffusion processes. *IEEE Transactions on Pattern Analysis and Machine Intelligence*, *25*(8), 1020–1036. doi:10.1109/TPAMI.2004.47

Goodman, J. W. (1996). *Introduction to Fourier optics* (2nd ed.). San Francisco, CA: McGraw-Hill.

Jain, A. K. (2006). *Fundamentals of digital image processing*. India: PHI.

Kuan, D. T., & Sawchuk, A. A. (1987). Adaptive restoration of images with speckle. *IEEE Transactions on Acoustics, Speech, and Signal Processing*, *35*, 373–383. doi:10.1109/TASSP.1987.1165131

Lee, J. S. (1981). Speckle analysis and smoothing of synthetic aperture radar images. *Computer Graphics and Image Processing*, *17*, 24–32. doi:10.1016/S0146-664X(81)80005-6

Lee, J. S. (1983). Digital image smoothing and the sigma filter. *Computer Vision Graphics and Image Processing*, *24*, 255–269. doi:10.1016/0734-189X(83)90047-6

Monroy, F. A., & Garcia-Sucerquia, J. (2010). Increment of lateral resolution in digital holography by speckle noise removal. *Optik (Stuttgart)*, *121*(22), 2049–2052. doi:10.1016/j.ijleo.2009.06.011

Perona, P., & Malik, J. (1990). Scale space and edge detection using anisotropic diffusion. *IEEE Transactions on Pattern Analysis and Machine Intelligence*, *12*, 629–639. doi:10.1109/34.56205

Romeny, B. ter H. (Ed.). (1994). *Geometry driven diffusion in computer vision*. Boston, MA: Kluwer.

Sharma, A., Sheoran, G., & Jaffery, Z. A., & Moinuddin. (2008). Improvement of signal-to-noise ratio in digital holography using wavelet transform. *Optics and Lasers in Engineering*, *46*(1), 42–47. doi:10.1016/j.optlaseng.2007.07.004

Srivastava, R., Gupta, J. R. P., & Parthasarathy, H. (2010). Comparison of PDE based and other techniques for speckle reduction from digitally reconstructed holographic images. *Optics and Lasers in Engineering*, *48*(5), 626–635. doi:10.1016/j.optlaseng.2009.09.012

Yaroslavskii, L. P., & Merzlyakov, N. S. (1989). *Methods of digital holography* (Parsons, D., Trans.). New York, NY: Consultants Bureau.

You, Y. L., & Kaveh, M. (2000). Fourth order partial differential equations for noise removal. *IEEE Transactions on Image Processing*, *9*, 1723–1730. doi:10.1109/83.869184

Yu, Y., & Acton, S. T. (2002). Speckle reducing anisotropic diffusion. *IEEE Transactions on Image Processing*, *11*(11), 1260–1270. doi:10.1109/TIP.2002.804276

KEY TERMS AND DEFINITIONS

Digital Holography: In digital holography, both the amplitude and the phase distribution of light coming from object being imaged can be recorded in any plane between the object and the observer producing the complete realistic field of view as originally observed.

Enhancement: This is one of the digital image processing tasks that enhances or highlights the visual quality of the image from its previous version.

Homomorphic Filter: This filter is used for speckle reduction from digital images using filters defined for additive noise removal in logarithmic domain. In logarithmic domain, the multiplicative noise converts in to additive noise.

Mathematical Hologram: In digital holography, the object's beam complex amplitude is called the mathematical hologram.

PDE-Based Filters: Partial differential equation based filters derived using variational calculus by minimizing the energy functional of the image defined in terms of gradient norm of the image.

Restoration: This is one of the digital image processing tasks that removes or reduces the noise from images to improve its visual quality.

Speckle Index: One of the performance measures for speckle noise. It can be regarded as an average reciprocal signal-to noise ratio (SNR).

Speckle Noise: A multiplicative noise that appears in digital holographic images. It degrades the visual quality of the image. This noise is introduced during the formation of image.

Speckle Reduction: It deals with the methods to reduce speckle noise.

Chapter 7
High-Speed, High-Resolution 3D Imaging Using Projector Defocusing

Song Zhang
Iowa State University, USA

Yuanzheng Gong
Iowa State University, USA

ABSTRACT

With the advance of software and hardware, three-dimensional (3D) scene digitization becomes increasingly important. Over the years, numerous 3D imaging techniques have been developed. Among these techniques, the methods based on analyzing sinusoidal structured (fringe) patterns stand out due to their achievable speed and resolution. With the development of digital video display technologies, digital fringe projection techniques emerge as a mainstream for 3D imaging. However, developing such a system is not easy especially when an off-the-shelf projector is used. The major challenging problems are: (1) the projection system nonlinearity; (2) the precise synchronization requirement; and (3) the projection system speed limit. This chapter will present an alternative route for 3D imaging while reducing these problems. The fundamentals of the proposed technique will be introduced, the analytical and experimental results will be shown, and its advantages and limitations will be addressed.

INTRODUCTION

With the release of Avatar and others three-dimensional (3D) movies, and the emergence of 3D TVs and monitors, 3D imaging technology started penetrating into our daily lives. Thus, 3D imaging has become unprecedentedly important and close to ordinary people.

DOI: 10.4018/978-1-61350-326-3.ch007

3D imaging is essential to represent the physical object with 3D contents either in a digital format or by an analog means. However, an enormous amount of effort has been put to represent the scene digitally because it is easier to manipulate in this manner. To digitalize a 3D scene, there are mainly two approaches: passive and active. The passive methods (e.g., stereo vision) are to recover the 3D information from natural 2D photographs. They essentially capture photographs of the object from different viewing angles and obtain the depth by finding the correspondences between the image pairs, and by establishing the triangulation between the object point and the camera sensor locations. These methods work well for applications where the accuracy is not the primary focus, such as the entertainment. However, hinging on identifying the correspondences between image pairs, the measurement accuracy is not high if the object does not have very strong texture information (Zhang, 2010b).

Active methods, on the other hand, recover depth information by actively placing some vivid features on the object surface to assist the correspondence establishment. Typically, an active light is used because of its surface noncontact nature. The active light can be a single wavelength laser, a range of color spectrum, or a broadband white light. The active features can be dots, lines, and area structured patterns (Salvi et al, 2010). For high-speed applications, the whole area structured patterns are usually desired. There are many ways to generate the structured patterns, such as laser interference, gratings, slide projectors, etc. However, the most convenient means is to use a digital video projector. The patterns can be different in terms of shape and structures, binary, sinusoidal, narrow and wide, etc. Among these techniques, the sinusoidal structured (fringe) patterns based methods stand out because it is most close to the natural light propagation (in sinusoidal way). The phase-shifting techniques have been studied over the past decades and have been used broadly in numerous applications (Gorthi & Rastogi, 2010). Conventionally, the fringe patterns are generated by laser interference, which is good in terms of measurement accuracy and stability. Digital fringe projection techniques, where a digital video projector is used, start expanding its use because of its simplicity.

There is some success and advancement in the technological development of digital fringe projection techniques for 3D imaging, which has been thoroughly reviewed in reference (Zhang, 2010a). The commercial digital video display systems are designed for the purposes other than 3D imaging. There are a number of challenges in order to use them for high-speed and high-accuracy 3D imaging. These include handling the problems of (1) the projection system nonlinearity, (2) the precise synchronization requirement, and (3) the projection system speed limit (Lei & Zhang, 2010).

The objective of this chapter is to present an alternative route for 3D imaging technique using a digital fringe projection and phase shifting technique. This new technique has the potential to significantly reduce the problems of the existing digital fringe projection technique, to drastically simplify the system development for non-experts, and thus to speed up its use in our daily lives. In particular, we will present some of our most recent research in high-speed 3D imaging area that uses this technique.

BACKGROUND

Over the past decades, a number of 3D imaging techniques have been developed including some with real-time capability (Huang et al, 2005; Li et al, 2010; Pawlowski et al, 2002; Quan et al, 1995; Takeda & Mutoh, 1983; Zhang & Huang, 2006a). With recent advancement in computational and shape analysis techniques, high-speed 3D imaging has become unprecedentedly important. Over the years, a number of techniques have been developed to reach real-time capability, including spacetime stereo (Zhang, et al, 2004, Davis, et al, 2005), structured light (Rusinkiewicz et al, 2002), and fringe projection (Zhang & Huang 2006). Among these techniques, fringe analysis stands out because of its numerous advantages (Gorthi & Rastogi, 2010).

A fringe pattern is essentially a special case of a structured pattern in that the stripe intensity varies sinusoidally. The Fourier profilometry method (Takeda & Mutoh, 1983) that uses a single fringe pattern could reach the fastest speed thus it has been applied to many areas (Su & Chen, 2001; Su & Zhang, 2010), and has been used to measure numerous extreme phenomena (Takeda, 2010). This method is basically to obtain the phase information by applying Fourier transform of the fringe pattern, and 3D information will be extracted from the phase. This technique is very suitable for measuring smooth surfaces, while has the limitations if it is used to measure complex shapes. Windowed Fourier transform technique endeavors to alleviate some problems of the conventional Fourier method by processing the fringe pattern patch by patch (Qian, K, 2004). However, challenges still remain for the single fringe pattern based Fourier technique to measure complicated objects.

To address the limitations of single fringe pattern based 3D imaging techniques, phase-shifting methods are proposed. The phase shifting methods use more fringe patterns with spatial or temporal shift in phase to perform better measurement. Instead of performing the measurement referring to the whole image, a phase-shifting method is to provide the measurement through a point-by-point operation. Therefore, this technique gives better spatial resolution (Zhang, 2010b). Various phase-shifting methods have been proposed including three-step, four-step, and least-square algorithms (Schreiber & Bruning, 2007). Since it requires a minimum number of three phase-shifted fringe images to allow for point by point 3D imaging, this technique requires capture three individual images, or a single color image for each measurement. Because only three fringe patterns are required to perform one 3D measurement at pixel-level spatial resolution, this technique has the potential to realize high-speed and high-resolution 3D imaging.

Conventionally, the fringe patterns are either generated by a mechanical grating or by laser interference. They have been very successfully applied to numerous industrial applications in optical metrology. However, it is typically not very easy for them to adjust fringe pitches (periods), nor accurately shift them in phase domain.

The technique of projecting sinusoidal fringe patterns with a digital video projector is called the digital fringe projection technique. It recently emerged as a mainstream in 3D imaging, and has the advantages of generating and controlling the fringe patterns accurately and easily. There is some success and advancement in using this technology for real-time 3D imaging (Li et al, 2010; Zhang & Huang, 2006a; Zhang & Yau, 2006, Zhang et al, 2006). However, developing a high-speed 3D imaging system with an off-the-shelf projector remains difficult, which will be explained in the next section.

To address the challenges of the existing 3D imaging technologies, this chapter will present a new 3D imaging approach that was recently developed in our research group. Because of some of its advantageous features of this new technique and also because of the fundamental limitations of the current off-the-shelf hardware components, this technique shows great potential to lead some breakthroughs in the field of high-speed 3D imaging. In particular, we will focus on the following three major pieces of work that we have recently developed: (1) improved the existing real-time 3D shape measurement speed without significantly increases its hardware costs (Gong & Zhang, 2010a); (2) reached a kHz 3D shape measurement speed with a simple and inexpensive digital-light-processing (DLP) projector (Gong & Zhang, 2010b); and (3) achieved a superfast phase-shifting method for unprecedentedly high-speed 3D imaging: 667 Hz (Zhang et al, 2010). In particular, we have developed a system that has doubled our real-time 3D imaging speed and reached the maximum rate of 120 Hz if a DLP projector and a three-step phase-shifting method are used. The second technique we have developed was essentially to convert a DLP projector to be a flexible sinusoidal grating system for fast motion capture. Implementing this technique to a recently innovated DLP Discovery projection platform, a potential tens-of-kHz rate 3D imaging is feasible. All these could not be realized without the defocusing technique that we have recently developed.

MAIN FOCUS OF THE CHAPTER

Principles of Digital Fringe-Projection and Phase-Shifting Technique

Figure 1 shows the setup of a digital fringe projection system. This is a typical triangulation based system. A computer generated sinusoidal fringe pattern is projected by a projector onto an object surface, a camera, from another viewing angle, captures the scattered fringe images by the object. The computer software is then used to process and recover the 3D shape. Since this is a triangulation based system, the correspondences between the projector projected image and the camera captured image must be identified. In a fringe projection system, this correspondence is established in phase domain. In other words, a point on the camera corresponds to the point projected by the projector only if both points have the same phase value. Because the structured pattern contains vertical stripes, each phase value corresponds to a vertical line on the projected image. Therefore, epipolar geometry is needed in order to identify the unique correspondences (Zhang & Huang, 2006b). Once the correspondence is identified, the depth information can be recovered based on triangulation. In this technique, the correspondence was established through the phase, which will be explained in the next Subsection.

Fundamentals of Phase-Shifting Algorithm

Phase-shifting techniques have been widely used in optical metrology. Over the years, a number of phase shifting algorithms have been developed including three-step, four-step, least square algorithms etc. (Schreiber & Bruning, 2007). All these algorithms differ in terms of the number of fringe images required, the amount of phase shift, but they are all the same in terms of phase calculation, pixel by pixel.

A three-step phase-shifting algorithm is very commonly used in high-speed applications since it requires the least number of fringe patterns. Three fringe images of a three-step phase-shifting algorithm can be represented as

$$I_1(x,y) = I'(x,y) + I''(x,y)\cos[\phi(x,y) - 2\pi / 3], \tag{1}$$

$$I_2(x,y) = I'(x,y) + I''(x,y)\cos[\phi(x,y)], \tag{2}$$

$$I_3(x,y) = I'(x,y) + I''(x,y)\cos[\phi(x,y) + 2\pi / 3]. \tag{3}$$

Here, $I'(x,y)$ is the average intensity, $I''(x,y)$ the intensity modulation, and $\phi(x,y)$ the phase to be solved for. Solving Equations (1)-(3) simultaneously, the phase can be obtained as:

$$\phi(x,y) = \tan^{-1}\left\{\frac{\sqrt{3}\left[I_1(x,y) - I_3(x,y)\right]}{2I_2(x,y) - I_1(x,y) - I_3(x,y)}\right\}. \tag{4}$$

The phase value provided by this equation only ranges from $-\pi$ to $+\pi$, which will result in 2π phase discontinuities. To obtain a continuous phase map, a phase unwrapping algorithm is usually needed (Ghialia & Pritt, 1998). Once the continuous phase map is obtained, 3D information can be recovered if the system is calibrated (Zhang & Huang, 2006b). To recover one 3D shape,

Figure 1. Setup of a digital fringe-projection and phase-shifting technique

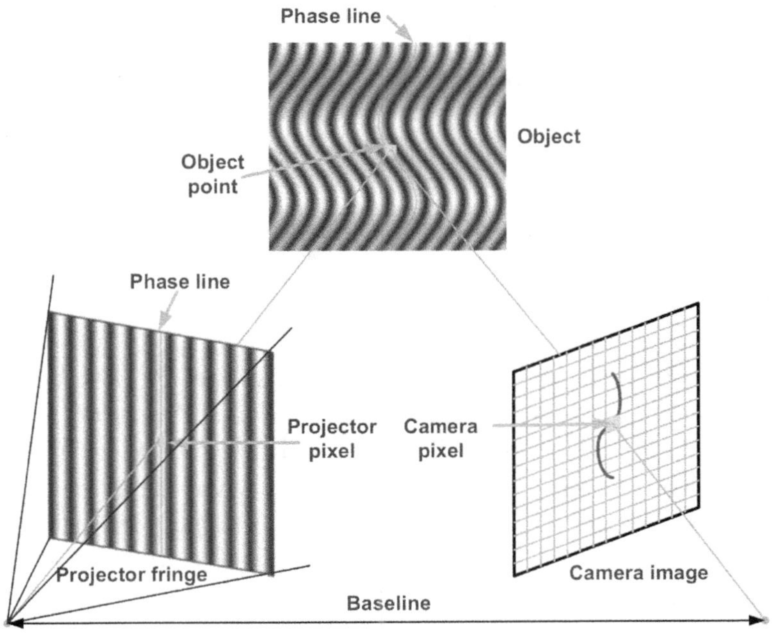

the system based on binary patterns uses codeword to establish correspondence between the captured images and the projected images. The codeword, a unique value, is formed by a sequence of stripes composed of 0s (purely black) or 1s (purely white) and can be determined from captured binary structured images. As explained previously, a method using binary structured patterns cannot reach pixel-level resolution spatially because the stripe width must be larger than one projector's pixel.

In this phase-shifting technique, the phase value is regarded to as the codeword that is used in the binary structured light technique because they are unique for each line on the projected fringe patterns. Therefore, once the phase is obtained from the captured fringe images, the codeword can be determined, 3D information can then be recovered from the phase. As can be seen in Equation (4), the phase here is calculated pixel by pixel, thus, the 3D information can be obtained pixel by pixel, which is advantageous over most other 3D imaging techniques. Therefore, this technique allows for pixel-level spatial resolution. Since only three images are required, it is possible to achieve high-speed (Zhang, 2010a).

In the meantime, from Equations (1) - (3), we can obtain the texture shown in Equation 5.

Because the texture is obtained point-by-point, and precisely aligned with the 3D geometry, this is another advantage of 3D imaging with a phase-shifting technique.

Box 1.

$$I_t(x,y) = I'(x,y) + I''(x,y) = \frac{I_1 + I_2 + I_3}{3} + \frac{\sqrt{3(I_1 - I_3)^2 + (2I_2 - I_1 - I_3)^2}}{3}. \tag{5}$$

Major Concerns of a Conventional Digital Fringe Projection System

Despite the success of using a digital fringe projection technique for real-time 3D imaging, challenges still remain, and problems still exist. There are three major problems of the existing digital fringe projection system for accurate 3D imaging, and for further improving its speed.

Challenge *#1: The nonlinearity of the projection system.* The first major challenge about this technique is the nonlinear effect of a projector. To perform high quality 3D imaging using a digital fringe projection and phase-shifting method, the projector nonlinearity calibration is usually mandatory. This is because the commercial video projector is usually a nonlinear device that is purposely designed to compensate for human vision. However, for 3D imaging, this nonlinear effect increases the complexity of the system development, and induces measurement errors. A variety of techniques have been studied including the methods that actively changing the fringe to be projected (Huang et al., 2002; Kakunai et al., 1999), and those that passively compensating for the phase errors (Guo et al., 2004; Pan et al., 2009; Zhang & Huang, 2007; Zhang & Yau, 2007a). Moreover, because the output light intensity does not change much when the input intensity is close to be 0 or/and 255 (Huang et al, 2002), it is impossible to generate sinusoidal fringe images with full intensity range (0-255). In addition, our experiments found that the projection nonlinear gamma actually changes over time, thus the system needs to be re-calibrated frequently.

Challenge *#2: The precise synchronization requirement for the whole system.* Because the projector is a digital device that generates the full grayscale image at a certain frequency, which is typically the refresh rate of the image. Therefore, in order to capture the grayscale images accurately and correctly, the camera must capture at least one full refresh cycle. For high speed 3D imaging, it is desirable to capture only one cycle. Therefore, the camera must be precisely synchronized with the projector, i.e., the camera must start its exposure when the image starts refresh and must stop its exposure when the refresh finishes. For instance, the DLP projector generates the grayscale fringe images by time modulation (Hornbeck, 1997), thus the camera exposure time cannot be shorter than the single channel projection time (1/360 sec). This limits its application to measure very fast phenomena when a very short exposure is required.

Challenge *#3: The speed limit of the projection system.* Because of its digital fringe generation nature, the 3D imaging speed is ultimately determined by the fringe projection rate: 120 Hz for a typical DLP projector. In order to capture a fast motion, a "solid-state" fringe pattern is usually needed. The solid-state fringe pattern can be generated by a mechanical grating, or by a laser interfering. However, as addressed earlier, the digital fringe projection technology usually cannot produce solid-state fringe pattern. To take advantage of the merits of digital fringe generation techniques, we need to circumvent the associate problems to achieve fast 3D imaging speed.

All these problems and challenges hinder the 3D imaging applications especially for precision measurement. On the contrast, if a technique can generate ideal sinusoidal fringe images without worrying about the problems introduced above, it would significantly simplify the 3D imaging system development, and drastically speed up its use in our daily life.

Proposed Technique

To address the problems of the current digital fringe projection system, we recently proposed a technique that is to generate sinusoidal fringe patterns based on defocusing effect (Lei & Zhang, 2009). This technique allows for "solid-state" fringe generation without requiring nonlinear gamma calibration. Instead of using 8-bit grayscale fringe images, this technique only uses binary (0s or 255s) structured patterns. The idea came from our two observations: (1) seemingly sinusoidal fringe patterns often appear on the ground when the light shines through an open window blind; and (2) the sharp features of an object are blended together in a blurring image that was captured by an out-of-focus camera. The former gives the insight that an ideal sinusoidal fringe image could be produced from a binary structured pattern, and the latter provides the hint that if the projector is defocused, the binary structured pattern might become ideal sinusoidal. Because only binary patterns are needed, the nonlinear response of the projector would not be a problem since only 0 and 255 intensity values are used. Moreover, phase shifting can be introduced by spatially moving the binary structured patterns. For instance, a $2\pi / 3$ phase shift can be realized by shifting the binary structured patterns spatially by 1/3 of its period. Therefore, a 3D shape measurement system based on a digital fringe projection technique can be developed without nonlinear gamma calibration.

This binary status coincides with the DLP technology that operates the digital micro mirrors ON/OFF, in binary stage. Therefore, theoretically, if a micro mirror is set to be a value of 0 or 255, it should stay OFF or ON all the time. By this means, the micro mirror will act as "solid state" (does not refresh). Therefore, the solid-state structured light will be realized. Because the structured patterns are generated in solid state and any segment of time can represent the signal, there is no precise synchronization requirement between the projector and the camera. In the meantime, the exposure time can be shorter than the channel projection time, namely, less than 1/120 second for a 120 Hz projector. Therefore, this technique allows for capturing extremely fast phenomena with a relatively inexpensive off-the-shelf DLP projector.

Because of the architecture of the digital cameras, the capture and data transfer usually cannot happen simultaneously if an external triggering mode is utilized. This limits the 3D imaging speed to 60 Hz for a 120 Hz projector (Zhang & Yau, 2007b; Li et al, 2010). With this new technology, it may enable the 120 Hz 3D imaging rate with the same hardware components. In addition, because only binary structured patterns are used, it actually allows for even faster fringe pattern switching rate since smaller size of data rate is needed. This, in turns, potentially allows for much faster 3D imaging rate.

Generating Sinusoidal Fringe Patterns with Projector Defocusing

Because of the advantages of a phase-shifting based technique, it is desirable to use sinusoidal fringe patterns for 3D imaging. However, as explained earlier, the existing techniques to generate sinusoidal fringe patterns have some challenges to tackle with. In the meantime, if only binary structured patterns are used, those problems can be significantly alleviated or eliminated.

To illustrate the viability of generating sinusoidal fringe patterns with a projector defocusing, we performed an experiment. In this experiment, a DLP projector (Dell M109S, Texas) is used to project the computer generated binary patterns onto a white board. A camera (The Imaging Source DMK 21BU04, North Carolina), with the board on its focal plane, is to capture the reflected fringe patterns by the board. The projector's focus is adjusted gradually from approximately perfectly focused to severely defocused. Figure 2 shows some frames of the captured fringe patterns. Figure 2(a) -(e) shows the progress of the binary structured pattern sent to a DLP projector with different degrees of defocusing. It clearly shows that the binary structured pattern becomes seemingly more and more sinusoidal with the degree of defocusing increases. At certain point, the fringe patterns become approximately sinusoidal. Of course, when the projector is defocused too much, all the patterns are blended together, and there are no obvious structured patterns. Figure

Figure 2. Binary structured pattern becomes more and more sinusoidal with the increase of degree of defocusing. (a)-(e) shows the progress of the fringe patterns; (f)-(j) shows their corresponding cross sections.

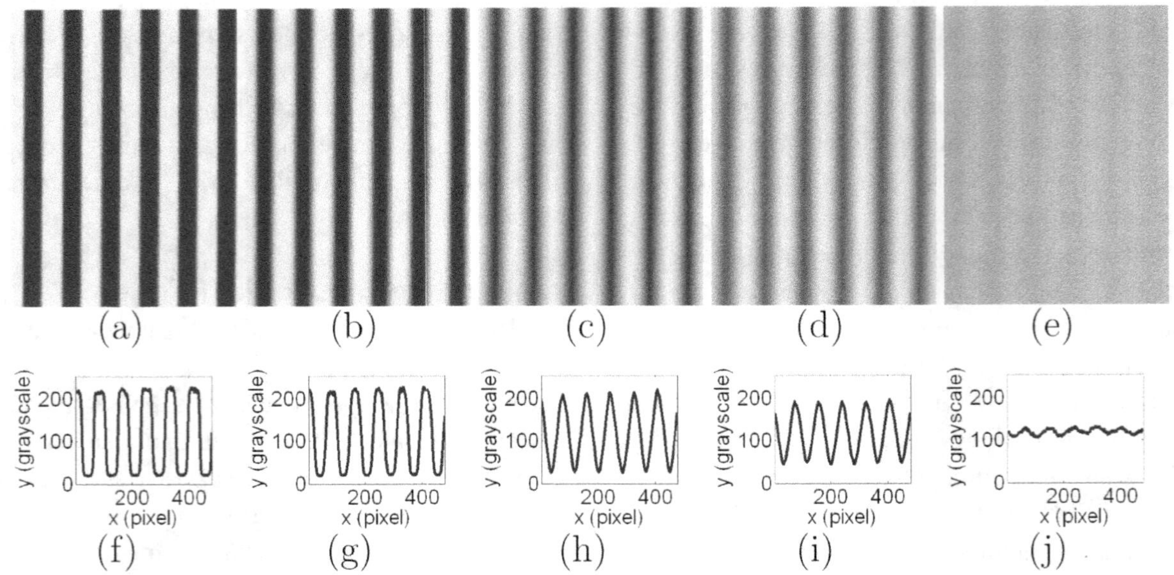

2(f)-(j) shows the one of the cross sections for each fringe image. This experiment shows that it seems feasible to generate approximately sinusoidal fringe patterns by properly defocusing the binary one.

Theoretical Analysis

Because the structured pattern contains vertical stripes with exactly the same structures, to understand how the pattern changes when the projector is defocused, we only need to understand its one horizontal cross section. The cross section is actually a square wave, which can be represented as

$$y(x) = \begin{cases} 0 & x \in [nT - T/2, nT) \\ 1 & x \in [nT,\ nT + T/2) \end{cases}. \tag{6}$$

Here n is an integer, and T the period of the signal. The square wave can be written in Fourier series as

$$y(x) = 0.5 + \sum_{k=0}^{k=\infty} \frac{2}{(2k+1)\pi} \sin\left[\frac{2\pi(2k+1)}{T} t\right]. \tag{7}$$

The imaging (or projection) system can be regarded as a point spread function (PSF). The defocusing effect is essentially to blur the images. The degree of blur can be modeled as applying different size of PSF. Since the square wave only contains odd harmonics without even ones, it is easier for a filter to suppress the higher frequency components. This indicates the feasibility of generating sinusoidal fringe patterns by defocusing binary structured ones. If the binary patterns are moved horizontally, phase-shifted fringe patterns will be generated after defocusing. Therefore, this technique can be used for 3D imaging using a digital fringe projection and phase-shifting technique.

Experiments

To demonstrate the viability of the proposed 3D imaging technology, a hardware system was developed. The system is composed of a DLP projector (Dell M109S, Texas), and a CCD camera (The Imaging Source DMK 21BU04, North Carolina). The projector is projecting binary fringe patterns with a period of 12 pixels per period, and a phase shift of 120 degrees (or 4 pixels in this case). The projector is properly defocused so that sinusoidal fringe patterns will be generated. Figure 3 (a)-(c) shows three captured fringe images. Equation (4) is applied to compute the wrapped phase map, as shown in Figure 3(d). This phase map is then unwrapped to obtain the continuous phase map by adopting a phase unwrapping algorithm (Zhang et al, 2007). Figure 3 (e) shows the unwrapped phase map. The unwrapped phase map can then be used to recover the 3D shape of the object (Zhang & Huang, 2006b). Figure 3(f)-(g) shows the rendered result of the 3D shape. Figure 3 (h) shows the zoom in view of the object, it clearly shows that the system can capture very fine details with a relatively low resolution camera (640 X 480).

Some Recent Advances

The defocusing technology has recently led some breakthroughs in high-speed 3D imaging field. In particular, three major technological improvements have been achieved, they are:

1. The improvement of the real-time 3D imaging speed.
2. The realization of a high-speed 3D imaging with an off-the-shelf inexpensive projector.
3. The achievement of a superfast phase-shifting method for 3D imaging.

In this section, we will elucidate these technologies.

Improve the Real-Time 3D Imaging Speed

Conventionally, a real-time 3D imaging technique based on a digital fringe projection and phase-shifting method requires sinusoidal fringe patterns to be sent to a focused DLP projector (Zhang & Huang, 2006a; Li et al, 2010). However, due to its digital fringe generation nature, the camera and the projector must be precisely synchronized. Modern projectors usually have no time gap between channels. Therefore, in order to reach the projection speed, the camera must be able to readout the data simultaneously while it exposures. However, when the external triggering mode is used, a relative inexpensive camera usually takes some time to readout the data asynchronously: usually 1 / (max frame rate) to readout the image before it takes another one. Moreover, a typical DLP projector has different time duration for different color channels to balance its output color. This means that the camera must be able to change its exposure time from frame to frame. In reality, this is not an easy task, especially when the external trigger mode is in use. Therefore, it is usually very difficult for an ordinary system to achieve the maximum 3D imaging speed: the projector's refresh rate. As a result, Only 60 Hz 3D imaging rate is achieved for a modified120 Hz projector (Zhang & Yau, 2007b, Li et al 2010). In order to solve this problem, a conventional approach is to employ a high-end camera so that it can capture images when the data is reading out, and it allows for precise timing changes from frame to frame. However, this type of camera is usually extremely expensive.

On the contrast, if a binary structured pattern is used, it does not require the camera capture the full projection channel. This is because any portion of the signal can represent the projected image and the capture can happen any time and with any exposure time during the image projection. Therefore, it allows the use of a relatively inexpensive camera to reach the maximum speed.

Figure 3. Example 3D shape measurement result using projector defocusing. (a)-(c) Three phase shifted fringe patterns; (d) Wrapped phase map; (e) Unwrapped phase map; (f) 3D rendering in shaded mode; (g) 3D shape with texture mapping; (h) Zoom-in view of the 3D shape.

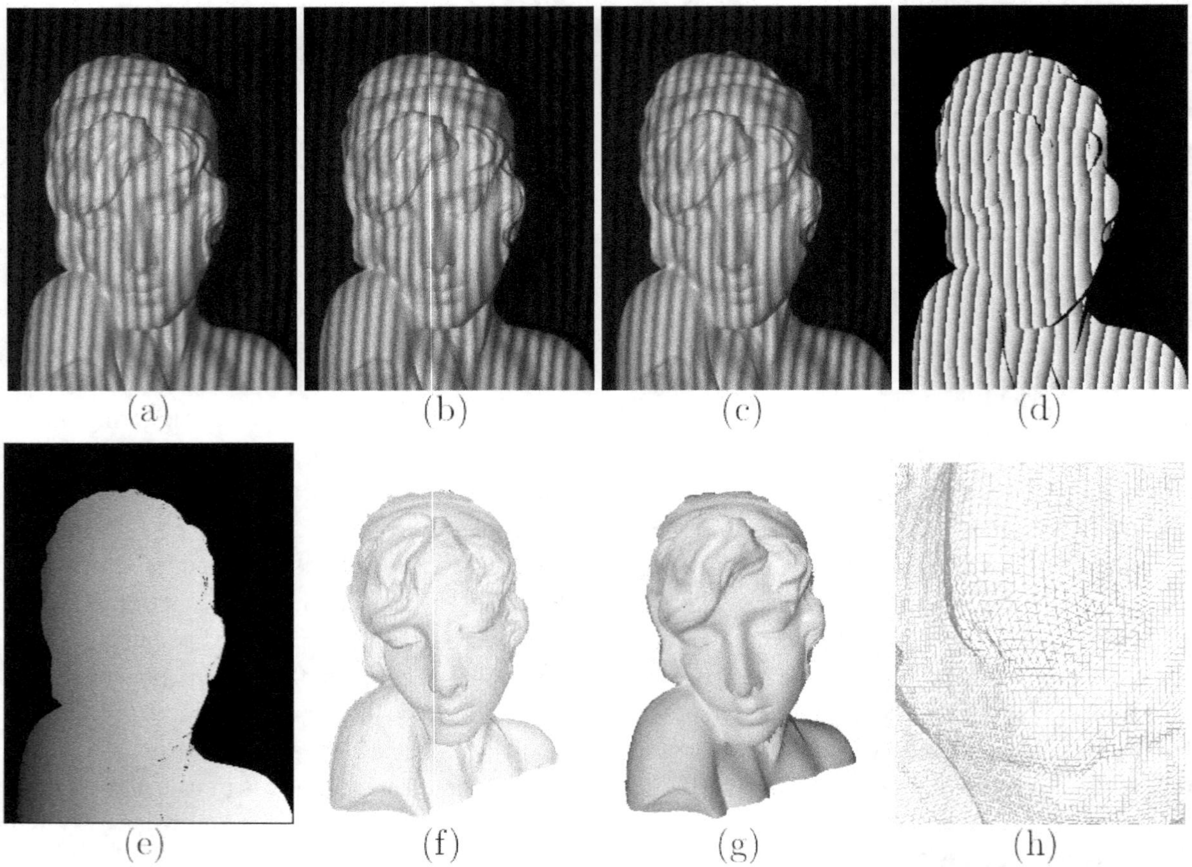

By this means, with a relatively inexpensive camera, the 3D imaging system can double the previously maximum achievable speed and reach the refreshing speed of a DLP projector: 120 Hz.

This technique was verified by implementing it into our previously developed real-time 3D imaging system (Zhang et al, 2007). In this system, a modified DLP projector (PLUS U5-632h, Japan) is used to switch the structured patterns at 120 Hz (RGB). Three color channels are encoded as three phase-shifted fringe patterns, and a camera (Jai Pulnix TM-6740CL, California) that is precisely synchronized with the projector is used to capture three color channels separately. We achieved 60 Hz 3D imaging rate with a conventional method, and the exposure time of the camera is precisely 2.78 ms since each projection channel lasts this amount of time. In this experiment, we chose the exposure time of the camera to be 0.78 ms and the camera capturing speed to be 360 Hz. If a conventional technique is used, the measurement cannot be correctly performed, as shown in Figure 4 top row.

On the contrast, if the method introduced in this chapter is used, 3D imaging can be performed accurately even if the exposure time is 0.78 ms, which is less than the channel projection time, 2.78 ms. Figure 4 shows the experimental result.

Figure 4. Experimental results of the 120 Hz 3D imaging technique with an exposure time of 0.78 ms (much less than the projection time 2.78 ms). The top row shows the measurement result using a conventional method. (a)-(c) Three phase-shifted fringe patterns. It can be seen that the fringe patterns lost its sinusoidal structures. From these fringe patterns, the phase map can be obtained as shown in (d). The phase map shows irregular structures. (e) The recovered 3D profile that is not correctly captured. The bottom row shows the measurement result using the proposed defocusing technique. The sinusoidal fringe patterns look normal and the 3D profile can be correctly captured. The resolution of the camera used is 224×480.

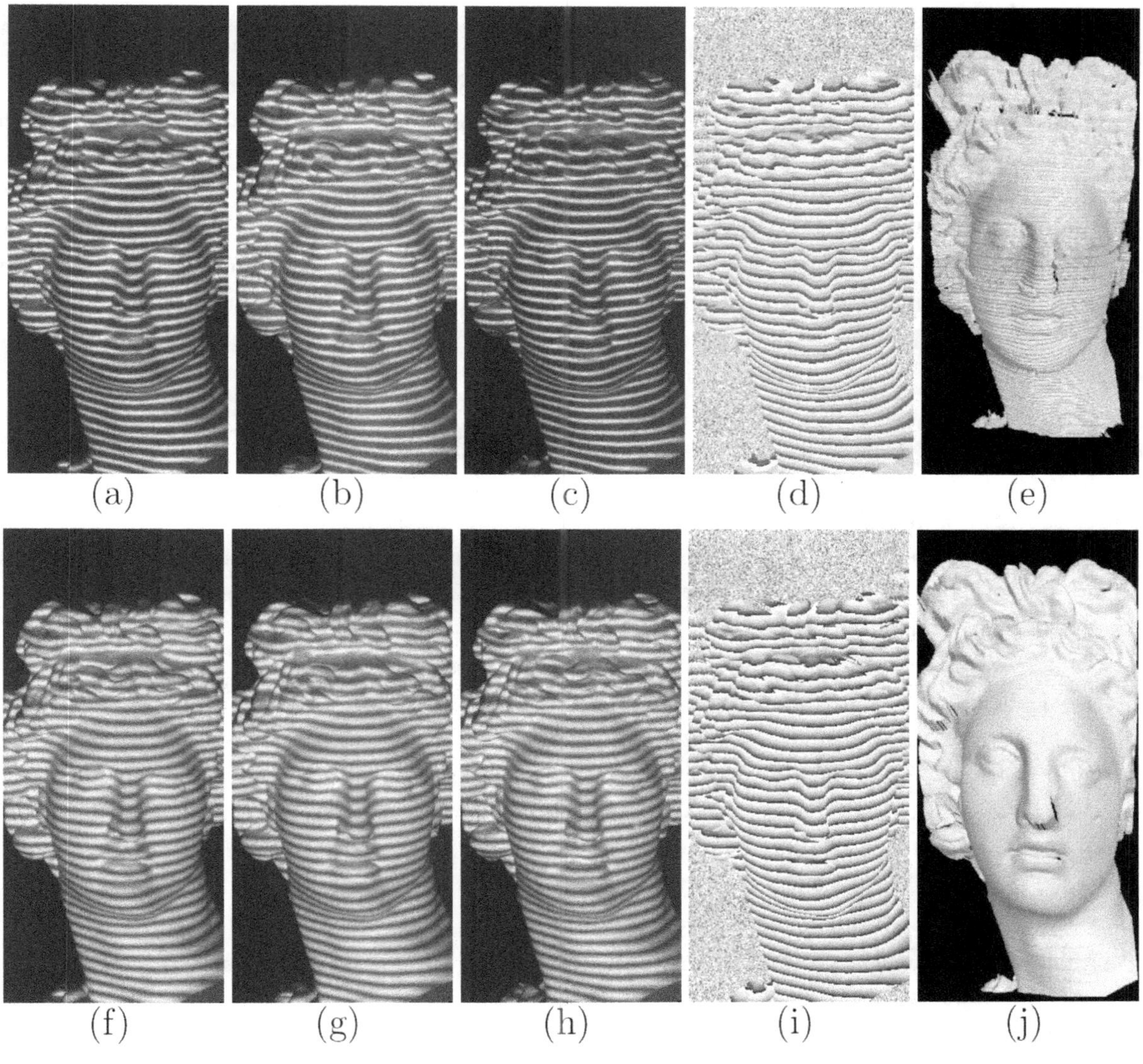

Realize High-Speed 3D Imaging with an Off-the-Shelf Inexpensive Projector

Because of its digital fringe pattern generation nature, the 3D imaging speed is ultimately determined by the fringe projection rate: 120 Hz for a typical DLP projector. Moreover, because the DLP projector generates the grayscale fringe images by time modulation, the camera exposure time cannot be shorter than the projection time. This limits its application to measure very fast motion when a very short exposure time is required. On the contrast, if we use binary structured patterns (0s and 255s), each micro-mirror always being one stage (either OFF or ON), thus identical structured patterns can be captured during any time interval. Again sinusoidal fringe patterns are generated by properly defocusing binary ones. As introduced earlier, from a single fringe pattern, 3D imaging can be performed through Fourier analysis. By this means, the 3D imaging speed can go beyond 120 Hz, and the exposure time can be shorter than the projector refreshing time (1/120 sec).

Experimentally, we used a very inexpensive DLP projector (Dell M109S) to project the fringe patterns at 60 Hz, and a high-speed CMOS camera (Phantom V9.1) to capture the projected fringe patterns. The camera captures the fringe images at 4000 Hz with exposure time of 240 μs. If a conventional fringe generation technique is used, where the 8-bit grayscale values are all used, the sinusoidal fringe patterns cannot be correctly captured. Figure 5 top row shows some typical frames of the captured fringe patterns. It clearly shows that the sinusoidal structure of the pattern is not very obvious and the fringe pattern cannot be captured correctly. This is because when the exposure time of the camera is much shorter than the channel projection time, the DLP projector cannot completely produce the full 8-bit grayscale image during the time period.

On the contrast, if a binary structured pattern and the defocusing technique are used, the fringe patterns can be captured with high quality. This is because each micro-mirror remains constant thus any time period will produce the same signal. Figure 5 bottom row shows the captured fringe patterns. It can be seen from these fringe images that even though the intensity of the fringe patterns are different, the sinusoidal structure is well preserved. The intensity variations are caused by the color of the projector. Even though a white structured pattern is used, the projector produces the white pattern by combining red, green, and blue channels in time sequence. Because the camera has different sensitivity to different spectrum of light, and the color light output of the projector has different intensity, the captured fringe pattern will be different in brightness during different projection timing.

Fourier method (Takeda & Mutoh, 1983) is one of the 3D imaging techniques that only require use of one single fringe pattern. Theoretically, the single fringe pattern can be written as shown in Equation (8).

If a Fourier transform is applied to the image and the conjugate frequency component and the DC component is filtered out, the resultant can be represented as the following signal in complex format

$$I^f(x,y) = \frac{1}{2} I''(x,y) e^{j\phi(x,y)}. \tag{9}$$

Box 2.

$$I(x,y) = I'(x,y) + I''(x,y)\cos[\phi(x,y)] = I'(x,y) + \frac{1}{2} I''(x,y)\left[e^{j\phi(x,y)} + e^{-j\phi(x,y)}\right]. \tag{8}$$

Figure 5. Comparison between captured fringe images with a high-speed camera and an off-the-shelf inexpensive projector when the exposure time is much shorter than each individual projection channel time. The projector refreshes at 60 Hz, and the camera captures at 4,000 Hz, the exposure time used for the camera is 240 μs. The top row shows some example fringe images if a conventional fringe projection technique is use. The sinusoidal structure is not very obvious in these images. The bottom row shows some example fringe images if the defocusing technique is used. The fringe pattern still maintains high-quality sinusoidal structures. Image resolution is 480×480.

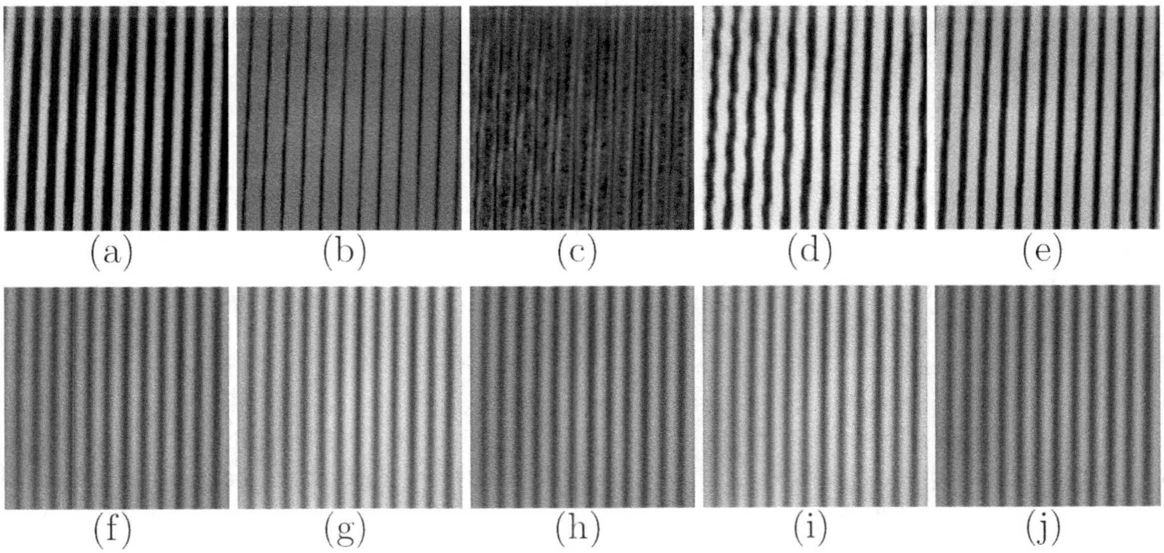

From this equation, the phase can be calculated by

$$\phi(x,y) = \tan^{-1}\left\{\frac{\mathrm{Im}[I^f(x,y)]}{\mathrm{Re}[I^f(x,y)]}\right\}. \tag{10}$$

Here, Im(x) is to obtain the imaginary part of the complex value x, and Re(x) is to obtain the real part of the complex value x. Once the phase is obtained, 3D information can be recovered from the phase following similar procedures as the phase-shifting technique.

To verify the performance of the proposed technique with the fast capturing rate, a rotating fan blade was measured. The fan is rotating at 1,793 revolutions per minute (rpm) during the experiment. Figure 6 shows one of the measurement results. The fringe pattern is captured at 4,000 Hz and exposure time of 80 μs. The image resolution is 480×480. This experiment shows that with a relatively inexpensive projector, the system can be used to measure very fast phenomena (rotating fan blade).

Achieve Superfast Phase-Shifting Method for 3D Imaging

Previously, we demonstrated that it was feasible to image very fast phenomena with the proposed fringe generation technique with a relatively inexpensive projector and the single fringe analysis technique. The Fourier method is very good to measure smooth surface at very fast speed (Karpinsky & Zhang, 2010). In the meantime, in order to measure very complex 3D structures, a phase-shifting technique is necessary. However, a typical digital video projector cannot change

Figure 6. 3D imaging result of a rotating fan blade with the proposed technique and a very inexpensive projector. The data is captured at 4000 Hz with an exposure time of 80 μs. (a) Photograph of the fan blade; (b) Fringe image; (c) Fourier spectrum; (d) Phase map; (e) 3D profile. The image resolution is 480×480.

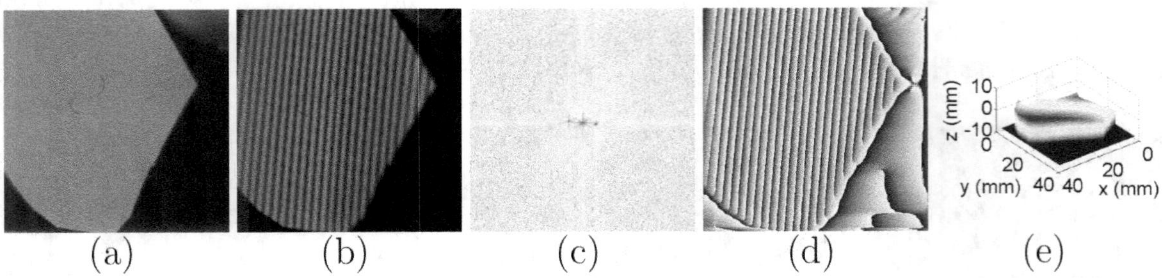

the phase-shifted fringe patterns at such a high frame rate.

The most recently developed DLP Discovery (Texas Instruments, Texas) technology has enabled 1-bit image switching rate at tens of kHz. This innovation shows great potential for 3D optical metrology because of its flexibility to control the projected light accurately (Hoefling, 2004a, Hoefling 2004b; Hoefling & Aswendt, 2009). We have verified the feasibility of using the DLP Discovery technology for superfast 3D imaging with a digital fringe projection and sinusoidal phase-shifting method (Zhang et al, 2010). In our experiments, we used a DLP Discovery D4000 with a 0.55" digital micro-mirror device (DMD) chip. It can switch binary images up to 32,550 Hz with a resolution of 1,024×768. On the contrast, if the same projection system is used to switch 8-bit grayscale images, it can reach approximately 291 Hz (or 97 Hz 3D imaging rate). The binary structured pattern combined with the defocusing technique thus can drastically increase the 3D imaging rate.

In this work (Zhang et al, 2010a), we used the DLP Discovery projection system that includes a DLP Discovery board (D4000) (Texas Instruments, Texas), an ALP High Speed (Digital Light Innovations, Texas), and an optical module (S3X) (Visitech, Norway). With a Phantom V9.1 (Vision research, NJ) digital camera, we successfully developed a system that can achieve fringe image acquisition at 2000 Hz at an image resolution of 576 X 576 with decent quality for white surfaces due to the low intensity of the light source. Because a three-step phase-shifting algorithm is used, the 3D imaging speed is actually 667 Hz. This research has proved the success of using such a platform to achieve an unprecedentedly fast 3D imaging rate.

Due to the low surface reflectivity of live rabbit heart surfaces, the previous system only achieved 333 Hz 3D imaging rate (Zhang et al, 2010b). To further increase the 3D imaging quality for the live hearts, we replaced the light source of the DLP Discovery projection system with a bright LED light (CBT-90-W, Luminus Devices, MA). This LED light has the potential to reach 2200 lumens output light. The same camera is used to perform the measurement at even higher frame rate. We have successfully achieved a speed of 2000 Hz 2-D imaging, or 667 Hz 3D imaging for live rabbit hearts measurement. The exposure time used for the data capture is 490 μs. Figure 7 shows the some typical frames of the 3D heart surfaces when it is beating. This experiment was performed in Prof. Igor R. Efimov's laboratory at Washington University in Staint Louis.

Figure 7. Measurement results of a live rabbit heart with the superfast 3D imaging system. The 3D image rate is 667 Hz, and the image resolution 576 X 576.

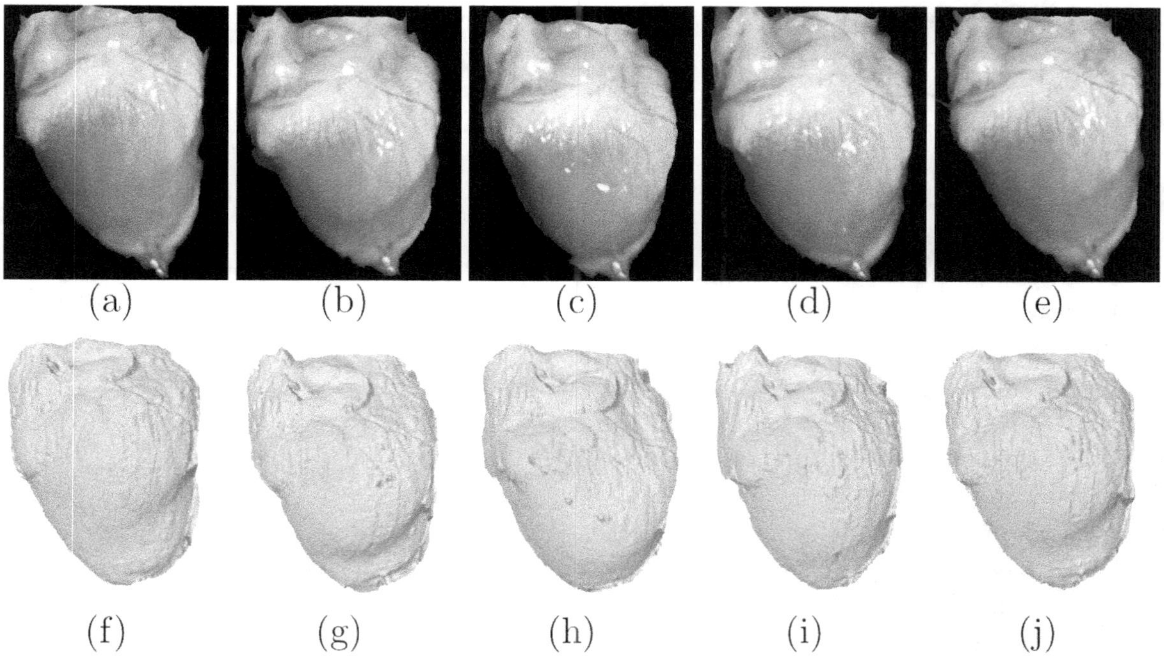

FUTURE RESEARCH DIRECTIONS

The digital fringe projection technology that uses a defocused projector has significantly simplified the 3D imaging system development, and has drastically advanced the area of high-speed imaging with off-the-shelf hardware components. However, this new technology is not trouble free: there are a number of challenging issues to explore in the future:

1. **The high-frequency harmonics phase errors.** The defocusing essentially is to filter out the high-frequency harmonics of the binary structured patterns. However, our experiments found that the high-frequency harmonics still exists for high-quality fringe patterns. In order to realize high-quality 3D imaging, the errors induced by the harmonics need to be reduced through either software approaches or hardware means.
2. **The depth range limitations.** For a conventional fringe generation technique, where fringe pattern is always sinusoidal if the nonlinearity of the projector is corrected. However, for this technique, the close-to-be-ideally sinusoidal fringe pattern can be generated within a small depth range. As addressed earlier, the high-quality fringe patterns are needed to perform good measurement. Therefore, the current technique can only perform high-quality measurement within relatively smaller depth range. Future research needs to be conducted to increase the depth range without sacrificing the merits of the proposed technique.
3. **The defocused projector calibration.** All existing calibration technologies assume that the projector is in focus, which is a natural way to develop a 3D imaging system. However, this proposed technology requires the projector be defocused, which makes the

projector (thus the system) calibration more complicated. In the future, new calibration methodologies have to be developed in order to perform high-accurate 3D imaging with the proposed technology.

CONCLUSION

We have presented a recently developed sinusoidal fringe generation technique, defocusing binary structured patterns, to realize high-speed, high-resolution 3D imaging. By this means, the 3D imaging system development has been significantly simplified since the projector nonlinearity does not bring any problems into the imaging system. Moreover, because this new technique coincides with the DLP projector's projection mechanism (binary operation), it permits some breakthroughs in the field of high-speed 3D imaging. We have presented three major studies that took advantage of this technology.

Of course, this technology is not trouble free: there are still a number of problems to be solved, and some challenges to be tackled with. Future research needs to be conducted to improve this technology. We believe that this technology will bring a lot new breakthroughs in the field of high-speed 3D imaging because of its simplicity and its closeness to nature.

ACKNOWLEDGMENT

We would like to thank Dr. Aamir Saeed Malik for inviting us to contribute to this exciting book, all the students in 3D machine vision laboratory for their assistance and great contributions. We also would like to thank Prof. Igor R. Efimov and his student Jacob I. Laughner from Washington University in St. Louis for their help on testing our superfast 3D imaging technology. Dr. Zhang would like to give his special thanks to Prof. Shing-Tung Yau from Harvard University, and Prof. Peisen S. Huang from Stony Brook University for their supervision and consistent support over the past years.

REFERENCES

Davis, J., Nehab, D., Ramamoorthi, R., & Rusinkiewicz, S. (2005). Spacetime stereo: A unifying framework for depth from triangulation. *IEEE Transactions on Pattern Analysis and Machine Intelligence*, *27*(2), 1–7. doi:10.1109/TPAMI.2005.37

Ghiglia, D. C., & Pritt, M. D. (1998). *Two-dimensional phase unwrapping: Theory, algorithms, and software*. New York, NY: John Willey & Sons.

Gong, Y., & Zhang, S. (2010a). *Improve real-time 3-D shape measurement speed by using projector defocusing*. In SPIE Optics & Photonics Conference: Interferometry XV: Applications, San Diego, California.

Gong, Y., & Zhang, S. (2010b). Ultrafast 3-D shape measurement with an off-the-shelf DLP projector. *Optics Express*, *18*(19), 19743–19754. doi:10.1364/OE.18.019743

Gorthi, S., & Rastogi, P. (2010). Fringe projection techniques: Whither we are? *Optics and Lasers in Engineering*, *48*(2), 133–140. doi:10.1016/j.optlaseng.2009.09.001

Guo, H., He, H., & Chen, M. (2004). Gamma correction for digital fringe projection profilometry. *Applied Optics*, *43*, 2906–2914. doi:10.1364/AO.43.002906

Hoefling, R. H. (2004a). High-speed 3D imaging by DMD technology. In J. R. Price, & F. Meriaudeau (Eds.), *Proceedings of SPIE Vol. 5303. Machine Vision Applications in Industrial Inspection XII* (pp. 188-194). San Jose, CA.

Hoefling, R. H. (2004b). ALP: Universal DMD controller for metrology and testing. In L.-C. Chien & M. H. Wu (Eds.), *Proceedings of SPIE Vol. 5303. Liquid Crystal Materials, Devices, and Applications X and Projection Displays X* (pp. 322-329). San Jose, CA.

Hoefling, R. H., & Aswendt, P. (2009). Real time 3D shape recording by DLP-based all-digital surface encoding. In L. J. Hornbeck, & M. R. Douglass (Eds.), *Proceedings of SPIE Vol. 7210. Emerging Digital Micromirror Device based Systems and Applications* (pp. 72100E). San Jose, CA.

Hornbeck, L. J. (1997). Digital light processing for high-brightness, high-resolution applications. In M. H. Wu (Ed.), *Proceedings of SPIE Vol. 3013. Projection Displays III* (pp. 27–40). Bellingham, WA.

Huang, P. S., Zhang, C., & Chiang, F.-P. (2002). High-speed 3-D shape measurement based on digital fringe projection. *Optical Engineering (Redondo Beach, Calif.)*, *42*(1), 163–168. doi:10.1117/1.1525272

Huang, Y., Quan, C., Jay, C. J., & Chen, L. J. (2005). Shape measurement by the use of digital image correlation. *Optical Engineering (Redondo Beach, Calif.)*, *44*(8), 1552–1559. doi:10.1117/1.2012202

Kakunai, S., Sakamoto, T., & Iwata, K. (2005). Profile measurement taken with liquid-crystal grating. *Applied Optics*, *38*(13), 2824–2828. doi:10.1364/AO.38.002824

Lei, S., & Zhang, S. (2009). Flexible 3-D shape measurement using projector defocusing. *Optics Letters*, *34*(20), 3080–3082. doi:10.1364/OL.34.003080

Lei, S., & Zhang, S. (2010). Digital sinusoidal fringe generation: Defocusing binary patterns VS focusing sinusoidal patterns. *Optics and Lasers in Engineering*, *48*, 561–569. doi:10.1016/j.optlaseng.2009.12.002

Li, Y., Jin, K., Jin, H., & Wang, H. (2010). High-resolution, high-speed 3D measurement based on absolute phase measurement. In P. K. Rastogi & E. Hack (Eds.) *AIP Conference Proceedings Vol. 1236. International Conference on Advanced Phase Measurement Methods in Optics and Imaging* (pp. 237-288). Monte Verita (Azcona), Switzerland.

Pan, B., Qian, K., Huang, L., & Asundi, A. (2009). Phase error analysis and compensation for non-sinusoidal waveforms in phase-shifting digital fringe projection profilometry. *Optics Letters*, *34*(4), 2906–2914. doi:10.1364/OL.34.000416

Pawlowski, M. E., Kujawinska, M., & Wgiel, M. G. (2002). Shape and motion measurement of time- varying three-dimensional objects based on spatiotemporal fringe-pattern analysis. *Optical Engineering (Redondo Beach, Calif.)*, *41*(2), 450–459. doi:10.1117/1.1430423

Qian, K. (2004). Windowed Fourier transform for fringe pattern analysis. *Applied Optics*, *43*(13), 2695–2706. doi:10.1364/AO.43.002695

Quan, C., Jay, C. J., Shang, H. M., & Bryanston-Cross, P. J. (1995). Contour measurement by fibre optics fringe projection and Fourier transform analysis. *Optics Communications*, *119*, 479–483. doi:10.1016/0030-4018(95)00287-I

Rusinkiewicz, S., Hall-Holt, O., & Levoy, M. (2002). Real-time 3D model acquisition. *ACM Transactions on Graphics*, *21*(3), 438–446. doi:10.1145/566570.566600

Salvi, J., Fernandez, S., Pribanic, T., & Llado, X. (2010). A state of the art in structured light patterns for surface profilometry. *Pattern Recognition*, *43*, 2666–2680. doi:10.1016/j.patcog.2010.03.004

Schreiber, H., & Bruning, J. H. (2007). Phase shifting interferometry. In Malacara, D. (Ed.), *Optical shop testing* (pp. 547–666). New York, NY: John Willey & Sons. doi:10.1002/9780470135976.ch14

Su, X., & Chen, W. (2001). Fourier transform profilometry: A review. *Optics and Lasers in Engineering, 35*(5), 263–284. doi:10.1016/S0143-8166(01)00023-9

Su, X., & Zhang, Q. (2010). Dynamic 3-D shape measurement: A review. *Optics and Lasers in Engineering, 48*(2), 191–204. doi:10.1016/j.optlaseng.2009.03.012

Takeda, M. (2010). Measurements of extreme physical phenomena by Fourier fringe analysis. In P. K. Rastogi & E. Hack (Eds.) *AIP Conference Proceedings Vol. 1236. International Conference on Advanced Phase Measurement Methods in Optics and Imaging* (pp. 445-448). Monte Verita (Azcona), Switzerland.

Zhang, L., Snavely, N., Curless, B., & Seitz, S. (2004). Spacetime faces: High resolution capture for modeling and animation. *ACM Transactions on Graphics, 23*(3), 548–558. doi:10.1145/1015706.1015759

Zhang, S. (2010a). Recent progresses on real-time 3-D shape measurement using digital fringe projection techniques. *Optics and Lasers in Engineering, 48*(2), 149–158. doi:10.1016/j.optlaseng.2009.03.008

Zhang, S. (2010b). High-resolution, high-speed 3D dynamically deformable shape measurement using digital fringe projection techniques. In M. K. Sharma (Ed.), *Advances in measurement systems* (pp. 29-50). Vukovar, Croatia: In-tech.

Zhang, S., & Gong, Y. Laughner, Lou, Q., Efimov, I. R., & van der Weide, D. (2010b). *High-resolution, superfast 3-D imaging using a phase-shifting method.* OSA Topical Meeting on Digital Image Processing and Analysis (DIPA). Tucson, AZ.

Zhang, S., & Huang, P. S. (2006a). High-resolution, real-time three-dimensional shape measurement. *Optical Engineering (Redondo Beach, Calif.), 45*(12), 123601. doi:10.1117/1.2402128

Zhang, S., & Huang, P. S. (2006b). Novel method for structured light system calibration. *Optical Engineering (Redondo Beach, Calif.), 45*(8), 083601. doi:10.1117/1.2336196

Zhang, S., & Huang, P. S. (2007). Phase error compensation for a three-dimensional shape measurement system based on the phase shifting method. *Optical Engineering (Redondo Beach, Calif.), 46*(6), 063601. doi:10.1117/1.2746814

Zhang, S., Li, X., & Yau, S.-T. (2007). Multilevel quality-guided phase unwrapping algorithm for real-time three-dimensional shape reconstruction. *Applied Optics, 46*(1), 50–57. doi:10.1364/AO.46.000050

Zhang, S., van der Weide, D., & Oliver, J. (2010a). Superfast phase-shifting method for 3-D shape measurement. *Optics Express, 18*(9), 9684–9689. doi:10.1364/OE.18.009684

Zhang, S., & Yau, S.-T. (2006). High-resolution, real-time 3-D absolute coordinate measurement based on a phase-shifting method. *Optics Express, 14*(11), 2644–2649. doi:10.1364/OE.14.002644

Zhang, S., & Yau, S.-T. (2007a). Generic nonsinusoidal phase error correction for three-dimensional shape measurement using a digital video projector. *Applied Optics, 46*(1), 36–43. doi:10.1364/AO.46.000036

Zhang, S., & Yau, S.-T. (2007b). High-speed three-dimensional shape measurement system using a modified two-plus-one phase-shifting algorithm. *Optical Engineering (Redondo Beach, Calif.), 46*(11), 113603. doi:10.1117/1.2802546

ADDITIONAL READING

Cheng, Y.-Y., & Wyant, J. C. (1984). Two-wavelength phase shifting interferometry. *Applied Optics, 23*, 4539–4543. doi:10.1364/AO.23.004539

Cheng, Y.-Y., & Wyant, J. C. (1985). Multiple-wavelength phase shifting interferometry. *Applied Optics*, *24*, 804–807. doi:10.1364/AO.24.000804

Cheng, Y.-Y., & Wyant, J. C. (1985). Multiple-wavelength phase shifting interferometry. *Applied Optics*, *24*, 804–807. doi:10.1364/AO.24.000804

Creath, K. (1987). Step height measurement using two-wavelength phase-shifting interferometry. *Applied Optics*, *26*, 2810–2816. doi:10.1364/AO.26.002810

Dhond, U., & Aggarwal, J. (1989). Structure from stereo—A review. *IEEE Transactions on Systems, Man, and Cybernetics*, *19*(6), 1489–1510. doi:10.1109/21.44067

Geng, Z. J. (1996). Rainbow 3-D camera: New concept of high-speed three vision system. *Optical Engineering (Redondo Beach, Calif.)*, *35*, 376–383. doi:10.1117/1.601023

Guo, H., & Huang, P. S. (2009). Absolute phase technique for the Fourier transform method. *Optical Engineering (Redondo Beach, Calif.)*, *48*, 043609. doi:10.1117/1.3122370

Huang, P. S., Hu, Q., Jin, F., & Chiang, F. P. (1999). Color-encoded digital fringe projection technique for high-speed three-dimensional surface contouring. *Optical Engineering (Redondo Beach, Calif.)*, *38*, 1065–1071. doi:10.1117/1.602151

Huang, P. S., & Zhang, S. (2006). Fast three-step phase-shifting algorithm. *Applied Optics*, *45*(21), 5086–5091. doi:10.1364/AO.45.005086

Huang, P. S., Zhang, S., & Chiang, F.-P. (2005). Trapezoidal phase-shifting method for three-dimensional shape measurement. *Optical Engineering (Redondo Beach, Calif.)*, *44*(12), 123601. doi:10.1117/1.2147311

Jia, P., Kofman, J., & English, C. (2007). Two-step triangular-pattern phase-shifting method for three-dimensional object-shape measurement. *Optical Engineering (Redondo Beach, Calif.)*, *46*(8), 083201. doi:10.1117/1.2768616

Mehta, R. P., Zhang, S., & Hadlock, T. A. (2008). Novel 3-D video for quantification of facial movement. *Otolaryngology - Head and Neck Surgery*, *138*(4), 468–472. doi:10.1016/j.otohns.2007.12.017

Pan, J., Huang, P. S., & Chiang, F.-P. (2006). Color phase-shifting technique for three-dimensional shape measurement. *Optical Engineering (Redondo Beach, Calif.)*, *45*, 013602. doi:10.1117/1.2151160

Polhemus, C. (1973). Two-wavelength interferometry. *Applied Optics*, *12*, 2071–2074. doi:10.1364/AO.12.002071

Sansoni, G., Carocci, M., & Rodella, R. (1999). Three-dimensional vision based on a combination of gray-code and phase-shift light projection: analysis and compensation of the systematic errors. *Applied Optics*, *38*(31). doi:10.1364/AO.38.006565

Su, X. Y., Zhou, W. S., Von Bally, G., & Vukicevic, D. (1992). Automated phase-measuring profilometry using defocused projection of a Ronchi grating. *Optics Communications*, *94*(6), 561–573. doi:10.1016/0030-4018(92)90606-R

Towers, D. P., Jones, J. D. C., & Towers, C. E. (2003). Optimum frequency selection in multi-frequency interferometry. *Optics Letters*, *28*, 1–3. doi:10.1364/OL.28.000887

Wang, Y., Gupta, M., Zhang, S., Wang, S., Gu, X., Samaras, D., & Huang, P. (2008). High resolution tracking of non-rigid 3D motion of densely sampled data using harmonic map. *International Journal of Computer Vision*, *76*(3), 283–300. doi:10.1007/s11263-007-0063-y

Wang, Y., Huang, X., Lee, C.-S., Zhang, S., Li, Z., & Samaras, D. (2004). High-resolution Acquisition, Learning and Transfer Dynamic 3D Facial Expression. *Computer Graphics Forum*, *23*(3), 677–686. doi:10.1111/j.1467-8659.2004.00800.x

Yelin, D., Bouma, B. E., Iftimia, N., & Tearney, G. J. (2003). Three-dimensional spectrally encoded imaging. *Optics Letters*, *28*, 2321–2323. doi:10.1364/OL.28.002321

Yelin, D., Bouma, B. E., Rosowsky, J. J., & Tearney, G. J. (2008). Doppler imaging using specturally-encoded endoscopy. *Optics Express*, *16*, 14831–14844.

Yelin, D., White, W. M., Motz, J. T., Yun, S. H., Bouma, B. E., & Tearney, G. J. (2007). Spectral-domain spectrally-encoded endoscopy. *Optics Express*, *15*, 2432–2444. doi:10.1364/OE.15.002432

Zhang, S. (2010). Flexible 3-D shape measurement using projector defocusing: Extended measurement range. *Optics Letters*, *35*, 931–933.

Zhang, S., Royer, D., & Yau, S.-T. (2006). GPU-Assisted high-resolution, real-time 3-D shape measurement. *Optics Express*, *14*(20), 9120–9129. doi:10.1364/OE.14.009120

Zhang, S., & Yau, S.-T. (2008a). Three-dimensional data merging using Holoimage. *Optical Engineering (Redondo Beach, Calif.)*, *47*(3), 033608. doi:10.1117/1.2898902

Zhang, S., & Yau, S.-T. (2008b). Absolute phase assisted three-dimensional data registration for a dual-camera structured light system. *Applied Optics*, *47*(17), 3134–3142. doi:10.1364/AO.47.003134

Zhang, Z. (2000). A flexible new technique for camera calibration. *IEEE Transactions on Pattern Analysis and Machine Intelligence*, *22*(11), 1330–1334. doi:10.1109/34.888718

KEY TERMS AND DEFINITIONS

Binary Structured Light: Light that is structured into a pattern, which can be used to encode a scene and only two intensity values are used for all the patterns.

Fringe Projection: Projecting sinusoidally varying fringe patterns (structured light) onto an object.

High Speed and High Resolution: The technique that could resolve both temporal and spatial resolution at very high details.

Phase Shifting: Process of taking multiple captured fringe patterns and performing phase wrapping and unwrapping to get an absolute phase map, which can then be used to acquire 3D coordinates.

Phase Unwrapping: Finding and removing 2π discontinuities resulting from the arctangent function used in phase wrapping. This can be done using a spatial phase unwrapping algorithm or using an encoded stair function.

Phase Wrapping: Taking multiple fringe patterns and wrapping them into a wrapped phase map. Typically, it is calculated by an arctangent function, which yields a phase map containing 2π discontinuities.

Projector Defocusing: An object is placed in a position where the projector is out of focus.

Section 2
Shape From X:
Algorithms & Techniques

Chapter 8
Three-Dimensional Scene Reconstruction:
A Review of Approaches

Dimitrios Chrysostomou
Democritus University of Thrace, Greece

Antonios Gasteratos
Democritus University of Thrace, Greece

ABSTRACT

The production of 3D models has been a popular research topic already for a long time, and important progress has been made since the early days. During the last decades, vision systems have established to become the standard and one of the most efficient sensorial assets in industrial and everyday applications. Due to the fact that vision provides several vital attributes, many applications tend to use novel vision systems into domestic, working, industrial, and any other environments. To achieve such goals, a vision system should robustly and effectively reconstruct the 3D surface and the working space. This chapter discusses different methods for capturing the three-dimensional surface of a scene. Geometric approaches to three-dimensional scene reconstruction are generally based on the knowledge of the scene structure from the camera's internal and external parameters. Another class of methods encompasses the photometric approaches, which evaluate the pixels' intensity to understand the three-dimensional scene structure. The third and final category of approaches, the so-called real aperture approaches, includes methods that use the physical properties of the visual sensors for image acquisition in order to reproduce the depth information of a scene.

DOI: 10.4018/978-1-61350-326-3.ch008

INTRODUCTION

Three-dimensional object and surface reconstruction from images is an important topic in various application areas, such as quality inspection, clinical photography, robotics, agriculture, and archaeology. In the domain of quality inspection, a large number of inspection tasks depend on three-dimensional reconstruction techniques, such as the surface measurement applied to high-precision engineered products such as aircraft wings (J. Xu, Xi, Zhang, & Shi, 2009; J. Xu, Xi, Zhang, Shi, & Gregory, 2010). Tasks of this kind usually require the accurate measurement of depth on small surfaces. Other tasks depend on the precise measurement of a sparse set of well defined points, for example to determine if an assembly process has been completed with the required accuracy, or measurement of the relative movement between important parts during a crash test. Three-dimensional clinical photographs have the potential to provide quantitative measurements that reduce subjectivity in assessing the surface anatomy of the subject before and after a surgical intervention by providing numeric scores for the shape, symmetry and longitudinal change of anatomic structures (Tepper et al., 2008; Janoos et al., 2009). Furthermore, the vast majority of nowadays mobile robots are equipped with one, two or more cameras in order to provide visual feedback in applications like maze exploration, map navigation and obstacle avoidance. (DeCubber, Nalpantidis, Sirakoulis, & Gasteratos, 2008; Nevado, Garcia-Bermejo, & Casanova, 2004; Nalpantidis, Kostavelis, & Gasteratos, 2009; Nalpantidis, Chrysostomou, & Gasteratos, 2009). Besides, in the field of agriculture, new applications emerged recently, such as a mobile robotic system use cameras to reconstruct the surface of the plants to find parasites and report them (Šeatović, 2008; Zhu, Lu, Luo, Tao, & Cheng, 2009). One more interesting application area includes the archaeological excavations and historic objects, where three-dimensional surface reconstruction is applied to many archaeological sites in order to preserve crucial details of the site and use them afterwards for 3D presentation and tourist attraction (El-Hakim, Beraldin, Picard, & Cournoyer, 2008; Remondino, El-hakim, Baltsavias, Picard, & Grammatikopoulos, 2008).

The key feature of a vision system is its capability to see or capture portions of the world and to obtain a density of sampling in space and time. This sampling density is essential in several applications. In surveillance, dense sampling of space and time might allow us to track a single pedestrian throughout a complex of buildings. Multiple views of a geometry can be used to obtain 3D reconstructions with appropriate assumptions, camera location and calibration information. Currently, they are used to produce reconstructions of quite complex geometries from a moving camera or from a static multi-camera system inside a room. For example, one might drive a camera through a city and build a geometric model from the resulting video or one might watch a hanging robot inside a seminar room while tracking it from the reconstructed data. Once multiple views establish correspondences, their observations might be used to recover a geometric model as well as a model of the cameras' current locations. The various application scenarios set different requirements on the reconstruction. In several tasks, it is sufficient to produce a sparse set of 3D points, where 3D information is available only for a very small number of pixels in the input images, while others require a dense reconstruction, with 3D information available for every pixel in the input images. Other important factors include the size, shape, and material of the objects, the number of required images, requirements on positions of the cameras or light sources, and the time allowed for image capture and reconstruction.

This chapter discusses different methods for capturing the three-dimensional surface of a scene. A first classification distinguishes geometric, photometric, and real-aperture approaches. Geometric approaches to three-dimensional scene reconstruction are generally based on the knowledge of the scene structure from the intrinsic and extrinsic camera parameters and are based on the modeling of the geometric aspects of image creation. These methods originate from the first photogrammetric methods and exploit the perspective projection of a three-dimensional scene into a two-dimensional image plane. Another class of three-dimensional scene reconstruction methods encompasses the photometric approaches, which evaluate the distribution of the pixel intensity in the image to infer the three-dimensional scene structure and are primarily based on photometric modeling. As long as sufficient information about the illumination conditions and the surface reflectance properties is available, these methods may provide dense depth maps of object surfaces. The third and final category of approaches includes methods, the so-called real aperture approaches, where they use the physical properties of the optical system of the visual sensors for image acquisition, in order to produce depth information about the scene. The rest of this chapter is organized as follows: The three aforementioned classes for 3D surface generation are reported in the next three, respective, sections. Future research challenges for scene reconstruction are discussed in the next section. Last concluding remarks are made in the last section of this chapter.

GEOMETRIC APPROACHES

Image Formation and Pinhole Camera Model

Two-dimensional image analysis has been the foundation for three-dimensional surface reconstruction since Laussedat and Meydenbauer developed the first photogrammetric methods back in 19th century (Laussedat, 1898; Meydenbauer, 1867). Now, that is used for mapping and reconstruction of buildings (Luhmann, 2003). The early photogrammetric methods were based on the geometric modeling of the image formation, exploiting the perspective projection of the 3D scene onto a flat 2D image plane.

The camera model used by most photogrammetric and computer vision approaches is the pinhole camera, as shown below in Figure 1. The projection of a 3D point given in the camera coordinate system C, $X^C = [x_l, y_l, z_l]$ into $X^I = [u_l, v_l]$ in image coordinates can be denoted by the projection function P:

$$X^I = P(K, X^c) \tag{1}$$

The parameter K defines the internal (focal length, lens distortion parameters) camera orientation. The projection function of a pinhole camera is defined as:

$$u_1 = -f\frac{x_1}{z_1} \tag{2}$$

$$v_1 = -f\frac{y_1}{z_1} \tag{3}$$

where f is the distance between pinhole and image plane. Once multiple cameras are considered, it is practical to introduce a world coordinate system W, and specify the orientation T_i of each camera relative to this world coordinate system. Then the projection function of a point in the world coordinate system needs to be transformed into the camera coordinate system of the ith camera, C_i using a camera orientation T_i. In this case the projection function depends on both internal orientation K_i and external orientation T_i:

$$X^I = P(T_i, K_i, X^C) \tag{4}$$

In the computer vision community, the internal camera orientation parameters are known as intrinsic camera parameters, while the external orientation parameters as extrinsic camera parameters. The projected points are then captured by a light sensitive device, typically a film or digital sensor.

Figure 1. Pinhole projection

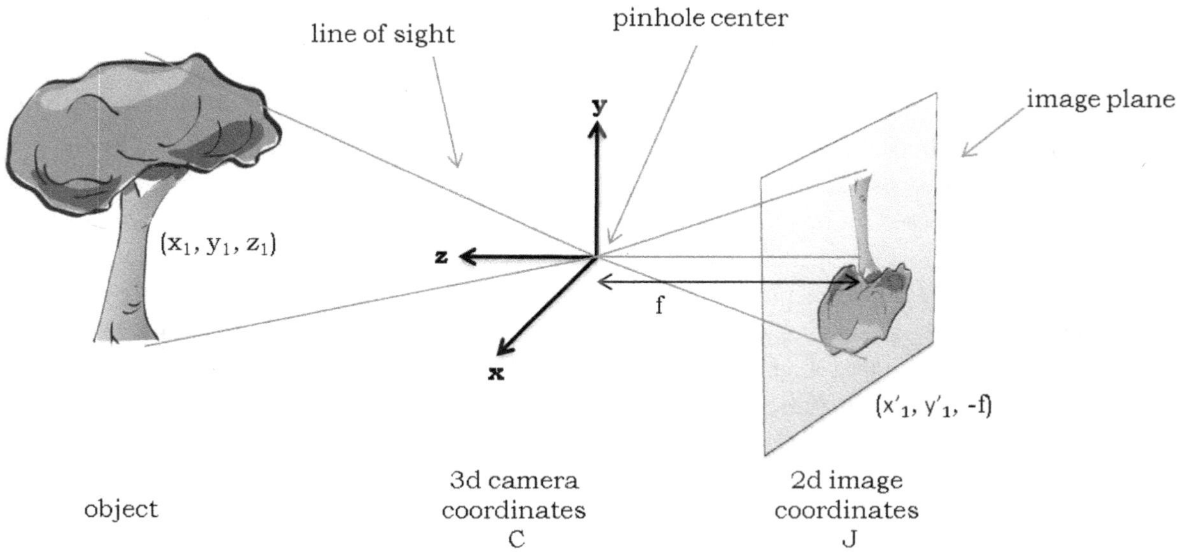

In the case of a digital sensor, the light sensitive area is sampled and the light intensity is measured at each sample point (Luhmann, 2003). Note that the 3D point in camera coordinates X^c cannot be determined uniquely given camera parameters K and image point X^J, since they only define a ray in C on which X^C is located.

Camera calibration is an emerging problem in the computer vision society and several researchers across the years have accomplished numerous solutions to a variety of occasions. To begin with, Barreto developed the first method for calibrating multiple cameras located across a room without using non-linear minimization and using a moving LED instead (Barreto & Daniilidis, 2004). Moreover, Sturm proposed a more generic concept for camera calibration by removing the parametric nature of the problem and adopted a more general projection model (Sturm & Ramalingam, 2004) and lately expanded this general imaging geometry to include central catadioptric cameras as well (Sturm & Barreto, 2008). Kanalla suggest a new generic camera model and calibration method for applications that need wide angle or fisheye lenses estimating all the parameters needed for a state-of-the-art accuracy (Kannala & Brandt, 2006). A very wide field of view and especially fisheye lenses are commonly used for space exploration and robotic applications and thus camera calibration methods for these applications using fisheye lenses were developed from Gennery and Courbon respectively (Gennery, 2006; Courbon, 2007). One of the most recent camera calibration methods was proposed by Furukawa, who achieve great pixel accuracy using multi-view stereo and bundle adjustment techniques to calibrate high resolution digital cameras (Furukawa & Ponce, 2009).

Bundle Adjustment

Most geometric methods for three-dimensional scene reconstruction from multiple images are based on establishing corresponding points in the images. For a single scene point X^W observed in *N images*, the corresponding points X^{J_i} in each image *i,* where $i = 1, \ldots, N$, can be determined manually or by automatic correspondence search methods. Given the extrinsic and intrinsic camera parameters, each image point defines a ray in three-dimensional space, and in the absence of measurement errors, all N rays intersect in the scene point X^W. Automatic detection of corresponding points in arbitrary scenes is a challenging problem and an active research area (Moreels & Perona, 2007; Li & Allinson, 2008; Tuytelaars & Mikolajczyk, 2008). First general scene reconstruction methods based on images acquired from different views were developed by (Kruppa, 1913). An overview of these early methods is given by (Luhmann, 2003). They aim to determine the intrinsic and extrinsic camera parameters and the three-dimensional coordinates of the scene points. Kruppa (1913) presented an analytical solution for the scene structure and extrinsic camera parameters from a minimal set of five corresponding image points. In classical bundle adjustment (D. Brown, 1958; Triggs, McLauchlan, Hartley, & Fitzgibbon, 2000; Lourakis & Argyros, 2009), scene points (structure) and camera orientation (motion) are recovered jointly and optimally from corresponding image points. The bundle adjustment error term:

$$E_B(\{T_i\},\{X_j\}) = \sum_{i=1}^{N} \sum_{j=1}^{M} [P(T_i, K_i, X_j) - x_{ji}]^2 \quad (5)$$

can be used to minimize the re-projection error E_B with respect to the unknown N internal camera orientations K_i, external camera orientation T_i and the M scene points Xj. Here, x_{ji} denotes the given 2D pixel coordinates (u_{ji}, v_{ji}) of feature j in image i. Bundle adjustment is a very flexible method, depending on the reconstruction task, and even values for all or some of parameters K_i, T_i and X_j might be unknown. By minimizing the bundle adjustment equation with respect to the unknown parameters, the bundle adjustment method can be used for calibration of internal and/or external camera parameters as well as pose estimation of objects. The method can be applied to image sequences acquired by the same camera, or to images acquired simultaneously by multiple cameras. It is also possible to use cameras with different projection functions P, for example pinhole and fish-eye cameras, in the same reconstruction task. If additional information about the scene is available, such as the position of some 3D points in world coordinates, additional terms can be added to the equation. Measurement uncertainties of the known variables can be used to compute the uncertainty of the estimated parameters. As the bundle adjustment equation is a nonlinear one, it is minimized using the Levenberg-Marqardt or Gauss-Newton algorithm. Even bundle adjustment tasks with many unknowns can be optimized efficiently, since the re-projection error of the jth point in view i only influences T_i, K_i in frames where the point j could be tracked as well as X_j. This leads to a sparse set of equations, which can be exploited by the optimization algorithm (Lourakis & Argyros, 2004).

In general, bundle adjustment provides accurate reconstruction of scene points for which correspondences could be established. Problems also occur when the correspondences contain outliers that do not comply with the assumption of a Gaussian re-projection error distribution. In that case the estimated parameters can contain gross errors that are not directly apparent in the statistics of the estimated parameters. Ways to work around outliers are based on screening the data for outliers, for example using RANSAC (Fischler & Bolles, 1981) together with a minimal case five point algorithm (Nister, 2004), or using a M-Estimator while minimizing the bundle adjustment equation. Usually correspondences can only reliably be extracted in high contrast image areas, resulting in a sparse 3D reconstruction, where areas with uniform or repetitive texture cannot be reconstructed.

Stereo Vision

In the cases where two cameras with known internal and external orientation observe the examined scene, a geometric limitation such as the epipolar constraint is used. This setup is exploited in the stereo vision approach for 3D reconstruction. The epipolar constraint simplifies the correspondence search problem, as it limits the correspondence search region for a given point in one image to a single line in the other one. Additionally, each 3D point can be calculated directly through triangulation, such that no bundle adjustment is required. Due to these simplifications, stereo vision is a widely used technique in close range 3D reconstruction. In most stereo systems, two views with known internal and external camera orientation are used. In a typical stereo application, a scene is simultaneously and continuously monitored by a pair of cameras whose centers of projection are located on a horizontal baseline. In many practical applications, the optical axes of the two cameras are parallel, and the images are taken with the same focal length. This is often called the standard stereo geometry and leads to epipolar lines oriented parallel to image rows or columns, where the correspondences can be estimated efficiently. It is possible to transform images from an arbitrary camera setup into images with horizontal or vertical epipolar lines, using a process known as stereo rectification (Forsyth & Ponce, 2002; Kruger, Wohler, Wurz-Wessel, & Stein, 2004).

Several surveys ((Barnard & Fischler, 1982; Dhond & Aggarwal, 1989; DeSouza & Kak, 2002; Scharstein & Szeliski, 2002; Brown, Burschka, & Hager, 2003; Lemaire, Berger, Jung, & Lacroix, 2007; Nalpantidis, Sirakoulis, & Gasteratos, 2008) provide an exhaustive overview of the different stereo methods. Given the internal parameters of the cameras, i.e. the focal length, the principal point and the distortion parameters and external ones, i.e. the position and the orientation of the cameras in the 3D space, the distance of an arbitrary object in the scene may be computed via disparity. The latter is the offset between the pixels in both images of the stereo pair. The collection of all the disparity points, in image coordinates, is the so called disparity map.

Robust determination of the corresponding points and, consequently, of disparity is the central problem to be solved by stereovision algorithms. An early survey by Barnard and Fischler (1982) reports the use of block and feature matching. Block matching approaches compare a small area in one image with potentially matching areas in the other one. Often cross correlation and the sum of squared differences are used as matching criteria. This assumes structures parallel to the image plane, known as fronto-parallel. At depth discontinuities or tilted areas, a block will contain pixels from different depths leading to less reliable matching results. Additionally uniform image areas cannot be matched reliably. Feature matching approaches extract suitable features like edges or curves (Mikolajczyk et al., 2005) and match these by computing suitable similarity measures. Since these features are usually well localized, feature based methods handle depth discontinuities better, but might provide a sparse disparity map, compared to block matching.

The robustness and accuracy of the disparity estimates can be improved by considering additional constraints during the matching process. For example, the smoothness constraint states that the disparity should vary smoothly, this is indeed useful for uniform and/or untextured areas where no correspondences can be established. The ordering constraint states that for opaque surfaces the order of correspondences is always preserved. Many stereo algorithms (Masrani & MacLean, 2006; Ben-Ari & Sochen, 2007) use dynamic programming (Z. Liu & Klette, 2008; MacLean, Sabihuddin, & Islam, 2010) to efficiently and optimally calculate the disparity values of a complete scanline while considering the ordering constraint. Constraints over the whole image, across several scanlines, are hard to integrate into the dynamic programming framework. Algorithms based on graph cuts (Boykov & Kolmogorov, 2004; Vogiatzis, Esteban, Torr, & Cipolla, 2007) can use the

Figure 2. Volumetric octree reconstruction from binary silhouettes (Szeliski 1993) © 1993 Elsevier (a) octree representations and its corresponding (b) tree structure; (c) input image of an object on a turntable; (d) computed 3D volumetric octree model.

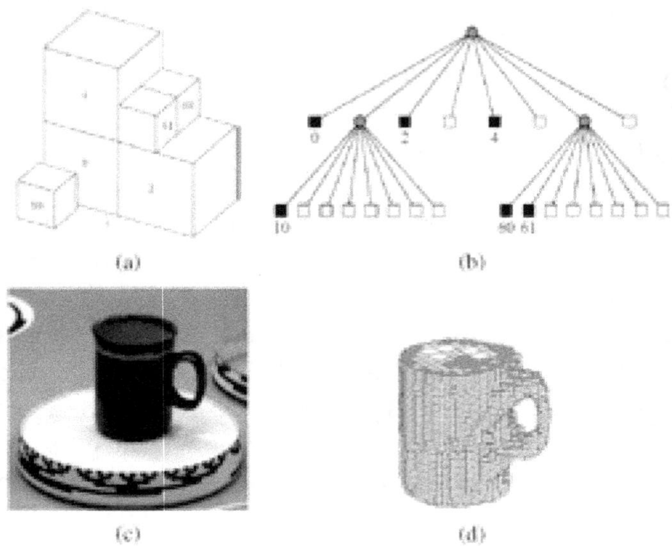

constraints globally during the reconstruction and are among the best performing stereo algorithms, in terms of reconstruction quality (Nalpantidis et al., 2008).

Shape from Silhouettes

In many situations, performing a foreground/background segmentation of the object of interest is a good way to initialize or fit a 3D model (Grauman, Shakhnarovich, & Darrell, 2003; Vlasic, Baran, Matusik, & Popović, 2008) or to impose a convex set of constraints on multi-view stereo (Kolev & Cremers, 2008). Over the years, a number of techniques have been developed to reconstruct a 3D volumetric model from the intersection of the binary silhouettes projected into 3D. The resulting model is called a visual hull, in analogy to the convex hull of a set of points, since the volume is maximal with respect to the visual silhouettes and since surface elements are tangent to the viewing rays (lines) along the silhouette boundaries (Boyer & Franco, 2003).

Some techniques first approximate each silhouette with a polygonal representation, and then intersect the resulting faceted conical regions in three dimensional space to produce polyhedral models (Aganj, Pons, Segonne, & Kerive, 2007; X. Liu, Yao, Chen, & Gao, 2008), which can later be fused using stereo methods (Esteban & Schmitt, 2004) or range data (Yemez & Wetherilt, 2007). Other approaches use voxel-based representations, usually encoded as octrees (Zhang & Smith, 2009), because of the resulting space-time efficiency. Figures 2a and b show an example of a 3D octree model and its associated colored tree, where black nodes are interior to the model, white nodes are exterior, and gray nodes are of mixed occupancy. Examples of octree-based reconstruction approaches include (Ladikos, Benhimane, & Navab, 2008; Azevedo, Tavares, & Vaz, 2010; Zhou, Gong, Huang, & Guo, 2010).

The most recent work on visual hull computation borrows ideas from image-based rendering and, hence, it is called image-based visual hull (Matusik, Buehler, & McMillan, 2001). Instead

Figure 3. Synthetic example of shape from shading application. (Zhang et al. 1999) © 1999 IEEE

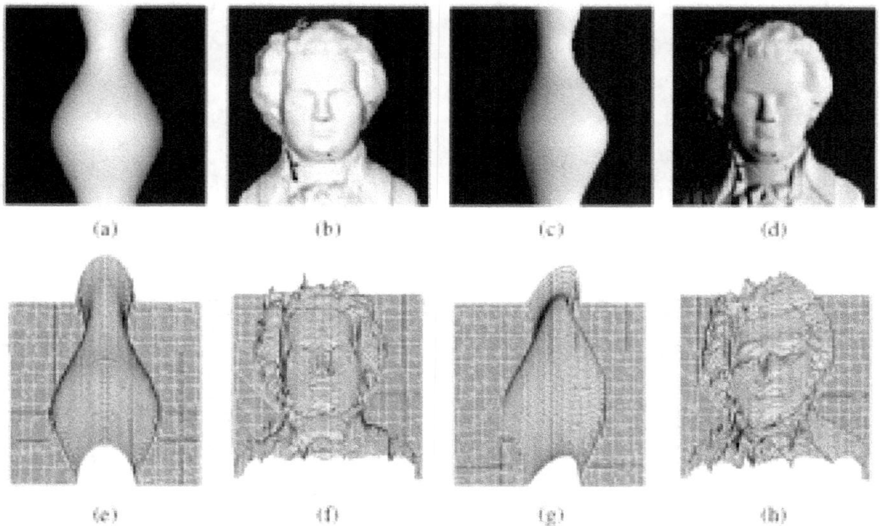

of pre-calculate a global 3D model, an image-based visual hull is recomputed for each new viewpoint, by successively intersecting viewing ray segments with the binary silhouettes in each image. This not only leads to a fast computation algorithm, but also enables fast texturing of the recovered model with color values from the input images. This approach can also be combined with high-quality deformable templates to capture and re-animate whole body motion (Vlasic et al., 2008).

PHOTOMETRIC APPROACHES

Shape from Shading

The problem of recovering the shape of a surface from this intensity variation is known as 'Shape-from-shading' (Horn, 1989) and typically handles smooth, non-textured surfaces. The images of smooth shaded objects, such as the ones shown in Figure 3, show clearly the shape of the object from just the shading variation. This is possible as the surface normal changes across the object, the apparent brightness changes as a function of the angle between the local surface orientation and the incident illumination.

Most shape from shading algorithms assume that the surface under consideration is of a uniform albedo and reflectance, and that the light source directions are either known or can be calibrated by means of a reference object. Under the assumptions of distant light sources and observer, the variation in intensity *(irradiance equation)* become purely a function of the local surface orientation,

$$I(x,y) = R(p(x,y),q(x,y)) \tag{6}$$

where $(p,q) = (z_x, z_y)$ are the depth map derivatives and $R(p,q)$ is called the reflectance map. For example, a diffuse (Lambertian) surface has a reflectance map that is the (non-negative) dot product between the surface normal $n = (p,q,1)) \,/\, \sqrt{1+p^2+q^2}$ and the light source direction $v = (v_x, v_y, v_z)$,

$$R(p,q) = \max\left(0, p\frac{pv_x + qv_y + v_z}{\sqrt{1+p^2+q^2}}\right) \tag{7}$$

where p is the surface reflectance factor (albedo). In principle, the two equations above can be used to estimate *(p,q)* using non-linear least squares or some other method. Unfortunately, unless additional constraints are imposed, there are more unknowns per pixel *(p,q)* than the measurements *(I)*. One commonly used constraint is the smoothness constraint,

$$\varepsilon_s = \int p_x^2 + p_y^2 + q_x^2 + q_y^2 dxdy = \int \| \nabla p \|^2 + \| \nabla q \|^2 \; dxdy, \tag{8}$$

and the *integrability constraint*,

$$\varepsilon_i \int (p_y - q_x)^2 dxdy, \tag{9}$$

which results straightforwardly, as for a valid depth map $z(x,y)$ with $(p,q) = (z_x, z_y)$, we have $p_y = z_{xy} = z_{yx} = q_x$.

Instead of first recovering the orientation fields *(p,q)* and then integrating these to obtain a surface, it is also possible to directly minimize the discrepancy in the image formation equation while finding the optimal depth map $z(x,y)$ (Horn, 1990). Unfortunately, shape from shading is both susceptible to local minima in the search space, and, similar to other variational problems that involve the simultaneous estimation of many variables, it can also suffer from slow convergence. Tsai and Shah utilized a preliminary step to improve the performance of shape from shading algorithms in all types of surfaces using linear approximation (Tsai & Shah, 1994). Other approaches, by using multi-resolution techniques (Szeliski, 1991) can help accelerate the convergence, while by using more sophisticated optimization techniques (Wilhelmy & Kruger, 2009) can help avoid local minima.

In practice, surfaces, other than plaster casts, are rarely of a single uniform albedo. Shape from shading therefore needs to be combined with some other technique or extended in some way to become useful. One way to do this is to combine it with stereo matching (Jin, Soatto, & Yezzi, 2000; Chow & Yuen, 2009) or known texture (surface patterns) (White & Forsyth, 2006). The stereo and/or texture components provide information in textured regions, while shape from shading helps fill in the information across uniformly colored regions and also provides finer information about surface shape. Another method is to combine the strength of graph cuts with the simplicity of shape from shading to produce accurate results (Chang, Lee, & Lee, 2008). The survey by (Durou, Falcone, & Sagona, 2008) not only reviews more recent techniques, but it also provides some comparative results.

Photometric Stereo

Another way to make shape from shading more reliable is to use multiple light sources that can be selectively turned on and off. This technique is called photometric stereo, since the light sources behave similarly to the cameras located at different locations in traditional stereo (Woodham, 1981; Basri, Jacobs, & Kemelmacher, 2007; Hernandez, Vogiatzis, & Cipolla, 2008). A different reflectance map, $R_1(p, q)$, $R_2(p, q)$, etc, corresponds to each individual light source. Given the corresponding intensities I_1, I_2, etc. of a pixel, in principle both an unknown albedo p and a surface orientation estimate *(p, q)* can be recovered. If the local orientation by n is parameterized, diffused surfaces, for non shadowed pixels, might be described by a set of linear equations:

$$I_k = pn \cdot v_k \tag{10}$$

from which we can recover pn using linear least squares method.

Once the surface normals or gradients have been recovered at each pixel, they can be integrated into a depth map using a variant of regularized surface fitting. (Nehab, Rusinkiewicz, Davis, &

Ramamoorthi, 2005) and (Harker & O'Leary, 2008) have presented some interesting results lately. When surfaces are specular, more than three light directions may be required. In fact, the irradiance equation not only requires that the light sources and camera be distant from the surface, it also neglects inter-reflections, which can be a significant source of the shading observed on object surfaces, e.g., the darkening seen inside concave structures such as grooves and crevasses (Nayar, Ikeuchi, & Kanade, 1991; Y. Xu & Aliaga, 2008).

Shape from Texture

Local anomalies of the imaged texture (e.g. anisotropy in the statistics of edge orientations for an isotropic texture, or deviations from assumed periodicity) are regarded as the result of projection and can also provide useful information about local surface orientation. Surface orientations which allow the original texture to be maximally isotropic or periodic are selected. Shape from texture algorithms require a number of processing steps, including the extraction of repeated patterns or the measurement of local frequencies in order to compute local affine deformations, and a subsequent stage to infer local surface orientation.

Details on these various stages can be found in the research literature (Loh & Hartley, 2005; Lobay & Forsyth, 2006; Galasso & Lasenby, 2007; Todd, Thaler, Dijkstra, Koenderink, & Kappers, 2007; Grossberg, Kuhlmann, & Mingolla, 2007; Jacques, De Vito, Bagnato, & Vandergheynst, 2008). When the pattern is more regular, it is possible to fit a regular but slightly deformed grid to the image, and to then use this grid for a variety of image replacement or analysis tasks (Y. Liu, Collins, & Tsin, 2004; Y. Liu, Lin, & Hays, 2004; Hays, Leordeanu, Efros, & Liu, 2006; Park, Brocklehurst, Collins, & Liu, 2009). This process becomes even easier if specially printed textured cloth patterns are used (White & Forsyth, 2006; White, Crane, & Forsyth, 2007).

REAL APERTURE APPROACHES

The geometric methods described in the previous section are all based on an ideal camera, which projects scene points into image points perfectly. However, a real camera system uses a lens of finite aperture, which results in images with a limited depth of field. The depth dependent blurring is not considered by the geometric methods and usually decreases the accuracy of the correspondence search methods.

The depth dependent defocusing is illustrated in Figure 4, where a scene point at distance d_0 is in focus (projected onto a single point in the image plane located at distance v), while points at other distances from the camera are spread onto a larger area, leading to a blurred image. If the light rays are traced geometrically, object points that are out of focus will be imaged to a circular disk. This disk is known to the photographers' community as the circle of confusion. Using the lens law:

$$\frac{1}{v} + \frac{1}{d} = \frac{1}{f} \tag{11}$$

its diameter C can be approximated by

$$C = Dv\left|\frac{1}{f} - \frac{1}{v} - \frac{1}{d}\right| \tag{12}$$

where f is the focal length and D is the diameter of the lens aperture. The depth dependent term $1/d$ approaches zero for larger values of d while the other terms stay constant, resulting in little change of the blur radius for objects at a large distance d. This limits the real aperture methods to close range scenarios, where two different depth values result in a measurable change of C.

Figure 4. Real aperture lens model used for depth from defocus

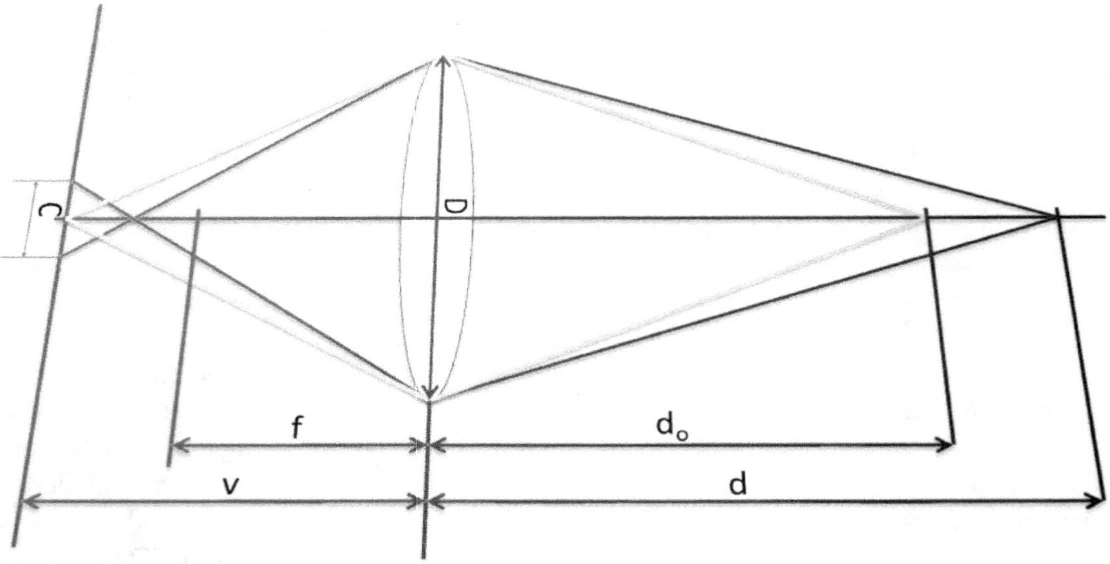

Shape from Focus

When the image is focused, knowledge of the camera parameters *f* and *v* can be used to calculate the depth *d* of the object. In *Depth from Focus (DfF)*, a sequence of images of a scene is obtained by continuously varying the distance v between the lens and the image detector (Shim & Choi, 2010). This leads to a series of differently blurred images. For each image, a sharpness measure is computed at each pixel in a local window and for each pixel position the image with the maximum focus measure is determined.

The main difference between the different *Depth from Focus* methods proposed in the literature is the choice of the focus measures, common measures are based on the strength of high frequency components in the amplitude spectrum. A particularly simple way is to use the image intensity variance of a local region. With a suitable criterion, the maximum of the sharpness measure can be interpolated, resulting in an improved depth resolution (Ramnath & Rajagopalan, 2009).

Depth *from Focus* is a comparably simple method, only one camera position is involved and the computational cost for depth recovery is quite low. Additionally, there is no correspondence problem and the accuracy of the method is relatively high. As the *Depth from Focus* method relies on high frequency image content it can only estimate the depth for surfaces with image texture. Like any other method based on real aperture effects, it is only applicable to close range scenarios, where the depth of field is small, compared to the object depth range. Sources of measurement errors include edge bleeding and the assumption of a constant depth of each window. A fundamental drawback is the requirement of a whole image focus series, a non-interpolating approach requires one image for each desired distance *D*.

Shape from Defocus

The main drawback of *Depth from Focus* is the necessity of an image series captured with multiple camera focus settings that scans the whole depth measurement range. *Depth from Focus* uses the camera parameters of the sharpest frame to determine the object depth. However, according to equation above, the radius of the circle of con-

Figure 5. Real time depth from defocus (Nayar et al. 1996) © 1996 IEEE: (a) the real-time focus range sensor, which includes a half-silvered mirror between the two telecentric lenses (lower right), a prism that splits the image into two CCD sensors (lower left), and an edged checkerboard pattern illuminated by a Xenon lamp (top); (b–c) input video frames from the two cameras along with (d) the corresponding depth map; (e–f) two different frames (you can see the texture if you zoom in) and (g) the corresponding 3D mesh model.

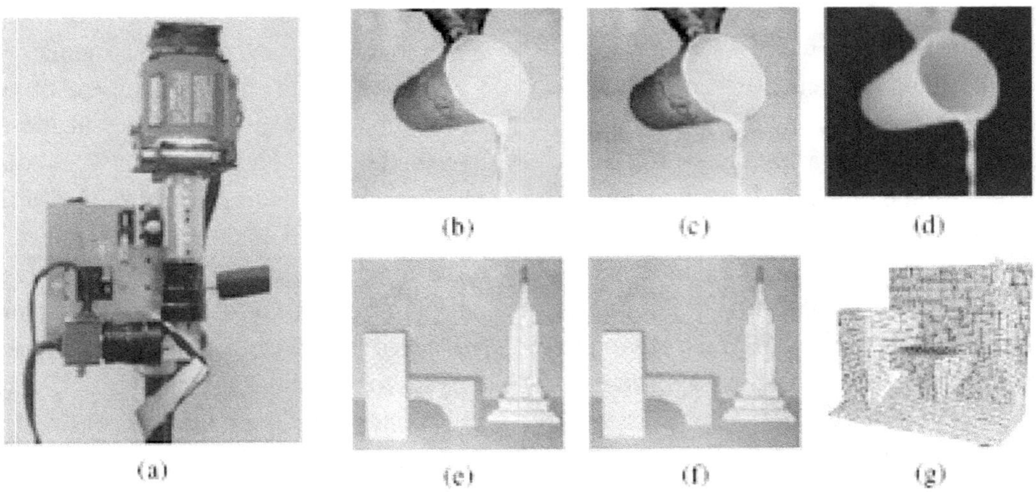

fusion is a function of the camera parameters and the depth of a scene point. Hence, the amount of blur observable in a defocussed image contains information about the depth of a scene point. A number of techniques have been developed to estimate depth from the amount of defocus (Favaro & Soatto, 2005; Pradeep & Rajagopalan, 2007; Favaro, 2007; Favaro, Soatto, Burger, & Osher, 2007).

In order to make such a technique practical numerous assumptions should be made as that the amount of blur increases in both directions as one moves away from the focus plane and therefore it is necessary to use two or more images captured with different focus distance settings (Favaro, 2007), or to translate the object in depth and to look for the point of maximum sharpness (Lou, Favaro, Bertozzi, & Soatto, 2007). Moreover, the magnification of the object can vary as the focus distance is changed or the object is moved. This can either be modeled explicitly (making correspondence more difficult), or telecentric optics and axial stereo, which approximate an orthographic camera and which require an aperture in front of the lens, can be used (Sahay & Rajagopalan, 2009). Besides, the amount of defocus must be reliably estimated. A simple approach is to simply average the squared gradient in a region, but this suffers from several problems, including the image magnification problem mentioned above.

Figure 5 shows an example of a real-time depth from defocus sensor, which employs two imaging chips at slightly different depths sharing a common optical path, as well as an active illumination system that projects a checkerboard pattern from the same direction. As you can see in Figure 5b–g, the system produces high-accuracy real-time depth maps for both static and dynamic scenes.

FUTURE RESEARCH CHALLENGES

The production of 3D models has been a popular research topic since a long time, and important progress has indeed been made since the early days. Nonetheless, the research community is well-aware of the fact that still much remains to be done. In this section, we list some of these challenges.

As seen in the previous section, there is a wide variety of techniques for creating 3D models, but depending on the geometry and material characteristics of the object or scene, one technique may be much better suited than another. For example, non-textured objects are a nightmare for traditional stereo, but too much texture may interfere with the patterns of structured-light techniques. Hence, one would seem to need a range of systems to deal with the variability of objects - e.g. in a museum - to be modeled. In fact, having to model the entire collection of diverse museums is a useful application area to think about, as it poses many of the pending challenges, often several at once. Another area is 3D city modeling, which has quickly grown in importance over the last years. It is another extreme in terms of conditions under which data have to be captured, in that cities represent an absolutely uncontrolled and large-scale environment. Many problems remain to be resolved, in that application area.

One of the most difficult challenges is the presence of many objects with an intricate shape, the scanning of which requires great precision combined with great agility of the scanner to capture narrow cavities and protrusions, deal with self-occlusions or fine carvings. The types of objects and materials that potentially have to be handled are very diverse, ranging from metal coins to woven textiles; stone or wooden sculptures; ceramics; gems in jewellery and glass. No single technology can deal with all these surface types and for some of these types of artifacts there are no satisfactory techniques yet. Also, apart from the 3D shape the material characteristics may need to be captured as well. The objects to be scanned range from needle size to an entire landscape or city. Ideally, one would handle this range of scales with the same techniques and similar protocols.

For many applications, data collection may have to be undertaken on-site under potentially adverse conditions or implying transportation of equipment to remote sites. Objects are sometimes too fragile or valuable to be touched and need to be scanned without human intervention. The scanner needs to be moved around the object, without it being touched, using portable systems.

Masses of data often need to be captured, like in city modeling examples. Efficient data capture and model building is essential if this is to be practical. Those undertaking the digitization may be technically trained. Not all applications are to be found in industry, and technically trained personnel might not exist. This raises the need for intelligent devices that ensure high quality data through (semi-) automation, self-diagnosis, and strong operator guidance. Also, precision is a moving target in many applications and as higher precisions are obtained, new applications emerge, that push for even higher precision.

An application where combined extraction of shape and surface reflectance occurs would be highly innovative. Increasingly, 3D scanning technology is aimed at also extracting high-quality surface reflectance information. Yet, there is an appreciable way to achieve high-precision geometry if it is combined with detailed surface characteristics like full-fledged BRDF (Bidirectional Reflectance Distribution Function) or BTF (Bidirectional Texture Function) information.

Besides, in-hand-scanning, i.e. scanning using portable devices is available. However, the choice is still restricted, especially when also surface reflectance information is required and when the method ought to perform well with any material, including metals. On-line scanning is another promising technique. Nowadays, the physical action of scanning and the actual processing of the data are still two separate steps. This may cause problems in the completeness and quality of the data as they can only be inspected after the scanning session is over and the data are analyzed and combined elsewhere. It may then be too late or too cumbersome to take corrective actions, such as acquiring a few additional scans. It would be very desirable if the system would extract the 3D data on the fly, and would give immediate visual feedback. This should ideally include steps like the integration and remising of partial scans. This would also be a great help in planning where to take the next scan during scanning.

Scanning using multi modal sensors may not only combine geometry and visual characteristics. Additional features like non-visible wavelengths (UV,(N)IR) could have to be captured, as well as haptic impressions. The latter would then also allow for a full replay to the public, where audiences can hold even the most precious objects virtually in their hands and explore them with all their senses.

Gradually computer vision is getting at a point where scene understanding becomes feasible. Out of 2D images, objects and scene types can be recognized. This will in turn have a drastic effect on the way in which 'low'- level processes can be carried out. If high-level, semantic interpretations can be fed back into 'low'- level processes like motion and depth extraction, these can benefit greatly. This strategy ties in with the opportunistic scanning idea. Recognizing what it is to be reconstructed in 3D (e.g. a car), would assist a system to decide how to perform, resulting in increased speed, robustness and accuracy. It can provide strong priors about the expected shape.

Finally, the real challenge is to make the technology available to mainstream audience using off-the-shelf components. In order to keep 3D modeling cheap, one would ideally construct the 3D reconstruction systems based on off-the-shelf, consumer products. This does not only reduce the price, but also lets the systems go along with fast-evolving, mass-market products. For instance, the resolution of still, digital cameras is steadily on the rise, so a system based on such cameras can be upgraded to higher quality without much effort or investment. Moreover, as most users will be acquainted with such components, the learning curve to use the system is probably not as steep as with a totally novel, dedicated technology.

Given the above considerations, 3D reconstruction of shapes and surfaces from multiple, un-calibrated images is one of the most promising 3D techniques. Objects or scenes are relatively small or large, depending on the appropriate optics and amount of camera data. These methods also give direct access to both shape and surface reflectance information, where both can be aligned without special alignment techniques.

CONCLUSION

Efficient implementations of several 3D reconstruction algorithms have been proposed lately. In this chapter, we described three different categories of methods: The first one calculates shape of objects and surfaces using pure geometric tools; the second one evaluates the intensity of the pixel neighborhoods inside the image space to examine the three-dimensional scene structure and, last, the third category examined in this chapter refers to the real – aperture approaches that take advantage of the physical properties of the visual sensors for image acquisition to reproduce depth and estimate shapes of the objects.

Concluding, it can be said that 3D visual reconstruction technology was accelerated significantly during the last decades, following the resolution booming in imaging devices and the extreme advance in computing power. We already are capable of reconstructing famous landmarks using everyday tourist photos from social web pages and finding our way on the map using triangulated image data from satellites.

Yet, 3D reconstruction is a laborious process which involves many limitations and difficulties to overcome. There is a large gap between computer algorithms and cognitive performance. One of the recent trends is using vast amount of image data to train machine learning models to act and think like real humans do. The complexity of the vision problems can be so high that smarter and more intelligent algorithms should be examined. The promising outcome of these potential research directions will assure us that three dimensional reconstructions will remain a very exciting area of study for the researchers in the future.

REFERENCES

Aganj, E., Pons, J.-P., S'egonne, F., & Kerive, R. (2007, October 14-20). Spatio-temporal shape from silhouette using four-dimensional delaunay meshing. In *ICCV'07: Proceedings of the 11th IEEE International Conference on Computer Vision* (pp. 1–8). Los Alamitos, CA: IEEE Computer Society.

Azevedo, T., Tavares, J., & Vaz, M. (2010). Three-dimensional reconstruction and characterization of human external shapes from two-dimensional images using volumetric methods. *Computer Methods in Biomechanics and Biomedical Engineering, 13*(3), 359–369. doi:10.1080/10255840903251288

Barnard, S. T., & Fischler, M. A. (1982). Computational stereo. [CSUR]. *ACM Computing Surveys, 14*(4), 553–572. doi:10.1145/356893.356896

Barreto, J. A. P., & Daniilidis, K. (2004). *Wide area multiple camera calibration and estimation of radial distortion.* In The Fifth Workshop on Omnidirectional Vision, Camera Networks and Non-Classical Cameras, Prague

Basri, R., Jacobs, D., & Kemelmacher, I. (2007). Photometric stereo with general, unknown lighting. *International Journal of Computer Vision, 72*(3), 239–257. doi:10.1007/s11263-006-8815-7

Ben-Ari, R., & Sochen, N. (2007). Variational stereo vision with sharp discontinuities and occlusion handling. In *ICCV'07: Proceedings of the IEEE International Conference on Computer Vision* (pp. 1-7). Los Alamitos, CA, USA: IEEE Computer Society.

Boyer, E., & Franco, J.-S. (2003, June 16-22). A hybrid approach for computing visual hulls of complex objects. In *CVPR'03* []. Los Alamitos, CA: IEEE Computer Society.]. *Proceedings of the IEEE Computer Society Conference on Computer Vision and Pattern Recognition, 1*, 695–701.

Boykov, Y., & Kolmogorov, V. (2004). An experimental comparison of min-cut/max-flow algorithms for energy minimization in vision. *IEEE Transactions on Pattern Analysis and Machine Intelligence, 26*(9), 1124–1137. doi:10.1109/TPAMI.2004.60

Brown, D. (1958). *A solution to the general problem of multiple station analytical stereo triangulation* (Tech. Rep. No. Technical Report No. 43 (or AFMTC TR 58-8)). Patrick Airforce Base, Florida: Technical Report RCA-MTP Data Reduction.

Brown, M. Z., Burschka, D., & Hager, G. D. (2003). Advances in computational stereo. *IEEE Transactions on Pattern Analysis and Machine Intelligence, 25*(8), 993–1008. doi:10.1109/TPAMI.2003.1217603

Chang, J., Lee, K., & Lee, S. (2008). Shape from shading using graph cuts. *Pattern Recognition, 41*(12), 3749–3757. doi:10.1016/j.patcog.2008.05.020

Chow, C. K., & Yuen, S. Y. (2009). Recovering shape by shading and stereo under Lambertian shading model. *International Journal of Computer Vision, 85*(1), 58–100. doi:10.1007/s11263-009-0240-2

Courbon, J., Mezouar, Y., Eck, L., & Martinet, P. (2007). A generic fisheye camera model for robotic applications. In *IROS'07: Proceedings of the IEEE/RSJ International Conference on Intelligent Robots and Systems,* (pp. 1683 -1688) San Diego, California, USA.

DeCubber, G., Nalpantidis, L., Sirakoulis, G. C., & Gasteratos, A. (2008). Intelligent robots need intelligent vision: Visual 3D perception. In *RISE'08: Proceedings of the EURON/IARP International Workshop on Robotics for Risky Interventions and Surveillance of the Environment*. Benicassim, Spain.

DeSouza, G., & Kak, A. (2002). Vision for mobile robot navigation: A survey. *IEEE Transactions on Pattern Analysis and Machine Intelligence, 24*(2), 237–267. doi:10.1109/34.982903

Dhond, U., & Aggarwal, J. (1989). Structure from stereo-A review. *IEEE Transactions on Systems, Man, and Cybernetics, 19*(6), 1489–1510. doi:10.1109/21.44067

Durou, J., Falcone, M., & Sagona, M. (2008). Numerical methods for shape-from-shading: A new survey with benchmarks. *Computer Vision and Image Understanding, 109*(1), 22–43. doi:10.1016/j.cviu.2007.09.003

El-Hakim, S., Beraldin, J., Picard, M., & Cournoyer, L. (2008). Surface reconstruction of large complex structures from mixed range data-the erechtheion experience. In *ISPRS'08: Proceedings of the XXI Congress of the International Society for Photogrammetry and Remote Sensing* (vol. 37, pp. 1077-1082). Lemmer, The Netherlands: Reed Business.

Esteban, C. H., & Schmitt, F. (2004). Silhouette and stereo fusion for 3D object modeling. *Computer Vision and Image Understanding, 96*(3), 367–392. doi:10.1016/j.cviu.2004.03.016

Favaro, P. (2007). Shape from focus and defocus: Convexity, quasiconvexity and defocus- invariant textures. In *ICCV'07: Proceedings of the 11th IEEE International Conference on Computer Vision* (pp. 1–7). Los Alamitos, CA: IEEE Computer Society.

Favaro, P., & Soatto, S. (2005). A geometric approach to shape from defocus. *IEEE Transactions on Pattern Analysis and Machine Intelligence, 27*(3), 406–417. doi:10.1109/TPAMI.2005.43

Favaro, P., Soatto, S., Burger, M., & Osher, S. (2008, March). Shape from defocus via diffusion. *IEEE Transactions on Pattern Analysis and Machine Intelligence, 30*(3), 518–531. doi:10.1109/TPAMI.2007.1175

Fischler, M. A., & Bolles, R. C. (1981). Random sample consensus: A paradigm for model fitting with applications to image analysis and automated cartography. *Communications of the ACM, 24*(6), 381–395. doi:10.1145/358669.358692

Forsyth, D. A., & Ponce, J. (2002). *Computer vision: A modern approach* (US ed.). Prentice Hall Professional Technical Reference.

Galasso, F., & Lasenby, J. (2007). Shape from texture of developable surfaces via Fourier analysis. In G. Bebis, et al. (Eds.), In *ISVC'07: Proceedings of the 3rd International Symposium on Advances in Visual Computing* (vol. 4841, pp. 702-713). Heidelberg, Germany: Springer.

Gennery, D. (2006). Generalized camera calibration including fish-eye lenses. *International Journal of Computer Vision, 68*, 239–266. doi:10.1007/s11263-006-5168-1

Grauman, K., Shakhnarovich, G., & Darrell, T. (2003). Inferring 3D structure with a statistical image-based shape model. In *ICCV'03: Proceedings of the 9th IEEE International Conference on Computer Vision* (vol. 1, pp. 641–647). Los Alamitos, CA: IEEE Computer Society.

Grossberg, S., Kuhlmann, L., & Mingolla, E. (2007). A neural model of 3D shape-from-texture: Multiple-scale filtering, boundary grouping, and surface filling-in. *Vision Research, 47*(5), 634–672. doi:10.1016/j.visres.2006.10.024

Harker, M., & O'Leary, P. (2008). Least squares surface reconstruction from measured gradient fields. In *CVPR'03: Proceedings of the IEEE Computer Society Conference on Computer Vision and Pattern Recognition*, (pp.1-7). Los Alamitos, CA: IEEE Computer Society.

Hays, J., Leordeanu, M., Efros, A. A., & Liu, Y. (2006). Discovering texture regularity as a higher-order correspondence problem. In A. Leonardis, H. Bischof, & A. Pinz (Eds.), *Eccv'06: Proceedings of the 9th European Conference on Computer Vision* (vol. 3952, pp. 522–535). Graz, Austria: Springer.

Hernandez Esteban, C., Vogiatzis, G., & Cipolla, R. (2008). Multiview photometric stereo. *IEEE Transactions on Pattern Analysis and Machine Intelligence*, *30*(3), 548–554. doi:10.1109/TPAMI.2007.70820

Horn, B. K. P. (1989). *Obtaining shape from shading information. Shape from shading* (pp. 123–171). Mit Press Series Of Artificial Intelligence Series.

Horn, B. K. P. (1990). Height and gradient from shading. *International Journal of Computer Vision*, *5*(1), 37–75. doi:10.1007/BF00056771

Jacques, L., De Vito, E., Bagnato, L., & Vandergheynst, P. (2008). Shape from texture for omnidirectional images. In *EUSIPCO'08: Proceedings of the 16th European Signal Processing Conference*.

Janoos, F., Mosaliganti, K., Xu, X., Machiraju, R., Huang, K., & Wong, S. (2009). Robust 3D reconstruction and identification of dendritic spines from optical microscopy imaging. *Medical Image Analysis*, *13*(1), 167–179. doi:10.1016/j.media.2008.06.019

Jin, H., Soatto, S., & Yezzi, A. J. (2000). Stereoscopic shading: Integrating multi-frame shape cues in a variational framework. In *CVPR'00* []. Los Alamitos, CA: IEEE Computer Society.]. *Proceedings of the IEEE Conference on Computer Vision and Pattern Recognition*, *1*, 1169.

Kannala, J., & Brandt, S. S. (2006). A generic camera model and calibration method for conventional, wide-angle, and fisheye lenses. *IEEE Transactions on Pattern Analysis and Machine Intelligence*, *28*, 1335–1340. doi:10.1109/TPAMI.2006.153

Kolev, K., & Cremers, D. (2008). Integration of multiview stereo and silhouettes via convex functionals on convex domains. In D. A. Forsyth, P. H. S. Torr, & A. Zisserman (Eds.), *ECCV'08: Proceedings of the 10th European Conference on Computer Vision* (vol. 5302, pp. 752–765). Heidelberg, Germany: Springer.

Kru¨ger, L. E., Wohler, C., Wurz-Wessel, A., & Stein, F. (2004). In-factory calibration of multiocular camera systems . In Osten, W., & Takeda, M. (Eds.), *Optical metrology in production engineering* (*Vol. 5457*, pp. 126–137). SPIE.

Kruppa, E. (1913). Zur ermittlung eines objektes aus zwei perspektiven mit innerer orientierung. *Sitzungsberichte der Mathematisch Naturwissenschaftlichen Kaiserlichen Akademie der Wissenschaften*, *122*, 1939–1948.

Ladikos, A., Benhimane, S., & Navab, N. (2008). Efficient visual hull computation for real- time 3d reconstruction using Cuda. In *Proceedings of the 2008 Conference on Computer Vision and Pattern Recognition Workshops* (pp. 1–8). Los Alamitos, CA: IEEE Computer Society.

Laussedat, A. (1898). *Recherches sur les instruments: Les méthodes et le dessin topographiques*. Gauthier-Villars.

Lemaire, T., Berger, C., Jung, I.-K., & Lacroix, S. (2007). Vision-based slam: Stereo and monocular approaches. *International Journal of Computer Vision*, *74*(3), 343–364. doi:10.1007/s11263-007-0042-3

Li, J., & Allinson, N. M. (2008). A comprehensive review of current local features for computer vision. *Neurocomputing*, *71*(10-12), 1771–1787. doi:10.1016/j.neucom.2007.11.032

Liu, X., Yao, H., Chen, X., & Gao, W. (2008). Shape from silhouettes based on a centripetal pentahedron model. *Graphical Models*, *70*(6), 133–148. doi:10.1016/j.gmod.2006.06.003

Liu, Y., Collins, R., & Tsin, Y. (2004). A computational model for periodic pattern perception based on frieze and wallpaper groups. *IEEE Transactions on Pattern Analysis and Machine Intelligence*, *26*(3), 354–371. doi:10.1109/TPAMI.2004.1262332

Liu, Y., Lin, W.-C., & Hays, J. (2004, August). Near-regular texture analysis and manipulation. *ACM Transactions on Graphics*, *23*(3), 368–376. doi:10.1145/1015706.1015731

Liu, Z., & Klette, R. (2008, November 25-28). Dynamic programming stereo on real-world sequences. In *ICONIP'08: Proceedings of the 15th International Conference on Advances in Neuro-Information Processing* (vol. 5507, pp. 527–534). Auckland, New Zealand: Springer.

Lobay, A., & Forsyth, D. (2006). Shape from texture without boundaries. *International Journal of Computer Vision*, *67*(1), 71–91. doi:10.1007/s11263-006-4068-8

Loh, A., & Hartley, R. (2005). Shape from non-homogeneous, non-stationary, anisotropic, perspective texture. In *BMCV'05: Proceedings of the British Machine Vision Conference* (pp. 69–78).

Lou, Y., Favaro, P., Bertozzi, A. L., & Soatto, S. (2007, 18-23 June). Autocalibration and uncalibrated reconstruction of shape from defocus. In *CVPR'07: Proceedings of the IEEE Conference on Computer Vision and Pattern Recognition* (pp. 1–8). Los Alamitos, CA: IEEE Computer Society.

Lourakis, M. I. A., & Argyros, A. A. (2004). *The design and implementation of a generic sparse bundle adjustment software package based on the Levenberg-Marquardt algorithm* (Technical Report 340). FORTH, Heraklion Crete, Greece, Institute of Computer Science.

Lourakis, M. I. A., & Argyros, A. A. (2009). SBA: A software package for generic sparse bundle adjustment. *ACM Transactions on Mathematical Software*, *36*(1), 1–30. doi:10.1145/1486525.1486527

Luhmann, T. (2003). *Nahbereichsphotogrammetrie Grundlagen, Methoden, Anwendungen. 2*. Heidelberg: Auflage, Wichmann Verlag.

MacLean, W., Sabihuddin, S., & Islam, J. (2010). Leveraging cost matrix structure for hardware implementation of stereo disparity computation using dynamic programming. *Computer Vision and Image Understanding*, *114*(11). doi:10.1016/j.cviu.2010.03.011

Masrani, D. K., & MacLean, W. J. (2006). A real-time large disparity range stereo system using fpgas. In *ICVS'06: Proceedings of the International Conference on Computer Vision Systems* (p. 13). Los Alamitos, CA, USA: IEEE Computer Society.

Matusik, W., Buehler, C., & McMillan, L. (2001). Polyhedral visual hulls for real-time rendering. In S. J. Gortler & K. Myszkowski (Eds.), *Proceedings of the 12th Eurographics Workshop on Rendering Techniques* (vol. 1, pp. 115–126). Springer.

Meydenbauer, A. (1867). Ueber die Anwendung der Photographie zur Architektur-und Terrain-Aufnahme. *Zeitschrift für Bauwesen*, *17*, 61–70.

Mikolajczyk, K., Tuytelaars, T., Schmid, C., Zisserman, A., Matas, J., & Schaffalitzky, F. (2005). A comparison of affine region detectors. *International Journal of Computer Vision*, *65*(1-2), 43–72. doi:10.1007/s11263-005-3848-x

Moreels, P., & Perona, P. (2007). Evaluation of features detectors and descriptors based on 3S objects. *International Journal of Computer Vision*, *73*(3), 263–284. doi:10.1007/s11263-006-9967-1

Nalpantidis, L., Chrysostomou, D., & Gasteratos, A. (2009). Obtaining reliable depth maps for robotic applications from a quad-camera system. In M. Xie, Y. Xiong, C. Xiong, H. Liu, & Z. Hu (Eds.), *ICIRA '09: Proceedings of the 2nd International Conference on Intelligent Robotics and Applications* (vol. 5928, pp. 906–916). Berlin, Germany: Springer.

Nalpantidis, L., Kostavelis, I., & Gasteratos, A. (2009). Stereovision-based algorithm for obstacle avoidance. In M. Xie, Y. Xiong, C. Xiong, H. Liu, & Z. Hu (Eds.), *ICIRA '09: Proceedings of the 2nd International Conference on Intelligent Robotics and Applications* (vol. 5928, pp. 195–204). Berlin, Germany: Springer.

Nalpantidis, L., Sirakoulis, G., & Gasteratos, A. (2008). Review of stereo vision algorithms: From software to hardware. *International Journal of Optomechatronics*, *2*(4), 435–462. doi:10.1080/15599610802438680

Nayar, S., Ikeuchi, K., & Kanade, T. (1991). Shape from interreflections. *International Journal of Computer Vision*, *6*(3), 173–195. doi:10.1007/BF00115695

Nayar, S., Watanabe, M., & Noguchi, M. (1996). Real-time focus range sensor. *IEEE Transactions on Pattern Analysis and Machine Intelligence*, *18*(12), 1186–1198. doi:10.1109/34.546256

Nehab, D., Rusinkiewicz, S., Davis, J., & Ramamoorthi, R. (2005). Efficiently combining positions and normals for precise 3D geometry. [TOG]. *ACM Transactions on Graphics*, *24*(3), 543. doi:10.1145/1073204.1073226

Nevado, M. M., Garcia-Bermejo, J. G., & Casanova, E. Z. (2004). Obtaining 3D models of indoor environments with a mobile robot by estimating local surface directions. *Robotics and Autonomous Systems*, *48*(2-3), 131–143.

Nister, D. (2004, June). An efficient solution to the five-point relative pose problem. *IEEE Transactions on Pattern Analysis and Machine Intelligence*, *26*(6), 756–777. doi:10.1109/TPAMI.2004.17

Park, M., Brocklehurst, K., Collins, R., & Liu, Y. (2009). Deformed lattice detection in real- world images using mean-shift belief propagation. *IEEE Transactions on Pattern Analysis and Machine Intelligence*, *31*(10), 1804–1816. doi:10.1109/TPAMI.2009.73

Pradeep, K., & Rajagopalan, A. (2007). Improving shape from focus using defocus cue. *IEEE Transactions on Image Processing*, *16*(7), 1920–1925. doi:10.1109/TIP.2007.899188

Ramnath, K., & Rajagopalan, A. N. (2009). Discontinuity-adaptive shape from focus using a non-convex prior. In J. Denzler, G. Notni, & H. Su¨ße (Eds.), *DAGM'09: Proceedings of the 31st DAGM Symposium* (vol. 5748, pp. 181–190). Heidelberg, Germany: Springer.

Remondino, F., El-Hakim, S., Baltsavias, E., Picard, M., & Grammatikopoulos, L. (2008). Image-based 3D modeling of the Erechteion, Acropolis of Athens. In *ISPRS'08: Proceedings of the XXI Congress of the International Society for Photogrammetry and Remote Sensing* (vol. 37, pp. 1083–1091). Lemmer, The Netherlands: Reed Business.

Sahay, R. R., & Rajagopalan, A. N. (2009). Real aperture axial stereo: Solving for correspondences in blur. In J. Denzler, G. Notni, & H. Su¨ße (Eds.), *DAGM'09: Proceedings of the 31st DAGM Symposium* (vol. 5748, pp. 362–371). Springer.

Scharstein, D., & Szeliski, R. (2002). A taxonomy and evaluation of dense two-frame stereo correspondence algorithms. *International Journal of Computer Vision*, *47*(1-3), 7–42. doi:10.1023/A:1014573219977

Sˇeatovi'c, D. (2008). A segmentation approach in novel real time 3d plant recognition system. In A. Gasteratos, M. Vincze, & J. K. Tsotsos (Eds.), *Proceedings6th International Computer Vision Systems Conference,* Santorini, Greece, May 12-15, 2008, (vol. 5008, pp. 363-372). Heidelberg, Germany: Springer.

Shim, S.-O., & Choi, T.-S. (2010). A novel iterative shape from focus algorithm based on combinatorial optimization. *Pattern Recognition*, *43*(10), 3338–3347. doi:10.1016/j.patcog.2010.05.029

Sturm, P., & Barreto, J. A. (2008). General imaging geometry for central catadioptric cameras. In D. A. Forsyth, P. H. S. Torr, & A. Zisserman (Eds.), *ECCV'08: Proceedings of the 10th European Conference on Computer Vision* (vol. 5305, pp. 609-622). Berlin, Germany: Springer.

Sturm, P., & Ramalingam, S. (2004). A generic concept for camera calibration. In T. Pajdla & J. Matas (Eds.), *ECCV'04: Proceedings of the 8th European Conference on Computer Vision* (vol. 3022, pp. 1-13). Berlin, Germany: Springer.

Szeliski, R. (1991). Fast shape from shading. *CVGIP: Image Understanding, 53*(2), 129–153. doi:10.1016/1049-9660(91)90023-I

Tepper, O., Karp, N., Small, K., Unger, J., Rudolph, L., & Pritchard, A. (2008). Three- dimensional imaging provides valuable clinical data to aid in unilateral tissue expander- implant breast reconstruction. *The Breast Journal, 14*(6), 543–550. doi:10.1111/j.1524-4741.2008.00645.x

Todd, J., Thaler, L., Dijkstra, T., Koenderink, J., & Kappers, A. (2007). The effects of viewing angle, camera angle, and sign of surface curvature on the perception of three-dimensional shape from texture. *Journal of Vision (Charlottesville, Va.), 7*(12). doi:10.1167/7.12.9

Triggs, B., McLauchlan, P. F., Hartley, R. I., & Fitzgibbon, A. W. (2000). Bundle adjustment - A modern synthesis. In B. Triggs, A. Zisserman, & R. Szeliski (Eds.), *ICCV '99: Proceedings of the International Workshop on Vision Algorithms: Theory and Practice* (vol. 1883, pp. 298–372). London, UK: Springer-Verlag.

Tsai, P.-S., & Shah, M. (1994). Shape from shading using linear approximation. *Image and Vision Computing, 12*(8), 487–498. doi:10.1016/0262-8856(94)90002-7

Tuytelaars, T., & Mikolajczyk, K. (2008). Local invariant feature detectors: A survey. *Foundations and Trends in Computer Graphics and Vision, 3*(3), 177–280. doi:10.1561/0600000017

Vlasic, D., Baran, I., Matusik, W., & Popovic, J. (2008). Articulated mesh animation from multi-view silhouettes. *ACM Transactions on Graphics, 27*(3), 1–9. doi:10.1145/1360612.1360696

Vogiatzis, G., Esteban, C. H., Torr, P. H. S., & Cipolla, R. (2007). Multiview stereo via volumetric graph-cuts and occlusion robust photo-consistency. *IEEE Transactions on Pattern Analysis and Machine Intelligence, 29*(12), 2241–2246. doi:10.1109/TPAMI.2007.70712

White, R., Crane, K., & Forsyth, D. A. (2007). Capturing and animating occluded cloth. [TOG]. *ACM Transactions on Graphics, 26*(3), 34. doi:10.1145/1276377.1276420

White, R., & Forsyth, D. A. (2006). Combining cues: Shape from shading and texture. *CVPR'06 . Proceedings of the IEEE Computer Vision and Pattern Recognition, 2*, 1809–1816.

Wilhelmy, J., & Kru¨ger, J. (2009). Shape from shading using probability functions and belief propagation. *International Journal of Computer Vision, 84*(3), 269–287. doi:10.1007/s11263-009-0236-y

Woodham, R. (1981). Analysing images of curved surfaces. *Artificial Intelligence, 17*(1-3), 117–140. doi:10.1016/0004-3702(81)90022-9

Xu, J., Xi, N., Zhang, C., & Shi, Q. (2009). Windshield shape inspection using structured light patterns from two diffuse planar light sources. *In IEEE/RSJ International Conference on Intelligent Robots and Systems* (pp. 721–726). Piscataway, NJ: IEEE Press.

Xu, J., Xi, N., Zhang, C., Shi, Q., & Gregory, J. (2010). Real-time 3D shape inspection system of automotive parts based on structured light pattern. *Optics & Laser Technology, 43*(1), 1–8. doi:10.1016/j.optlastec.2010.04.008

Xu, Y., & Aliaga, D. G. (2009). An adaptive correspondence algorithm for modeling scenes with strong interreflections. *IEEE Transactions on Visualization and Computer Graphics, 15*, 465–480. doi:10.1109/TVCG.2008.97

Yemez, Y., & Wetherilt, C. J. (2007). A volumetric fusion technique for surface reconstruction from silhouettes and range data. *Computer Vision and Image Understanding*, *105*(1), 30–41. doi:10.1016/j.cviu.2006.07.008

Zhang, J., & Smith, S. (2009). Shape similarity matching with octree representations. *Journal of Computing and Information Science in Engineering*, *9*(3), 034503. doi:10.1115/1.3197846

Zhang, R., Tsai, P.-S., Cryer, J. E., & Shah, M. (1999). Shape from shading: A survey. *IEEE Transactions on Pattern Analysis and Machine Intelligence*, *21*(8), 690–706. doi:10.1109/34.784284

Zhou, K., Gong, M., Huang, X., & Guo, B. (2010). Data-parallel octrees for surface reconstruction. [PrePrints]. *IEEE Transactions on Visualization and Computer Graphics*, 99.

Zhu, B., Lu, J., Luo, Y., Tao, Y., & Cheng, X. (2009). 3D surface reconstruction and analysis in automated apple stem-end/calyx identification. *International Journal of Pattern Recognition and Artificial Intelligence*, *52*(5), 1775–1784.

ADDITIONAL READING

Cyganek, B. (2007). *An introduction to 3d computer vision techniques and algorithms*. John Wiley & Sons.

Davies, E. R. (2004). *Machine vision: Theory, algorithms, and practicalities*. San Francisco, CA, USA: Morgan Kaufmann Publishers Inc.

Faugeras, Olivier, Luong, Quang-Tuan, and Papadopoulou, T. (2001). *The Geometry of Multiple Images: The Laws That Govern The Formation of Images of A Scene and Some of Their Applications*. MIT Press, Cambridge, MA, USA.

Forsyth, D. A., & Ponce, J. (2002). *Computer vision: A modern approach (US ed.)*. Prentice Hall Professional Technical Reference

Hartley, R. I., & Zisserman, A. (2004). *Multiple view geometry in computer vision* (2nd ed.). Cambridge University Press. doi:10.1017/CBO9780511811685

Paragios, N., Chen, Y., & Faugeras, O. (2005). *Handbook of mathematical models in computer vision*. Secaucus, NJ, USA: Springer-Verlag New York, Inc.

Steger, C., Ulrich, M., & Wiedemann, C. (2008). *Machine vision algorithms and Applications*. Wiley VCH.

KEY TERMS AND DEFINITIONS

Bundle Adjustment: The method to acquire accurate reconstruction of points by minimizing the reprojection error using Levenberg-Marqardt or Gauss-Newton algorithms.

Camera Calibration: The method to estimate the internal and external camera parameters.

PhotometricStereo: The method to estimate shapes using shading variations and multiple light sources.

Pinhole Camera: The fundamental camera model for projection estimation.

Shape from Defocus: The method to estimate shapes estimating the amount of blur.

Shape from Focus: The shape reconstruction approach to measure depth using the camera parameters like focus.

Shape from Shading: The problem of recovering the shape of a surface using shading variations.

Shape from Silhouettes: The method to estimate shape of objects calculating the silhouette boundaries from images.

Shape from Texture: The method to estimate shapes measuring texture anomalies.

Stereo Vision: The use of two cameras to simulate human vision system and estimate depth from the camera system.

Chapter 9
Comparison of Focus Measures under the Influence of Various Factors Effecting their Performance

Aamir Saeed Malik
Universiti Teknologi Petronas, Malaysia

ABSTRACT

This chapter presents a comparison of eleven focus measures which are categorized in four main classes or groups. The performance of focus measures is evaluated by considering various factors that might hinder their smooth operation. These factors include illumination variation, texture reflectance, object distance variation, distance variation in between consecutive frames, and various types of noise including Gaussian, Shot, and Speckle noise. The focus measures are tested for depth estimation for 3D shape recovery using Shape From Focus (SFF) techniques. Three measures are used to compare the performance of the focus measures, namely, visual inspection as a qualitative measure and root mean square error and correlation as quantitative measures.

INTRODUCTION

Depth map estimation for three-dimensional shape recovery from one or multiple observations is a challenging problem of computer vision. This depth map can subsequently be used in interpolation and approximation techniques and algorithms leading to the recovery of a three dimensional structure of the object, a requirement of a number of high level vision applications. However, the basic problem of imaging systems, such as the digital-camera, is that depth information is lost while projecting a 3D scene onto 2D image plane.

DOI: 10.4018/978-1-61350-326-3.ch009

Figure 1. Image formation of a 3D object

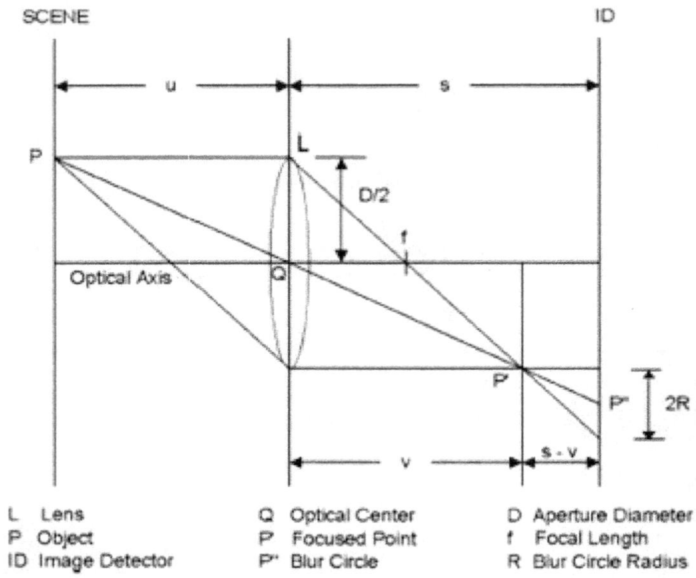

Therefore, one fundamental problem in computer vision is the reconstruction of a geometric object from one or several observations.

There are a variety of 3D Shape estimation methods that try to address this problem. They include Shape From Focus, Defocus, Texture, Motion etc. They are generally referred to as Shape From X and are classified as optical passive methods. In this chapter, we limit our discussion to Shape From Focus (SFF). SFF is based on focus which is an accommodation cue (Mennucciy, 1999) that can be measured from blurring in the image, which increases with the distance of imaging system from the plane of focus. Techniques that retrieve spatial information, by looking at multiple images of the same scene, taken with different geometry or position of imaging devices, are classified as Shape From Focus (SFF).

The objective of Shape From Focus (SFF) is to find out the depth of every point of the object from the camera lens. Hence, finally we get a depth map which contains the depth of all points of the object from the camera lens where they are best focused or in other words, where they show maximum sharpness.

The basic image formation geometry is shown in Figure 1. In Figure 1, the parameters related to the camera are already known. We need to calculate 'u', i.e., depth of object from the lens. We make a depth map by calculating 'u' for every pixel. We can use the lens formula to calculate 'u'. If the image detector (ID) is placed exactly at a distance v, sharp image P' of the point P is formed at v (see Figure 1). Then the relationship between the object distance u, focal distance of the lens f, and the image distance v is given by the Gaussian lens law:

$$\frac{1}{f} = \frac{1}{u} + \frac{1}{v} \tag{1}$$

Therefore, in SFF, a sequence of images that correspond to different levels of object focus is obtained. A sharp image and the relative depth can be retrieved by collecting the best focused points in each image. The absolute depth of object surface patches can be calculated from the focal length and the position of lens that gave the sharpest image of the surface patches. The depth or best focus is obtained by using some focus measure. A Focus Measure operator is one that calculates the best focused point in the image, i.e., focus measure is defined as a quantity to evaluate the sharpness of a pixel locally (Helmli, 2001).

EVOLUTION OF FOCUSING METHODS

1960s-1970s

One of the earliest focusing relationships was provided by (Horn, 1968). He described the application of Fourier transform method to the focusing problem. His analysis included errors due to resolution limits, noise, lens positioning, diffraction, servo inaccuracy and lens motion. Aperture-plane distortion was considered by (Buffington, 1974) because they analyzed the sharpness of the image. They proposed that sharpness of the image reaches a maximum value only for an undistorted image. What they didn't know was that they were additionally describing one of the earliest focus measure operators. In 1976, (Erteza, 1976) related the sharpness to the focus control. He derived a sharpness index function from the intensity distribution in an image and used it for correctness of focus.

1980s

(Greon, 1982) presented a comparison of autofocus algorithms in 1982. They found that the best algorithms are based upon squared gradient of the image and normalized image standard deviation. In 1983, (Jarvis, 1983) surveyed a variety of approaches to generalize range finding including, depth from texture, focusing, stereo, motion, ultrasonic, laser etc. He discussed entropy, variance and sum modulus difference as focus measures for estimating range using focus. (Krotkov, 1987) discussed how to best determine the focus motor position providing sharpest focus on an object point and how to compute the distance to a sharply focused point. Also, focus related research was published by (Pentland, 1987; Gillespie, 1989).

1990s

First major SFF system was demonstrated by (Nayar, 1990). He also described a focus measure based on gradient as well as approximation method based on Gaussian interpolation for the 3D microscope based imaging system. (Dillion, 1992) defined a hybrid range scheme based on shape from stereo and shape from focus in 1992. In 1993, various focusing methods were implemented on a prototype camera system called SPARCS (Subbarao, 1993). This was one of the first implementation of a 3D camera system. (Choi & Subbarao, 1995) for the first time proposed the calculation of focus measure over a surface in 3D plane instead of a 2D image plane. (Asada, 1998) proposed a method eliminating windowing in 1998. This resulted in improvement in computational speed. References (Nair, 1992, Xiong, 1993, Bove, 1993, Nayar, 1994, Noguchi, 1994, Nayar, 1996, Subbarao, 1998, Choi, Lee & Ho, 1999, Yun, 1999) also contributed valuably for this era.

Since 2000

Various new focus measures have been proposed in the literature since 2000. (Zhang, 2000) described a new focus measure based on 2nd/4th order central moment in 2000. This type provides explicit expression of point spread function. In 2001, (Helmi, 2001) proposed three new focus measures based on mean, curvature and point focus methods. Wavelet found its way to focus measure too in 2002 (Kautsky, 2002) just as it did in so many other research fields lately. Another focus measure based on Chebyshev moments was proposed by (Yap, 2004). Dynamic programming

(Mozerov, 2000; Milios, 2000), DCT (Shen, 2006) and few others (Pradeep, 2006, Tsai, 2003, Takahashi, 2004) were also proposed as one of the possible solutions.

CLASSIFICATION OF FOCUS MEASURES

A Focus Measure operator is one that calculates the best focused point in the image, i.e., focus measure is defined as a quantity to evaluate the sharpness of a pixel locally. The success of any focus measure depends on how accurate the sharpness in image pixels could be detected. Hence, algorithms and techniques based on calculating sharpness and edges in an image automatically become potential candidates for the selection of focus measure. The focus measures can be classified into four main classes or groups.

Derivative-Based Focus Measures

Laplacian-Based Focus Measure

Laplacian, being a point and symmetric operator, is suitable for accurate shape recovery. This focus measure is obtained by adding second derivatives in the x and y directions.

$$Laplacian = \frac{\partial^2 g(x,y)}{\partial x^2} + \frac{\partial^2 g(x,y)}{\partial y^2} \qquad (2)$$

In case of textured images, x and y components of Laplacian operator may cancel out ($g_{xx} = -g_{yy}$) and yield no response. Therefore, Modified Laplacian (ML) is computed by adding squared 2nd derivates. For discrete model, self convolution of Sobel operator is applied.

$$ML = \left(\frac{\partial^2 g(x,y)}{\partial x^2}\right)^2 + \left(\frac{\partial^2 g(x,y)}{\partial y^2}\right)^2 \qquad (3)$$

If the image has rich textures with high variability at each pixel, focus measure can be calculated considering single pixel. In order to improve robustness for weak-texture images, (Nayar, 1990, 1994) presented focus measure at (x,y) as Sum of ML (SML) values in a local window.

$$SML = \sum_{p(x,y)\in U(x_0,y_0)} \left(\frac{\partial^2 g(x,y)}{\partial^2 x}\right)^2 + \left(\frac{\partial^2 g(x,y)}{\partial^2 y}\right)^2 \qquad (4)$$

Tenenbaum Focus Measure

It is gradient magnitude maximization method that measures the sum of squared responses of horizontal and vertical Sobel masks. For robustness, it is also summed in a local window.

$$TEN = \sum_{p(x,y)\in U(x_0,y_0)} \left(G_x(x,y)^2 + G_y(x,y)^2\right)^2 \qquad (5)$$

M_2 Focus Measure

Various focus measures were proposed by (Subbarao, 1993). The focus measures proposed were based on image grey level variance (M_1), energy of image gradient (M_2) and energy of image Laplacian (M_3).These focus measures are similar to those described above, i.e., M_1 is similar to Gray Level Variance (GLV) Focus Measure, M_2 is similar to Tenenbaum Focus Measure and M_3 is similar to Laplacian focus measure. M_2 was further modified by (Xiong, 1993). They employed M_2 method using Fibonacci search followed by exhaustive search. They mentioned to use either M_2 or Tenegrad (Krotkov, 1987) with zero threshold. M_2 is computed as:

$$M2 = \sum_{x=i-N}^{i+N} \sum_{y=j-N}^{j+N} (g_x^2 + g_y^2) \qquad (6)$$

where: $g_x(x,y) = g_i(x+1,y) - g_i(x,y)$ & $g_y(x,y) = g_i(x,y+1) - g_i(x,y)$

Statistics-Based Focus Measures

Gray Level Variance (GLV) Focus Measure

In case of a sharp image, the variance of gray-level is higher than that in a blur image.

$$GLV = \frac{1}{N-1} \sum_{p(x,y)\in U(x_0,y_0)} \left(g(x,y) - \mu_{U(x_0,y_0)}\right)^2 \quad (7)$$

With $\mu_{U(x_0,y_0)}$ the mean of the gray values in the neighborhood $U(x_0,y_0)$

Mean Method (MM) Focus Measure (Helmli, 2001)

The ratio of mean grey value to the center grey value in the neighborhood can also be used as a focus measure. The ratio of one shows a constant grey-level or absence of texture. Ratio is different in case of high variations. It is also summed in local window. Let U(x,y) be the neighborhood region, g(x,y) be the center gray value and μ be the mean value, then mathematically this focus measure is given as:

$$MM = \begin{cases} \dfrac{\mu(U(x,y))}{g(x,y)}, & \mu(U(x,y)) > g(x,y) \\ \dfrac{g(x,y)}{\mu(U(x,y))}, & else \end{cases} \quad (8)$$

Curvature Focus Measure (Helmli, 2001)

The curvature in a sharp image is expected to be higher than that in a blur image. First, the surface is approximated using a quadratic equation $f(x,y) = ax + by + cx^2 + dy^2$. The coefficients (a, b, c, d) are calculated using a least squares approximation technique (Nayar, 1996). Then these coefficients are combined to obtain a focus measure.

Moment-Based Focus Measures

2nd order and 4th order central moments are used to obtain the focus information from a sequence of images (Zhang, 2000). Finally, a curve is obtained that is used to express the blur property of the imaging system. Another way is to represent the low and high spatial frequency components as low and high order Chebyshev moments (Yap, 2004). Focus measure is defined as the ratio of the norm of the high order moments to that of low order moments.

Energy-Based Focus Measures

Signal Power (SP) Focus Measure (Ligthart, 1982)

Square of the intensity value is taken and it is summed in a 2D window. The maximum value of SP is the sharpest value.

$$SP = \sum_{x=1}^{N} \sum_{y=1}^{N} g(x,y)^2 \quad (9)$$

Histogram Entropy (HE) Focus Measure (Jarvis, 1983, Krotkov, 1987, Gillespie, 1989)

Generally, a sharply focused edge has two spikes in histogram corresponding to each side of the edge. Hence, the result is a bimodal intensity histogram whereas a blurred edge does not exhibit this behavior and hence it is different from that of a sharp edge. Let I be the grey level and P(I) be the frequency of occurrence of I then:

$$HE = -\sum_i P(I)\ln[P(I)],\ P(I) \neq 0 \qquad (10)$$

HE is minimum when P(I) is zero for all except one value of I and it is maximum when all P(I) are equal. Therefore, the sharp edge will have less entropy compared to the blurred edge.

Transform-Based Focus Measures

Some of the following focus measures can also be classified as energy based focus measures. However, we have put them under this section to emphasize the role of transform domain.

DCT-Based Focus Measures

The AC energy component can be used as a focus measure (Baina, 1995). This provides information about variance component of the luminance. The best focus is where AC energy component is maximum. Let G(u,v) be the image after DCT transformation then this focus measure is given as:

$$AC = \sum_{u=0}^{N}\sum_{v=0}^{N} G(u,v)^2 \qquad (11)$$

Another focus measure (Shen, 2006) is calculated as a ratio between the energy of AC part of the DCT image to that of DC part, i.e., by dividing the energy of high frequency band by that of low frequency band. This focus measure is given as:

$$CR = \frac{\textit{Sum of square of AC Coefficients}}{\textit{Sum of square of DC Coefficient}} \qquad (12)$$

Wavelet Based Focus Measures (Kautsky, 2002, Yang, 2003, Xie, 2006)

Ratio of high pass band and low pass band norms can be taken as a focus measure (Kautsky, 2002; Xie, 2006). The measure exhibits monotonic behavior with respect to degree of defocusing. Another way is to exploit the properties of the wavelet transform coefficients in high frequency subbands. Let the wavelet transform images at level-1 LH, HL and HH subbands be denoted as G_{LH}, G_{HL} and G_{HH} respectively. Let μ_{LH}, μ_{HL} and μ_{HH} be the expectation of wavelet coefficients and U_{LH}, U_{HL} and U_{HH} be the operator windows in each subband with *w* and *l* as width and length respectively. Daubechies orthogonal wavelet basis D6 is used for computing following:

$$WAV = \frac{1}{wl}\left[\sum_{(i,j)\in E_{LH}} \left(G_{LH}(i,j) - \mu_{LH}\right)^2 + \sum_{(i,j)\in E_{HL}} \left(G_{HL}(i,j) - \mu_{HL}\right)^2 + \sum_{(i,j)\in E_{HH}} \left(G_{HH}(i,j) - \mu_{HH}\right)^2\right] \qquad (13)$$

Fourier-Based Focus Measure (Malik, 2008)

This focus measure is based on an optical transfer function implemented in the Fourier domain, and it is denoted as OM. The theory of this focus measure is based on bipolar incoherent image processing (Poon, 2001). Let g(x,y) be input image frames, F & F^{-1} be Fourier and inverse Fourier transform, K_x and K_y be spatial frequencies, σ_1 and σ_2 be filtering parameters, then mathematically, this focus measure is represented as:

$$\text{OM (i,j)} = \sum_{x=i-N}^{i+N}\sum_{y=j-N}^{j+N} \text{Real}\left[F^{-1}\left\{F\left|g(x,y)\right|^2(\exp\{-\sigma_1(k_x^2+k_y^2)\} - \exp\{-\sigma_2(k_x^2+k_y^2)\})\right\}\right] \qquad (14)$$

NOTE: For all simulations, the calculations are done by summing in a local window like SML (see equation 4). In addition, we have taken into consideration the optimum window size (Malik, 2007) and we have used window of 3×3.

APPROXIMATION TECHNIQUES

After obtaining a robust focus measure, some approximation technique (like interpolation, surface estimation, polynomial fitting etc) can be applied in order to construct a more accurate depth range image. Most of the approximation techniques for SFF mentioned in the literature (Subbarao, 1995; Nayar, 1996; Yun, 1999; Choi, Asif & Yun, 1999; Asif, 2001; Ahmad, 2005; Malik, 2007) use some kind of focus measure. Some of the approximation techniques include Focus Image Surface Method (Subbarao, 1995), piecewise curved surface method (Yun, 1999), SFF using Neural Network (Asif, 2001) and SFF using Dynamic Programming (Ahmad, 2005).

EXPERIMENTAL DESIGN

Test Images

We have used a "Test" image, a sequence of 97 simulated cone images, 97 real cone images, 87 real planar object images and 68 real microscopic object (Lincoln head on US penny) images for noise analysis. For other types of analysis, we used only microscopic objects, namely, Lincoln head part on US penny, TFT-LCD cell, V-groove and micro-sphere. Figure 2 shows one of the frames for these images.

Focus Measures Selected for Comparison

We selected focus measures from each of the four categories. The focus measures selected include Sum of Modified Laplacian (SML, eq. 4), Tenenbaum (TEN, eq. 5), M2 (eq. 6), Gray Level Variance (GLV, eq. 7), Mean Method (MM, eq. 8), Signal Power (SP, eq. 9), Histogram Entropy (HE, eq. 10), DCT based AC energy (AC, eq. 11), DCT based ratio of AC energy to DC energy (CR, eq. 12), wavelet based (WAV, eq. 13) and Fourier based (OM, eq. 14). The description of these focus measures is given in section III.

Image Quality Metric

An image quality metric can be derived as a measure of the perceived difference from a reference image. The fundamental assumption is that any

Figure 2. Test images

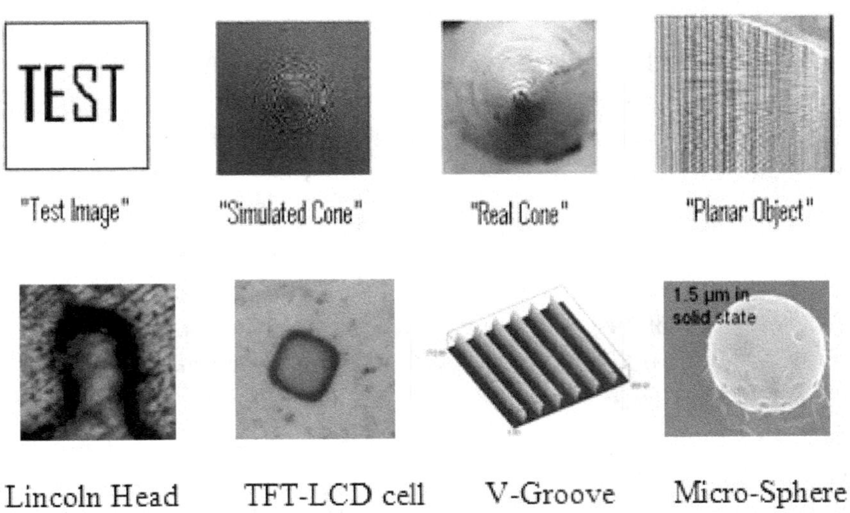

reduction in quality is caused by some perceived difference. If no differences can be perceived, then the reproduced image is indistinguishable from the original and the image quality is at its maximum. We take the ground truth depth map for simulated cone and all microscopic objects as a reference.

Various metrics have been proposed to deal with both general and specific aspects of image quality. We use visual inspection, Root Mean Square Error (RMSE) and Correlation as image quality metrics.

Design of Experiments

Noise

We compared the focus measures by adding Gaussian, speckle and shot noise to the "Test" image, simulated cone sequence, real cone sequence, planar object sequence and microscopic coin sequence. Gaussian and speckle noise are added at five noise levels having variance = 0.5, 0.05, 0.005, 0.0005 and 0.00005 at each level. Shot noise is also added at five noise levels with noise densities = 0.5, 0.05, 0.005, 0.0005 and 0.00005 at each level.

Pre-Filtering

We pre-filtered the noisy images using Weiner filter, Frost filter and Median filter for Gaussian noise, speckle noise and shot noise respectively. Then we compared the above sequence of test images using various focus measures.

Illumination

We study the effects of illumination by changing the microscope source illumination levels. 50 W (1000 Lumens) halogen lamp is used. Illumination is controlled with various steps. We select 3 illumination levels, i.e., low (~20% of source illumination), medium (~50% of source illumination) and high (~100% of source illumination).

Texture Reflectance

We study the effect of texture reflectance by selecting 4 microscopic objects of different material and texture, i.e., copper alloy (coin), transparent glass (TFT-LCD cell), reflective silicon (V-Groove) and transparent plastic (micro-sphere).

Distance Variation within Image Frames

We perform this experiment by acquiring different number of images for same z-distance. We select 3 distance variation levels for this experiment.

Distance Variation with Different Object Positions

We study the effect of placing object at different distances from the CCD camera. We perform this experiment for 4 different distances for coin and V-Groove while 3 different object positions for the TFT-LCD cell.

NOISE ANALYSIS

Qualitative Analysis

Figure 3(a) shows the "Test" image with uniform background of white color and "TEST" written in black over it. The result is converted to binary images so that actual edges extracted can be viewed. Figures 3(b) ~ (e) show the sharp variations in pixel values calculated by four focus measures, one each from the four groups. As can be seen from the images in Figure 3, all the methods are able to estimate the sharp variations clearly. There is no noise in the images in Figure 3. In Figure 4, we have added the Gaussian noise with zero mean and variance is 0.1. The effects of noise can now be seen on the results of the focus measures.

Figure 3. Applying focus measure operators on the test image

(a) Original (b) TEN (c) GLV (d) HE (e) OM

Figure 4. Results with Gaussian noise (mean=0 & var=0.1) addition

(a) Gaussian noise (b) TEN (c) GLV (d) HE (e) OM

In Figure 5, we changed the value of variance to 0.5 for the Gaussian noise. As can be seen, the results of focus measures deteriorate as the noise in increased.

Now we calculate the depth map for sequence of 97 simulated cone images. Figure 6 shows some of the frames of the simulated cone images. We obtain depth maps using all the eleven focus measures. These depth maps are obtained without any addition of noise to the sequence of images. Figure 7(a) shows the ground truth depth map while Figures 7(b) to 7(l) show the depth maps calculated using the eleven focus measures while no noise is added to the images. As can be seen from the figures, the 3D depth map obtained using all focus measures is comparable. However, the depth maps of GLV and OM appear to be smoother than the rest of the focus measures.

Now consider Figure 8. Noise is now added to the sequence of the images of simulated cone. Noise added is Gaussian with zero mean and variance equal to 0.05. Figures 8(a) to 8(k) show

Figure 5. Gaussian noise with zero mean and variance = 0.5

(a) Gaussian noise (b) TEN (c) GLV (d) HE (e) OM

Figure 6. Some frames for the simulated cone

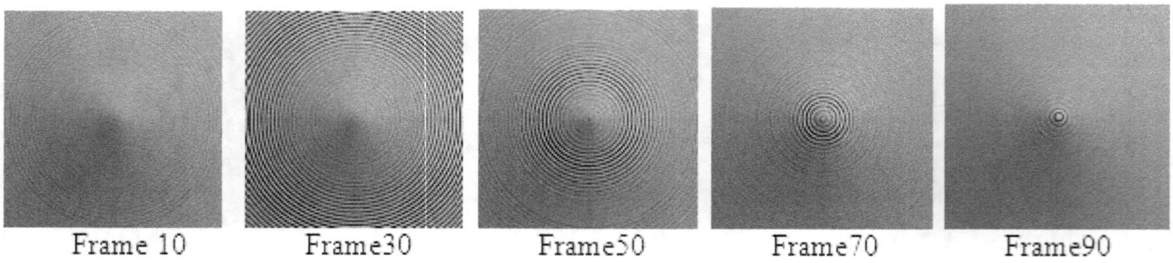

the depth maps calculated using the eleven focus measures. As can be seen from the figures, the 3D depth map obtained using OM is recognizable followed by GLV and TEN. The depth maps using other focus measures have degraded significantly. Infact, the noise added to the pixel values is enhanced in the depth map for others and hence it results in spikes originating from pixels all over the image.

Now we calculate the depth map for sequence of 97 real cone images. Figure 9 shows some of the frames of the real cone images. Figure 10 shows the depth maps for real cone with Gaussian noise added. Now Gaussian noise with zero mean and variance equal to 0.005 is added. This time we have decreased the noise variance from 0.05 to 0.005. SML, SP, HE and WAV results deterio-

Figure 7. Depth maps for the simulated cone object

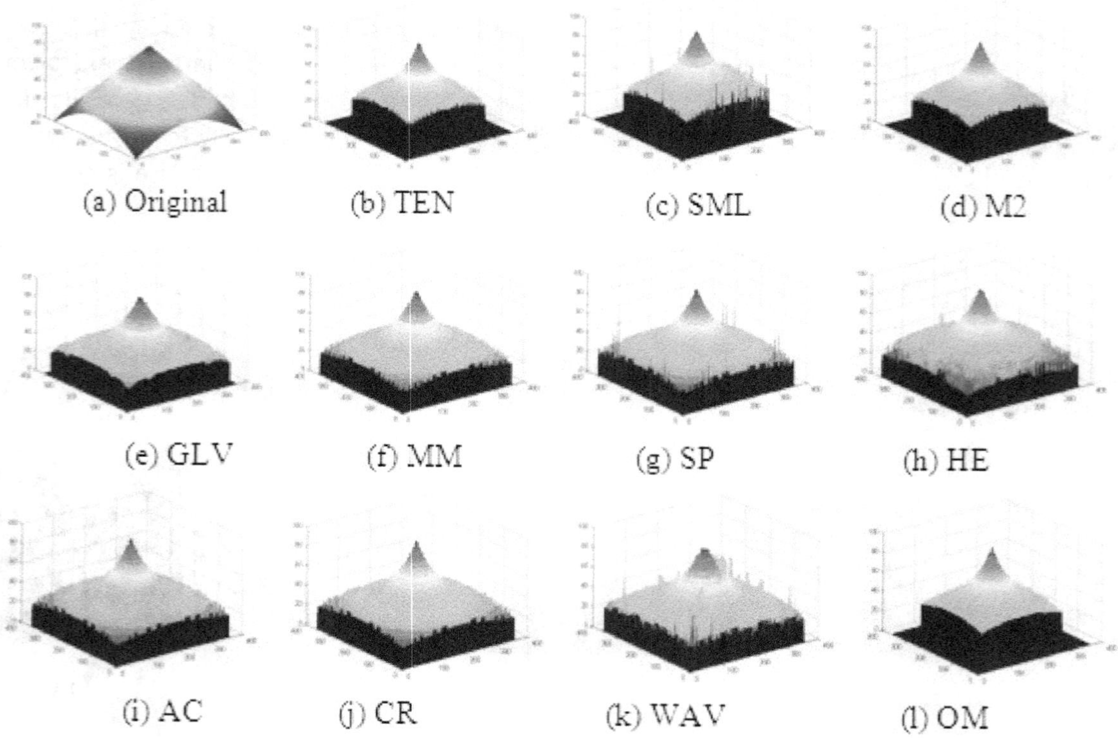

Figure 8. Depth maps for the simulated cone object when Gaussian noise is added

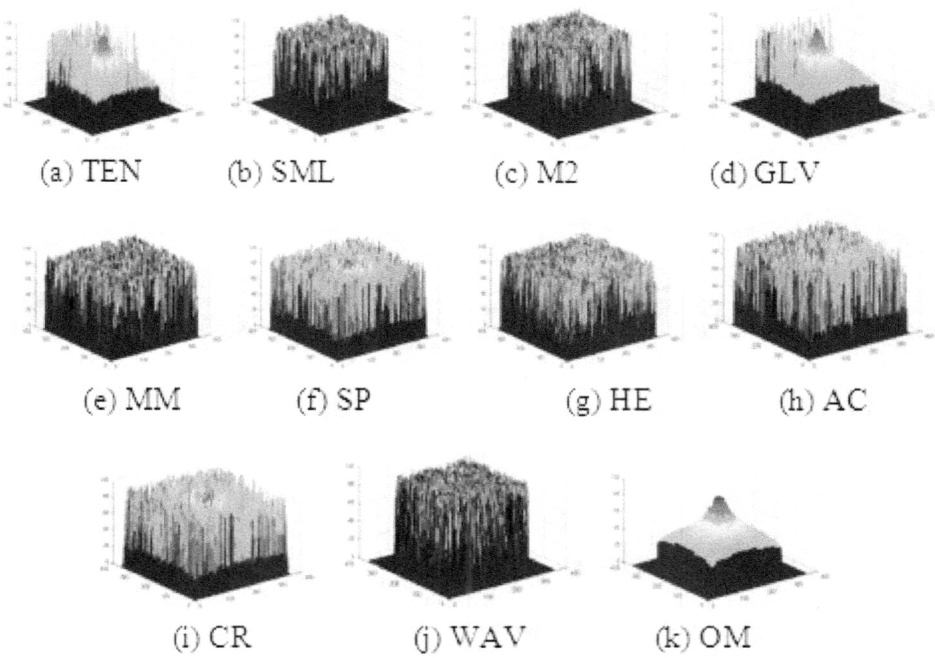

rate significantly while rest of the focus measures show comparable behavior.

Till now all the results shown are for Gaussian noise only. Now we consider two more types of noise, i.e., Shot noise and Speckle noise. We add the bipolar shot noise to the sequence of images of the planar object. Consider Figure 11 where Figure 11 (a) to (k) shows the depth maps for all eleven focus measures when the bipolar shot noise is added to the planar sequence of images. The noise density used is 0.0005. As can be seen from the images, the depth maps for SML, M2, MM, SP, HE, AC, CR and WAV are again degraded with spikes originating from the pixels all over the image hence making the shape of the planar object unrecognizable. However, it can also be seen from the depth maps that the results of TEN and GLV are better than the others except OM. OM shows exceptionally good results for the planar object.

Figure 9. Some frames for the real cone

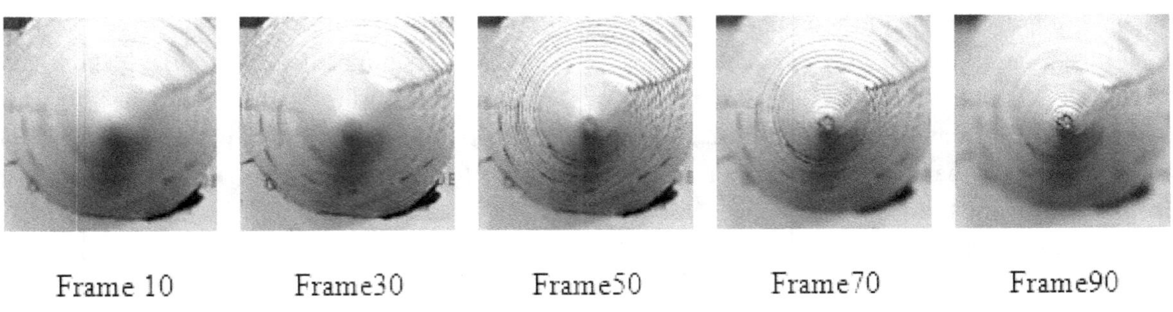

Figure 10. Depth maps for the real cone object when Gaussian noise is added

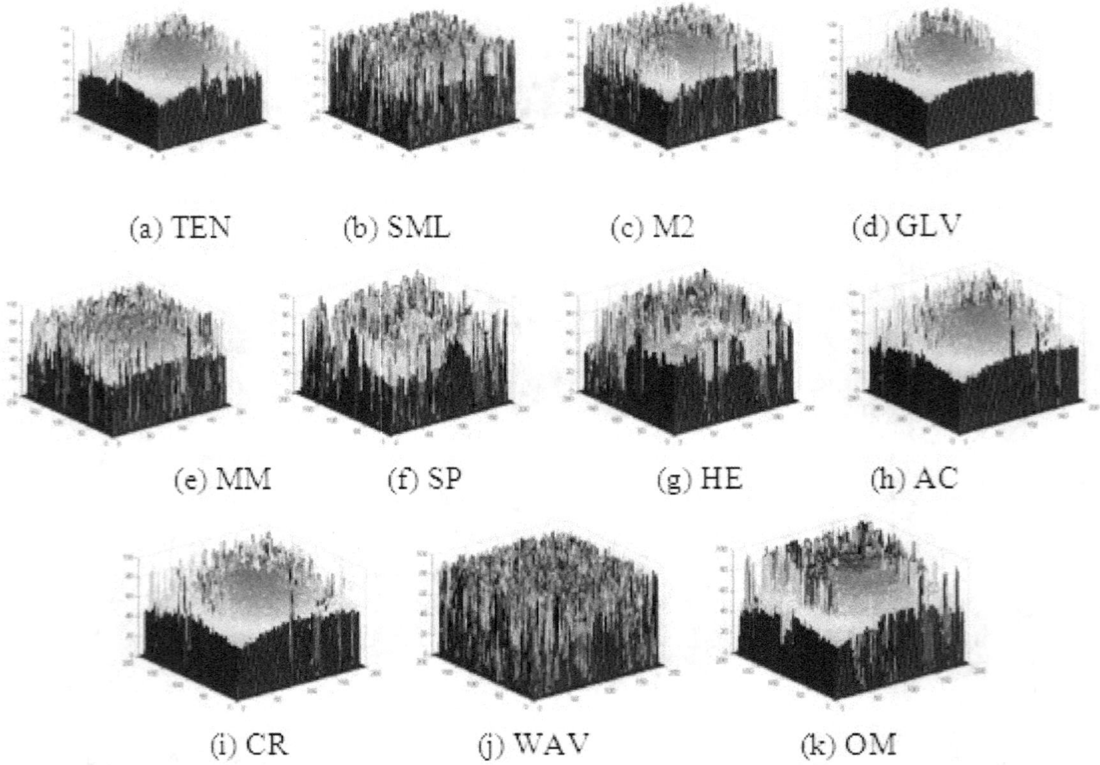

Similarly, Figure 12(a) to 12(k) shows the depth maps for all the eleven focus measures when the multiplicative speckle noise is added to the sequence of images of microscopic Lincoln head. The noise variance is 0.00005.

Quantitative Analysis

We used two metric measures to compare these focus measures. The metric measures used are Root Mean Square Error (RMSE) and Correlation. The reference image for comparison is the ground truth depth map for the simulated cone. For real objects like real cone etc such data was not available. Hence, we obtained more accurate depth maps (as compared to using the focus measure alone) by using the approximation technique (Fuzzy using Neural Network).

For simulated cone data, we have 97 numbers of steps corresponding to 97 images. Each step is of equal distance. On the other hand, if the depth maps contain the object distances from the camera, then the RMS error is obtained in term of object distances. The object distances from the lens can be easily computed using eq (1), if we know the depth map in terms of image number and the parameters of the camera. For example in Table 1, the RMS error for TEN is about 19 lens steps out of 97 steps. Same is true for other object sequences too.

For the simulated cone, Table 1 shows the comparison of all the focus measures in the absence of noise. Corresponding Figure 13 depicts the results for RMSE and correlation. It is seen from these tables that the performance of all focus measures is comparable with the exception of OM

Figure 11. Depth maps for the planar object when shot noise is added to the images

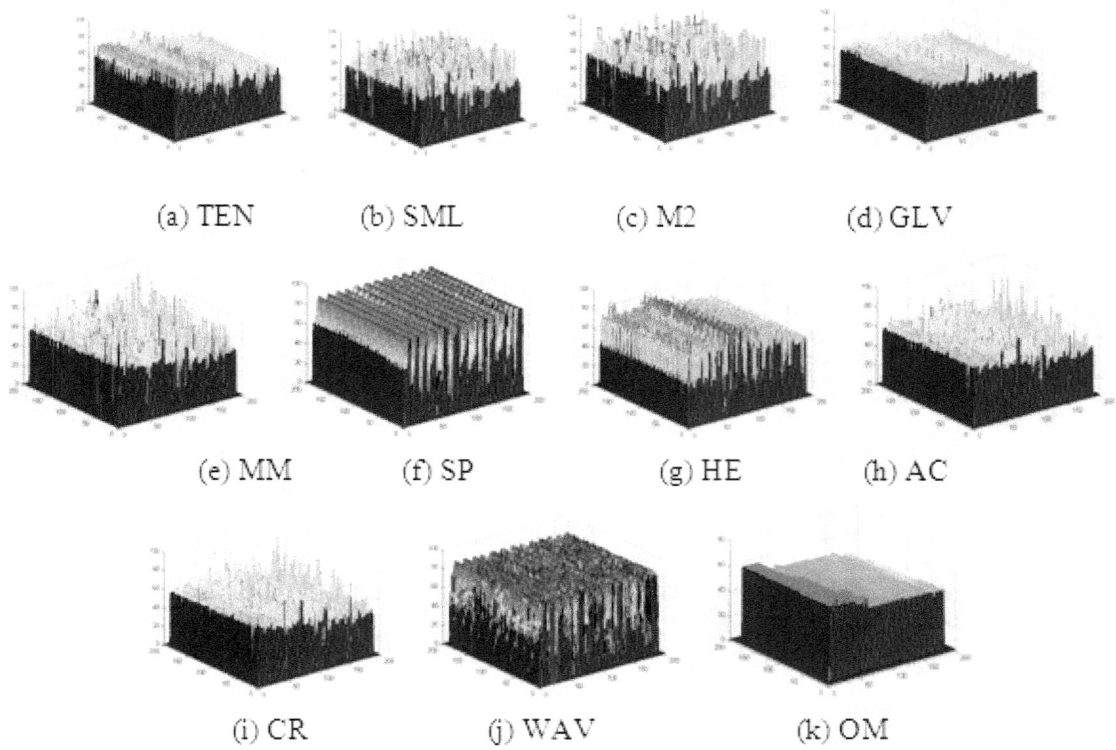

and WAV. OM performs better than the rest while WAV performs worst than the rest.

Figures 14 to 16 show the graphs for RMSE and correlation for all three types of noise. The results in Figure 14 are shown for data with Gaussian noise of zero mean and varying variance values, i.e., variance = 0.5, 0.05, 0.005, 0.0005, 0.00005. OM shows robustness to Gaussian

Table 1. Performance in the absence of noise (simulated cone)

Focus Measure	RMSE	Correlation
TEN	19.7017	0.8991
SML	19.6557	0.8999
M2	19.6535	0.9005
GLV	19.6816	0.8985
MM	19.5736	0.8988
SP	19.6916	0.8961
HE	19.5673	0.9001
AC	19.6768	0.8997
CR	19.6065	0.9009
WAV	20.891	0.8603
OM	14.2114	0.9119

Figure 12. Depth maps for microscopic object when speckle noise is added to the images

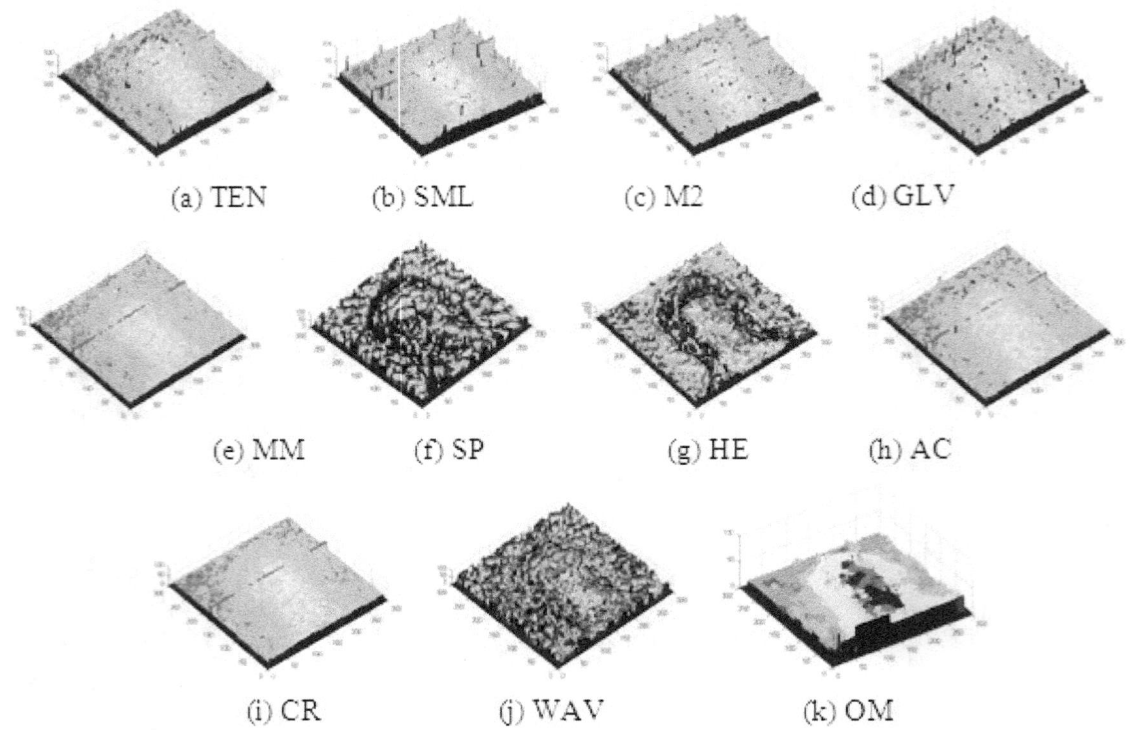

(a) TEN (b) SML (c) M2 (d) GLV

(e) MM (f) SP (g) HE (h) AC

(i) CR (j) WAV (k) OM

Figure 13. Comparison in absence of noise

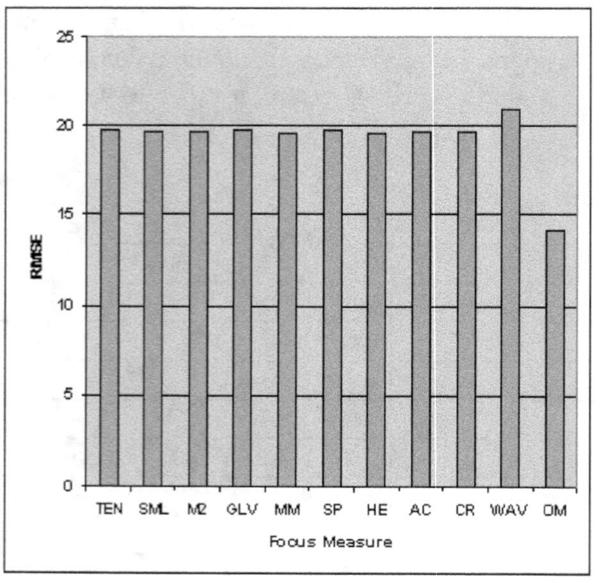

(a) RMSE

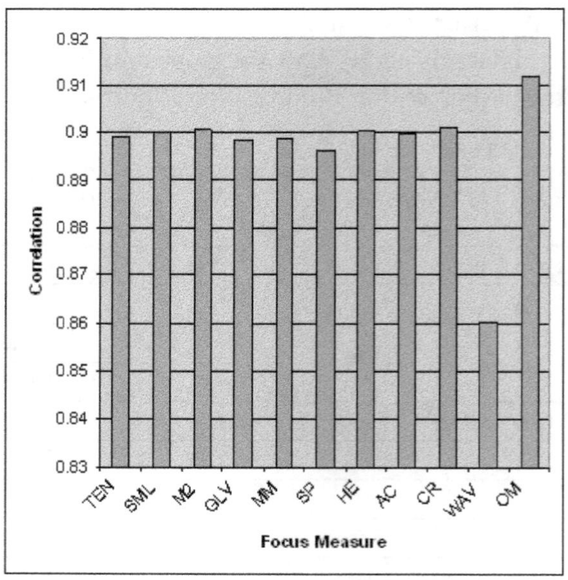

(b) Correlation

Figure 14. Comparison of focus measures (Gaussian Noise)

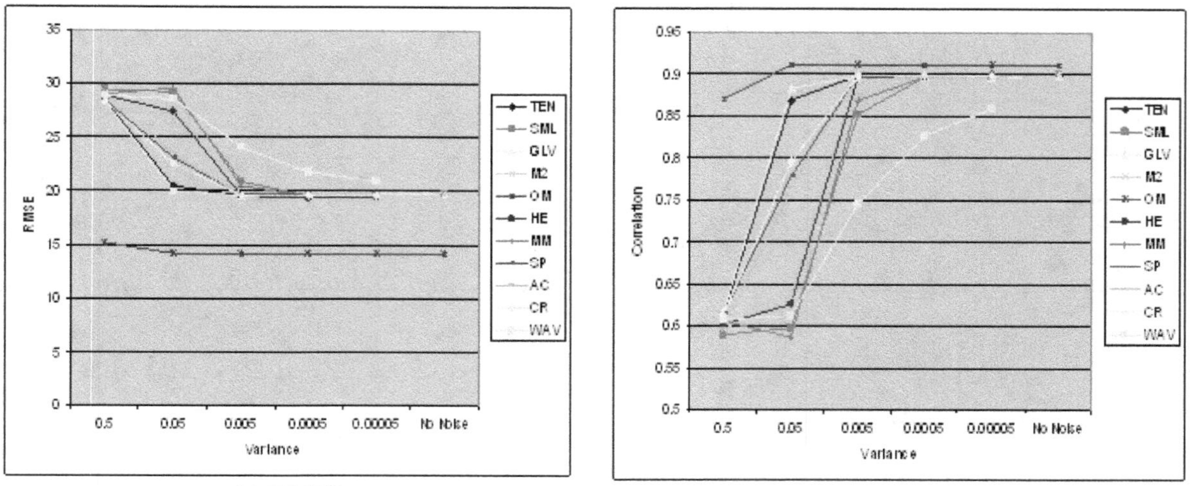

(a) RMSE (b) Correlation

noise and is not affected while GLV and TEN performance degrades for upper most noise level while their performance increases considerably for the other noise levels. The rest of the focus measures deteriorate at medium and high noise levels while showing comparable performance at low noise levels.

Now consider Figure 15 which shows the results of the eleven focus measures in the presence of Shot noise of various densities, i.e., values of noise density from 0.5 to 0.00005. OM and

Figure 15. Comparison of focus measures (Shot Noise)

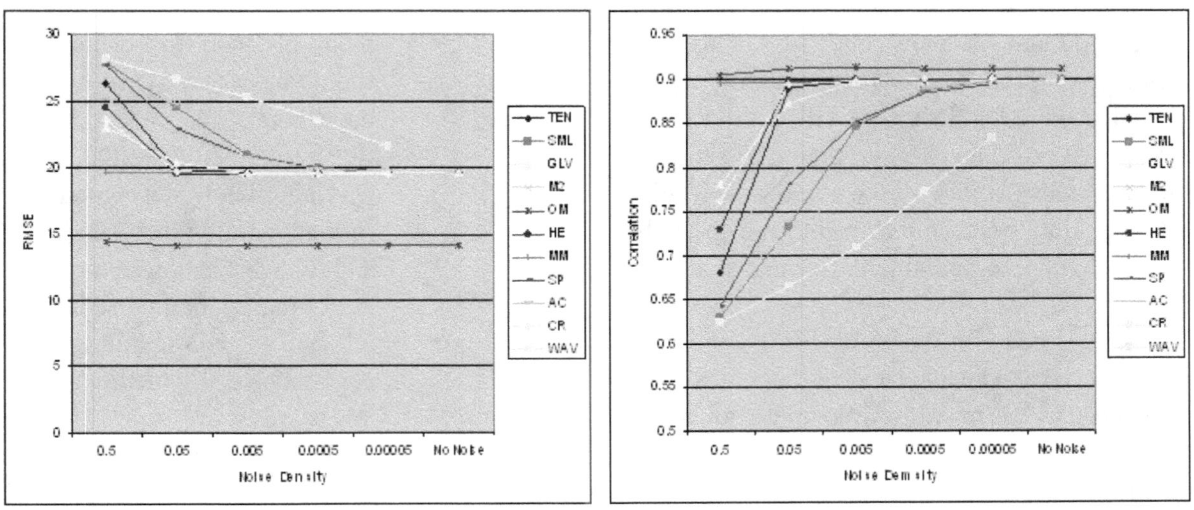

(a) RMSE (b) Correlation

Figure 16. Comparison of focus measures (Speckle Noise)

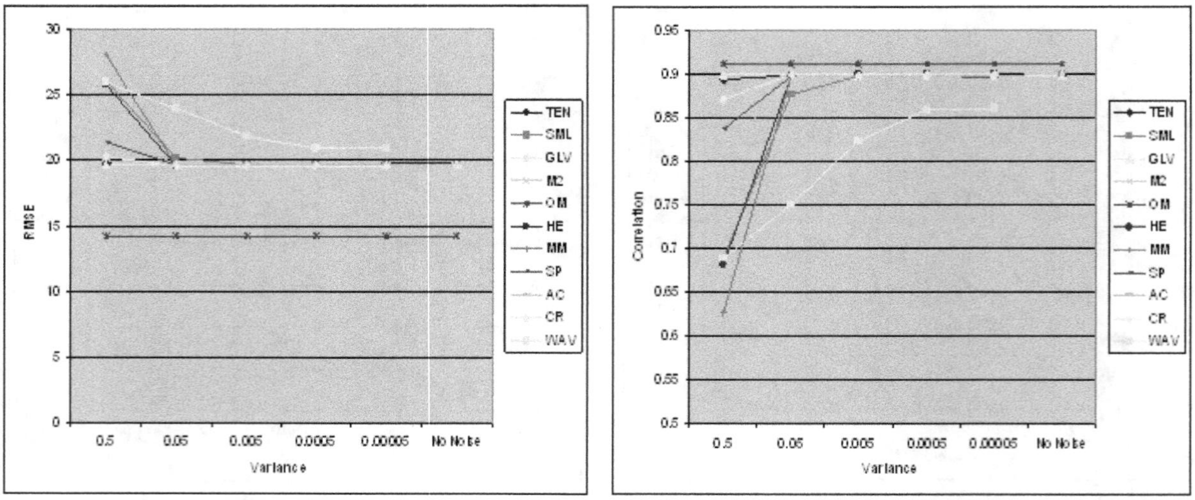

(a) RMSE (b) Correlation

MM show robustness to shot noise. Among rest of four focus measures, performance of GLV, CR, HE and TEN degrades for upper most noise level while their performance increases considerably for the other noise levels. The rest of the focus measures deteriorate at medium and high noise levels while showing comparable performance at low noise levels.

Now consider Figure 16 which shows the results of the eleven focus measures in the presence of Speckle noise with varying variances, i.e., 0.5, 0.05, 0.005, 0.0005, 0.00005. OM, TEN, GLV, CR, SP, M2 and AC show robustness to speckle noise. Among rest of four focus measures, performance of MM, SML and HE degrades for upper most noise level while their performance increases considerably for the other noise levels. WAV deteriorate at medium and high noise levels while showing comparable performance at low noise levels.

Till now we have shown the results for simulated cone. Hence, keeping in view the results of all the objects, we can make the following evaluation.

- Overall Performance:
  - Gaussian Noise:
    - OM shows good performance followed by GLV, CR, TEN and HE
    - AC, M2 and MM should be avoided at high and medium noise levels
    - WAV, SP and SML should be avoided
  - Shot Noise:
    - OM, TEN, CR and GLV show better performance at all noise levels
    - Rest of the focus measures should be avoided except at low noise level
  - Speckle Noise:
    - OM and TEN show good performance followed by GLV, CR and AC which show comparable performance at all noise levels
    - HE, M2 and MM exhibits better performance for medium noise levels
    - SML, WAV and SP should be avoided

NOISE PRE-FILTERING

We have used the following filters before applying focus measures:

1. **Wiener Filer is used for Gaussian noise.** It filters an intensity image that has been degraded by constant power additive noise. Since we already know that this additive noise is Gaussian noise, therefore, we use this information for implementing this filter for Gaussian noise.
2. **Median filter is used for Shot noise.** It is the most commonly used filter for shot noise and it's very effective.
3. **Frost filter (Frost, 1982) is used for Speckle noise.** The Frost filter is based on the multiplicative speckle model and the local statistics.

There is little improvement in the results of focus measures after the usage of Wiener and Frost filter at high noise levels. However, improvement can be observed at the medium and low noise levels. However, median filter improves the result remarkably. Most of the focus measures are not affected by the shot noise at medium and low noise levels. They are only affected at high noise level. The results of the focus measures improve significantly showing almost no affect of shot noise at medium and low noise levels.

TEXTURE REFLECTANCE AND ILLUMINATION

Reflectance is defined as a physical quantity that measures how well a surface reflects light or other electromagnetic radiations. In other words, reflectance is a measure of the percentage of light reflected from a surface (the rest is absorbed.). Reflectance is a dimensionless quantity, comprised in the range 0 to 1. A bright white surface will reflect almost all of the light falling on it, perhaps having a reflectance of over 99%, while a very deep black will reflect less than 1% of the incident light. A mid-gray surface reflects around 18% of the light falling on it - and thus absorbs 82%. The amount of light reflected by a surface element also depends on the material. For example, reflectance value for snow is 0.93 while that for stainless steel is 0.80.

Let g(x,y) be the captured image frame, $i_s(x,y)$ be the amount of source illumination incident on the object and r(x,y) be the reflectance. Then, g(x,y) is defined as a product of illumination and reflectance (Gonzalez, 2002):

$$g(x,y) = i_s(x,y)r(x,y)$$

where, $0 < i_s(x,y) < \infty$ and $0 < r(x,y) < 1$

Hence, by changing the source illumination and object material, the resultant image g(x,y) changes. Therefore, we selected the following for our experiments:

Objects: US Penny (copper plated), TFT-LCD Cell (transparent glass), V-Groove (reflective silicon surface), Micro-Sphere(transparent plastic)

Source Illumination: 50 W (1000 Lumens) halogen lamp is used. Illumination is controlled with various steps. We select 3 illumination levels, i.e., low (~20% of source illumination), medium (~50% of source illumination) and high (~100% of source illumination). Figure 17 shows frame# 30 at different illumination levels.

Table 2 shows the comparison of various focus measures. This comparison is on scale 1 to 3 where this value simply implies the order with 1 as highest order (best result) while 3 as the lowest order number (worst result). This numbering is relative based on RMSE, correlation and visual inspection of depth map. The 'f' indicates that the focus measure fails and 'o' means ONLY that focus measure showed any performance.

Medium Lamp Level: Overall, the best results are obtained at this level. TEN, M2, GLV, AC and CR focus measures did not fail for any object. Best performance is shown by OM followed by

Figure 17. Frame# 30 at different illumination levels

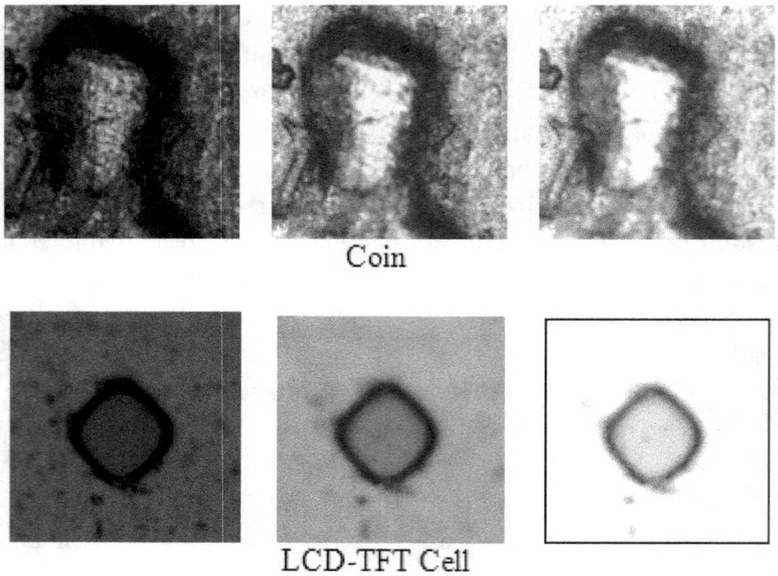

CR, AC and M2. However, OM fails for V-Groove at all lamp levels.

Low & High Lamp Levels: It was observed that focus measures generally fail at low or high lamp levels. At low lamp level, all focus measures fail for at least two objects with best performance shown by CR followed by OM, AC and M2. CR is also the focus measure that fails only for one object at high lamp level while rest fails for two or more objects.

One thing is clear from Table 2 that the performance of focus measures varies for varying textures as well as source illumination levels. No one focus measure performs satisfactorily

Table 2. Study of reflectance & illumination

Lamp	**Low Lamp Level**				**Medium Lamp Level**				**High Lamp Level**			
Object	**C**	**L**	**V**	**S**	**C**	**L**	**V**	**S**	**C**	**L**	**V**	**S**
TEN	2	2	f	f	3	2	2	2	f	2	2	f
SML	1	3	f	f	2	2	f	2	f	1	f	f
M2	1	2	f	f	3	2	2	1	f	2	2	f
GLV	3	2	f	f	3	2	2	2	f	2	1	f
MM	3	f	f	f	f	3	f	2	3	3	f	f
SP	f	f	o	f	f	f	3	1	f	f	1	o
HE	f	f	f	f	3	f	f	2	f	f	f	f
AC	1	2	f	f	3	2	2	1	f	2	1	f
CR	1	1	f	f	3	2	1	1	f	2	1	o
WAV	f	f	o	f	f	f	3	2	f	f	3	f
OM	2	1	f	f	1	1	f	1	3	1	3	f

C=Coin, L=LCD, V=V-Groove, S=Micro-Sphere

for all textures at varying illumination levels. Therefore, Table 2 is a guideline for selection of focus measures based on known texture properties (copper, transparent glass, reflecting silicon, transparent plastic) and known illumination level. Table 2 can be expanded to a very meaningful lookup table for the selection of focus measures based on object and environment properties by the addition of more materials with varying textures.

DISTANCE VARIATION

Within Frames

The performance of SFF algorithms depends on various parameters. One of them is the number of image frames acquired for depth estimation. Small number of frames implies larger distance in between the frames resulting in more error introduced due to non-continuous behavior. Large number of images means lesser distance in between the frames but more error due to the factors associated with the image acquisition equipment. Hence, we study the effects of this distance variation within the frames on the focus measures. Using the microscopic setup described earlier, we acquire the images for US Penny, TFT-LCD Cell, V-Groove and Micro-Sphere at a resolution of 300x300. Table 3 shows the three distance levels within the frames.

Table 3. Three distance level variations

S. No.	Number of Images Acquired	Distance in between Frames (μm)
1	45	0.72
2	75	0.43
3	105	0.3

Table 4 shows the comparison of focus measures on scale 1 to 3 where this value simply implies the performance level with 1 as highest order (best performance) while 3 as the lowest order number. This numbering is relative based on RMSE, correlation and visual inspection of depth map. In addition, 'f' indicates that the focus measure fails in that case and 'o' means only that focus measure showed any performance.

For small distance in between frames, the focus measures generally tend to fail for all objects.

Table 4. Study of distance variation within frames

Distance b/w Frames	Large Distance (45 Images)				Medium Distance (75 Images)				Small Distance (105 images)			
Object	**C**	**L**	**V**	**S**	**C**	**L**	**V**	**S**	**C**	**L**	**V**	**S**
TEN	2	1	f	f	2	1	2	2	f	2	f	f
SML	2	2	f	f	1	2	f	2	o	2	f	f
M2	2	1	f	f	1	1	3	3	f	2	f	f
GLV	2	1	f	o	1	1	2	2	f	2	f	f
MM	f	f	f	f	2	f	f	2	f	3	f	f
SP	f	3	o	o	f	f	f	1	f	f	f	o
HE	f	3	f	f	f	3	f	1	f	f	f	f
AC	2	1	f	f	1	1	2	2	f	1	f	f
CR	2	1	f	f	1	1	1	2	f	1	o	f
WAV	3	3	o	f	3	f	f	3	f	f	f	f
OM	1	1	f	f	1	1	f	2	f	1	o	f

C=Coin, L=LCD, V=V-Groove, S=Micro-Sphere

Table 5. Object placed at different positions

S.No.	Coin (μm)	TFT-LCD Cell (μm)	V-Groove (μm)
1	0	0	0
2	16	8	8
3	32	16	16
4	48	x	24

The vibration error of the microscopic system increase as more images are acquired within same range which results in decrease in distance within frames. SML, CR and OM show some performance for most number of objects at this level. Large distance in between frames results in smaller number of images being acquired. Hence, some information is lost in between frames. OM performs well at this level too. However, GLV, SP and WAV show some level of performance for most number of the objects. Almost all focus measures give some level of performance for medium distance level. Best performance is shown by CR followed by GLV, AC and OM.

Different Object Positions

Another parameter affecting the performance of focus measures is the distance of the object from the imaging device. The best results are obtained where the object is best focused. Hence, we study the effects of object distance variation from the imaging device. Figure 18 shows frame# 30 at different distances. Using the microscopic setup, we acquire the images for US Penny, TFT-LCD Cell and V-Groove at a resolution of 300x300. Table 5 shows the different object positions for each of the image sequences. For different objects, the distance variation is different because of the different sizes of the objects. In Table 5, 0 means the first position of the object. The object is moved away from the imaging device at a constant distance rate.

Figure 18. Frame# 30 when objects are placed at different distances

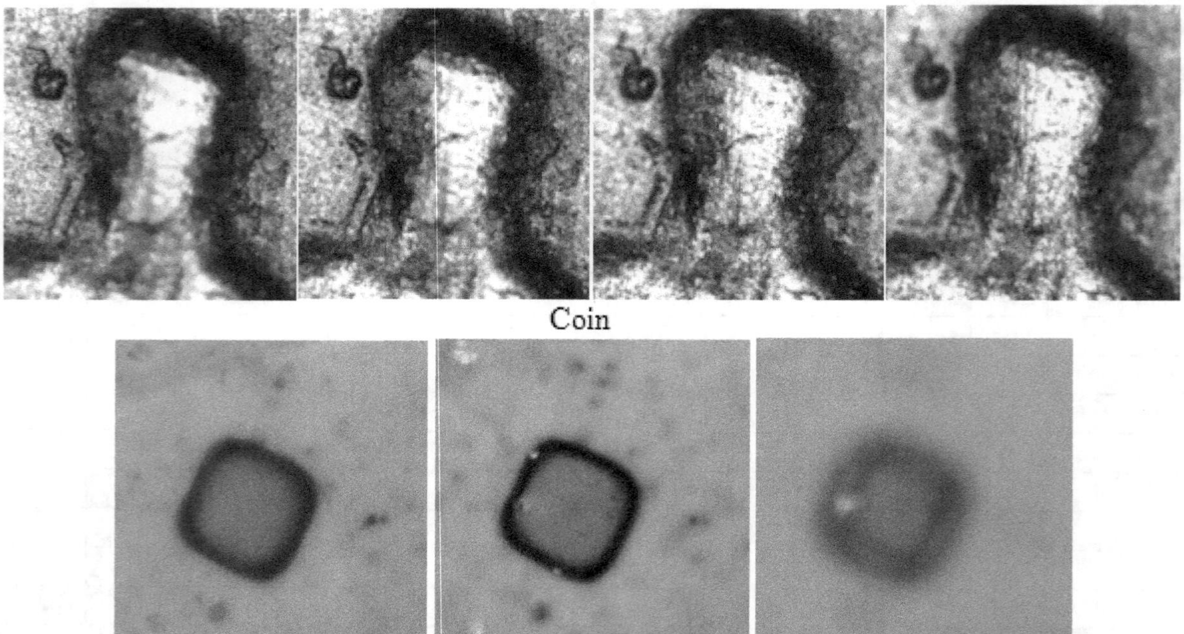

Table 6 shows the comparison of various focus measures. The comparison is on scale 1 to 3 where this value simply means 1 as highest order (best performance) while 3 as the lowest order number. The 'f' indicates that the focus measure fails in that case and 'o' means only that focus measure showed any performance.

At distance of 0 µm, objects are generally out of focus and then they cone in focus and finally they become defocused again at maximum distances as shown in Table 6. Almost all focus measures fail when the object is defocused. However, some level of performance is shown by SML, MM and OM. At distances where the object comes in focus, OM shows best performance for coin and TFT-LCD cell followed by SML and CR. For V-Groove, best performance is shown by GLV followed by CR and TEN. Most of the focus measures fail for this object as discussed in previous sections.

Table 6. Study of distance variation with reference to object position

(a) Coin				
Object Position	**0 µm**	**16 µm**	**32 µm**	**48 µm**
TEN	f	2	3	f
SML	f	1	2	o
M2	f	1	2	f
GLV	f	2	2	f
MM	o	3	3	o
SP	f	f	2	f
HE	f	f	f	f
AC	f	1	2	f
CR	f	1	2	f
WAV	f	f	f	f
OM	o	1	1	f

(b) TFT-LCD Cell			
Object Position	**0 µm**	**8 µm**	**16 µm**
TEN	3	2	f
SML	2	1	3
M2	3	2	f
GLV	3	2	f
MM	3	3	f
SP	3	f	f
HE	f	3	f
AC	3	2	f
CR	2	2	f
WAV	3	f	f
OM	1	1	2

(c) V-Groove				
Object Position	**0 µm**	**8 µm**	**16 µm**	**24 µm**
TEN	f	f	2	2
SML	f	f	f	f
M2	f	f	3	3
GLV	f	f	1	2
MM	o	f	f	f
SP	f	3	f	f
HE	f	f	f	f
AC	f	f	2	3
CR	f	f	2	1
WAV	f	3	f	f
OM	f	3	f	f

FUTURE RESEARCH DIRECTIONS

Focus measures are generally computed on individual 2D frames in the 3D space. In future, new focus measures need to be developed that can be applied directly in the 3D space and that can exploit the 3D nature of the object. Additionally, considerable time is required for the computation of focus measures for all the frames because the number of frames is large. Therefore, parallel processing and distributed processing algorithms need to be exploited for reducing the computational complexity of the focus measures algorithms.

CONCLUSION

In this chapter, focus measures are classified into four main groups, namely, derivative, statistics, energy and transform based. Total of eleven focus measures are selected with at least two from each of the group. We tested and compared these

Table 7. Lookup table for the selection of focus measure

Focus Measures		Factors effecting the Focus Measures: Noise				Texture Reflectance				Source Illumination			Distance Variation with respect to: Frames			Objects		
		General robustness to noise	Robustness to Gaussian noise	Robustness to Shot noise	Robustness to Speckle noise	Metal Alloys like Copper+Zinc	Transparent Glass Types	Reflective Silicon Type	Transparent Polymer Type	Bright	Average/Normal	Dim	Very Large	Average	Very Small	In-Focus	De-Focus	**Low Computational Complexity**
Derivative Based	TEN	X	X	X	X		X	X			X		X	X		X		X
	SML					X	X				X			X	X	X	X	X
	M2						X	X	X		X	X	X	X		X		X
Statistics Based	GLV		X	X			X	X		X	X		X	X		X		
	MM										X			X		X	X	X
Energy Based	SP								X		X			X		X		X
	HE		X								X			X		X		
Transform Based	AC						X	X	X	X	X	X	X	X		X		
	CR		X	X			X	X	X	X	X	X	X	X	X	X		
	WAV										X			X		X		
	OM	X	X	X	X	X	X		X	X	X	X	X	X	X	X	X	

focus measure using 'TEST' image (Table 7 and 8), simulated cone images, real cone images, slanted planar object images and three microscopic objects. The experiments were conducted for noise, pre-filtering the noisy images, effects of illumination and texture reflectance and effects of distance variation within the frames as well as with respect to various object positions. We used visual inspection of depth maps as a qualitative measure while RMSE and Correlation as quantitative metric measures to compare the performance of the focus measures. After extensive and wide base of experimentation, we found that OM, CR, GLV and TEN provide good results in most of the conditions. However, it was concluded that no one focus measure can be used for every condition. The performance of focus measures vary with change in various above-mentioned factors. Hence, we provide our recommendations in Table 7 that can be used as a lookup table for the selection of focus measures.

ACKNOWLEDGMENT

This work is supported by the E-Science grant funded by the Ministry of Science, Technology and Innovation (MOSTI), Government of Malaysia (No: 01-02-02-SF0064).

REFERENCES

Ahmad, M. B., & Choi, T.-S. (2005). A heuristic approach for finding best focused shape. *IEEE Transactions on Circuits and Systems for Video Technology*, *15*(4), 566–574. doi:10.1109/TCSVT.2005.844450

Asada, N., Fujiwara, H., & Matsuyama, T. (1998). Edge and depth from focus. *International Journal of Computer Vision, 26*(2), 153–163. doi:10.1023/A:1007996810301

Asif, M., & Choi, T.-S. (2001). Shape from focus using multilayer feedforward neural network. *IEEE Transactions on Image Processing, 10*(11), 1670–1675. doi:10.1109/83.967395

Baina, J., & Dublet, J. (1995). Automatic focus and iris control for video cameras. In *IEE Conference on Image Processing and its Application* (pp. 232-235).

Bove, V. M. (1993). Entropy based depth from focus. *Optical Society of America, 10*(4), 561–566. doi:10.1364/JOSAA.10.000561

Choi, K. S., Lee, J.-S., & Ko, S.-J. (1999). New autofocusing technique using the frequency selective weighted median filter for video cameras. *IEEE Transactions on Consumer Electronics, 45*(3).

Choi, T.-S., Asif, M., & Yun, J. (1999). Three-dimensional shape recovery from focused image surface. In *IEEE International Conference on Acoustics, Speech and Signal Processing*, vol. 6 (pp. 3269-3272).

Dillion, C., & Caelli, T. (1992). Generating complete depth maps in passive vision systems. *IEEE Pattern Recognition, 1*, 562–566.

Erteza, A. (1976). Sharpness index and its application to focus control. *Applied Optics, 15*(4), 877–881. doi:10.1364/AO.15.000877

Frost, V. S., Stiles, J. A., Schanmugan, K. S., & Holzman, J. C. (1982). A model for radar images and its application to adaptive digital filtering of multiplicative noise. *IEEE Transactions on Pattern Analysis and Machine Intelligence, 4*, 157–166. doi:10.1109/TPAMI.1982.4767223

Gillespie, J., & King, R. (1989). The use of self-entropy as a focus measure in digital holography. *Pattern Recognition Letters, 9*, 19–25. doi:10.1016/0167-8655(89)90024-X

Gonzalez, R. C., & Woods, R. E. (2002). *Digital image processing* (2nd ed.). Prentice Hall.

Helmli, F. S., & Scherer, S. (2001). *Adaptive shape from focus with an error estimation in light microscopy*. In 2nd International Symposium on Image and Signal Processing and Analysis (ISPA01).

Horn, B. K. P. (1968). *Focusing*. M.I.T., Project MAC, AI Memo. 160.

Jarvis, R. A. (1983). A perspective on range finding techniques for computer vision. *IEEE Transactions on Pattern Analysis and Machine Intelligence, 5*(2), 122–139. doi:10.1109/TPAMI.1983.4767365

Kautsky, J., Flusser, J., Zitova, B., & Simberova, S. (2001). A new wavelet based measure of image focus. *Pattern Recognition Letters, 23*, 1785–1794. doi:10.1016/S0167-8655(02)00152-6

Krotkov, E. (1987). Focusing. *International Journal of Computer Vision, 1*, 223–237. doi:10.1007/BF00127822

Ligthart, G., & Greon, F. C. A. (1982). A comparison of different autofocus algorithms. In *IEEE International Conference on Pattern Recognition* (pp. 597-600).

Malik, A. S., & Choi, T. S. (2007). Consideration of illumination effects and optimization of window size for accurate calculation of depth map for 3D shape recovery. *Pattern Recognition, 40*(1), 154–170. doi:10.1016/j.patcog.2006.05.032

Malik, A. S., & Choi, T. S. (2007). Application of passive techniques for three dimensional cameras. *IEEE Transactions on Consumer Electronics, 53*(2), 258–264. doi:10.1109/TCE.2007.381683

Malik, A. S., & Choi, T. S. (2008). A novel algorithm for estimation of depth map using image focus for 3D shape recovery in the presence of noise. *Pattern Recognition*, *41*(7), 2200–2225. doi:10.1016/j.patcog.2007.12.014

Mennucciy, A., & Soatto, S. (1999). On observing shape from defocused images. In *International Conference on Image Analysis and Processing* (pp. 550-555).

Milios, E., & Petrakis, E. (2000). Shape retrieval based on dynamic programming. *IEEE Transactions on Image Processing*, *9*(1), 141–147. doi:10.1109/83.817606

Mozerov, M., Kober, V., & Choi, T.-S. (2000). 3D stereo matching based on modified dynamic programming. *IEEE Pattern Recognition and Image Analysis*, *10*(1), 90–96.

Muller, R. A., & Buffington, A. (1974). Real time correction of atmospherically degraded telescope images through image sharpening. *Journal of the Optical Society of America*, *64*(9), 1200–1210. doi:10.1364/JOSA.64.001200

Nair, H. N., & Stewart, C. V. (1992). Robust focus ranging. In *IEEE Conference on Computer Vision and Pattern Recognition* (pp. 309-314).

Nayar, S., Noguchi, M., Watanabe, M., & Nakagawa, Y. (1996). Real time focus range sensors. *IEEE Transactions on Pattern Analysis and Machine Intelligence*, *18*(12), 1186–1198. doi:10.1109/34.546256

Nayar, S. K., & Nakagawa, Y. (1990). Shape from focus: An effective approach for rough surfaces. In *IEEE CRA9* (pp. 218-225).

Nayar, S. K., & Nakagawa, Y. (1994). Shape from focus. *IEEE Transactions on Pattern Analysis and Machine Intelligence*, *16*(8), 824–831. doi:10.1109/34.308479

Noguchi, M., & Nayar, S. (1994). Microscopic shape from focus using active illumination. *IEEE Pattern Recognition*, *1*, 147–152.

Pentland, A. P. (1987). A new sense for depth of field. *IEEE Transactions on Pattern Analysis and Machine Intelligence*, *9*(4), 523–531. doi:10.1109/TPAMI.1987.4767940

Poon, T.-C., & Banerjee, P. P. (2001). *Contemporary optical image processing* (1st ed.). New York, NY: Elsevier Science Ltd.

Pradeep, K. S., & Rajagopalan, A. N. (2006). Improving shape from focus using defocus information. In *Proceedings of the 18th International Conference on Pattern Recognition (ICPR 06)*, vol. 1, (pp. 731-734). Hong Kong.

Shen, C.-H., & Chen, H. H. (2006). Robust focus measure for low contrast images. In *IEEE International Conference on Consumer Electronics* (pp. 69-70).

Subbarao, M., & Choi, T.-S. (1995). Accurate recovery of 3D shape from image focus. *IEEE Transactions on Pattern Analysis and Machine Intelligence*, *17*(3), 266–274. doi:10.1109/34.368191

Subbarao, M., Choi, T.-S., & Nikzad, A. (1993). Focusing techniques. *Optical Engineering (Redondo Beach, Calif.)*, *32*(11), 2824–2836. doi:10.1117/12.147706

Subbarao, M., & Tyan, J. K. (1998). Selecting the optimal focus measure for autofocusing and depth from focus. *IEEE Transactions on Pattern Analysis and Machine Intelligence*, *20*(8), 864–870. doi:10.1109/34.709612

Takahashi, K., Kubota, A., & Naemura, T. (2004). A focus measure for light field rendering. In *IEEE International Conference on Image Processing* (pp. 2475-2478).

Tsai, D., & Chou, C. (2003). A fast focus measure for video display inspection. *Machine Vision and Applications*, *14*, 192–196.

Xie, H., Rong, W., & Sun, L. (2006). Wavelet-based focus measure and 3D surface reconstruction method for microscopy images. In *International Conference on Intelligent robots and Systems* (pp. 229-234). Beijing.

Xiong, Y., & Shafer, S. A. (1993). Depth from focusing and defocusing. In *IEEE Conference on Computer Vision and Pattern Recognition* (pp. 68-73).

Yang, G., & Nelson, B. J. (2003). Wavelet-based autofocusing and unsupervised segmentation of microscopic images. In *International Conference on Intelligent robots and Systems* (pp. 2143-2148), Las Vegas.

Yap, P. T., & Raveendran, P. (2004). Image focus measure based on Chebyshev moments. In *IEE ISP, 151*(2), 128-136.

Yun, J., & Choi, T.-S. (1999). Accurate 3-D shape recovery using curved window focus measure. In *IEEE International Conference on Image Processing,* vol. 3 (pp. 910-914).

Zhang, Y., Zhang, Y., & Wen, C. (2000). A new focus measure method using moments. *Image and Vision Computing, 18*, 959–965. doi:10.1016/S0262-8856(00)00038-X

ADDITIONAL READING

Malik, A. S. (2010). Selection of Window Size for Focus Measure Processing. In *2010 IEEE International Conference on Imaging Systems and Techniques (pp. 426-430),* Thessaloniki, Greece.

Malik, A. S., & Choi, T. S. (2009). Comparison of Polymers: A New Application of Shape From Focus. *IEEE Transactions on Systems, Man and Cybernetics. Part C, Applications and Reviews, 39*(2), 246–250. doi:10.1109/TSMCC.2008.2001714

Malik, A. S., & Choi, T. S. (2009). Analysis of Effects of Texture Reflectance and Source Illumination on Focus Measures for Microscopic Images. In *2009 International Conference on Computer and Electrical Engineering: Vol. II (pp. 529-532),* Dubai, UAE.

Malik, A. S., & Choi, T. S. (in press). Depth Map Estimation based on Linear Regression using Image Focus. *International Journal of Imaging Systems and Technology.*

Malik, A. S., Nisar, H., & Choi, T.-S. (2011). A Fuzzy-Neural Approach for Estimation of Depth Map using Focus. *Applied Soft Computing, 11*(2), 1837–1850. doi:10.1016/j.asoc.2010.05.030

Shim, S.-O., Malik, A. S., & Choi, T.-S. (2009). Accurate Shape from Focus based on Focus Adjustment in Optical Microscopy. *Microscopy Research and Technique, 72*(5), 362–370. doi:10.1002/jemt.20662

Shim, S.-O., Malik, A. S., & Choi, T.-S. (2010). Pre-Processing for noise reduction in depth estimation. In *2nd International Conference on Digital Image Processing, SPIE: Vol. 7546 (pp. 754625-2~7),* Singapore.

Shim, S.-O., Malik, A. S., & Choi, T.-S. (in press). Noise Reduction using Mean Shift Algorithm for Estimating 3D Shape. *Imaging Science Journal.*

KEY TERMS AND DEFINITIONS

3D Shape Recovery: To completely reconstruct the 3D shape of an object in x, y and z-planes.

All-in-Focus Image: An all-in-focus image consist of the sharpest pixels values corresponding to best focus.

Depth Map: A Depth map represents the 3D information of an object in the z-plane. Focus measure: An operator that measures the sharpness of the pixel values, i.e. the focus quality of the image.

Metric Measures: These are the quality measures like RMSE, correlation etc that are used to assess the 3D reconstruction of the object.

Shape from Focus: This is one of the shape from X methods based on the focus settings to capture images and then reconstruct the 3D shape.

Shape From X: These are the optical passive methods to recover the 3D shape. The X represents focus, defocus, texture, motion, stereo, shading.

Chapter 10
Image Focus Measure Based on Energy of High Frequency Components in S-Transform

Muhammad Tariq Mahmood
Korea University of Technology and Education, Korea

Tae-Sun Choi
Gwangju Institute of Science and Technology, Korea

ABSTRACT

Focus measure computes sharpness or high frequency contents in an image. It plays an important role in many image processing and computer vision applications such as shape from focus (SFF) techniques and multi-focus image fusion algorithms. In this chapter, we discuss different focus measures in spatial as well as in the transform domains. In addition, we suggest a novel focus measure in S-transform domain, which is based on the energy of high frequency components. A localized spectrum, by using variable window size, provides a more accurate method of measuring image sharpness as compared to other focus measures proposed in spectral domains. An optimal focus measure is obtained by selecting an appropriate frequency dependent window width. The performance of the proposed focus measure is compared with those of existing focus measures in terms of three dimensional shape recovery and all-in-focus image generation. Experimental results demonstrate the effectiveness of the proposed focus measure.

INTRODUCTION

Imaging devices, particularly those with lenses of long focal lengths, usually suffer from limited depth-of-field. Therefore, in the acquired images, some parts of the object are well-focused while the other parts are defocused with a degree of blur. Usually, a focus measure is used to compute the image focus quality that plays an important role in many image processing and computer vision applications. For example, the performance of the shape from focus (SFF) techniques and multi-focus fusion algorithms depend on accurate focus

DOI: 10.4018/978-1-61350-326-3.ch010

measurements (Ahmad & Choi, 2005; Simonov & Rombach, 2009; Wang, Ma, & Gu, 2010). In multi-focus fusion, an all-in-focus image is obtained from two or more blurred images. In the stack of images, the well-focused regions are distinguished from the blurred regions by commuting the focus quality. The fused image is obtained by combining the well-focused regions. On the other hand, SFF is a passive optical technique, which uses focus as a cue for depth estimating. In this technique, a sequence of images is acquired at different focus levels by translating object along the optical axis. First, focus value is computed for each pixel in the image sequence by using a focus measure. The best-focused pixels among the sequence provide the depth information. A rough depth map is then obtained by maximizing the focus measure along the optical axis. In the second step, an approximation technique is used to refine the initial depth map (Ahmad & Choi, 2005; Simonov & Rombach, 2009; Wang, et al., 2010). These depth estimation techniques generally rely on the accuracy of the focus measure. Thus, in SFF techniques, a robust and accurate focus measure is of fundamental importance. In addition, many focus measure based techniques have been successfully utilized in many industrial applications. Using the variations of the focus, it is possible to measure the surface roughness and metrology(Kyte, 2010; Malik & Choi, 2009). It can also be employed in surface characterization, evaluation of tolerances and wear analysis in 3D, accurate 3D measurement of micro-gear wheels (Kyte, 2010).

Focus measure computes sharpness or high frequency contents in an image. Acquired images through the camera aperture are result of convolution of actual image and low pass filter i.e. point spread function (PSF). Therefore, ideally, a focus measure is a high pass filter that should response to the high frequency contents in an image. In the literature, many focus measures have been reported in spatial and frequency domains. In spatial domain, derivative and statistical analysis of image intensities commonly used to compute the sharpness (Krotkov, 1988; Malik & Choi, 2007; Nayar & Nakagawa, 1994b). In frequency domains, focus measures usually compute total energy of high frequency components. Some focus measures in transform domain calculate the ratio of the high frequency components to the low frequency components (Kautsky, Flusser, Zitov, & Simberov, 2002; Sang-Yong, Kumar, Ji-Man, Sang-Won, & Soo-Won, 2008; Xie, Rong, & Sun, 2007). The studies of these focus measures have revealed that frequency components of different energies affect the focus measurement. For example, in discrete wavelet transform (DWT) based focus measures, high frequency components at the second level have a higher effect on image sharpness (Mahmood, Shim, & Choi, 2009). Similarly, in discrete cosine transform (DCT) based focus measures, frequency components in the middle are of greater interest regarding focus measurement (Mahmood, et al., 2009). The use of optical transfer function (OTF) i.e. low pass filter in frequency domain increases the robustness of the focus measure (Malik & Choi, 2008). In other words, the quality of focus measurements depends upon the frequency spectrum of the image.

Due to a variable window size, a recently proposed S-transform (ST) has certain advantages over DWT and other time-frequency analysis tools, and it has gained considerable attention in signal and image processing (Brown, Zhu, & Mitchell, 2005; Stockwell, Mansinha, & Lowe, 1996). In this chapter, we suggest the use of ST, with modified window width scheme, to compute the image focus. The window width affects the energy of the transformed components. In the proposed method, the window width depends on the variation in frequency along with two adjustable parameters. The optimal values of these adjustable parameters are chosen in such a way that the energy concentration be maximized. The energy of localized spectrum is taken as a criterion to compute the focus quality. Experimental results demonstrate the effectiveness of the proposed method.

Figure 1. Basic image formation in convex lens

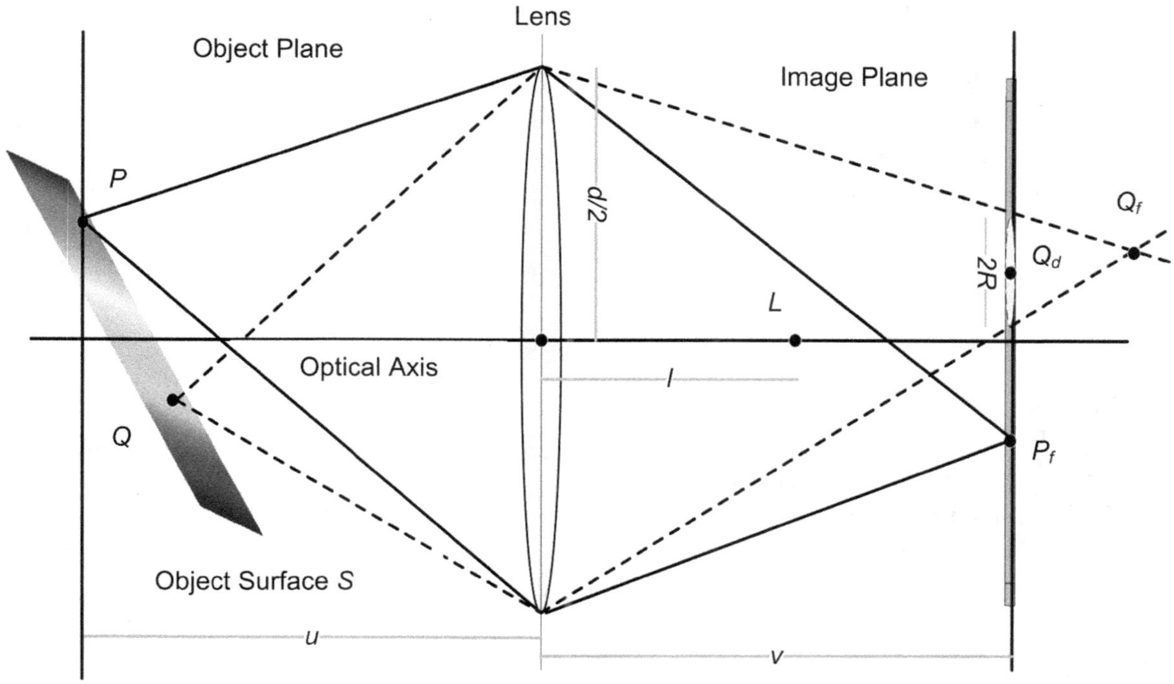

In the remaining chapter, we start with the formation of image in convex lens and define focus measure. Different focus measures in spatial and frequency domains are explained. Then, details about the focus measure in ST are given. Later sections of this chapter explain experimental setup and comparative analysis.

BACKGROUND

Focus Measure

In order to illustrate the theoretical aspects of the focus measure, we consider the paraxial geometric optics model that is circularly symmetric around the optical axis (Brown, Lauzon, & Frayne, 2010; Sejdic, Djurovic, & Jiang, 2008; Stockwell, et al., 1996). Figure 1 shows the basic geometry of image formation through a thin convex lens. Let P and Q are two points on a visible surface of an object and all light rays, which are radiated from these points of the object, are intercepted by the lens and converged at the image detector. A point P has its well-focused image P_f on the image plane while a defocused image Q_d of the point Q is obtained on the image detector. Well-focused points satisfy the lens law:

$$\frac{1}{f} = \frac{1}{u} + \frac{1}{v} \qquad (1)$$

where f is the focal length of the lens, u is the distance of the object point P from the lens, and v is the distance between lens and image plane. At any other distance $u' \neq u$ of the object point from the lens will not be well focused on the image plane. According to geometric optics, the aperture defines the shape of the blurred image of the point P. Since, we have considered the aperture to be circular, the blurred image is also a circle of radius R with uniform brightness inside the circle and zero outside it. In practice, the image of a point is not a crisp circular patch with constant brightness. Due to diffraction, polychromatic illumination, lens aberrations etc., it will be a circular blob with the brightness falling of gradually at the border. Therefore, this blurring effect is usually modeled by two-dimensional Gaussian function, which is defined by:

$$h(x,y)=\frac{1}{2\pi r^2}\exp\left(-\frac{x^2+y^2}{2r^2}\right) \tag{2}$$

This function is also known as Point Spread Function (PSF) where, r is a spread parameter corresponding to the standard deviation of the distribution of the PSF. Hence, a sensed image is the convolution of the actual image and a Gaussian function, i.e.

$$I(x,y) = I_a(x,y)*h(x,y) \tag{3}$$

where $I(x,y)$ is the sensed image, $I_a(x,y)$ is the actual image, * is the convolution operator and $h(x,y)$ is the PSF. The radius of the blurred circle R and the width of the Gaussian function r are related by $r = cR$. Where c is a constant and it can be approximated through camera calibration (Horn, 1986). Further, the Optical Transfer Function (OTF) is the corresponding PSF in frequency domain. Convolution in the spatial domain corresponds to the multiplication in the Fourier domain, the OTF is written as

$$G(u,v) = H(u,v).G_a(u,v) \tag{4}$$

where $G(u,v)$, $H(u,v)$, and $G_a(u,v)$ are the Fourier transformations of the functions $I(x,y)$, $h(x,y)$, and $I_a(x,y)$ respectively. The OTF can be expressed as

$$H(u,v)=\exp\left(-\frac{u^2+v^2}{2}r^2\right) \tag{5}$$

It is notable that OTF exhibits the characteristics of a low pass filter as low frequencies are passed unattenuated, while higher frequencies are reduced in magnitude. Therefore, a focus measure will be a high pass filter that is capable of computing high frequency components effectively in the image. The focus measure increases with the increase of focus quality and it attains maximum value at well-focused frame number. Hence, a well-focused image will have larger amount of high frequency contents as compared to de-focused image of the same scene.

Focus Measures in Spatial Domain

The focus measures in spatial domain are usually computed locally. Let $U(x,y)$ be the neighborhood of size $d \times d$ of a point (x,y) in an image $I(x,y)$. It is defined as,

$$U(x,y)=\left\{(\xi,\eta)\,|\,|\xi-x|\le d \wedge |\eta-y|\le d\right\} \tag{6}$$

One of the famous categories of focus measures in spatial domain is based on image derivatives. These focus measures are based on the idea that the larger difference in intensity values of neighboring pixels analogous to the sharper edges. Broadly, they can be divided into two sub-categories: first and second derivative based methods. A method based on gradient energy is investigated by (Pentland, 1987; Subbarao, Choi, & Nikzad, 1993a) that uses the Sobel operators to estimate the gradient of the image. The focus measure is defined as,

$$F_{TEN}(x,y)=\sum_{(\xi,\eta)\in U(x,y)}\left(\left(G_x*I(\xi,\eta)\right)^2+\left(G_y*I(\xi,\eta)\right)^2\right) \tag{7}$$

Where G_x and G_y are Sobel operators in x and y directions and can be written in kernel form as:

$$G_x = \begin{pmatrix} -1 & 0 & 1 \\ -2 & 0 & 2 \\ -1 & 0 & 1 \end{pmatrix}, \quad G_y = \begin{pmatrix} 1 & 2 & 1 \\ 0 & 0 & 0 \\ -1 & -2 & -1 \end{pmatrix}$$

Some variants of gradient based methods like threshold absolute gradient (TAG), squared gradient (SG), and Brenner gradient (BG) (Tenenbaum, 1970) have also been studied. Second order derivative operator (Laplacian), being a point and symmetric operator, is suitable for measuring image sharpness. It has capability to suppress the lower frequencies more strongly than a first order derivative. The Laplacian of an image in defined as

$$\nabla^2 I = \frac{\partial^2 I(x,y)}{\partial x^2} + \frac{\partial^2 I(x,y)}{\partial y^2} \tag{8}$$

The partial derivatives of an image in horizontal and vertical directions are approximated as;

$$\begin{aligned} \frac{\partial^2 I(x,y)}{\partial x^2} &= I(x+1,y) + I(x-1,y) - 2I(x,y) \\ \frac{\partial^2 I(x,y)}{\partial y^2} &= I(x,y-1) + I(x,y-1) - 2I(x,y) \end{aligned} \tag{9}$$

Several focus measures have been proposed by modifying the Laplacian operator (Brenner et al., 1976; Santos et al., 1997; Tian, Shieh, & Wildsoet, 2007). Among these, sum modified Laplacian (SML) focus measure based on second derivative has gain considerable attention (Helmli & Scherer, 2001; Subbarao, Choi, & Nikzad, 1993b; Subbarao & Tyan, 1998; Thelen, Frey, Hirsch, & Hering, 2009). In this focus measure, first, an image is convolved with Laplacian operator. The x and y components may have opposite sign and can yield zero response by canceling the effect of each other. To overcome this limitation, it is modified by taking the energy of the Laplacian. In order to improve robustness for weak textured image, the resultant values are summed up within a small window. The focus value for each pixel is thus computed as;

$$F_{SML}(x,y) = \sum_{(\xi,\eta)\in U(x,y)} \begin{bmatrix} \left|I(\xi+1,\eta) + I(\xi-1,\eta) - 2I(\xi,\eta)\right| + \\ \left|I(\xi,\eta+1) + I(\xi,\eta-1) - 2I(\xi,\eta)\right| \end{bmatrix} \tag{10}$$

The second derivative based focus measures provide more accurate results as compared to the first derivative based methods. However, these methods are more sensitive to noise.

Many focus measures have been reported based on the statistical analysis of image intensities (Nayar & Nakagawa, 1994a). Among these, the gray level variance (GLV) is the most famous. The larger variance of intensity values within a small window corresponds to the sharper image and vice versa. The focus value for the central pixel of a small neighborhood is computed by calculating the variance of intensity values as:

$$F_{GLV}(x,y) = \frac{1}{d^2} \sum_{(\xi,\eta)\in U(x,y)} \left[I(\xi,\eta) - \mu\right]^2 \tag{11}$$

where μ is the mean gray level value within the window of size $d \times d$.

$$\mu = \frac{1}{d^2} \sum_{(\xi,\eta)\in U(x,y)} I(\xi,\eta) \tag{12}$$

Recently, (Groen, Young, & Ligthart, 1985; Mahmood, Choi, & Choi, 2008b; Mahmood, et al., 2009; Wee & Paramesran, 2007; Yap & Raveendran, 2004; Zhang, Zhang, & Wen, 2000) suggested a focus measure by applying a robust band-pass filter defined in the frequency domain and based on bipolar incoherent image processing. The squared values of the responses of real in the spatial domain are then summed over a neighborhood.

$$F_{OTF} = \sum_{(\xi,\eta)\in U(x,y)} \tilde{I}\left(\xi,\eta\right) \tag{13}$$

where $\tilde{I}\left(\xi,\eta\right)$ is the response of the filtered image in frequency domain and represented as

$$\tilde{I}\left(x,y\right) = \mathrm{Re}\left[\Gamma^{-1}\left\{\Gamma\left(\left|I\left(x,y\right)\right|^{2}\right)h_{OTF}\left(x,y\right)\right\}\right] \tag{14}$$

where Γ is Fourier transform, Γ^{-1} is its inverse Fourier transform, $h_{OTF}(x,y)$ is the optical transform function and Re indicates the real part of the transformed components.

Focus Measures in Discrete Cosine Transform

The DCT of a signal is a real valued transform, which represents data in the frequency domain. For a given image $I(x,y)$ of size $N \times N$, two dimensional DCT coefficients $C(u,v)$ can be computed by with several variants, the following is one of the most commonly used (Malik & Choi, 2008).

$$F(u,v) = \omega(u)\omega(v)\sum_{x=0}^{N-1}\sum_{y=0}^{N-1} I(x,y)\cos\left[\frac{2x+1}{2N}\pi u\right]\cos\left[\frac{2y+1}{2N}\pi v\right], \quad u,v = 0,1,2,\ldots,N-1 \tag{15}$$

where

$$\omega(u) = \omega(v) = \begin{cases} \sqrt{1/N} & for \quad u,v=0 \\ \sqrt{2/N} & for \quad u,v \neq 0 \end{cases} \tag{16}$$

For $(u,v=0)$, the coefficient $F(0,0)$ represents average of the data and it is known as DC part. The values for $C(u \neq 0, v \neq 0)$ are known AC components. Most variations in the data are collected by AC part. In literature, some focus measures have also been proposed in the DCT domain by using the energy of its coefficients. (Ahmed, Natarajan, & Rao, 1974) proposed the energy of the AC part of DCT as a focus measure.

$$F_{DCT1} = \sum_{u=0}^{N-1}\sum_{v=0}^{N-1} F\left(u,v\right)^{2}, \quad (u,v) \neq (0,0) \tag{17}$$

It is observed that the mid frequency components have more influence on the focus quality as compared to the very low and very high frequency components (Baina & Dublet, 1995). The focus measure is thus computed by square of the convolution with 4 × 4 image block **B** and a DCT operator.

$$F_{DCT2} = \sum_{x}\sum_{y}\left[\mathbf{B}_{x,y} * \mathbf{O}_{DCT}\right]^{2} \tag{18}$$

where

$$\mathbf{O}_{DCT} = \begin{pmatrix} 1 & 1 & -1 & -1 \\ 1 & 1 & -1 & -1 \\ -1 & -1 & 1 & 1 \\ -1 & -1 & 1 & 1 \end{pmatrix} \tag{19}$$

For the images taken with the low illumination, the ratio between energies of AC and DC parts of the DCT is a better choice for measuring the focus quality (Sang-Yong, et al., 2008).

$$F_{DCT3} = \frac{E_{AC}}{E_{DC}} \tag{20}$$

In this instance, E_{AC} and E_{DC} are the energies of the AC and DC parts of DCT of an image block. Strengthening the idea of selective frequency component for image focus, (Shen & Chen, 2006) proposed another focus measure by using Bayes spectral entropy function. It is defined as

$$F_{DCT4} = 1 - \sum_{u=0}^{N-1} \sum_{v=0}^{N-1} \tilde{F}(u,v)^2 \quad (21)$$

where $\tilde{F}(u,v)$ stands for normalized DCT coefficients.

Focus Measures in Discrete Wavelet Transform

DWT is the decomposition of the signal into approximations and details coefficients obtained by expanding the signal in terms of the scaling function and basis function. Two-dimensional discrete wavelet transform of an image *I(x,y)* of size *N* × *M* is give as:

$$W_{\phi}(j_0, m, n) = \frac{1}{\sqrt{MN}} \sum_{x=0}^{M-1} \sum_{y=0}^{N-1} I(x,y) \varphi_{j_0,m,n}(x,y) \quad (22)$$

$$W_{\psi}^{i}(j, m, n) = \frac{1}{\sqrt{MN}} \sum_{x=0}^{M-1} \sum_{y=0}^{N-1} I(x,y) \psi_{j,m,n}^{i}(x,y) \quad i = \{H, V, D\} \quad (23)$$

where $W_{\phi}(j_0, m, n)$ and $W_{\psi}^{i}(j, m, n)$ represent approximation and detailed coefficients respectively. The approximation coefficients are computed by inner products of original image with scaling function $\varphi_{j_0,m,n}(x,y)$ whereas, detailed or wavelet coefficients are obtained through the basis functions $\psi_{j,m,n}^{i}(x,y)$. These scale and basic functions are defined as

$$\varphi_{j_0,m,n}(x,y) = 2^{j/2} \varphi(2^j x - m, 2^j y - n) \quad (24)$$

$$\psi_{j,m,n}^{i}(x,y) = 2^{j/2} \psi^{i}(2^j x - m, 2^j y - n) \quad i = \{H, V, D\} \quad (25)$$

where index *i* is a superscripts and indicates the directional wavelets i.e., horizontal, vertical, and diagonal wavelets. Approximations components represent low pass bands or smoothness whereas detailed components contain high frequency contents in the image.

In wavelet domain, first, (Kristan, Pers, Perse, & Kovacic, 2006) proposed a focus measure based on energy of wavelet coefficients. It is the ratio of Euclidean norms of high pass bands and low pass bands.

$$F_{DWT1} = \frac{\|h_w(I)\|}{\|l_w(I)\|} \quad (26)$$

Where $h_w(I)$ and $l_w(I)$ are total number of coefficients in high pass bands and low pass bands respectively, of the input image *I*. The $\|\cdot\|$ operator is Euclidean norm. (Kautsky, et al., 2002) proposed two focus measures in wavelet domain. These focus measures are very similar to the first and second order moments of the high frequency components. Considering a small window *U(x,y)* of size *d* × *d*, Daubechies orthogonal wavelets with basis 6 are applied and only the first level of decomposition is considered.

$$F_{DWT2} = \frac{1}{d^2} \left(\sum_{(x,y)\in U_H} \left| W_{\psi}^{H}(x,y) \right| + \sum_{(x,y)\in U_V} \left| W_{\psi}^{V}(x,y) \right| + \sum_{(x,y)\in U_D} \left| W_{\psi}^{D}(x,y) \right| \right) \quad (27)$$

$$F_{DWT3} = \frac{1}{d^2} \left(\sum_{(x,y)\in U_H} \left(W_{\psi}^{H}(x,y) - \mu^{H} \right)^2 + \sum_{(x,y)\in U_V} \left(W_{\psi}^{V}(x,y) - \mu^{V} \right)^2 + \sum_{(x,y)\in U_D} \left(W_{\psi}^{D}(x,y) - \mu^{D} \right)^2 \right) \quad (28)$$

where μ^H, μ^V, and μ^D are mean of horizontal, vertical, and diagonal detailed coefficients respectively. In other words, the focus values are determined by calculating L1-norm and L2-norm divided by total no of coefficients, of high frequency components. (Ge & Nelson, 2003) concluded that the detailed components, obtained at the second level, have stronger discriminating properties with respect to focus quality. Hence, the absolute sum of high frequency wavelet coefficients is taken as a focus measure.

$$F_{DWT4} = \sum_{(x,y)\in U_H} \left|W_\psi^H(x,y)\right| + \sum_{(x,y)\in U_V} \left|W_\psi^V(x,y)\right| + \sum_{(x,y)\in U_D} \left|W_\psi^D(x,y)\right| \quad (29)$$

(Jui-Ting, Chun-Hung, See-May, & Homer, 2005) proposed another focus measure in wavelet domain. The ratio of energies of the high frequency components to the low frequency components is taken as focus quality measure.

$$F_{DWT5} = \frac{E_H^2}{E_L^2} \quad (30)$$

where E_H^2 and E_L^2 are defined as

$$E_H^2 = \sum_{(x,y)\in U_H} \left(W_\psi^H(x,y)\right)^2 + \sum_{(x,y)\in U_V} \left(W_\psi^V(x,y)\right)^2 + \sum_{(x,y)\in U_D} \left(W_\psi^D(x,y)\right)^2 \quad (31)$$

$$E_H^2 = \sum_{(x,y)\in U_H} \left(W_\phi(x,y)\right)^2 \quad (32)$$

FOCUS MEASURE IN S-TRANSFORM

S-Transform

Fourier transform (FT) is one of the fundamental tools for frequency analysis of signals. However, spatial information is lost in this frequency domain, which is essential in many cases. To overcome this problem, several transforms have been reported including short time Fourier transform (STFT) and wavelet transforms (WT). S-transform has been proposed by (Xie, et al., 2007) and is being widely utilized in image and single processing. Let us start with the Fourier analysis of a signal *h*(*t*) that can be written as:

$$H(\omega) = \int_{-\infty}^{+\infty} h(t)\, e^{-i2\pi\omega t}\, dt \quad (33)$$

where *w* denotes the frequency. This spectrum can be referred to as the time-averaged spectrum. Now, if we multiply time series with a window function g(t) point by point then the Fourier spectrum will be given as:

$$H(\omega) = \int_{-\infty}^{+\infty} h(t)\, g(t)\, e^{-i2\pi\omega t}\, dt \quad (34)$$

Due to the compact form in the spatial as well as in the frequency domains, Gaussian window is used. So setting the Gaussian window as under;

$$g(t) = \frac{1}{\sigma\sqrt{2\pi}} \exp\left(-\frac{(t-\tau)^2}{2\sigma^2}\right) \quad (35)$$

This window is a function of both translation τ and dilation (window width σ). If the window width σ is fixed, a special case of ST, which is equivalent to the short-time Fourier transform (STFT). However, in ST, the window size varies at each point. By setting the width of Gaussian window as a function of frequency i.e.;

$$\sigma(\omega) = 1/|\omega| \quad (36)$$

In other words, the window function becomes a function of time and frequency. Width of the window is determined by the frequency. In the time domain, window is wider for lower, and narrower for higher frequencies. In conclusion, the window provides good localization in the frequency domain for low frequencies, while it provides good localization in the time domain for higher frequencies. It is clear that the time–frequency atoms for the S-transform are arranged in the same way as for the wavelet transform. The ST can be written as:

$$\chi(\tau,\omega) = \frac{|\omega|}{\sqrt{2\pi}} \int_{-\infty}^{+\infty} h(t) \exp\left(-\frac{(t-\tau)^2 \omega^2}{2} \right) e^{-i2\pi\omega t} dt. \tag{37}$$

Energy Concentration

Energy concentration measure concerns with the representation of the signals in time-frequency domains. Maximum energy concentration measure provides the sharpness or peakedness of the transformed signal. In literature, several energy concentration measures have been reported (Stockwell, et al., 1996). We adopted one of them as defined by (Jones & Parks, 1990; Sejdic, et al., 2008).

$$E_c = \frac{\int_{-\infty}^{+\infty}\int_{-\infty}^{+\infty} \left|\chi(\tau,\omega)\right|^4 d\tau d\omega}{\left(\int_{-\infty}^{+\infty}\int_{-\infty}^{+\infty} \left|\chi(\tau,\omega)\right|^2 d\tau d\omega\right)^2} \tag{38}$$

This measure is similar to the kurtosis in statistics. It is L_4 – norm divided by the square of L_2 – norm of the transformed coefficients. It should be clear that energy concentration should be the maximum for a good representation of a signal in S-domain. The energy concentration or kurtosis of the frequency components in the transform domain may directly be used for measuring the focus quality of an image (Jones & Parks, 1990). It is worthy to mention here that kurtosis of ST coefficients can be used as an energy concentration measure.

Modified S-Transform

It is inferred from the discussion in previous sections that the window size affects the transformed components and their energy. The total energy of the high frequency components further affects the focus quality and we obtain an improved focus measure by modifying the standard deviation of the Gaussian window as:

$$\sigma(\omega) = \frac{\alpha}{|\omega|^{\beta}}, \quad \alpha, \beta > 0 \tag{39}$$

The above modification is very similar to that proposed in (Caviedes & Oberti, 2004; Feng, Han, & Zhu, 2006) for a higher energy concentration. However, instead of using a single parameter, we use two parameters namely α and β that tune the width of the window. The modified ST is given as:

$$\chi'(\tau,\omega) = \frac{|\omega|^{\beta}}{\alpha\sqrt{2\pi}} \int_{-\infty}^{+\infty} h(t) \exp\left(-\frac{(t-\tau)^2 \left(|\omega|^{\beta}\right)^2}{2\alpha^2} \right) e^{-i2\pi\omega t} dt \tag{40}$$

In the modified ST, the window width corresponds to $\alpha = 1$ and $\beta = 1$ representing the standard ST. The value of α determines the intercept for the maximum window size. For $\alpha < 1$, the maximum window size becomes less than one and the window becomes narrower in the time domain. However, in the case of $\alpha > 1$, the window becomes wider. Increasing the value of β shortens the window width and vice versa. An effective focus measure can be obtained by providing appropriate values of α and β. To determine the optimal values of α and β, the grid search algorithm is applied. The main steps of the algorithm are listed in Table 1.

To illustrate the effect of parameters on window width and energy concentration, some experiments have been conducted using a synthetic cross chirp

Table 1. Parameters optimization

1	Initialize parametrs α and β and step size.
2	Compute ST $\chi'(\tau,\omega)$ using (41).
3	Calculate E_c using (38) for each value of α and β.
4	Determine the parameters α and β that provide optimal energy concentration measure. $\alpha,\beta = \max_{\alpha,\beta}\left(E_c\left(\alpha,\beta\right)\right)$
5	Use these optimal values for computing modified ST.

time series (Sejdic, et al., 2008). Figure 2 (a) shows the variations in window size with respect to frequency for different values of β and α = 1. It can be observed that in standard S-transform, a wider window is used for lower frequencies and on the other hand, a narrower window is used for higher frequencies. This choice of window width function leads to a disadvantage which is the same assignment of the standard deviation for all signal components at all frequencies because of the fact that σ is always defined as a reciprocal of the frequency. Some signals would benefit from different value of the standard deviation for the window function. The effect of parameters α and β on energy concentration is shown in Fig. 2 (b). It can be observed that the standard ST does not provide maximum energy concentration. If some optimal values for α and β are determined for a given signal, an improved time-frequency localization becomes possible. Figure 2 (c)-(d) shows the contour of the chirp time series obtained after applying standard ST and modified ST respectively. The modified ST components have been computed using the values α = 0.7 and β = 0.7. In other words, energy concentration in the time-frequency representation of a signal can be improved by adjusting these parameters.

In the case of a discrete signal $h[pT]$, ST can be written as $\left\{\tau = n/_{NT}, \omega = pT\right\}$.

$$\chi'\left(pT,\frac{n}{NT}\right) = \sum_{n'=0}^{N-1} H\left(\frac{n'+n}{NT}\right)\exp\left(-\frac{2\pi^2 n'^2}{n^2}\right)\exp\left(\frac{i2\pi n'p}{N}\right) \tag{41}$$

where $n,p = 0,1,2,\ldots,N\text{-}1$ and $H(\cdot)$ denotes the Fourier transform of the signal.

Computing Depth and All-in-Focus Image

In SFF, an image sequence $I_z(x,y)$ is acquired through a charge-coupled device (CCD) camera by varying the focus level. In the obtained sequence, the total number of images is Z. Considering a small image of size $N \times M$ around each pixel (x,y) of the sequence, we use 2D ST for measuring the focus quality. The discrete form of 2D ST is written as

$$\chi'\left(pT_x, qT_y, \frac{n}{NT_x}, \frac{m}{MT_y}\right) = \sum_{n'=0}^{N-1}\sum_{m'=0}^{M-1} H\left(\frac{n'+n}{NT_x}, \frac{m'+m}{MT_y}\right) \exp\left(-\frac{2\pi^2 n'^2}{n^2}\right)\exp\left(\frac{i2\pi n'p}{N}\right) \exp\left(-\frac{2\pi^2 m'^2}{m^2}\right)\exp\left(\frac{i2\pi m'q}{M}\right) \tag{42}$$

where NT_x and MT_y denote frequency coordinates, T_x and T_y represent time samples, and p,q,n and

Figure 2. (a) Effect of parameters on window width, (b) energy concentration measure for varying parameters alpha and beta, (c) contour plot of Chirp function for standard ST, (d) contour plot of Chirp function for modified ST.

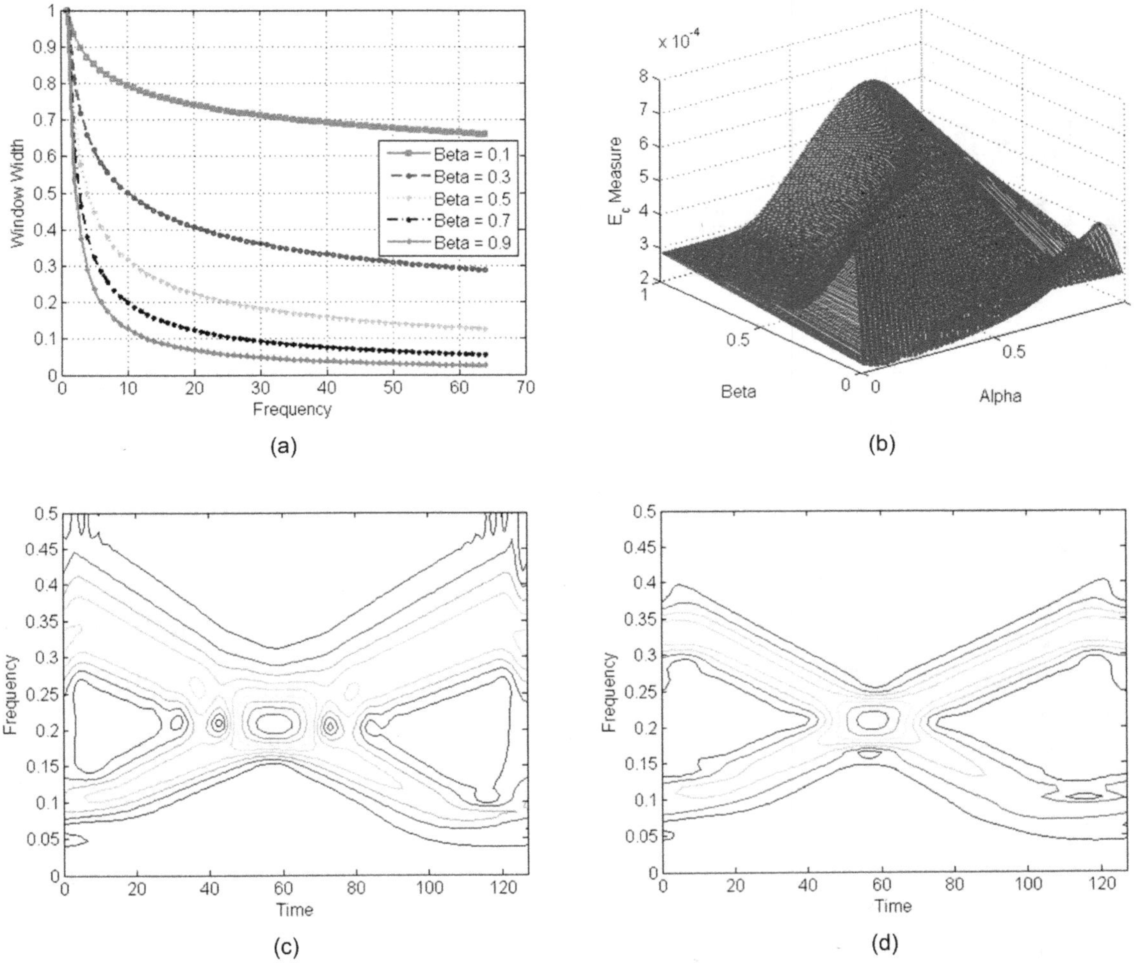

m are indices in the spatial and transform domains. At zero frequencies, the transformed components $\chi'\left(pT_x, qT_y, 0, 0\right)$ represent the average values of the data. Thus, we exclude these components to compute the sharpness. The proposed focus measure is computed as,

$$F_{ST} = \sum_{p=0}^{N-1}\sum_{q=0}^{M-1}\sum_{n=0}^{N-1}\sum_{m=0}^{M-1}\left|\chi'\left(pT_x, qT_y, \frac{n}{NT_x}, \frac{m}{MT_y}\right)\right|^2, (n,m) \neq (0,0). \tag{43}$$

By applying the above focus measure for each pixel in the sequence, we determine the focus volume $I_z'\left(x,y\right)$ as,

$$I_z'\ (x,y) = F_{ST}(I_z(x,y)), \quad z = 1,2,\cdots,Z. \tag{44}$$

To improve the robustness, the initial focus values are accumulated within a small 3D neighborhood $U_z(x,y)$ around each point $(x,y)_z$ of size $(d \times d \times d)$ and a refined focus volume is obtained as,

Table 2. Summary of depth estimation and recovering all-in-focus image through proposed focus measure

1	For each pixel (x,y) in the image sequence $Iz(x,y)$, a small image of size $N \times M$ is taken. Initialize $n, m, p, q, T_x, NT_x, T_y, MT_y$.
2	Compute the Fourier transform using fast Fourier transform (FFT) method.
3	Calculate 2D Gaussian window using optimal window width at the current frequency $\left(\frac{n}{NT_x}, \frac{m}{MT_y}\right)$.
4	Shift the Fourier spectrum $H(n,m)$ to $H(n'+n,m'+m)$.
5	Compute point wise multiplication of shifted spectrum and Gaussian window.
6	Repeat steps 3 to 5 for $N \times M$ times.
7	Compute initial focus value using (43).
8	Repeat steps 2 to 7 for each pixel in the image sequence.
9	Recover depth map and restore all-in-focus image through (46) and (47) respectively.

$$I''_z(x,y) = \sum_{(x,y,z) \in U_z(x,y)} I'_z(x,y) \quad (45)$$

For each object point, the sharpest pixel provides the depth information. The depth map $D(x,y)$ is computed by maximizing the focus measure along the optical axis, as shown below:

$$D(x,y) = \underset{z}{\operatorname{argmax}}(I''_z(x,y)),\ z = 1,2,\cdots,Z. \quad (46)$$

On the other hand, the all-in-focus image I_{aif} is computed by taking gray level values at the maximum focus as

$$I_{aif}(x,y) = I_D(x,y) \quad (47)$$

For brevity, the steps to compute depth map and all-in-focus image using the ST based focus measure are listed in Table 2.

RESULTS AND DISSCUSION

To report the effectiveness of the proposed focus measure, we conducted experiments using image sequences of three real objects. A sequence of 97 images of a real cone was obtained by using simple charge-coupled device CCD camera. The dimensions of each image are 200 × 200 pixels. Its actual depth map was estimated by using the cone parameters. Images for the real test objects, a statue on the one cent coin and letter *I* engraved into the surface, were obtained using microscope control system (Shim & Choi, 2010) shown in Figure 3. The system comprises of a personal computer, a frame grabber board, a CCD camera, motor driver with 2.5 *nm* (nano meter) step length, a microscope and image capturing software. The CCD camera is mounted on the microscope and the images were acquired by controlling the lens position. Keeping the illumination conditions constant, sequences of 60 images for the both test objects obtained under 50× magnifications.

We have actual depth map and corresponding all-in-focus image of real cone. Therefore, it is

Figure 3. Microscope control system

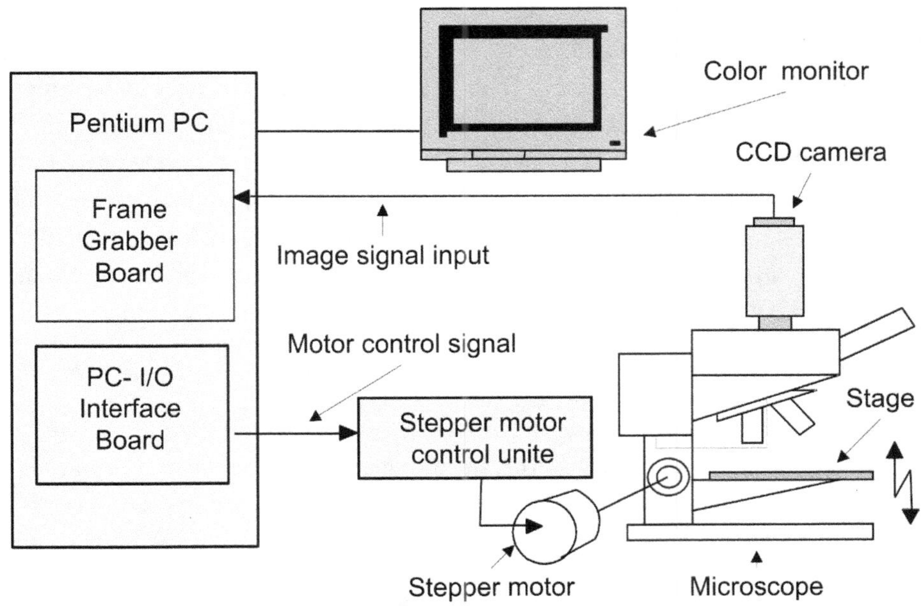

possible to evaluate different focus measures quantitatively. We choose three quantitative measures: root mean square error (RMSE), universal image quality index (UIQI), and structural similarity index measure (SSIM). RMSE measures the average of square of the error or distortion between actual and estimated depth maps or all-in-focus images.

$$RMSE = \sqrt{\frac{1}{XY}\sum_{x=1}^{X}\sum_{y=1}^{Y}\left(D(x,y) - D'(x,y)\right)^2} \tag{48}$$

where $D(x,y)$ and $D'(x,y)$ are actaul and estimated depth maps respectively. UIQI measure the similarity between actual and a distorted images. The distortion is modled by combining three factors: correlation,mean luminance, and contrast (Stockwell, et al., 1996). Mathematically, it is expressed as,

$$UIQI = \frac{4\sigma_{DD'}\bar{D}\bar{D}'}{\left(\sigma_D^2 + \sigma_{D'}^2\right)\left(\left(\bar{D}\right)^2 + \left(\bar{D}'\right)^2\right)} \tag{49}$$

where $\bar{D}$ and $\bar{D}'$ represent mean of the actual and estimated depth maps respectivly. Extending their work on distortion measure, (Zhou & Bovik, 2002) suggested another quality measure that is based on strutural information. This measure is defined as,

$$SSIM = \frac{\left(2\bar{D}\bar{D}' + c_1\right)\left(2\sigma_D\sigma_{D'} + c_2\right)}{\left(\left(\bar{D}\right)^2 + \left(\bar{D}'\right)^2 + c_1\right)\left(\sigma_D^2 + \sigma_{D'}^2 + c_2\right)} \tag{50}$$

where, σ_D is the variance of orignal depth map, $\sigma_{D'}$ is the variance of estimated depth map, and $\sigma_{DD'}$ is the covariance between original and compted depth maps. c_1 and c_2 are constants.

To find the appropriate values of parameters α and β, we used well known grid search technique as explained in Table 1. In this technique, we assign grid range and the step size for each parameter. Energy concentration is computed using modified ST for a small image patch of size 15x15 by varying the values of α and β. To make computationally effective search, we use the concept of coarse search by selecting large size of grid range with large step size then we refine the grid using small ranges [1,5] and [.5,2] for α and β, respectively, with the step size 0.1. The overall average of the optimal values from randomly selected fifty image patches are found to α=3 and β=1.7. After that, same values have been used for computing focus measure for all image sequences. It is difficult to find generic optimal values for these parameters that are valid for all images. Apart from the image distribution, parameters values also depend upon the length of the signal (patch size) and maximum frequency present. These two factors determine the frequency samples at each time instance. For image patches of same sizes the variations in parameters α and β is comparatively small. However, if the size of the image patch largely differs then these parameters have significantly different optimal values.

Figure 4. First row, frame number 20 extracted from each image sequence: (a) real cone, (b) coin, and (c) letter I. Second row: frame number 50 extracted from each image sequence: (d) real cone, (e) coin, and (f) letter I. (g)–(i) all-in-focus images obtained through the proposed focus measure F_{ST}

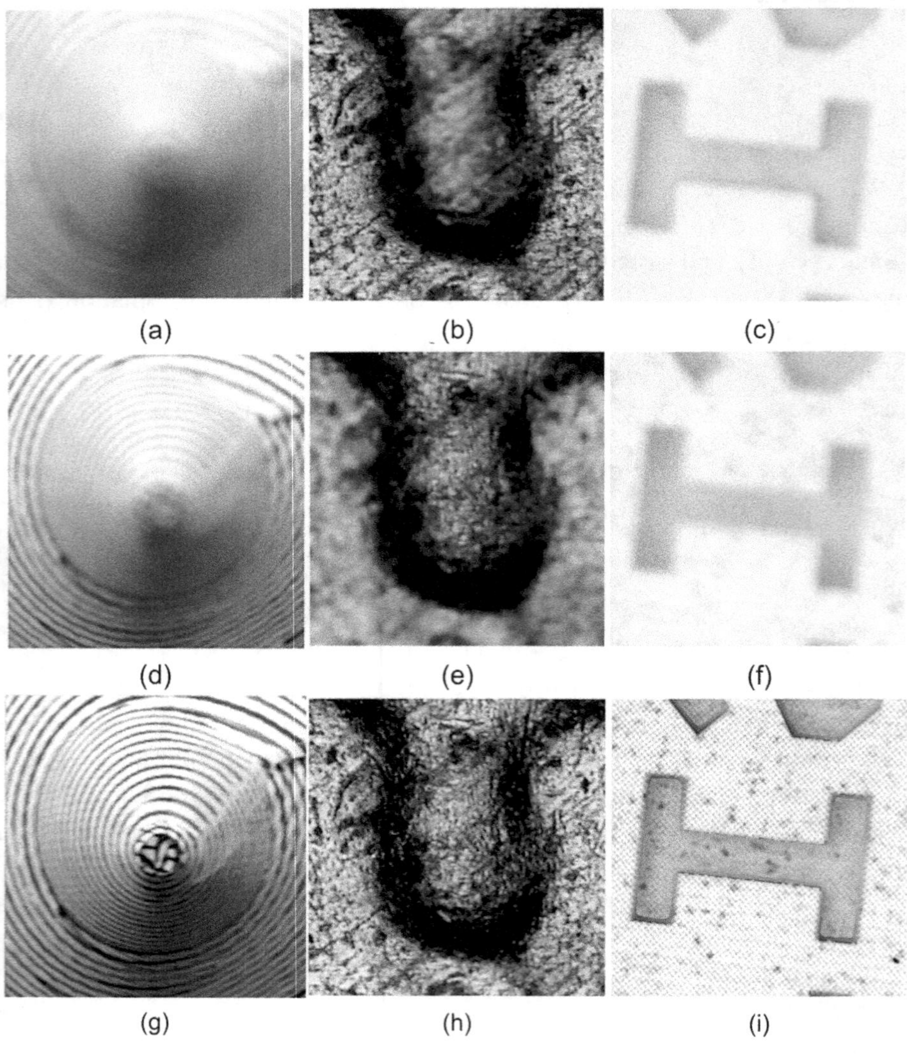

Table 3. Performance comparison in terms of depth map recovery and all-in-focus image generation

Measure	Depth map			All-in-focus image		
	RMSE	UIQI	SSIM	MSE	UIQI	SSIM
F_{SML}	10.3063	0.7734	0.9906	26.1593	0.9580	0.9695
F_{TEN}	14.2214	0.6994	0.9759	78.4633	0.9218	0.9390
F_{GLV}	10.3947	0.8151	0.9910	26.5559	0.9654	0.9807
F_{DCT}	13.1364	0.8371	0.9891	21.8088	0.9665	0.9752
F_{DWT}	9.5982	0.8152	0.9951	16.9414	0.9679	0.9775
F_{ST}	9.4518	0.8537	0.9966	14.6745	0.9724	0.9812

The performance of the proposed focus measure is compared with the existing focus measures F_{SML}, F_{TEN}, F_{GLV}, F_{DCT}, and F_{DWT}. The focus measures in frequency domain F_{DCT} and F_{DWT} compute ratio of the energies of high frequency components to low frequency components. First two rows of Figure 4 show frames numbers 20 and 50 extracted from each of image sequences of test objects real cone, coin, and letter *I*. It can be observed that in frame 20 some parts of the objects are well-focused and others are defocused and not clearly visible. Similarly, the well-focused parts in frame 20 are defocused in frame 60 and defocused parts in frames 20 are becomes well focused. Third row of the Figure 3 shows the all-in-focus image computed through the proposed method. In restored images, the objects are well focused the quality of the image is improved significantly.

Table 3 shows the performance comparisons in terms of depth map and all-in-focus image generation using the quality measures RMSE, UIQI, and SSIM. The numerical values for these metrics are computed using true depth map / all-in-focus image and the estimated depth maps / and all-in-focus images of real cone. It can be observed that RMSE, UIQI, and SSIM values for the proposed focus measure are better among others. The results computed using F_{DWT} are comparable with the proposed method. However, the performance of F_{TEN} has provided the lowest performance.

Figure 5 shows the depth maps computed by using existing focus measures F_{SML}, F_{GLV}, F_{DCT} and the proposed focus measure F_{ST} for three test objects real cone, coin, and letter *I*. For lucid comparisons, after applying focus measures, 3D window of size 5 × 5 × 5 has been used for summation of focus values. This step helps to suppress the noisy focus measurements. A significant performance difference can be observed between the 3D shapes recovered using the proposed and existing focus measures. It can be observed that 3D shapes recovered through the proposed scheme are better than the conventional methods. On the other hand, 3D shapes reconstructed through the traditional methods generated coarse surfaces due to their limited capability in computing focus measurements accurately. Moreover, noisy focus values introduce error in depth values and relatively more spikes can be observed in the constructed depth maps of the objects. It is notable that the F_{SML} focus measure poorly recovered the depth of the lower right corner of the real cone. This is due to the low illumination and weak textured area of the real cone images. However, our proposed method successfully recovered the depth from this weak textured area.

FUTURE RESEARCH DIRECTIONS

From various studies, it is concluded that the initial focus measurements contain noise that result in

Figure 5. Reconstructed 3D shapes of test objects (left-most column) real cone, (central column) coin, and (right-most column) letter I using (first row) F_{SML}*, (second row)* F_{GLV}*, (third row)* F_{DCT}*, and (fourth row)* F_{ST}*.*

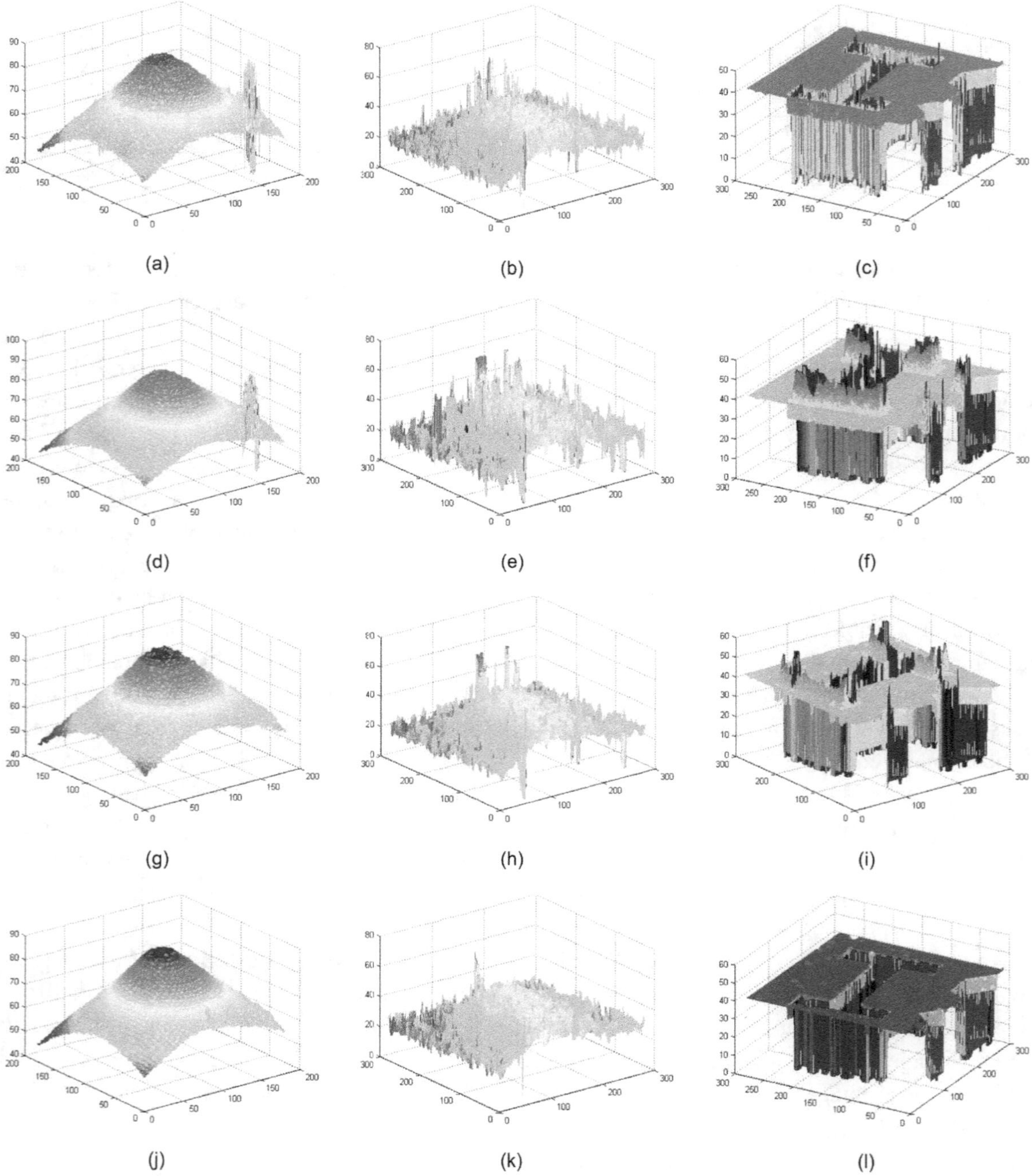

inaccurate depth map and all-in-focus image. The filtering of initial focus volume may provide better focus measurements. This fact is also highlighted in earlier work (Mahmood, Choi, & Choi, 2008a; Malik & Choi, 2008; Zhou, Bovik, Sheikh, & Simoncelli, 2004). The use of OTF in frequency domain acts as low pass filtering whereas PCA discriminate the focus measurements in frequency domains. The focus measurements with ST may also be enhanced through a similar operation. However, instead of applying linear filtering i.e., uniformly estimating focus measurements using a fixed window, adaptive nonlinear filtering will provide better results. In addition, we found that the computations of ST are expensive. This is perhaps the main reason that limits the utility of ST in many applications. However, recently some fast algorithms have been proposed to improve its efficiency (Brown, Lauzon, & Frayne, 2009).

CONCLUSION

In this chapter, we have explored the image focus measure that has many important applications in image processing and machine vision applications. A comprehensive discussion has been done regarding various focus measures that have been reported in spatial and transform domains. In addition, we have proposed a new focus measure employing the energy of high frequency components in ST. A modification for the window width function is suggested that affects the energy as well as focus measure. The optimal parameters are obtained using the energy concentration criterion. The proposed focus measure is then tested for real images with respect to multi-focus restoration and depth map extraction. The comparative analysis has shown the effectiveness of the proposed focus measure.

REFERENCES

Ahmad, M. B., & Choi, T. S. (2005). A heuristic approach for finding best focused shape. *IEEE Transactions on Circuits and Systems for Video Technology, 15*(4), 566–574. doi:10.1109/TCSVT.2005.844450

Ahmed, N., Natarajan, T., & Rao, K. R. (1974). Discrete cosine transfom. *IEEE Transactions on Computers, C-23*(1), 90–93. doi:10.1109/T-C.1974.223784

Baina, J., & Dublet, J. (1995). Automatic focus and iris control for video cameras.*Fifth International Conference on Image Processing and its Applications,* (pp. 232-235).

Brenner, J. F., Dew, B. S., Horton, J. B., King, T., Neurath, P. W., & Selles, W. D. (1976). An automated microscope for cytologic research: A preliminary evaluation. *The Journal of Histochemistry and Cytochemistry, 24*(1), 100–111. doi:10.1177/24.1.1254907

Brown, R. A., Lauzon, M. L., & Frayne, R. (2009). A general description of linear time-frequency transforms and formulation of a fast, invertible transform that samples the continuous s-transform spectrum non-redundantly. *IEEE Transactions on Signal Processing: Accepted for future publication.*

Brown, R. A., Lauzon, M. L., & Frayne, R. (2010). A general description of linear time-frequency transforms and formulation of a fast, invertible transform that samples the continuous s-transform spectrum nonredundantly. *Signal Processing . IEEE Transactions on, 58*(1), 281–290. doi:10.1109/TSP.2009.2028972

Brown, R. A., Zhu, H., & Mitchell, J. R. (2005). Distributed vector processing of a new local multiscale Fourier transform for medical imaging applications. *IEEE Transactions on Medical Imaging, 24*(5), 689–691. doi:10.1109/TMI.2005.845320

Caviedes, J., & Oberti, F. (2004). A new sharpness metric based on local kurtosis, edge and energy information. *Signal Processing Image Communication, 19*(2), 147–161. doi:10.1016/j.image.2003.08.002

Feng, J., Han, Z., & Zhu, M. (2006). Adaptive kurtosis optimization autofocus algorithm. *Journal of Electronics (China), 23*(4), 532–534. doi:10.1007/s11767-004-0174-3

Ge, Y., & Nelson, B. J. (2003, 27-31 Oct. 2003). Wavelet-based autofocusing and unsupervised segmentation of microscopic images. *IEEE/RSJ International Conference on Intelligent Robots and Systems,* (vol. 3 pp. 2143-2148).

Groen, F. C. A., Young, I. T., & Ligthart, G. (1985). A comparison of different focus functions for use in autofocus algorithms. *Cytometry, 6*(2), 81–91. doi:10.1002/cyto.990060202

Helmli, F. S., & Scherer, S. (2001). *Adaptive shape from focus with an error estimation in lightmicroscopy* (pp. 188–193).

Horn, B. K. (1986). *Robot vision*. McGraw-Hill Higher Education.

Jones, D. L., & Parks, T. W. (1990). A high resolution data-adaptive time-frequency representation. *IEEE Transactions on Acoustics, Speech, and Signal Processing, 38*(12), 2127–2135. doi:10.1109/29.61539

Jui-Ting, H., Chun-Hung, S., See-May, P., & Homer, C. (2005, 13-16 December). Robust measure of image focus in the wavelet domain. *Proceedings of 2005 International Symposium on Intelligent Signal Processing and Communication Systems,* (pp. 157-160).

Kautsky, J., Flusser, J., Zitov, B., & Simberov, S. (2002). A new wavelet-based measure of image focus. *Pattern Recognition Letters, 23*(14), 1785–1794. doi:10.1016/S0167-8655(02)00152-6

Kristan, M., Pers, J., Perse, M., & Kovacic, S. (2006). A Bayes-spectral-entropy-based measure of camera focus using a discrete cosine transform. *Pattern Recognition Letters, 27*(13), 1431–1439. doi:10.1016/j.patrec.2006.01.016

Krotkov, E. (1988). Focusing. *International Journal of Computer Vision, 1*(3), 223–237. doi:10.1007/BF00127822

Kyte, B. (2010). *Optical 3D micro metrology*. Retrieved November 15, 2010, from http://www.alicona.com/

Mahmood, M. T., Choi, W. J., & Choi, T. S. (2008a). PCA-based method for 3D shape recovery of microscopic objects from image focus using discrete cosine transform. *Microscopy Research and Technique, 71*(12), 897–907. doi:10.1002/jemt.20635

Mahmood, M. T., Choi, W. J., & Choi, T. S. (2008b). PCA-based method for 3D shape recovery of microscopic objects from image focus using discrete cosine transform. *Microscopy Research and Technique, 71*(12), 897–907. doi:10.1002/jemt.20635

Mahmood, M. T., Shim, S. O., & Choi, T. S. (2009). Shape from focus using principal component analysis in discrete wavelet transform. *Optical Engineering (Redondo Beach, Calif.), 48*, 057203. doi:10.1117/1.3130232

Malik, A. S., & Choi, T. S. (2007). Consideration of illumination effects and optimization of window size for accurate calculation of depth map for 3D shape recovery. *Pattern Recognition, 40*(1), 154–170. doi:10.1016/j.patcog.2006.05.032

Malik, A. S., & Choi, T. S. (2008). A novel algorithm for estimation of depth map using image focus for 3D shape recovery in the presence of noise. *Pattern Recognition, 41*(7), 2200–2225. doi:10.1016/j.patcog.2007.12.014

Malik, A. S., & Choi, T. S. (2009). Comparison of polymers: A new application of shape from focus. *IEEE Transactions on Systems, Man and Cybernetics. Part C, Applications and Reviews*, *39*(2), 246–250. doi:10.1109/TSMCC.2008.2001714

Nayar, S. K., & Nakagawa, Y. (1994a). Shape from focus. *IEEE Transactions on Pattern Analysis and Machine Intelligence*, *16*(8), 824–831. doi:10.1109/34.308479

Nayar, S. K., & Nakagawa, Y. (1994b). Shape from focus. *IEEE Transactions on Pattern Analysis and Machine Intelligence*, *16*(8), 824–831. doi:10.1109/34.308479

Pentland, A. P. (1987). A new sense for depth of field. *IEEE Transactions on Pattern Analysis and Machine Intelligence*, *9*(4), 523–531. doi:10.1109/TPAMI.1987.4767940

Sang-Yong, L., Kumar, Y., Ji-Man, C., Sang-Won, L., & Soo-Won, K. (2008). Enhanced autofocus algorithm using robust focus measure and fuzzy reasoning. *IEEE Transactions on Circuits and Systems for Video Technology*, *18*(9), 1237–1246. doi:10.1109/TCSVT.2008.924105

Santos, A., Solorzano, O. D., Vaquero, J. J., Pena, J. M., Malpica, N., & Del, P. (1997). Evaluation of autofocus functions in molecular cytogenetic analysis. *Journal of Microscopy*, *188*(3), 264. doi:10.1046/j.1365-2818.1997.2630819.x

Sejdic, E., Djurovic, I., & Jiang, J. (2008). A window width optimized s-transform. *EURASIP Journal on Advances in Signal Processing, ID 672941*, 13 pages.

Shen, C. H., & Chen, H. H. (2006). *Robust focus measure for low-contrast images.*

Shim, S., & Choi, T. S. (2010). A novel iterative shape from focus algorithm based on combinatorial optimization. *Pattern Recognition*, *43*(10), 3338–3347. doi:10.1016/j.patcog.2010.05.029

Simonov, A. N., & Rombach, M. C. (2009). Sharp-focus image restoration from defocused images. *Optics Letters*, *34*(14), 2111–2113. doi:10.1364/OL.34.002111

Stockwell, R. G., Mansinha, L., & Lowe, R. P. (1996). Localization of the complex spectrum: The S transform. *IEEE Transactions on Signal Processing*, *44*(4), 998–1001. doi:10.1109/78.492555

Subbarao, M., Choi, T. S., & Nikzad, A. (1993a). Focusing techniques. *Optical Engineering (Redondo Beach, Calif.)*, *31*(11), 2824–2836. doi:10.1117/12.147706

Subbarao, M., Choi, T. S., & Nikzad, A. (1993b). Focusing techniques. *Optical Engineering (Redondo Beach, Calif.)*, *31*(11), 2824–2836. doi:10.1117/12.147706

Subbarao, M., & Tyan, J. K. (1998). Selecting the optimal focus measure for autofocusing and depth-from-focus. *IEEE Transactions on Pattern Analysis and Machine Intelligence*, *20*(8), 864–870. doi:10.1109/34.709612

Tenenbaum, T. M. (1970). *Accommodation in computer vision.* Stanford University.

Thelen, A., Frey, S., Hirsch, S., & Hering, P. (2009). Improvements in shape-from-focus for holographic reconstructions with regard to focus operators, neighborhood-size, and height value interpolation. *IEEE Transactions on Image Processing*, *18*(1), 151–157. doi:10.1109/TIP.2008.2007049

Tian, Y., Shieh, K., & Wildsoet, C. F. (2007). Performance of focus measures in the presence of nondefocus aberrations. *Journal of the Optical Society of America. A, Optics, Image Science, and Vision*, *24*(12), 165–173. doi:10.1364/JOSAA.24.00B165

Wang, Z., Ma, Y., & Gu, J. (2010). Multi-focus image fusion using PCNN. *Pattern Recognition*, *43*(6), 2003–2016. doi:10.1016/j.patcog.2010.01.011

Wee, C. Y., & Paramesran, R. (2007). Measure of image sharpness using eigenvalues. *Information Sciences*, *177*(12), 2533–2552. doi:10.1016/j.ins.2006.12.023

Xie, H., Rong, W., & Sun, L. (2007). Construction and evaluation of a wavelet-based focus measure for microscopy imaging. *Microscopy Research and Technique*, *70*(11), 987–995. doi:10.1002/jemt.20506

Yap, P. T., & Raveendran, P. (2004). Image focus measure based on Chebyshev moments. *IEE Proceedings. Vision Image and Signal Processing*, *151*(2), 128–136. doi:10.1049/ip-vis:20040395

Zhang, Y., Zhang, Y., & Wen, C. (2000). A new focus measure method using moments. *Image and Vision Computing*, *18*(12), 959–965. doi:10.1016/S0262-8856(00)00038-X

Zhou, W., & Bovik, A. C. (2002). A universal image quality index. *IEEE Signal Processing Letters*, *9*(3), 81–84. doi:10.1109/97.995823

Zhou, W., Bovik, A. C., Sheikh, H. R., & Simoncelli, E. P. (2004). Image quality assessment: From error visibility to structural similarity. *IEEE Transactions on Image Processing*, *13*(4), 600–612. doi:10.1109/TIP.2003.819861

KEY TERMS AND DEFINITIONS

All-in-Focus Image: An all-in-focus image comprises of best focus parts that are extracted from a stack of visual observation.

Depth Map: A Depth map represents depth information instead of intensity values corresponding to two- dimensional array of an image.

Discrete Cosine Transform: The discrete cosine transform (DCT) of a signal is a real valued transform that expresses a sequence of finitely many data points in terms of a sum of cosine functions oscillating at different frequencies.

Discrete Wavelet Transform: The discrete wavelet transform (DWT) is the decomposition of the signal into approximations and details coefficients obtained by expanding the signal in terms of the scaling function and basis function.

Energy Concentration Measure: The Energy concentration measure concerns with the representation of the signals in time-frequency domains. Maximum energy concentration measure provides the sharpness or peakedness of the transformed signal.

Focus Measure: An operator that measures the focus quality of the image by utilizing the high frequency contents in the image.

Shape from Focus: A technique that retrieves 3D structure of an object from a sequence of its images with different focus settings.

S-Transform: The S transform is a generalization of the Short-time Fourier transform (STFT), which uses a variable window instead of a constant window that depends on the frequency.

Chapter 11

Combining Focus Measures for Three Dimensional Shape Estimation Using Genetic Programming

Muhammad Tariq Mahmood
Korea University of Technology and Education, Korea

Tae-Sun Choi
Gwangju Institute of Science and Technology, Korea

ABSTRACT

Three-dimensional (3D) shape reconstruction is a fundamental problem in machine vision applications. Shape from focus (SFF) is one of the passive optical methods for 3D shape recovery, which uses degree of focus as a cue to estimate 3D shape. In this approach, usually a single focus measure operator is applied to measure the focus quality of each pixel in image sequence. However, the applicability of a single focus measure is limited to estimate accurately the depth map for diverse type of real objects. To address this problem, we introduce the development of optimal composite depth (OCD) function through genetic programming (GP) for accurate depth estimation. The OCD function is developed through optimally combining the primary information extracted using one (homogeneous features) or more focus measures (heterogeneous features). The genetically developed composite function is then used to compute the optimal depth map of objects. The performance of this function is investigated using both synthetic and real world image sequences. Experimental results demonstrate that the proposed estimator is more accurate than existing SFF methods. Further, it is found that heterogeneous function is more effective than homogeneous function.

DOI: 10.4018/978-1-61350-326-3.ch011

INTRODUCTION

Inferring three-dimensional (3D) shape of an object from two-dimensional (2D) images is a fundamental problem in computer vision (Ahmad & Choi, 2005; Nayar & Nakagawa, 1994; Thelen, Frey, Hirsch, & Hering, 2009). Broadly, 3D shape recovery algorithms based on optical reflective model can be categorized into active and passive techniques. In active techniques, depth of the object of interest is computed by investigating transmission or reflection of signals such as ultrasound or infrared rays. While, passive methods infer the depth of the object by analyzing information from the captured images. The shape from focus (SFF) is one of the passive methods to estimate 3D structure of the object based on image focus analysis. It is a famous one in the paradigm of shape from X, where X denotes the cue used to infer the shape as stereo, motion, shading, de-focus, and focus. The SFF technique has been successfully utilized in many industrial applications, i.e. microelectronics (Niederost, Niederost, & Scucka, 2003), industrial inspection (S. O. Shim, Malik, & Choi, 2009), medical diagnostics (Boissenin et al., 2007), 3D cameras (Malik & Choi, 2007a), TFT-LCD color filter manufacturing (Ahmad & Choi, 2007), and roughness comparison of polymers (Malik & Choi, 2009). In addition, it has also been employed to measure roughness, and geometry of large components, such as engine blocks and aircraft turbines (Kyte, 2010).

In SFF, an image sequence is acquired by translating object along the optical axis. It is important to note that acquired images from lenses with limited depth of field have both the areas in and out of focus. However, it is possible to compute the well-focused image from the image sequence taken at different focus levels by computing the high frequency image contents. A criterion, usually known as focus measure, is used to compute focus quality of each pixel in the image sequence. Focus quality is computed for each pixel in the image sequence and an initial depth map is obtained by maximizing the focus measure along the optical axis. In the literature, many focus measure operators are reported in spatial (Helmli & Scherer, 2001; Krotkov, 1988; Subbarao & Tyan, 1998) and transform domains (Mahmood, Choi, & Choi, 2008; Mahmood, Shim, & Choi, 2009; Malik & Choi, 2008; Sun, Duthaler, & Nelson, 2004; Xie, Rong, & Sun, 2007). Once an initial depth map is computed, some approximation technique is applied to further refine these results (Malik & Choi, 2007b; Nayar & Nakagawa, 1994; Subbarao & Choi, 1995). Most of these techniques use a single focus measure to estimate initial depth map. Due to the diverse nature of real images, it is not possible for a single focus measure to perform equally well under different scenarios. Therefore, it is difficult to choose a suitable focus measure for specific conditions. Another drawback with existing techniques is that the error introduced in computing initial depth map is propagated to the approximation step. In such scenario, there is a demand of a new generalized optimal depth estimator that may effectively incorporate useful information from more than one focus measures.

In this connection, we propose a novel idea of combining initial depth and focus values extracted from various focus measures. Using this concept, the advantages of one focus measure can overcome the shortcomings of others. However, the problem is how to combine in a best possible way. Under such circumstances, we introduce genetic programming (GP) based technique that optimally combines the initial information extracted from one or more focus measures. GP approach works on the principles of natural selection and recombination to search the space for possible solutions under a fitness criterion. Due to the flexibility of adjustable parameters, GP optimization technique (dos Santos, Ferreira, Torres, Gonlves, & Lamparelli, 2010; Kouchakpour, Zaknich, & Brnl, 2009; Koza, Streeter, & Keane, 2008; Langdon, 2000; Mallipeddi, Mallipeddi, & Suganthan, 2010) has been widely used in the applications of image processing (Petrovic & Crnojevic, 2008), pattern recognition (Majid, 2006), and computer vision (Majid, Khan, & Mirza, 2006). In the proposed scheme, GP based optimal composite depth (OCD) functions are developed using homogenous and heterogeneous feature sets. In the first step, set of features, consisting of initial focus and depth values that are computed through existing focus measures, is obtained. The useful features information and random constant values are combined through arithmetic operators to develop the OCD function. The composite function is then used to compute the optimal depth map. The improved performance of the developed function is investigated using synthetic and real images. Experimental results demonstrate the superiority of the proposed GP based scheme over the conventional focus measures.

In the remainder of this chapter, we describe the SFF scheme and present a brief summary of existing focus measures and approximation techniques. Then, after providing some experimental results describing the effect of focus measures on the depth map, we are motivated to suggest GP based scheme. Later sections of this chapter explain experimental setup and comparative analysis.

BACKGROUND

Shape from Focus

Techniques that retrieve spatial information from multiple images of the same scene, taken at different focus levels, are classified as Shape From Focus (SFF). In SFF, the objective is to find out the depth by measuring the distance of well-focused position of every object point from the camera lens. Once, distances for all points of the object are found, the 3D shape can easily be recovered. Figure 1 shows the schematic of SFF technique. Initially, an object of unknown depth is kept on reference plane and then translated in the optical direction in fixed finite steps of δd with respect to a real aperture camera. At every step, an image is captured and a stack of visual observations is obtained. Due to the limited depth-of-field of the camera and the 3D nature of the object, the captured images are space-variantly blurred such that some parts of the object come into focus in each frame. The distances between the focus plane and reference plane are known. A point *P* on the surface of the object becomes focused gradually and at one stage, it will be in sharp focus. We are interested at this sharp focus stage as it provides information about the depth of this point. All light rays, which are radiated from the object point *P*, are intercepted by the lens and converged at the point *P'* on image plane. Well-focused point *P* satisfies the lens law:

$$\frac{1}{f} = \frac{1}{u} + \frac{1}{v} \tag{1}$$

where f is the focal length of the lens, u is the distance of the object point from the lens, and v is the distance between lens and image plane. At any other distance $u \neq u'$ of the object point from the lens will not be well-focused on image plane. According to geometric optics, the aperture defines the shape of the blurred image of the point *P*. Since, we have considered the aperture to be circular, the blurred image is also a circle of radius *R* with uniform brightness inside the circle and zero outside it. In practice, the image of a point is not a crisp circular patch with constant brightness. Due to diffraction, polychromatic illumination, lens aberrations etc., it will be a circular blob with the brightness falling of gradually at the border. Therefore, this blurring effect is usually modeled by two-dimensional Gaussian function which is defined by:

$$h(i,j) = \frac{1}{2\pi r^2} \exp\left(-\frac{i^2 + j^2}{2r^2}\right) \tag{2}$$

Figure 1. Schematic of shape from focus

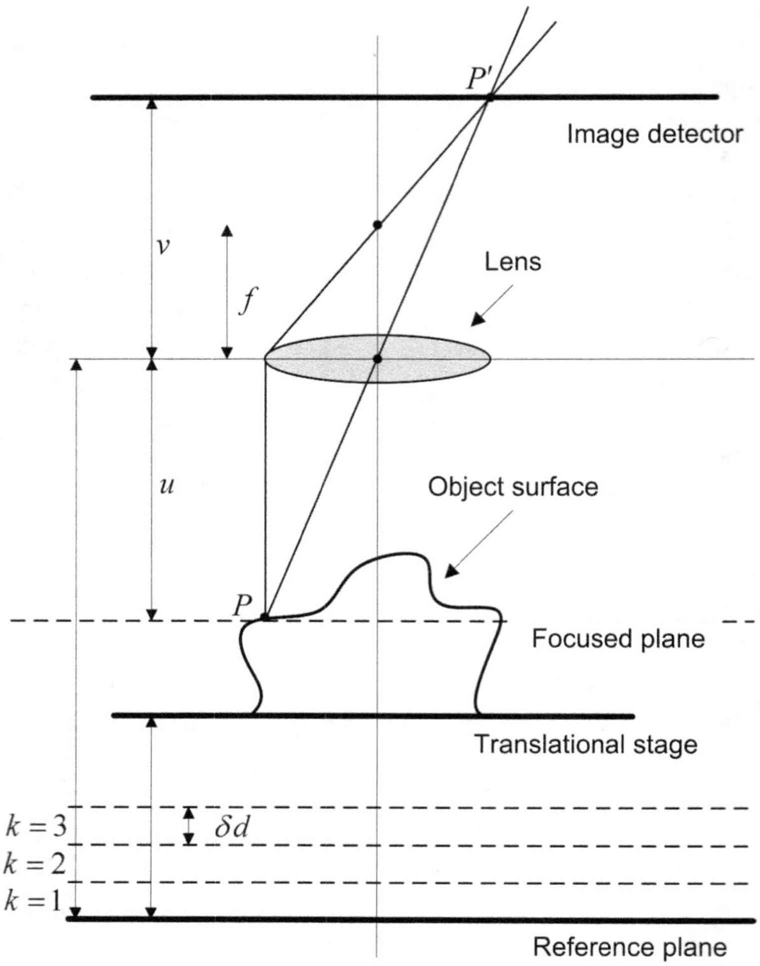

This function is also known as Point Spread Function (PSF) where, r is a spread parameter corresponding to the standard deviation of the distribution of the PSF. Hence, a sensed image is the convolution of the actual image and a Gaussian function, i.e.

$$g(i,j) = g_a(i,j) * h(i,j) \tag{3}$$

where $g(i,j)$ is the sensed image, $g_a(i,j)$ is the actual image, * is the convolution operator and $h(i,j)$ is the PSF. The radius of the blurred circle R and the width of the Gaussian function r are related by $r = cR$. Where c is a constant and it can be approximated through camera calibration (Pentland, 1987). The main issue in SFF scheme is to determine the particular distances of all object points from the camera for which they are well focused at image plane. The focus measure increases with the increase of focus quality and it attains maximum value at well-focused frame number. Therefore, a well-focused image will have larger amount of high frequency contents as compared to de-focused image of the same scene. The main problem in the construction of an accurate depth map is to locate the best-focused pixel using the obtained image frames at each object point.

Focus Measures

The basic step in SFF is to compute the focus quality or sharpness for each pixel in the image sequence. One of the famous categories of focus measures is based on image derivatives. These measures are based on the idea that the larger difference in intensity values of neighboring pixels is analogous to the sharper edges. Broadly, they can be divided into two sub-categories: first and second derivative based methods. In gradient based methods include Threshold Absolute Gradient (TAG) (Santos et al., 1997), Squared Gradient (SG) (Tian, Shieh, & Wildsoet, 2007), Brenner Gradient (BG) (Brenner et al., 1976), and Tenenbaum focus measure (TEN) (Tenenbaum, 1970) . Among these, TEN is the most famous. In this method, horizontal and vertical Sobel operators are applied. The focus quality is measured by computing the magnitude of the gradient vector components. In the remaining text, it is referred as F_2. Second derivative based operator Laplacian, being a point and symmetric operator, is suitable for measuring image sharpness. Several focus measures have been proposed by modifying the Laplacian operator (Helmli & Scherer, 2001; Subbarao, Choi, & Nikzad, 1993; Subbarao & Tyan, 1998; Thelen, et al., 2009). Nayar and Nakagawa proposed sum modified Laplacian (SML) focus measure (Nayar & Nakagawa, 1994) that is denoted by F_1 in the remaining text. In this measure, first, an image is convolved with the Laplacian operator. The components obtained through the Laplacian operator may have opposite sign and can yield zero response by canceling the effect of each other. To overcome this limitation, it is modified by taking the energy of the Laplacian operator. In order to improve robustness, the resultant values are summed up within a small window. The second derivative based focus measures provide more accurate results as compared to the first derivative based measures. However, these measures are more sensitive to noise. Among the statistic based focus measures (Groen, Young, & Ligthart, 1985; Wee & Paramesran, 2007; Yap & Raveendran, 2004; Zhang, Zhang, & Wen, 2000), the Gray Level Variance (GLV) (Krotkov, 1988) focus measure has gained the most attention. The main concept is that the larger variance of intensity values within a small window corresponds to the sharper image and vice versa. The focus value is computed by calculating the variance of intensity values. In the remainder text, this focus measure is denoted by F_3.

Some focus measures have also been proposed in transform domains. The main concept is the energy of high frequency components remains analogous to the image sharpness. In discrete cosine transform (DCT) domain, the energy of high frequency components or ratio of energies of high frequency and low frequency components is taken as focus measure. The entropy of the normalized DCT coefficients has been suggested as a focus measure by (Kristan, Perv, Pervse, & Kovavcic, 2006). In discrete wavelet transform (DWT) domain, the ratio of energies of the high and low frequency components is considered as focus measure (Xie, et al., 2007). Another robust focus measure based on optical transfer function (OTF) has been proposed by (Malik & Choi, 2008). Recently, a focus measure based on the energy of high frequency components in S-transform is suggested by (Mahmood & Choi, 2010). Focus measures in transform domains also provide comparable accuracy. However, these methods are rather expensive to compute.

Depth Refinement Techniques

Once an initial depth is extracted by applying a focus measure, it is further refined to obtain an accurate depth map. In literature, many approximation and machine learning based methods have been suggested. Some of them are briefly discussed here.

Gaussian Interpolation

Gaussian interpolation has been suggested to compute the accurate depth position by (Nayar & Nakagawa, 1994). First, focus values are computed through F_1 focus measure. The Gaussian model is fitted to three focus values near the peak of the focus curve. The initial depth is replaced with the mean value of the fitted curve. However, parametric interpolation methods such as Gaussian curve fitting may not yield optimal depth, as the focus values (focus curve) may not follow any specific distribution. Moreover, in reality objects have complex geometry, so the focus values computed over a single frame may not capture the effect of focus values from the neighboring frames.

Focused Image Surface

Traditional methods do not consider the fact that the focused image of the 3D object being in 3D space too. The concept of Focused Image Surface (FIS) is introduced by (Subbarao & Choi, 1995). Based on this concept, they also proposed a method SFF-FIS to recover an accurate 3D shape. FIS is defined as the surface formed by the set of points at which the object points are focused by the camera lens. First, an initial shape is computed using a focus measure F_1. Then, the initial shape is refined by approximating focus measure over 3D space. A planar window in the image volume instead of a single frame is used to approximate the optimal focus. The SFF-FIS provides better results; however, this method is computationally expensive as it searches the plane that provides optimal focus measure from a huge number of possible planes.

Focused Image Surface through Curved Window

Further extending the work on SFF-FIS, (Choi & Yun, 2000) suggested the estimation of FIS through piecewise curved surface approximation instead of planar window. For objects with complex geometry, the planar window may not yield accurate results. Moreover, small neighborhood around the initial depth estimate may not provide sufficient information for optimal focus measure. Therefore, (Choi & Yun, 2000) proposed higher order approximation. the initial shape estimate is computed through the focus measure F_3. The piecewise curved surface around each pixel is estimated by using second order Lagrangian polynomials with nine control points. The central pixel of the approximated curve provides optimal focus measure. It provides better results, however, the related computational cost is increased.

Neural Network-Based SFF

Neural networks are capable of learning any arbitrary nonlinear function from a set of observations. An approach based on neural networks to obtain an optimal FIS has been suggested by (Asif & Choi, 2001). In this method, initial focus values are computed through the focus measure F_3. The focus measurements from 3D neighborhood of each point are provided to the input layer of the neural network. The optimal focus measure is learned by maximizing the focus value at the output layer. It provides better 3D shape, however, it is difficult to get a generalized function for optimal focus measure for arbitrary objects.

Dynamic Programming-Based SFF

In another work, (Ahmad & Choi, 2005) proposed the use of dynamic programming (DP) to obtain an accurate shape of the object. An initial FIS is estimated using a traditional focus measure. Through DP, a refined FIS is then obtained by optimizing focus measure in 3D space. DP optimization is based on Bellman's principal of optimality which states that the optimal path between two given points is also optimal between any two points lying on that path. In this method, the problem of shape recovery is split into a series of small problems.

Thus computationally this approach is effective than traditional methods. The shape of the object is searched in the whole image focus volume.

Combinatorial Optimization for SFF

In this approach, the SFF is modeled as combinatorial optimization problem (S. Shim & Choi, 2010). To reduce the computational complexity, a local search algorithm is proposed. First, the initial estimate is obtained by applying a conventional focus measure and then initial depth is obtained by maximizing the focus measure along the optical axis. At each point, the neighborhood is defined from the initial depth by taking several preceding and following frames with respect to the initial depth. The intermediate image volume is obtained by collecting the pixels values of neighborhood at each point. The updated solution is retrieved from the intermediate image volume. This update process continues until the convergence criterion is fulfilled. The process to obtain temporary image volume has the effect of aligning the curved object patch, corresponding to the focused image surface perpendicular to the optical axis. Therefore, applying the focus measure on the intermediate image volume gives more accurate focus level at each pixel.

The above-discussed methods provide good results but their performance relies upon the initial estimate obtained from a single focus measure. Another drawback of these traditional approaches is their noise sensitivity due to the gradient-based focus measures. That is why; the performance of these SFF methods is deteriorated significantly under diverse conditions.

FOCUS MEASURE VS. DEPTH MAP

In this section, we will discuss the effect of focus measure on depth map. Comparative studies of focus measure operators (Groen, et al., 1985; Subbarao & Tyan, 1998; Sun, et al., 2004; Tian, et al., 2007) revealed that different focus measures provide different focus values and thus result in different depth values. Practically, it is hard to predict the suitable focus measure among a large list. Many factors including window size, mask size, noise level, illumination, contrast, affect the accuracy of the computed focus values and depth map. Some focus measure work well in normal conditions however, others perform poorly in weak textured and low illumination conditions. The performance of derivative based focus measures is relativity poor relating to the parts of images with less texture. Moreover, in noise-free environment, the second derivative based focus measure is more accurate as compared to the first derivative based focus measure. However, its performance is degraded with the increase of noise variance. Similarly, higher order moments provide good focus measures but are more sensitive to noise and complex in computation.

To illustrate the effect of window size and accuracy of focus measurements for various focus measures, we carried out some experiments by retrieving image sequence of simulated cone. The image acquiring procedure and experimental setup are explained in sub section "Implementation details". Since, we have true depth map for the simulated cone object, it is possible to compute discrepancies between estimated and true depth maps through a quantitative measure. Figure 2 (a) shows the root mean square error (RMSE) computed for depth maps obtained through the focus measures F_1, F_2, and F_3. The accuracies of estimated depth maps are differing from each other. We can observe that F_1 provides better results than F_2 and F_3 It is also notable, that as we increase the window size, the depth map become smoother and performance difference between focus measure operators becomes smaller. Increasing the window size actually makes the object surface smooth and spikes (wrong depth estimates) are suppressed. A comprehensive study about the effect of neighborhood size and illumination in SFF techniques is done by (Malik & Choi, 2007b). Malik and Choi

Figure 2. (a) Effect of window size on depth map accuracy, and (b) focus curves for the object point (80,80) of the simulated cone obtained through different focus measures

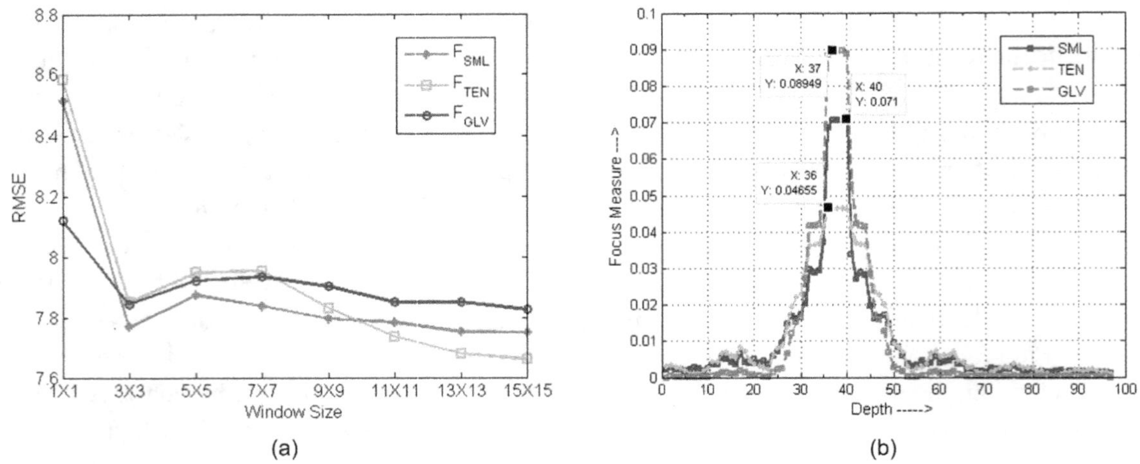

concluded that a larger window size deteriorates the depth map accuracy. In addition, the uniform averaging (summation of focus values within the window) is another source of inaccuracies in depth map computation. During this process, many noiseless focus measurements are altered.

We computed focus curve for the sequence of the object point (80, 80) of simulated cone by using different focus measures. For all focus measure, window size 5 × 5 has been used to aggregate focus values. The true depth value for this point is 38.97. From Fig. 2 (b), it can be observed that the three focus measures F_1, F_2 and F_3 have provided different depth values. The focus measure F_1 has provided the best focus pixel at position 40, F_3 has provided depth value 36, and depth value 37 is computed through the focus measure F_3

Summarizing the above discussion, we conclude that it is difficult to compute accurate depth map using a single focus measure. Moreover, real images with diverse types of illumination and contrast lead to the erroneous depth map estimation through a single focus measure. Therefore, we propose GP based optimization technique that automatically combines the useful information to develop an optimal function for accurate depth estimation. Through GP evolution cycle, the most satisfactory solution in the shape of a numerical expression is developed. This composite function might compensate the shortcomings of one focus measure while taking advantages of the other focus measure for accurate depth map estimation. In the subsequent section, we will describe the GP based approach to develop OCD function.

Proposed Scheme based on Genetic Programming

Our aim is to develop GP based depth estimation function, $F': x \rightarrow y$, that maps the useful input information x to the target depth y values. The proposed scheme is divided into *Preparing training data, Function development* and *Depth estimation*. In first module, training data is formed by computing features from by applying some focus measure. During GP process, optimal depth estimation function is developed using the training data. *Depth estimation* module is used to estimate the optimal depth map of the object. The block diagram of the proposed GP based scheme is shown in Figure 3.

Figure 3. GP based optimal depth map estimation scheme

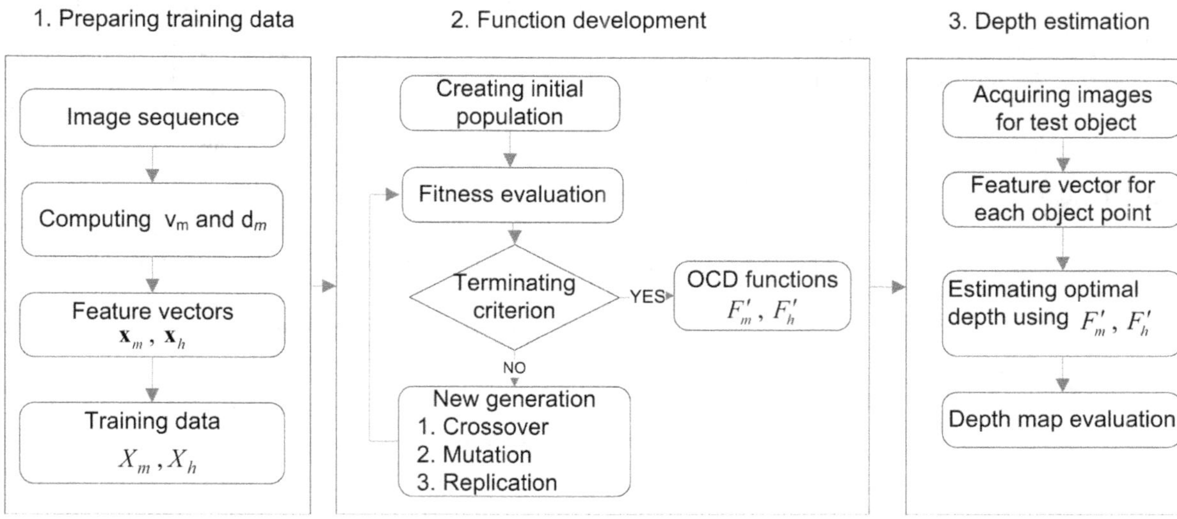

Preparing Training Data

In the proposed scheme, the first step is to construct a set of informative feature vectors. The feature vector x consists of initial depth and focus values computed by applying some focus measure operator on an image sequence of the object. An image sequence $g^{(k)}(i,j)$ is acquired through a CCD camera by translating the object along the optical axis, where $i = 1,2,\ldots,I$ and $j = 1,2,\ldots,J$ indicate the number of rows and columns of each image in the sequence. Here $k = 1,2,\ldots,K$ denotes the frame number in the image sequence. In order to obtain a focus volume $g_m^{\prime(k)}(i,j)$, m^{th} focus measure operator F_m is applied on each image in the sequence, i.e.,

$$g_m^{\prime(k)}(i,j) = F_m(g^{(k)}(i,j)), \quad k = 1,2,\cdots,K. \quad (4)$$

For each object point, the sharpest pixel provides depth information. Therefore, the initial depth map is constructed by selecting the maximum values along the optical axis as:

$$d_m(i,j) = \operatorname*{argmax}_k \, (g_m^{\prime(k)}(i,j)), \quad k = 1,2,\cdots,K. \quad (5)$$

Corresponding to each initial depth $d_m(i,j)$, the best focus values $v_m(i,j)$ are computed from the image focus volume, i.e.

$$v_m(i,j) = g_m^{\prime(d_m)}(i,j). \quad (6)$$

The training dataset $X = \{(\mathbf{x}^{(n)}, y^{(n)})\}_{n=1}^{N}$, is constructed by selecting N data point out of total $L = I \times J$ points. Each data point consists of input feature vector and corresponding target value. The feature vector x consists of initial depth and focus values. The initial depth d_m and focus value v_m are computed using Eq. (7) and Eq. (8) respectively. In this work, we construct homogeneous and heterogeneous feature vectors. The homogeneous feature vector x_m is constructed using F_m focus measure, whereas the heterogeneous feature vector x_h is formed from various focus measures. Using these two types of feature vectors, homogeneous function $F'_m : \mathbf{x}_m \rightarrow y$ and heterogeneous function $F'_h : \mathbf{x}_h \rightarrow y$ are developed.

To develop homogenous feature vector x_m we use the initial depth and 3D neighborhood around best-focused pixel. Neighboring focus values of the best-focused pixel have greater influence on the depth value. There are many approaches to construct 3D neighborhood (Bernd, 2005). However, we use the six-pixel neighborhood around each best focused pixel in the focus volume $g_m'^{(k)}(i,j)$. Thus, a set s_m of seven focus values is obtained, i.e.,

$$s_m = \begin{Bmatrix} v_m^{(1)} = g_m'^{(d_m)}(i,j), v_m^{(2)} = g_m'^{(d_m)}(i-1,j), \\ v_m^{(3)} = g_m'^{(d_m)}(i+1,j), v_m^{(4)} = g_m'^{(d_m)}(i,j-1), \\ v_m^{(5)} = g_m'^{(d_m)}(i,j+1), v_m^{(6)} = g_m'^{(d_m-1)}(i,j), \\ v_m^{(7)} = g_m'^{(d_m+1)}(i,j) \end{Bmatrix} \quad (7)$$

Thus, for each point (i,j), an eight-dimensional homogenous feature vector x_m is constructed, i.e.,

$$x_m = (d_m, s_m) \quad (8)$$

From the object with known depth data, the training dataset is formed by randomly selecting N pairs of data points i.e., $X_m = \left\{\left(\mathbf{x}_m^{(n)}, y^{(n)}\right)\right\}_{n=1}^{N}$.

We found that the heterogeneous combination of prediction models is more informative than the homogenous combination of prediction models (Khan, Abdul, & Anwar, 2005; Majid, 2006; Majid, et al., 2006). Applying this concept, we construct heterogeneous feature vector using various focus measures to develop heterogeneous function $F_h' : \mathbf{x}_h \to y$. The heterogeneous feature vector x_h for each object point (i,j) is constructed by including the initial depth d_m and best focused pixel v_m from each of M focus measures, i.e.,

$$x_h = (d_1, \ldots, d_M, v_1, \ldots, v_M) \quad (9)$$

Currently, we use three most commonly used focus measures F_1, F_2, and F_3 for $m = 1,2,3$ respectively, however, other focus measures in spatial and transform domains can also be incorporated. In this way, a six-dimensional homogeneous feature vector $x_h = (d_1, d_2, d_3, v_1, v_2, v_3)$ is formed. From the simulated object with known depth, we prepare training set $X_h = \left\{\left(\mathbf{x}_h^{(n)}, y^{(n)}\right)\right\}_{n=1}^{N}$ of N observations, where x_h is the nth input vector and $y^{(n)}$ is corresponding scalar target value, respectively. For brevity, we use separate notations for the homogeneous vector, heterogeneous vector and corresponding OCD functions.

Function Development

In this module, the main task is the adequate representation of desired solutions in tree-like data structure. To develop homogeneous function F_m', we provide eight-dimensional homogenous feature vector x_m as input variables for GP tree. Whereas, six-dimensional heterogeneous feature vector x_h gives six input variables for the development of heterogeneous function F_h'. Further, randomly generated numbers in the range of [0-1] are used as constant terminals. On the other hand, the set of non-terminals in GP tree comprises of simple arithmetic and trigonometry operators i.e., *plus, minus, times, divide, sin, cos, log, power,* and *exp*. Table 1 shows all necessary parameter setting for GP simulation.

In GP evolution process, first, an initial population of size Z is generated using ramped half-and-half method (Majid, et al., 2006). In a population, possible candidate solution V is a constituent of randomly selected variables, constants and mathematical functions. In second step, the fitness of each individual tree in the population is assessed using mean square error (MSE) as a fitness criterion. This minimum value of fitness measure $Fit(V)$ indicates how effectively a GP individual moves towards the optimal solution (Majid, 2006).

$$Fit(V) = \frac{1}{N}\sum_{n=1}^{N}\left(eval\left(V^{(n)}\right) - y^{(n)}\right)^2, \quad n = 1,2,\cdots,N. \quad (10)$$

Table 1. GP parameters setting to develop OCD function

GP parameters	**Set values**
Terminals set	• A homogenous feature vector x_m to develop homogenous function $F'_m : \mathbf{x}_m \to y$; $\mathbf{x}_m = \left(d_m, v_m^{(1)}, v_m^{(2)}, v_m^{(3)}, v_m^{(4)}, v_m^{(5)}, v_m^{(6)}, v_m^{(7)}\right)$ • A heterogeneous feature vector x_h to develop heterogeneous function $F'_h : \mathbf{x}_h \to y$; $x_h = (d_1, d_2, d_3, v_1, v_2, v_3)$ • Random constants in the range of [0-1]
Functions set	plus, minus, times, divide, log, sin, cos, exp, power
Fitness criterion	Mean square error
Population size and No. of generations	50 and 300, respectively
Population initialization	Ramped half and half
Initial tree depth	5
Expected offspring	Rank85
Operators probabilities	Variable crossover/mutation ratio
Population sampling	Tournament
Survival	Keep the best individuals

where $y^{(n)}$ is the nth target depth value corresponding to the nth training pattern. In the third step of GP process, based on the survival of fittest rule, i.e. the best candidates ranked and selected from the population. The probability of individual to be selected within the population is computed as (Majid, 2006):

$$\Pr\left(V_z\right) = \frac{Fit\left(V_z\right)}{Total\ Fitness}, \tag{11}$$

where

$$Total\ Fitness = \sum_{z=1}^{Z} Fit\left(V_z\right). \tag{12}$$

The selected candidates are used for the creation of next generation. Crossover, mutation, and replication operators are applied on the selected individuals to generate new population. Crossover operator creates offspring by exchanging genetic material between two individual parents. To obtain good results through crossover, we used tournament selection method (Koza, et al., 2008). This selection works by selecting trees at random from the current generation. Two trees with the highest fitness values are exchanged sub-trees resulting in two new possible solutions. Crossover helps in converging to optimal/near-optimal solution. However, in mutation process, a small part of individual often brings diversity in the solution space. For GP simulation, a ratio of crossover/mutation is automatically adapted. During simulation, each new generation has a slightly higher average fitness score. In this way, the solution space is refined and converges to the optimal/near optimal solution. The simulation is stopped if either the number of generations reaches the maximum limit or the fitness value (MSE) approaches the minimum set value.

Depth Estimation

Once, OCD function is developed through the GP module, the optimal depth map of an object can easily be computed. It is to be noted that for OCD function development, we prepare training data from simulated object with known depth. Moreover, randomly selected data points are used not only to reduce the training time but also to improve the generalization. During the development of OCD functions, we found that the order of the feature vectors is insignificant.

In order to obtain a complete depth map of an object, we need to provide input feature vector for each object point to the depth estimator. Consider again, an image sequence $g^{(k)}(i,j)$ of the test object that is acquired using CCD camera by displacing object toward the camera lens. For total K images, each of size $I \times J = L$ in the sequence, we are interested to estimate depth $\hat{d}(i,j)$ for each object point (i,j) First, lexicographically arranged vectors are represented in a matrix $X = \left\{\left(\mathbf{x}^{(n)}\right)\right\}_{n=1}^{L}$ of size $L \times R$, where x represents a R dimensional feature vector and L indicates the total feature vectors respectively. The optimal depth value $\hat{d}$ is estimated using the OCD function F', i.e.;

$$\hat{d} = F'\left(\mathbf{x}\right) \tag{13}$$

The function F' may be homogeneous or heterogeneous function. In case of homogeneous function F'_m, the feature vector x_m obtained through m^{th} focus measure will be used to estimate depth. Similarly, heterogeneous feature vector x_h will be used to estimate depth from heterogeneous function F'_h.

RESULTS AND DISSCUSSION

Implementation Details

For the development of OCD function, we prepared training data from the simulated object. A sequence of 97 images, each of size 360 × 360, of a simulated cone was generated synthetically by using simulation software (Subbarao & Choi, 1995). For each pixel, a homogeneous feature vector x_1 is formed by applying F_1 focus measure. In this way, feature vectors along with true depth values are used to construct training data of 129600 (=360 × 360) points. To avoid the over fitting problem and to reduce the training time, we randomly selected 40,000 data points.

Figure 4 (a) shows the curve of the best fit individual, in each generation, during the development of homogeneous function F_1'. The complexity is expressed as a function of tree depth level and the number of nodes. During GP evolution, some constructive blocks are created, which try to minimize the destruction of useful building blocks. As a result, the size of GP individual grows exponentially without appreciable improvement in performance curve of the best individual. This behavior can be clearly observed in the middle region of the accuracy curve in Figure 4 (a). This might be due to the occurrence of bloating phenomenon during GP evolution (Langdon, 2000; Majid, 2006). Many branches may not contribute in improving the performance. As a result, the best genome's total number of nodes increases and average tree depth becomes very large. Therefore, with the increase of complexity, the performance curve of the best individual approaches towards the optimal solution. During the evolution, the selected parameters are optimally combined to develop homogenous OCD function. Figure 4 (b) shows the graphical representation of the best homogeneous OCD function F_1' obtained at the end of simulation. Similarly, we obtain numerical expressions for the homogeneous

Figure 4. (a) Improvement in accuracy/fitness score exhibited by the best individual in each generation; and (b) Tree representation of the best individual F_1' obtained at the end of GP simulation

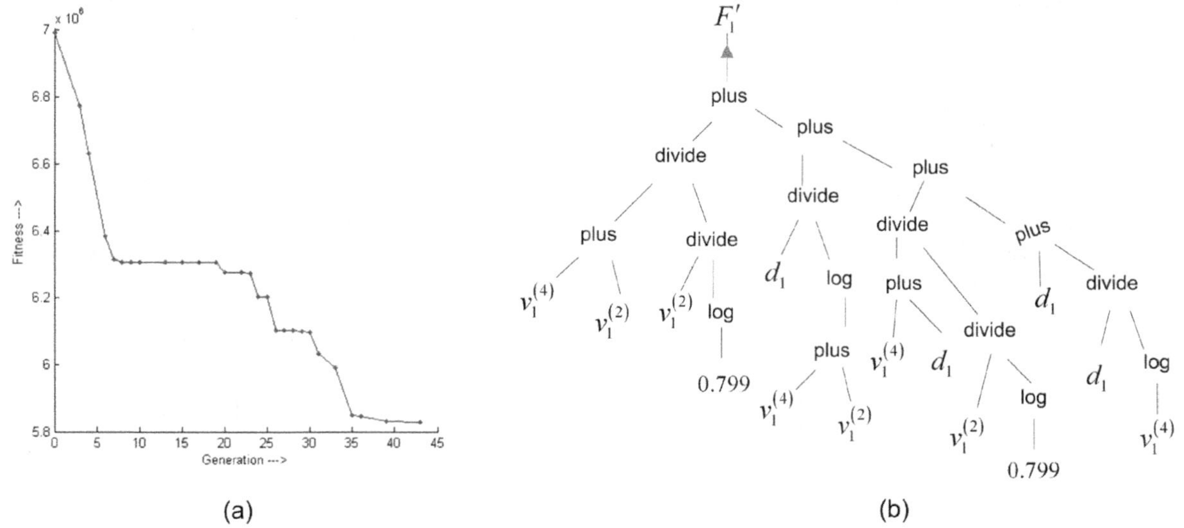

functions F_2' and F_3' through the GP evolutionary cycle for homogeneous feature vectors x_2 and x_3 respectively. These expressions, in prefix forms, are given as follows:

$$F_1'(\mathbf{x}_1) = \mathit{plus}(\mathit{divide}(\mathit{divide}(v_1^{(2)}, \log(0.799)), \mathit{plus}(v_1^{(4)}, v_1^{(2)})),\ \mathit{plus}(\mathit{divide}(d_1, \log(\mathit{plus}(v_1^{(4)}, v_1^{(2)}))), \mathit{plus}(\mathit{divide}(\mathit{divide}(v_1^{(2)}, \log(0.799)), \mathit{plus}(v_1^{(4)}, d_1)), \mathit{plus}(\mathit{divide}(d_1, \log(v_1^{(4)})), d_1)))). \tag{14}$$

$$F_2'(\mathbf{x}_2) = \mathit{plus}(d_2, \mathit{times}(\log(v_2^{(3)}), \sin(\mathit{times}(\mathit{times}(\log(v_2^{(5)}), \sin(\mathit{times}(\log(d_2), \log(0.40753)))), 0.4638)))). \tag{15}$$

$$F_3'(\mathbf{x}_3) = \mathit{plus}(\mathit{divide}(v_3^{(3)}, v_3^{(1)}), \mathit{plus}(\mathit{divide}(\sin(\mathit{plus}(0.44722, \mathit{plus}(v_3^{(3)}, v_3^{(2)}))), \mathit{plus}(\mathit{divide}(v_3^{(3)}, v_3^{(4)}), \mathit{plus}(\mathit{divide}(\sin(\mathit{plus}(0.44722, \mathit{plus}(v_3^{(3)}, v_3^{(2)}))), \mathit{power}(0.39891, \exp(\mathit{divide}(v_3^{(5)}, v_3^{(6)})))), \mathit{plus}(\mathit{divide}(\sin(\mathit{plus}(0.44722, \mathit{plus}(v_3^{(3)}, v_3^{(2)}))), \mathit{power}(0.39891, \exp(\mathit{divide}(v_3^{(5)}, v_3^{(6)})))), d_3)))), \mathit{plus}(\mathit{divide}(v_3^{(3)}, v_3^{(1)}), d_3))). \tag{16}$$

The heterogeneous function $F_h'(\mathbf{x}_h)$ is developed using heterogeneous feature vectors x_h where x_h is formed by applying focus measures F_1, F_2, and F_3. The best obtained numerical expression is given as:

$$F_h'(\mathbf{x}_h) = \mathit{plus}(\mathit{plus}(\log(\sin(\mathit{plus}(\log(0.7116), \log(\mathit{times}(\log(\log(d_2)), d_2))))), \mathit{plus}(\log(\sin(\log(d_2))), \mathit{plus}(\log(\log(d_1)), \mathit{plus}(\mathit{power}(v_1, \mathit{divide}(\mathit{plus}(0.4306, d_1), v_3)), \mathit{plus}(\log(\sin(\log(d_2))), \mathit{plus}(\log(\log(d_1), \mathit{plus}(\mathit{power}(v_1, \mathit{divide}(\mathit{plus}(0.4306, d_1), v_3)), \mathit{plus}(\log(\sin(\log(d_2))), \mathit{plus}(\log(\sin(\mathit{plus}(\log(0.7116), \log(\mathit{times}(\log(\log(d_2)), \mathit{plus}(\log(\sin(\log(d_1))), \mathit{plus}(\log(\log(d_1)), \mathit{plus}(\mathit{power}(v_1, \mathit{divide}(\mathit{plus}(0.4306, d_1), v_3)), \mathit{plus}(\log(\sin(\log(d_2))), \mathit{plus}(\log(\sin(\mathit{plus}(\log(0.7116), \log(\mathit{times}(\log(\log(d_2)), d_2))))), \mathit{plus}(\log(\sin(\log(d_1))), \mathit{plus}(\log(\log(d_1)), \mathit{plus}(\log(\sin(\log(d_2))), \mathit{plus}(\log(\sin(\log(d_2))), \mathit{plus}(\log(d_2), d_1))))))))))))))), \mathit{plus}(\log(\sin(\log(d_1))), \mathit{plus}(\log(\log(d_1)), \mathit{plus}(\log(\sin(\log(d_2))), \mathit{plus}(\log(\sin(\log(d_2))), \mathit{plus}(\log(\log(v_3)), d_1)))))))))))))), \mathit{plus}(\mathit{plus}(\log(0.7116), \log(\sin(\log(d_2)))), \mathit{plus}(\log(\sin(\log(d_2))), \log(d_2)))). \tag{17}$$

The improved performance of the developed functions is highly dependent on the optimal combination of initial depth and focus parameters along with arithmetic functions and some random constants. These homogeneous and heterogeneous functions are employed to estimate the 3D structure of any arbitrary object.

Experiments with Synthetic Images

To report the improved performance of OCD functions, we carried out several experiments using both synthetic and real image sequences. In case of synthetic object, it is possible to compute qualitatively discrepancies between the true and the estimated depth maps. We use three metrics: Mean Square Error (MSE), correlation, and Peak Signal-to-Noise Ratio (PSNR). MSE quantifies the amount by which a processed depth map differs from the ground-truth depth map. It measures the average of square of the error or distortion between the true and the estimated depth maps as:

$$MSE = \frac{1}{L}\sum_{i=1}^{I}\sum_{j=1}^{J}\left(\hat{d}\left(i,j\right) - d_t\left(i,j\right)\right)^2 \quad (18)$$

where $d_t(i,j)$ and $\hat{d}\left(i,j\right)$ are true and estimated depth maps respectively. The smaller the RMSE is, better the result will be. The correlation metric provides the similarity between two depth maps. Correlation coefficient indicates the strength and direction of a linear relationship between the actual and approximated depth maps. It can be computed as:

$$Correlation = \frac{\sum_{i=1}^{I}\sum_{j=1}^{J}\left(d_t\left(i,j\right) - \overline{d_t\left(i,j\right)}\right)\left(\hat{d}\left(i,j\right) - \overline{\hat{d}\left(i,j\right)}\right)}{\sqrt{\left(\sum_{i=1}^{I}\sum_{j=1}^{J}\left(d_t\left(i,j\right) - \overline{d_t\left(i,j\right)}\right)^2\right)\left(\sum_{i=1}^{I}\sum_{j=1}^{J}\left(\hat{d}\left(i,j\right) - \overline{\hat{d}\left(i,j\right)}\right)^2\right)}}. \quad (19)$$

where $\overline{d_t\left(i,j\right)}$ and $\overline{\hat{d}\left(i,j\right)}$ are the means of the original depth map and estimated depth map respectively. The PSNR is one of the most commonly used as a measure of quality of reconstruction. PSNR is the ratio between the maximum possible power of a signal and the power of corrupting noise. It is usually expressed in terms of the logarithmic decibel scale:

$$PSNR = 10\log_{10}\left(\frac{max_d^2}{MSE}\right) \quad (20)$$

where $\max_d$ is the maximum depth value of the object. A higher PSNR indicates the reconstruction of higher quality.

Figure 5 (first row) shows different frames extracted from the image sequence of simulated cone. From these images, it can be observed that some parts of the cone are well focused and some parts are defocused. Second row in the Figure 5 shows depth maps of simulated cone object computed through the traditional focus measures F_1, F_2, F_3 and heterogeneous F_h' function. For the traditional focus measures, first, focus volumes are computed by using window size $5 \times 5$and then depth maps are obtained by maximizing the focus measure along the optical direction. To estimate the depth map from heterogeneous F_h' function, the feature vector x_h for each object point is computed as described in Section 4.1. It can be observed that 3D shapes recovered through the proposed approach are better than conventional methods. The heterogeneous F_h' function, based on the optimal combination of conventional focus measures, has provided more accurate depth map.

Table 2 shows the performance comparisons in terms of MSE, correlation, and PSNR. The numerical values for these metrics are computed using true depth map of simulated cone and estimated depth maps using traditional focus measures, homogeneous, and heterogeneous functions. Homogeneous feature vectors x_1, x_2, and x_3 for each object point are computed using focus measures F_1, F_2, and F_3 respectively. Depth maps of simulated cone are estimated using homogeneous functions F_1', F_2', and F_3'. It can be observed that the estimated depth maps through the homogeneous and heterogeneous functions are closer to the actual depth map of simulated cone. The correlation value for the heterogeneous function F_h' is increased by 3.46%, 3.21%, and 3.37% than the conventional measures respectively. GP based heterogeneous function exhibits improvement 13.69%, 114.81%, and 114.45% as compared to traditional methods F_1, F_2, and F_3 respectively. Similarly, improvement in terms of PSNR measure can be observed from the Table 4. Figure 9 shows the performance comparison in pictorial form. It

Figure 5. (First row) sample frames extracted from image sequence of synthetic object simulated cone (a) frame number 20, (b) frame number 40, (c) frame number 60 (d) frame number 80. (Second row) depth map reconstructed using (e) F_1, (f) F_2, (g) F_3, and (h) F_h'.

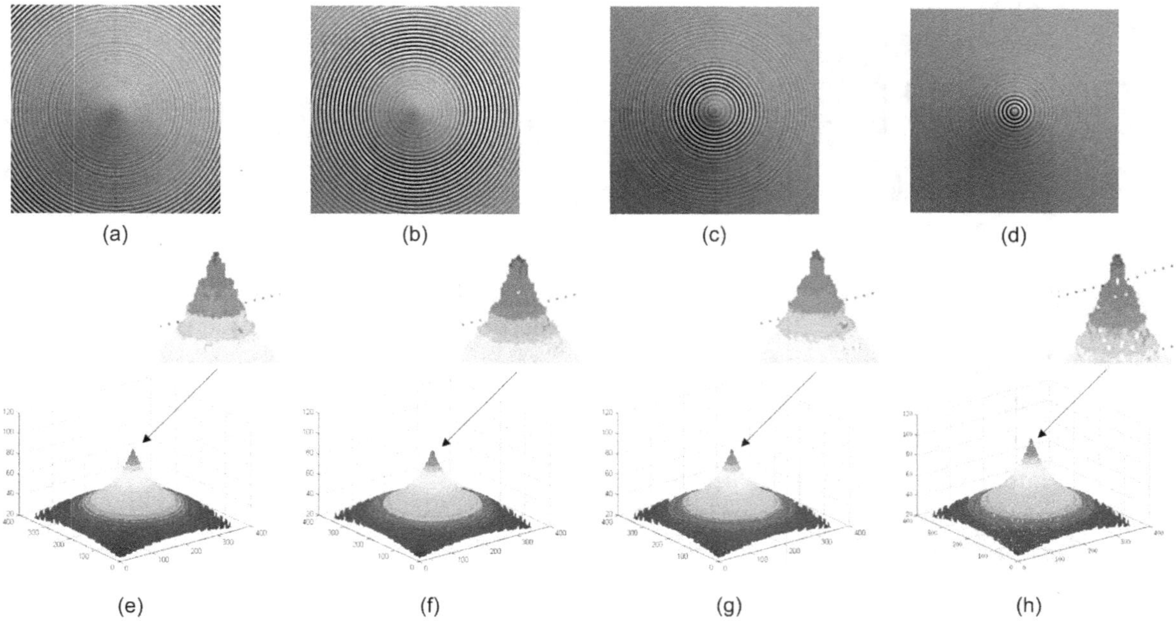

is observed that the GP based functions F_1', F_2', F_3', and F_h' provided considerable improvement as compared to the conventional measures in terms of MSE, correlation and PSNR. Further, it is inferred that the heterogeneous function F_h' is more effective than homogeneous functions F_1', F_2', and F_3'.

Experiments with Real Images

Images for real objects have been acquired through the microscope control system (MCS) (Ahmad & Choi, 2007; Malik & Choi, 2007a). The MCS consists of a charge-coupled device (CCD) camera mounted on microscope, a frame grabber card integrated with computer, a motor driver with step size 2.5 *nm* to move the object plane along the optical axis, and software for capturing images by controlling the step size and number of images. The second object taken was TFT-LCD color filter. These filters have size in microns and are used to develop thin and bright displays. Sequence of 90 images, each of size 300 × 300, has been considered for experiments. The third object is a slanted planar, while the fourth object is a real cone whose 97 images, each of size 200 × 200, were taken using CCD camera. The real cone object was made of hardboard with black and white stripes drawn on its surface for dense texture. First row of Figure 6 shows the sample images extracted from the image sequences of these objects.

Figure 6 shows 3D shapes reconstructed for test objects TFT-LCD color filter, real cone, and planar object. The depth maps of these objects are computed using traditional methods F_1, F_2, F_3, and the heterogeneous function F_h'. In case of traditional methods, first, focus volumes are computed using focus measures and focus value are aggregated by using a window size 5 × 5. Then the depth maps are obtained by maximizing the focus measure along the optical axis. For the function F_h', feature vector x_h for each object point

Table 2. Performance comparisons of different methods

Measure	Conventional			Homogeneous			Heterogeneous
	F_1	F_2	F_3	F_1'	F_2'	F_3'	F_h'
MSE	62.0373	63.0752	62.7372	33.3168	26.7515	50.4217	17.0687
Correlation	0.9367	0.09391	0.9376	0.9491	0.9632	0.9484	0.9697
PSNR	20.9631	20.8911	20.9144	23.9144	24.6491	21.8635	26.5677

is computed as described in Section "Preparing training data. It is to be noted that the data generated for depth estimation from real objects is entirely different from the training data generated by the simulated cone. For each test object, the complete depth map is computed using F_h'. Figure 6 shows the improved performance of the proposed GP based scheme. It can be observed that 3D shapes recovered through the proposed scheme are better than conventional methods. On the other hand, 3D shapes reconstructed through the traditional methods generated coarse surfaces due to their limited capability in computing focus measurements accurately. Noisy focus values introduce error in depth values and relatively more spikes can be observed in the constructed depth maps of the objects. However, F_h' has significantly reduced the effect of noisy-focus measurements and generated accurate depth maps.

FUTURE RESEARCH DIRECTIONS

In proposed GP based scheme, the main concept is to estimate a nonlinear function for computing optimal depth of the object. In probability theory and mathematics, function estimation is also known as regression. This kind of function estimation is possible through many ways such as Gaussian process regression (GPR), support vector regression (SVR), and generalized regression neural networks (GRNN). For such learning algorithms, the features (input data) play vital role to learn a generalized nonlinear function. Therefore, the input feature vector can be enriched by including additional useful information. The accuracy of the estimated depth map may be enhanced by including information from more robust focus measures.

CONCLUSION

In this chapter, we explain the SFF, i.e. one famous passive method to estimate the structure of an object from a sequence of images. A comprehensive discussion is made on existing approximation and machine learning based SFF approaches. Besides this, we developed an improved performance composite function for optimal depth estimation of real objects through GP based technique. The main advantage of the proposed scheme is that useful information of individual focus measures is automatically selected and combined during GP evolution cycle. Another benefit is that this generic method does not depend on a specific focus measure. Moreover, the capability of proposed depth estimator can be enhanced by adding more useful information through focus measures. The performance of this method is also investigated for homogenous and heterogeneous combination using single and multiple focus measures respectively. Through various experiments, it is found that heterogeneous combination is more informative. Experimental results have demonstrated that our generalized depth estimator has provided more accurate depth maps than existing SFF methods.

Figure 6. (First row) Sample frames extracted from image sequence of real objects (a) TFT-LCD color filter, (b) real cone, (c) planar object. 3D shapes reconstructed for TFT-LCD color filter, real cone, and planar object using (second row) (d-f) F_1, (third row) (g-i) F_2, (forth row) (j-l) F_3, and (fifth row) (m-o) F_h'.

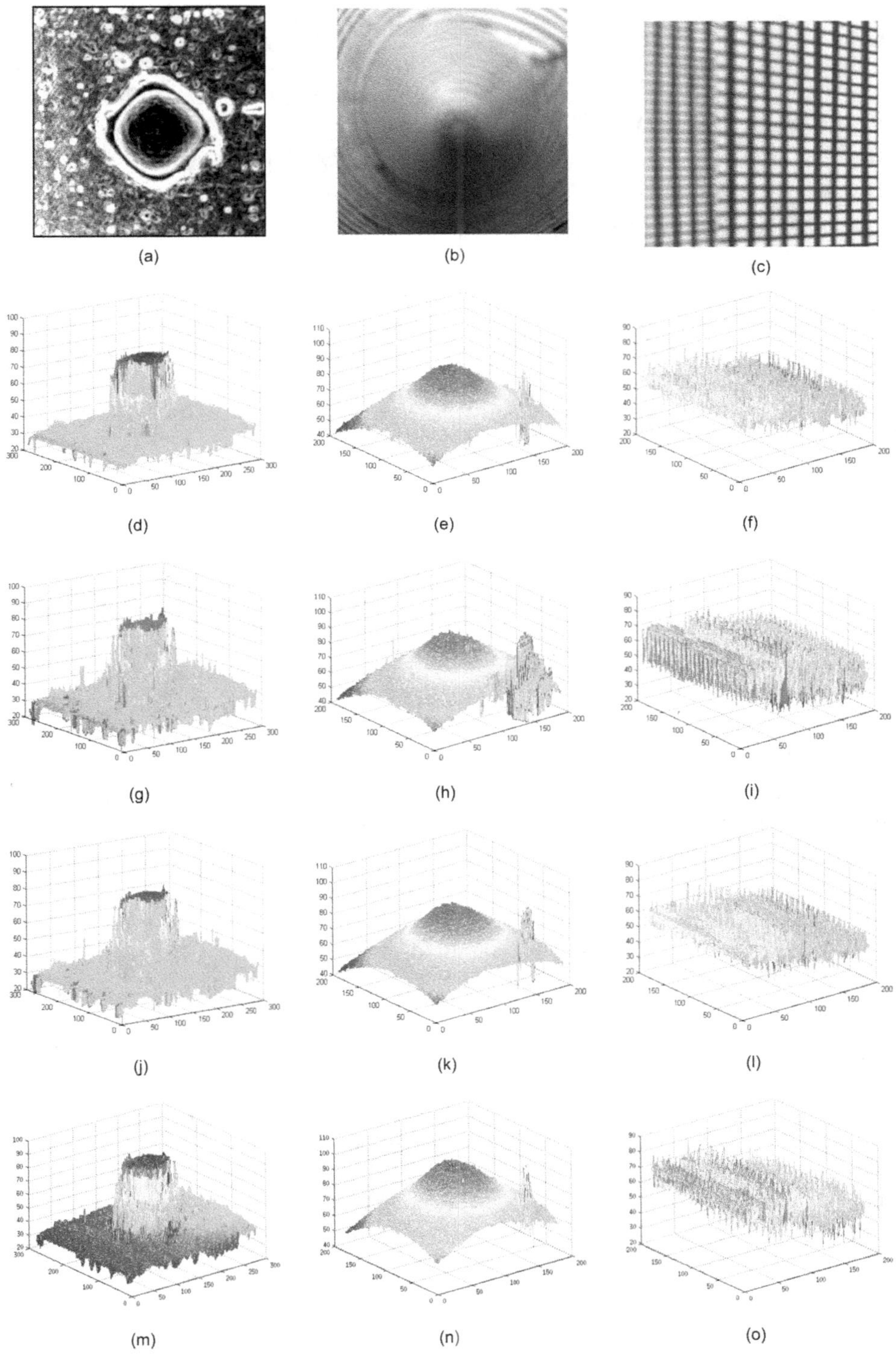

REFERENCES

Ahmad, M. B., & Choi, T. S. (2005). A heuristic approach for finding best focused shape. *IEEE Transactions on Circuits and Systems for Video Technology*, *15*(4), 566–574. doi:10.1109/TCSVT.2005.844450

Ahmad, M. B., & Choi, T. S. (2007). Application of three dimensional shape from image focus in LCD/TFT displays manufacturing. *IEEE Transactions on Consumer Electronics*, *53*(1), 1–4. doi:10.1109/TCE.2007.339492

Asif, M., & Choi, T. S. (2001). Shape from focus using multilayer feedforward neural networks. *IEEE Transactions on Image Processing*, *10*(11), 1670–1675. doi:10.1109/83.967395

Bernd, J. (2005). *Digital image processing* (6th ed.). Heidelberg, Germany: Springer-Verlag.

Boissenin, M., Wedekind, J., Selvan, A. N., Amavasai, B. P., Caparrelli, F., & Travis, J. R. (2007). Computer vision methods for optical microscopes. *Image and Vision Computing*, *25*(7), 1107–1116. doi:10.1016/j.imavis.2006.03.009

Brenner, J. F., Dew, B. S., Horton, J. B., King, T., Neurath, P. W., & Selles, W. D. (1976). An automated microscope for cytologic research a preliminary evaluation. *The Journal of Histochemistry and Cytochemistry*, *24*(1), 100–111. doi:10.1177/24.1.1254907

Choi, T. S., & Yun, J. (2000). Three-dimensional shape recovery from the focused-image surface. *Optical Engineering (Redondo Beach, Calif.)*, *39*(5), 1321–1326. doi:10.1117/1.602498

dos Santos, J. A., Ferreira, C. D., & Torres, R. S. (in press). Gon?lves, M. A., & Lamparelli, R. A. C. (2010). A relevance feedback method based on genetic programming for classification of remote sensing images. [*Corrected Proof.*]. *Information Sciences*.

Groen, F. C. A., Young, I. T., & Ligthart, G. (1985). A comparison of different focus functions for use in autofocus algorithms. *Cytometry*, *6*(2), 81–91. doi:10.1002/cyto.990060202

Helmli, F. S., & Scherer, S. (2001). Adaptive shape from focus with an error estimation in light-microscopy.*Proceeding of the 2nd International Symposium on Image and Signal Processing and Analysis* (pp. 188-193).

Khan, A., Abdul, M., & Anwar, M. M. (2005). Combination and optimization of classifiers in gender classification using genetic programming. *International Journal of Knowledge-Based and Intelligent Engineering Systems*, *9*(1), 1–11.

Kouchakpour, P., Zaknich, A., & Brnl, T. (2009). Dynamic population variation in genetic programming. *Information Sciences*, *179*(8), 1078–1091. doi:10.1016/j.ins.2008.12.009

Koza, J. R., Streeter, M. J., & Keane, M. A. (2008). Routine high-return human-competitive automated problem-solving by means of genetic programming. *Information Sciences*, *178*(23), 4434–4452. doi:10.1016/j.ins.2008.07.028

Kristan, M., Perv, J., Pervse, M., & Kovavcic, S. (2006). A Bayes-spectral-entropy-based measure of camera focus using a discrete cosine transform. *Pattern Recognition Letters*, *27*(13), 1431–1439. doi:10.1016/j.patrec.2006.01.016

Krotkov, E. (1988). Focusing. *International Journal of Computer Vision*, *1*(3), 223–237. doi:10.1007/BF00127822

Kyte, B. (2010). Optical 3D micro metrology. Retrieved November 15, 2010, from http://www.alicona.com/

Langdon, W. B. (2000). Size fair and homologous tree genetic programming crossovers. *Genetic Programming and Evolvable Machines*, *1*(1/2), 95–119. doi:10.1023/A:1010024515191

Mahmood, M. T., & Choi, T. S. (2010). Focus measure based on the energy of high-frequency components in the S transform. *Optics Letters*, *35*(8), 1272–1274. doi:10.1364/OL.35.001272

Mahmood, M. T., Choi, W. J., & Choi, T. S. (2008). PCA-based method for 3D shape recovery of microscopic objects from image focus using discrete cosine transform. *Microscopy Research and Technique*, *71*(12), 897–907. doi:10.1002/jemt.20635

Mahmood, M. T., Shim, S. O., & Choi, T. S. (2009). Shape from focus using principal component analysis in discrete wavelet transform. *Optical Engineering (Redondo Beach, Calif.)*, *48*, 057203. doi:10.1117/1.3130232

Majid, A. (2006). *Optimization and combination of classifiers using Genetic Programming. Faculty of Computer Science*. Pakistan: GIK Institute.

Majid, A., Khan, A., & Mirza, A. M. (2006). Combination of support vector machines using genetic programming. *International Journal of Hybrid Intelligent Systems*, *3*(2), 109–125.

Malik, A. S., & Choi, T. S. (2007a). Application of passive techniques for three dimensional cameras. *IEEE Transactions on Consumer Electronics*, *53*(2), 258–264. doi:10.1109/TCE.2007.381683

Malik, A. S., & Choi, T. S. (2007b). Consideration of illumination effects and optimization of window size for accurate calculation of depth map for 3D shape recovery. *Pattern Recognition*, *40*(1), 154–170. doi:10.1016/j.patcog.2006.05.032

Malik, A. S., & Choi, T. S. (2008). A novel algorithm for estimation of depth map using image focus for 3D shape recovery in the presence of noise. *Pattern Recognition*, *41*(7), 2200–2225. doi:10.1016/j.patcog.2007.12.014

Malik, A. S., & Choi, T. S. (2009). Comparison of polymers: A new application of shape from focus. *IEEE Transactions on Systems, Man and Cybernetics. Part C, Applications and Reviews*, *39*(2), 246–250. doi:10.1109/TSMCC.2008.2001714

Mallipeddi, R., Mallipeddi, S., & Suganthan, P. N. (2010). Ensemble strategies with adaptive evolutionary programming. *Information Sciences*, *180*(9), 1571–1581. doi:10.1016/j.ins.2010.01.007

Nayar, S. K., & Nakagawa, Y. (1994). Shape from focus. *IEEE Transactions on Pattern Analysis and Machine Intelligence*, *16*(8), 824–831. doi:10.1109/34.308479

Niederost, M., Niederost, J., & Scucka, J. (2003). *Automatic 3D reconstruction and visualization of microscopic objects from a monoscopic multifocus image sequence*. International Archives of Photogrammetry, Remote Sensing and Spatial Information Sciences.

Pentland, A. P. (1987). A new sense for depth of field. *IEEE Transactions on Pattern Analysis and Machine Intelligence*, *9*(4), 523–531. doi:10.1109/TPAMI.1987.4767940

Petrovic, N. I., & Crnojevic, V. (2008). Universal impulse noise filter based on genetic programming. *IEEE Transactions on Image Processing*, *17*(7), 1109–1120. doi:10.1109/TIP.2008.924388

Santos, A., Solorzano, O. D., Vaquero, J. J., Pena, J. M., Malpica, N., & Del, P. (1997). Evaluation of autofocus functions in molecular cytogenetic analysis. *Journal of Microscopy*, *188*(3), 264–272. doi:10.1046/j.1365-2818.1997.2630819.x

Shim, S., & Choi, T. S. (2010). A novel iterative shape from focus algorithm based on combinatorial optimization. *Pattern Recognition*, *43*(10), 3338–3347. doi:10.1016/j.patcog.2010.05.029

Shim, S. O., Malik, A. S., & Choi, T. S. (2009). Accurate shape from focus based on focus adjustment in optical microscopy. *Microscopy Research and Technique*, *72*(5), 362–370. doi:10.1002/jemt.20662

Subbarao, M., & Choi, T. S. (1995). Accurate recovery of three-dimensional shape from image focus. *IEEE Transactions on Pattern Analysis and Machine Intelligence*, *17*(3), 266–274. doi:10.1109/34.368191

Subbarao, M., Choi, T. S., & Nikzad, A. (1993). Focusing techniques. *Optical Engineering (Redondo Beach, Calif.)*, *31*(11), 2824–2836. doi:10.1117/12.147706

Subbarao, M., & Tyan, J. K. (1998). Selecting the optimal focus measure for autofocusing and depth-from-focus. *IEEE Transactions on Pattern Analysis and Machine Intelligence*, *20*(8), 864–870. doi:10.1109/34.709612

Sun, Y., Duthaler, S., & Nelson, B. J. (2004). Autofocusing in computer microscopy: Selecting the optimal focus algorithm. *Microscopy Research and Technique*, *65*(3), 139–149. doi:10.1002/jemt.20118

Tenenbaum, T. M. (1970). *Accommodation in computer vision*. Stanford University.

Thelen, A., Frey, S., Hirsch, S., & Hering, P. (2009). Improvements in shape-from-focus for holographic reconstructions with regard to focus operators, neighborhood-size, and height value interpolation. *IEEE Transactions on Image Processing*, *18*(1), 151–157. doi:10.1109/TIP.2008.2007049

Tian, Y., Shieh, K., & Wildsoet, C. F. (2007). Performance of focus measures in the presence of nondefocus aberrations. *Journal of the Optical Society of America. A, Optics, Image Science, and Vision*, *24*(12), 165–173. doi:10.1364/JOSAA.24.00B165

Wee, C. Y., & Paramesran, R. (2007). Measure of image sharpness using eigenvalues. *Information Sciences*, *177*(12), 2533–2552. doi:10.1016/j.ins.2006.12.023

Xie, H., Rong, W., & Sun, L. (2007). Construction and evaluation of a wavelet-based focus measure for microscopy imaging. *Microscopy Research and Technique*, *70*(11), 987–995. doi:10.1002/jemt.20506

Yap, P. T., & Raveendran, P. (2004). Image focus measure based on Chebyshev moments. *IEE Proceedings. Vision Image and Signal Processing*, *151*(2), 128–136. doi:10.1049/ip-vis:20040395

Zhang, Y., Zhang, Y., & Wen, C. (2000). A new focus measure method using moments. *Image and Vision Computing*, *18*(12), 959–965. doi:10.1016/S0262-8856(00)00038-X

KEY TERMS AND DEFINITIONS

Depth Map: A Depth map represents depth information instead of intensity values corresponding to two- dimensional array of an image.

Fitness Measure: It evaluates the performance of each individual (solution) in the population.

Focus Measure: An operator that measures the focus quality of the image by utilizing the high frequency contents in the image.

Focused Image Surface: The focused image surface (FIS) is formed by the set of points at which the object points are focused by the camera lens.

Genetic Programming: It is a well-known genetic operators based machine learning technique that can combine optimally arithmetic operators and data features.

Shape From Focus: A technique that retrieves 3D structure of an object from a sequence of its images with different focus setting.

Training Data: The training dataset consists of a set of input feature vectors with their corresponding target values.

Chapter 12
"Scanning from Heating" and "Shape from Fluorescence": Two Non-Conventional Imaging Systems for 3D Digitization of Transparent Objects

Fabrice Mériaudeau
Université de Bourgogne, France

R. Rantoson
Université de Bourgogne, France

G. Eren
Université de Bourgogne, France

L. Sanchez-Sécades
Université de Bourgogne, France

O. Aubreton
Université de Bourgogne, France

A. Bajard
Université de Bourgogne, France

D. Fofi
Université de Bourgogne, France

I. Mohammed
Université de Bourgogne, France

O. Morel
Université de Bourgogne, France

C. Stolz
Université de Bourgogne, France

F. Truchetet
Université de Bourgogne, France

ABSTRACT

3D surface acquisition is a subject which has been studied to a large extent. A significant number of techniques for acquiring shape have been proposed, and a wide range of commercial solutions are available. Nevertheless, today's systems still have difficulties when digitizing objects with non-Lambertian surfaces in the visible light spectrum, as is the case of transparent, semi-transparent or highly reflective materials (e.g. glass, crystals, some plastics and shiny metals). In this chapter, some of the issues of traditional scanning systems are addressed by considering various approaches using the radioactive properties of materials, the polarization information of the reflected light as well as the generated fluorescence applied

DOI: 10.4018/978-1-61350-326-3.ch012

to the digitization of transparent object These approaches led to three recent techniques which can be referred as shape from polarization, shape from fluorescence as well as shape from heating (SFH). The two latest approaches will be exposed throughout this chapter.

INTRODUCTION

The computer vision community has extensively developed techniques to determine the shape of objects. Laser light based scanning systems and structured lighting systems are probably the most commonly used techniques for acquiring the 3D shape of objects, however, reliable solutions are still lacking for non-opaque materials (specular or transparent objects). To overcome this problem, powder is usually spread onto the object prior digitization. This supplementary step is troublesome (the object has to be cleaned afterwards), and the accuracy is dependent on the powder thickness and homogeneousness. Various attempts have been proposed over the last few years for 3D surface acquisition of transparent objects and an exhaustive review can be found in (Ihrke, 2008) but the presented methods require a priori about the object or assumptions about the interactions of the light with the object surface.

This chapter presents two new approaches which are an extension of the well known structured lighting method to the thermal infrared range as well as to the UV range with induced fluorescence.

BACKGROUND

In this chapter, we present a new technique for 3D range scanning of transparent objects. 3D range scanning has been investigated for several decades and most of the proposed approaches assume a diffuse reflectance of the object's surface. The broad literature on the subject is usually divided into active and passive techniques. Active light techniques, whose recent review is proposed by Blais (2004), include laser range scanning, coded structured light systems (Salvi, 2004) and time-of-flight scanners (Bokhabrine 2010) whereas passive techniques are mainly stereovision (Horn, 1986), "shape from optical flow", shape from shading.... or multiview acquisition system (Harvent, 2010).

The further a surface deviates from the Lambertian reflectance assumption, the less accurate standard 3D range scanning techniques become. Figure 1 is an example of a glass bottle scanned by a commercial scanner without any preparation of the sample surface (powder spray) prior digitization.

Coating the object with a powder prior digitization might solve the problem (see Figure 2), on the other hand, this cannot be done in many applications because it involves additional handlings of the objects (coating, cleaning) which include higher processing costs.

The literature survey (Ihrke, 2008), (Ihrke, 2010) pinpoints several techniques to partially overcome this problem. For instance, in the computer graphics community Goesele et al., (Goesele, 2004) proposed a method for determining the scattering behaviour of translucent objects by using a laser, but the geometry was initially acquired by covering the object with a white coating. Similarly, Matusik et al., (Matuzik, 2002) presented an acquisition and rendering system for transparent and refractive objects from arbitrary viewpoints using a novel illumination, but the recovered geometry is just the visual hull (i.e. a very rough approximation of the object's shape). Morris and Kutulakos (Morris, 2007) proposed a method based on scatter-trace photography that provides good results for complex objects with an inhomogeneous interior. Ohara *et al.* (Ohara, 2003) estimated the depth of the edge of a trans-

Figure 1. (a) glass bottle; and (b) the 3D rendering obtained with a Vi 910 Minolta scanner without prior preparation of the surface

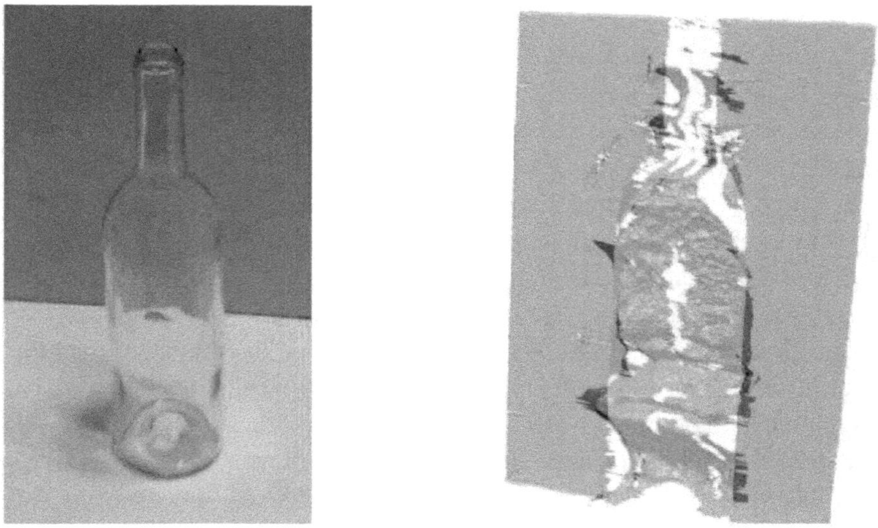

(a) (b)

Figure 2. From left to right: (a) specular object, same object after being coated with a powder; (b) transparent object

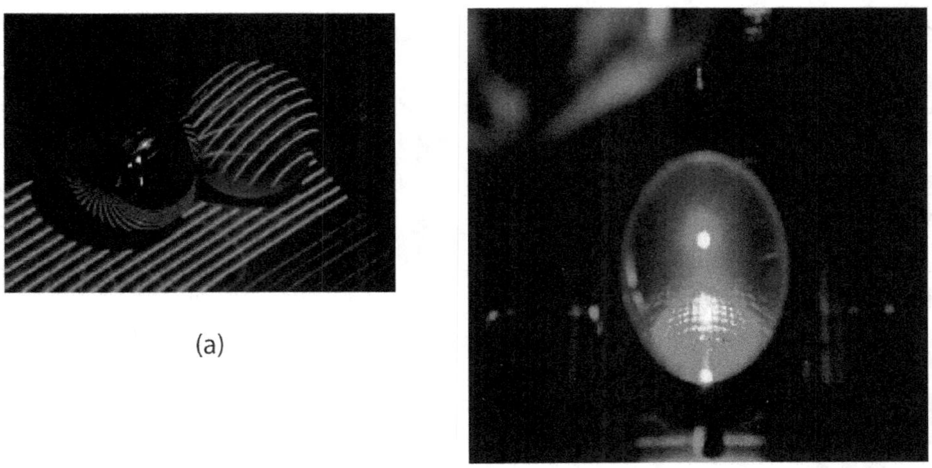

(a) (b)

parent object by using the shape from focus. Ben-Ezra and Nayar (Ben Ezra 2003) used the structure from motion to estimate the parameterized surface shape of transparent objects. These methods, however, do not estimate the arbitrary shapes of transparent objects. Trifonov *et al.* (Trifonov, 2006) have recently estimated the shape of a transparent object through visible tomography. However, their technique requires immersing in a liquid the object to be digitized and assumes perfect index matching (to avoid spurious interreflections and refractions) between the fluid and the object. Hullin *et al.* (Hullin, 2008) have developed a similar technique by immersing the object to be digitized in fluorescent liquid.

Myasaki and coworkers (Miyazaki 2002), (Miyazaki, 2004), (Saito, 1999) have developed a technique relying based on "shape from polarization" which was later on applied to metallic specular surfaces (Morel 2006). A recent extension of this method with multispectral information has been developed for transparent object (Ferraton 2009). However, these techniques required prior information on the refraction index of the object, or its estimation during the process, and ambiguities (zenithal and azimutal angles) which need to be solved or inferred from the experimental set-up. Recent works based on polarization coupled with inverse raytracing (D. Miyazaki, Ikeuchi (2005) led to interesting results which cope with some of the cons of the previous methods by taking into account the inter-reflections and not suffering from the ambiguity of the incident angle. However, this technique is still based on numerous assumptions such as a known back surface shape, known refractive index and a homogenous lighting system which cannot be guaranteed for any object shape.

Other methods have been developed over the last decade and are thoroughly presented by Ihrke (Ihrke, 2008) in its broad and complete survey on recent developed techniques for 3D digitization of specular and transparent surface. This chapter presents two new approaches for addressing some of the issues of visible light laser scanners by considering the heat pattern released by the object and the fluorescence induced by UV irradiation.

Regarding, the information carried by the heat released by samples, similar techniques based on an *IR* sensor to infer 3-D information have recently been developed: Pelletier and Maldague (Pelletier 1997) were among the pioneers to work on this idea. Their technique ("shape from heating") requires a presegmentation of the image to isolate linear patches and nonlinear patches, which are, afterward, used to lead to "extraction of relative depth" and "extraction of surface orientation". The technique is restricted to simple shapes such as cylinders, and accurate measurements for more complex objects have not yet been achieved. An extension of this technique, called the "shape from amplitude", has recently been proposed (Liu 2006). The technique uses amplitude images that are obtained in pulse phase thermography.

Ming *et al.* (Ming, 2005; Ming, 2007) calibrated an *IR* camera that acquires a sequence of 2-D thermographic images and then reconstructed the 3-D temperature distribution from the captured 2-D thermal images by the Octree carving technique. This technique is based on imaging the heat released by a surface that is being mechanically processed and is, by its principle, close to our system; however, the 3-D reconstruction principle based on the shape from silhouette is totally different from the technique presented in this chapter.

Sadjadi (Sadjadi, 2007) and Prakash *et al.*(Prakash, 2006) proposed a passive stereoscopic system. Both techniques suffer from the lack of texture on *IR* images leading to a sparse 3-D representation. To cope with this lack of information, our system relies on an *IR* pattern that is being simply projected onto the object (Eren, 2009), (Meriaudeau, 2010), and the heat released by the object (which has been heated by the *IR* radiation) is then imaged by a spatially calibrated *IR* sensor; the technique relies, therefore, on the observation of the emitted pattern.

Regarding shape from fluorescence or 3D from Fluorescence generated by UV radiations, to the best of our knowledge, only one approach is related to this (Hullin, 2008) but requires the immersion of the sample and is therefore not suitable for numerous applications.

The rest of this chapter is organized as follows: The first part is related to a 3-D laser scanning method called Scanning from Heating, the second part presents recent results obtained with the Shape from Fluorescence technique. The last part is dedicated to new research directions and the chapter ends with a short conclusion.

"SHAPE FROM HEATING" FOR 3D DIGITIZATION OF TRANSPARENT OBJECT

Shape From Heating (SFH) is a new method for 3D shape estimation based on local surface heating by a laser source and observation of it by a thermal camera. This technique is based on imaging the heat released by the object (which has been heated by an IR radiation) by a spatially calibrated IR sensor; the technique relies, therefore, on the observation of the emitted pattern.

Two approaches were recently developed: a point scanning approach (Enen, 2009) and a structured lighting system approach (Meriaudeau, 2010). In the point scanning approach (See Figure 3), the system is calibrated using the Zhang method

Figure 3. Schematic of the "shape from heating" point approach

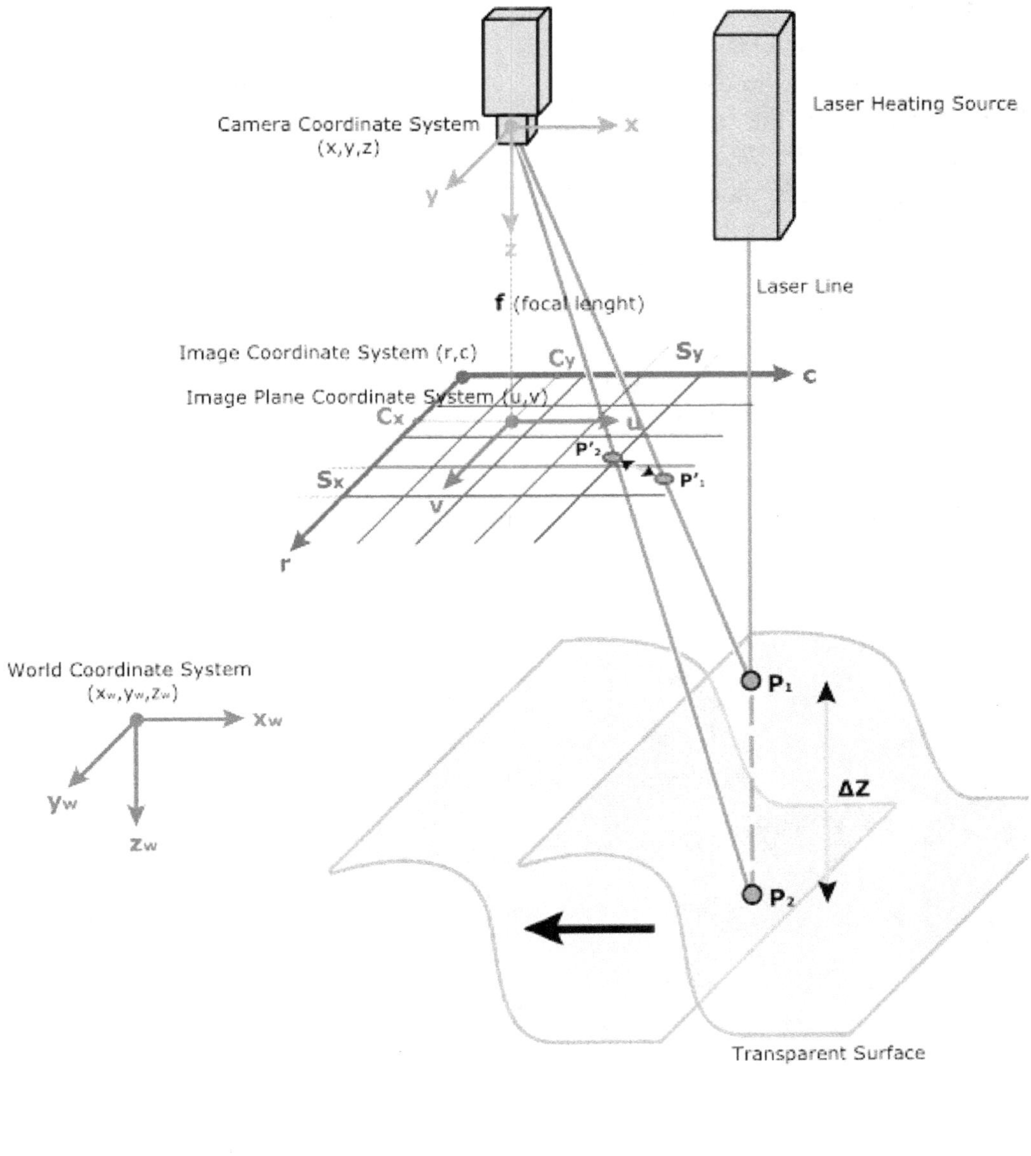

Figure 4. Calibration grid

(Zhang, 1999) with a thermal calibrating grid (See Figure 4) specially designed for this method.

Once the camera is calibrated, various "Z positions" are scanned so as to establish a corresponding table between the position in the image reference frame and the Z position in the world coordinate system. The transparent object is placed on a moving platform. The laser heat source, which is fixed, can be a point (Eren, 2009) or structured with a hemi-spherical lens (Meriaudeau, 2010) to generate a laser sheet, reducing the scanning time. For the structured lighting condition, the quadrangle techniques was used, so as to simultaneously calibrate the camera as well as the laser related to the camera (Figure 5).

The thermal camera (3 to 5 mm or 8-12 mm) is fixed and imaged the released heat pattern (see figure 6).

A CO_2 laser was chosen for the irradiation because at a wavelength of 10.6 µm, the absorption rate of most of glass or plastic materials is very high at the surface level, enabling therefore to get an accurate 3D digitization of the object surface without having any volume effect.

The SFH concept is as follows:

Figure 5. Calibration using the quadrangle technique

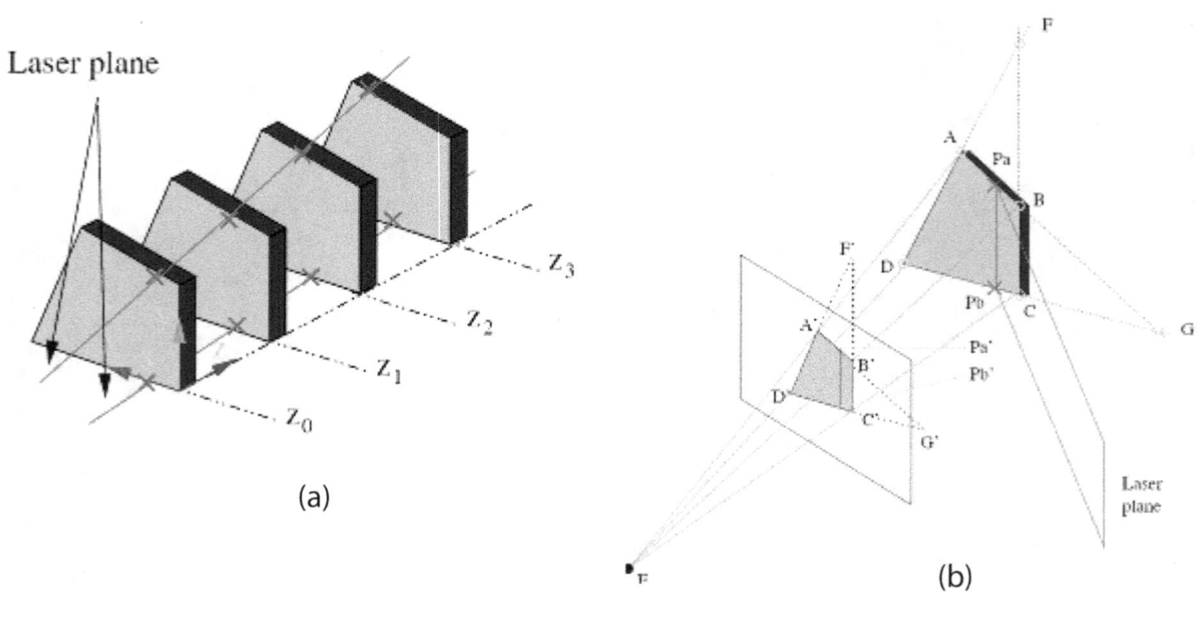

Figure 6. Experimental configuration for the structured lighting system configuration

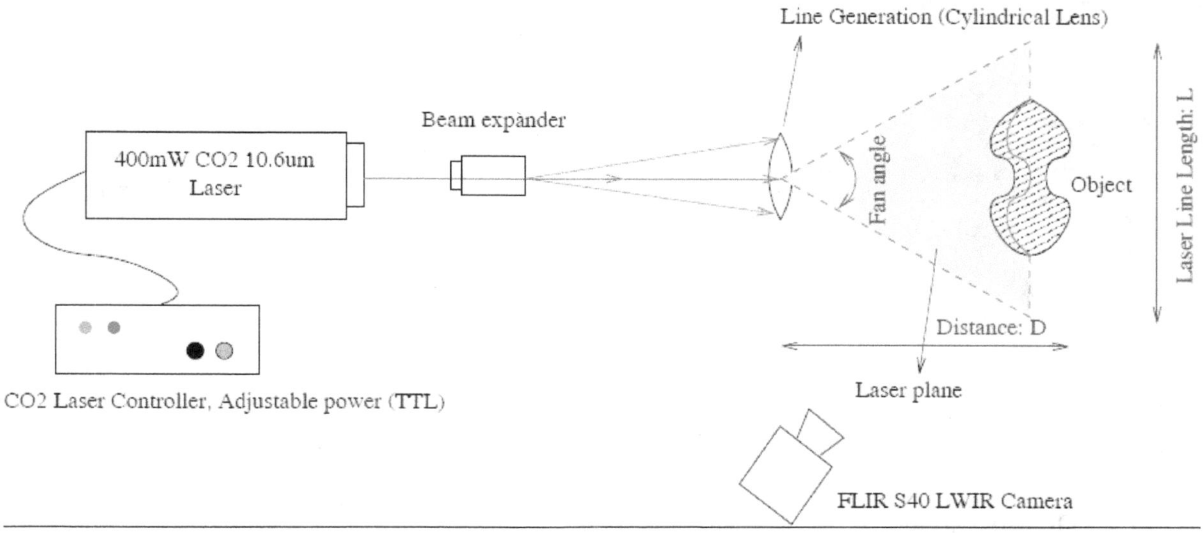

- When the laser is on, the sample is being heated.
- When the laser is turned off, the heat which is released by the sample is then imaged by the IR camera (see Figure 6) which has been previously calibrated so as to infer the 3D coordinates of the objects (Eren, 2009), (Meriaudeau, 2010).
- The method relies on few assumptions which are most of the time fulfilled:
    - The object surface is opaque to the laser heating source, and laser energy is

Figure 7 (a) Images of a heat pattern released in the case of a glass; (b) scanned using the Scanning from heating: Line configuration (Meriaudeau, 2010)

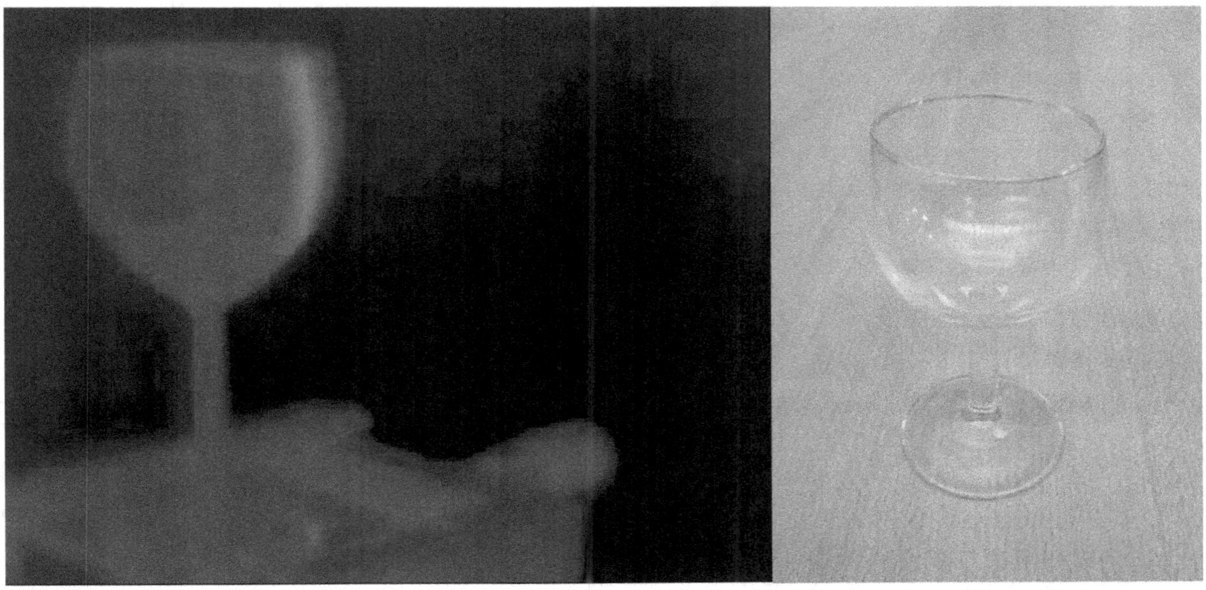

(a) (b)

Figure 8. From left to right: Glass container and its 3D representation obtained from SFH, Glass and its 3D representation obtained from SFH

absorbed by the surface at the object (without penetrating into the object).

- ◦ Once the surface is heated, the emission of thermal radiation is omnidirectional, so that it can be observed by the thermal camera.

- Calibration procedures involve either the quadrangle principle (Meriaudeau, 2010) realized with a glass quadrangle for the laser line projection technique or the general technique (Zhang, 1999) using a chessboard (Eren, 2009) which has to be realized with a graphite paint to enable the localization of landmarks in the IR spectrum after heating.

Many experiments were carried out on glass samples or plastic samples. Figures 8, 9, and 10 illustrate these results with some error map obtained by computing the normal difference between a reference object (obtained with a prior coating of the object with a Minolta@ scanner).

"SHAPE FROM FLUORESCENCE" FOR 3D DIGITIZATION OF TRANSPARENT OBJECT

In this experiment, a classical triangulation system based on the use of UV laser source to estimate transparent objects shape (see Figure 11) was developed. The choice of the environment range is motivated by the specificity of the application materials (glasses and some plastics) which exhibit fluorescence when irradiated with UV.

Our stereoscopic system is composed of two classical cameras working in visible range associated with a low cost UV laser source (Rantoson, 2010). Under the effect of ultraviolet irradiation, the fluorescence generated at the object surface is emitted in a diffuse way and captured with our calibrated stereoscopic rig (Figure 11). As in classical active system, the emitted light spots detected by the stereoscopic sensors are used to simplify stereo matching step exploiting epipolar

Figure 9. From left to right: a plastic bottle, its 3D digitization using SFH and the error map with a ground truth providing a commercial scanner Vi910 from Minolta@

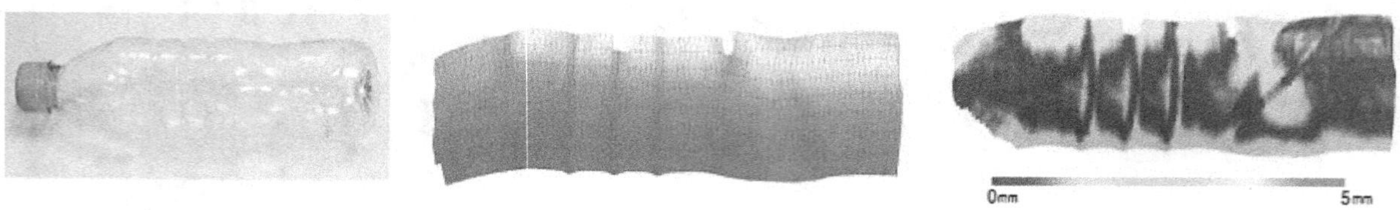

Figure 10. Error map between the 3D digitization of the glass (of Figure 3) using the SFH with line projection and a reference obtained with a commercial scanner Vi910 from Minolta@

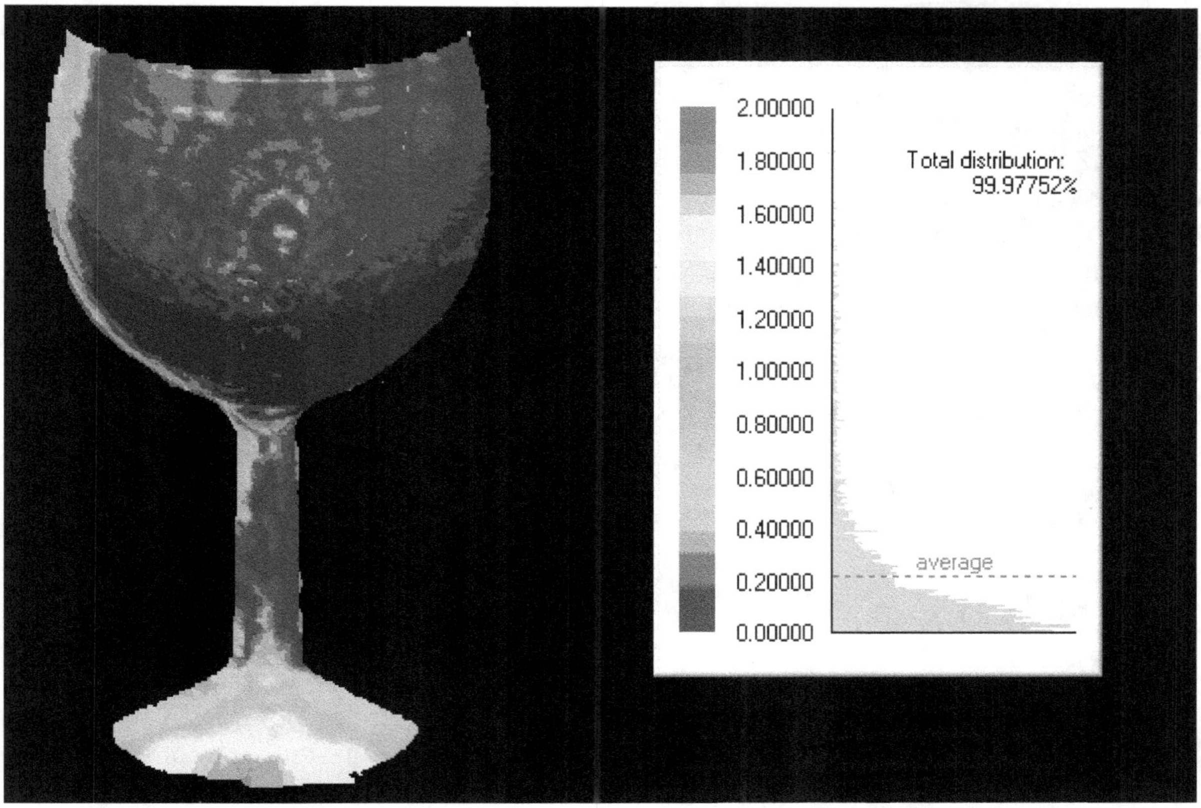

Figure 11. From left to right, experimental set-up of our shape from fluorescence system: a UV laser beam generates fluorescence (right image) on a glass sample (middle image) which is then imaged by a calibrated stereo–rig

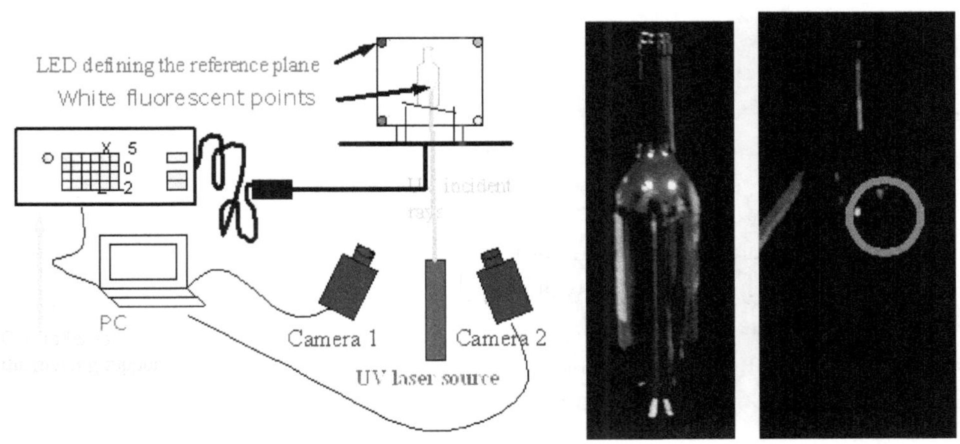

constraint (Faugeras, 1993). The experiments are run in dark room environment.

The shape from Fluorescence method is as follows:

Let us briefly recall the main equation representing the image forming process. Considering $M = (X, Y, Z, 1)^T$ as the 3D homogenous coordinate of an object point in world coordinate system and $m = (u, v, 1)^T$ its corresponding 2D homogeneous coordinate image point in the image frame, with a scale factor $s \neq 0$, the equation is as follows:

$$s \cdot m = A[R\ t]M \quad (1)$$

where,

$$A = \begin{pmatrix} \alpha_u & 0 & u_0 \\ 0 & \alpha_v & v_0 \\ 0 & 0 & 1 \end{pmatrix}$$

A is the matrix of intrinsic parameters with (α_u, α_v) as the focal length and $(u_0, v_0)^T$ the coordinate of the camera center in the image system. Specific for each camera, intrinsic parameters characterize the internal geometrical and optical parameters of the camera. R and t define the matrix of extrinsic parameters reduced to rotation and translation geometric transformations.

$$R = [r_1 r_2 r_3] \quad (2)$$

where, $r_1 = (r_{1i}\ r_{2i}\ r_{3i})^T$, i = {1,2,3}

And

$$t = (t_1 t_2 t_3)^T \quad (3)$$

Extrinsic parameters characterize the position and the orientation between the camera system and the world system. The developed reconstruction process behind the acquisition system could be summarized by the four main steps described below:

- Calibration step.

We preliminarily calibrated each camera (intrinsic and extrinsic parameters) separately using homography (Rantoson 2010) from captured images of a planar square chessboard in different orientations to determine the intrinsic parameters (*A* in *eq.1*). The spatial relationship between both cameras is thereby established from the extrinsic parameters.

- Matching step.

The matching step criterion is based on the epipolar constraint. For each given pair of images acquired by the stereoscopic cameras, the centroid of a fluorescent point in the left image is matched with the closest centroïd of a fluorescent point of the right image to the epipolar line. Let E, be the space of the fluorescent points m'_i in the right image, $i=1,\dots,n$ with n as the number of elements in E, and Δ, the epipolar line whose equation is represented by $au' + bv' + c = 0$, where a, b, c are the constants calculated from the image coordinate of a given point m in the left image. The corresponding point of m is the closest point $m' \in E$ whose distance to the epipolar line is the minimum. This minimization problem is represented by the equations presented below:

$$m' = \min_{m_i' \in E} d(m'_i, \Delta) \quad (4)$$

where,

$$d(m_i', \Delta) = \frac{au' + bv' + c}{\sqrt{a^2 + b^2}}$$

where d is the orthogonal distance between a point $m_i' = (u', v')^T$ and the epipolar line Δ.

- 3D Reconstruction step in a camera system.

From each matched pair of points (*m, m'*), the 3D homogenous camera coordinate $M_c = (x,y,z,1)^T$ of the object is estimated by least square methods knowing the intrinsic and extrinsic parameters of each camera. The computation is done in a chosen camera system, the left in this case, assuming the equations below:

For the left camera:

$$s \cdot m = A[I\ 0]M \tag{5}$$

For the right camera:

$$s \cdot m' = A'\begin{bmatrix} R_{rl} & t_{rl} \end{bmatrix} M_c \tag{6}$$

where I is a 3×3 identity matrix, A' is the matrix of intrinsic parameters related to the right camera and $(R_{rl}\ t_{rl})$ is the spatial relationship between the calibrated stereoscopic cameras.

Due to our experimental configuration for which the two cameras are fixed as well as the UV laser providing a fixed direction of incident rays, the measured image points of the object loose their actual spatial distribution in the camera coordinate system, a supplementary step was required to get the 3D model in the world coordinate system

The following part describes the computation of the definitive 3D data in the world coordinate system in accordance with the object's structure.

- 3D Reconstruction step in the world system.

The spatial relationship, denoted by S_i in eq.7, between the world system (represented by the reference plane) and the left camera system is required to be known for each position i of the object to project its 3D coordinate points of the camera system onto the world system (eq.7). Therefore,

$M_i = S_i M_{c_i}$ with

$$S_i = \begin{pmatrix} R_i & t_i \\ 0 & 1 \end{pmatrix}^{-1} \tag{7}$$

The process operates in three steps:

Step 1: Location of the plane in the left image by an automatic extraction algorithm of the four LED spots: a global threshold is first applied on one or more of the RGB components of the image in accordance with the color of the LED to be tracked. The threshold pixels' RGB components are then used to verify the Mahalanobis distance criterion as in eq.8. The intensities of pixels representing a LED spot are modeled by a Gaussian distribution. There are subsequently four ($k=1...4$) distributions that represent the LED spots. Considering RGB space, each distribution (k) is characterized by a mean vector μ_k and a covariance matrix Σ_k related to the intensities of a given class of spot. Gaussian parameters are learned from few training images. Hence, to track an LED spot (k) whose Gaussian parameters are defined, we verified for each given pixel if its intensity I satisfies the inequality expressed below:

$$D_M(I) = \sqrt{(I - \mu_k)^T \Sigma_k^{-1} (I - \mu_k)} \leq T_k \tag{8}$$

where D_M is the Mahalanobis distance for a given intensity I towards a considered class of distribution (k) and T_k a threshold defined by the maximum Mahalanobis distance calculated from the training data used to estimate the Gaussian distribution parameters. Consequently, all pixels satisfying the inequality (8) formed the LED spot (k) and the spot centroid is computed to select the representative point used for defining the plane. The four led spots are tracked in a fixed sequence to correctly define the plane orientation that must be maintained for each analyzed image.

Figure 12. From left to right, Glass sample, 3D sample points obtained from our technique, Error map between our results and those obtained with a Vi 910 Minolta scanner @ on a powdered sample

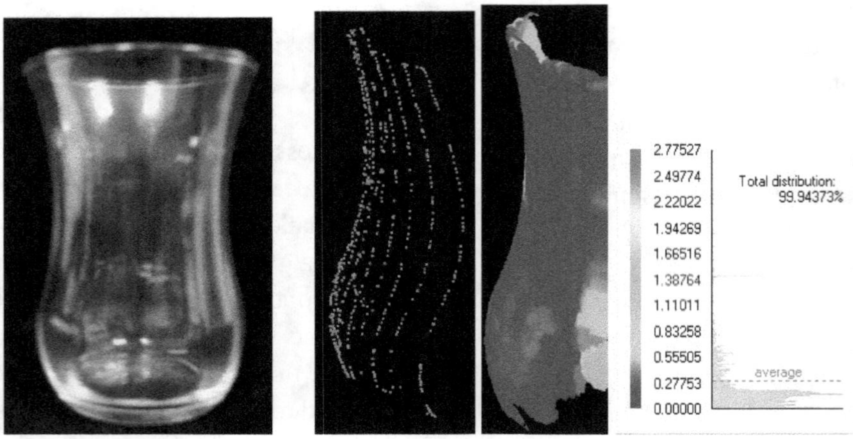

Step 2: Computation of the extrinsic parameters S_i in eq.7 for a position i by homography, that relate the coordinates of the reference plane in world system and its camera system.

Step 3: Determination of the object 3D coordinates in the world system (eq.7) using the extrinsic parameters derived from step 2.

Figure 12 to 13 present some of the results recently obtained with our method.

FUTURE RESEARCH DIRECTIONS

Shape from Heating

This technique provides fast and reliable results and is currently implemented in the industry (Eren, 2010) to control various glass objects. We are also investigating, in the case of a laser point interaction, the shape of the heated pattern

Figure 13. Glass bottle reconstruction (original presented on figure 11) and the associated error map

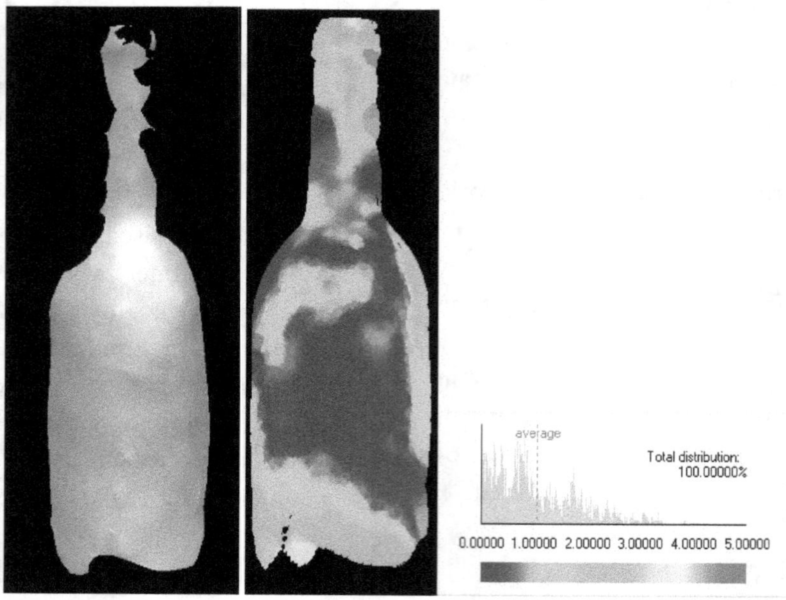

(Gaussian shape) being released (Eren, 2010), which also provides useful information about the local normal direction of the sample surface.

The process is further extended in the case of metallic specular objects, the principle relies on the same concept but the heating process is carried out at a lower wavelength range to enable a better absorption of the metallic surfaces. First results are promising and should soon be published.

This only drawback of the technique is its high cost because it involves Infrared Camera which, depending on the NETD can be quite expensive. However, based on the market trend, prototype should soon be available for less than 50,000 USD.

Shape from Fluorescence

We showed the feasibility of transparent objects shape estimation by a system relying on the fluorescence generated by UV irradiations onto the sample surface. Thanks to the transportability of the system and the accuracy provided, our approach is perfectly adaptable for industrial applications such as glass inspection in quality control and 3D modeling of transparent objects. All objects reacting by fluorescence with an adequate UV laser source are measurable with the proposed system. With our UV laser source almost all transparent objects (glasses and plastics) are measurable (3D surface acquisition) irrespective of their color, their shape complexity and the thickness of the glass, except for very thin plastics. We are currently working on a deeper analysis of the physical phenomenon occurring at the interface in order to understand and compensate if any volume effects appear.

CONCLUSION

This chapter has presented two recent techniques (2009, 2010) for digitization of transparent objects. These two techniques: shape from Heating (Eren, 2009), (Meriaudeau, 2010) and shape from Fluorescence (Rantoson, 2010) do not need any a priori on the sample (index of refraction) or on the interaction (multiple reflexions, refraction) whereas all other techniques dedicated to the 3D digitization of transparent objects need strong a priori or complicated experimental set-up and assumptions on the interaction process.

The only drawback of the shape from heating technique is the high cost involved in the experimental set-up whereas the drawback for the Shape from Fluorescence is linked to volume effect which might occur during the interaction.

REFERENCES

Agronik, G., Sirat, G. Y., Paz, F., & Wilner, K. (2005). Conoscopic holography. In *Proc. SPIE*, vol 5972.

Ben-Ezra, M., & Nayar, S. K. (2003). What does motion reveal about transparency? *Proc. IEEE Int'l Conf. Computer Vision*, (pp. 1025-1032).

Blais, F. (2004). Review of 20 years of range sensor development. *Journal of Electronic Imaging*, *13*(1), 231–243. doi:10.1117/1.1631921

Bokhabrine, Y. (2010). *Application des techniques de numérisation tridimensionnelle au contrôle de process de pièces de forge.* Doctoral Thesis, Université de Bourgogne.

Eren, G., Aubreton, O., Meriaudeau, F., Sanchez Secades, L. A., Fofi, D., Truchetet, F., & Erçil, A. (2009). Scanning from heating: 3D shape estimation of transparent objects from local surface heating. *Opt. Express, 17*(14), 11 457–11 468.

Faugeras, O. (1993). *Three dimensional computer vision: A geometric viewpoint.* The MIT Press.

Ferraton, M., Stolz, C., & Meriaudeau, F. (2009). Optimization of a polarization imaging system for 3D measurements of transparent objects. *Optics Express*, *17*(23), 21077–21082. doi:10.1364/OE.17.021077

Goesele, M. (2004). Disco: Acquisition of translucent objects. *ACM Transactions on Graphics, 23*(3), 835–844. doi:10.1145/1015706.1015807

Harvent, J. (2010). *Mesures de formes par corrélation muli-images: Application à l'inspection de pièces aéronautiques à l'aide d'un système multi-caméra.* Thése de l'Université Toulouse 3 Paul Sabatier.

Hata, S., Saitoh, Y., Kumamura, S., & Kaida, K. (1996). Shape extraction of transparent object using genetic algorithm. *Proc. Int'l Conf. Pattern Recognition*, (pp. 684-688).

Horn, B. K. P. (1986). *Robot vision* (p. 509). Cambridge, MA: MIT Press.

Hullin, M. B., Fuchs, M., Ihrke, I., Seidel, H.-P., & Lensch, H. P. A. (2008). Fluorescent immersion range scanning. *ACM Trans. Graphics, 27*(3), 87:1–87:10.

Ihrke, I., Kutulakos, K. N., Hendrik, P. A., Lensch, M. M., & Heidrich, W. (2008). State of the art in transparent and specular object reconstruction. *Proc. of EUROGRAPHICS 2008.*

Ihrke, I., Kutulakos, K. N., Hendrik, P. A., Lensch, M. M., & Heidrich, W. (2010). Transparent and specular object reconstruction. *Computer Graphics Forum, 29*(8), 2400–2426. doi:10.1111/j.1467-8659.2010.01753.x

Liu, C., Czuban, L., Bison, P., Grinzato, E., Marinetti, S., & Maldague, X. (2006). Complex -surfaced objects: Effects on phase and amplitude images in pulsed phase thermography. *Proceedings of the 12th A-PCNDT 2006 – Asia-Pacific Conference on NDT*, 5th – 10th Nov 2006, Auckland, New Zealand.

Matusik, W., et al. (2002). Acquisition and rendering of transparent and refractive objects. *Eurographics Workshop on Rendering*, (pp. 267-278).

Meriaudeau, F., Sanchez-Secades, L. A., Eren, B. G., Erçil, B. A., Truchetet, F., Aubreton, A. O., & Fofi, A. D. (2010). 3D scanning of non-opaque objects by means of infrared imaging. *IEEE Transactions on Instrumentation and Measurement, 59*(11), 2898–2906. doi:10.1109/TIM.2010.2046694

Ming, Y., Ng, H., & Du, R. (2005). Acquisition of 3D surface temperature distribution of a car body. In *Proc. IEEE Int. Conf. Inf. Acquisition*, Hong Kong.

Ming, Y., Ng, H., Yu, M., Huang, Y., & Du, R. (2007). Diagnosis of sheet metal stamping processes based on 3-D thermal energy distribution. *IEEE Transactions on Automation Science and Engineering, 4*(1), 22–31. doi:10.1109/TASE.2006.873227

Miyazaki, D., & Ikeuchi, K. (2005). Inverse polarization raytracing: Estimating surface shape of transparent object. *Proceedings of International Conference on Computer Vision and Pattern Recognition*, (pp. 10-917).

Miyazaki, D., Kagesawa, M., & Ikeuchi, K. (2004). Transparent surface modeling from a pair of polarization images. *IEEE Transactions on Pattern Analysis and Machine Intelligence, 26*(1), 73–82. doi:10.1109/TPAMI.2004.1261080

Miyazaki, D., Saito, M., Sato, Y., & Ikeuchi, K. (2002). Determining surface orientations of transparent objects based on polarization degrees in visible and infrared wavelength. *Journal of the Optical Society of America. A, Optics, Image Science, and Vision, 19*(4), 687–694. doi:10.1364/JOSAA.19.000687

Morel, O. (2006). Active lighting applied to three-dimensional reconstruction of specular metallic surfaces by polarization imaging. *Applied Optics, 45*(17), 4062–4068. doi:10.1364/AO.45.004062

Morris, N. J. W., & Kutulakos, K. N. (2007). Reconstructing the surface of inhomogeneous transparent scenes by scatter-trace photography. In *Proc. IEEE 11th ICCV*, (pp. 1–8).

Ng, Y.-M. H., & Du, R. (2005). Acquisition of 3D surface temperature distribution of a car body. *Proceedings of the 2005 IEEE International Conference on Information Acquisition,* June 27 - July 3, 2005, Hong Kong and Macau, China.

Ng, Y.-M. H., Yu, M., Huang, Y., & Du, R. (2007). Diagnosis of sheet metal stamping processes based on 3-D thermal energy distribution. *IEEE Transactions on Automation Science and Engineering*, *4*(1), 22–31. doi:10.1109/TASE.2006.873227

Ohara, K., Mizukawa, M., Ohba, K., & Taki, K. (2003). 3D modeling of micro transparent object with integrated vision. *Proc. IEEE Conf. Multisensor Fusion and Integration for Intelligent Systems,* (pp. 107-112).

Pelletier, J. F., & Maldague, X. (1997). Shape from heating: A two-dimensional approach for shape extraction in infrared images. *Optical Engineering (Redondo Beach, Calif.)*, *36*, 371–375. doi:10.1117/1.601210

Prakash, S., Lee, P. Y., & Caelli, T. (2006). 3D mapping of surface temperature using thermal stereo. In *Proc. IEEE ICARCV*, (pp. 1–4).

Rantoson, R., Stolz, C., Fofi, D., & Meriaudeau, F. (2010). *3D reconstruction of transparent objects exploiting surface fluorescence caused by UV radiation*. ICPR Conference, Istanbul Turkey.

Sadjadi, F. A. (2007). Passive 3D imaging using polarimetric diversity. *Optics Letters*, *32*(3), 229–231. doi:10.1364/OL.32.000229

Sadjadi, F. A. (2007). Extraction of surface normal and index of refraction using a pair of passive infrared polarimetric sensors. *Proceedings of the IEEE International Workshop on Object Tracking and Classification beyond the Visible Spectrum*, Minneapolis, Minnesota, June 2007.

Saito, M., Sato, Y., Ikeuchi, K., & Kashiwagi, H. (1999). Measurement of surface orientations of transparent objects by use of polarization in highlight. *Journal of the Optical Society of America. A, Optics, Image Science, and Vision*, *16*(9), 2286–2293. doi:10.1364/JOSAA.16.002286

Salvi, J., Pagès, J., & Batle, J. (2004). Pattern codification strategies in structured light systems. *Pattern Recognition*, *37*(4), 827–849. doi:10.1016/j.patcog.2003.10.002

Trifonov, B., Bradley, D., & Heidrich, W. (2006). Tomographic reconstruction of transparent objects. In T. Akenine-Möller & W. Heidrich (Eds.), *Eurographics Symposium on Rendering*.

Yamazaki, M., Iwata, S., & Xu, G. (2007). Dense 3D reconstruction of specular and transparent objects using stereo cameras and phase-shift method . In Yagi, Y. (Eds.), *ACCV 2007, Part II, LNCS 4844* (pp. 570–579).

Zhang, Z. (1999). Flexible camera calibration by viewing a plane from unknown orientations. In *Proceedings of ICCV*.

Zhang, Z. (2000). A flexible new technique for camera calibration. *IEEE Transactions on Pattern Analysis and Machine Intelligence*, 1330–1334. doi:10.1109/34.888718

KEY TERMS AND DEFINITIONS

Active Methods: The structured light or texture/pattern or both projected on the scene.

Passive Methods: Light reflected from the object is captured by cameras for 3D shape reconstruction.

IR: InfraRed radiations.

UV: UltraViolet radiations.

Section 3
Stereoscopy & Autostereoscopy

Chapter 13

Modular Stereo Vision: Model and Implementation

Ng Oon-Ee
Monash University Sunway Campus, Malaysia

Velappa Ganapathy
University of Malaya, Malaysia

S.G. Ponnambalam
Monash University Sunway Campus, Malaysia

ABSTRACT

The two-frame stereo vision algorithm is typically conceived of and implemented as a single process. The standard practice is to categorize individual algorithms according to the 'type' of process used. Evaluation is done based on the quality of the depth map produced. In this chapter, we demonstrate that the stereo vision process is actually composed of a number of inter-linked processes. Stereo vision is shown to be modular in nature; algorithms implementing it typically implement distinct stages of the entire process. The modularity of stereo vision implies that the specific methods used in different algorithms can be combined to produce new algorithms. We present a model describing stereo vision in a modular manner. We also provide examples of the stereo vision process being implemented in a modular manner, with practical example code. The purpose of this chapter is to present this model and implementation for the use of researchers in the field of computational stereo vision.

INTRODUCTION

"Stereo vision" is a specific method for 3-dimensional imaging. Other methods include multi-view vision, optical flow, and photometric methods relying on known light sources or shadows. Stereo vision is quite robust in application, not relying on any fixed external constraints (besides the scene being lighted). Stereo vision setups do not require controlled illumination which may disturb the scene being observed. They are relatively cheap in cost and power consumption relative to multi-view methods and laser rangefinder imaging.

DOI: 10.4018/978-1-61350-326-3.ch013

The disadvantages of stereo vision are high complexity and ambiguity in the stereo vision process itself (this is expanded on in the next section). 'Pure' two-frame stereo vision is much more difficult than multi-view methods which can rely on many different observations; it is therefore more vulnerable to noise and lack of scene texture. Much research has been done into solving the stereo vision problem; some of that work is referenced in this chapter.

In this chapter, we examine the stereo vision process as a system composed of distinct processes. The stereo vision process is typically viewed as a kind of black box which describes a single process applied on two input images to produce a depth map. This chapter describes the processes which make up the typical stereo vision algorithm. Based on these processes, a model for modular stereo vision is presented.

This model is not meant merely as an academic description, but as a practical guide to implementing stereo vision algorithms. Algorithms developed based on this model will have component parts (called modules) which are inter-changeable with other implementations of the same process. In this way, a newly developed approach for one process in the model can potentially be used in any number of existing algorithms. To this end, this chapter also presents a practical implementation of the model in C++.

To summarize, this chapter presents a model describing the stereo vision process as a series of inter-linked processes. The practical advantages of this model over the standard 'black box' view of stereo vision are discussed. An implementation of this model is included to demonstrate those advantages.

COMPUTATIONAL STEREO VISION

This section provides a brief introduction to computational stereo vision. Readers who are familiar with stereo vision and its implementation on computers should feel free to skip to the next section on Modular Stereo Vision.

The first sub-section describes the challenges inherent in stereo vision, and gives a general summary of how these challenges are tackled by existing algorithms. The second sub-section discusses how the typical stereo vision algorithm is considered to be a black box when compared against a competing algorithm and the background behind this understanding. The final sub-section briefly discusses the quantitative measurement of stereo vision result quality.

The Stereo Vision Challenge

Stereo vision is the process of converting two views of a scene to its depth map. For computational stereo vision, typically two cameras are used in the epipolar configuration with parallel visual axes (Figure 1). This is analogous to the physical arrangement of our eyes (Ganapathy and Ng, 2008, Figure 1). In general, the term 'stereo vision' should only be used to describe the two-view case, since multi-view vision is a very different field of research. Also, some hybrid depth-perception methods rely on a combination of visual data with other non-visual cues; these are not sufficiently related to stereo vision to warrant inclusion in this discussion.

The depth of a scene element is measured from the baseline between the camera apertures. The disparity of a scene element is the difference between the position of that element in each view. For example, the disparity d of Object 1 in Figure 1 is $xl_1 - xr_1$. Given a baseline b between the two cameras and a common focal length f, the relationship between depth z and disparity is:

$$z = \frac{b \bullet f}{d} \quad (0.1)$$

Due to this inverse relationship, if we know the disparity of an object, we can calculate its depth and therefore its position relative to the cameras. The real challenge in stereo vision is matching

Figure 1. Standard epipolar stereo geometry

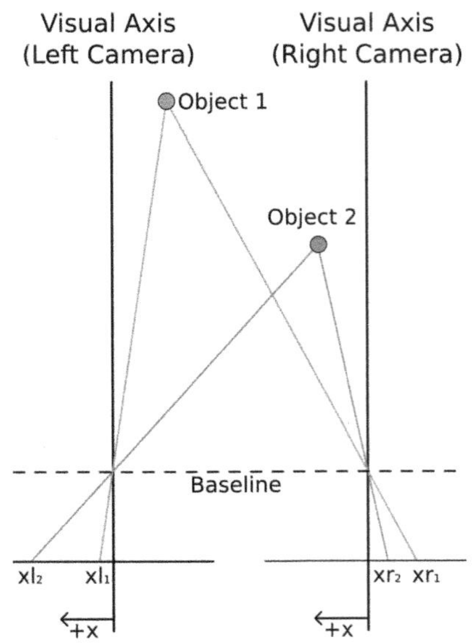

the pixels at xl_1 and xr_1 while avoiding wrong matches, for example between pixel xl_1 and xr_2.

Computational stereo vision can therefore be classified as a matching problem. This matching problem is non-trivial, since for natural noisy images any pixel can theoretically be a good match for many other pixels, only one of which would actually be the correct match. The two challenges faced in stereo matching are matching ambiguity and occlusion.

Stereo matching can be ambiguous due to repetitive texture (this is sometimes referred to as the 'picket fence' problem), lack of texture, and differences in the cameras used for both views. Baker et al. (2001) analyze the effect of lighting and surface reflectivity on stereo vision and conclude that the stereo matching problem is inherently ambiguous even in the noiseless case.

Occlusion in stereo vision happens when a scene element is visible in one view but is blocked by another (nearer) scene element in the other view. Occlusion occurs in any scene with elements at different depths. Šára and Bajcsy (1997) state that it is not possible to "overcome the lack of information needed to resolve it" and that at best it is only possible to "reduce the *expectancy* of errors".

Despite the twin challenges of matching ambiguity and occlusion, many algorithms exist to solve the stereo matching problem with varying degrees of robustness and success. Scharstein and Szeliski (2002) provide a good taxonomy of such algorithms; the same authors maintain an up-to-date website[1] for comparing the results of recent best-performing stereo matching algorithms. The exact method used differs widely. Yoon and Kweon (2006) and other aggregation-based methods average nearby or related pixels to reduce ambiguity. Tola et al. (2010) and other feature-matching methods are based on matching local feature descriptors. Recent research activity has primarily been focused on graph-theory based methods, with the disparity map being represented as a connected graph to be optimized; for example Yang et al. (2006) use belief propagation while Deng et al. (2007) use graph cuts. For efficiency purposes, Criminisi et al. (2007) and related methods optimize in one dimension at a time rather than as a fully connected graph.

In general, existing stereo algorithms work by providing 'support' for the assigning of a particular disparity to a pixel. This 'support' is normally taken from nearby pixels, either directly through aggregation or by using those pixels to construct a feature descriptor. Graph-theory based methods utilize this same concept of 'support' by implicitly making every node's value dependent on other nodes. The problem of occlusion is handled by attempting to make sure occluded pixels do not form part of the 'support' for non-occluded pixels, based on assumptions on image content. The most common assumption used for this purpose is that disparity boundaries (where occluded pixels appear) correlate strongly with intensity boundaries in the original image.

In this discussion, we have ignored the problem of image rectification. Image rectification is necessary due to the differences in camera configurations between both views of the stereo scene, but is generally well understood and only needs to be calibrated once. Methods such as those proposed by Fusiello et al. (2000) and Zhou and Li (2008) can be used for this purpose.

In short, ambiguity and occlusion are the two main challenges to be overcome when performing stereo matching. Tackling these problems is typically accomplished using the concept of 'support' for a particular pixel-disparity combination; many methods are available for doing this.

Stereo Algorithms as Black Boxes

All stereo vision algorithms take two views of a scene and produce a depth map of the scene. The simplest way to evaluate a stereo vision algorithm is to see it as a black box which applies a certain function to the input views to produce a depth map. This allows comparisons to be made between differing stereo vision algorithms by simply comparing the output depth maps when both algorithms are provided with the same input views.

The pattern of viewing stereo vision algorithms as black boxes is well-established in the literature. Banks and Corke (2001) compare the effects of radiometric differences on various algorithms, Scharstein and Szeliski (2002) present the largest comparison of unique stereo vision algorithms available in the literature, and Brown et al. (2003) evaluate the performance of stereo algorithms specifically in the occluded regions of a scene. In these and other similar works, the common denominator is that each compares multiple stereo vision algorithms based solely on the final depth map.

While black box evaluation is intuitive and simple, it by necessity ignores the complexities of modern stereo vision algorithms. For example, the comparisons in Banks and Corke (2001) are between similar algorithms, while the breadth of algorithms covered by Scharstein and Szeliski (2002) means that wildly differing methods are compared. In recognition of this, Scharstein and Szeliski (2002) classify stereo algorithms based on a taxonomy similar to the general one provided in the previous sub-section. Recently published stereo vision algorithms typically classify themselves according to this taxonomy for evaluation purposes rather than attempting to compare their performance with the performance of totally unrelated algorithms.

Recent comparative papers such as Hirschmüller and Scharstein (2007, 2009) recognize the inherent problems of apples-to-oranges comparison and choose to focus specifically on comparing similar methods. In the process, the authors are able to draw specific conclusions on the effects of the specific elements examined in the paper (cost calculation methods) on stereo matching performance over different scenes and with radiometric differences.

Black box evaluation is reflective of real world performance, because it is the depth map which is the ultimate product of any stereo vision algorithm. However, more in-depth analysis is necessary to understand the contribution of the different elements of the modern stereo algorithm, a task made difficult if algorithms are only compared based on black box evaluation.

Measuring Stereo Vision Output Quality

Quantitative evaluations of results are necessary in any scientific endeavour. In the stereo vision research field, almost all publications utilize a simple quality measure based on the percentage of correctly-matched pixels or its complement. Scharstein and Szeliski (2002) were not the first to utilize this measure, but their website[i] makes extensive use of it in its comparative study of stereo vision algorithms. The results of new algorithms are regularly added to this website by researchers.

The quality measure used by Scharstein and Szeliski (2002) also includes specifying the error rate (percentage of wrongly-matched pixels) in two regions of the image, the non-occluded region and the discontinuity region (pixels near discontinuities). As we have previously mentioned, stereo algorithms work by obtaining support for individual pixel-disparity assignments. This leads to the hypotheses that error rates would be higher not just at occluded points but also at pixels near to those occluded points (discontinuities in the depth map). The submitted measurements agree with this hypotheses; all algorithms submitted to this website[i] exhibit much higher error rates in the discontinuity region than in the overall image.

All algorithms submitted to the website[i] above are required to utilize the same parameters across all test images. This applies both to generic parameters such as disparity range searched and noise threshold level as well as algorithm-specific parameters such as occlusion penalties and *apriori* standard deviation value. This is meant to favour more robust algorithms which work well over a multitude of images compared to algorithms which specialize in particular image types.

An alternative method of quantifying quality is the method of Kostlivá et al. (2007), known as Receiver Operating Characteristics (ROC) analysis. ROC analysis is based on the false positive and false negative rates given by a particular run of an algorithm. Instead of focusing on 'average' performance over varying images, evaluation using ROC analysis is done by repeatedly evaluating the output of the algorithm with a range of parameters and noting the best achievable performance and plotting the relationship between false positive and false negative rates. This plot gives the 'best case' performance of the algorithm under varying parameters, and provides more discriminability between algorithms than the standard method. It is also more complicated and time-consuming to measure.

In general, the standard method is sufficient to indicate the performance of a stereo vision algorithm. While the requirement for constant parameter settings is understandable, it results in algorithms being evaluated on their 'average' performance rather than 'best' performance, which may not be the desired effect depending on the intentions of the reviewer. ROC analysis provides a good alternative, though a time-consuming one. There remains a need for a generic evaluation method for stereo vision algorithms similar to the evaluation methods which exist for other computer vision fields (Cardoso and Corte-Real, 2005).

MODULAR STEREO VISION: BEYOND CLASSIFICATION

Classifying stereo vision algorithms by methodology does not adequately reflect the relationships between different stereo vision algorithms. The quick survey of the literature above shows that algorithm classification is necessary for the purposes of relative algorithm evaluation. A comparison of existing stereo vision algorithms shows that the stereo vision process itself is modular in nature. This section describes the modularity of computational stereo vision and provides a model for this modularity. The Modular Stereo Vision model was first presented by Ng and Ganapathy (2008), an updated version is presented here (Figure 2).

The Modular Stereo Vision Model (MSVM) consists of four stages. Each stage consists of one or more processes. Data flow is indicated by arrows, some processes have feedback arrows indicating *in situ* processing or iteration.

The central concept of the MSVM is the cost volume. It contains the estimated cost of assigning a disparity value to each pixel. This matching cost is commonly defined as the probability of such an assignment.

The pre-processing stage operates on the rectified input stereo images. Pre-processing stage processes are easily identified since they only operate on a single input image at a time without using information from the other image. Pre-processing stage processes range from common mean filters to complex segmentation algorithms. One pre-processing filter which is often mistakenly identified as a cost calculation

Figure 2. Modular stereo vision model

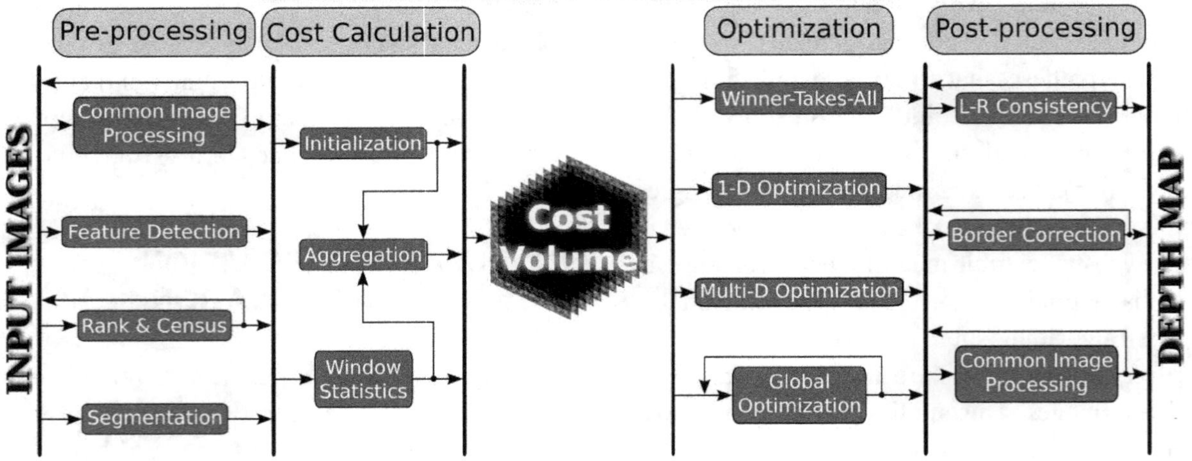

process is the rank and census transform of Zabih and Woodfill (1994).

The cost calculation stage operates on a pair of stereo images. Processes from this stage generate the cost volume, either through initialization based on pixel values or through feature/window analysis. Aggregation can optionally be applied. In this stage, the aggregation process has been heavily focused on in research, with interesting and varied works such as Yoon and Kweon (2008) and Adhyapak et al. (2007) being significantly more advanced than simple fixed-window aggregation techniques.

The cost volume is optimized by optimization stage processes. These range from the simple 'winner-takes-all' algorithm to complex global algorithms. In recent research, global optimization algorithms such as the belief propagation algorithm of Felzenszwalb and Huttenlocher (2004) and the graph cut method of Hong and Chen (2004) have been heavily studied, as evidenced by the number of such algorithms in the top rankings of the Middlesbury site[i].

The post-processing stage is similar to the pre-processing stage, but operates on the depth map rather than on the input images. The left-right consistency check is widely used, and border correction methods such as those presented in Hirschmüller et al. (2002) are also possible. Post-processing is not a popular field of research in stereo vision, largely because more improvement can be seen from working on the cost calculation and optimization stages.

Using the MSVM, a stereo vision algorithm is defined by the combination of processes used to implement it. There are many ways to implement each process. Each separate method is a single module which implements one or more processes. Modules should be drop-in replacements for each other within the same process.

As an example of module-based representation of a current stereo vision algorithm, take the 3-step Dynamic Programming method of Cox et al. (1996). No pre-processing is specified, so the first stage is skipped. For the second stage, the initialization process is implemented by the Absolute Difference module. For the third stage, the 3-step Dynamic Programming module implements the 1-D optimization process. Finally, no post-processing is specified.

Having specified the modules, it is then possible to consider changing those modules. One of the simplest changes would be substituting another initialization module. The sampling-insensitive version of the Absolute Difference method developed by Birchfield and Tomasi (1998) can also implement the initialization stage. Alternatively, an aggregation module could be added, such as the adaptive support-weight aggregation of Yoon and Kweon (2006). The resulting algorithm can be evaluated against the initial algorithm, revealing the specific effect of the module which was changed or added.

The MSVM can be seen as a generalization of the taxonomy of Scharstein and Szeliski (2002). Where the taxonomy focuses on classifying algorithms, the MSVM focuses on identifying the components (processes) which make up those algorithms. This provides the following benefits:

1. Encourages process improvement at a finer level by making explicit the multiple processes each algorithm implements.
2. Allows like-to-like comparison of processes rather than apples-to-oranges comparison of algorithms.
3. Suitable for direct translation to computational implementation.

Item 1 has been covered in the previous example. Item 2 indicates that the stage/process model of the MSVM lends itself to the comparison of processes. For example, a new initialization method should be evaluated against other initialization methods maintaining the same aggregation, optimization, and pre/post-processing modules. For more informative analysis, a range of modules can be used. Hirschmüller and Scharstein (2009) and others already do comparisons in this manner, the MSVM provides a model justifying the selection of modules to be compared.

As indicated by item 3, the MSVM can be implemented directly to computer code. This facilitates the comparisons mentioned in the previous paragraph, as well as code re-use (a new segmentation module can be substituted in place of an old one without disturbing the second to fourth stage of the algorithm). While there are efficiency costs to pay when implementing stereo vision algorithm costs in this manner (due to additional memory requirements and data copying) the resulting modules can be rapidly substituted, allowing mix-and-match algorithms to be constructed for any particular application.

The MSVM goes beyond the classification of stereo vision algorithms to explicitly defining these algorithms as a product of their component processes. Implementing stereo vision algorithms in a modular manner according to the MSVM brings benefits in focused analysis, fair evaluation of stereo process methods, and actual implementation for easy deployment.

IMPLEMENTING MODULAR STEREO VISION

This section discusses a practical implementation of the MSVM. First we discuss the selection of object-oriented programming as the programming paradigm. This is followed by the specification of classes along with associated fields and behaviours. Finally, some abbreviated example code is provided for well-known stereo vision-related algorithms.

MSVM Programming Paradigm

Programming paradigms define a certain way to program solutions to a computational problem. Programming paradigms provide instructions and defined boundaries to the coding process. This results in more uniform and easily understandable code, at the cost of limiting the programmer's freedom. Procedural programming, functional programming, and object-oriented programming are examples of programming paradigms.

Programming paradigms are not mutually exclusive, and deciding on the 'best' programming paradigm for a computational problem is not an exact science. For the MSVM, we believe that the object-oriented programming (OOP) paradigm is the most suitable. In general terms, OOP is implemented using 'objects' containing data and methods. Since the term 'method' has a defined meaning in the MSVM, we substitute the term 'function' to mean 'OOP method'. Each object should theoretically be self-contained, with the behaviour of an object defined by its functions; an object's data should be self-contained and non-accessible by external objects/functions.

OOP is a good fit for the MSVM because each stereo vision module can be implemented as a single object. Furthermore, OOP concepts such as inheritance and polymorphism are useful in creating portable and re-usable modules. Inheritance in OOP refers to the ability of objects to inherit from other objects, making the second object a specialization of the first object. The resulting relationship is referred to as an 'is-a' relationship. An ancestor object 'car' could have a derived object 'BMW', with the obvious relationship that a 'BMW' is a 'car'. Polymorphism in OOP refers to how different objects can be used in place of each other; this is typically implemented through inheritance. A 'BMW' object can thus be used anywhere a 'car' object is required, though the inverse is not usually allowed.

In an OOP implementation of the MSVM, the basic object is the stereo vision module. The stereo vision algorithm consists of several modules which operate sequentially, with the main program itself operating as a simple shell passing image and operational data between the modules. Polymorphism is applied such that each module can be treated almost identically by the main program; a basic 'module' object is inherited by the various algorithms which implement each module. The behaviour of a stereo vision module is defined as follows:

- Takes image(s) as input
- Produces image(s) as output
- Takes a 'settings' object as input
- Provides information on the implemented algorithm

Every implemented module should adhere to the behaviour listed above. The 'settings' object is analogous to a notepaper being passed from person to person in a team project with instructions for each person. Obviously, stereo vision modules operate on images, the exact number of images depends on the process being implemented. Finally, the module should provide some information to the main program on how exactly it should be handled (the format of input/output images, the process being implemented etc.).

OOP is not the only paradigm which can or should be used for implementing the MSVM. However, for the reasons stated above, we believe it is a good fit for the MSVM problem. We have described the objects which constitute an OOP implementation of the MSVM.

MSVM C++ Class Specifications

The MSVM implementation described in this section is implemented using C++. As with the selection of the programming paradigm, there is no 'best' programming language for implementing MSVM. One of the obvious requirements would be OOP capability, though almost all popular languages allow implementation of the OOP paradigm to some degree. In general, low-level programming languages are preferable for efficiency reasons, though this can be somewhat offset by efficient compilers for higher-level languages. Another consideration would be the availability of helper libraries for standard tasks which should not need to be programmed from scratch by the researcher.

C++ fulfils all the requirements above. The language enables and encourages the use of the OOP paradigm through its implementation of classes with public/private inheritance. Polymorphism is also available through the base class. C++ is one of the most efficient programming languages, being relatively low-level. The OpenCV[2] computer vision library is available for C++ (also in C and Python), and provides many useful helper func-

Algorithm 1.

```
class alg_method {
public:
  virtual ~alg_method();
  virtual void run(asMat* const inImage,stereo_opt* const myOpt);
  virtual void run(asMat* const inLeft,asMat* const inRight,
      stereo_opt* const myOpt);
  asMat* getResult();
  virtual int requestedInputs() = 0;
  virtual int requestedOutputs() = 0;
  virtual bool returnsNew() = 0;
  virtual int stageImplemented() = 0;
  ...
}
```

tions to deal with digital images; the software library received a major update in October 2009 which enhanced type-safety and ease-of-use specifically in the C++ interface. A final benefit of using C++ is its backward-compatibility with C. Cox et al. (1996), Scharstein and Szeliski (2002) and other researchers have released the source code for their stereo matching algorithms; this code is typically written in either C or C++. Using C++ as the language for implementing the MSVM allows re-use of such code, whether it was written in C or C++.

We provide here a specification for an implementation of the MSVM which we call 'anystereo'. The basic 'module' object of the MSVM is implemented as an abstract base class containing unimplemented functions. In accordance with the OOP paradigm, an object is defined by its behaviour. In C++, the abstract base class has abstract (virtual) functions, at least one of which is 'pure virtual' and must be implemented by a derived class. The virtual functions provide a defined interface for interacting with the object – the *behaviour* of the object (also known as the Application Programming Interface). The module object 'alg_method' can be implemented with the following public functions (Algorithm 1).

The functions above describe the behaviour detailed previously for generic stereo vision modules. There is an overloaded run() function which takes pointers to one or two asMat objects (described later in this section) and one options object. getResult() returns an asMat object pointer. The remaining four functions are pure virtual and allow the caller to determine what inputs and outputs the algorithm requires, whether the algorithm creates a new asMat object, and what stage of stereo vision is implemented by the algorithm.

Besides the object interface, each alg_method object also has several defined data fields. These are defined as protected members which are directly accessible by derived classes but not from the caller (Algorithm 2).

The inImage, inLeft, and inRight asMat objects and the stereo_opt object are recommended for use by derived classes. However, the result and result2 asMat objects and the various FLAG boolean fields have a specific meaning. The asMat objects store the result(s) of the module's algorithm, while the FLAG fields specify the current state of the algorithm and should be set to true when the algorithm has finished running and has populated the appropriate output objects.

Algorithm 2.

```
protected:
  asMat *inImage_, *inLeft_, *inRight_;
  stereo_opt *options_;
  bool FLAG_complete;
  asMat* result;
  bool FLAG_result_valid;
  asMat* result2;
  bool FLAG_result2_valid;
```

Finally, the output of the API functions indicating the stage implemented and the type of input/output expected is specified using the following enum defines (Algorithm 3).

These enums are included in the class definition. Derived classes would need to return the values indicated above. Details on how this is done will be provided later in this section.

To complete our specification, we provide the definition of the asMat and stereo_opt objects. The asMat class inherits from cv::Mat, the basic matrix object used by the OpenCV[ii] computer vision library. The asMat class is a thin wrapper to cv::Mat, meaning it can be used wherever a cv::Mat is required (for example by built-in OpenCV[ii] functions); in this regard it is similar to the cv::Mat_ class. The asMat class adds a 'planes' data field so that it can be used to hold a 3-dimensional cost volume (a d-by-w-by-h volume is simply a matrix with width w and height d x h). The relevant portion of asMat.h is shown here, more documentation on behaviour inherited from cv::Mat should be obtained from the OpenCV wiki[3] (Algorithm 4).

Algorithm 3.

```
enum svStage {
  svStageERROR,
  svPRE,
  svCOST,
  svOPT,
  svPOST,
  svCOMBINED
};
enum svInputOutput {
  svInputOutputERROR,
  sv2IM,
  svVOL
};
```

The stereo_opt object provides an interface to set and retrieve various options, which can be textual or numerical. A helper getdepth() function is also provided to convert numerical depth (32-bit, for example) to the appropriate OpenCV[ii] value. The options are stored separately based on data-type in a std::map (from the C++ Standard Template Library) and indexed/accessed by their name. The setopt() function is overloaded for the different option data-types (Algorithm 5).

This section has provided the reasons for selecting the C++ language for implementing the MSVM using the OOP paradigm. Also provided are public and protected definitions for the stereo_alg, asMat, and stereo_opt classes. This API forms the basis for the 'anystereo' MSVM implementation.

Example Module Code

To demonstrate the utility of the API specified above, we present some examples of common stereo vision algorithm modules. These examples are meant to illustrate how various technical

Algorithm 4.

```
class asMat: public cv::Mat {
public:
  int planes;
  asMat(): cv::Mat(), planes(1) { }
  asMat(int _rows, int _cols, int _type)
  ...
};
```

Algorithm 5.

```
class stereo_opt {
private:
  std::map<std::string,int> intopt;
  std::map<std::string,double> doubleopt;
  std::map<std::string,std::string> stropt;
public:
  void setopt(const std::string& optname, const int& opt);
  void setopt(const std::string& optname, const double& opt);
  void setopt(const std::string& optname, const std::string& opt);
  const int getint(const std::string& optname) const;
  const double getdouble(const std::string& optname) const;
  const std::string getstr(const std::string& optname) const;
  const int getdepth(const int& depth, const bool& colour=false) const;
};
```

issues relating to modularity are handled. The actual mathematical algorithms themselves are not implemented fully; instead they are represented as C++ comments to save space. Error-checking and option-handling code is not shown for the same reason unless crucial and specific to the algorithm being implemented.

Initialization Modules

Initialization algorithms produce a pixel-wise difference volume between the two input images, which serves as the cost volume. The exact difference measure used varies between algorithms. We take advantage of polymorphism and inheritance to speed implementation of various initialization algorithms. Since the mathematical difference for this type of initialization can be separated from the algorithm implementation, an init_hash class can be used to implement all initialization algorithms which use a pixel-wise difference. One pure virtual function is defined which needs to be implemented by derived classes, the create_hash() function (Algorithm 6).

Derived functions implement the specifics for the create_hash() function. The hashtable itself is a 256-by-256 integer array indicating the cost to be assigned to each pair of input pixel values. For example, the most common pixel-difference measure is the absolute difference measure (Algorithm 7).

Algorithm 6.

```
class init_hash: public alg_method {
protected:
  template<class T> void runTemplate();
  int* hashtable;
  init_hash() { hashtable = 0; }
  virtual void create_hash() = 0;
public:
  virtual ~init_hash() { delete [] hashtable; }
  void run(asMat* const inLeft,asMat* const inRight,
      stereo_opt*  const myOpt);
  bool returnsNew() { return 1; }
  int requestedInputs() { return sv2IM; }
  int requestedOutputs() { return svVOL; }
  int stageImplemented() { return svCOST; }
};
void init_hash::run(asMat* const inLeft,
    asMat* const inRight,stereo_opt* const myOpt) {
  inLeft_ = inLeft; inRight_ = inRight; options_ = myOpt;
  /* call runTemplate with the appropriate data-type specialization */
}
template<class T>
void init_hash::runTemplate() {
  create_hash();
  /* implement initialization using the values in the hashtable array */
}
```

Algorithm 7.

```
class cost_init_ad: public init_hash {
public:
  void create_hash() {
    hashtable = new int[256*256];
    for (int i=0 ; i<=255 ; ++i)
      for (int j=0 ; j<=255 ; ++j)
        if (i>j) { hashtable[i*256+j] = i-j; }
        else { hashtable[i*256+j] = j-i; }
  }
};
```

The 'heavy-lifting' algorithm-wise is done in the init_hash class, with the derived init_ad class only required to implement the mathematical equation $| l_{xy} - r_{xy} |$. The same init_hash class can be used to implement the closely-related squared difference and truncated absolute difference measures. Additionally, a cost_init_file class can be created which loads an image file to be used as the hashtable. This allows arbitrary pixel-value relations to be implemented separately and loaded into the stereo algorithm without changing any code (Algorithm 8).

The derived classes of init_hash are not only limited to simple measures such as those shown before. Ng and Ganapathy (2009) describe an initialization algorithm based on modeling the intensity relationships between the input images. This algorithm is specifically targeted at handling exposure differences between input images efficiently. The implementation is shown below; specifics (especially the mathematical background) should be obtained from the referenced paper (Algorithm 9).

The initialization process algorithms shown here demonstrate the advantages of polymorphism and inheritance in implementing similar algorithms without having to re-implement code.

Optimization Modules

The optimization stage produces a depth map from a cost volume. Optimization algorithms attempt to find the depth map which best satisfies some criteria with regards to the cost volume. Depending on the initialization method used, the objective may be to minimize the cost (with standard pixel-difference measures) or to maximize the cost (for example when using normalized cross-correlation).

It is possible to simply specify that all cost stage algorithms generate cost volumes where lower costs are better. In this implementation of the MSVM, we take a different approach which allows the same optimization modules to be used regardless of whether the cost volume is 'lower-is-better' or 'higher-is-better'. An additional class alg_method_opt is created from which all optimization algorithms are derived (Algorithm 10).

Optimization algorithms should then implement the runMin() function rather than the run() function. These algorithms almost invariably perform cost minimization by default, deriving from alg_method_opt ensures that no changes need to be made to the algorithms to handle the case where higher costs are better. Instead, the responsibility for indicating that higher costs are better is placed on the cost stage algorithm, which would need to set the _findmax option appropriately.

Since most of the API functions have already been implemented in alg_method_opt, optimization methods can be very brief, basically consisting only of the implementation of runMin(). For example, the following shows the implementation of the winner-takes-all module, which just creates the output depth map based on the lowest cost for each pixel (Algorithm 11).

For comparison, we also present our implementation of the 3-step dynamic programming algorithm by Cox et al. (1996). An extra level of data templatization is used since a temporary array is required for this algorithm (Algorithm 12).

In both the opt_wta and opt_dp3 classes, inheriting alg_method_opt allowed the same algorithm to be used for both lower-is-better and higher-is-better type costs. This demonstrates the inherent flexibility in implementing stereo algorithms based on the MSVM.

FUTURE RESEARCH DIRECTIONS

3D Imaging is a wide field of research. The recent advent of 3D movies, animated or otherwise, has demonstrated the maturity of the relatively simple task of 3D projection. 3D interpretation, however, will continue to be a very open field for continuous research. Stereo vision research can be expected

Algorithm 8.

```
class cost_init_sd: public init_hash {
public:
  void create_hash() {
    hashtable = new int[256*256];
    for (int i=0 ; i<=255 ; ++i)
      for (int j=0 ; j<=255 ; ++j)
        hashtable[i*256+j] = (i-j)*(i-j);
  }
};
class cost_init_tad: public init_hash {
public:
  void create_hash() {
    hashtable = new int[256*256];
    int temp, trunc = options_->getint("truncation");
    for (int i=0 ; i<=255 ; ++i) {
      for (int j=0 ; j<=255 ; ++j) {
        if (i>j)
          if (i-j > trunc) { hashtable[i*256+j] = trunc; }
          else { hashtable[i*256+j] = i-j; }
        else
          if (j-i > trunc) { hashtable[i*256+j] = trunc; }
          else { hashtable[i*256+j] = j-i; }
  }
};
class cost_init_file: public init_hash {
public:
  void create_hash() {
    asMat filehash = cv::imread("filehash.png",0);
    unsigned char* filehashPtr = (unsigned char*) filehash.data;
    hashtable = new int[256*256];
    for (int i=0 ; i<=255 ; ++i)
      for (int j=0 ; j<=255 ; ++j)
        hashtable[i*256+j] = filehashPtr[i*step+j];
    filehash.release();
  }
};
```

to lead the way, as results in this field are directly applicable to applications in other fields such as multi-view vision.

As stereo vision algorithms become more complicated, we expect that research will shift from being focused on complete algorithms to being focused on specific processes as defined

Algorithm 9

```
class cost_init_quartile_match: public init_hash {
private:
  const std::vector<int> quartile(const asMat * const inImage);
  void init_hash(const std::vector<int> leftQuartiles,
      const std::vector<int> rightQuartiles,
      const alg_math_monotone_cubic *const mc_L,
      const alg_math_monotone_cubic *const mc_R);
  void fill_hash(const int n);
public:
  void create_hash();
};
void cost_init_quartile_match::create_hash() {
  const std::vector<int> leftQuartiles = quartile(inLeft_);
  const std::vector<int> rightQuartiles = quartile(inRight_);
  hashtable = new int[256 * 256];
  /* mc_L and mc_R are functions to calculate the monotone cubic */
  /* interpolation of the quartiles previously calculated        */
  init_hash(leftQuartiles, rightQuartiles, mc_L, mc_R);
  fill_hash(256);
}
const std::vector<int> cost_init_quartile_match::quartile(
    const asMat * const inImage) {
  /* calculate image quartiles based on entire input image */
}
void cost_init_quartile_match::init_hash(
    std::vector<int> const leftQuartiles,
    std::vector<int> const rightQuartiles,
    const alg_math_monotone_cubic * const mc_L,
    const alg_math_monotone_cubic * const mc_R) {
  /* Fill in the intensity curve based on the monotone cubic */
  /* interpolated values.                                    */
}
void cost_init_quartile_match::fill_hash(const int n) {
  /* Find the Manhattan distance from the interpolated intensity */
  /* curve, which is used as the matching cost.                  */
}
```

in the Modular Stereo Vision Model. Current research is heavily focused on optimization processes. As potential for improvement decreases in that process, more research will be conducted into the cost calculation stage as well as 'guided' pre/post-processing methods.

The increasing availability of high-grade consumer graphics hardware is also a potential

Algorithm 10.

```
class alg_method_opt: public alg_method {
private:
  asMat* flipImage;
  template<class T>
  void runTemplate(asMat* const inImage,stereo_opt* const myOpt);
protected:
  alg_method_opt() { flipImage = 0; };
  virtual void runMin(asMat* const inImage,stereo_opt* const myOpt) = 0;
public:
  virtual ~alg_method_opt() { if(flipImage) { flipImage->release(); } };
  void run(asMat* const inImage,stereo_opt* const myOpt);
  int requestedInputs() { return svVOL; }
  int requestedOutputs() { return sv1IM; }
  bool returnsNew() { return 1; }
  int stageImplemented() { return svOPT; }
};
template<class T>
void alg_method_opt::runTemplate(asMat* const inImage,
    stereo_opt* const myOpt) {
  flipImage = new asMat(h,w,inImage->depth());
  T *inPtr = (T*) inImage->data;
  T *flipPtr = (T*) flipImage->data;
  for (int r = 0;r < h;++r)
    for (int c = 0;c < w;++c)
      flipPtr[r*step+c] = 0 - inPtr[r*step+c];
  runMin(flipImage,myOpt);
}
void alg_method_opt::run(asMat* const inImage, stereo_opt* const myOpt) {
  if (myOpt->getint("_findmax"))
    /* call runTemplate with the appropriate data-type specialization */
  else { runMin(inImage,myOpt); }
}
```

game-changer for the 3D Imaging field, as it has been for the previous decade. A consumer purchasing a latest-generation desktop machine would own a graphics card which surpasses the capabilities of the expensive purpose-built FPGAs previously favoured for image processing. Mairal et al. (2006), Labatut et al. (2006) and Yang et al. (2006) are examples of new stereo vision algorithms with the computational burden shifted to consumer graphics cards.

The gradual shift from CPU-based to GPU-based programming will have tremendous implications in the stereo vision field, allowing for speed increases of an order of magnitude or more. Some algorithms which cannot be efficiently implemented on the GPU (such as the more complex

Algorithm 11.

```
class opt_wta: public alg_method_opt {
private:
  template<class T> void runTemplate();
public:
  void runMin(asMat* inImage,stereo_opt* myOpt);
};
template<class T> inline void opt_wta::runTemplate() {
  result = new asMat(h,w,CV_8UC1);
  unsigned char *outPtr = (unsigned char*) result->data;
  int match;
  for (int r = 0;r < h;++r) {
    for (int c = 0;c < w;++c) {
      minimum = abs_max_val;  match = 0;
      for (int d=0;d<disp;++d) {
        if (sumPtr[d*sum_size+r*sum_step+c] < minimum) {
          match = d;
          minimum = sumPtr[d*sum_size+r*sum_step+c];
        }
      }
      outPtr[r*step+c] = match + d_min;
    }
  }
}
void opt_wta::runMin(asMat* const inImage, stereo_opt* const myOpt) {
  /* call runTemplate with the appropriate data-type specialization */
}
```

dynamic programming techniques) will see little to no development. The MSVM implementation presented in this chapter would need to be replaced by one which reduces data transfer, since that is the main bottleneck in GPU programming.

A big open question in stereo vision research currently is that of measuring the quality of stereo vision algorithms. Earlier in this chapter a summary of the current methodology has been presented. Research into better quality measures complete with realistic image sets would be of tremendous usefulness to the field of stereo vision, and indeed to 3D Imaging research as a whole. Hirschmüller and Scharstein (2007) and Scharstein and Pal (2007) are examples of such initiatives.

CONCLUSION

This chapter provided an overview on computational stereo vision. The stereo vision field was defined, as well as the major challenges faced were discussed. A short description of the current model for stereo vision was presented followed by a description of stereo vision quality measures. After that, the Modular Stereo Vision Model was presented. This model described the processes

Algorithm 12.

```
class opt_dp3: public alg_method_opt {
private:
  template<class IN_C> void run2();
  template<class IN_C, class OUT_C> void runTemplate();
  asMat *cost;
  bool FLAG_cost_valid;
  asMat *match;
public:
  void runMin(asMat* vol_cost,stereo_opt* myOpt);
};
template<class IN_C, class OUT_C>
inline void opt_dp3::runTemplate() {
  match = new asMat(h*(d_max-d_min+3),w,CV_8UC1);
  result = new asMat(h,w,CV_8UC1);
  cost = new asMat(h*(d_max-d_min+3),w,
      options_->getdepth(options_->getint("out_depth")));
  /* Processing is row-by-row */
  for (int r = 0; r <= h-1; ++r) {
    /* Initialize boundary pixels to maximum allowable value        */
    /* Loop through pixels, assigning lowest cost of three possible */
    /* 'moves' - same disparity, increase disparity, or decrease    */
    /* disparity. A change in disparity necessitates and occlusion  */
    /* cost.                                                        */

    /* Finally, use a reverse-traversal loop to trace the minimum  */
    /* cost path back to the original pixel (at 0,0). Depth map is */
    /* assigned based on this minimum cost path.                   */
  }
  FLAG_complete = true;
}
void opt_dp3::runMin(asMat* const vol_cost,stereo_opt* const myOpt) {
  /* call run2 with the input data-type specialization */
}
template<class IN_C> void opt_dp3::run2() {
  /* call runTemplate with the output data-type specialization */
}
```

which make up the typical stereo vision algorithm. The excellent work of Scharstein and Szeliski (2002) formed the basis for this generalized model. The validity of this model was demonstrated using a few examples of well-known stereo vision algorithms.

An implementation of the MSVM was presented. This implementation was presented sequentially. First, the programming paradigm used was discussed; Object-oriented programming was chosen as the most suitable paradigm. Next, the Application Programming Interface was specified. The language used was C++ for this particular implementation. Finally, examples were given on module implementation, complete with abbreviated example code. The code focused on MSVM-specific functionality rather than the algorithms being implemented due to space concerns. Some comments on the future of 3D imaging in general and Modular Stereo Vision in particular were also included to round up the chapter.

REFERENCES

Adhyapak, S. A., Kehtarnavaz, N., & Nadin, M. (2007). Stereo matching via selective multiple windows. *Journal of Electronic Imaging*, *16*(1), 013012–1–14. doi:10.1117/1.2711817

Baker, S., Sim, T., & Kanade, T. (2001). A characterization of inherent stereo ambiguities. In *Proceedings Eighth IEEE International Conference on Computer Vision*, (vol. 1, pp. 428–435)

Banks, J., & Corke, P. (2001). Quantitative evaluation of matching methods and validity measures for stereo vision. *The International Journal of Robotics Research*, *20*(7), 512–532. doi:10.1177/02783640122067525

Birchfield, S., & Tomasi, C. (1998). A pixel dissimilarity measure that is insensitive to image sampling. *IEEE Transactions on Pattern Analysis and Machine Intelligence*, *20*(4), 401–406. doi:10.1109/34.677269

Brown, M., Burschka, D., & Hager, G. (2003). Advances in computational stereo. *IEEE Transactions on Pattern Analysis and Machine Intelligence*, *25*(8), 993–1008. doi:10.1109/TPAMI.2003.1217603

Cardoso, J., & Corte-Real, L. (2005). Toward a generic evaluation of image segmentation. *IEEE Transactions on Image Processing*, *14*(11), 1773–1782. doi:10.1109/TIP.2005.854491

Cox, I. J., Hingorani, S. L., Rao, S. B., & Maggs, B. M. (1996). A maximum likelihood stereo algorithm. *Computer Vision and Image Understanding*, *63*(3), 542–567. doi:10.1006/cviu.1996.0040

Criminisi, A., Blake, A., Rother, C., Shotton, J., & Torr, P. H. S. (2007). Efficient dense stereo with occlusions for new view-synthesis by four-state dynamic programming. *International Journal of Computer Vision*, *71*(1), 89–110. doi:10.1007/s11263-006-8525-1

Deng, Y., Yang, Q., Lin, X., & Tang, X. (2007). Stereo correspondence with occlusion handling in a symmetric patch-based graph-cuts model. *IEEE Transactions on Pattern Analysis and Machine Intelligence*, *29*(6), 1068–1079. doi:10.1109/TPAMI.2007.1043

Felzenszwalb, P., & Huttenlocher, D. (2004). Efficient belief propagation for early vision. In *Proceedings of the 2004 IEEE Computer Society Conference on Computer Vision and Pattern Recognition,* (vol. 1, pp. I-261–I-268).

Fusiello, A., Trucco, E., & Verri, A. (2000). A compact algorithm for rectification of stereo pairs. *Machine Vision and Applications*, *12*(1), 16–22. doi:10.1007/s001380050120

Ganapathy, V., & Ng, O. (2008). Stereo vision based robot controller. In *IEEE International Conference on Systems, Man and Cybernetics,* (pp. 1849–1854).

Hirschmüller, H., Innocent, P. R., & Garibaldi, J. (2002). Real-time correlation-based stereo vision with reduced border errors. *International Journal of Computer Vision*, *47*(1), 229–246. doi:10.1023/A:1014554110407

Hirschmüller, H., & Scharstein, D. (2007). Evaluation of cost functions for stereo matching. In *IEEE Conference on Computer Vision and Pattern Recognition,* (pp. 1–8).

Hirschmüller, H., & Scharstein, D. (2009). Evaluation of stereo matching costs on images with radiometric differences. *IEEE Transactions on Pattern Analysis and Machine Intelligence, 31*(9), 1582–1599. doi:10.1109/TPAMI.2008.221

Hong, L., & Chen, G. (2004). Segment-based stereo matching using graph cuts. In *Proceedings of the 2004 IEEE Computer Society Conference oComputer Vision and Pattern Recognition,* (vol. 1, pp. I-74–I-81).

Kostlivá, J., Čech, J., & Šára, R. (2007). Feasibility boundary in dense and semi-dense stereo matching. In *IEEE Conference on Computer Vision and Pattern Recognition,* (pp. 1–8).

Labatut, P., Keriven, R., & Pons, J. (2006). Fast level set multi-view stereo on graphics hardware. In *Third International Symposium on 3D Data Processing, Visualization, and Transmission,* (pp. 774–781).

Mairal, J., Keriven, R., & Chariot, A. (2006). Fast and efficient dense variational stereo on GPU. In *Third International Symposium on 3D Data Processing, Visualization, and Transmission,* (pp. 97–104).

Ng, O., & Ganapathy, V. (2008). A novel modular framework for stereo vision. In *2009 IEEE/ASME International Conference on Advanced Intelligent Mechatronics*, (pp. 857–862).

Ng, O., & Ganapathy, V. (2009). Efficient handling of over/under-exposure in stereo vision. In *International Conference of Soft Computing and Pattern Recognition,* (pp. 569–574).

Šára, R., & Bajcsy, R. (1997). On occluding contour artifacts in stereo vision. In *Proceedings IEEE Computer Society Conference on Computer Vision and Pattern Recognition* (pp. 852–857).

Scharstein, D., & Pal, C. (2007). Learning conditional random fields for stereo. In *IEEE Conference on Computer Vision and Pattern Recognition,* (pp. 1–8).

Scharstein, D., & Szeliski, R. (2002). A taxonomy and evaluation of dense two-frame stereo correspondence algorithms. *International Journal of Computer Vision, 47*(1), 7–42. doi:10.1023/A:1014573219977

Tola, E., Lepetit, V., & Fua, P. (2010). DAISY: An efficient dense descriptor applied to wide-baseline stereo. *IEEE Transactions on Pattern Analysis and Machine Intelligence, 32*(5), 815–830. doi:10.1109/TPAMI.2009.77

Yang, Q., Wang, L., Yang, R., Wang, S., Liao, M., & Nister, D. (2006). Real-time global stereo matching using hierarchical belief propagation. In *Proceedings of the British Machine Vision Conference (BMVC 2006)*, (pp. 989–998).

Yoon, K., & Kweon, I. S. (2006). Adaptive support-weight approach for correspondence search. *IEEE Transactions on Pattern Analysis and Machine Intelligence, 28*(4), 650–656. doi:10.1109/TPAMI.2006.70

Yoon, K., & Kweon, I. S. (2008). Distinctive similarity measure for stereo matching under point ambiguity. *Computer Vision and Image Understanding, 112*(2), 173–183. doi:10.1016/j.cviu.2008.02.003

Zabih, R., & Woodfill, J. (1994). Non-parametric local transforms for computing visual correspondence . In *Computer Vision* (pp. 151–158). ECCV.

Zhou, J., & Li, B. (2008). Image rectification for stereoscopic visualization. *Journal of the Optical Society of America. A, Optics, Image Science, and Vision, 25*(11), 2721–2733. doi:10.1364/JOSAA.25.002721

ADDITIONAL READING

Agarwal, A., & Blake, A. (2010). Dense Stereo Matching over the Panum Band. *Pattern Analysis and Machine Intelligence . IEEE Transactions on, 32*(3), 416–430.

Baker, S., Sim, T., & Kanade, T. (2003). When is the shape of a scene unique given its light-field: a fundamental theorem of 3D vision? *Pattern Analysis and Machine Intelligence . IEEE Transactions on, 25*(1), 100–109.

Čech, J., & Šára, R. (2005). Complex Correlation Statistic for Dense Stereoscopic Matching. In *Image Analysis* []. Springer Berlin / Heidelberg.]. *Lecture Notes in Computer Science, 3450*, 598–608. doi:10.1007/11499145_61

Delon, J., & Rougé, B. (2007). Small Baseline Stereovision. *Journal of Mathematical Imaging and Vision, 28*(3), 209–223. doi:10.1007/s10851-007-0001-1

Deriche, R. (1990). Fast algorithms for low-level vision. *Pattern Analysis and Machine Intelligence . IEEE Transactions on, 12*(1), 78–87.

Gallup, D., Frahm, J., & Mordohai, P. Qingxiong Yang, & Pollefeys, M. (2007). Real-Time Plane-Sweeping Stereo with Multiple Sweeping Directions. In *Computer Vision and Pattern Recognition, 2007. CVPR '07. IEEE Conference on* (pp. 1-8).

Habbecke, M., & Kobbelt, L. (2007). A Surface-Growing Approach to Multi-View Stereo Reconstruction. In *Computer Vision and Pattern Recognition, 2007. CVPR '07. IEEE Conference on* (pp. 1-8).

Kim, J. C., Lee, K. M., Choi, B. T., & Lee, S. U. (2005). A dense stereo matching using two-pass dynamic programming with generalized ground control points. In *Computer Vision and Pattern Recognition, 2005. CVPR 2005. IEEE Computer Society Conference on* (Vol. 2, pp. 1075-1082 vol. 2).

Kostkova, J., & Šára, R. (2003). Stratified Dense Matching for Stereopsis in Complex Scenes. In *Proceedings of the British Machine Vision Conference (BMVC2003)* (pp. 339-348).

Leung, C., Appleton, B., & Sun, C. (2008). Iterated dynamic programming and quadtree subregioning for fast stereo matching. *Image and Vision Computing, 26*(10), 1371–1383. doi:10.1016/j.imavis.2007.11.013

Loop, C., & Zhengyou Zhang. (1999). Computing rectifying homographies for stereo vision. In *Computer Vision and Pattern Recognition, 1999. IEEE Computer Society Conference on*. (Vol. 1, p. 131).

Ma, X., & Zha, H. (2008). Graph-based Stereo Matching by Incorporating Monocular Cues. In *3D Data Processing, Visualization, and Transmission, Fourth International Symposium on* (pp. 153-158).

Mikolajczyk, K., & Schmid, C. (2005). A performance evaluation of local descriptors. *Pattern Analysis and Machine Intelligence . IEEE Transactions on, 27*(10), 1615–1630.

Monaco, J. P., Bovik, A. C., & Cormack, L. K. (2008). Nonlinearities in Stereoscopic Phase-Differencing. *Image Processing . IEEE Transactions on, 17*(9), 1672–1684.

Moravec, K., Harvey, R., & Bangharn, J. (2000). Scale trees for stereo vision. *Vision, Image and Signal Processing . IEEE Proceedings*, *147*(4), 363–370.

Pajares, G., & de la Cruz, J. (2002). The non-parametric Parzen's window in stereo vision matching. *Systems, Man, and Cybernetics, Part B, IEEE Transactions on*, 32(2), 225-230.

Papadimitriou, D., & Dennis, T. (1996). Epipolar line estimation and rectification for stereo image pairs. *Image Processing . IEEE Transactions on*, *5*(4), 672–676.

Rogowitz, B. E., Frese, T., Smith, J. R., Bouman, C. A., & Kalin, E. B. (1998). Perceptual image similarity experiments. In *Human Vision and Electronic Imaging III* (Vol. 3299, pp. 576-590).

Scharstein, D., & Szeliski, R. (2002). A Taxonomy and Evaluation of Dense Two-Frame Stereo Correspondence Algorithms. *International Journal of Computer Vision*, *47*(1), 7–42. doi:10.1023/A:1014573219977

Scharstein, D., & Szeliski, R. (2007, August 29). Stereo. vision.middlebury.edu. Retrieved January 12, 2009, from http://vision.middlebury.edu/stereo/

Van Meerbergen, G., Vergauwen, M., Pollefeys, M., & Van Gool, L. (2002). A Hierarchical Symmetric Stereo Algorithm Using Dynamic Programming. *International Journal of Computer Vision*, *47*(1), 275–285. doi:10.1023/A:1014562312225

Viola, P., & Wells, W. M. III. (1997). Alignment by Maximization of Mutual Information. *International Journal of Computer Vision*, *24*(2), 137–154. doi:10.1023/A:1007958904918

We recommend the following publications by Daniel Scharstein to all who are interested in stereo vision research; these provide a good background on the area. Scharstein, D. (1999). *View synthesis using stereo vision*. Springer.

We recommend the following publications describing interesting stereo vision algorithms. Some of the more recent publications are also indexed in the Middlebury stereo vision website previously recommended.

We recommend the following publications describing lower-level vision issues which directly affect the stereo vision field.

Wu, C., & Wang, Z. (2006). Stereo Correspondence Using Stripe Adjacency Graph. *In Pattern Recognition, 2006. ICPR 2006. 18th International Conference on* (Vol. 1, pp. 123-126).

Yoon, K., & Kweon, I. S. (2008). Distinctive Similarity Measure for stereo matching under point ambiguity. *Computer Vision and Image Understanding*, *112*(2), 173–183. doi:10.1016/j.cviu.2008.02.003

Zabulis, X., & Kordelas, G. (2006). Efficient, Precise, and Accurate Utilization of the Uniqueness Constraint in Multi-View Stereo. In *3D Data Processing, Visualization, and Transmission, Third International Symposium on* (pp. 137-144).

Zhou, J., & Li, B. (2008). Image rectification for stereoscopic visualization. *Journal of the Optical Society of America. A, Optics, Image Science, and Vision*, *25*(11), 2721–2733. doi:10.1364/JOSAA.25.002721

KEY TERMS AND DEFINITIONS

Application Programming Interface (API): Conceptually – the description of the behavior of a class. Practically: The functions and data members of a class which are publicly available to all users of the class.

Disparity: Difference between the positions of a scene element in both views, inversely correlated to scene depth.

Modular Stereo Vision: Concept which defines stereo vision algorithms as a combination of inter-linked modules which taken together implement stereo vision.

Object-Oriented Programming: A programming paradigm incorporating the concepts of inheritance and polymorphism, a suitable paradigm for the implementation of modular stereo vision.

Occlusion: The condition where a scene element from one view is not visible in the other view due to being covered by another scene element.

Stereo Algorithm: A computational process which implements stereo vision.

Stereo Method: An algorithm which implements some section or stage of a stereo algorithm.

Stereo Module: A computational implementation of a stereo method which allows its use as part of a modular stereo vision algorithm.

Stereo Vision: The process of combining two distinct views of a scene into a unified map of scene depth.

ENDNOTES

1 http://vision.middlebury.edu/stereo/

2 http://opencv.willowgarage.com/

3 http://opencv.willowgarage.com/wiki

Chapter 14
Stereoscopic Vision for Off-Road Intelligent Vehicles

Francisco Rovira-Más
Polytechnic University of Valencia, Spain

ABSTRACT

After mechanization, the next disruptive technology in agriculture will probably be robotization. The introduction of information technology and automation in farm fields started in the eighties with the advent of the Global Positioning System (GPS) and the subsequent development of Precision Agriculture. While being indispensable for many innovative applications, global positioning is not sufficient for all situations encountered in the field, where local sensing is essential if accurate and updated, information has to control automated vehicles. Safeguarding, high resolution mapping, and real time monitoring can only be achieved with local perception sensors such as cameras, lasers, and sonar rangers. However, machine vision offers multiple advantages over other sensing alternatives, and among imaging sensors, stereo vision provides the richest source of information for real time actuation. This chapter presents an overview of current and future applications of 3D stereo vision to off-road intelligent vehicles, with special emphasis in real problems found in agricultural environments and practical solutions devised to cope with them, as image noise, system configuration, and 3D data management. Several examples of stereo perception engines implemented in robotized off-road vehicles illustrate the concepts introduced along the chapter.

INTRODUCTION

An intelligent off-road vehicle is a vehicle that, in addition to perform optimally in off-road environments, has been endowed with certain degree of artificial intelligence (AI). Typical equipment designed to work in off-highway conditions include agricultural machinery, forestry machines, construction vehicles, military trucks, and planetary rovers. The intelligent and automatic tasks typically demanded from these vehicles are directly related to the purpose or main activity for which they have been designed, although most of the in-

DOI: 10.4018/978-1-61350-326-3.ch014

strumentation and techniques employed are common among off-road vehicles. This chapter focuses on off-road vehicles used for civilian applications, especially those related with agricultural production systems. In this particular case, the objective is usually to assist operators so that vehicles can perform their tasks in *semi-autonomous mode*; that is, the driver sits in the cabin of the vehicle for supervision and security reasons while several tasks are carried out simultaneously, and some of them automatically. Unlike planetary rovers that cope with totally unstructured terrains, farm-based equipment often navigates through fields orderly arranged and partially structured by crop rows, tree lanes, guiding trellis, or greenhouse walls. Even those operations occurring in barren fields are limited by field boundaries, natural streams, or irrigation canals.

The presence of solid structures sharing the vital space used by intelligent vehicles in their regular motion is both an advantage and a disadvantage. The former is due to the fact that structures of known characteristics provide additional information to the vehicle and create visual features that can be tracked for navigation and localization. The latter, however, poses a critical problem for automated vehicles as most of the obstacles found in agricultural fields are not traversable and the possibility of having an accident is always present. In fact, this risk is most likely the hardest impediment to automation in agricultural fields, where intelligent machines –even in semi-autonomous mode– are assumed to outperform humans. The existence of semi-structured environments results in the need of reliable perception capabilities, and among the range of practical possibilities, machine vision occupies a preeminent position. It is, precisely, under these circumstances of partially structured terrains where stereoscopic vision finds its privileged niche as it provides three-dimensional (3D) representations of field scenes in the vicinity of intelligent vehicles, at high rates, and with a resolution never reached by satellite or airborne imagery. This chapter explains how to configure the stereo perception engine of an off-road vehicle, discusses the main issues and difficulties involved with real time 3D perception, proposes practical solutions to deal with common problems encountered in the field, and finally analyzes two popular applications within off-road agricultural equipment.

BACKGROUND

Although the principles of stereoscopy have been known since the nineteen century, the availability of commercial stereo cameras only dates from the turn of this century. The processing speed of current computers allows the execution of algorithms that can correlate two stereo images and generate a depth map in real time. The level of detail and amount of information supplied by stereoscopic perception has placed stereo-based devices in a privileged position among other sensors used in field robotics. Mars exploration (Olson et al., 2003) and defense mobile robots like Urbie rely on stereo cameras to acquire critical information of remote and often hazardous environments. The application of 3D vision technology to agricultural vehicles, in spite of having a high potential (Rovira-Más, 2003), is still in its infancy. Some timid efforts have been made to apply the idea of stereoscopy to automatically locate fruits in plants (Kondo et al., 1996), but human intervention has been normally required to assist in pixel matching. Real time stereo-based perception for mobile robots is relatively recent, and although some solutions have been successfully developed for small indoor robots (Herath et al., 2006) and on-highway vehicles (Kato et al., 1996), the scenarios typically perceived in these applications are substantially different from those encountered by off-road vehicles; therefore, the latter demand specific solutions motivated by very distinctive needs. Even the off-road prototypes that participated in the DARPA Grand Challenge competition, organized by the United

States Department of Defense, were set to fulfill elaborated missions that nothing have to do with habitual agronomical tasks (Kogler et al., 2006).

Conventional stereo cameras provide perceptual information at three levels: the original (2D) images that comprise the stereo pairs (left image and right image), the disparity (2D) image that holds the basic depth information, and the 3D point cloud that recreates the perceived scene in a discrete set of points. Critical information may be retrieved at any stage, so, for example, color data in RGB format is obtainable from the raw images of the stereo pairs when at least one of the imagers of the camera supports RGB color. The disparity image, in spite of storing spatial and depth information, is actually a two-dimensional image to which conventional image processing techniques can be applied. However, the fact that (disparity) image pixels can retrieve the 3D location of critical features selected from the scene, allowed Rovira-Más et al. (2004) to segment crop rows in disparity images with the purpose of determining the trajectory of an automatically steered tractor. The availability of the three-dimensional coordinates for each point of an image-determined trajectory was advantageous in the transformation from image domain to real world, which typically represents a delicate stage for monocular cameras working outdoors. The appearance of a vanishing point when a camera captures images of parallel rows from a ground vehicle is often a difficult challenge for monocular vision, but it is correctly handled by stereo cameras. The main complexity for this disparity-based row tracking, though, rested on the segmentation of the images, which were set in such a way that background soil was saturated and consequently filtered by the correlation algorithm, only remaining the pixels representing the crop rows providing navigation information. A varying ambient illumination, common in outdoor conditions, complicated the permanent and robust discrimination of crop rows, especially with alternating vehicle heading at sunrise and sunset.

Although the previous example describes a technique that uses disparity images as source data for guiding a vehicle, the majority of stereo applications employ 3D point clouds as initial data from which critical perceptual information can be extracted. Point clouds will be, therefore, the main data to be processed in stereo perception. There are three fundamental problems that need to be solved before an intelligent vehicle can make use of the perception information acquired with a stereo camera: *noise reduction*, *data conditioning*, and the *extraction of critical information*. The presence of noise is a common issue for real time stereo correlation, and as a result, it has led to the development of diverse filtering algorithms usually incorporated in commercial off-the-shelf stereo cameras. Nevertheless, there is always uncontrollable noise that remains unfiltered by the proprietary software loaded in the camera, and consequently ends up forming part of the disparity image, leading to 3D points with wrong –and often impossible– coordinates. This sort of noise is dangerous because it can be hard to detect and could destabilize an automated vehicle very easily. The first place where correlation mismatches can be noticed is in the disparity image, and therefore noise filtering can start at this stage. As a matter of fact, high quality disparity images often result in rich and meaningful 3D point clouds. Bailey et al. (2007) applied a Gaussian filter to blur both images of a stereo pair and eliminate pixel noise. Salt and pepper noise has also been efficiently eliminated from disparity images (Wong & Jarvis, 2004), but Rovira-Más et al. (2009) found that the application of spectral analysis to palliate the effect of noise in disparity images was not helpful for orchard scenes, where noisy blobs composed of several miscorrelated pixels could not be completely eradicated. This type of consistent mismatches is, precisely, the one posing the biggest challenges for mapping and navigation. A wrongly-detected branch, for instance, can halt a vehicle before finishing a task, or more dangerously, it can force a vehicle to deviate from its safest trajectory to avoid inexistent obstacles.

The *Validity Box* approach, explained in the next section, has been successfully applied to lessen the effect of noise causing coordinates to point at unrealistic locations. After the reduction of noise, the following stage in the processing of stereo information is data conditioning for the extraction of key information. This step comprises the set of operations that prepare massive and unstructured point clouds for an effective extraction of information in real time. A point cloud can be considered a large number of disconnected points that only have coherence when appropriately treated. The utilization of 3D data through the idea of *evidence grids* (Moravec, 1996) has been a popular resource among the robotics community dealing with indoor applications. The non-probabilistic approach of *density grids* (Rovira-Más et al., 2006) has worked efficiently outdoors, and therefore results convenient for processing the 3D perceptual information acquired by automatic off-road vehicles. The third and last essential matter to deal with when analyzing 3D point clouds is the identification of vital information for the pursuit of the assigned mission. These data will depend on each particular application and so will be the degree of precision and reliability required. The resolution needed in the reconstruction of agricultural scenes will also be related to the size of potential objects being perceived. Sometimes, a vehicle does not need to identify the obstacles that invade its surroundings; rather, it is more practical to assess if those obstacles can be traversed by the robotic vehicle. Because point clouds constitute unstructured data for which no standard modeling exists (Marr, 1982), each case needs to be treated with the required degree of specificity. The analysis of terrain according to its *traversability* has resulted appealing for planetary rovers (Singh et al., 2000), and has been explored as well for robotized agricultural equipment by Rovira-Más (2009). Regardless of the specific application developed, the best guarantee for successful stereoscopic perception entails a favorable configuration of the system, the efficient reduction of noise, and a proficient method to handle and process massive point clouds. The following sections try to bring some light on these crucial steps, and the last one describes two agricultural applications that illustrate most of the ideas, concepts and approaches presented along the chapter.

THE WEIGHT OF NOISE IN REAL-TIME FIELD APPLICATIONS

Issues with Stereo Mismatches in Outdoor Environments

Television watchers remember, before the popularization of digital technology, how annoying it was the presence of noise in broadcast images. It was a nuisance but there were no further consequences besides the obvious exasperation. When the final outcomes of vision-based intelligent systems are navigation commands, safeguarding alerts, yield-production estimations, or any other electronic signal serving as the basic control input of an automatic operation, the consequences of image degradation are far deeper than just irritation. Noise can mask important phenomena occurring in the sensed scene or involve a vehicle in a dangerous situation. Figure 1 shows problematic noise found in typical field scenes. Cloudy skies, for example, are a common source of error when the matching algorithm is misled and clouds are positioned within few meters from the ground. In general, field scenes are well illuminated and full of texture, what is propitious for good stereo correlation and quality disparity images. Sometimes, however, ambiguous areas in the scene, i. e. those portions with pixels of equal intensity, originate mismatches and erroneous blobs appear in disparity maps. Once a pixel is associated with a disparity value –either correct or incorrect–, it will lead to a given 3D location defined by three Cartesian coordinates. A noisy blob in the disparity image will result in a misplaced set of points in the 3D point cloud; its size and 3D position will determine the level of corruption introduced in the point cloud. Rain is another potential source of errors as it might confuse the correlation algorithm, although it is not very important in practice as most off-road vehicles do not operate with stormy weather. Muddy terrains complicate maneuverability of heavy off-road vehicles, and humidity can jeopardize certain tasks such as harvesting.

Practical Solutions and Recommendations

When a group of noisy points skip the firmware-embedded correlation filters and appear in the final 3D point cloud as outliers, it is very difficult to isolate and remove them. In this situation, the information known beforehand as a consequence of operating in semi-structured scenarios can be helpful, in addition to other logic assumptions such as the suppression of those points located under the reference ground. An effective way to remove harmful outliers is by defining a 3D box indicating a reasonable placement for scene points, called the *Validity Box*. All points falling outside this box do not need to be considered in further processing. The dimensions of the box depend on each application and are limited by intersecting planes parallel to the Cartesian planes defined by the ground coordinates indicated in Figure 3b. Negative heights, for example, can be neglected in common orchard scenes as the ground is usually flat without deep holes. Maximum ranges –distances from the camera to the objects– depend on the configuration of the camera, i.e. baseline and lenses chosen, and therefore it makes no sense to process points that are too far to be detected reliably with a given configuration. If the average height of the trees in the field is less than 4 m, all the points above 5 m from the ground have a high probability of being noise such as the clouds in the sky of Figure 1.

When vehicles navigate between tree rows, there is no interest in considering points that belong to adjacent rows, which will likely represent outliers because an onboard camera cannot perceive through the dense vegetation of surrounding

Figure 1. Noise in stereo images of field scenes

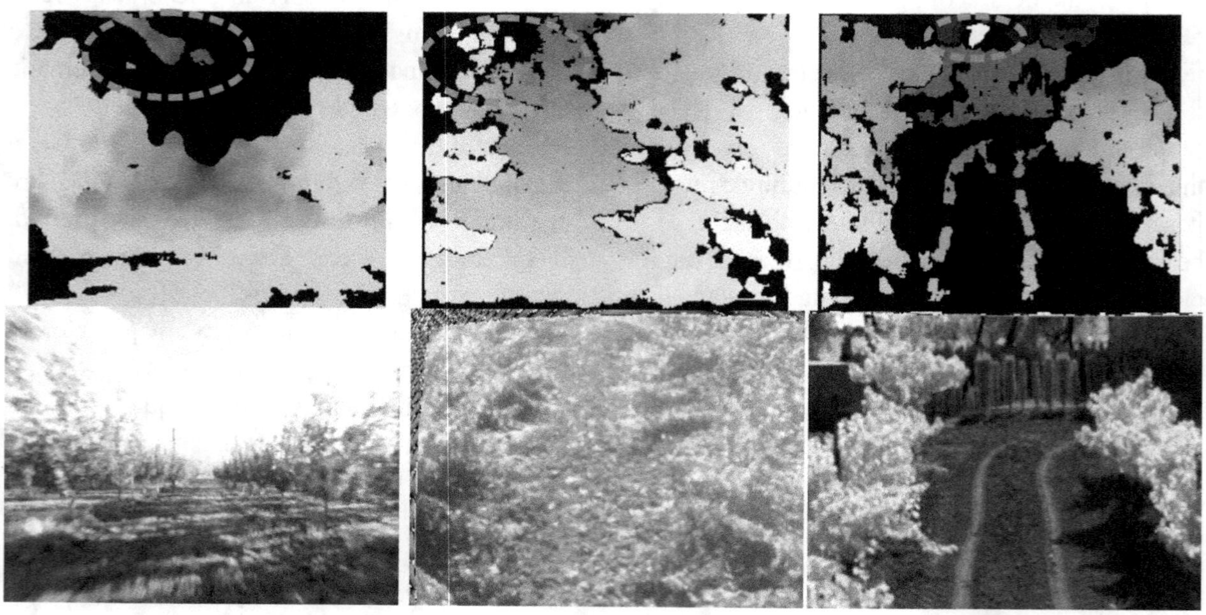

trees. Following a rationale of this kind, we can limit the volume where the camera is expected to retrieve 3D information. The points outside the box (*"outboxers"*) tend to be small in number but large in error (*outliers*). As an example, a robotic tractor equipped with a stereo camera of 20 cm baseline and 8 mm lenses was set to perceive 3D information of a vineyard, composed of plants 1.5 m high and disposed in rows separated 3 m. The onboard stereo system used a box 5 m high (Z), 6 m width (X), and 15 m long (Y). The camera was mounted on the top-center of the cabin, approximately 3 m (Z) from ground level. The effects of salt and pepper noise, as well as other electronic noise, resulting in isolated outliers inside the Validity Box were palliated with data condensation through density grids.

CONFIGURATION OF 3D PERCEPTION ENGINES FOR OFF-ROAD VEHICLES

In-Field Needs for Automated Vehicles

The tasks performed by off-road vehicles and the environments in which they operate are very different from those related to other sort of vehicles; on-highway trucks and automobiles navigate at higher speeds and may be continuously surrounded by other light vehicles, many small robots are designed to work indoors and find guiding features in doors or walls, planetary rovers roam around unstructured terrain where no man-made structures are ever found. Agricultural off-road vehicles usually move around crop or tree rows that leave small tolerances for the vehicle to go through the field without causing damage to the vegetation. Many times these tolerances are in the order of a few centimeters. With these precision requirements, GPS-based navigation can greatly benefit from small local adjustments only possible with perception sensors such as stereo cameras. However, the targeted field of view has to be finely determined in order to reach high levels of detail in the reconstructed 3D view. Precision needs strongly depend on both the vehicle used and the task performed. Nonetheless, some general requirements can be enunciated for the average off-road intelligent vehicle. First of all, the perceptive capabilities of the vehicle should assure real-time awareness, and therefore we cannot afford such a wide field of view that data processing has to be carried out off-line. In general, two target distances (or ranges) need to be adequately covered: *medium* distances between 10 and 30 m, and *short* ranges in the vehicle's vicinity. Medium distances include crop-based feature detection for automatic guidance, as for example trajectory target points or cut-uncut harvesting edges. Perception in the surrounding area of vehicles is important for safeguarding reasons, since it is essential to be aware of any nearby obstacle that might become a potential hazard for automated operations. Given the importance of adjusting the field of view of the camera to the sensed scene, it is essential to optimally select the three main design parameters of stereo cameras: *baseline*, *lenses*, and targeted *range interval*. The following section describes a procedure to find the best configuration of a binocular stereo rig.

Determination of Optimal Baselines and Lenses

The objective of this procedure is to find the fundamental camera parameters that yield high quality 3D perception for a determined visual field. These parameters are the *baseline*, or distance between the optical center of both lenses, and the *focal length* of the lenses. Every time a lens is changed or the baseline modified, the camera needs to be calibrated with a chess-like board. For this reason, it is more convenient to use precalibrated cameras with fixed optics and permanent baseline, but before choosing definitive lenses and baseline, we need to make sure that the needed field of view is appropriately covered by the stereo sensor and the requested ranges are accurately estimated. As mentioned before, a rich and noiseless disparity image is a good indicator of 3D perception quality, but there is a need for a quantitative evaluation in terms of objective quality indicators. A way to conduct this evaluation consists of placing target objects of size similar to that of potential objects to detect separated from the camera the approximated expected range, and capturing a stereo image of the test scene with the camera under evaluation.

Once the stereo image has been acquired, its corresponding 3D point cloud let us compare between the stereo-estimated distances and the actual ones along the three coordinates x_1, x_2, x_3 in such a way that three *relative errors* can be calculated according to Equation 1, where x_1, x_2, and x_3 are the real distances directly measured with a standard tape, and x_1', x_2', and x_3' are the estimated distances directly read from the 3D point cloud. The distances under comparison need to be replicated for multiple features randomly sampled from the testing scene and then averaged before proceeding with the calculation of *relative errors* $\{\varepsilon_{x1}=\varepsilon_x, \varepsilon_{x2}=\varepsilon_z, \varepsilon_{x3}=\varepsilon_y\}$. Once the relative errors have been determined for the three Cartesian axes, those defining a plane parallel to the camera and therefore containing points of equal range, for example x_1 and x_2, will give the planar efficiency η_{2D} according to Equation 2. If, in addition, the relative error in ranges, say x_3, is also taken into account following Equation 3, we will have an estimate of the 3D quality achieved for that camera configuration. The assessment of 3D perception quality with efficiencies η_{2D} and η_{3D} helps to compare different camera configurations before choosing the ideal setup for a given application. After the selection of lenses and baseline has been made according to the highest values of η_{2D} and η_{3D}, a stereo camera with fixed optics can be ordered to avoid the problems caused by a loss of accuracy in the calibration parameters as a consequence of vehicle vibrations or accidental drops of the camera. The chart of Figure 2 plots the η_{2D} and η_{3D} efficiencies estimated with various combinations of lenses and baselines for a set of flat targets situated at 12 m from the camera. The target consisted of a matrix of nine boxes equally spaced with a front pattern combining black and white solid squares. This test considered ranges of around 12 m as the critical medium distances under study. In the chart, each point represents a combination of baseline (*B*) and focal length (*f*) analyzed in the experiment. The position of each square in the two-dimensional plot of Figure 2 indicates a particular perception quality for the combination *B-f* tried for the detection of 12-m ranges, as the abscissas give the planar efficiency (η_{2D}) while the ordinates represent the stereo efficiency (η_{3D}). In particular, Figure 2 highlights three combinations that are preferable (η_{2D} and η_{3D} over 90%) for sensing at 12 m ranges: {*B*=19 cm; *f*=8 mm}, { *B*=19 cm; *f*=12 mm }, { *B*=15 cm; *f*=16 mm }.

$$\varepsilon_{x_i} = \left| \frac{\min(x_i, x_i')}{\max(x_i, x_i')} - 1 \right| \cdot 100 \qquad i = \{1,2,3\} \tag{1}$$

$$\eta_{2D} = (1 - 0.01 \cdot \varepsilon_{x_1}) \cdot (1 - 0.01 \cdot \varepsilon_{x_2}) \cdot 100 \tag{2}$$

$$\eta_{3D} = \eta_{2D} \cdot (1 - 0.01 \cdot \varepsilon_{x_3}) \tag{3}$$

PROCESSING 3D DATA IN REAL TIME

Raw Data from Compact Binocular Cameras and Coordinate Systems

After finding the best possible configuration of the camera given by baseline and lenses, as well as the optimum location for the sensor in the vehicle, we are ready to perceive the world in three dimensions. The output of the camera is formed by a set of points precisely located in space by their three Cartesian coordinates. Two elemental checks need

Figure 2. Assessment of 3D perception quality for target ranges of 12 m

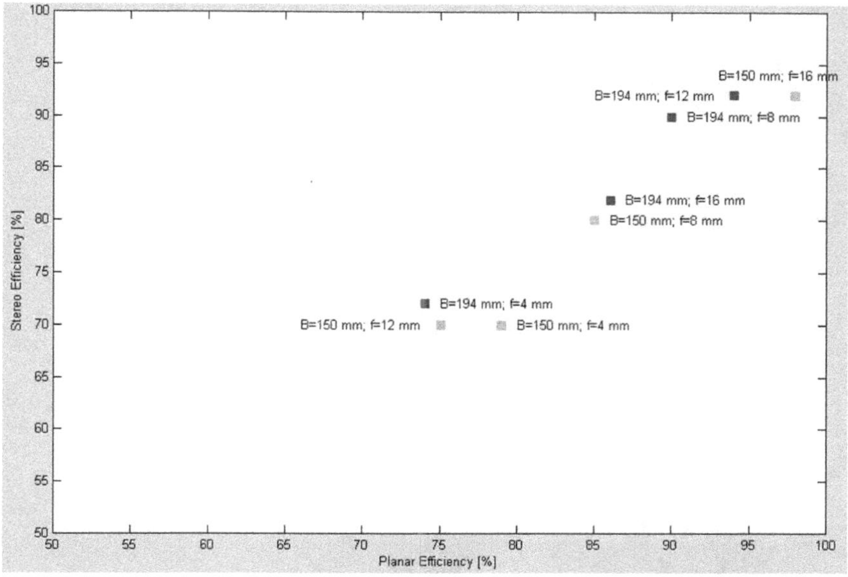

to be made at this point: the definition of camera coordinates and the units in which the coordinates are expressed. This information is provided by the camera manufacturer, and without it all the data acquired in the field is useless, as objects and distances need to be properly referenced and measured. Typical units for camera coordinates are meters and millimeters. The definition of *camera coordinates* includes the position of the origin and the description of the three Cartesian axes intersecting in the origin. Origins are normally set at the optical center of one of the two lenses, hereafter denominated the reference lens. Figure 3a illustrates a usual coordinate frame for stereo cameras, where the reference lens is the left one (seen from behind the camera).

Following a convention adopted by the machine vision community, the X_cY_c plane is coincident with the image plane, the Y_c coordinates follow the vertical dimension of the image growing downwards, and the X_c axis coincides with the horizontal dimension of the image with increasing values from left to right. The third dimension, Z_c, indicates the ranges or distances from the optical center of the reference lens to the detected objects. The camera coordinates portrayed in Figure 3a, while being comfortable for a permanent camera mounted with the image plane perpendicular to the ground, are not convenient for cameras on board off-road vehicles. To begin with, vehicle-fixed cameras are often tilted in such a way that ranges are not parallel to the ground. Furthermore, this inclination angle may change with time, making impossible the fusion of coordinates generated when the camera is set under different inclination angles. Off-road intelligent vehicles are, after all, ground vehicles, and any information related to them needs to be properly referenced to them and, by extension, to the ground. The *ground coordinates* represented in Figure 3b are very practical and convenient for off-road machines to perceive in the open field. They keep the origin at ground level so that heights are coincident with the Z coordinates. The XY plane is therefore coplanar with the ground where the vehicle rests on. This definition establishes the center of coordinates at the intersection of the Z axis with the XY plane. The Y axis points at the forward direction and indicates the distance between objects and cameras, i. e. the ranges, and

Figure 3. Fundamental systems of coordinates for stereo vision perception. (a) Camera coordinates (b) Ground coordinates.

the X axis is perpendicular to the forward direction as represented in Figure 3b. Given that the locations of the points in the 3D cloud will be initially given in camera coordinates, they will have to be transformed to ground coordinates according to Equation 4, where (x_c, y_c, z_c) are the camera coordinates, (x, y, z) are the ground coordinates, h_c is the camera height taken at the optical center of the reference lens, and ϕ is the camera inclination angle. Figure 4 illustrates this transformation process for a generic point P.

$$\begin{bmatrix} x \\ y \\ z \end{bmatrix} = \begin{bmatrix} 1 & 0 & 0 \\ 0 & -\cos\varphi & \sin\varphi \\ 0 & -\sin\varphi & -\cos\varphi \end{bmatrix} \cdot \begin{bmatrix} x_c \\ y_c \\ z_c \end{bmatrix} + h_c \cdot \begin{bmatrix} 0 \\ 0 \\ 1 \end{bmatrix} \quad (4)$$

Figure 4. Transformation from camera coordinates to ground coordinates

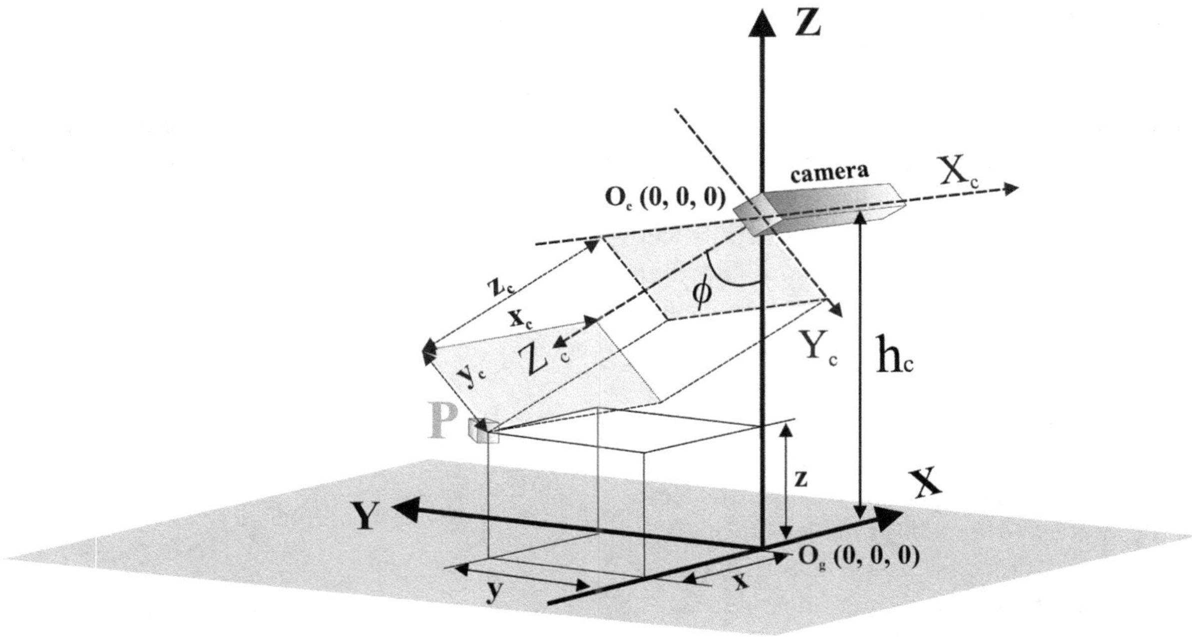

Density Grids and 3D Density

After the coordinate transformation has been conducted for every valid pixel of the disparity map, the stereo information is still in the form of a 3D point cloud, an unstructured set of points with no apparent relation among them. The point clouds can be massive, and even a moderate image resolution of 320 x 240 easily yields disparity images of over 50000 correlated pixels. The perception computer on the vehicle needs to process these data and make sense of the discrete amount of points forming the cloud. Each point provides the 3-coordinate location of a determined feature belonging to the scene captured by the camera; therefore, the accumulation of points in certain region of space is a good indicator of a potential solid object. This idea led to the concept of *3D density* (Rovira-Más et al., 2006), which is defined as the *number of stereo-correlated points per volume unit*. The definition of 3D density (d_{3D}) avoids the strong dependency of the number of stereo-matched points on the image resolution. The presence or absence of an object must never depend on the resolution of the images used. As consecutive images may be exposed to different illumination patterns, the amount of valid pixels can differ significantly from one image to the next one; therefore, the calculation of densities in any image has to be normalized by the maximum value of the 3D density reached in that image before comparisons can be made among images. With this adjustment, regardless of image resolution, and even of the total number of pixels correlated, the high density portions of the reconstructed space will represent solid objects in the actual scene.

A practical way of applying the concept of 3D density is through *density grids*, which can be constructed by dividing the space ahead of the vehicle in a regular grid and calculating the 3D density for every cell. In spite of the positive results obtained with density grids to process 3D data in real time, there is still a concern remaining: the distribution of d_{3D} along the Y axis decreases quadratically. This means that the 3D density depends on the range of the cell evaluated, phenomenon that can be easily explained by the physical fact that the terrain closer to the camera is described by a greater number of pixels since the field of view opens as the range grows. As a result, further targets will be represented by fewer points and the 3D point cloud will get scarcer as it separates from the camera. A less populated point cloud is still valid to discern between solid objects and empty space, but the 3D density threshold used to discriminate objects from emptiness cannot be uniformly applied to the entire image unless a *range-based correction* is applied. This correction normalizes the densities within the grid according to a reference range, usually near the camera where the d_{3D} has not dropped too much. The general form for the corrective formula is that of Equation 5, where $[d_{3D}]_c$ is the compensated density, d_{3D} is the original density, Y_{cell} is the range measured at the center of the corrected cell, and K is a constant that depends on the magnitude and units of the reference range as well as the quadratic fit curve. Figure 5 shows a typical orchard scene (a) and two density grids associated to it: a *frontal grid* composed of square cells of size 50 mm resulting in a grid of resolution 200 x 100 (b); and a *top view grid* (c) made of square cells of 120 mm side that lead to a grid resolution of 47 x 125. The top grid of Figure 5c clearly traces the lane free of obstacles between the two rows detected. The separation between the adjacent rows is about 20 cells, which approximately corresponds to 2.5 meters. Notice that the maximum range represented spans 125 cells or the equivalent length of 15 m ahead of the camera. The color code assigned to every cell represents a particular value of 3D density, obtained by counting the number of correlated points inside the cell, normalizing it, and compensating it according to its range. The grids of Figure 5 do not portray the ground of the traversable inter-row lane in spite of being represented in the original image of the scene (Figure 5a). As a matter of fact, there were many correlated pixels coming from the ground, but in order to enhance non-traversable obstacles –trees in this case–, all the points located under 0.8 m were suppressed in the final display of the grid. Although grid cells are represented by squares, 3D density is defined as points per volume unit, and therefore the actual cells considered are long prisms of square section, where the cross section is precisely the square cell and the main length is limited by the layer gathering the critical information. The top grid of Figure 5c, for example, uses prismatic cells with a cross section of 0.12 x 0.12 (m^2), and a main length of 4.2 m from Z = 0.8 m to Z = 5 m.

$$\left[d_{3D}\right]_c = K \cdot Y_{cell}^2 \cdot d_{3D} \tag{5}$$

AGRICULTURAL APPLICATIONS

Global 3D Terrain Mapping

The pervasive diffusion of satellite localization systems such as GPS has motivated the development of new concepts and disciplines like *precision agriculture*. The core idea behind precision farming is to endow agricultural operations with higher levels of precision, those levels never reached before, what in a nutshell can be understood and summarized as *applying the right quantity of input exactly where it is needed and just at the right time*. This procedure entails handling large amounts of data, high updating rates, and instant actuation. The information exchanged in precision agriculture applications is typically managed and expressed in the form of globally referenced maps, where satellite imagery blends with locally acquired information to compose useful maps for the producers. However, these maps lack high resolution when they are based on satellite images, and seldom can they be updated very often because the final user has no free access to the source of information. A stereo-based terrain map can offer the high degree of detail typical of local perception

Figure 5. Density grids for stereo perception in off-road environments: (a) Field scene; (b) Frontal density grid; (c) Top view density grid

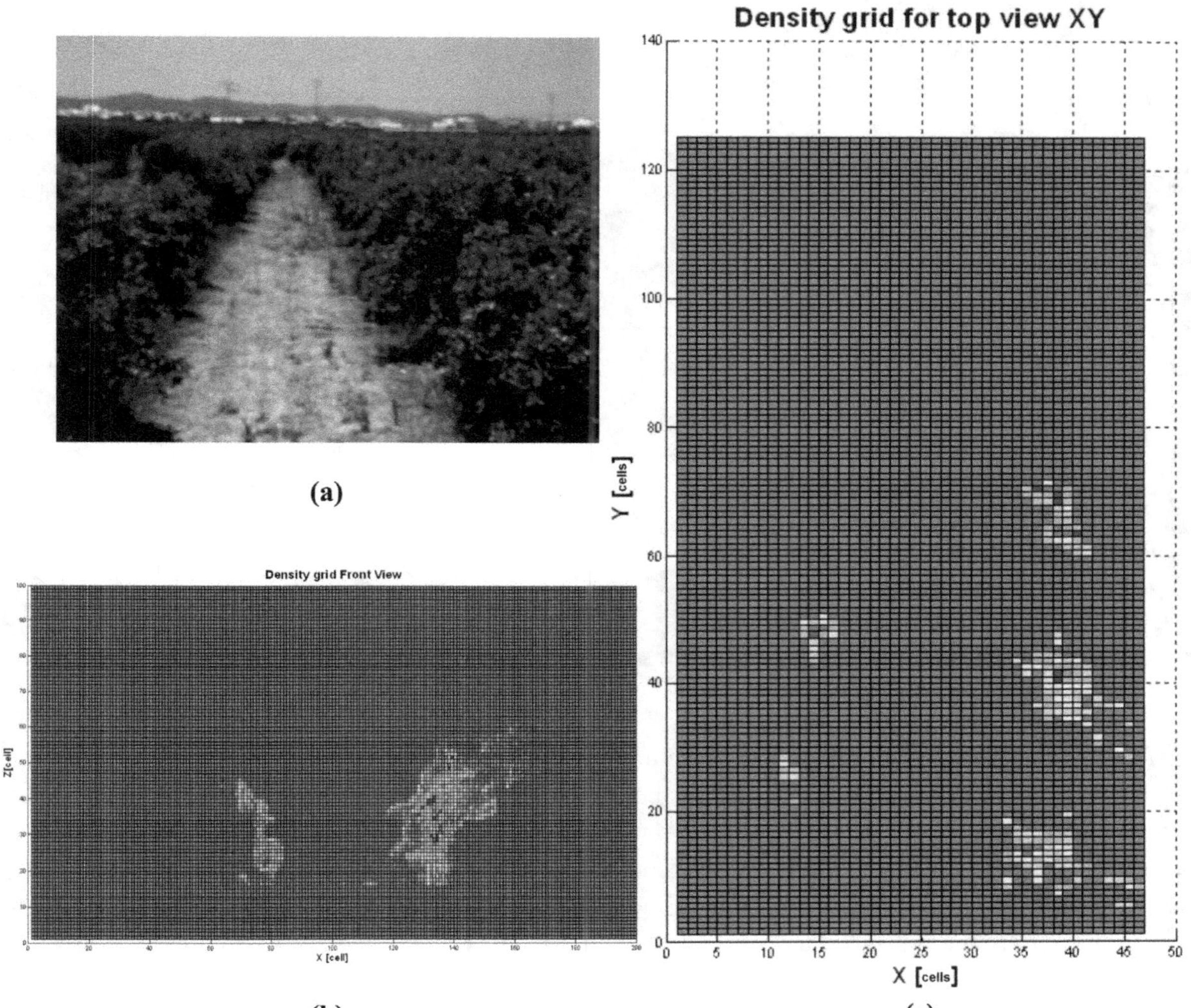

(a) (b) (c)

plus all the advantages of global positioning. In addition, the visual information gathered in the field is available in three dimensions, which allows for multiple views easily adaptable to any application pursued. *Global 3D terrain mapping* requires three different sources of information: 3D perception with a *stereo camera*, vehicle attitude estimated with an *inertial measurement unit*, and global positioning data acquired by a *GPS receiver*. Apart from the 3D location of each point in the cloud, the stereo camera can also provide the true color associated to each point in the scene so that realistic 3D maps can be rendered as shown in Figure 6.

The main difficulties encountered when creating 3D global maps are caused by the complexity of managing large amounts of data accumulated in massive point clouds, and the strong dependency of the map consistency on the accuracy of localization and attitude sensors. The procedure to assemble a global map starts by merging the individual local maps obtained from

Figure 6. Global 3D mapping of agricultural environments

every pair of images grabbed by the stereo camera. This superposition of point clouds can be carried out either on-the-fly or offline after all the data have been acquired in the field. Given that mapping does not interfere with other farming operations, terrain maps can be built while other production tasks are being performed by the intelligent vehicle. Global three-dimensional maps have multiple uses. They can provide safeguarding warnings and navigational assistance, from them we may estimate the bulk of canopies or the growth rate of crops, and any object represented in them is immediately situated in a Cartesian frame for future applications. Figure 6 illustrates the concept of 3D field mapping applied to an orchard in full production. The degree of detail achieved with ground-based 3D perception allows the identification of a person lying down between two rows of apple trees.

Automatic Guidance of Tractors and Harvesters

The major advantage of stereoscopic vision over monocular vision is the availability of ranges. Depth assessment is always ambiguous with single two-dimensional images because any pixel in these images may belong to different points in the real world, depending on the relative position and orientation of the camera with respect to the sensed scene; generally speaking, the presence of vanishing points induces such effect. This advantage of stereo vision has been used to detect obstacles interfering with a vehicle's course, either to find the optimum path for autonomous navigation or as a safety tool to avoid collisions. Given that the information acquired with a stereo camera is far richer than that obtained with alternative systems such as nodding lasers and ultrasonic devices, the practical applicability of stereoscopic vision to mobile robotics is relentlessly increasing.

The strategy planned to guide tractors and combines depends on each application and on the architecture available in the vehicle. Quite likely, camera coordinates will have to be transformed to ground coordinates, but their further conversion to a global frame will not always be necessary. Very often, as the cases illustrated in Figure 7, it is more convenient to use a vehicle-fixed coordinate system; other times, as the case of using 3D global maps to trace a path reference, East-North coordinates are essential. In any case, and whatever the option chosen, the procedure to follow entirely fits the methodology outlined along this chapter: noise treatment, camera optimal configuration, coordinate transformation, and density grid representation. What rest are, of course, the interpretation of every density grid and the calculation of their subsequent decision making commands, which will involve different operations according to each particular application developed. The top row of Figure 7 provides the basic information related to the automated guidance of a corn harvester: the position and orientation of the camera on the harvester head, strategically offset from the head by an extension rod to better sense the cut-uncut edge, selected as the main guidance feature; a sample image of the stereo pair taken from the chosen camera position; and finally the corresponding top-view density grid signaling the machine-detected position of the guiding edge. Notice that the reconstructed corn row is straight and vertical regardless of the camera position and attitude, what indicates a correct scene perception and coordinate transformation. The bottom row of Figure 7 shows a different way to guide the same vehicle with the same stereo camera. The major difference between both approaches is caused by the new position of the camera, which results in a completely different morphology of visual scenes. This time, the cut-uncut edge is not properly sensed within the available field of view, as shown in the sample image included in the figure.

For that reason, rather than trying to find the dividing edge, the objective is to identify the corn rows in front of the harvester as main tracking features for guiding the vehicle. This change in strategy is favored by the capability of stereo cameras mounted on top of the harvester cabin to situate the rows ahead with respect to the vehicle-fixed system of coordinates. Placing the camera on the cabin provides a more compact solution than offset locations at the outermost end of extension bars, as protruding linkages are always problematic in dense environments. Any plant stem or leaf might hit the camera and alter the optimum orientation angle. The density grid resulting from processing two stereo images captured from the cabin roof clearly identifies the position and dimensions of the five rows perceived in the field of view of the camera. Note that the 3D density (d_{3D}) represented in the final grid was properly range-compensated according to the approach explained before and practically executed with Equation 5. The density grid of Figure 7 (bottom right) representing the five rows of corn is composed of square cells of 10 cm side and 1.5 m depth, which give a grid resolution of 60 x 150 cells, i. e. a field of view 6 m wide and 15 m long. The computer resources required to deal with density grids are significantly less demanding than those involved with 3D renderization, reconstruction, and data processing. Grid resolutions such as those used in the examples of Figure 7 are much easier to handle than the original 51480 correlated points comprising the disparity image that led to the five-row grid. The fact that the five rows of corn are parallel in the density grid, while there is a vanishing point in the original gray-level image (bottom center in Figure 7), indicates that coordinate transformations and scene reconstruction were correctly executed. These tests were conducted in the USA Midwest where habitual corn spacing is 76 cm (30 inches).

Figure 7. Automatic steering of a corn harvester based on stereoscopic vision

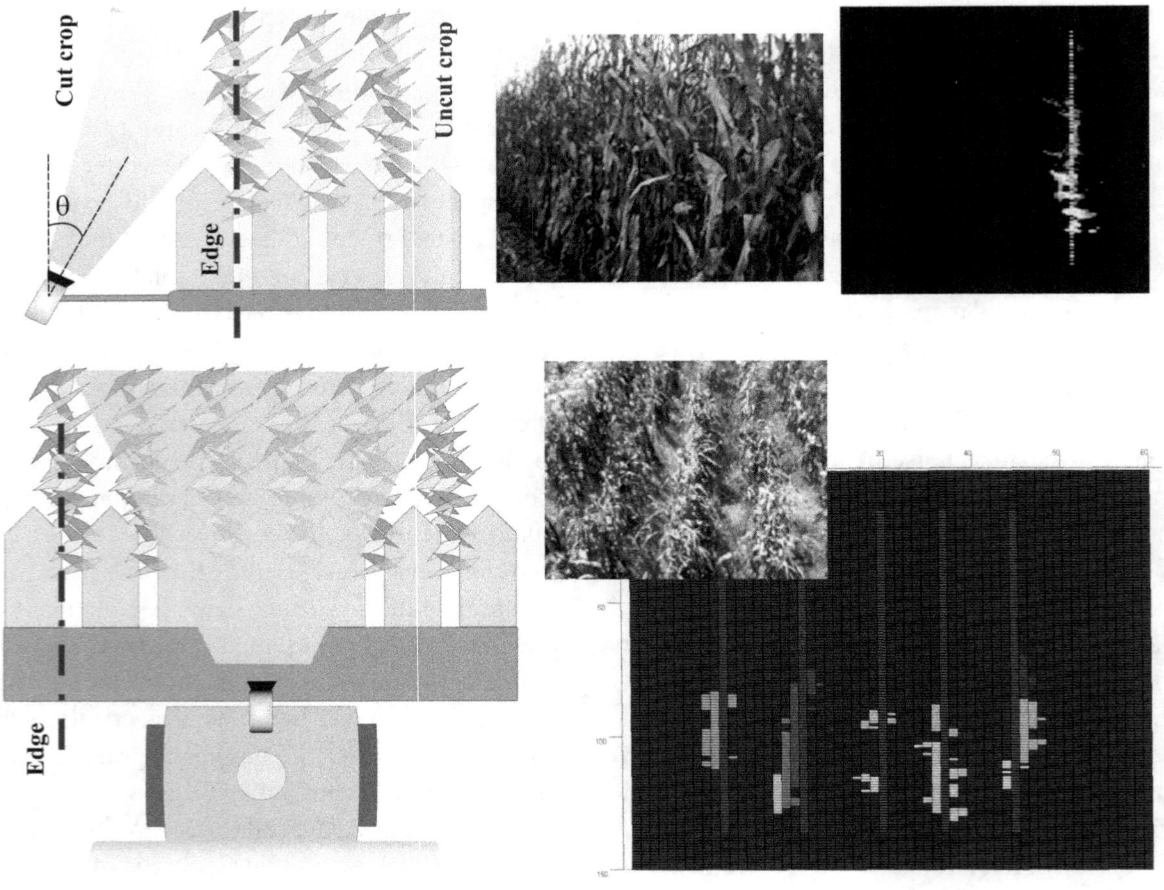

FUTURE RESEARCH DIRECTIONS

Stereoscopic cameras provide a wealth of perceptual information at such a high rate that no other commercial sensor can currently match them. Yet, no sensor is perfect and therefore redundancy with other sensors will most likely be necessary in the forthcoming years, but as technology advances, the capabilities of these cameras will be higher and higher. At the beginning, only cutting-edge high technology projects, such as Mars Pathfinder, implemented them as key perception sensors. However, the popularization of binocular compact stereo rigs has extended their use to a wide variety of engineering fields. Agriculture is one of them. However, in spite of being a field where robotization can provide great benefits, mostly due to the need of performing repetitive tasks in semi-structured and harsh environments, it has been to a great extent overlooked. Yet, potential is prominent and interest high. The deployment of intelligent vehicles for off-road applications will surely induce research, projects, and prototypes with stereoscopic cameras over the next decades.

CONCLUSION

This chapter gives a general view of the applicability of stereo vision perception to the design of intelligent off-road vehicles, and provides a framework to integrate stereo cameras in intelligent equipment performing common tasks in agricultural systems. The methodology proposed starts finding the most favorable configuration of the sensor in relation to lenses and baselines, and continues setting the basic steps in the processing of 3D data by reducing the impact of noise, transforming the initial coordinates to a convenient frame, and conditioning the perceptual information through the concepts of 3D density and density grids. All these consecutive stages constitute the necessary preparation onboard for the execution of decision-making routines in real time. The whole process has been illustrated along two particular applications of high interest in agricultural robotics: global 3D terrain mapping and automated driving. The novelty of real time stereo technology, and the elevated requirements of safety and precision demanded by manufacturers of agricultural vehicles, place the practical realization of intelligent vehicles –and derived commercial expansion– in its infancy. However, the inexorable flow of technology towards vehicle automation and agricultural robotics in the upcoming years presages a very active role for stereoscopic vision as the chief provider of 3D perception.

REFERENCES

Bailey, M., Chanler, A., Maxwell, B., Micire, M., Tsui, K., & Yanco, H. (2007). Development of vision-based navigation for a robotic wheelchair. In Proc. *10th International Conference on Rehabilitation Robotics* (pp.951-957). IEEE.

Herath, D. C., Kodagoda, K. R. S., & Dissanayake, G. (2006). Modeling errors in small baseline stereo for SLAM. In Proc. *9th International Conference on Control, Automation, Robotics, and Vision* (pp. 1-6). IEEE.

Kato, S., Tomita, K., & Tsugawa, S. (1996). Visual navigation along reference lines and collision avoidance for autonomous vehicles. In Proc. *Intelligent Vehicles Symposium* (pp. 385-390). IEEE.

Kogler, J., Hemetsberger, H., Alefs, B., Kubinger, W., & Travis, W. (2006). Embedded stereo vision system for intelligent autonomous vehicles. In Proc. *Intelligent Vehicles Symposium* (pp. 64-69). IEEE.

Kondo, N., Nishitsuji, Y., Ling, P. P., & Ting, K. C. (1996). Visual feedback guided robotic cherry tomato harvesting. *Transactions of the ASABE*, *39*(6), 2331–2338.

Marr, D. (1982). *Vision: a computational investigation into the human representation and processing of visual information*. San Francisco, CA: W. H. Freeman & Co.

Moravec, H. P. (1996). Robot spatial perception by stereoscopic vision and 3D evidence grids. *Tech. Report CMU-RI-TR-96-34*. Pittsburgh, PA: Carnegie Mellon University.

Rovira-Más, F. (2003). *Applications of stereoscopic vision to agriculture*. Unpublished doctoral dissertation, University of Illinois at Urbana-Champaign.

Rovira-Más, F. (2009). 3D vision solutions for robotic vehicles navigating in common agricultural scenarios. In Proc. *4th IFAC International Workshop on Bio-Robotics, Information Technology, and Intelligent Control for Bioproduction Systems*. IFAC.

Rovira-Más, F., Reid, J. F., & Zhang, Q. (2006). Stereovision data processing with 3D density maps for agricultural vehicles. *Transactions of the ASABE*, *49*(4), 1213–1222.

Rovira-Más, F., Wang, Q., & Zhang, Q. (2009). Noise reduction in stereo disparity images based on spectral analysis. *ASABE Paper 096258*, St. Joseph, MI.

Rovira-Más, F., Zhang, Q., & Reid, J. F. (2004). Automated agricultural equipment navigation using stereo disparity images. *Transactions of the ASABE*, *47*(4), 1289–1300.

Singh, S., Simmons, R., Smith, T., Stentz, A., Verma, V., Yahja, A., & Schwehr, K. (2000). Recent progress in local and global traversability for planetary rovers. In Proc. *International Conference on Robotics and Automation* (pp. 1194-1200). IEEE.

Wong, E. T. P., & Jarvis, R. (2004). Real time obstacle detection and navigation planning for a humanoid robot in an indoor environment. In Proc. *Conference on Robotics, Automation and Mechatronics* (pp. 693-698). IEEE.

ADDITIONAL READING

Gibbs, W. W. (2008). Innovations from a robot rally. *Scientific American* Reports . *Special Edition on Robotics*, *18*(May), 80–88.

Han, S., Reid, J. F., & Rovira-Más, F. (2009). Vision-aided system and method for guiding a vehicle. US Patent 7,610,123 B2.

Konolige, K. (1997). Small Vision Systems: hardware and implementation. In Proc. *International Symposium on Robotic Research* (pp. 111-116).

Ma, Y., Soatto, S., Kosecká, J., & Sastry, S. S. (2006). *An invitation to 3-D vision. From images to geometric models*. New York, NY: Springer.

Martin, M. C., & Moravec, H. P. (1996). Robot evidence grids. *Tech. Report CMU-RI-TR-96-06*. Pittsburgh, PA: Carnegie Mellon University.

McCorduck, P. (2004). *Machines who think*. Natick, MA: A. K. Peters, Ltd.

Olson, C. F., Abi-Rached, H., Ye, M., & Hendrich, J. P. (2003). Wide-baseline stereo vision for Mars rovers. In Proc. *International Conference on Intelligent Robots and Systems, 2* (pp. 1302-1307). IEEE.

Rovira-Más, F. (2009). Recent innovations in off-road intelligent vehicles: in-field automatic navigation. *Recent Patents on Mechanical Engineering*, *2*, 169–178. doi:10.2174/1874477X10902030169

Rovira-Más, F., & Han, S. (2006). Kalman filter for sensor fusion of GPS and machine vision. *ASABE Paper 063034*, St. Joseph, MI.

Rovira-Más, F., Han, S., Wei, J., & Reid, J. F. (2005). Fuzzy logic model for sensor fusion of machine vision and GPS in autonomous navigation. *ASABE Paper 051156*, St. Joseph, MI.

Rovira-Más, F., Wang, Q., & Zhang, Q. (2009). Bifocal stereoscopic vision for intelligent vehicles. *International Journal of Vehicular Technology*. ID 123231.

Rovira-Más, F., Wang, Q., & Zhang, Q. (2009). Design parameters for adjusting the visual field of binocular stereo cameras. *Biosystems Engineering*, *105*, 59–70. doi:10.1016/j.biosystemseng.2009.09.013

Rovira-Más, F., Zhang, Q., & Hansen, A. (2007). Dynamic behavior of an electrohydraulic valve: typology of characteristic curves. *Mechatronics*, *17*, 551–561. doi:10.1016/j.mechatronics.2007.07.003

Rovira-Más, F., Zhang, Q., & Hansen, A. (2010). *Mechatronics and intelligent systems for off-road vehicles*. London: Springer.

Rovira-Más, F., Zhang, Q., & Reid, J. F. (2005). Creation of three-dimensional crop maps based on aerial stereoimages. *Biosystems Engineering*, *90*(3), 251–259.

Rovira-Más, F., Zhang, Q., & Reid, J. F. (2008). Stereo vision three-dimensional terrain maps for precision agriculture. *Computers and Electronics in Agriculture*, *60*, 133–143. doi:10.1016/j.compag.2007.07.007

Rovira-Más, F., Zhang, Q., & Reid, J. F. (2009). Method for processing stereo vision data using image density. US Patent 7,587,081 B2.

Zabih, R., & Woodfill, J. (1994). Non-parametric local transform for computing visual correspondence. In Proc. *European Conference of Computer Vision*, Stockholm (pp. 151-158).

KEY TERMS AND DEFINITIONS

3D Density: Number of stereo-correlated points per volume unit, represented as d_{3D}.

3D Point Clouds: Discrete set of points providing a three-dimensional representation of the scene. Every valid pixel in a disparity image corresponds to a unique point in its corresponding 3D point cloud.

Density Grid: Regular grid covering the space of interest in front of the camera whose cells carry 3D density information. It can be either two-dimensional or three-dimensional.

Disparity Images or Disparity Maps: Images with the same resolution of the left and right stereo images, but carrying the disparity value for every pixel matched. The intensity value of each valid pixel is proportional to its disparity value. They provide the depth, or ranges, of all the objects detected.

Disparity: Horizontal distance measured in pixels between the position of the same point of the scene detected in both left and right stereo images. Only left-right correlated pixels will have an associated disparity value and therefore 3D information.

Intelligent Vehicle: Conventional or concept vehicle equipped with artificial intelligence techniques that automate some of its basic functions.

Stereo Baseline: Horizontal distance comprised between the optical centers of the two lenses mounted on a binocular camera. The baseline is a fundamental parameter to determine the field of view of the camera.

Validity Box: Portion of volume delimited by intersecting orthogonal planes where 3D locations have high likelihood of occurrence. Points of the cloud located outside the Validity Box are discarded and considered stereo mismatches.

Chapter 15
Effectiveness of New Technology to Compose Stereoscopic Movies[1]

Hiroki Takada
University of Fukui, Japan

Yasuyuki Matsuura
Nagoya University, Japan

Masaru Miyao
Nagoya University, Japan

ABSTRACT

The most widely known theory of motion sickness and asthenopia are based on the concept of sensory conflict, a disagreement between vergence and visual accommodation while viewing stereoscopic images. Visually induced motion sickness (VIMS) can be measured by using psychological and physiological methods. We quantitatively measured vergence, visual accommodation, head acceleration, and body sway before and during exposure to conventional and new stereoscopic movies. Sickness symptoms appeared with exposure to stereoscopic images. We found that some analytical index for stabilograms increased significantly when the subjects viewed a 3D movie. VIMS could be detected by using these indices. While lateral sway is dependent on the transverse component of head movement while watching the conventional stereoscopic movie, we examine whether this tendency is reduced by Power 3D.

INTRODUCTION: DEPTH ESTIMATION AND VIEW SYNTHESIS

With rapidly growing market for three-dimensional (3D) movies, 3D TVs, and 3D gaming, we may now be entering a time called the "Era of 3D". The general public has also started to become comfortable with stereoscopic vision. However, there are concerns about the effects on the human body from continuously watching 3D images, for examples, visually induced motion sickness (VIMS), visual fatigue, and asthenopia. Although their mechanisms are still not fully understood, there is a great need for more knowledge about the effects of those products on users and guidelines

DOI: 10.4018/978-1-61350-326-3.ch015

Figure 1. Binocular parallax. (a) near vision and (b) far vision

(a) (b)

for safety watching 3D images. We herein show the effects of stereoscopic images that may cause the VIMS or the simulator adaptation syndrome (SAS) in human. The goal of this chapter is to present a new technology to counter the causes of the VIMS.

On the other hand, an increasing number of people need to perform near-visual tasks such as operations on video display terminals (VDTs) with the development of computers and the widespread use of the Internet. Working under such conditions for several hours induces the contraction of the muscles involved in focus adjustment around the eyeball, such as the ciliary muscles. The abnormal contraction of ciliary muscles due to the performance of a near-visual task for several hours causes various vision problems such as asthenopia and visual loss. Further, this contraction has been reported to induce the cervicobrachial and psychoneurotic syndromes (Gomzi, 1994; Nakazawa *et al.*, 2002).

For persons afflicted with pseudomyopia, performing stretching exercises of the ciliary muscles alleviates strain and temporarily improves the myopic condition. These exercises can be performed by alternately repeating the negative and the positive accommodation of the eye. Miyao *et al.* (1996) experimentally demonstrated that the accommodation of the eye was possible by gazing at stereoscopic images displayed on a liquid crystal display (LCD) or a cathode ray tube (CRT).

Human beings perceive three-dimensional (3D) objects by the simultaneous vergence and lens accommodation in natural binocular vision. The depth of vergence and accommodation agreed under natural viewing conditions. They also perceive virtual images by using the same mechanism. A general stereoscopic view is obtained by using the binocular parallax (Figure 1).

It has been commonly explained that lens accommodation makes us focus on the surface of a display although the *optical axes of lens* are crossed at the virtual image (Figure 2) while viewing stereoscopic images (Cruz-Neira *et al.*, 1993). There is discrepancy between vergence and accommodative focus. That is, there is contradictory depth information between vergence and accommodation, called discordance, in the visual system. According to previous textbooks on 3D imaging, the VIMS and asthenopia are caused by this discordance. However, it seems to be an incorrect explanation. It has been shown that our focus is not always fixed on the surface of a display while viewing a stereoscopic image as follows:

Figure 2. Vergence and lens accommodation

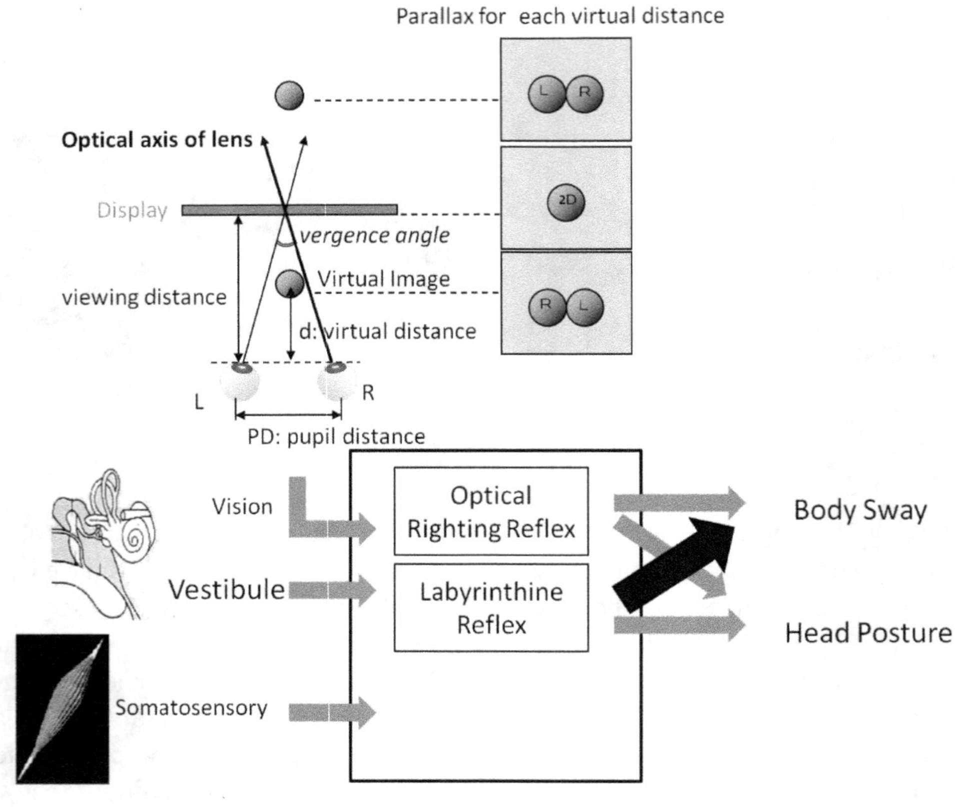

Under natural viewing conditions with binocular vision, we measured lens accommodation for 40 s (Hasegawa *et al.*, 2009; Omori *et al.*, 2009), while spherical images moved virtually toward and away from the subject on a head-mounted display (HMD), a liquid crystal display (LCD), and a cathode ray tube (CRT). Displays were positioned such that an image appeared in the upper portion of a dichroic mirror placed in front of the subject's eyes. 2D and 3D moving images were observed through reflection in the dichroic mirror, and refraction could be measured at the same time by transmitting infrared rays through the dichroic mirror. The refractive index of the right lens accommodation was measured by using a modified version of an original apparatus with an accommodo-refractometer (Nidek AR-1100) when the subjects gazed at the presented image via a small mirror with both eyes (Miyao *et al.*, 1992). The refraction in the case of the subjects was less than +0.5 Diopter (D), so both eyes were emmetropic. As a result, accommodation was set to approximately 3 (D) in front of the eyes even when the stereoscopic sphere reached the nearest point. Immediately after the sphere flew across the distant sky, the accommodation was approximately 1 (D). The synchronization of the accommodation with the movement process of the sphere is shown only in the 3D movie. Hence, the ciliary muscle is repeatedly strained and relaxed while the vision contains virtual movement of 3D images. Moreover, focal accommodation in the near-vision condition did not differ greatly with the different types of display. It was also shown, irrespective of whether the liquid crystal shutter glasses were used, that accommodation was easy and comfortable when focusing on virtually distant movements on the considered displays.

Patterson and Martin (1992) reviewed stereopsis and pointed out that the perceived depth for a crossed disparity follows predictions derived from constancy in most cases, whereas the perceived depth for an uncrossed disparity is frequently less than the predicted value. They reported that among several possible distance cues related to the computation of the perceived depth, one set of cues involves proprioceptive information from accommodation, vergence, or both.

Depending on the audiovisual condition, stereoscopic videos that use binocular stereoscopic vision often induce the unpleasant symptoms of asthenopia, headache, difficulty in focusing, dizziness, disorientation, and nausea (Ukai *et al.*, 2008). Ataxia in stereoscopic video-induced sickness has been reported previously. The influence of video-induced motion sickness on the body has been measured by employing subjective scales such as the simulator sickness questionnaire (SSQ) (Kennedy *et al.*, 1993). Further, it is also measured by quantitatively investigating the relationship between external factors and internal conditions using physiological indices such as respiratory functions, electrocardiograms, skin electrical activity, fluctuation of the center of gravity, and electrogastrograms (Holomes *et al.*, 2001; Himi *et al.*, 2004; Yokota *et al.*, 2005). However, there is no established methodology detecting the VIMS due to 3D movie in an early stage. An objective index are required to measure degree of the VIMS, which is also useful in examining whether a developing 3D movie can be regarded as a safety product.

Recent studies suggest that maintaining postural stability is a major goal of animals (Stoffregon *et al.*, 2000) and that they experience sickness symptoms in circumstances where they have not acquired strategies for maintaining their balance (Riccio & Stoffregon, 1991). In the next section, backgrounds involved in the VIMS and stabilometry are reviewed as a preparation to introduce our methodology evaluating VIMS.

BACKGROUND

Visually Induced Motion Sickness (VIMS)

Historical chronicles of the human experience with motion sickness-like symptoms date back at least to Hippocrates, and while Julius Caesar, Lawrence of Arabia, and Admiral Nelson suffered bouts of sickness (Money, 1972), adaptation and repeated exposure minimized these adverse effects (Kennedy & Kennedy, 2007).

More recent human encounters with motion environments, including simulators, virtual environments and even some commercially available video games that create the illusion of motion, demonstrate the general rule that motion sickness adversely affects operational efficiency among susceptible individuals (Benson, 1978). Although the most widely known theory of motion sickness is based on the concept of sensory conflict (Oman, 1982; Reson, 1978), Riccio and Stoffregen (1991) argued that motion sickness is not caused by sensory conflict but by postural instability. The VIMS has been attributed to a disagreement between vergence and visual accommodation while viewing 3D images (Okuyama et al., 1996). Stoffregen and Smart (1999) reported that the onset of motion sickness may be preceded by significant increases in the postural sway (Stoffregen *et al.*, 1999). The equilibrium function in humans deteriorates when viewing 3D movies (Takada *et al.*, 2007).

Nowadays, liquid crystal displays (LCDs) are extensively used as general visual display terminals. They have several features such as large display size, reduction in weight and size because of miniaturization, and low power consumption. However, users viewing movies on LCDs often complain of the blurring and bleeding of images and experience VIMS. Typical LCDs are said to be inferior to cathode-ray tube displays with regard to motion picture display. This is because while the later is an impulse-based display and the temporal waveform of each pixel is a luminance impulse that is only a few milliseconds long, a typical LCD (a conventional LCD) is a voltage-hold-type display, which implies that the voltages across the pixels are held during the entire frame period (16.7 ms). A voltage-hold-type display has a blur in its motion-frame picture because while human eyes track the movement of the picture, the picture is fixed for a certain period (field period) and a time gap is generated in its display. These problems can be avoided in LCD displays by using the pseudo-impulse driving method to realize a higher performance. The blurred images on the LCDs sometimes induced "image sickness" in viewers, which is an unpleasant feeling that is similar to motion sickness. On the other hand, optokinetic stimulation is known to trigger motion sickness (Lestienne et al., 1977). In particular, anterior displacement of the centre of gravity remarkably increased during the body sway when random dots were rotated vertically at a speed of 40-60 deg/s as optokinetic stimulation to the subjects. The conventional LCD might aggravate symptom of the motion sickness that was caused by some sensory conflicts. Furthermore, newly developed optically compensated bend display could suppress the symptom of the motion sickness (Fujikake et al., 2007).

VIMS can be measured by psychological and physiological methods, and the simulator sickness questionnaire (SSQ) is a well-known psychological method for measuring the extent of motion sickness (Kennedy *et al.*, 1993). The SSQ is used herein for verifying the occurrence of VIMS. The following parameters of autonomic nervous activity are appropriate for the physiological method: heart rate variability, blood pressure, electrogastrography, and galvanic skin reaction (Holomes and Griffin, 2001; Himi *et al.*, 2004; Yokota *et al.*, 2005). A wide stance (with midlines of the heels 17–30 cm apart) reportedly results in a significant increase in the total locus length in the stabilograms for individuals with high SSQ scores, while the length in those of the individuals with low scores is less affected by such a stance (Scibora *et al.*, 2007).

Stabilometry

The human standing posture is maintained by the body's balance function, which is an involuntary physiological adjustment mechanism termed the righting reflex (Okawa *et al.*, 1995). To maintain a standing posture when locomotion is absent, the righting reflex, centered in the nucleus ruber, is essential. Sensory signals such as visual inputs and auditory and vestibular inputs as well as proprioceptive inputs from the skin, muscles, and joints are involved in the body's balance function (Kaga, 1992). The evaluation of this function is indispensable for diagnosing equilibrium disturbances such as cerebellar degenerations, basal ganglia disorders, and Parkinson's disease in patients (Okawa *et al.*, 1996). Stabilometry has been used for evaluating this equilibrium function qualitatively and quantitatively. The stabilometry is useful not only for medical diagnosis but also for achieving control of upright standing by two-legged robots and for preventing elderly people from falling (Fujiwara and Toyama, 1993).

Even when a young, healthy individual attempts to stand still, the centre of gravity of his/her body and the centre of pressure (COP) under his/her feet move relative to a global coordinate system (Collins & De Luca 1993), which is induced by the complex sensorimotor control system. A plot of time-varying coordinates of the COP is known as a stabilogram. The COP could be measured in accordance with stabilometry in which many of the earlier studies limited the analysis of the plots to summary statistics, i.e., calculation of the length of sway path (total locus length, L), average radial area (area of sway, A), locus length per unit area (L/A) etc. (Suzuki *et al.*, 1996). These parameters have been proposed to quantify the instability involved in the standing posture, and such parameters are widely used in clinical studies. In particular, the last parameter (L/A) depends on the fine variations involved in posture control (Okawa *et al.*, 1995). This index is then regarded as a gauge for evaluating the function of the proprioceptive control of standing in human beings. However, it is difficult to clinically diagnose the disorders of the balance function and to identify the decline in the equilibrium function by utilizing the abovementioned indices and measuring patterns in the stabilogram. Large interindividual differences might make it difficult to understand the results of such a comparison. Thus, Collins and De Luca (1993) introduced another method known as stabilogram diffusion analysis that provides a quantitative statistical measure of the apparently random variations of the COP trajectories recorded during quiet upright stance in humans. This analysis generates a stabilogram diffusion function (SDF) that summarizes the mean square COP displacement as a function of the time interval between COP comparisons. SDFs have a characteristic two-part form that suggests the presence of two different control regimes: a short-term open-loop control behavior and a longer-term closed-loop behavior (Peterka, 2000).

Mathematically, the sway in the centre of pressure (COP) is described by a stochastic process (Emmerrik *et al.*, 1993; Newell *et al.*, 1997). The anterior-posterior direction y was considered to be independent of the mediallateral direction x (Goldie *et al.*, 1989). We examined the adequacy of using stochastic differential equations (SDEs) on the Euclid space $\mathbf{E}^2 \ni (x, y)$

$$\frac{\partial x}{\partial t} = -\frac{\partial}{\partial x} U_x(x) + w_x(t)$$
$$\frac{\partial y}{\partial t} = -\frac{\partial}{\partial y} U_y(y) + w_y(t)$$

and investigated the most adequate equation as mathematical models that generate the stabilograms. Here, pseudorandom numbers were generated by the white noise terms $w_x(t)$ and $w_y(t)$. Constructing the nonlinear SDEs from the stabilograms in accordance with the following equations (Takada *et al.*, 2001):

$$U_x(x) = -\frac{1}{2}\ln G_x(x) + const.,$$
$$U_y(y) = -\frac{1}{2}\ln G_y(y) + const., \quad (1)$$

we observed that their temporally averaged potential functions U_x and U_y have several minimal points, where G(z), the distribution of the observed point z, is related in the following manner to $U_z(z)$, the (temporal averaged) potential function, in the stochastic differential equation (SDE), which has been considered a mathematical model of the sway. In the vicinity of these points, a local stable movement with a high-frequency component can be generated as a numerical solution to the SDE. Hence, fluctuations could be observed in the neighborhood of the minimal points. A high density of the observed COP can be expected in this area on the stabilogram; the sparse density (SPD) is regarded as an index for its measurement (Takada *et al.*, 2003).

The correlation between head movement and the movement of the center of gravity has been investigated in general, and a corporative effect was seen in their relationship (Sakaguchi *et al.*, 1995). By showing a stereoscopic movie to the subjects, Takeda *et al.* (1995) verified that there is a corporative correlation between the head movement and the sway. In the control theory, the transfer function analysis is widely used for investigating a system. We denote the Fourier transform by a capital letter corresponding to the letter of the function being transformed (such as y(t) and Y(f)). The transfer function H(f) is defined as a Fourier transform of the impulse response h(f). In our experiments, we cannot observe the output signal of the transfer system but only the signal added to the noise n(t). On the basis of a theorem (Winner-Khinchine):

$$W_{xx} = |X(f)|^2 = \sigma_x^{\,2}\, F(R_{xx}), \qquad (2)$$

we can easily estimate a power spectrum W_{xx}. On the right-hand side of Equation (2), σ_x expresses the standard deviation, and $F(R_{xx})$ indicates the Fourier transform of the auto-correlation function with respect to the signal x(t) (Kido, 2007).

In the next section, our researches are herein reviewed. By using the SSQ and stabilometry (body sway), we examined whether the VIMS was induced by a stereoscopic movie. The aim of our study is to propose a methodology to measure the effect of 3D images on the equilibrium function. Moreover, we wondered if the noise terms vanished from the mathematical model of the body sway. Using our Double-Wayland algorithm (Takada *et al.*, 2006), we evaluate the degree of visible determinism for the dynamics of the sway. We also investigate the relationship between the body sway and head acceleration by using a transfer function analysis.

METHODOLOGY TO EVALUATE VISUALLY INDUCED MOTION SICKNESS (VIMS)

Problems

There have been VIMS and eye-strain issues in stereoscopic movies. Why are stereoscopic images unnatural for human vision? According to a common view, these issues are caused by a certain sensory conflict while viewing stereoscopic images. A clue to solve this difficult problem can be obtained by using psychological and physiological methods. However, there is no established method that can measure the degree of VIMS. Herein, we assume that the input signal, x(t), is the head acceleration in the transfer system to control the body sway (or maintain the upright posture). The transfer function that controls the sway is estimated as discussed in the following paragraphs (Takada *et al.*, 2009b). In this section, we support our hypothesis: VIMS changes the system to control the body sway.

Subjects

Ten healthy volunteers (age: 23.6 ± 2.2 years) participated in our study. All of them were Japanese and lived in Nagoya and its surrounding areas. They provided informed consent prior to participation. The following subjects were excluded from the study: subjects working in the night shift, those dependent on alcohol, those who consumed alcohol and caffeine-containing beverages after waking up and less than 2 h after meals, those who had been using prescribed drugs, and those who may have suffered from an otorhinolaryngologic or neurological disease in the past (except for conductive hearing impairment, which is commonly found in the elderly). In addition, the subjects must have experienced motion sickness at some time during their lives.

We ensured that the body sway was not affected by environmental conditions. By using an air conditioner, we adjusted the room temperature to 25 °C and kept the room dark. All subjects were tested from 10 a.m. to 5 p.m. in the room. The subjects wore an HMD (iWear AV920; Vuzix Co. Ltd.) on which two types of images were presented in a random order: (I) a visual target (circle) whose diameter was 3 cm and (II) a conventional 3D movie that shows a sphere approaching and moving away from the subjects irregularly.

Design

The subjects stood without moving on the detection stand of a stabilometer (G5500; Anima Co. Ltd.) in the Romberg posture with their feet together for 1 min before the sway was recorded. Each sway of the COP was then recorded at a sampling frequency of 20 Hz during the measurement, while the head acceleration was simultaneously recorded by the active tracer (AC-301A; GMS Co. Ltd.) at 50 Hz. The subjects were instructed to maintain the Romberg posture for the first 60 s and a wide stance (with the midlines of heels 20 cm apart) for the next 60 s. The subjects viewed one of the images, i.e., (I) or (II), on the HMD from the beginning to the end.

Simulator Sickness Questioner (SSQ)

The SSQ was filled before and after stabilometry. After the exposure to a conventional 3D movie (II), scores for SSQ-N (nausea), SSQ-OD (eyestrain), SSQ-D (disorientation), and SSQ-TS (total score) were 11.4 ± 3.7, 18.2 ± 4.1, 23.7 ± 8.8, and 19.8 ± 5.3, respectively. Sickness symptoms seemed to appear with the exposure to the stereoscopic images although there were large individual differences.

Stabilograms

We have shown typical stabilograms (Fujikake *et al.*, 2009). In these stabilograms, the vertical axis shows the anterior and posterior movements of the COP, and the horizontal axis shows the right and left movements of the COP. The amplitudes of the sway that were observed during exposure to the movies tended to be larger than those of the control sway. Although a high density of COP was observed in the stabilograms, the density decreased in stabilograms during exposure to the conventional stereoscopic movie. Furthermore, stabilograms measured in an open leg posture with the midlines of heels 20 cm apart were compared with those measured in the Romberg posture. COP was not isotropically dispersed but was characterized by the considerable movement in the anterior-posterior (y) direction (Fujikake *et al.* 2009). During exposure to 3D movie, the diffusion of COP was larger in the lateral (x) direction and had spread to the extent that it was equivalent to the control stabilograms. Moreover, we calculated several indices that are commonly used in the clinical field (Suzuki *et al.*, 1996) for stabilograms, such as "area of sway," "total locus length," and "total locus length per unit area." The new quantification indices were termed "SPD" and "total locus length of chain" (Takada *et al.*, 2003). According to the two-way analysis of variance (ANOVA) with repeated measures, there was no interaction between the factors of posture (Romberg posture or standing posture with their feet wide apart) and images (I or II). With respect to the total locus length and the sparse density, there were main effects in response to both factors ($p < 0.01$). Multiple comparisons revealed that these indices significantly increased when the subjects viewed the 3D movie (II) on the HMD with the Romberg posture. A similar result was statistically obtained with the comparison of images (I or II) on a LCD (Takada *et al.*, 2008). VIMS could be detected by these indices for stabilograms.

Transfer Function Analysis

When the subjects stood with their feet close together (Romberg posture), the coherence function between the head acceleration x(i) and the movement of the centre of gravity y(j) was estimated as

$$coh_{x(i)y(j)}(f) = |W_{x(i)y(j)}|^2 / (W_{x(i)x(i)} W_{y(j)y(j)}), \quad (3)$$

where i and j expressed the component (1: lateral and 2: anterior/posterior). By using the Fast Fourier transform algorithm, we estimated the power spectrums $W_{x(i)x(i)}$, $W_{y(j)y(j)}$. On the basis of Equation (3), we calculated cross spectrums $W_{x(i)y(j)}$. The coherence indicates an index for the degree of the linear correlation between the input and the output signals ($0 \leq coh \leq 1$). There exists a completely linear correlation between these signals when coh = 1. We assumed that a linear system intervenes between the head and the body sway only if coh ³ 0.12 (significant correlation coefficient for $N = 512$, $p < 0.01$). Moreover, we estimated the transfer function as follows:

$$H(f) = W_{x(i)y(j)} / W_{x(i)x(i)}, \tag{4}$$

and the transfer function gain (TFG) $|H(f)|$. When the subjects stood with the Romberg posture, the transfer function analysis was implemented with the head acceleration (input) and the body sway (output). We estimated the coherence function (3), i.e., $coh_{x(1)y(1)}(f)$, $coh_{x(1)y(2)}(f)$, $coh_{x(2)y(1)}(f)$, and $coh_{x(2)y(2)}(f)$. For any frequency, $coh_{x(1)y(1)}(f)$ and $coh_{x(1)y(2)}(f)$ were less than 0.12 (significant correlation coefficient for $N = 512$, $p < 0.01$). On the other hand, $coh_{x(2)y(2)}(0.51)$ was more than 0.12. $coh_{x(2)y(j)}(0.51)$ and $coh_{x(2)y(j)}(7)$ during the exposure to the 3D movie (II) remarkably increased for $j = 1, 2$ (See Figure 3).

Complex System Analysis

By estimating the translation errors, we mathematically measured the degree of determinism

Figure 3. Significant coherence

images / Coh x(i) y(j)	(I)	(II)
Coh x(1) y(1)		
Coh x(1) y(2)		
Coh x(2) y(1)		+
Coh x(2) y(2)	+	+

The Symbol "+" indicates that significant coherence has been observed for any frequency. i and j expressed the component.

in the dynamics of the sway of COP. Representative results of the Double-Wayland algorithm are derived from the lateral sway x as shown in Figure 3. Whether subjects were exposed to the 3D movies or not, E_{trans} derived from the temporal differences of the time series x, y was approximately 1 (Figure 4). These translation errors in each embedding space were not significantly different from the translation errors derived from the time series x, y. $E_{trans} > 0.5$ was obtained by the Wayland algorithm, which implies that the time series could be generated by a stochastic process in accordance with a previous standard (Matsumoto *et al.*, 2002). The threshold 0.5 is half of the translation error resulting from a random walk. The body sway has been described previously by stochastic processes (Collons and De Luca, 1993; Emmerrik *et al.*, 1993; Newell *et al.*, 1997), which was shown with the Double-Wayland algorithm (Takada *et al.*, 2006). Moreover, $0.8 < E_{trans} < 1$ obtained from the temporal differences of these time series exceeded the translation errors estimated by the Wayland algorithm. The exposure to 3D movie would not change it into a deterministic one. Mechanical variations were not observed in the locomotion of the COP. We assumed that the COP was controlled by a stationary process, and the sway during exposure to the static control image (I) could be compared with that when the subject viewed 3D movies. Indices for stabilograms might reflect the coefficients in stochastic processes although the translation error did not exhibit a significant difference between the stabilograms measured during exposure to the static control image (I) and the conventional 3D movie (II).

Controversies

Scibora *et al.* (2007) concluded that the total locus length of subjects with prior experience of motion sickness increases with exposure to a virtual environment when they stood with their feet wide apart, whereas, in our study, the degree of sway was found to be reduced significantly when the subjects stood with their feet wide apart than when they stood with the Romberg posture. A clear change in the form of the potential function (1) occurs when the feet are wide apart. Irrespective of posture, the indicators involved in the stabilogram (total locus length and SPD) during exposure to the conventional 3D movie (II) were greater than that during exposure to the control image (I). Moreover, the total locus length of chain tended to increase when the subjects were exposed to the conventional 3D images (II) compared that when they were exposed to (I). Hence, we noted postural instability with the exposure to the conventional stereoscopic images (II) by using these indicators.

The variance in the stabilogram depends on the form of the potential function in the mathematical model of the body sway (SDEs); therefore, it is important to focus on the nonlinearity of the potential function. The total locus length was increased during the exposure to the conventional 3D images (II), which might be caused by the diminution of the gradient in the bottom of the potential function. We herein note that it is possible to estimate the decrease in the gradient of the potential function by using the SPD by performing a one-way analysis of variance.

SOLUTIONS AND RECOMMENDATIONS

New Processing of Stereoscopic Images

Recently, a novel 3D video construction method has been developed to prevent video-induced motion sickness (Yasui *et al.*, 2006; Kakeya, 2007). Humans perceive actual objects by simultaneous vergence and accommodation of the lens, but stereoscopic videos generally consist of the unnatural images perceived along a fixed

Figure 4. Results of Double-Wayland algorithm

Images	Direction	classification (rows) / dimension in embedding space (columns)		1	2	3	4	5	6	7	8	9	10
(I)	Lateral	time series	mean	1.04	0.93	0.89	0.86	0.80	0.76	0.71	0.66	0.63	0.58
			SD***	0.41	0.31	0.28	0.26	0.23	0.21	0.18	0.17	0.16	0.14
	Lateral	Differences	Mean	0.79	0.80	0.82	0.80	0.81	0.81	0.82	0.83	0.81	0.81
			SD	0.13	0.10	0.11	0.09	0.09	0.09	0.10	0.12	0.13	0.14
	Anterior /Posterior	time series	mean	0.92	0.87	0.86	0.82	0.78	0.74	0.69	0.66	0.61	0.57
			SD	0.30	0.29	0.30	0.29	0.26	0.25	0.22	0.21	0.20	0.18
	Anterior /Posterior	Differences	mean	0.93	0.92	0.91	0.92	0.92	0.92	0.92	0.91	0.90	0.90
			SD	0.18	0.12	0.09	0.10	0.11	0.08	0.11	0.12	0.11	0.10
(II)	Lateral	time series	mean	1.15	1.07	1.05	1.00	0.95	0.88	0.82	0.76	0.70	0.66
			SD	0.21	0.18	0.18	0.19	0.17	0.15	0.15	0.14	0.13	0.13
	Lateral	Differences	mean	0.85	0.87	0.85	0.86	0.87	0.88	0.86	0.85	0.83	0.83
			SD	0.17	0.12	0.09	0.11	0.13	0.11	0.11	0.12	0.13	0.15
	Anterior /Posterior	time series	mean	1.12	1.01	0.96	0.91	0.89	0.81	0.78	0.72	0.68	0.66
			SD	0.20	0.21	0.20	0.20	0.18	0.14	0.15	0.13	0.12	0.13
	Anterior /Posterior	Differences	mean	0.98	0.97	0.96	0.92	0.95	0.92	0.90	0.89	0.87	0.86
			SD	0.15	0.11	0.13	0.08	0.08	0.09	0.08	0.08	0.07	0.07

***SD means a standard deviation of translation errors estimated in each embedding space by the Double-Wayland algorithm.

optical axis of lens, negating such vergence and accommodation. Stereoscopic images that are prepared using the Power 3D method reduce the inconsistency between experienced and actual senses (Nishihira & Tahara, 2001). This is why the Power 3D approach counters the cause of VIMS.

With respect to a general stereoscopic view, the images are composed of photographs taken by two cameras or by computer graphics (CG). Camera axes are fixed and crossed at the point of the virtual image at which the creator expects viewers to gaze. That is, viewers would suffer from finding the anomalous vergence if they looked at the other elements in a stereoscopic flame. According to International Workshop Agreement 3 (IWA3), International Organization for Standardization (ISO) suggested that the popping-out effect not be used considerably (Figure 2). We herein introduce a new technology to construct stereoscopic movies (Power 3D). The new technology (Nihihara & Tahara, 2001) sets each camera axis as well as human beings that change the vergence angle corresponding to the visual distance of subjects for photography (Figure 5). Moreover, camera axes are also set to be parallel as well as natural binocular vision in order to construct the background (Figure 5a). These elements of far/near visions are superimposed on the flame (Figure 5c). Viewers might not feel a sense of incongruity if they gaze at any elements in the flame. This technology has already been applied to the "Eyesight-Recovering Stereoscopic Movie System" produced by Olympus Visual Communications Co., Ltd. Parallax images for each eye are alternately presented at 75 Hz for the prevention of photosensitive seizures.

We assume that the high density of observed COP decreases during exposure to stereoscopic images. In this section, we showed that the sparse density (SPD) was a useful index in stabilometry to measure VIMS. Our previous studies have shown that the degree of video-induced motion sickness is reduced in body sway by viewing stereoscopic videos prepared by using this method on the HMD (Takada *et al.*, 2009a) and on an LCD (Takada *et al.*, 2009d), respectively. In the next section, we present an application of this new technology, Power 3D, to health promotion.

Figure 5. A stereoscopic image constructed by Power 3D method

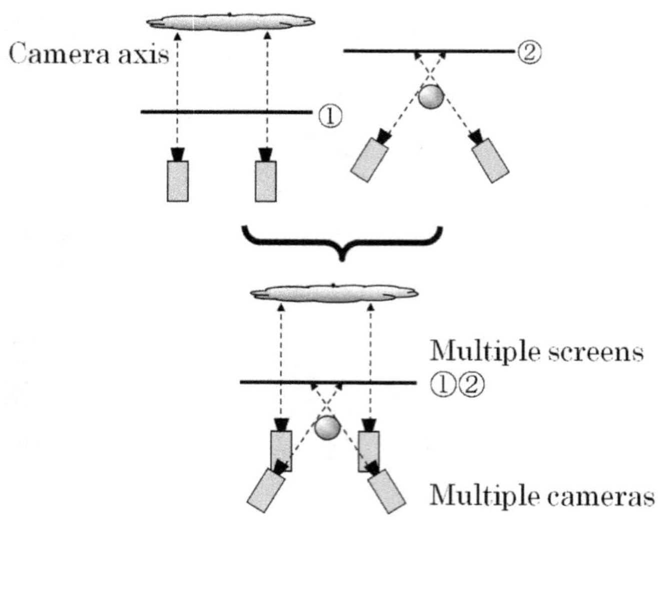

FUTURE RESEARCH DIRECTIONS

Applications

We assume that it is possible to improve an abnormal accommodative function of the lens by activating the muscles by alternately repeating negative and positive accommodation. By improving the abnormal accommodative function, we can improve or prevent these vision problems. We call this operation "accommodation training (AT)." In Japan, an apparatus called MD-SS was developed by Kobayashi (1994). The objective of this apparatus is to recover visual acuity. This apparatus works by using a Landolt ring drawn on a flat plate that moves back and forth over a distance of 2 m in order to encourage alternately repeating negative and positive accommodation in the observers. However, a large space was required to employ the AT with the use of this MD-SS. On the other hand, the Eyesight-Recovering Stereoscopic Movie System is commonly known as "Dr. Rex." In the Dr. Rex, an LCD displaying stereoscopic videos and the visual acuity recovery device equipped with liquid crystal shutter eyeglasses. Dr. Rex Eye Care Program contains some stereoscopic video contents that simulate near and distant visual conditions (Figure 6). Alternately displaying the videos on the LCD and device at appropriate intervals is expected to improve and prevent myopia, presbiopia, and visual fatigue. The abnormal contraction of ciliary muscles due to the performance of a near-visual task for several hours causes various vision problems such as asthenopia and visual loss. However, these problems can be resolved by activating the muscles by alternately repeating negative and positive accommodation. In this study, we have verified the effect of accommodation training that uses the strategy of presenting a stereoscopic movie to myopic youth and measuring the spherical diopter (SPH), visual acuity (far/near vision), and subjective index of asthenopia obtained using a visual analog scale (VAS). The results of subjective evaluations before (pre) and after (post) viewing the stereoscopic videos were compared using the Wilcoxon signed ranks test, where the significance level p was set to be 0.05.

a) **Myopic population** will be increasing in the near future with the rapid development of industry or other social factors in some countries. Thirty two myopic students aged 20 ± 1 years (16 males and 16 females) were chosen as the subjects. One group performed the AT for 6 min, and the other group underwent a near-visual task during the same period as the control group. The uncorrected distant visual acuity increased in 17 of the 32 subjects (53.1%). The visual acuity on day 11 was considerably higher in the AT group than in the control group ($p < 0.05$). The visual acuity improved in the AT group. This result suggests that the AT has a cumulative positive effect on eyesight. The AT would prevent the deterioration of visual acuity even though there was no significant difference in the SPH between the groups. We considered that the AT using the stereoscopic movie did not deform the lens, thus not improving myopia fundamentally. However, the visual acuity and the near-point accommodation function were enhanced by this accommodation training, which also led to a decrease in asthenopia (Sugiura *et al.*, 2010).
b) **Visual inspection workers** suffered from eye fatigue after their work. We have investigated the visual acuity improving effect of the device using stereoscopic videos for 22 visual inspection workers aged 37 ± 6 years. These subjects were also divided into two groups. One group underwent the Dr. Rex treatment, in which they viewed a stereoscopic video for 6 min after the visual inspection work, and the other group was not given any task to perform during the first three consecutive days. Thereafter, the

Figure 6. Idea involved in the Dr. Rex Eye Care Program

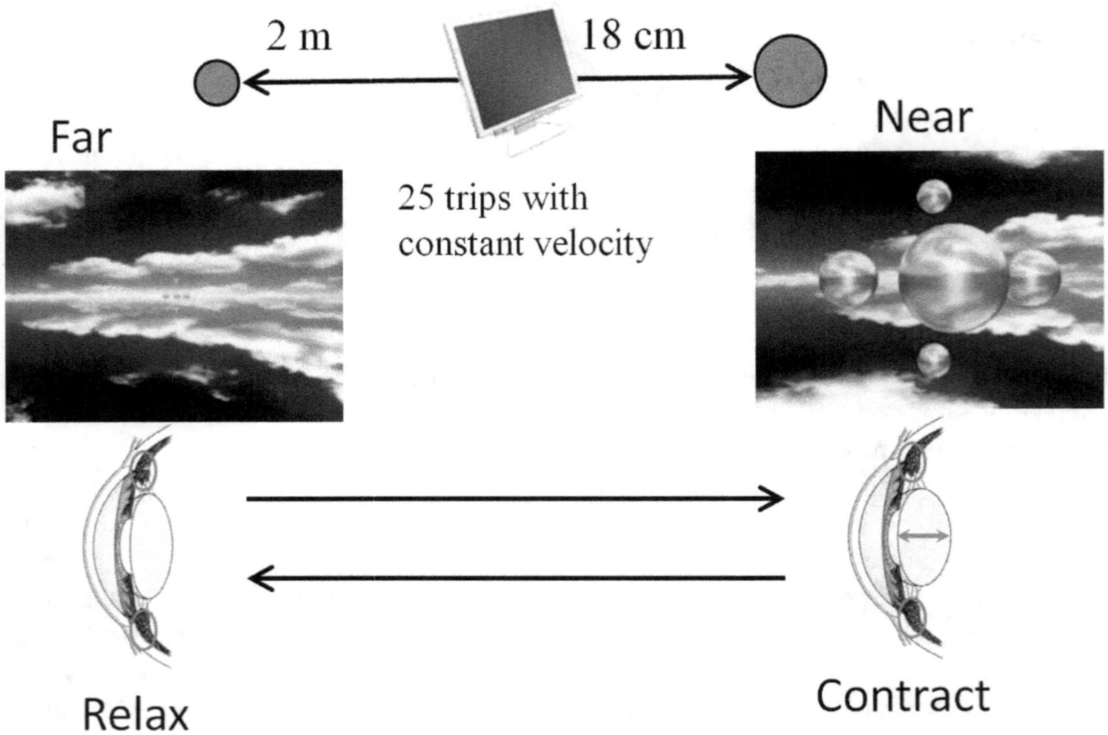

groups switched tasks, and the experiment was performed in a similar manner to collect data without the influence of task order. The above-mentioned items were performed before the visual inspection work on the morning of the first day and after the treatment for the six experimental days. Although the dioptric comparison between the control and the Dr. Rex treatment groups showed that there was no significant difference between the values for the groups ($p < 0.05$), the binocular BVA increased in 13 of the 22 visual inspection workers (59.1%). The visual acuity of the control group without the Dr. Rex treatment showed an improvement. The myopic tendency increased because of the visual inspection work. Moreover, it was possible that the subjects became skilled in the vision test. However, the results obtained from the Wilcoxon signed ranks test showed that the distant visual acuity in the Dr. Rex treatment group increased considerably as compared to that in the control group ($p < 0.05$). As compared to the near-visual acuity in the control group, that in the Dr. Rex treatment group had increased significantly on day 3 ($p < 0.05$). The VAS in the Dr. Rex treatment group also increased significantly on day 3 as compared to that in the control group. There seemed to be not only an visual acuity improving effect but also a reduction of the visual fatigue by the Dr. Rex treatment for more than 3 consecutive days (M. Takada *et al.*, 2010).

c) **Presbiopic population** will be also increase in the near future with an aging society. After the age of 40 in most people, and by the age of 45 in virtually everyone, a clear, comfortable focus at a near distance becomes more difficult with eyes that see clearly at a far distance. This normal condition is known as presbyopia, and it is due both to a lessening of flexibility of the crystalline lens and to a generalized weakening of the ciliary muscle. We have reported that both the uncorrected binocular near- and distant-visual acuities improved in *middle-aged* subjects, suggesting that viewing stereoscopic videos reduced strain and increased the flexibility of the ciliary muscles, which temporarily recovered the visual acuity. In contrast, dioptric measurements did not change in the eyes of members of either group. The duration of treatment may have been too short to modify the eyeball (lens) structure. It is suggested that the short-term repeated use of the Dr. Rex visual acuity recovery device would increase the near-visual acuity. We expect that this may assist the improvement in and prevention of presbyopia (Takada *et al.*, 2009c).

Based on the Dr. Rex visual acuity recovery device, a Web-based system will be developed to diffuse the health promotion. The frequency at which parallax images for each eye are alternatively presented must be tuned to fit the frequency on household displays. Through the Web-based system, we will be able to obtain considerable data to elucidate the Dr. Rex visual acuity recovery device. In the future, we suggest that the apparatus be used for several months in order to verify its long-term effects on visual acuity and asthenopia.

Theory

Human beings perceive actual objects with simultaneous vergence and lens accommodation in binocular vision. Virtual images are perceived via the same mechanism, although, as we previously reported, the focus is not always fixed on the surface of a display when stereoscopic images are being viewed. We should investigate the effect of stereoscopic images on the visual functions with careful deliberation. We have already developed a method to simultaneously measure accommodation and vergence in order to provide further support for this theory. In the next step, we use "Power 3D" to test visual functions in a stereoscopic view with a very wide amplitude. We also measure the accommodation and vergence in natural vision to confirm that these measurements are correct. We found that both accommodation and vergence were consistent with the distance from the subject to the object using the Power 3D system (Figure 7).

CONCLUSION

In order to evaluate the VIMS, we performed the simultaneous recording of the center of gravity with the head acceleration during the exposure to a 2D image and a 3D movie in this study. According to the transfer function analysis, the anterior/posterior head acceleration could affect the lateral body sway during the VIMS caused by the 3D movie. In addition to the analysis of stabilograms, the transfer function between the head posture (input) and the body sway (output) is considered to be useful for the prediction/detections of the VIMS.

Power 3D was developed by Olympus Visual Communications Co. Ltd. as a 3D technology that does not induce 3D sickness (uncomfortable feeling of nausea when viewing unnatural stereoscopic movies). The Power 3D approach counters the cause of VIMS because the technology uses free-viewpoint binocular stereoscopic graphics. Using Power 3D, subjects can see very close stereoscopic images in front of their face as well as distant mountain views. Conventional 3D views are generated with fixed-viewpoint binocular stereoscopic graphics. When subjects view a close target (crossed view), far mountains cannot be fused. When they see far mountains, the close target (crossed view) is split, and two targets are seen.

Figure 7. Typical examples of the simultaneous measurement

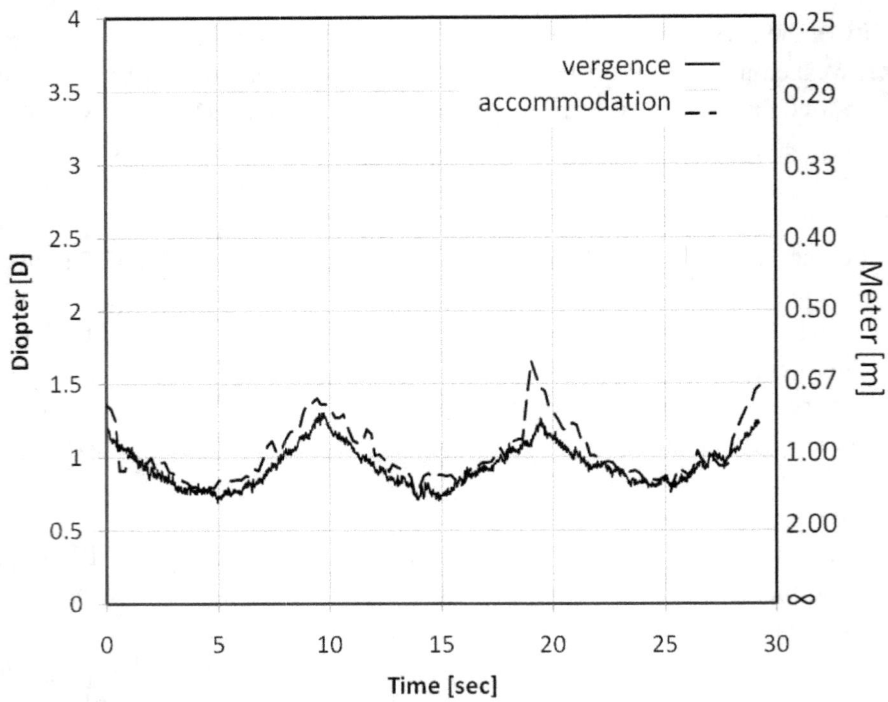

REFERENCES

Benson, A. J. (1978). *Spatial disorient: General aspects. Aviation medicine*. London, Uk: Tri-Med Books.

Collins, J. J., & De Luca, C. J. (1993). Open-loop and closed-loop control of posture: A random-walk analysis of center of pressure trajectories. *Experimental Brain Research*, *95*, 308–318. doi:10.1007/BF00229788

Cruz-Neira, C., Sandin, D. J., & DeFanti, T. A. (1993). Surround-screen projection-based virtual reality: The design and implementation of the CAVE. *Proceedings of the 20th Annual Conference on Computer Graphics and Interactive Techniques*. (pp. 135-142). Anaheim, CA: ACM SIGGRAPH.

Emmerrik, R. E. A., Van Sprague, R. L., & Newell, K. M. (1993). Assessment of sway dynamics in tardive dyskinesia and developmental disability: sway profile orientation and stereotypy. *Movement Disorders*, *8*, 305–314.

Fujikake, K., Miyao, M., Honda, R., Omori, M., Matsuura, Y., & Takada, H. (2007). Evaluation of high quality LCDs displaying moving pictures, on the basis of the form obtained from statokinesigrams. *Forma*, *22*(2), 199–206.

Fujikake, K., Miyao, M., Watanabe, T., Hasegawa, S., Omori, M., & Takada, H. (2009). Evaluation of body sway and the relevant dynamics while viewing a three-dimensional movie on a head-mounted display by using stabilograms . In Shumaker, R. (Ed.), *Virtual and mixed reality, LNCS 5622* (pp. 41–50). Berlin, Germany: Springer-Verlag. doi:10.1007/978-3-642-02771-0_5

Fujiwara, K., & Toyama, H. (1993). Analysis of dynamic balance and its training effect: Focusing on fall problem of elder persons. *Bulletin of the Physical Fitness Research Institute*, *83*, 123–134.

Goldie, P. A., Bach, T. M., & Evans, O. M. (1989). Force platform measures for evaluating postural control - Reliability and validity. *Archives of Physical Medicine and Rehabilitation*, *70*, 510–517.

Gomzi, M. (1994). Work environment and health in VDT use-An ergonomic approach. *Arhiv za Higijenu Rada i Toksikologiju, 45*, 327–334.

Hasehawa, S., Omori, M., Watanabe, T., Fujikake, K., & Miyao, M. (2009). Lens accommodation to the stereoscopic vision on HMD . In Shumaker, R. (Ed.), *Virtual and mixed reality, LNCS 5622* (pp. 439–444). Berlin, Germany: Springer-Verlag. doi:10.1007/978-3-642-02771-0_49

Himi, N., Koga, T., Nakamura, E., Kobashi, M., Yamane, M., & Tsujioka, K. (2004). Differences in autonomic responses between subjects with and without nausea while watching an irregularly oscillating video. *Autonomic Neuroscience: Basic and Clinical, 116*, 46–53. doi:10.1016/j.autneu.2004.08.008

Holomes, S. R., & Griffin, M. J. (2001). Correlation between heart rate and the severity of motion sickness caused by optokinetic stimulation. *Journal of Psychophysiology, 15*, 35–42. doi:10.1027//0269-8803.15.1.35

Kaga, K. (1992). *Memaino Kouzo (Structure of vertigo)*. Tokyo, Japan: Kanehara.

Kakeya, H. (2007). *MOEVision: Simple multiview display with clear floating image. SPIE Proceedings, 6490* (p. 64900J). Bellingham, WA: SPIE.

Kennedy, R. S., & Kennedy, R. C. (2007). The past, present and future of research in visually induced motion sickness. *Proceedings of VIMS2007: The first International Symposium on Visually Induced Motion Sickness, Fatigue, and Photosensitive Epileptic Seizures* (pp. 3-8). Hong Kong: Hong Kong University of Science and Technology Press.

Kennedy, R. S., & Lane, N., E., Berbaum, K. S., & Lilienthal, M. G. (1993). A simulator sickness questionnaire (SSQ) - A new method for quantifying simulator sickness. *The International Journal of Aviation Psychology, 3*, 203–220. doi:10.1207/s15327108ijap0303_3

Kido, K. (2007). *Digital Fourier transform (II)*. Tokyo, Japan: Corona Publishing.

Kobayashi, S. (1994). Eye sight recovering apparatus. *Japan Patent, 6*, 339501.

Matsumoto, T., Tokunaga, R., Miyano, T., & Tokuda, I. (2002). *Chaos and time series*. Tokyo, Japan: Baihukan.

Miyao, M., Ishihara, S., Saito, S., Kondo, T., Sakakibara, H., & Toyoshima, H. (1996). Visual accommodation and subject performance during a stereographic object task using liquid crystal shutters. *Ergonomics, 39*(11), 1294–1309. doi:10.1080/00140139608964549

Miyao, M., Otake, Y., & Ishihara, S. (1992). A newly developed device to measure objective amplitude of accommodation and papillary response in both binocular and natural viewing conditions. *Japanese Journal of Industrial Health, 34*, 148–149.

Money, K. E. (1972). Measurement of susceptibility to motion sickness. In M. P. Lansberg (Ed.), *AGARD Conference Proceedings No. 109: Predictability of Motion Sickness in the Selection of Pilots*. Nueilly –Sur-Seine, France: Advisory Group for Aerospace Research and Development.

Nakazawa, T., Okubo, Y., Suwazono, Y., Kobayashi, E., Komine, S., Kato, N., & Nogawa, K. (2002). Association between duration of daily VDT use and subjective symptoms. *American Journal of Industrial Medicine, 42*, 421–426. doi:10.1002/ajim.10133

Newell, K. M., Slobounov, S. M., Slobounova, E. S., & Molenaar, P. C. (1997). Stochastic processes in postural center-of-pressure profiles. *Experimental Brain Research, 113*, 158–164. doi:10.1007/BF02454152

Nishihara, T., & Tahara, H. (2001). *Apparatus for recovering eyesight utilizing stereoscopic video and method for displaying stereoscopic video*. (U.S. Patent, US7404693B2).

Okawa, T., Tokita, T., Shibata, Y., Ogawa, T., & Miyata, H. (1995). Stabilometry - Significance of locus length per unit area (L/A) in patients with equilibrium disturbances. *Heiko Shinkei Kagaku, 55*(3), 283–293. doi:10.3757/jser.55.283

Okawa, T., Tokita, T., Shibata, Y., Ogawa, T., & Miyata, H. (1996). Stabilometry-significance of locus length per unit area (L/A). *Heiko Shinkei Kagaku, 54*(3), 296–306. doi:10.3757/jser.54.296

Okuyama, F., Yana, K., Ikeda, T., & Oyamada, K. (1996). Accomodative response and vergence eye movement by stereoscopic image. *ITE Technical Report, 20*(24), 13–18.

Oman, C. (1982). A heuristic mathematical model for the dynamics of sensory conflict and motion sickness. *Acta Oto-Laryngologica*, (Supplement 392), 1–44.

Omori, M., Hasegawa, S., Watanabe, T., Fujikake, K., & Miyao, M. (2009). Comparison of measurement of accommodation between LCD and CRT at the stereoscopic vision gaze . In Shumaker, R. (Ed.), *Virtual and mixed reality, LNCS 5622* (pp. 90–96). Berlin, Germany: Springer-Verlag. doi:10.1007/978-3-642-02771-0_10

Patterson, R., & Martin, W. L. (1992). Human stereopsis. *Human Factors, 34*, 669–692.

Peterka, R. J. (2000). Postural control model interpretation of stabilogram diffusion analysis. *Biological Cybernetics, 82*(4), 335–343. doi:10.1007/s004220050587

Reason, J. (1978). Motion sickness adaptation: a neural mismatch model. *Journal of the Royal Society of Medicine, 71*, 819–829.

Riccio, G. E., & Stoffregen, T. A. (1991). An ecological theory of motion sickness and postural instability. *Ecological Physiology, 3*(3), 195–240. doi:10.1207/s15326969eco0303_2

Sakaguchi, M., Taguchi, K., Ixhiyama, T., Netsu, K., & Sato, K. (1995). Relationship between head sway and center of foot pressure sway. *Auris, Nasus, Larynx, 22*(3), 151–157.

Scibora, L. M., Villard, S., Bardy, B., & Stoffregen, T. A. (2007). Wider stance reduces body sway and motion sickness. *Proceedings of VIMS2007: The first International Symposium on Visually Induced Motion Sickness, Fatigue, and Photosensitive Epileptic Seizures* (pp. 18-23). Hong Kong: Hong Kong University of Science and Technology Press.

Stoffregen, T. A., Hettinger, L. J., Haas, M. W., Roe, M. M., & Smart, L. J. (2000). Postural instability and motion sickness in a fixed-base flight simulator. *Human Factors, 42*, 458–469. doi:10.1518/001872000779698097

Stoffregen, T. A., Smart, L. J., Bardy, B. J., & Pagulayan, R. J. (1999). Postural stabilization of looking. *Journal of Experimental Psychology. Human Perception and Performance, 25*, 1641–1658. doi:10.1037/0096-1523.25.6.1641

Sugiura, A., Takada, H., Yamamoto, T., & Miyao, M. (2010). *Effect of accommodation training by stereoscopic movie presentation on myopic youth. SPIE Proceedings, 7524(H)*. Bellingham, WA: SPIE.

Suzuki, J., Matsunaga, T., Tokumatsu, K., Taguchi, K., & Watanabe, Y. (1996). Q&A and a manual in stabilometry. *Heiko Shinkei Kagaku, 55*(1), 64–77. doi:10.3757/jser.55.64

Takada, H., Fujikake, K., & Miyao, M. (2009d). *On a qualitative method to evaluate motion sickness induced by stereoscopic images on liquid crystal displays. HCII 2009, Virtual and Mixed Reality* (*Vol. 5622*, pp. 254–262). Berlin, Germany: Springer-Verlag.

Takada, H., Fujikake, K., Miyao, M., & Matsuura, Y. (2007). Indices to detect visually induced motion sickness using stabilometry. *Proceedings of VIMS2007: The first International Symposium on Visually Induced Motion Sickness, Fatigue, and Photosensitive Epileptic Seizures* (pp. 178-183). Hong Kong: Hong Kong University of Science and Technology Press.

Takada, H., Fujikake, K., Omori, M., Hasegawa, S., Watanabe, T., & Miyao, M. (2008). Reduction of body sway can be evaluated by sparse density during exposure to movies on liquid crystal displays. *The International Federation for Medical and Biological Engineering (IFMBE)* [Berlin, Germany: Springer-Verlag.]. *Proceedings*, *23*, 987–991.

Takada, H., Fujikake, K., Watanabe, T., Hasegawa, S., Omori, M., & Miyao, M. (2009a). A method for evaluating motion sickness induced by watching stereoscopic images on head-mounted display. [Bellingham, WA: SPIE.]. *SPIE Proceedings*, *7237*, 1P–1.

Takada, H., Kitaoka, Y., Ichikawa, S., & Miyao, M. (2003). Physical meaning on geometrical index for stabilometry. *Heiko Shinkei Kagaku*, *62*(3), 168–180. doi:10.3757/jser.62.168

Takada, H., Kitaoka, Y., & Shimizu, Y. (2001). Mathematical index and model in stabilometry. *Forma*, *16*(1), 17–46.

Takada, H., Morimoto, T., Tsunashima, H., Yamazaki, T., Hoshina, H., & Miyao, M. (2006). Applications of double-Wayland algorithm to detect anomalous signals. *Forma*, *21*(2), 159–167.

Takada, H., Yamamoto, T., Miyao, M., Aoyama, T., Furuta, M., & Shiozawa, T. (2009b). Effect of a stereoscopic movie on the correlation between head acceleration and body sway . In Shumaker, R. (Ed.), *Virtual and mixed reality, LNCS 5622* (pp. 90–96). Berlin, GErmany: Springer-Verlag. doi:10.1007/978-3-642-02771-0_14

Takada, H., Yamamoto, T., & Sugiura, A. A., & Miyao, M. (2009c). Effect of an eyesight recovering stereoscopic movie system on visual acuity of middle-aged and myopic young people. *The International Federation for Medical and Biological Engineering (IFMBE) Proceedings, 25*(11), (pp. 331-334). Berlin, Germany: Springer-Verlag.

Takada, M., Miyao, M., Shiomi, T., Matsuura, Y., Omori, M., & Takada, H. (2010). Effect of eyesight-recovering stereoscopic movie system on visual acuity and fatigue of visual inspection workers. [Freiburg, Germany: IADIS Press.]. *Proceedings of the IADIS International Conferences, 2010*, 494–497.

Takeda, T., Izumi, S., & Sagawa, K. (1995). On the correlation between the head movement and the movement of the center of gravity using HMD. *Proceedings of the 1995 IEICE General Conference*, (pp. 203). Tokyo, japan: The Institute of Electronics, Information and Communication Engineers.

Ukai, K., & Howarth, P. A. (2008). Visual fatigue caused by viewing stereoscopic motion images. *Displays*, *29*, 106–116. doi:10.1016/j.displa.2007.09.004

Yasui, R., Matsuda, I., & Kakeya, H. (2006). *Combining volumetric edge display and multiview display for expression of natural 3D images. SPIE Proceedings, 6055 (pp.0Y1-0Y9)*. Bellingham, WA: SPIE.

Yokota, Y., Aoki, M., & Mizuta, K. (2005). Motion sickness susceptibility associated with visually induced postural instability and cardiac autonomic responses in healthy subjects. *Acta Oto-Laryngologica*, *125*, 280–285. doi:10.1080/00016480510003192

ADDITIONAL READING

Allison, R. S., Gillam, B. J., & Palmisano, S. A. (2009). Stereoscopic discrimination of the layout of ground surfaces. *Journal of Vision (Charlottesville, Va.)*, *9*(12), 1–11. .doi:10.1167/9.12.8

Baloh, R. E., & Honrubia, V. (1989). *Clinical neurophysiology of the vestibular system*. 2nd edition. Philadeiphia: FA Davis.

Barlow, H. B., Blakemore, C., & Pettigrew, J. D. (1967). The neural mechanism of binocular depth discrimination. *The Journal of Physiology*, *193*(2), 327–342.

Blakemore, C. (1970). A new kind of stereoscopic vision. *Vision Research*, *10*(11), 1181–1199. doi:10.1016/0042-6989(70)90036-2

Bowman, D. A., Kruijff, E., LaViola, J. J., & Poupyrev, I. (2004). *3D user interfaces-Theory and practice*. Redwood City, CA, USA: Addison Wesley Longman Publishing Co. Inc.

Campbell, F. W., & Robson, J. G. (1968). Application of Fourier analysis to the visibility of gratings. *The Journal of Physiology*, *197*(3), 551–566.

Cassin, B. (1993). *Dictionary of eye terminology. Gainesville*. Fl, USA: Triad Publishing Co.

Cumming, B. G., & Judge, S. J. (1986). Disparity-induced and blur-induced convergence eye movement and accommodation in the monkey. *Journal of Neurophysiology, May;55*(5), 896-914.

Fordell, H., Bodin, K., Bucht, G., & Malm, J. (2010, JUN). A virtual reality test battery for assessment and screening of spatial neglect. *Acta Neurologica Scandinavica*, *21*. doi:.doi:10.1111/j.1600-0404.2010.01390.x

Frisby, J. P., Buckley, D., Wishart, K. A., Porrill, J., Garding, J., & Mayhew, J. E. (1995). Interaction of stereo and texture cues in the perception of three-dimensional steps. *Vision Research*, *35*(10), 1463–1472. doi:10.1016/0042-6989(95)98726-P

Gibson, J. J., & Carel, W. (1952). Does motion perspective independently produce the impression of a receding surface? *Journal of Experimental Psychology*, *44*(1), 16–18. doi:10.1037/h0056030

Gillam, B., Chambers, D., & Russo, T. (1988). Postfusional latency in stereoscopic slant perception and the primitives of stereopsis. *Journal of Experimental Psychology. Human Perception and Performance*, *14*(2), 163–175. doi:10.1037/0096-1523.14.2.163

Gillam, B., Flagg, T., & Finlay, D. (1984). Evidence for disparity change as the primary stimulus for stereoscopic processing. *Perception & Psychophysics*, *36*(6), 559–564. doi:10.3758/BF03207516

Heuer, H., & Rapp, K. (2009). Pointing in stereoscopic space. *Perception*, *38*(11), 1663–1677. doi:10.1068/p6370

Howard, I. P., & Rogers, B. J. (1995). *Binocular vision and stereopsis*. New York: Oxford University Press.

Howland, H. C., Dobson, V., & Sayles, N. (1987). Accommodation in infants as measured by photorefraction. *Vision Research*, *27*(12), 2141–2152. doi:10.1016/0042-6989(87)90128-3

Hubel, D. H., & Wiesel, T. N. (1962). Receptive fields, binocular interaction and functional architecture in the cat's visual cortex. *The Journal of Physiology*, *160*, 106–154.

Julesz, B. (1964). Binocular depth perception without familiarity cues. *Science*, *145*, 356–362. doi:10.1126/science.145.3630.356

Ma, Y., Soatto, S., Kosecka, J., & Sastry, S. (2003). *An invitation to 3-D vision - From images to geometric models*. New York: Springer New York.

Marr, D., & Poggio, T. (1976). Cooperative computation of stereo disparity. *Science*, *194*(4262), 283–287. doi:10.1126/science.968482

Marr, D., & Poggio, T. (1979). A computational theory of human stereo vision. [London: Royal Society Publishing.]. *Proceedings of the Royal Society of London. Series B. Biological Sciences*, *204*(1156), 301–328. doi:10.1098/rspb.1979.0029

Morningstar, M. W., Pettibon, B. R., Schlappi, H., & Trevor, T. V. (2005). Reflex control of the spine and posture: a review of the literature from a chiropractic perspective. *Chiropractic & Osteopathy*, *13*, 13–16. doi:10.1186/1746-1340-13-16

Mowforth, P., Mayhew, J. E., & Frisby, J. P. (1981). Vergence eye movements made in response to spatial-frequency-filtered random-dot stereograms. *Perception*, *10*(3), 299–304. doi:10.1068/p100299

Nikara, T., Bishop, P. O., & Pettigrew, J. D. (1968). Analysis of retinal correspondence by studying receptive fields of binocular single units in cat striate cortex. *Experimental Brain Research*, *6*(4), 353–372. doi:10.1007/BF00233184

Ninio, J. (1985). Orientational versus horizontal disparity in the stereoscopic appreciation of slant. *Perception*, *14*(3), 305–314. doi:10.1068/p140305

Norcia, A. M., & Tyler, C. W. (1984). Temporal frequency limits for stereoscopic apparent motion processes. *Vision Research*, *24*(5), 395–401. doi:10.1016/0042-6989(84)90037-3

Rogers, B. (1988). Vision: perspectives on movement. *Nature*, *333*(6168), 16–17. doi:10.1038/333016a0

Rogers, B., & Cagenello, R. (1989). Disparity curvature and the perception of three-dimensional surfaces. *Nature*, *339*(6220), 135–137. doi:10.1038/339135a0

Sakaguchi, M., Taguchi, K., Ixhiyama, T., Netsu, K., & Sato, K. (1995). Relationship between head sway and center of foot pressure sway. *Auris, Nasus, Larynx*, *22*(3), 151–157.

Shin, H. K., Lee, J. H., Jin, H. J., Yoon, T. H., & Kim, J. C. (2010). Stereoscopic three-dimensional display based on polarization-switching device with low cross talk and high contrast ratio. *Optics Letters*, *35*(13), 2227–2229. .doi:10.1364/OL.35.002227

Van der Meer, H. C. (1978). Linear combinations of stereoscopic depth effects in dichoptic perception of gratings. *Vision Research*, *18*(6), 707–714. doi:10.1016/0042-6989(78)90149-9

Wheatstone, C. (1852). Contributions to the physiology of vision. Part the first. On some remarkable, and hitherto unobserved, phenomena of binocular vision. *Philosophical Magazine Series 4*, *3*(18), 241–267. doi:.doi:10.1080/14786445208646997

KEY TERMS AND DEFINITIONS

Accommodation: Accommodation is the process by which the eyes change optical power to maintain a clear image (focus) of an object as its distance changes. Human beings vary optical power by changing the form of the elastic lens using the ciliary body (<15 diopters). Accommodation is also closely related to vergence movements as follows. Under normal conditions, changing the focus of the eyes to look at an object at different distances will automatically cause vergence and accommodation.

Camera Axis: Camera axis is the optical axis of a camera. Each camera axis is shown as an imaginary line passing through the optical center of the lens system of a camera and perpendicular to the focal plane (Figure 5). Camera axes are fixed and crossed at the point of the virtual image at which the creator expects viewers to gaze.

Head Acceleration: Head acceleration is automatically caused by stabilization of the visual field in human beings. Righting reflexes can be subcategorized as follows: optical; neck; body-on-head; body-on-body; and labyrinthine, which plays a key role in the control of head posture. Some of the reflexes and neuroanatomy have been defined and illustrated separately. However, collective reflexes and their interactions have not been elucidated, although a cooperative effect was seen in the relationship between head movement and the movement of the center of gravity.

Motion Sickness: Often occurs because the vestibular information cannot be combined with the visual information. The visual field becomes unstable if there is a disagreement between the visual and vestibular information; this causes the vestibulo-ocular (VO) and optokinetic (OK) reflexes, respectively. The VO reflex cooperates with the OK reflex; this stabilizes the visual field in human beings.

Parallax: Parallax is an apparent displacement or difference in the apparent position of an object viewed along 2 different lines of sight. Exaggerated *convergence*, as mentioned below, is termed cross-eyed viewing (for example, focusing on the nose). As shown in Figure 1a, a double background is seen during cross-eyed viewing. When looking into the distance, the eyes diverge until their lines of sight are parallel, effectively fixating the same point at infinity. In this case, 2 fingers are seen (Figure 1b).

Stabilometry: Stabilometry has been used for the qualitative and quantitative evaluation of equilibrium. A projection of a subject's center of gravity onto a detection stand is measured as the average of the center of pressure (COP) of both feet. The COP is traced for each time step, and the time series of the projections is traced on an xy plane. The temporally vicinal points are connected to create a stabilogram. The body sway is complemented by the optical righting reflex (Figure 2).

Vergence: Vergence is the simultaneous movement of both eyes in opposite directions to obtain or maintain single binocular vision. When an organism with binocular vision looks at an object, the eyes must rotate around a vertical axis so that the image is projected at the centre of the retina in both eyes. To look at an object that is closer, the eyes rotate towards each other (*convergence*), whereas to look at an object that is farther away, the eyes rotate away from each other (*divergence*). Vergence is measured by the angle of inclination between these 2 lines (Figure 2).

ENDNOTE

1 This work was supported in part by the Hori Information Science Promotion Foundation.

Chapter 16
Low-Complexity Stereo Matching and Viewpoint Interpolation in Embedded Consumer Applications

Lu Zhang
IMEC, Belgium

Ke Zhang
IMEC, Belgium

Jiangbo Lu
Advanced Digital Sciences Center, Singapore

Tian-Sheuan Chang
National Chiao-Tung University, Taiwan

Gauthier Lafruit
IMEC, Belgium

ABSTRACT

Viewpoint interpolation is the process of synthesizing plausible in-between views - so-called virtual camera views - from a couple of surrounding fixed camera views. To make viewpoint interpolation possible for low/moderate-power consumer applications, a further quality/complexity trade-off study is required to conciliate algorithmic quality to architectural performance. In essence, the inter-dependencies between the different algorithmic steps in the processing chain are thoroughly analyzed, aiming at an overall quality-performance model that pinpoints which algorithmic functionalities can be simplified with minor global input-output quality degradation, while maximally reducing their implementation complexity w.r.t. arithmetic and line buffer requirements. Compared to state-of-the-art CPU and GPU platforms running at several GHz clock speed, our low-power 100 MHz FPGA implementation achieves speedups with one to two orders of magnitude, without impeding on the visual quality, reaching over 100 frames per second VGA high-quality, 64-disparity search range stereo matching and enabling viewpoint interpolation in low-power, embedded applications.

DOI: 10.4018/978-1-61350-326-3.ch016

Figure 1. Interpolation of Left/Right camera views into a rendered virtual viewpoint for eye-gaze correction in video teleconferencing (bottom), possibly augmented with auto-stereoscopic 3D displays where each pixel projects multi-directional viewing cones from which two are captured by the viewer's eyes (top)

INTRODUCTION

Figure 1 shows a typical eye-gaze correcting video conferencing application where virtual camera viewpoint interpolation restores straight eye contact to video tele-conference participants by interpolating surrounding views of the user/viewer/participant captured through cameras all around the display. This principle can be extended towards rendering multiple, adjacent, interpolated viewpoints for auto-stereoscopic, shutter-glasses-free 3D displays, where depth impression is obtained by rendering – for each pixel - up to ten different images in different viewing cones (see Figure 1 (top)), two of which being captured by the viewer's eyes. Ultimately, these dozens of views are calculated through viewpoint interpolation from a single pair of cameras, capturing stereoscopic content.

An essential DSP kernel in this process is the extraction of depth from the stereo cameras. Though we humans do not experience the difficulty of perceiving depth from our binocular view on the outside world, this depth extraction – also called stereo matching – is an incredibly complex processing step that only recently has been ported to embedded platforms (Woodfill, 2004; van der Horst, 2006) at the expense of the quality of the extracted depth image (also called dense depth map) in targeting near-to-real-time performances.

Figure 2 confirms we achieve competitive, real-time processing (over 100 frames per second at VGA resolution, including frame buffer access latency), while preserving high-quality standards, as confirmed by the very low Bad Pixel Error Rate (BPER) reported in Figure 2(d), following the definition of (Scharstein, 2002), i.e. the average difference between calculated and ground truth disparities over all pixels in the image (cfr. Figure 2(b)), using the test images of http://vision.middlebury.edu/stereo/. The black arrows refer

Figure 2. Frame rate (a) - (frames per second – fps) and quality (a,d) figures of merit (Bad Pixels Error Rate – BPER – cfr. definition in (b)) on different platforms (CPU, GPU and proposed FPGA implementation). The arrows compare implementations on FPGA (black arrow) and GPU (grey arrow) of the same/similar reference stereo matching code from two authors of this chapter.

(a)

	Image	Tsukuba	Venus	Teddy	Cones
	Resolution	384 x 288	434 x 383	450 x 375	450 x 375
	Disparity dmax	15	19	59	59
Frame	FPGA, 100 MHz	296	195	193	193
Rate	GeForce 8800 GPU	57	n/a	12	12
(fps)	PIV, 4 GHz CPU	7.14	3.84	1.21	1.19
Bad Pixel	FPGA, 100 MHz	3.84	1.2	**7.17**	**5.41**
Error Rate	GeForce 8800 GPU	**1.71**	**0.55**	9.9	6.66
(BPER %)	PIV, 4 GHz CPU	1.99	0.62	9.75	6.28

$$BPER = \frac{1}{N}\sum_{(x,y)}(|dC(x,y) - dT(x,y)| > 1.0)$$

dC = Calculated disparity (b)

dT = Ground Truth disparity

(c)

Frame Rate (FPS)					
Publication	Algorithm	Implementation	Disparity Range	Frame Rate	MDE/s
Proposed	MiniCensus + Adaptive	1 x FPGA	64	47 @ 1024 x 768	2365
Jin et al. 2010	Census + Fixed	1 x FPGA	64	230 @ 640 x 480	4521
Zhang et al. 2009	SAD + Adaptive	GPU GeForce8800 GTX	64	12 @ 450 x 375	129
Park et al. 2007	Trellis Based	1 x FPGA	128	30 @ 320 x 240	249
Wang et al. 2006	Dynamic Programming	GPU Radeon XL1800	16	43 @ 320 x 240	53
Kuhn et al. 2003	SSD + Fixed	ASIC	25	50 @ 256 x 192	61

(d)

BPER (%)													
Publication	*Tsukuba 384 x 288*			*Venus 434 x 383*			*Teddy 450 x 375*			*Cones 450 x 375*			Average Benchmark
	nocc.	*all*	*disc*	*nocc*	*all*	*disc*	*nocc*	*all*	*disc*	nocc	all	disc	
Proposed	3.84	4.34	14.2	1.2	1.68	5.62	7.17	12.6	17.4	5.41	11	13.9	**8.2**
Jin et al. 2010	9.79	11.56	20.29	3.59	5.27	36.82	12.5	21.5	30.57	7.34	17.58	21.01	17.24
Zhang et al. 2009	1.71	2.22	6.74	0.55	0.87	2.88	9.9	15	19.5	6.66	12.3	13.4	**7.65**
Park et al. 2007	2.63	n/a	n/a	3.34	n/a	n/a	n/a	n/a	n/a	n/a	n/a	n/a	n/a
Wang et al. 2006	2.05	4.22	10.6	1.92	2.98	20.3	7.23	14.4	17.6	6.41	13.7	16.5	9.82
Kuhn et al. 2003	n/a	n/a	n/a	n/a	n/a	n/a	n/a	n/a	n/a	n/a	n/a	n/a	n/a

to the presented FPGA solution, and the grey arrows correspond to a comparable solution on GPU from one of the co-authors, with one order of magnitude higher clock speed, though reaching a 15 times lower frame rate (or Million Disparity Estimations per second – MDE/s) at a marginally higher quality (lower BPER: 7.65% versus 8.2%).

STEREO MATCHING AND VIEWPOINT INTERPOLATION

The main principle of viewpoint interpolation, e.g. the mid-viewpoint extraction of Figure 3, is to shift pixels halfway between their corresponding positions in a Left/Right image pair. Since pixels attached to foreground objects show a larger displacement (called disparity, inversely proportional to depth) from the Left to the Right image, than background pixels (see Figure 3(b)), a dense (i.e. for each pixel in the image) depth map should first be extracted, see Figure 3(b'): dark-grey pixels correspond to the background, light-grey pixels to foreground objects. This process is called Stereo Matching, and shows some resemblance with Motion Estimation in video coding. However, in contrast to the rectangular windows used in motion estimation with Sum-of-Absolute-Differences (SAD) matching cost, high-quality stereo matching demands irregular shaped windows, so-called Adaptive Support Windows, with a more robust cost function: the Hamming distance in the Census Transform. This is an important difference that has major consequences on the hardware design choices. A large part of the chapter is devoted to these issues.

Figure 3. Stereo matching (a) extracts disparity (b) and its inverse, i.e. depth (b'), from left and right camera views. An interpolated view (halfway) is obtained by shifting all pixels over half their respective disparity (c). If not properly handled, occlusions create disturbing ghosting artifacts (d).

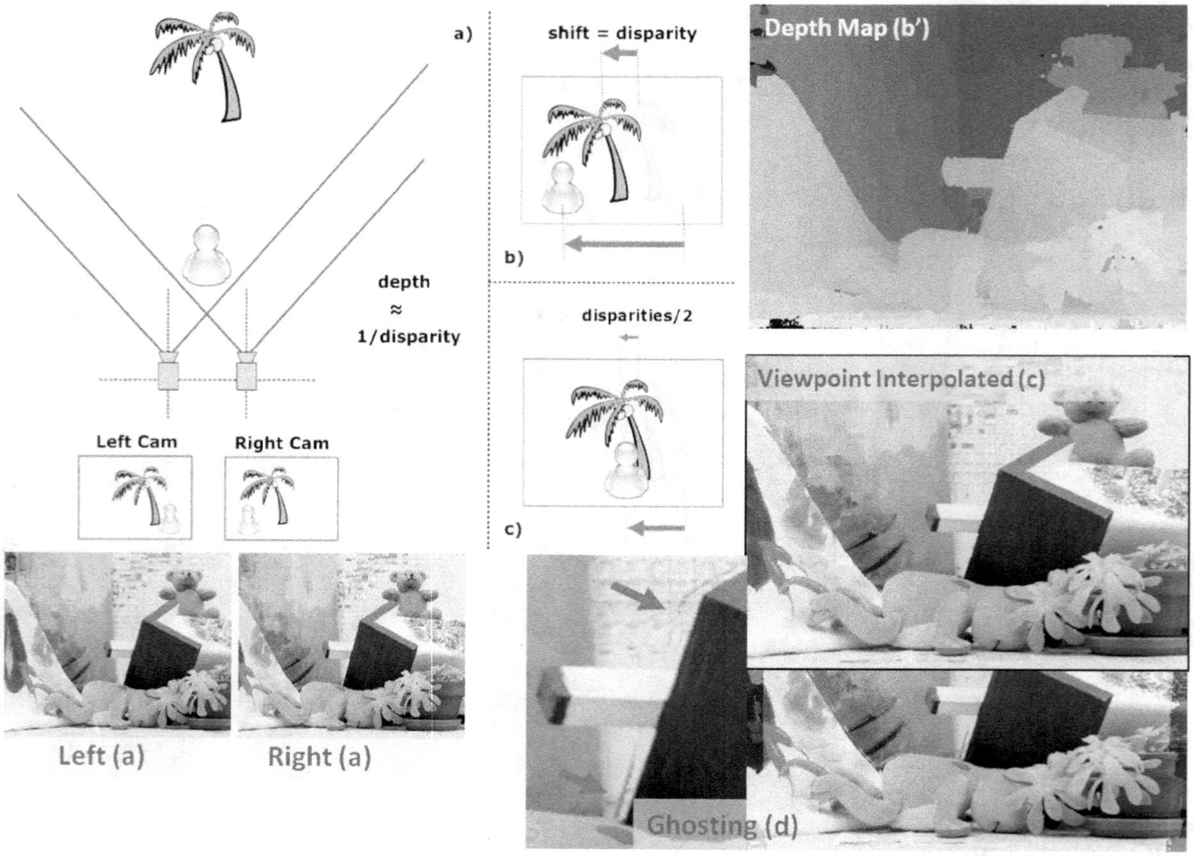

Additionally, in this stereo matching process, some pixels around the object boundaries are occluded in one of the two views, and hence it is a quite challenging process to faithfully extract depth information around object boundaries/ edges, and select in the succeeding viewpoint interpolation process the appropriate pixels from the left and right images to fill in the holes left by the occlusions after pixel shifting to an intermediate viewpoint. The present chapter will not dive into the details of viewpoint interpolation; instead, we pinpoint the intricate relationship between stereo matching and viewpoint interpolation, and how the final target of viewpoint interpolation influences the stereo matching specs. For more information about viewpoint interpolation, the reader is referred to (Lu, 2007; Lu, 2009b).

Though incorrect depth and occlusion estimations at the object edges lead to disturbing ghosting artifacts in the viewpoint interpolated image (see Figure 3(d)), the depth map does fortunately not have to be perfect for viewpoint interpolation applications (in contrast to metrological applications): boundary edge-preservation and the noise reduction of depth map specific noise are the main attention points, since they directly influence the viewpoint interpolation; all other aspects are less critical. Consequently, some freedom is left to make compromises between the stereo matching and associated viewpoint interpolation quality on one hand, and the hardware implementation complexity on the other hand. The result is that we achieve high-quality stereo matching and viewpoint interpolation, while reaching - at moderate implementation cost - one to two orders of magnitude frame rate performance increase over state-of-the-art solutions, see Figure 2.

ALGORITHM-HARDWARE CO-DESIGN IN STEREO MATCHING AND VIEWPOINT INTERPOLATION

Algorithmic/Visual Quality versus Hardware Complexity

This chapter serves the purpose of providing some insights on how aforementioned quality-complexity compromises are made. We here follow a tutorial approach giving a high-level overview of the design choices, heavily relying on visual results shown throughout the figures. We will guide the reader throughout three main stages of the design process, following a top-down pathfinding approach:

- A high-level algorithmic overview of the different processing steps in the full end-to-end pipeline, giving the reader a feeling of some important DSP kernels and identifying some important attention points.
- A "hardware-friendly" algorithmic refinement step with important algorithmic modifications that keep high-quality visual results while relaxing rebellious hardware constructs. Many images will be worth a thousand words to explain the design choices. The reader will gradually become familiar to concepts like Support Window, Line Buffers, arithmetic complexity reduction, … little by little being confronted to more hardware related aspects.
- A low-level hardware view with registers, line buffers and arithmetic parallelization constructs, clearly supporting our claims w.r.t. implementation complexity. We show final performance figures, comparing frame rate and MDE/s (Million Disparity Estimations per second), accompanied by RTL complexity figures (number of ALUs, RAM blocks, …).

Census Transform and Adaptive Support Regions

The target application being a symbiosis of Figure 1's eye gaze correction and auto-stereoscopic 3D display rendering for mobile video chatting, a Left/Right Image pair taken from webcams at each side of the mobile display (instead of one single top camera – see Figure 1(d)) should be DSP processed in order to create up to ten intermediate viewpoints of the user's picture, each rendered in a different direction through a lenticular sheet covering the display, see Figure 1(c). Even with moderate resolution images, high image data processing rates will be required to sustain such real-time, multiview rendering.

Figure 2 shows competing PIV CPUs, GeForce 8800 GPUs and recent FPGA/ASIC implementations, where the latter are the main candidates for low-power, mobile applications. Though providing acceptable frame rates and/or image sizes, they unfortunately suffer from a high Bad Pixel Error Rate (BPER- see Figure 2(b)) defined by Middlebury's Stereoscopy Best Practice Reference (http://vision.middlebury.edu/stereo/), exhibiting large pixel value differences between the calculated, interpolated viewpoint and its ground truth from Middlebury's test set (Scharstein, 2002). Additionally, for this prior art implementations, it is unclear how the quality degrades with radiometric variations in the Left/Right image pair, i.e. when the average luminance in left and right images are different.

Census Transform

In terms of robustness to radiometric differences, (Hirschmuller, 2007) evaluated different cost functions, e.g. rank, census, and etc. Census Transform is often preferred in hardware implementations (Chang, 2010), compared to other wide-spread cost functions, like Sum-of-Absolute-Differences (SAD) and Normalized Cross-Correlation (NCC): SAD is very sensitive to radiometric variations, while NCC (very robust to radiometric variations (Zhang, 2009c)) is too costly for hardware implementation due to its normalizing factor, i.e. the sqrt() function in its denominator.

The simplicity of the Census Transform with its local weights window and Hamming distance calculations (see Figure 4(a)) does not impede on its robustness to the stereo matching process in practical conditions where left and right image might exhibit very different luminance levels from mismatched cameras, as shown in Figure 5(a): with SAD an unrecognizable depth map with a disastrous 96% BPER is obtained for the Left1/Right1 image pair with mismatched luminance, while with the Census Transform a very reasonable depth map is obtained with a lower than 10% BPER, which might be regarded as a very good threshold of fidelity. We will come back on the Census Transform and its quality implications when touching the subject of viewpoint interpolation later in this chapter.

Adaptive Support Regions

For those familiar with video coding, we might say that stereo matching is very similar to motion estimation in video coding: a square region of pixels in a frame is compared to a large, rectangular region in another frame (in video coding this is another time frame; in stereo matching this is another camera) - called a search region - in a sliding approach. Figure 4(b) illustrates the concept: a square region of pixels around pixel x in the left frame is compared to the corresponding square region at position $x\text{-}d$ in the right frame, with d being the hypothesis disparity (motion vector in video coding terminology). All values of d from 0 to d_{max} (the disparity range) are traversed and the best matching position is selected through a Winner-Takes-All (WTA) approach.

The matching cost in comparing two square regions of pixels (one at position x in the left image, the other at position $x\text{-}d$ in the right image) is often taken (similar to video coding) as the Sum-of-Absolute-Difference (SAD) over all pixels taken two by two over the two square regions. However, in contrast to video coding where the average luminance over successive frames remains fairly constant since the frames are captured by the same, unique camera with fixed capturing parameters, the frames to match in stereo matching come from two different cameras and hence might exhibit very different luminances, incuring a severe matching cost bias to which the stereo matching is very sensitive. Hence, other approaches, already mentioned earlier – e.g. Normalized Cross Correlation, Census Transform - have been proposed, from which we have selected the latter, as explained in previous section.

Moreover, stereo matching differs substantially from motion estimation in video coding in the sense that stereo matching should deliver *physically correct* disparity vectors to create proper viewpoint interpolation results, while in video coding the motion vectors should be *good enough* as to minimize the bit coding entropy energy (i.e. the randomness of the sequence of bits representing the motion vectors), and sometimes this criterion does not coincide with the physical motion vectors fidelity. In practice, this deviation from physical reality occurs when square regions of pixels intersect image regions from two different objects in the scene. Therefore, another key aspect to obtain accurate depth maps in the stereo matching, is adaptive cost aggregation. Tombari et al. (Tombari, 2008) evaluated different methods of local cost aggregation. Among them, the adaptive weight method (Yoon, 2006) reaches top

Figure 4. The (mini-) Census Transform (a) compares the (example) pixel values with the window center pixel value to create Hamming Distance Vectors for each image (Left and Right Census Vector). Disparity estimation for fixed regions (b) and Adaptive Support Regions (c) for all candidate disparity values (d=0 to d=dmax). Left and Right image might have different support region shapes, and hence a Correlation Region (d) is extracted to correlate the results from the Left-to-Right and Right-to-Left disparity estimation.

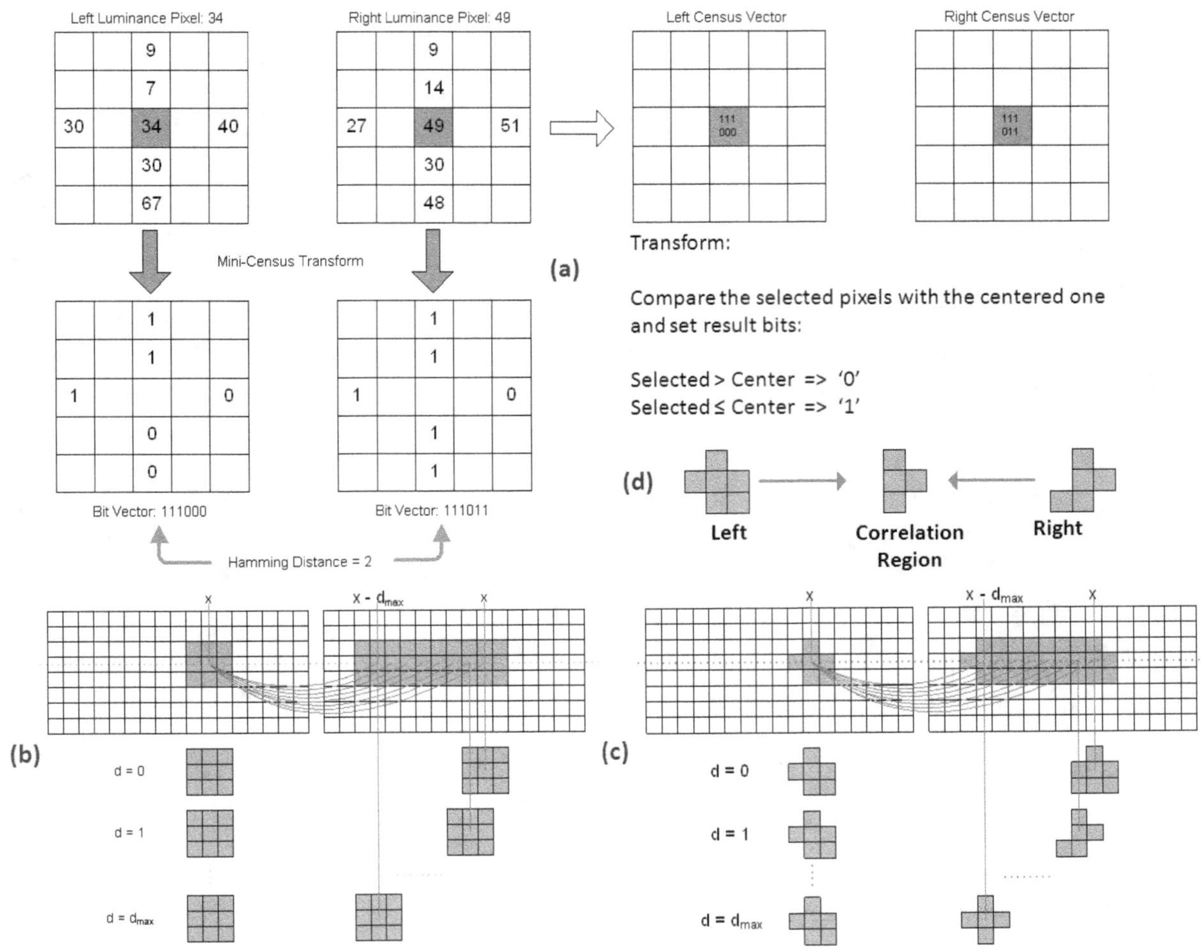

matching accuracy. It assigns an adaptive weight to each support pixel based on its spatial distance and color difference to the anchor pixel. Though producing accurate results, the adaptive weight method is computationally intensive and consumes a lot of memory, due to its pixel-wise adaptive weight. Another concept is found in Adaptive Support Regions (Zhang, 2009a; Zhang, 2009b; Zhang, 2009c; Zhang, 2009d; Lu, 2009a). These are irregularly-shaped regions (in contrast to the square regions with motion estimation in video coding) where all pixels surrounding pixel x have a similar pixel value, and hence are believed to be part of the same object in the image. Support Regions may hence be regarded as edge-preserving micro-segments in the image, similar to (Zitnick, 2004), with the slight nuance that Adaptive Support Regions partially overlap each other. The

Figure 5. (a) Depth map and its quality (BPER) for Adaptive Support Region (ASR) with Census Transform, compared to SAD with different luminance values for Left and Right images (Left$_1$, Right$_1$) (b) Depth map quality for different settings in the disparity estimation under matched luminance levels between Left and Right image (Left$_2$, Right$_2$): Rect = Rectangular window, SAD = Sum of Absolute Differences, ASR = Adaptive Support Region.

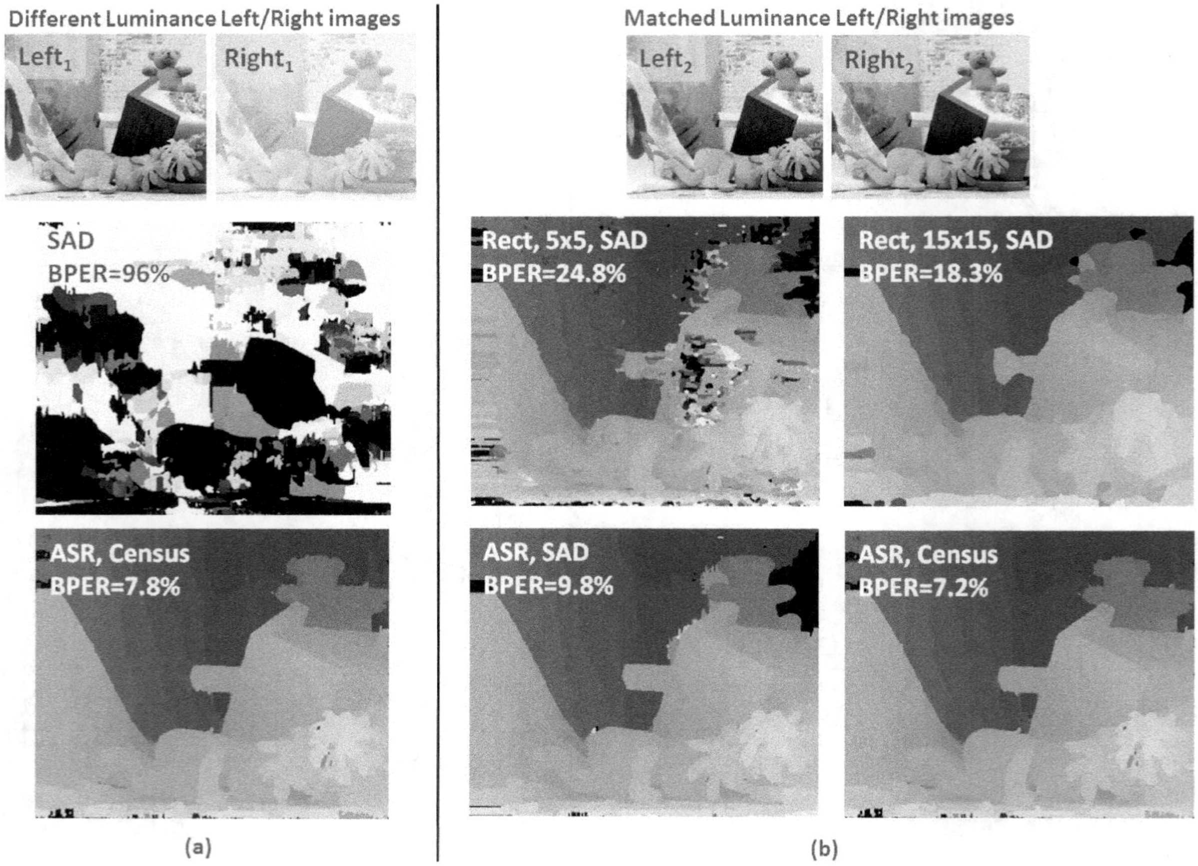

stereo matching (or motion estimation in video coding) of Figure 4(b) should then be adapted to the one shown in Figure 4(c), with the additional constraint that:

- The number of pixels may vary from one region to the other, which influences the matching cost criterion (and hence this pixel count per region has to be tracked), and
- Two differently shaped pixel regions to be compared should be restricted to their common Correlation Region, as shown in Figure 4(d). How these Adaptive Support Regions (ASR) are determined will be explained in a while. At this point in time, let's rather confirm the superiority of ASR over Rectangular (Rect.) regions, as shown in Figure 5(b): ASR with SAD is twice better in BPER than Rect. with SAD, but remains worse than ASR combined with the Census Transform, which exhibits a BPER of 7.2%, well below 10% and is hence to be considered as a high-quality depth map (Figure 5(b)-Bottom-Right). Additionally, as explained earlier, the Census Transform

is fairly insensitive to luminance variations (see Figure 5(a)-Bottom) and hence Census Transform combined with Adaptive Support Regions is the selected approach for performing stereo matching at high quality.

Line Buffers in Census Transform and Adaptive Support Regions

To reach a desirable quality-complexity trade-off it is important to have a closer look at the impact of the so far explained Census Transform and Adaptive Support Regions on the hardware implementation. In contrast to full-featured computer systems disposing of a wealth of RAM memory, embedded systems and in particular FPGAs and ASICs have only scarce memory resources, and every memory register and RAM block should be well-used in order to avoid slow off-chip memory accesses, which would impede on the final system performances. Consequently, unlike in classical image processing software engineering where full frames are kept in memory, our design reads the images in progressive scanline order and stores only a limited number of successive lines in a so-called Line Buffer, as shown in Figure 6. The most recently read pixels (e.g. values 0, 3 and 1 in the bottom part of Window W in Figure 6(a)) are pushed into the lowest Line Buffer SRAM L_2 of Figure 6(b) and remain available for processing unless they are pushed out of the upper Line Buffer SRAM L_1, as exemplified by the values 1, 3 and 4 in the upper part of window W in Figure 6(a). All values of the window W can then simultaneously be accessed for a (in this example) 3x3 window processing. When window W moves one pixel to the right, the corresponding underlying 3x3 pixels will be available in the WinReg registers of Figure 6(b). Obviously, the height in number of lines of window W determines the number of image lines to buffer: height-1 lines should be stored (see Figure 6(a): 2 lines L_1 and L_2, for a 3x3 window).

In FPGA and ASIC design, this is an important metric to be taken into account. In particular, the Census Transform of Figure 4(a) with a window size of 5x5 pixels, and the simplified Adaptive Support Region construction with limited vertical span (see Figure 7(c) - center), result in a limited Line Buffer cost.

Figure 7 shows the basic idea behind the construction of the Adaptive Support Regions: a two-by-two-armed cross (two horizontal arms, two vertical arms) is constructed around pixel x by successively adding pixels to the region as long as their luminance difference with x is below a certain threshold – the Luminance Difference Threshold (LDT) - see Figure 7(a). For each vertical position V_d below/above (and including) x (see Figure 7(b)) a horizontal arm is constructed following the same principles to the left h_l and the right h_r. A detailed description of the involved calculations can be found in (Zhang, 2009d).

The stereo matching quality variations for different horizontal and vertical arm lengths of the cross spanned over each pixel x is shown in Figure 7(c). We clearly see that an acceptable BPER lower than 10% is reached for the vertical arm lengths choice V=5 of Figure 7(c)-center. Making the vertical arm length adaptive below this number in a kind of Adaptive Vertical Support Region approach (see Figure 7(c) "V=adapt") does not modify the quality of the outcome, and does not provide any hardware implementation advantages.

The horizontal arm length has only a minor impact on memory requirements, but it will impact the involved arithmetic logic and the number of registers and parallel processing paths. In view of the quality results shown in Figure 7(c), a maximum horizontal arm length of 15 is selected. Note that the horizontal arm lengths remain adaptive (below a length of 15), otherwise there would be no Adaptive Support Region creation possible, since the vertical arm lengths are now fixed.

Figure 6. 2D local window processing based on extracting a square region from a Scan-Line Buffered image through Line Buffers L_1 and L_2 (a) implemented in SRAM-blocks on FPGA (b)

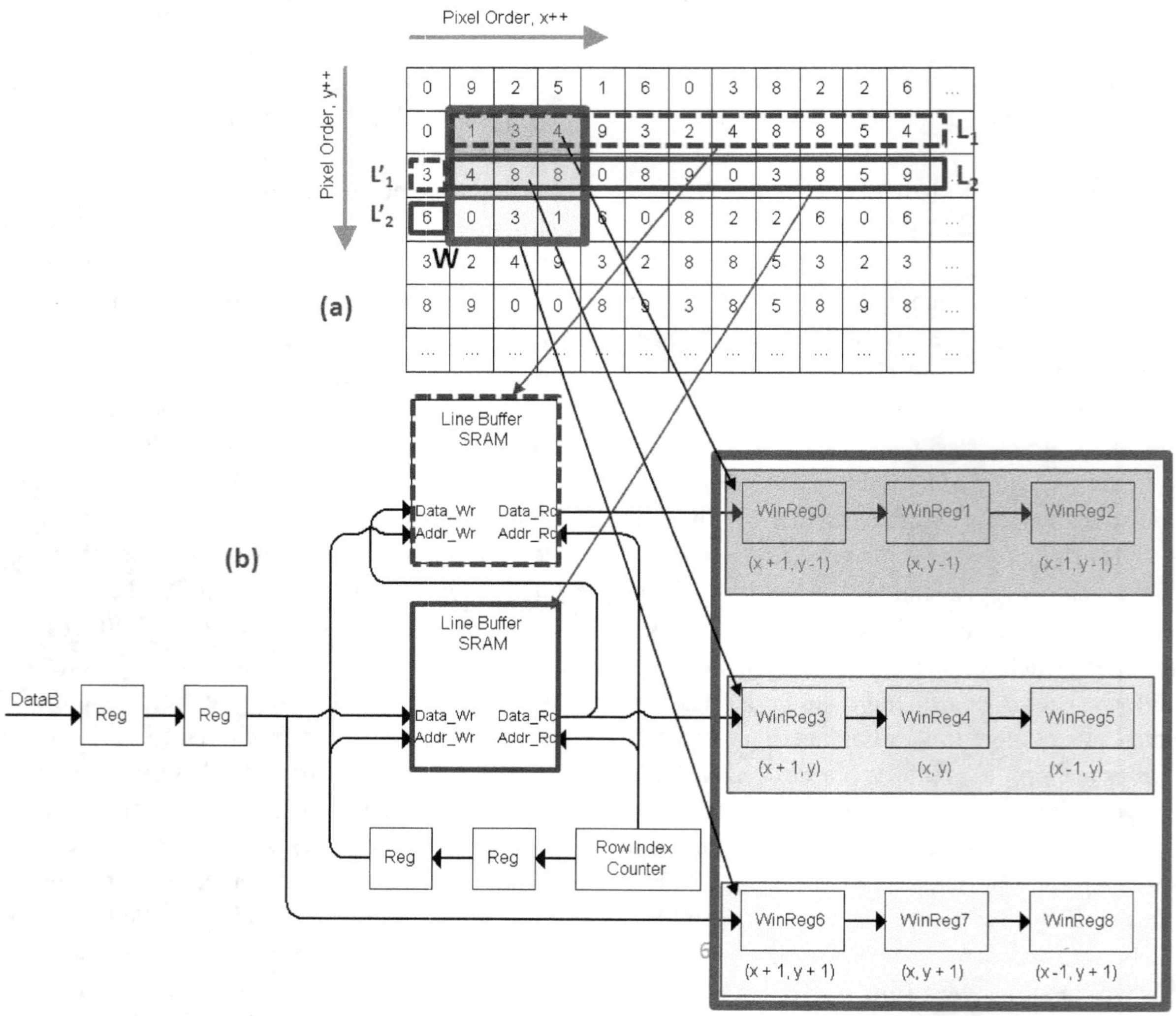

Run-Time Parameters for Viewpoint Interpolation

Thus far, we have made some algorithmic choices - the Census Transform and Adaptive Support Regions - and fixed their hardware-sensitive parameters in order to limit the Line Buffer memory requirements to a limited number of image lines to store. We also have verified their robustness against parametric variations, e.g. the luminance variation robustness, reported in Figure 5(a).

One algorithmic parameter (hardly influencing the hardware) remains to be studied: the Luminance Difference Threshold (LTD) of Figure 7, used in the two-by-two armed, cross-based Adaptive Support Region construction: the larger the allowed LTD, the larger the support regions will be, but also the higher the risk that edge preservation is endangered. Fortunately, according to Figure 8 showing the Normalized Error (i.e. the Bad Pixel Error Rate scaled between 0 and 1 for the most common test images of the Middlebury data set)

Figure 7. Cross-based (a) Adaptive Support Region extraction (b) and the depth map results (c) for different values of the Horizontal and Vertical arm lengths

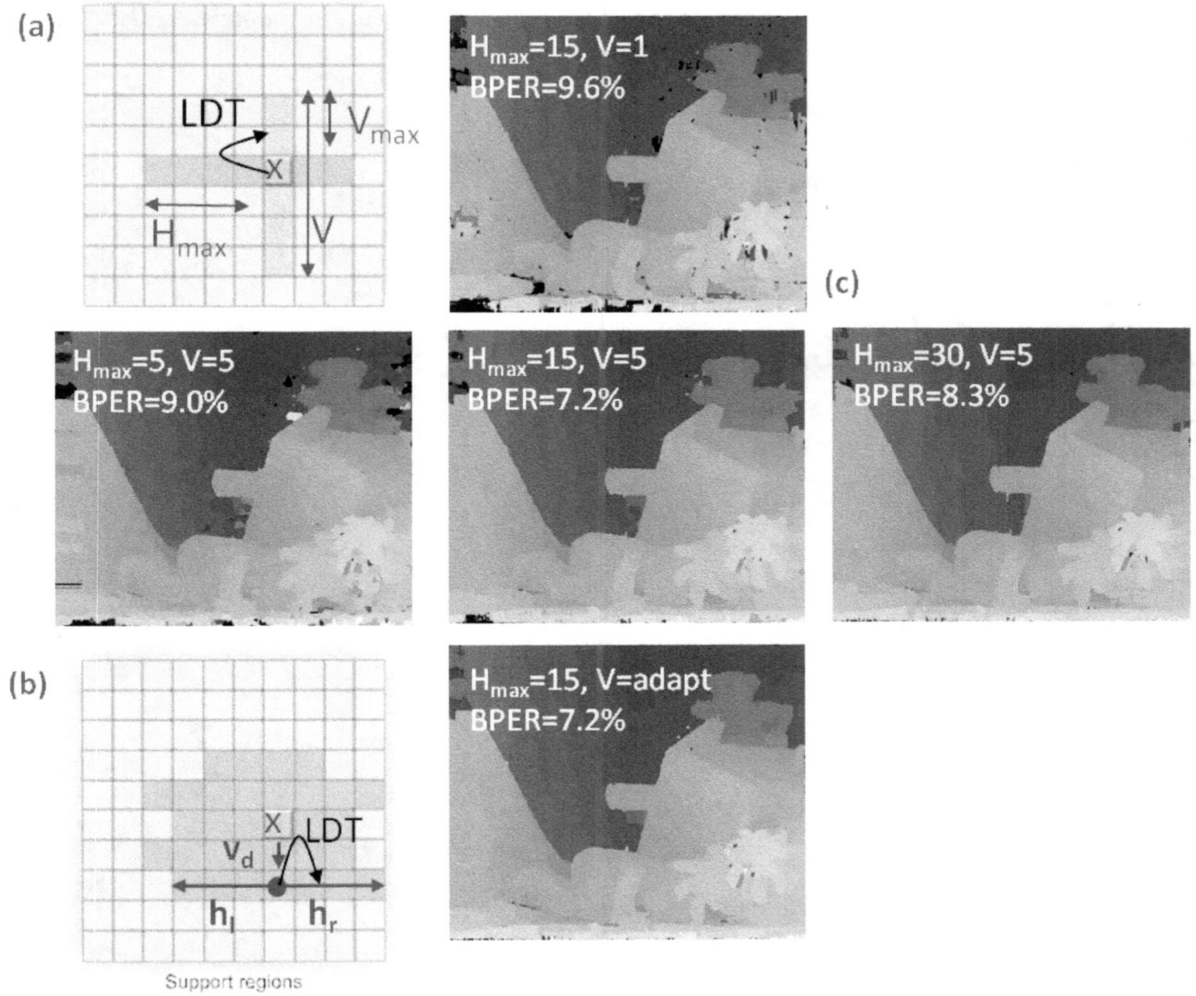

with sweeping LDT values, for image regions ranging from Non-Occluded regions (nonocc) to Depth Discontinuity (disc) regions (Scharstein, 2002), a comfortable LDT setting valley V exists in which the depth image fidelity is high (i.e. low Normalized Error). The Venus test image (dashed curves in Figure 8) from the Middlebury data set exhibits a more peculiar Normalized Error behavior (with scaling parameters following the average of the common image data set), extending the best LDT valley V to V', yet keeping the LDT parameter range acceptably small in most practical applications.

Hence, the final quality of the stereo matching and viewpoint interpolation – though being dependent on the LDT parameter - remains high in this valley V setting range, as confirmed by the depth maps (top) and viewpoint interpolation (bottom) results of Figure 9 for the Teddy test sequence; visual results for other Middlebury test sequences can be found in (Zhang, 2011). For extreme LDT values far away from the valley V in Figure 8, the stereo matching and viewpoint

Figure 8. Depth map quality (Normalized Error, scaling BPER between 0 and 1 for the most common images) over different regions of the image (Non-occluded, All, Discontinuous, as defined by the Middlebury benchmark) as a function of the Luminance Difference Threshold (LDT) parameter of Figure 7 (a,b), with best values in Valley V for the most common images, and V' when including exceptional cases.

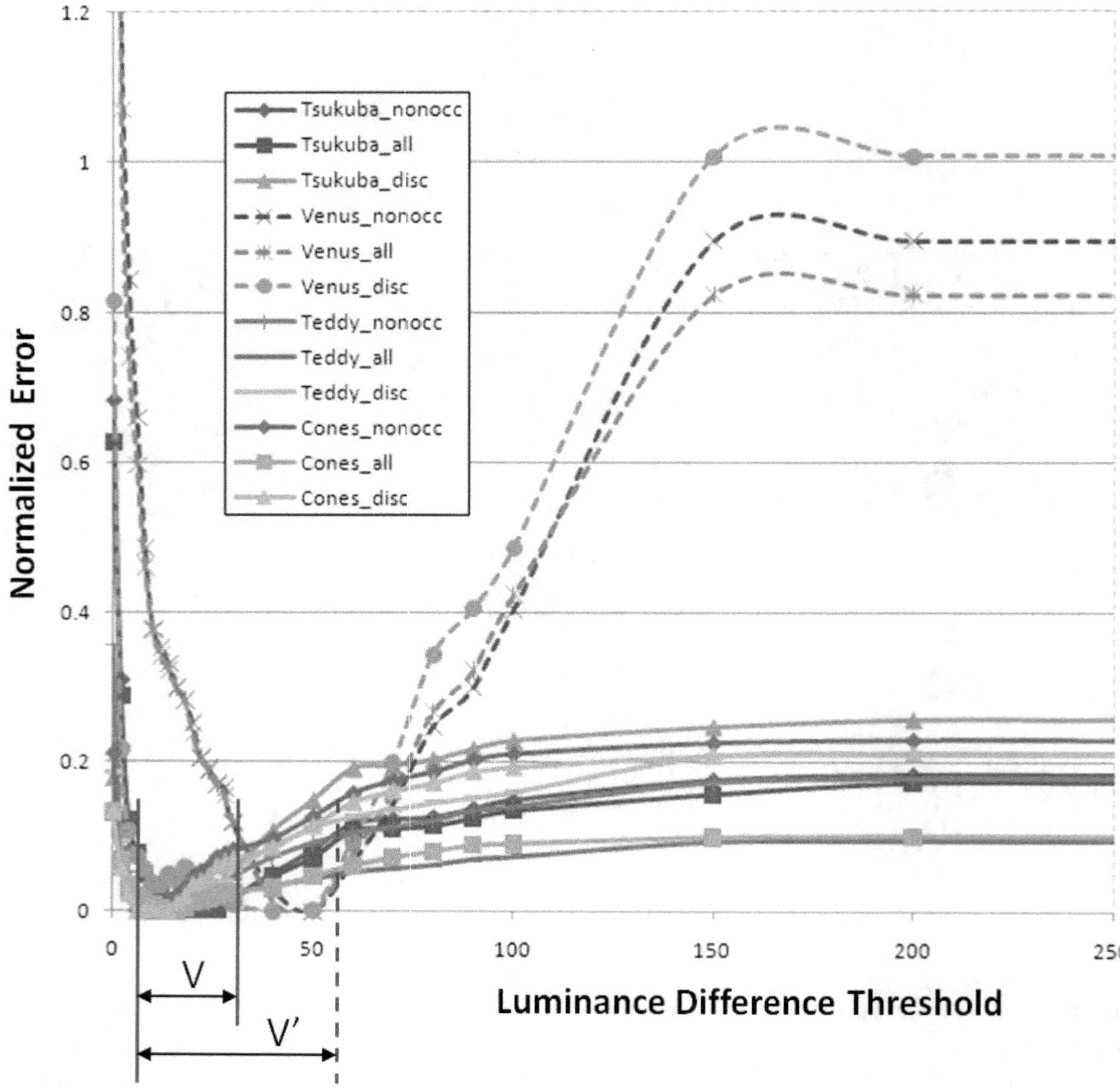

interpolation results might exceptionally deteriorate as in Figure 10, with the jagged roof of Figure10-right only appearing for LDT values in the right-most saturation region of Figure 8.

We may thus conclude that our stereo matching and viewpoint interpolation is robust against parametric variations, while at the same time being hardware-friendly w.r.t. memory requirements.

Figure 9. Depth map (top) and Viewpoint Interpolation (bottom) quality for two settings of the Luminance Difference Threshold (LDT) parameter inside region V (LDT=6 to 25) of Figure 8

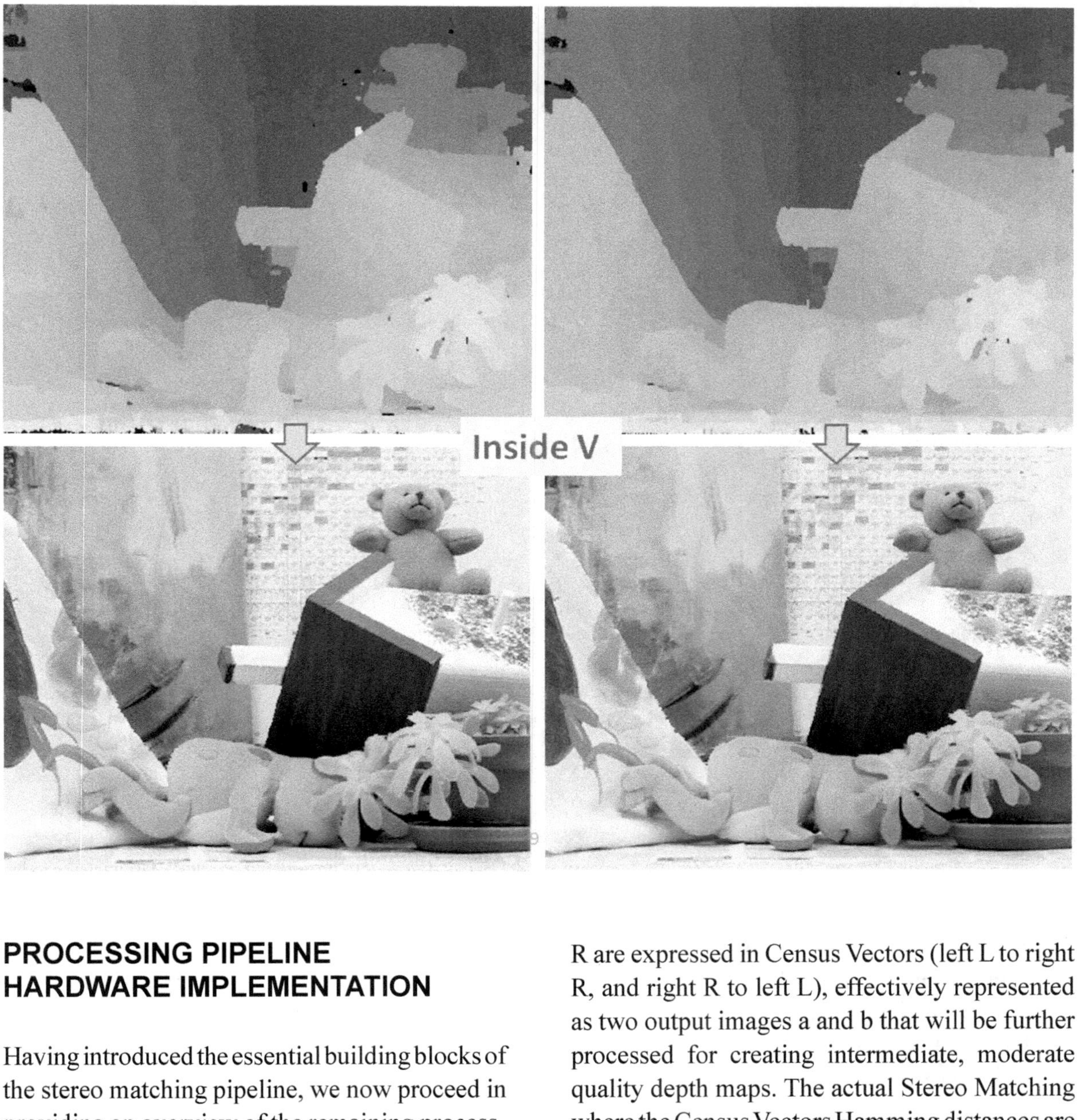

PROCESSING PIPELINE HARDWARE IMPLEMENTATION

Having introduced the essential building blocks of the stereo matching pipeline, we now proceed in providing an overview of the remaining processing steps related to the refinement of intermediate depth maps. Figure 11 inventories the three main processing steps as embedded on our Stratix-III EP3SL150 FPGA implementation:

A pre-processing step with the Census Transform and the Adaptive Support Region extraction. Hamming distances between input images L and R are expressed in Census Vectors (left L to right R, and right R to left L), effectively represented as two output images a and b that will be further processed for creating intermediate, moderate quality depth maps. The actual Stereo Matching where the Census Vectors Hamming distances are aggregated in parallel for each hypothetical disparity i.e. the matching from the Left to the Right image, is calculated over all possible disparities concurrently with the best match selected by a Winner-Takes-All step, in accordance to Figure 4(c) with the (equivalent to motion estimation in video coding) search window in the right image.

Figure 10. Depth map (top) and Viewpoint Interpolation (bottom) quality for two settings of the Luminance Difference Threshold (LDT) parameter outside region V (LDT=6 to 25) of Figure 8

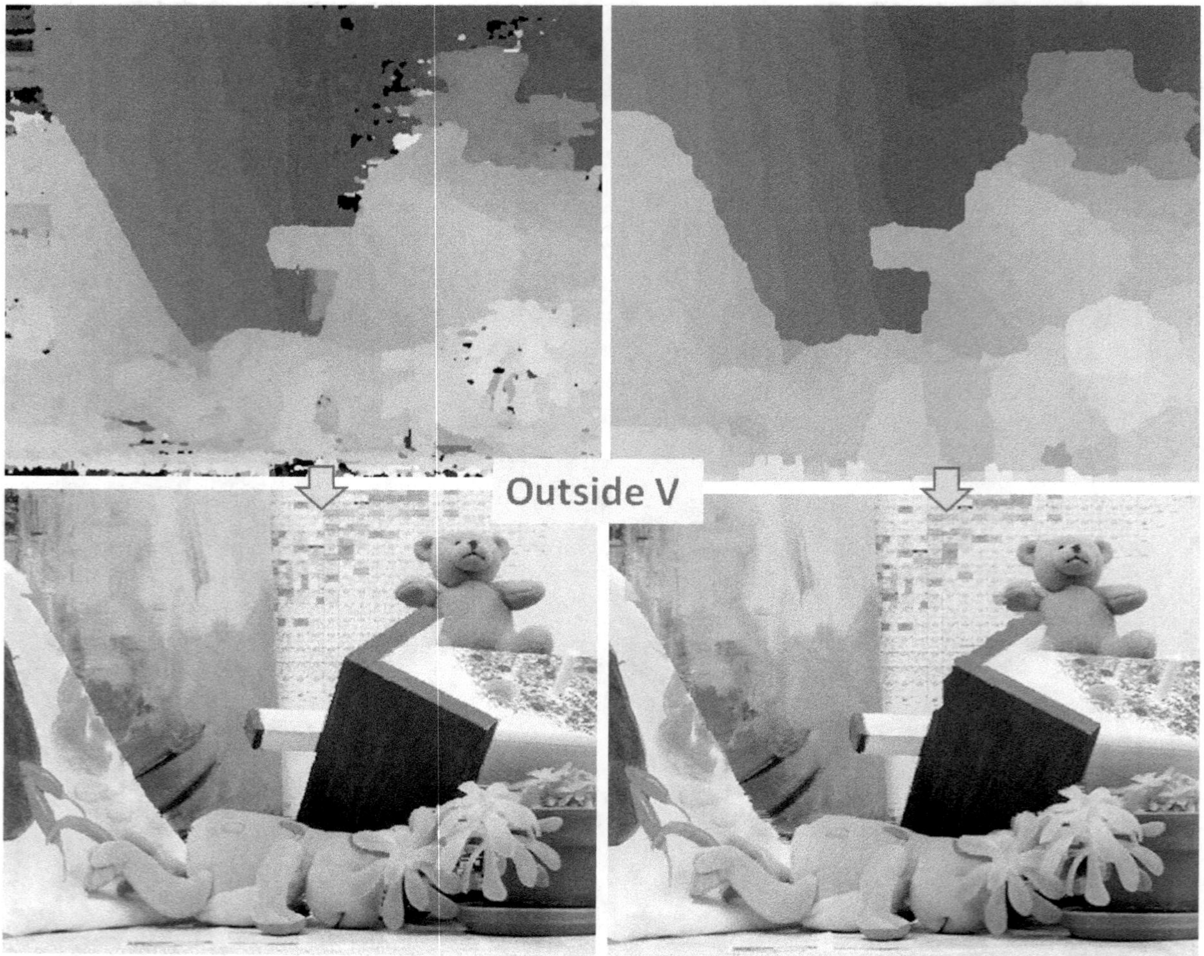

A similar procedure from Right to Left is calculated by reusing the same but time-shifted aggregation results, effectively creating two intermediate depth maps A and B

The post-processing combines these two depth maps A and B through a consistency check into a combined depth map C, which is further denoised to create the unique final depth map D.

Figure 11 also shows some FPGA specific I/O hardware fabrics: data is processed through Sources and Sinks for data I/O and Slaves (registers) for I/O control. Avalon Source, Avalon Sink and Avalon Slave in Figure 11 are industry-standard on-chip interconnection interfaces defined by Altera (Altera, 2009). The links to external DDR2 memory are used for providing video source and disparity map storage, and all stereo matching processing remains on-chip, with image data entering in scanline order. The FPGA has sufficient on-chip RAM to store a limited number of successive scanlines as to allow processing over small 2D windows, typically 5 up to a dozens of scanlines for SVGA resolution. Key to the best processing throughput is to fully pipeline all processing steps, i.e. except for the pipeline latency, a new income pixel gets its disparity at

Figure 11. Pre-processor, Stereo Matcher and Post-processor FPGA flowgraph, reading Left/Right images (L, R) from DDR2 SDRAM, creating Census Transformed images (a, b), Left-to-Right and Right-to-Left depth map estimations (A,B), combined to a single depth map (C), denoised to (D) and providing additional occlusion information (O) for viewpoint interpolation.

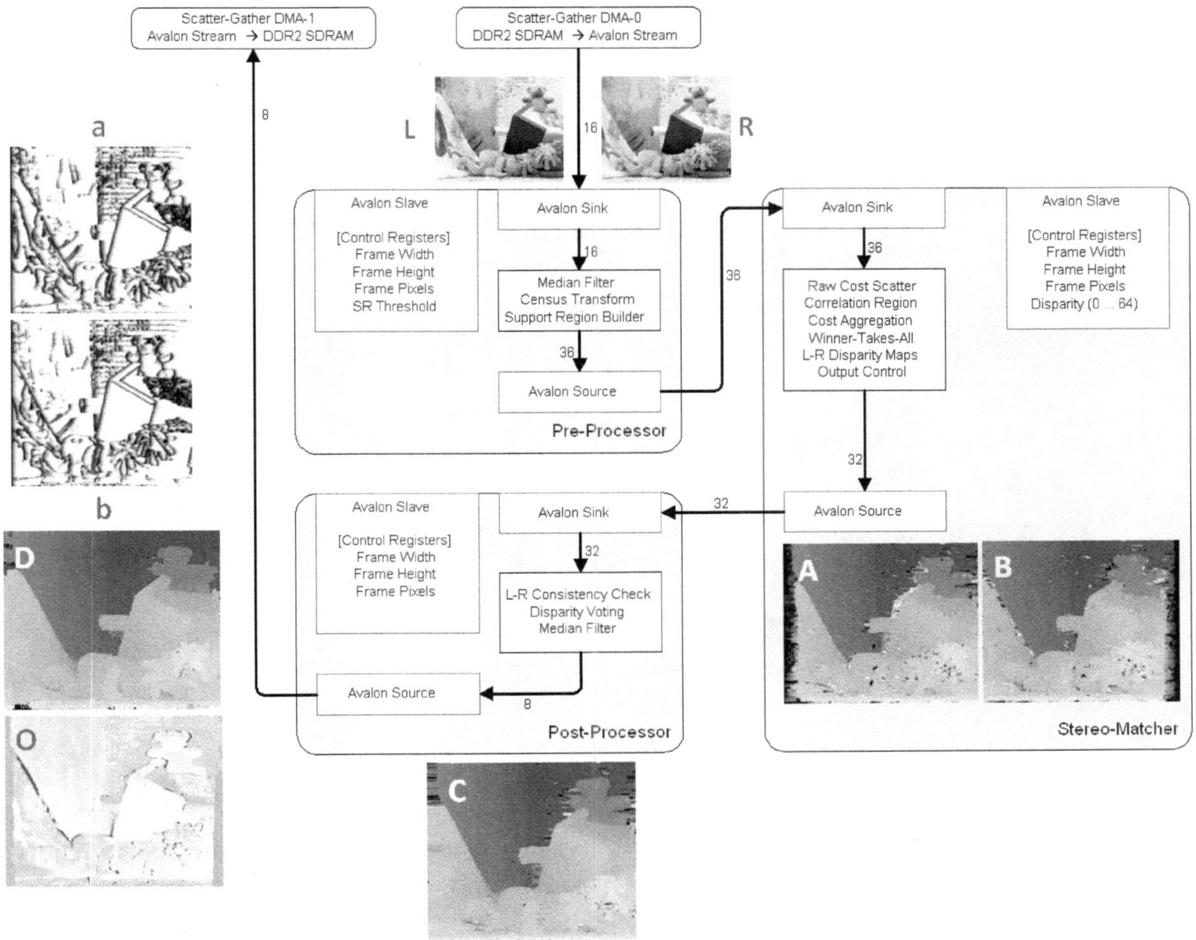

the end of the pipeline, and valid disparities come successively in scanline order, synchronized with the input pixel rate.

Pre-Processor

Besides the Census Transform and Adaptive Support Region calculation, minor other processing is performed, e.g. median filtering for denoising (also present in other steps of the processing pipeline), which we will not further elaborate on. Several transformed images (e.g. a and b of Figure 11) that contain census vectors and support region information are generated for the original stereo image pair. Important to know is that the output of the preprocessor does not contain disparity hypothesis, and involved processing is performed on the left and right image separately. The non-zero disparity hypothesis is left to the stereo matcher, which uses the census vectors and adaptive support regions calculated in the pre-processor.

Figure 12. Left-to-Right Aggregation Cost over Adaptive Support Regions (a) is decomposed over successive lines (b) with all disparity estimations in parallel (c)

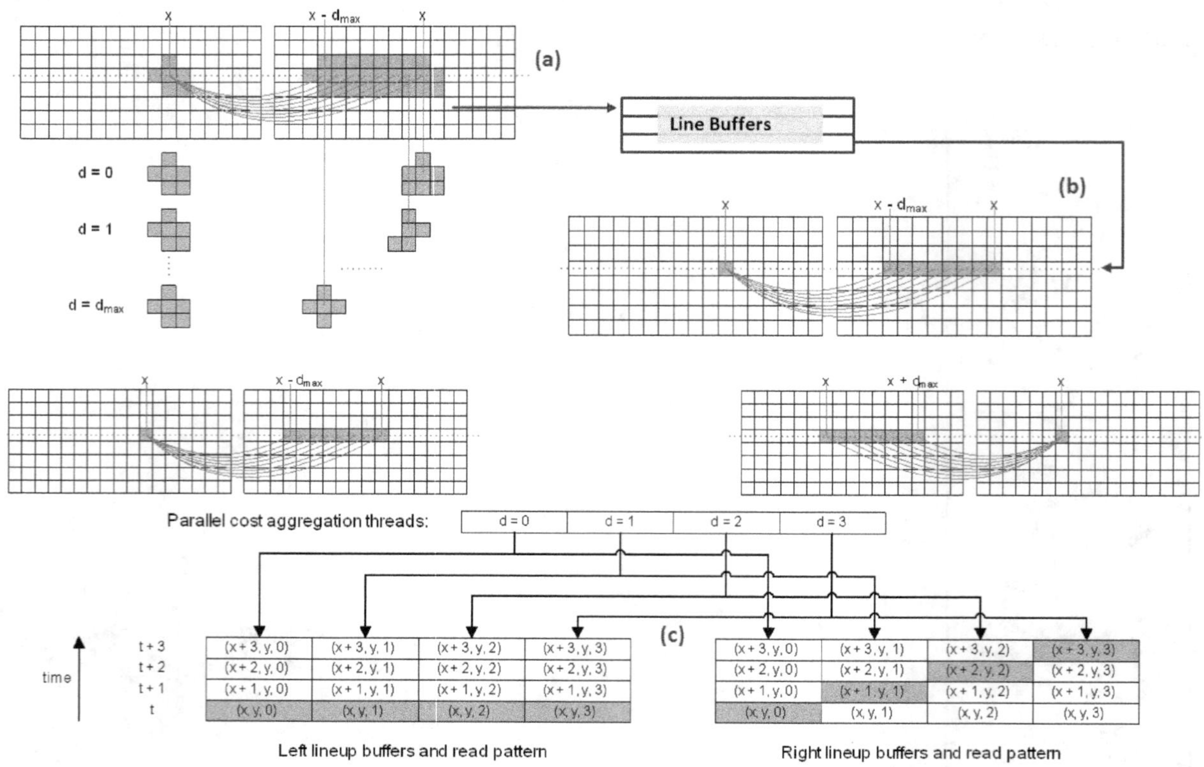

Stereo Matcher

Following the processing pattern explained in Figure 4(c) and (d), a Left-to-Right and Right-to-Left depth map estimation is performed, creating the images A and B of Figure 11. The Adaptive Support Regions of the Left and Right (Census Transformed) images are matched according to Figure 4(c-d) by decomposing the regions in successive lines and calculating over each line the matching or Aggregation Cost for all possible hypothesis disparities d (d=0→3 in Figure 12(c)). The decomposition in lines is taken care by Line Buffers in a similar way as explained in Figure 6, for each Cost Aggregation process, e.g. the Left-to-Right aggregation in Figure 12(a). Interestingly, the information for the Right-to-Left aggregation of Figure 12(b) is available at almost the same time instance; there are nevertheless slight time shifts, as suggested by Figure 12(c), compensated for by additional line-up buffers so that Left-to-Right and Right-to-Left results can be appropriately combined in the final decision taking.

Figure 13 shows the parallelized cost aggregation process for concurrently computing four disparity hypotheses (d=0→3) in more details. In essence, the process involves the Census Transformed images a and b from Figure 11 containing – as indicated in Figure 13 - Hamming distances (HammingDist) between Left and Right Census Transformed image frames. For each disparity value, a raw matching cost is calculated (a kind of difference between left and right pixels) for each pixel involved in the 2D correlation region of Figure 12(a). All raw matching costs in a 2D correlation region are also aggregated in parallel and a Winner-Takes-All (WTA) best match provides the answer to which disparity hypothesis is

Figure 13. Data path for calculating the final aggregation cost (best d), starting from the Left/right data of Figure 12(c)

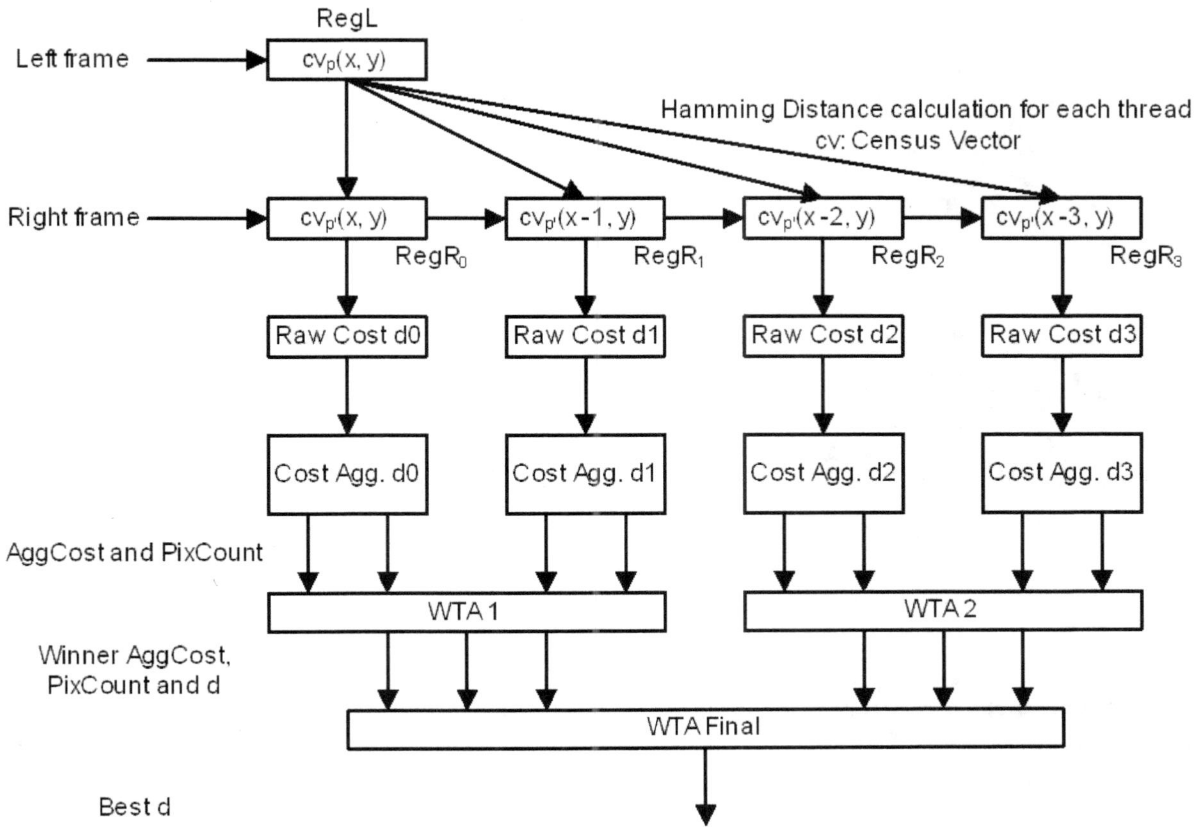

the one to be chosen as the final disparity for the given pixel *(x,y)*. This process is performed in scanline order with the appropriate line buffers pixel by pixel, such that for each new pixel that is read into the system, an output disparity value is provided with a minimal delay equal to the pipeline latency. Disparity values are hence also output in scanline order at the rate the images are input into the system.

For simple illustration, Figure 13 only shows 4 threads working in parallel, each associated with a disparity hypothesis. In principle, the proposed parallel architecture supports any number of parallel threads, depending on the maximum desired disparity. Our EP3SL150 FPGA implementation supports up to 64 parallel threads, which corresponds to evaluating 64 disparity hypotheses concurrently. In this case, the raw cost computing and cost aggregation modules are simply duplicated, but the WTA becomes a tree-like structure.

Figure 14 shows how the arithmetic complexity in the cost aggregation can be further reduced. First, instead of performing a pixel-by-pixel addition process over the 2D region around pixel *p* (Figure 14(a)), the 2D region is decomposed in a separable way into several horizontal stripes in which the summations are done according to Figure 14(b), followed by a vertical summation as shown in Figure 14(a)-(b-c). The horizontal summation is done with the help of integral computing, where the sum of all values from *a* to *b* in a horizontal stripe of Figure 14(b) can be calculated as the difference between the sum of values from 0 to *b* and the sum of values from 0 to *a*. These

Figure 14. Simplification of the Aggregation Cost calculation by (a) separable (2 x 1D) decomposition, (b) Integral Sum computations, and (c) division-free decision taking

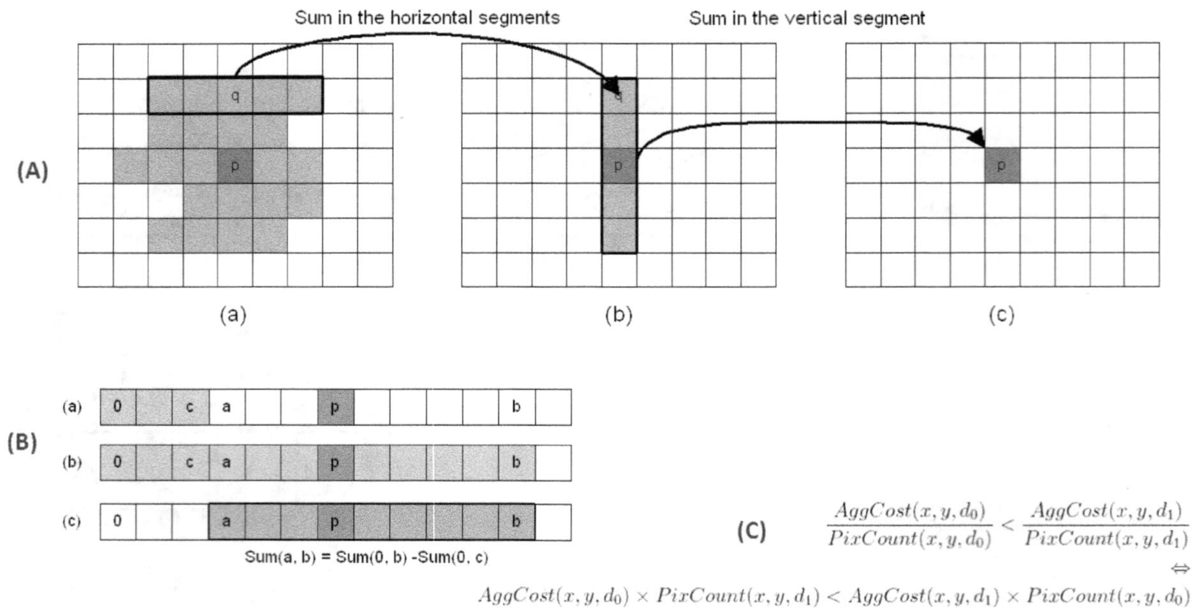

sums are pre-calculated, before actually starting the calculation process of Figure 14(b)-(c). More details can be found in (Zhang, 2009d).

A second trick to reduce the complexity consists in avoiding divisions, see Figure 14(c). Actually, since the support regions have an irregular shape (no square regions) which adapts to the region around the pixel under test (cfr. Adaptive Support Regions section), the number of involved pixels in the aggregation varies, and hence the calculation of the average aggregation cost involves in principle a division by the (varying) number of pixels. Fortunately, as observed in Figure 14(c), the mathematical relation for deciding between two candidate disparities d_0 and d_1 (with the averaging denominator PixCount) can be easily transformed in a way where the divisions are replaced by multiplications, hence avoiding complex arithmetic processes, difficult (or costly) to implement in hardware.

Finally, all DSP calculations of the proposed solution are floating-point operation free, which is a distinctive advantage over e.g. GPU implementations using floating-point operations by default.

Post-Processor

Starting from the output images of the stereo matcher (images a and b of Figure 15), a Left-Right (L-R) consistency check is performed validating the equality of the Left-to-Right and Right-to-Left disparities, and in case this is not satisfied, the corresponding disparity value is replaced by its closest valid disparity, using a 2D histogram voting scheme.

We so obtain the unique (combined Left-to-Right and Right-to-Left) depth map image c of Figure 15, which is already an acceptable depth map (less than 10% BPER), but still contains substantial amounts of horizontal stripes noise, which clearly reveals the (hardware-friendly) scanline processing nature of the algorithm. In order to achieve a higher quality depth map, a 2D histogram-based majority voting over the Adaptive Support Region of the pixel under test is

Figure 15. The Left-Right (L-R) consistency check (from (a,b) to (c)) and the simplified post-processing 2x 1D disparity voting with median filtering (d), compared to the full 2D disparity voting (d1 and d2)

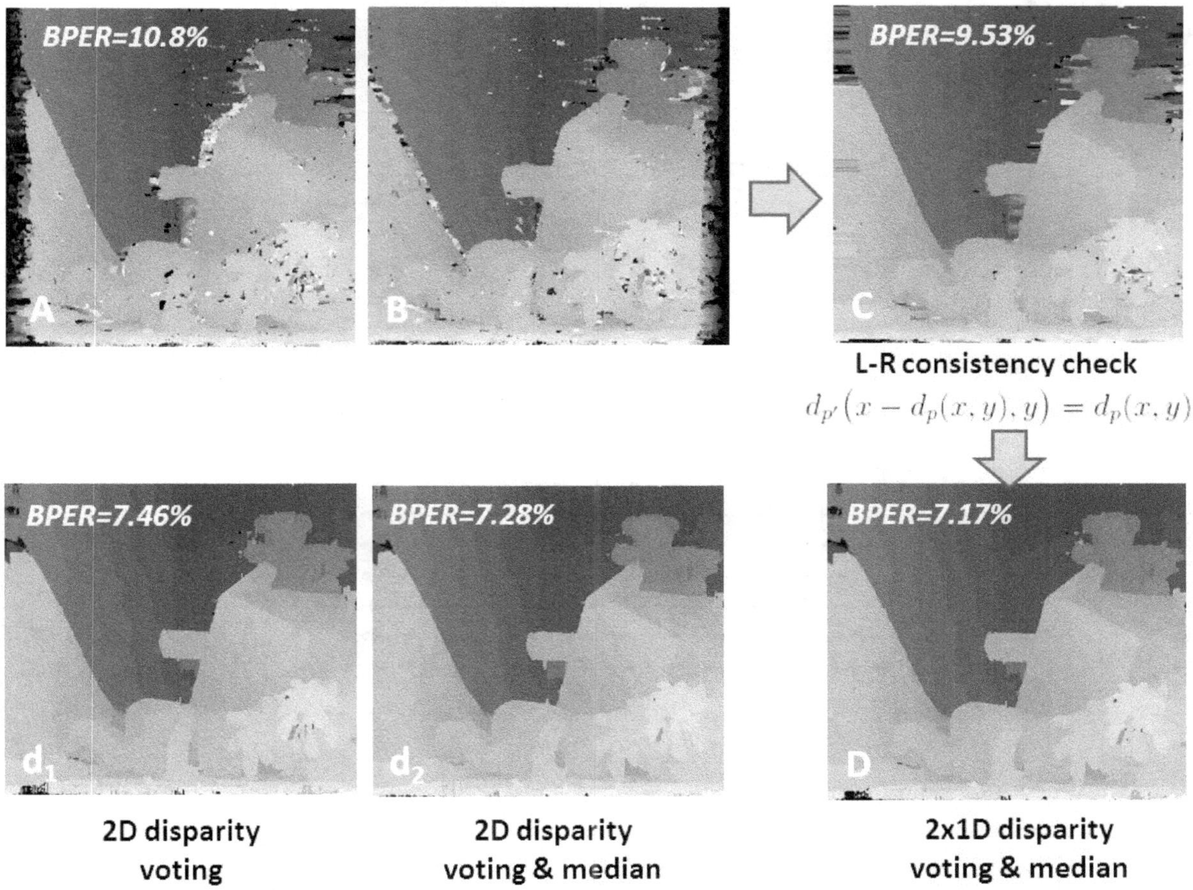

performed, replacing its disparity by the most often occurring disparity vector present in the support region of the pixel – see Figure 16 (a-b-c). An additional improvement can be obtained with median filtering, creating the image d_2 of Figure 15 from image d_1. However, in view of processing simplification, we have tested the idea of performing majority voting first on the horizontal lines in the support region, followed by an additional majority voting along the vertical line passing through the pixel under test – see Figure 16 (d-e-f) – providing an even better quality depth map, as illustrated in Figure 15(d). Since the quality is improved and the processing follows a separable pattern (horizontal versus vertical processing) similar to the one followed in the construction of the support region (cfr. the two-by-two armed, the cross-based pattern in Figure 7(a)), we have opted for this approach in reaching a better quality-complexity trade-off.

As a final remark, also observe that the original depth maps A and B from Figure 11 contain directional information (Left-to-Right vs. Right-to-Left), hence occlusion information is hidden in the differences between these depth maps: a pixel not visible in one of the views will create a different effect when comparing the Left/Right images in one or the other order. This is a wonderful opportunity to catch information about object edges in the image, and this information is

Figure 16. 2D histogram disparity voting (b) on 2D Adaptive Support Region (a) into the pixel under test (c). Its separable counterpart (d,e) does not necessarily provide the same histogram (f) and output results, but as shown in Figure 15, the final results are very competitive (even better) than with the direct 2D voting scheme.

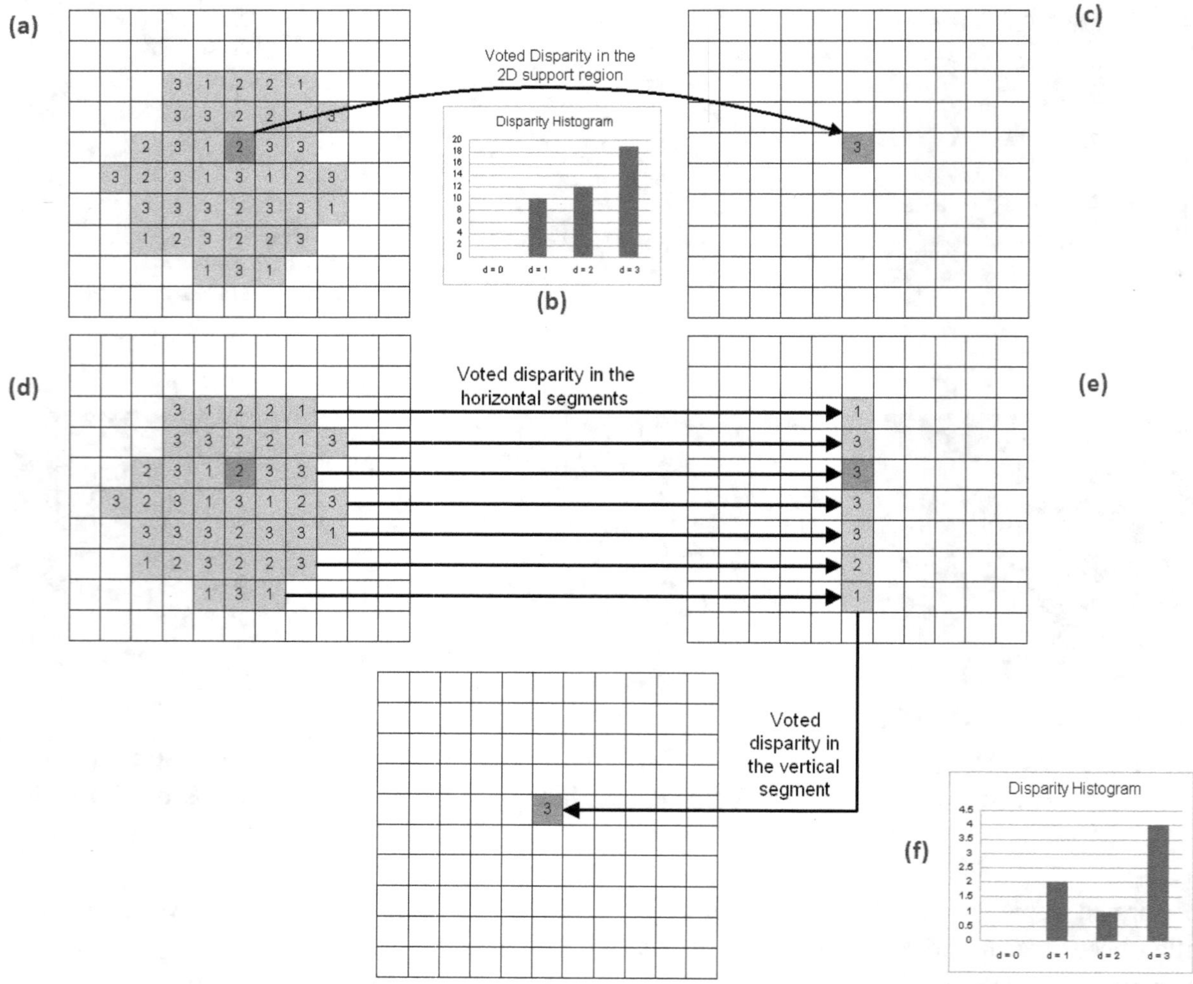

transmitted (after all post-processing steps) to the viewpoint interpolation module: pixels rendered in the viewpoint interpolated view, but occluded in one of the Left/Right image pair, should be extracted from the appropriate image, as suggested by the different colors in Figure 11(O). The reader is referred to (Lu, 2009b) for more information w.r.t. this occlusion holes filling process.

FPGA Implementation and Performance Figures

The proposed hardware architecture is scalable for implementing any number of parallel computing threads, each responsible for the computations involved in a specific disparity hypothesis d in the range ($0 \ldots d_{max}$). Over the processing pipeline shown in Figure 11, it is mainly the Stereo Matcher that can take benefit of the parallelization determined by the maximum allowed disparity

Figure 17. Hardware utilization of the FPGA resources on an Altera Stratix-III EP3SL150

	Combinational ALUTs		Memory ALUTs		Registers		DSP Blocks		SRAM Bits	
	Total: 113,600	Ratio	Total: 56,800	Ratio	Total: 113,600	Ratio	Total: 384	Ratio	Total: 5,630,976	Ratio
Pre-Processor	3,310	3%	288	1%	2,075	2%	0	0%	417,792	7%
Post-Processor	11,733	10%	536	1%	9,995	9%	0	0%	376,832	7%
Stereo-Matcher 16	8,477	7%	4,864	9%	18,743	16%	60	16%	589,824	10%
Stereo-Matcher 32	16,430	14%	9,728	17%	37,175	33%	124	32%	884,736	16%
Stereo-Matcher 64	32,329	28%	19,456	34%	74,039	65%	252	66%	1,474,560	26%
SoC 64	58,557	52%	20,288	36%	95,404	84%	256	67%	3,755,311	67%

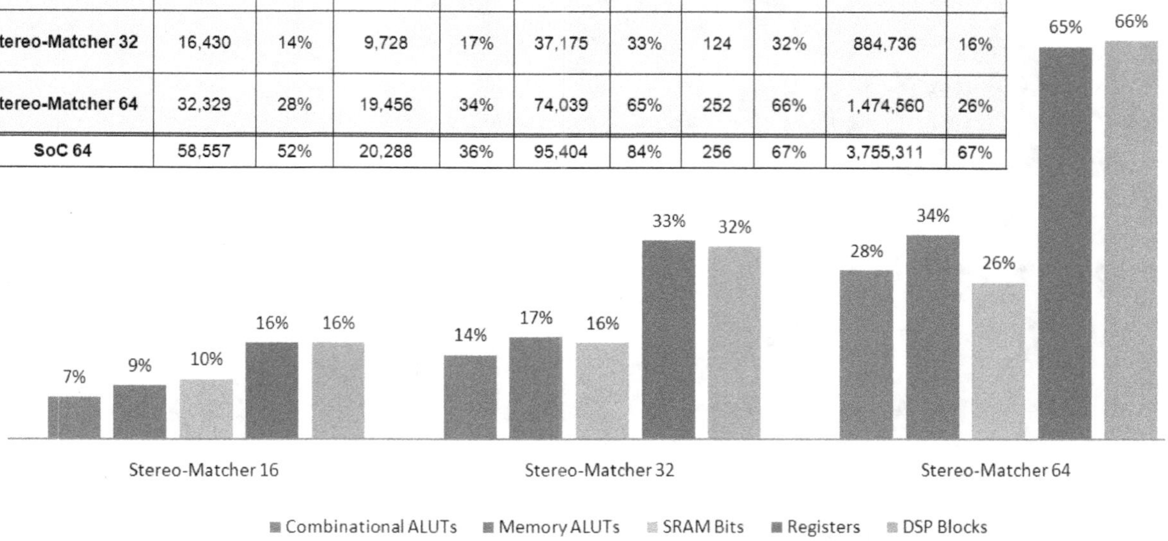

d_{max}; the Pre-Processor and Post-Processor are only slightly affected by d_{max}.

The actual parallelization is constrained by the available hardware resources on the FPGA. The EP3SL150 FPGA from Altera's Stratix III family was selected as the implementation and evaluation platform, on which we have successfully implemented 16, 32 and 64 parallel disparity computing threads, targeting different application scenarios.

Figure 17 gives an overview of the number of LUTs, SRAMs, etc. used by different processing kernels on the Stratix-III EP3SL150 FPGA, as well as the scalability of the Stereo-Matcher in different settings (d_{max} = 16, 32, 64) . The limiting factors in the current implementation are the required registers and hardware DSP blocks available on the FPGA. Balancing optimizations are possible by replacing some dedicated hardware DSPs with LUTs, and reducing the usage of pipeline registers.

To evaluate the performance of the complete stereo matching pipeline, additional logic components were implemented, creating a complete programmable system on FPGA, including the Nios-II soft processor, DMAs, DDR2-SDRAM controller, timers and JTAG UART communication cores, etc. FPGA resources utilized by the complete System-on-Chip (SoC) with 64 parallel threads are shown in Figure 17. The Nios-II CPU and stereo matching cores are all clocked at 100MHz.

Figure 18 shows the final FPGA performances in frames per second for four different image sizes (388x288 to 1024x768). The frame rates have been measured in two different settings:

- On-FPGA Frame rate where the data access to external DDR2 memory is disregarded, and pure computing performance

Figure 18. Performance figures (kCycles and Frame rate) of the pipeline of Figure 11 on Stratix-III EP3SL150 FPGA

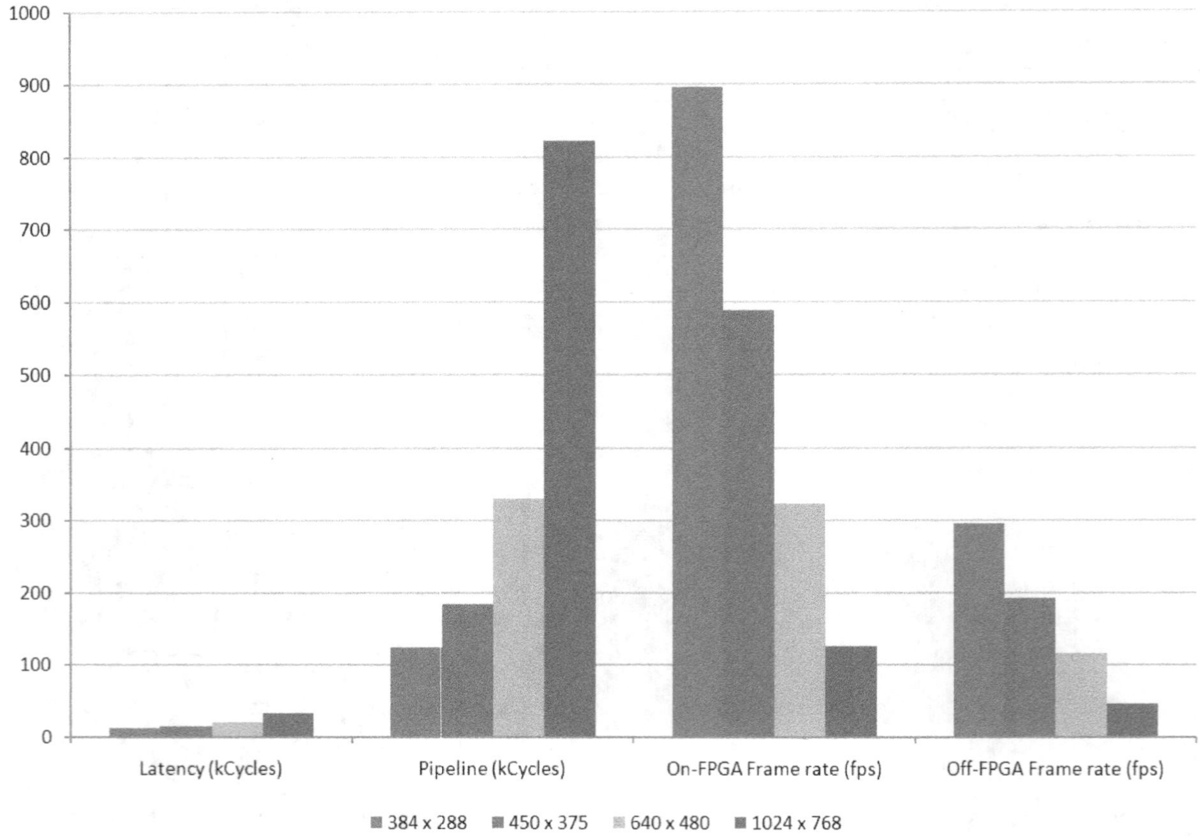

and associated frame rate is given by the relation 100MHz / (total pipeline cycles - pipeline latency cycles), corresponding to the first three regions of Figure 18. Frame rates up to 125 fps, resp. 900 fps for 1024x768 and 388x288 images are obtained.

- Off-FPGA Frame rate where the data access to external DDR2 memory is included (synchronization with on-board Nios-II processor), reducing the performances with by factor 2.5 to 3, but nevertheless still reaching 47 fps, resp. 296 fps on 1024x768 and 388x288 video streams.

The average power consumption over the full Stratix-III evaluation board was measured to be around 5-6W (including all peripheral drivers that could not be disabled/disconnected from the main FPGA board), which is in accordance with Altera's Quartus II tool power estimation, reporting – for the FPGA core only - a total thermal power dissipation of 4.8W (3.3W dynamic power, 0.7W static power, 0.8W I/O power). Based on (Kuon, 2007) providing power translation rules for logic, DSP and memory when migrating from FPGA to ASIC design, the power figures for an equivalent 90 nm ASIC design are estimated to be in the range of 370 mW to 510 mW. These power figures are comparable to state-of-the-art 90 nm stereo matching ASIC solutions, reporting e.g. 445 mW in (Liang, 2009) and 760mW to 1.2W in (Islam, 2008).

CONCLUSION

We have presented an algorithm-architecture co-design of viewpoint interpolation where a plausible in-between view is synthesized from the left/right image pairs of a stereo camera rig. A lot of attention has been devoted in fine-tuning the algorithms such that quality of the rendered images is conciliated with scarcity of the hardware resources, in combining floating-point operation-free DSP processing with limited scanline buffer memory. The stereo matching has been implemented and verified on an Altera EP3SL150 FPGA, processing over 100 fps stereo VGA video with DDR2 SDRAM frame access over DMA. With more dedicated video frame access protocols (e.g. direct camera data capture) overcoming some I/O synchronization issues, a performance increase with a factor 2.5 to 3 is expected. Compared to prior-art CPU and GPU implementations, a performance increase with one to two orders of magnitude is obtained, without the need of high clock speeds (100 MHz FPGA compared to a couple of GHz for CPU/GPU) and hence enabling low-power, embedded applications.

REFERENCES

Altera. (2009). Avalon interface specifications. Retrieved from http://www.altera.com/ literature/ manual/ mnl% 20avalon% 20spec.pdf

Chang, Y.-C., Tsai, T.-H., Hsu, P.-H., Chen, Y.-C., & Chang, T.-S. (2010). Algorithm and architecture of disparity estimation with mini-census adaptive support weight. *IEEE Transactions on Circuits and Systems for Video Technology*, *20*(6), 792–805. doi:10.1109/TCSVT.2010.2045814

Hirschmuller, H., & Scharstein, D. (2007), Evaluation of cost functions for stereo matching. *IEEE Conf. on Computer Vision and Pattern Recognition*.

Horst, J., Leeuwen, R., Broers, H., Kleihorst, R., & Jonker, P. (2006). A real-time stereo SmartCam, using FPGA, SIMD and VLIW. *Proc. 2nd Workshop on Applications of Computer Vision (Graz, May 12, Austria)*, (pp. 1-8).

Islam, J., Chun, P. W., MacLean, W. J., & Kirischian, L. (2008). Lowering power consumption using run-time reconfiguration for stereo rectification. *Canadian Conference on Electrical and Computer Engineering*, (pp. 1693-1698).

Kuon, I., & Rose, J. (2007). Measuring the gap between FPGAs and ASICs. *IEEE Transactions on Computer-Aided Design of Integrated Circuits and Systems*, *26*(2), 203–215. doi:10.1109/TCAD.2006.884574

Liang, C.-K., Chao-Chung Cheng, C.-C., Lai, Y.-C., Chen, L.-G., & Chen, H. H. (2009). Hardware-efficient belief propagation. *IEEE Conference on Computer Vision and Pattern Recognition*, (pp. 80-87).

Lu, J., Rogmans, S., Lafruit, G., & Catthoor, F. (2007). High-speed stream-centric dense stereo and view synthesis on graphics hardware. *IEEE International Workshop on Multimedia Signal Processing - MMSP*, (pp. 243-246).

Lu, J., Rogmans, S., Lafruit, G., & Catthoor, F. (2009b). Stream-centric stereo matching and view synthesis: A high-speed approach on GPUs. *IEEE Transactions on Circuits and Systems for Video Technology*, *19*(11), 1598–1611. doi:10.1109/TCSVT.2009.2026948

Lu, J., Zhang, K., Lafruit, G., & Catthoor, F. (2009a). Real-time stereo matching: A cross-based local approach. *IEEE International Conference on Acoustics, Speech, and Signal Processing - ICASSP*, (pp. 733-736).

Scharstein, D., & Szeliski, R. (2002). A taxonomy and evaluation of dense two-frame stereo correspondence algorithms. *International Journal of Computer Vision*, *47*(1), 7–42. doi:10.1023/A:1014573219977

Tombari, F., Mattoccia, S., & Stefano, L. (2008), Classification and evaluation of cost aggregation methods for stereo correspondence. *IEEE Conference on Computer Vision and Pattern Recognition*.

Woodfill, J., Gordon, G., & Buck, R. (2004). Tyzx DeepSea high speed stereo vision system. In *Proceedings of the Workshop on Real Time 3-D Sensors and Their Use, IEEE Conference on Computer Vision and Pattern Recognition*.

Yoon, K. J., & Kweon, S. (2006). Adaptive support-weight approach for correspondence search. *IEEE Transactions on Pattern Analysis and Machine Intelligence*, *28*(4), 650–656. doi:10.1109/TPAMI.2006.70

Zhang, K., Lu, J., & Lafruit, G. (2009d). Cross-based local stereo matching using orthogonal integral images. *IEEE Transactions on Circuits and Systems for Video Technology*, *19*(7), 1073–1079. doi:10.1109/TCSVT.2009.2020478

Zhang, K., Lu, J., Lafruit, G., Lauwereins, R., & Van Gool, L. (2009a). *Real-time accurate stereo with bitwise fast voting on CUDA*. 5th IEEE Workshop on Embedded Computer Vision.

Zhang, K., Lu, J., Lafruit, G., Lauwereins, R., & Van Gool, L. (2009b). Accurate and efficient stereo matching with robust piecewise voting. *IEEE International Conference on Multimedia & Expo - ICME*, (pp. 93-96).

Zhang, K., Lu, J., Lafruit, G., Lauwereins, R., & Van Gool, L. (2009c), Robust stereo matching with fast normalized cross-correlation over shape-adaptive regions. *IEEE International Conference on Image Processing - ICIP*, (pp 2357-2360).

Zhang, L., Zhang, K., Chang, T., Lafruit, G., Kuzmanov, G., & Verkest, D. (2011). *Real-time high-definition stereo matching on FPGA*. Nineteenth ACM/SIGDA International Symposium on Field-Programmable Gate Arrays.

Zitnick, C. L., Kang, S. B., Uyttendaele, M., Winder, S., & Szeliski, R. (2004). *High-quality video view interpolation using a layered representation*. SIGGRAPH.

KEY TERMS AND DEFINITIONS

ASIC: Application Specific Integrated Circuit.

ASR: Adaptive Support Regions.

BPER: Bad Pixel Error Rate.

Census Transform: Local Transform mapping pixel values within a square region to a bit string.

DDR: Double Data Rate – SDRAM with data transfer on rising and falling edges of the clock.

DDR2: next generation DDR with speed improvements thanks to technology tricks on the chip die itself.

DMA: Direct Memory Access.

DSP: Digital Signal Processing.

FPGA: Field Programmable Gate Array.

HD: High Definition.

LDT: Luminance Difference Threshold.

MDE/s: Million Disparity Estimations per second.

NCC: Normalized Cross-Correlation.

SAD: Sum of Absolute Differences.

SDRAM: Dynamic Random Access Memory.

SoC: System-on-Chip.

SRAM: Static Random Access Memory.

SVGA: Super Video Graphics Adapter, 800x600 pixels.

VGA: Video Graphics Adapter, 640x480 pixels.

WTA: Winner-Takes-All.

Chapter 17
The Use of Watermarking in Stereo Imaging

Dinu Coltuc
Valahia University Targoviste, Romania

ABSTRACT

The manipulation and processing of stereo image sequences demand higher costs in memory storage, transmission bandwidth, and computational complexity than of monoscopic images. This chapter investigates scenarios for cost reduction by using reversible watermarking. The basic principle is to embed some data by reversible watermarking instead of either computing or storing/transmitting it. Storage and/or bandwidth are reduced by embedding into one frame of a stereo pair the information needed to recover the other frame. Computational complexity is reduced by embedding the disparity map. The cost of extracting the embedded disparity map is considerably lower than the one of computing it. Experimental results are provided.

INTRODUCTION

Watermarking is the imperceptible hiding of information about a certain digital work (image, sound, text, etc.) within that work itself. Watermarking is notoriously used to increase data security (by copyrighting, fingerprinting, authentication). Besides the increase of security, watermarking is also used to provide a transmission channel associated with some given data. The watermark becomes annotation data whose insertion in a host is meant to increase the value of the host. The range of possible applications of annotation watermarking is large. For instance, information related to the content of images can be embedded to facilitate content-based indexing, retrieval and manipulation of digital images and image regions. As for any other kind of images, copyrighting, fingerprinting, authentication and annotation are of interest for stereo images, too.

DOI: 10.4018/978-1-61350-326-3.ch017

This chapter investigates the use of reversible watermarking in certain annotation applications meant to reduce costs associated to stereo image manipulation or processing. These costs refer to storage, bandwidth or computational effort. We remind that stereo images are sequences of two or more images of the same scene taken from slightly different viewpoints. Obviously, compared with monochrome images, stereo images demand more space for storing and a larger bandwidth for transmitting. Furthermore, three-dimensional (3D) information is usually recovered from two-dimensional (2D) images by matching features in scenes observed from different viewpoints. The cost of the 3D information, besides storage or transmission, demands computational effort as well.

The basic principle of our approach is to embed data, by reversible watermarking, into stereo images. First of all, a scheme of high capacity reversible watermarking is introduced. The use of reversible watermarking in stereo is motivated by two aspects: reversibility and capacity. The reversibility allows, after the extraction of the embedded data, the recovery of the original image without any distortion. Nowadays reversible watermarking algorithms provide enough embedding capacity to deal with real applications. Thus, for natural graylevel images, the embedding bit-rate is of 2-3 bits per pixel.

The reversible watermarking is fragile. This means that the embedded data is lost even in case of simple image processing tasks as histogram modification, contrast enhancement, filtering, image resizing, lossy compression, etc. On the other side, the embedded information adds some more value to the data and the user has no interest to destroy it. Therefore, if necessary, such image processing tasks should be performed only after the embedded data has been extracted.

As an immediate application of reversible watermarking in stereo imaging, the embedding of the disparity map into stereo images is discussed. Thus, ground truth disparity maps are embedded directly into stereo images, instead of being stored and transmitted as additional files. Furthermore, the disparity map is crucial for computing 3D information. Instead of either computing or storing/transmitting the disparity map each time the 3D information is needed, it is simply extracted from the embedded watermark. The mathematical complexity of reversible watermark embedding and extraction is considerably lower than the one of computing the disparity map. The quality of the embedded images is good. Furthermore, the reversible watermarking can be reverted in order to exactly recover the stereo images. The embedding of the disparity map by reversible watermarking is discussed also in Khan et al, 2009, Ali & Khan, 2009.

The main contribution of the chapter is storage/bandwidth reduction for stereo images. For a pair of stereo image frames, the basic principle is to embed into one frame the information needed to recover the other frame. Thus, only one image is stored or transmitted and its content is directly accessible. The bandwidth and the storage size are halved, i.e., they are reduced to the ones of monoscopic images. When the stereo context is needed, the embedded data is extracted and the second image of the pair is recovered. The reversible watermarking allows the exact recovery of one image frame. The quality of the other frame depends on two features: the size of the information needed to be embedded and the distortion introduced by the watermarking.

The chapter ends with a general discussion on the use of watermarking in stereo imaging. The strengths and weaknesses of reversible watermarking for complexity or storage/bandwidth reduction are presented. Further extensions are investigated, as well.

BACKGROUND

The major issue of this chapter is storage/bandwidth reduction for stereo image pairs. This problem can be directly approached by compression. There is, indeed, much research on efficient stereo image compression. The direct compression of each image frame does not exploit the high correlation between stereo frames. Most of the work is focused on disparity estimation and disparity compensation (DE/DC) schemes. The basic DE/DC principle is to use one of the images in the stereo pair as a reference and to estimate the other image by using disparity (Dinstein et al., 1989). The reference, the disparity and the residual error are further encoded. Besides DE/EC schemes, other schemes have been considered as well.

While size and bandwidth are considerably reduced, the compression-based methods have the drawback that, once compressed, the content of the stereo pair is not visible anymore. Hence, for any simple task (as the mere visualization) stereo images should be first decompressed. In order to eliminate this drawback, we proposed a watermarking based approach, (Coltuc, 2007), where the bandwidth and the size are halved, i.e., they are reduced to the ones of monoscopic images. The basic idea is to embed into one image, by reversible watermarking, the information needed to recover the other one. Thus, only one image is stored or transmitted and its content is directly accessible. When the stereo context is needed, the embedded data is extracted and the second image of the pair is recovered. By storing/transmitting only one image frame, one has stereo imagining at the same expense in memory or bandwidth as for monoscopic imaging.

As stated in the introduction, the range of annotation applications is large. The embedding of stereo by watermarking is such an annotation application. A rather similar application is the embedding of the chrominance components into image luminance. Only luminance is transmitted and the graylevel image is directly available (Campisi et al., 2002). Obviously, the chrominance information should be extracted before color displaying. Another application is the transmission of the audio data hidden within the video sequence (Swanson et al., 1997). Such applications where data hiding is used to improve coding efficiency are known as compressive data hiding (Campisi et al., 2002). Let us consider two more examples (Cox et al., 2008). Video and audio channels of a television signal are processed separately and synchronization may be lost. A typical example is when the motion of the lips is either ahead or behind the speech (lip-sync). Tektronix's digital watermark encoder for synchronizing audio and video signals embeds a highly compressed version of the audio signal into the video signal. When all signal processing is completed, the audio signal is compared to the embedded one. Thus, any time delay is detected and removed prior to broadcasting. Similarly, a solution to synchronize the display of the lyrics with the music for some MP3 players is to embed the lyrics directly into the audio signal by using watermarking.

For compressive data hiding, the stringent requirement is watermarking capacity. Robustness (usually demanded in watermarking) is not an issue. One can suppose that nobody is interested to destroy, remove or replace the embedded data. Furthermore, the users are aware that usual image processing (lossy compression, image enhancement, etc.,) can destroy the embedded information. Otherwise stated, if the user is interested to take advantage of the embedded information, no modifications (neither signal processing, nor intentional attacks) are expected. Otherwise, the embedded information can be lost.

An original aspect of our approach is the use of reversible watermarking for compressive data hiding. We remind that reversible watermarking appeared to extend the use of watermarking for special domains, like military, legal, medicine, etc., where no data distortion is admitted. For such domains, the imperceptibility of the embedding is not enough. Reversible watermarking allows at detection not only the extraction of the embedded data without any loss, but also the exact recovery of the original host image. Nowadays, the development of high capacity reversible watermarking schemes provides the context of a change of paradigm. In our opinion, the reversible watermarking can be used instead of the classical one in most of the applications. The advantage is that, in certain conditions, one can get reversibility as a byproduct of the marking.

Such an example is the distortion-free robust watermarking paradigm, (Friedrich et al., 2002), which joins the reversible and the robust watermarking. If the embedding bit-rate is high enough, almost any degree of robustness can be added to the reversible watermarking schemes by multiple marking (Coltuc & Chassery, 2007). Robust watermarking is the first one to be performed. Then, by reversible watermarking, the information needed to invert both the robust and the reversible watermarking is embedded. In case of no attacks, the robust watermark is detected and the authorized party exactly recovers the original. In case of attacks, one can suppose that the robust watermark can still be detected, but the reversibility is lost. The robust authentication problem can be approached in a similar way. To conclude, for the case of compressive hiding, the robustness is not a must. On the contrary, a high capacity is necessary. The capacity requirement is fulfilled by nowadays reversible watermarking schemes. For natural images, currently high capacity reversible watermarking algorithms provide bit-rates greater than 2 bits per pixel.

STEREO EMBEDDING

As said above, the basic idea of our approach is to embed data by reversible watermarking in order to save storage space, bandwidth or computational complexity. Before investigating the use of watermarking in stereo imaging, a high capacity reversible watermarking scheme is briefly introduced.

Reversible Watermarking Scheme

The embedding capacity is the main issue of the reversible watermarking scheme. The highest capacity reversible watermarking schemes proposed so far are the ones based on difference expansion, DE, (Tian, 2003). The basic idea of DE schemes is to create space for data embedding by expanding the difference either between pairs of pixels, or between pixels and their estimates. For instance, by expanding two times the difference, its least significant bit is freed and a bit of data can be embedded. In the following paragraphs we will focus on a version of the high capacity reversible watermarking scheme of Coltuc & Chassery, 2007. The scheme can provide more than 1 bpp in a single embedding level.

Let image pixels be indexed, for instance, on rows, from left to right and from top to bottom. Let x_i and x_{i+1} be the graylevels of two consecutive pixels. Let n be a fixed integer, $n \geq 2$, and let w be an integer in $[0, n-1]$. Let us further replace x_i by X_i:

$$X_i = x_i + (n-1)(x_i - x_{i+1}) + w \tag{1}$$

The replacement is done if no overflow or underflow appears, i.e., for 8 bit graylevel images, $0 \leq X_i \leq 255$. By replacing x_i with X_i, the difference D_i between the consecutive pixels located at i and $i+1$ becomes: $D_i = X_i - x_{i+1} = n(x_i - x_{i+1}) + w$. Since $D_i - w$ is divisible by n, the embedded codeword w can be simply recovered by taking:

$$w = D_i \bmod n \tag{2}$$

Next, once w is available, the original graylevel x_i follows as:

$$x_i = \frac{X_i + (n-1)x_{i+1} - w}{n} \tag{3}$$

The equations (1), (2) and (3) provide the framework for reversible data embedding. Once a pair of pixels is transformed and embedded with equation (1), the next pair is processed and so on. From equation (3), it clearly appears that, in order to recover the original graylevel x_i, the original graylevel of the consecutive pixel, x_{i+1}, is needed. Therefore, the detection and the data extraction should proceed in reverse scanning order.

At detection, for each pixel, one should know if the pixel was transformed or not. The classical solution is to use a lossless compressed location map. The drawback of using a location map is an increase in mathematical complexity because of the lossless compression stage. Lossless compression can be avoided by using the divisibility with n introduced by the transform. The idea is simple. An integer code r is reserved to indicate if a pixel was not transformed. The not transformed pixels are modified (by simply subtracting or adding some correction codes) to provide the result r for equation (2). Let us suppose that a pair (x_i, x_{i+1}) does not fulfill the conditions to be transformed. Then, a correction code c_i should be subtracted from x_i such that the result of equation (2) is r. It immediately appears that c_i should be computed as:

$$c_i = r - (x_i - x_{i+1}) \bmod n \tag{4}$$

In order to recover the original pixel values, the correction codes should be stored together with the payload. For simplicity, let $r = n-1$. Since $n-1$ is reserved, the watermark codewords are limited to [0, $n-2$]. The information provided by the location map is replaced by the result of equation (2): $n-1$ for a not transformed pixel or an integer code in [0, $n-2$], otherwise. The scheme does not need lossless compression, but there is a certain loss in embedding capacity by reducing the range of the watermark codewords.

The embedding capacity of the proposed scheme depends on the number of transformed pixels (i.e., on image statistics) and on the parameter n. The theoretical upper bound of the embedding capacity provided by such a scheme is $\log_2 n$ bpp (for location map based detection) or $\log_2(n-1)$ for divisibility based detection. For $n \le 4$ the scheme operates close to the theoretical upper bound. By increasing n, the number of pairs subject to overflow/underflow increases as well and the difference with respect to the theoretical upper bound increases. For natural images, the maximum embedding capacity is obtained, depending on image content, for $8 \le n \le 15$. The further increase of n does not provide any improvement: the decrease of the number of transformed pixels is more significant than the increase of the number of bits of the corresponding codewords.

In order to match the capacity of the scheme with the one demanded by the application at hand, a simple threshold control scheme can be used. Thus, the pixel at the location i is transformed not only if no overflow/underflow appears, but also if $x_i - x_{i+1}$ is less than a certain threshold. By limiting the difference, the distortion is limited, too.

Let us next consider the test stereo images *Art* and *Dolls* shown in Figure 1 (available at http://vision.middlebury.edu/stereo/data/). The test images consist of 7 rectified views taken from equidistant points along a line (Scharstein &Pal, 2005). Each stereo pair was composed by taking frames 2 and 6. The images are of high quality, full-color (24 bits), 1300 × 1100 pixels (cropped to the overlapping field of view).

The reversible watermarking scheme was introduced for graylevel images. For the case of color images in format RGB, a simple solution is to separately embed each color plane. As discussed above, the parameters n and the threshold control the capacity and the distortion of the watermarking. The experimental results for the left frame of the two test images, namely PSNR with respect to capacity, are plotted in Figure 2. The PSNR measures the distortion introduced by the watermarking. Greater the PSNR, higher the quality of the image. Together with the results for the three color planes, the results for the graylevel versions of the test images are provided as well. The graylevel version, $I(x,y)$, has been computed as:

$$I(x,y)=0.2989R(x,y)+0.5870G(x,y) +0.1140\, B(x,y) \tag{5}$$

Figure 1. Test image pairs: Art (top) and Dolls (bottom) of http://vision.middlebury.edu/stereo/data/

Figure 2. PSNR with respect to embedding capacity by threshold control reversible watermarking scheme on red, green, blue and graylevel of Art (a) and Dolls (b)

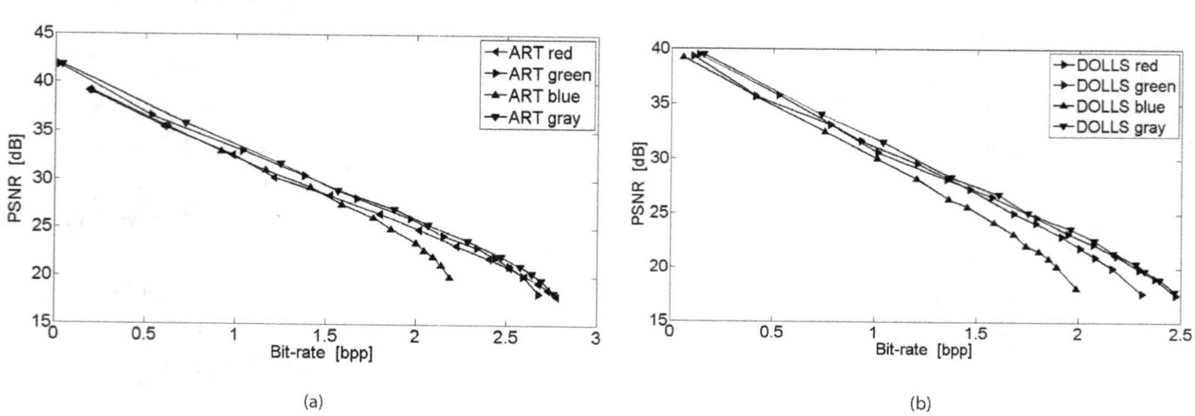

where $R(x,y)$, $G(x,y)$ and $B(x,y)$ are the red, green and blue plane, respectively.

As it can be seen from Figure 2, the shape of the curves is quite similar. For both images, the embedding bit-rates are greater than 2 bpp. The results are slightly better for the test image *Art*. This is due to the fact that there are less details (equivalently, larger uniform areas) in *Art* than in *Dolls*.

Embedding Disparity Maps

In some cases, disparity maps are provided together with stereo image pairs. For instance, for the test images of Figure 1, the authors also offer ground truth disparity maps obtained by using the structured lighting technique described in Scharstein & Szeliski, 2003. Ground truth data are of great interest in benchmarking the performance of different stereo matching techniques.

The ground truth disparity maps for *Art* and *Dolls* test images are shown in Figure 3. They are provided as additional image files. For instance, the maps of Figure 3 are given as PNG (Portable Network Graphics) of sizes 108,186 bytes (Art left map), 107,927 bytes (*Art*, right map), 158,234 bytes (*Dolls*, left map) and 152,410 bytes (*Dolls*, right map). Instead of keeping ground truth disparity maps in additional files, they can be directly embedded into the corresponding images.

Figure 2 clearly shows that the reversible watermarking provides enough capacity to embed the disparity maps. In order to embed the left disparity map of *Art* into its left frame one should have a bit-rate of 0.56 bits per color pixel, i.e., an embedding bit-rate of less than 0.2 bpp for each color frame. We have inserted the left ground truth disparity map by embedding 0.19 bpp into the red plane, 0.27 bpp into the green plane and 0.2 bpp into the blue plane. The PSNRs of the embedded color planes are 39.10 dB, 39.39 dB and 39.04 dB. The embedding is completely imperceptible. To conclude, the average PSNR is greater than 39 dB for an embedding bit-rate of 0.66 bits per color pixel. Besides the disparity map, there are more 18,66 Kbytes available for embedding additional data. The embedding of the right frame of *Art* image gives similar results.

The embedding of the disparity maps of *Dolls* in the corresponding image frames demands a slightly larger bit-rate. For instance, for the left disparity map one needs 0.82 bits per color pixel, i.e., less than 0.3 bpp for each color plane. One can easily obtain this global bit-rate by embedding 0.27 bpp into the red plane at a PSNR of 37.42 dB, 0.39 bpp at 36.21 dB into the green plane and 0.22 bpp at 37.32 dB into the blue plane. One gets a global bit-rate of 0.88 bits per color pixel at an average PSNR of 36.98 dB. The situation is similar for the embedding of the right ground truth disparity map.

The embedding of both ground truth disparity maps into their corresponding color frames is imperceptible. For graylevel images, since the entire embedding bit-rate should be ensured by a single image plane, the distortion becomes visible. In Figure 4 left, the result of the embedding of the entire ground truth disparity map into the graylevel version of *Dolls* image is presented. Compared with the original (center), the image looks noisy, but there are not any annoying artifacts. In order to improve the visibility, only a 256 x 256 region of *Dolls* is shown in Figure 4. Meantime, it should be stressed that the embedding was done by reversible watermarking. Therefore, the original image can be recovered at zero distortions from the marked copy.

If only one map is embedded into both image frames, the embedding bit-rate is halved and the distortion decreases accordingly. See in Figure 4 right, the result of inserting into the graylevel version at half of the embedding bit-rate. The resulted image is less noisy than the one of Figure 4, left. Even if the ground truth is not available, the disparity map can be computed and embedded. Then, every time the 3D information is needed, the disparity map is simply extracted.

Figure 3. Ground truth disparity maps for Art (top) and Dolls (bottom)

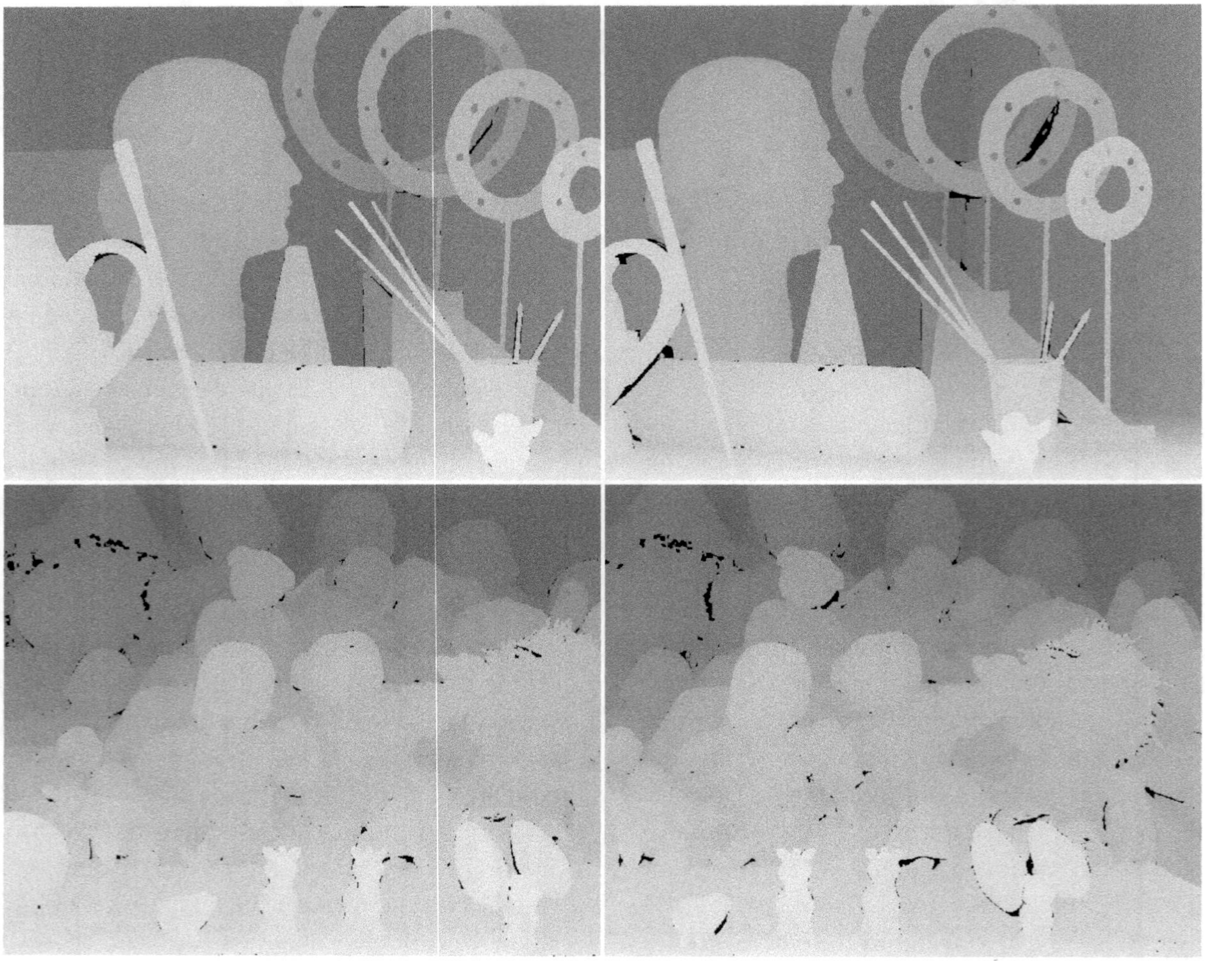

Figure 4. Original (center) and results of ground truth disparity map embedding into a single graylevel frame (left), into both frames (right)

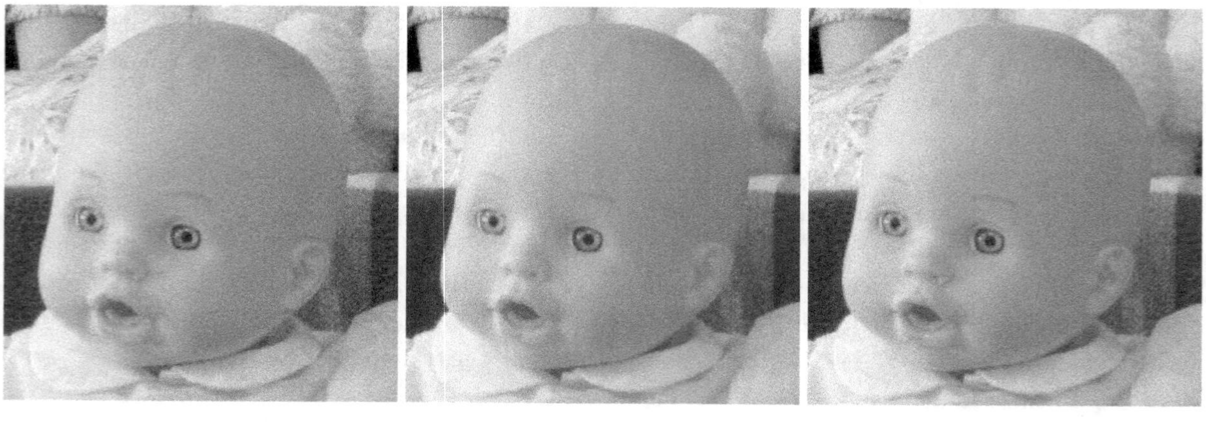

By inserting the ground truth disparity map into the corresponding image files, the need of supplementary files is eliminated. The disparity maps are linked with the corresponding image data. The management of the information is simplified and the risk of any data mix up disappears. Besides the elegance of the approach, it should be also mentioned that no major artifacts are introduced and, the reversibility of the embedding allows the exact recovery of the original stereo images.

The stereo matching problem is of high mathematical complexity. The embedding of the disparity map into the stereo images reduces the mathematical complexity of 3D computation without the burden of transmitting additional files. Obviously, the extraction of the embedded disparity map is by far less mathematically complex than its computation.

Stereo Embedding: Halving Storage/ Bandwidth for Stereo Images

Let S_l and S_r be the left and right frames of a stereo image pair. The main idea of the stereo embedding by reversible watermarking is to hide into one image (e.g., the left frame S_l), all the information needed to recover the other frame, S_r. Furthermore, S_l is stored or transmitted. The content of S_l is continuously available for displaying or for any other tasks. When the stereo content is needed, the information embedded into S_l is extracted and S_r is recovered. The size of the information to be hidden is crucial for this approach. A straightforward solution is to embed the compressed residual between the two stereo frames, $R=S_l - S_r$. Depending on the embedding capacity, the residual can be lossless or lossy compressed. This solution is simple, but of rather limited usefulness. We have tested it on generated stereo images (Coltuc, 2007).

An efficient solution is to consider a disparity compensation scheme. The disparity represents the difference in position between corresponding points in left and right frames. In the hypothesis that the stereo frames are rectified, the points should lie on the same row. By establishing the correspondence for all the pixels of a frame, a dense disparity map is obtained. There is a huge literature on disparity computation (Scharstein & Szeliski, 2002). In the sequel, we use the sum of absolute differences (*SAD*). For each pixel of the left frame, $S_l(x,y)$, the *SAD* is computed to the pixels of the left frame:

$$SAD(x,y,d) = \sum_{-w \le u,v \le w} |\, S_l(x+u, y+v) - S_r(x+u, y+v+d)\,| \tag{6}$$

where the search window is of size $(2w+1)$ x$(2w+1)$. Then, the disparity map is:

$$D(x,y) = \arg\min_d SAD(x,y,d) \tag{7}$$

Since equation (6) establishes the correspondence between the pixels $S_l(x,y)$ and $S_r(x,y+D(x,y))$, one can estimate the right frame as:

$$\hat{S}_r(x,y) = S_l(x, y + D(x,y)) \tag{8}$$

Finally, the estimation error is:

$$E(x,y) = S_r - \hat{S}_r(x,y) \tag{9}$$

The right frame is exactly recovered if the estimated right frame and the estimation error are available. Furthermore, if the left frame and the disparity are available, the right frame is estimated by equation (8). Therefore, as soon as one can embed into the left frame the disparity map and the residual error, one has enough information to recover exactly the left frame. We remind that the reversible watermarking is invertible, i.e., not only the embedded data is exactly extracted, but also the cover image is exactly recovered. The only problem is to ensure the embedding capacity needed to store the required additional information. In order to reduce the necessary embedding bit-rate, the disparity map and the estimation error are compressed. Depending on the embedding bit-rate provided by the reversible watermarking, either lossless or lossy compression is used.

The above scheme can be immediately extended to color images. In the case of color images, a single disparity map is computed. Each color plane of the right frame is estimated by using the corresponding color plane of the left frame and the disparity map. Then, the estimation error is computed. The information to be embedded into the left frame consists of the disparity map and the estimation errors for the three color planes. Therefore, into each color channel of the left frame, besides the corresponding estimation error, only approximately one third of the disparity map should be embedded. It should be observed that, for graylevel images, the entire disparity map and the estimation error should be embedded into one graylevel frame. Therefore, better performances should be expected for color images.

The embedding of disparity map and estimation errors could have been done by classical watermarking. For instance, the classical watermarking by LSB substitution ensures high embedding capacity at very low distortion. However, it should be observed that the exact recovery of left frame color channels is possible only if the corresponding color channels of the right frame and their estimation errors are available. The embedding by reversible watermarking allows, at detection, the exact recovery of the cover image.

The size of the disparity map and the three estimation error planes is too big to be embedded without lossy compression. We have used JPEG compression. The compression ratio is controlled by four parameters (one controls the quality of the disparity map and the others three control the quality of the estimation error planes). The parameters are tuned in order to match the available embedding capacity provided by the reversible watermarking stage. Higher the compression ratio, lower the quality of the recovered right frame. The PSNRs of the reconstructed left frame in function of the size of the data to be embedded (bits per color pixel) for *Art* and *Dolls* test images are plotted Figure 5. The PSNR of the color image is computed as the average of the PSNRs of the three color planes. The quality of the recovered frame is slightly better for *Dolls* image. Given the high values of the PSNRs, even at low embedding bit-rates, the recovered left frames are of very good quality. No visual artifacts are present.

In order to evaluate the quality of the embedded frame, an example is given in Figure 6. The left image represents a detail of *Dolls* test image embedded at 3 bits per color pixel (i.e., 1 bpp for each color plane) and the right image represents the same region embedded at 6 bits per color pixel. The original region is shown in center. As it can be seen, at low-embedding bit-rates the embedded image looks identical to the original. At high embedding bit-rates, an increase of contrast is visible and the image looks noisy. However, this is not a major problem since at detection the left frames can be exactly recovered.

The disparity compensation scheme offers at detection, as a byproduct, the disparity map. Sometimes, instead of recovering the right frame, extracting only of the disparity map may be sufficient. This is the case of extracting 3D information from stereo images. The direct recovery of the disparity map means a great saving in computational complexity. As discussed above, the extraction of the disparity map is by far less expensive than its computation.

Solutions and Recommendations

In order to match the embedding bit-rate provided by the image at hand, or to simply reduce the distortions of the embedded frame, lossy compression is used. A good policy is to compress at higher rates the estimation errors and to compress at lower rates or even lossless the disparity map. As

Figure 5. Experimental results: PSNR of the recovered right frame

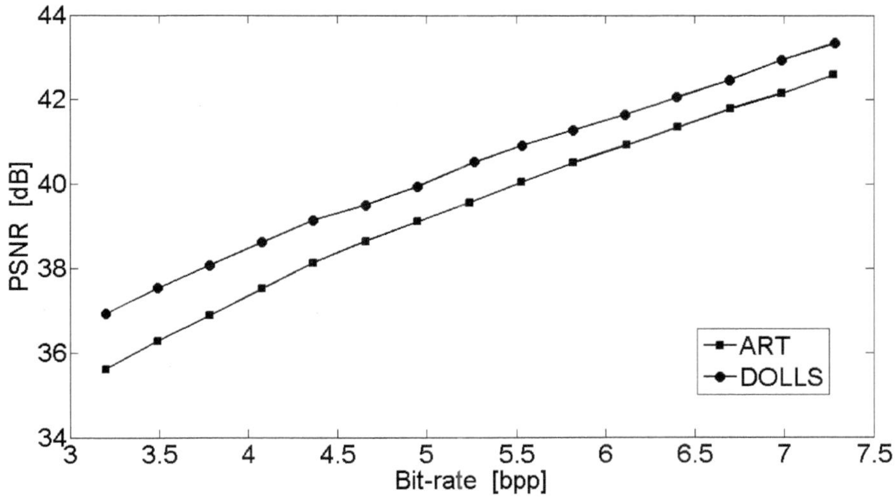

Figure 6. Detail of embedded left frame of Dolls at 3 bpp (left) and 6 bpp (right) and non-embedded detail (center)

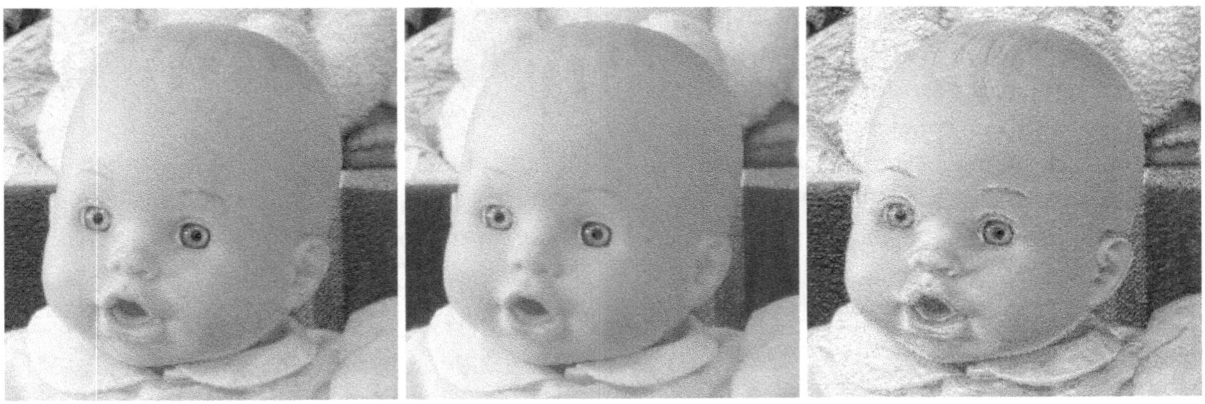

discussed above, the disparity map is of interest beyond the recovery of the right frame.

FUTURE RESEARCH DIRECTIONS

We have discussed the halving of storage or bandwidth for stereo images. An immediate extension of this approach is for stereo sequences. The embedding of left frame sequences with the data for recovering the right frames immediately reduces at half the amount of data, while keeping the sequence visible. An interesting extension for the case of image sequences is to exploit the high redundancy existing among the left frames to further reduce the size of the stored transmitted data keeping the sequence content visible.

We discussed the embedding of the ground truth disparity maps (or the computed disparity maps) into stereo pairs or the embedding into one frame of the information needed to recover the other frame. Obviously, the increasing of the amount of embedded data increases the distortions as well. Meantime, since the embedding was done by reversible watermarking, these distortions are removed at detection. In this context, it is interesting to investigate the jointly embedding of more data to extend the scope of the applications discussed above.

CONCLUSION

The embedding of stereo information by reversible watermarking has been investigated. First, the embedding of disparity maps has been discussed. This eliminates the need of storing and manipulating additional files. Meantime, at detection, the disparity map is provided at a low mathematical complexity cost, namely the cost of watermark extraction.

Next, by embedding more data, not only the disparity map, but also the estimation error, the bandwidth and storage requirements when operating with stereo images are halved, i.e., only the watermarked frame is stored and transmitted. Compared with the stereo image compression, the proposed approach has the advantage that image content remains available during image manipulation. When the stereo context is needed, the embedded data is extracted and the other frame is recovered. The quality of the recovered frame depends on the hiding bit-rate provided by the reversible watermarking. The proposed approach can be extended to stereo sequences.

Another original aspect of our approach is the use of reversible watermarking. This ensures the exact recovery of the watermarked frames. Nowadays reversible watermarking algorithms provide enough embedding capacity for this kind of applications.

ACKNOWLEDGMENT

This work is supported by UEFISCU, Romania, Grant ID 2200/2008.

REFERENCES

Ali, A., & Khan, A. (2009). Reversible watermarking for 3D cameras: Hidden depth maps . In Grgic, M. (Eds.), *Recent advances in multiple signal processes and communications* (pp. 495–521). Berlin, Germany: Springer-Verlag. doi:10.1007/978-3-642-02900-4_19

Campisi, P., Kundur, D., Hatzinakos, D., & Neri, A. (2002). Compressive data hiding: An unconventional approach for improved color image coding. *EURASIP Journal on Applied Signal Processing*, (2): 152–163. doi:10.1155/S1110865702000550

Coltuc, D. (2007). On stereo embedding by reversible watermarking. *International Symposium on Signals, Circuits and Systems, ISSCS'07,* (pp. 93-96).

Coltuc, D. (2007). Improved capacity reversible watermarking. *Proceedings of the IEEE International Conference on Image Processing,* (pp. 249-252).

Coltuc, D., & Caciula, I. (2009). Stereo embedding by reversible watermarking: Further results. *International Symposium on Signals, Circuits and Systems, ISSCS'09,* (pp. 121-124).

Coltuc, D., & Chassery, J.-M. (2007). Distortion-free robust watermarking: A case study. *Proceedings of SPIE: Security, Steganography, and Watermarking of Multimedia Contents IX, 6505*, 739–746.

Cox, I. J., Miller, M. L., Bloom, J. A., Fridrich, J., & Kalker, T. (2008). *Digital watermarking and steganography*. Elsevier, Morgan Kaufmann Publishers.

Dinstein, I., Guy, G., Rabany, J., Tzelgov, J., & Henik, A. (1989). On the compression of stereo images: Preliminary results. *Signal Processing, 17*(4), 373–382. doi:10.1016/0165-1684(89)90122-9

Ellinas, J. N. (2009). Reversible watermarking on stereo image sequences. *International Journal of Signal Processing, 5*(3), 210–215.

Fridrich, J., Goljan, M., & Du, R. (2002). Lossless data embedding - New paradigm in digital watermarking. *EURASIP Journal on Applied Signal Processing,* (2): 185–196. doi:10.1155/S1110865702000537

Khan, A., Mahmood, M. T., Ali, A., Usman, I., & Choi, T.-S. (2009, January). *Hiding depth map of an object in its 2D image: Reversible watermarking for 3D cameras*. Paper presented at 27th Inter. Conference on Consumer Electronics, Las Vegas, USA.

Scharstein, D., & Pal, C. (2007). *Learning conditional random fields for stereo*. IEEE Computer Society Conference on Computer Vision and Pattern Recognition, CVPR 2007.

Scharstein, D., & Szeliski, R., (2002). A taxonomy and evaluation of dense two-frame stereo correspondence algorithms. *International Journal for Computer Vision, 47*(1/2/3), 7–42.

Scharstein, D., & Szeliski, R. (2003). High-accuracy stereo depth maps using structured light. *IEEE Computer Society Conference on Computer Vision and Pattern Recognition,* (pp. 195–202).

Swanson, M. D., Zhu, B., & Tewfik, A. H. (1997). Data hiding for video in video. *Proceedings of the IEEE International Conference on Image Processing, ICIP, 97*, 676–679. doi:10.1109/ICIP.1997.638586

Tian, J. (2003). Reversible data embedding using a difference expansion. *IEEE Transactions on Circuits and Systems for Video Technology, 13*(8), 890–896. doi:10.1109/TCSVT.2003.815962

ADDITIONAL READING

Alattar, A. M. (2004). Reversible Watermark Using the Difference Expansion of a Generalized Integer Transform. *IEEE Transactions on Image Processing, 13*(8), 1147–1156. doi:10.1109/TIP.2004.828418

Aydinoglu, H., & Hayes, M. H. (1998). Stereo image coding: a projection approach . *IEEE Transactions on Image Processing, 7*(4), 506–516. doi:10.1109/83.663495

Bhatnagar, G., Kumar, S., Raman, B., & Sukavanam, N. (2009). Stereo image coding via digital watermarking . *Journal of Electronic Imaging, 18*. .doi:10.1117/1.3210015

Brown, M. Z., Burschka, D., & Hager, G. D. (2003). Advances in computational stereo. *Transactions on Pattern Analysis and Machine Inteligence, 25*(8), 993–1008. doi:10.1109/TPAMI.2003.1217603

Coltuc, D. (2007). Towards Distortion-Free Robust Image Authentication. *Journal of Physics: Conference Series*, IOP Publishing, vol. 77.

Coltuc, D. (2009). Modified Versions of Tian's Difference Expansion Reversible Watermarking. *Proceedings of the IEEE International Conference on Image Processing, ICIP, 09*(4225-4228).

Coltuc, D., & Chassery, J.-M. (2007). Very Fast Watermarking by Reversible Contrast Mapping. *IEEE Signal Processing Letters, 15*(5), 255–258. doi:10.1109/LSP.2006.884895

Duarte, M. H. V., Carvalho, M. B., da Silva, E. A. B., Pagliari, C. L., & Mendona, G. V. (2005). Multiscale Recurrent Patterns Applied to Stereo Image Coding. *IEEE Transactions on Circuits and Systems for Video Technology, 15*(11), 1434–1474. doi:10.1109/TCSVT.2005.856926

Ellinas, J. N. (2009). Reversible Watermarking on Stereo Image Sequences. *International Journal of Signal Processing, 5*(3), 210–215.

Frajka, T., & Zeger, K. (2003). Residual Image Coding for Stereo Image Compression. *Optical Engineering (Redondo Beach, Calif.)*, *42*(1), 182–189. doi:10.1117/1.1526492

Hartung, F., & Kutter, M. (1999). Multimedia watermarking techniques. *Proceedings of the IEEE*, *87*(7), 1079–1107. doi:10.1109/5.771066

Jiang, J., & Edirisinghe, E. A. (2002). A hybrid scheme for low bit-rate coding of stereo images. *IEEE Transactions on Image Processing*, *11*(2), 123–134. doi:10.1109/83.982820

Kamstra, L., & Heijmans, H. J. A. M. (2005). Reversible Data Embedding Into Images Using Wavelet Techniques and Sorting. *IEEE Transactions on Image Processing*, *14*, 2082–2090. doi:10.1109/TIP.2005.859373

Kim, H. J., Sachnev, V., Shi, Y. Q., Nam, J., & Choo, H.-G. (2008). A Novel Difference Expansion Transform for Reversible Data Embedding. *IEEE Transactions on Information Forensics and Security*, *3*, 456–465. doi:10.1109/TIFS.2008.924600

Kumar, S., & Raman, B. (2009). An Optimally Robust Digital Watermarking Algorithm for Stereo Image Coding, Recent Advances in Multimedia Signal Processing, Springer, 2009.

Lee, S., Yoo, C. D., & Kalker, T. (2007). Reversible Image Watermarking Based on Integer-to-Integer Wavelet Transform . *IEEE Transactions on Circuits and Systems for Video Technology*, *2*(3), 321–330.

Luo, L., Chen, Z., Chen, M., Zeng, X., & Xiong, Z. (2010). Reversible Image Watermarking Using Interpolation Technique, (2010). *IEEE Transactions on Information Forensics and Security*, *5*(1), 187–193. doi:10.1109/TIFS.2009.2035975

Mendona, V. (2005). Multiscale Recurrent Patterns Applied to Stereo Image Coding . *Transactions on Circuits and Systems for Videotechnology*, *15*(11), 1434–1474. doi:10.1109/TCSVT.2005.856926

Perkins, M. G. (1992). Data Compression of Stereopairs . *IEEE Transactions on Communications*, *40*(4), 684–696. doi:10.1109/26.141424

Scharstein, D., & Szeliski, R., (2002). A taxonomy and evaluation of dense two-frame stereo correspondence algorithms. *International Journal for Computer Vision*, 47(1/2/3), 7–42.

Servetto, S. D., Ramchandran, K., & Orchard, M. T. (1999). Image coding based on a morphological representation of wavelet data. *IEEE Transactions on Image Processing*, *8*(9), 1161–1174. doi:10.1109/83.784429

Strintzis, M. G., & Malassiotis, S. (1999). Object-based coding of stereoscopic and 3D image sequences: a review. *IEEE Signal Processing Magazine*, *16*(3), 14–28. doi:10.1109/79.768570

Thodi, D. M., & Rodriguez, J. J. (2006). Expansion Embedding Techniques for Reversible Watermarking. *IEEE Transactions on Image Processing*, *15*, 721–729.

Tian, J. (2003). Reversible Data Embedding Using a Difference Expansion . *IEEE Transactions on Circuits and Systems for Video Technology*, *13*(8), 890–896. doi:10.1109/TCSVT.2003.815962

Wang, X., Shao, C., Xu, X., & Niu, X. (2007). Reversible Data-Hiding Scheme for 2-D Vector Maps Based on Difference Expansion. *IEEE Transactions on Information Forensics and Security*, *2*, 311–319. doi:10.1109/TIFS.2007.902677

Woo, W., & Ortega, A. (2000). Overlapped block disparity compensation with adaptive windows for stereo image coding. *IEEE Trans. on Circuits and Systems for Videotechnology*, *10*(2), 194–200. doi:10.1109/76.825718

KEY TERMS AND DEFINITIONS

Disparity: The difference in position between the correspondence points of the stereo frames.

Fragile Watermarking: The embedded information fails to be extracted after the slightest modification of the embedded host data.

Peak Signal-to-Noise Ratio (PSNR): The ratio between the maximum possible power of a signal and the power of corrupting noise that affects the signal. PSNR is commonly used in image processing as a quality metric.

Reversible Watermarking: Besides the extraction of the embedded information, the host data is exactly recovered.

Robust Watermarking: The embedded information can be extracted after intentional or not intentional modifications (attacks) of the embedded host.

Stereo Images: Pairs of images (left and right frames) of the same scene taken from slightly different viewpoints.

Watermarking Capacity (Payload): The amount of information embedded into a host.

Watermarking: Imperceptible embedding of information into digital host data as images, videos, sound, text files, etc.

Chapter 18
Introduction to Autostereoscopic Displays

Armin Grasnick
Sunny Ocean Studios Pte. Ltd., Singapore

ABSTRACT

This chapter is an introduction to the principles of operation in autostereoscopic displays. It explains the most important autostereoscopic technologies and their principles, the image representation, and the resulting strengths and weaknesses. Beside the general principles, all necessary steps for a successful 3D display design are illustrated. This includes the fundamental dimensions, the generation of the screen images, as well as the creation of the 3D optics. To characterize and classify a certain 3D display, a display metric for autostereoscopic displays is proposed. Even though all parameters are explained for a static 3D system, the basic principles are also applicable for dynamic systems (i.e. 3D displays with head or eye tracking). In such cases, the described geometrics are only correct for a singular point in time.

INTRODUCTION

Since the early days of television and with the progress in television technology, there were consistent technical adaptations of various features to improve the presentation of realistic 3D. Even as the glasses based 3D technologies are quite popular nowadays, it was and is an everlasting dream of viewers, engineers and researchers to achieve the same impressive 3D image quality without glasses too. The formulas, equations and matrices used in this chapter deal with 3D technology without glasses and they are illustrated by examples. The matrix generation is explained by using the universal formula (Grasnick, 2010). In such cases the notation is according to Mathematica (Wolfram Mathematica). All samples are created with Mathematica 7.

DOI: 10.4018/978-1-61350-326-3.ch018

DEFINITION

An autostereoscopic display is a device for representing a 3D scene without the need of viewing aids. There are two main principles of spatial impression (Grasnick, 2010)]: Spatial Existence (used in volumetric displays) and Binocular Disparity (used in stereoscopic and autostereoscopic displays).

"Spatial Existence" means the real presence of spatial information within an observed volume. This kind of representation can be achieved by projecting a sequence of flat images or light points on rotating flat screens (i.e. the "100 million voxel display" (Favalora et al., 2002)), rotating curved screens (i.e. the "Felix 3D display" (Langhans et.al., 1998)), fog screens (DiVerdi et. al., 2006) or varifocal mirrors (Fuchs and Pizer, 1986). Other volumetric technologies are multi-layer displays (Sullivan, 2002), light excitation in a solid medium (Downing, 1997), multi-lens projection (Grasnick, 2001) and of course holographic displays (Spatial Imaging Group @ MIT, 2010). These and similar devices are also referred as "volumetric displays". Because of their complex and sophisticated technologies, volumetric displays have only a little commercial impact today.

The 3D effect in stereoscopic displays is principally caused by the difference in between the left-eye and right-eye images (Binocular Disparity). Usually, an observer has to wear 3D glasses to separate the stereoscopic image pair to the "destined" eyes. If this separation doesn't needs any additional 3D glasses, the device is typically called "autostereoscopic". Autostereoscopic displays are 3D devices at which the spatial impression is mainly based on the reproduction of a disparity in between the represented perspective images without the requirement of viewing aids.

GENERAL PRINCIPLE

As per definition, an autostereoscopic 3D impression is based on binocular disparity. The stereopsis is the determining principle, but all other monocular or binocular depth cues can be used to improve the 3D quality.

An autostereoscopic display has to contain at least two elements: A display device to represent the specific image data (screen image) and an optical modulator to separate parts of the screen image(s) into different parts of the viewing area.

A common display device will show the images as raster image, in which each pixel position can be described with two coordinates. The raster image is a combination of certain number of raster images, representing different perspective views. The combination rule for the screen image can be completely specified in a two dimensional matrix (Figure 1).

where

i, j position indices
i_0 first horizontal index
i_n last horizontal index
j_0 first vertical index
j_m last vertical index
V perspective view number at position i, j

The optical modulator (Figure 2) is an array of optical elements. For the most popular technologies, the optical elements are arranged in one layer. Similar to the screen image, the arrangement of the optical elements can be described also with a matrix. This matrix could be a transformation of the screen image matrix (and vice versa), at which the perspective view number has been replaced with the number of the optical element, the balance or the anchor point of the element. As this number represents now a certain characteristic and value of optical modulation, these values could be described as the "optical mode" for a specific pixel position.

where

k, l position indices
k_0 first horizontal index

Figure 1. Screen image matrix

	i_0	i_1	i_2	i_3	...	i_n
j_0	$V(i_0,j_0)$	$V(i_1,j_0)$	$V(i_2,j_0)$	$V(i_3,j_0)$	...	$V(i_n,j_0)$
j_1	$V(i_0,j_1)$	$V(i_1,j_1)$	$V(i_2,j_1)$	$V(i_3,j_1)$	...	$V(i_n,j_1)$
j_2	$V(i_0,j_2)$	$V(i_1,j_2)$	$V(i_2,j_2)$	$V(i_3,j_2)$	...	$V(i_n,j_2)$
j_3	$V(i_0,j_3)$	$V(i_1,j_3)$	$V(i_2,j_3)$	$V(i_3,j_3)$	...	$V(i_n,j_3)$
...	...	...	...	...	...	...
j_m	$V(i_0,j_m)$	$V(i_1,j_m)$	$V(i_2,j_m)$	$V(i_3,j_m)$	...	$V(i_n,j_m)$

Figure 2. Optical modulator matrix

	k_0	k_1	k_2	k_3	...	k_n
l_0	$W(k_0,l_0)$	$W(k_1,l_0)$	$W(k_2,l_0)$	$W(k_3,l_0)$	...	$W(k_n,l_0)$
l_1	$W(k_0,l_1)$	$W(k_1,l_1)$	$W(k_2,l_1)$	$W(k_3,l_1)$	...	$W(k_n,l_1)$
l_2	$W(k_0,l_2)$	$W(k_1,l_2)$	$W(k_2,l_2)$	$W(k_3,l_2)$	...	$W(k_n,l_2)$
l_3	$W(k_0,l_3)$	$W(k_1,l_3)$	$W(k_2,l_3)$	$W(k_3,l_3)$	...	$W(k_n,l_3)$
...	...	...	...	...	...	...
l_m	$W(k_0,l_m)$	$W(k_1,l_m)$	$W(k_2,l_m)$	$W(k_3,l_m)$	...	$W(k_n,l_m)$

k_n last horizontal index
l_0 first vertical index
l_m last vertical index
O optical mode at position k, l

The fundamental dimensions for an autostereoscopic display design are shown in Figure 3. In the illustrated case, the optical layer is placed in between the screen and the observer.

In certain circumstances it could be preferred to move the optical layer behind the screen, in between backlight and screen (Figure 4). As in geometric optics, there is an upside down image of the object in the Figure 3 (analog to real imag-

Figure 3. Optical layer as overlay

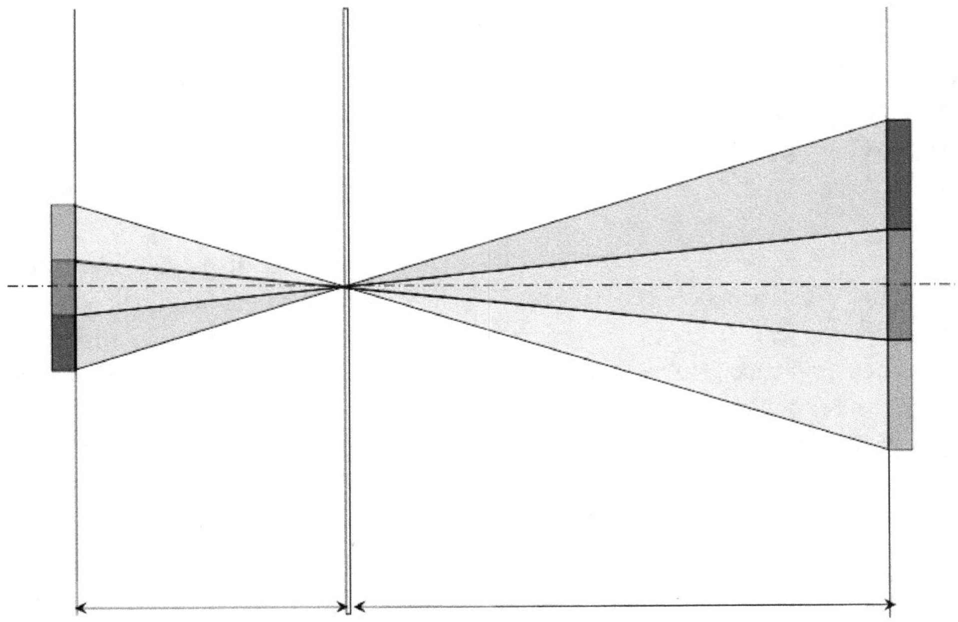

Figure 4. Optical layer as illumination

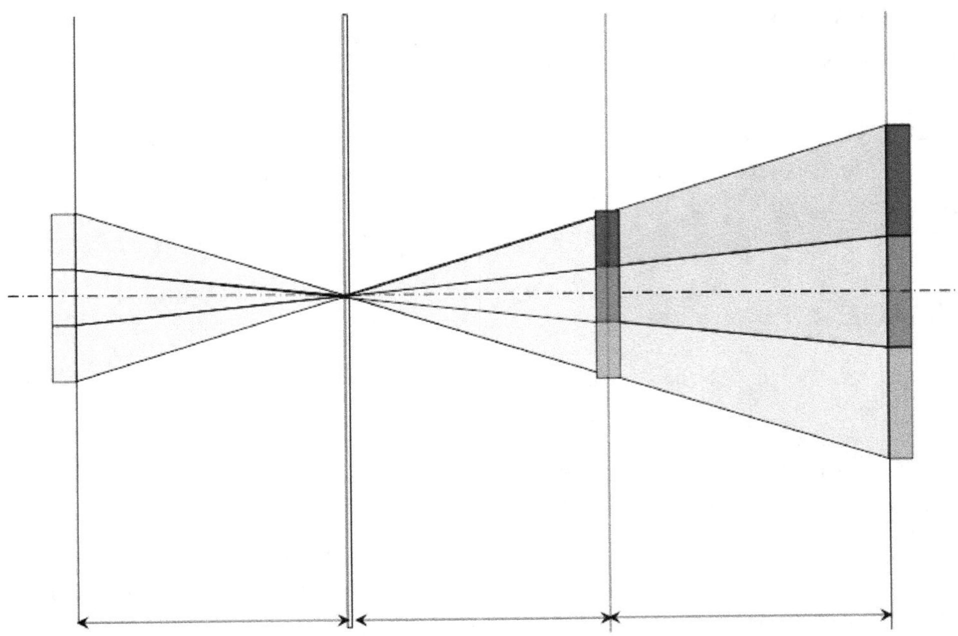

ing) and an upright image in Figure 4 (virtual imaging). The projected pixel size is identical with the eyes distance in the most cases, but it is possible to insert intermediate views to reduce the view flipping or the accommodation-convergence conflict. These principal arrangements works with parallax barriers, lenses or other optical layer based designs.

where

- a object distance (distance between screen layer and optical layer)
- a' image distance (distance between optical layer und observer position)
- A object size (pixel or sub-pixel size)
- A' image size (projected object size in image distance)

where

- a object distance (distance between backlight and optical layer)
- a' image distance (distance between optical layer und screen)
- a'' projection distance (distance between screen and observer position)
- A object size (backlight area in pixel or sub-pixel size)
- A' image size (pixel or sub-pixel size)
- A'' projection size (projected image size in projection distance)

SCREEN IMAGE

Multiplexing

Considering existing installations and sales numbers, the major group of 3D displays is based on flat panel displays (FPD) with only one screen layer. Autostereoscopic displays with more than one layer has been presented from time to time, but hasn't achieved the same level of relevance as single layer displays.

Figure 5. Sub-pixel multiplexing in a red-cyan anaglyph

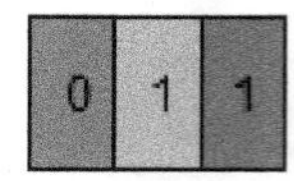

Spatial Multiplexing

The screen image displayed on a standard graphic display can be described using a 2dimensional data array. Anaglyph stereo might be one of the simplest examples for spatial multiplexing. In this example, the screen image is multiplexed in the display sub-pixels.

Sub-pixel multiplexing is a common technology for the propagation of flat panel displays for stereoscopic and autostereoscopic representations. As a pixel is build by three sub-pixel elements (red, green blue), a sub-pixel combination allows a higher density of views on the screen (Figure 5).

A way to achieve such kind of mixing with pixel-based algorithms is color channel mixing or –permutation.

Temporal Multiplexing

By introducing dynamic system, a time parameter is required to locate the pixel. Probably the best-known case of the time sequential mode is the stereo shutter. This mode can be used for projection as well as for flat panel displays. If the refresh rate of the playback device is fast enough to avoid any flickering, the implementation in applications is quite easy.

COMBINATION MATRICES

A combination matrix for an autostereoscopic display represents the arrangement of the perspective views in the screen image. The screen image matrix is interconnected with the optical modulator matrix, but it is possible to have many different screen image matrices for the same optical modulator matrix.

This could be beneficial for the adaption of a real autostereoscopic display for different viewing parameters (i.e. viewing distance or 3D depth), numbers of perspective views (i.e. fewer numbers for real-time applications) or other screen image adjustment (i.e. shifting, tilting, scaling) without any change in the mechanical design or the optics array.

Matrix Generation

Certainly it is possible to create the same matrices using different mathematic algorithms. In this chapter, a simple solution for matrix creation is shown. The administration and adaption for different autostereoscopic displays is easy through setting few parameters. The equations can be directly used in symbolic programming.

2-Dimensional

A 2dimensional array needs at least two coordinates $i, j = f(x, y)$. Two position modulation parameters are used to define the increments to the contiguous cell in x-direction (q_A) and y-direction (q_B). Two other variables determine the repetition factor. These are q_X and q_Y, at which the subscript explicated the direction of action.

$$V = FractionalPart\left[\frac{IntegerPart\left[\frac{i}{qX}\right]qA + IntegerPart\left[\frac{j}{qY}\right]qB}{n}\right]n \tag{1}$$

n-Dimensional

For various reasons it could be desirable to add more dimensions to the formula. This might be necessary if the design of the device needs wavelength separation or analysis of the polarization state. For such intentions, but not limited to them, a more universal equation could be used.

$$V = FractionalPart\left[\frac{\sum_{i=0}^{m}\left(q_{Di}IntegerPart\left[\frac{a_y}{qRi}\right]\right)}{n}\right]n \tag{2}$$

a = position parameter as function of the array index, q_D = Dimension (position) modulation parameter,

q_R = Repetition parameter

An example for a four-dimensional matrix generation is given with the following symbolic expression, shown in Equation (3). i=f(x), j=f(y), k=f(z), l=f(t).

Box 2.

$$MatrixForm\left[Table\left[Round\left[FractionalPart\left[\frac{IntegerPart\left[\frac{i}{qX}\right]qA + IntegerPart\left[\frac{j}{qY}\right]qB + IntegerPart\left[\frac{k}{qZ}\right]qC + IntegerPart\left[\frac{l}{qT}\right]qD}{n}\right]n\right], \{y, y_{\min}, y_{\max}\}, \{x, x_{\min}, x_{\max}\}, \{z, z_{\min}, z_{\max}\}, \{t, t_{\min}, t_{\max}\}\right]\right] \tag{3}$$

Figure 6. Four states of the spatiotemporal stereoscope

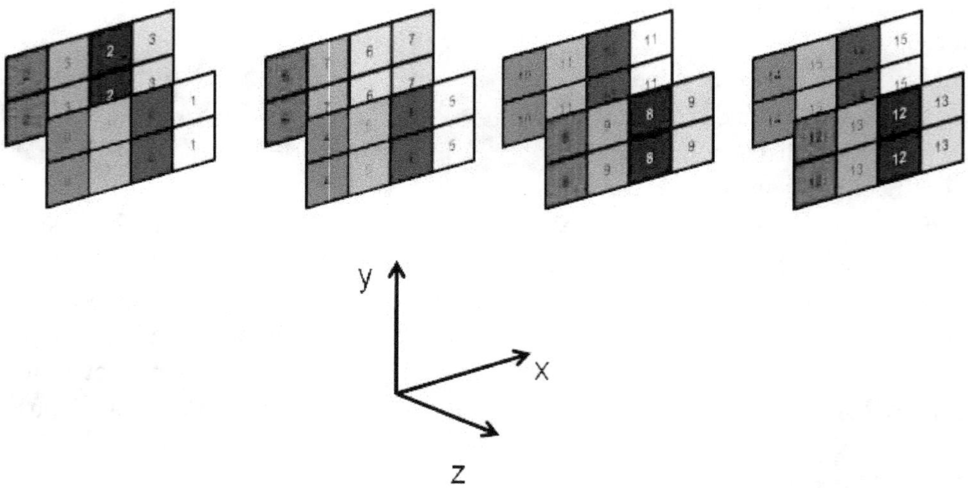

Table 1. Image pairs and switching state of on representation cycle

	State 1 (t = 0)	**State 2 (t = 1)**	**State 3 (t = 2)**	**State 4 (t = 3)**
Front display (z = 1)	0/1 (on)	4/5 (on)	8/9 (off)	12/13 (off)
Rear display (z = 0)	2/3 (off)	6/7 (off)	10/11 (on)	14/15 (on)

Example: Spatiotemporal Stereoscope

Spatiotemporal is a description of space and time depending phenomena. This description has been used for the characterization of image multiplexing procedures (Jang et. al., 2004).

The spatiotemporal stereoscope was introduced as a virtual device or "Gedankenexperiment" (Mach, 1905) to explain a 4dimensial multiplexing. This virtual spatiotemporal stereoscope contains two display layers with addressable pixels in columns and rows. Only the top layer is switchable to a transparent or translucent mode. Each layer shows a combination of two images, interlaced in columns. Any appropriate technique is used to separate the images and provide a binocular vision.

Figure 6 shows a multiplexing of 16 different perspective images (views). The views are addressable by their numbers. Each pixel position needs now four dimensions to become traceable; x,y,z and t.

In this example, the front display has to become transparent/translucent in off-state. As is not defined, if the front panel could be switched to a fully opaque mode, both displays are always addressed, even in off state. Setting some dimension parameters, the matrix can be calculated with the following code in Equation (4).

Setting the parameters, the output matrix shows the exact states (Figure 6).

(n=16, q_A=1, q_B=1, q_C=2, q_D=4, q_X=1, q_Y=1,q_Z=1, q_T=1, i=FractionalPart[x/2]2, k=1-z, l=t)

$$\begin{pmatrix} \begin{pmatrix} 2 & 6 & 10 & 14 \\ 0 & 4 & 8 & 12 \end{pmatrix} & \begin{pmatrix} 3 & 7 & 11 & 15 \\ 1 & 5 & 9 & 13 \end{pmatrix} & \begin{pmatrix} 2 & 6 & 10 & 14 \\ 0 & 4 & 8 & 12 \end{pmatrix} & \begin{pmatrix} 3 & 7 & 11 & 15 \\ 1 & 5 & 9 & 13 \end{pmatrix} \\ \begin{pmatrix} 2 & 6 & 10 & 14 \\ 0 & 4 & 8 & 12 \end{pmatrix} & \begin{pmatrix} 3 & 7 & 11 & 15 \\ 1 & 5 & 9 & 13 \end{pmatrix} & \begin{pmatrix} 2 & 6 & 10 & 14 \\ 0 & 4 & 8 & 12 \end{pmatrix} & \begin{pmatrix} 3 & 7 & 11 & 15 \\ 1 & 5 & 9 & 13 \end{pmatrix} \end{pmatrix}$$

Table 2. Dependent views on x, y, z and t

	X0				X1				X2				x3				
z0	2	6	10	14	3	7	11	15	2	6	10	14	3	7	11	15	y0
z1	0	4	8	12	1	5	9	13	0	4	8	12	1	5	9	13	
z0	2	6	10	14	3	7	11	15	2	6	10	14	3	7	11	15	y1
z1	0	4	8	12	1	5	9	13	0	4	8	12	1	5	9	13	
	t0	t1	t2	t3	t0	t1	t2	t3	t0	t1	t2	t3	t0	t1	t2	t3	

Box 3.

$$FractionalPart\left[\frac{IntegerPart\left[\frac{i}{qX}\right]qZ + IntegerPart\left[\frac{j}{qY}\right]qB + IntegerPart\left[\frac{k}{qZ}\right]qC + IntegerPart\left[\frac{l}{qT}\right]qD}{n}\right]n \tag{4}$$

Table 2 shows the number of view dependent on x, y, z and t.

OPTICAL LAYER

Matrices

An optical layer for a 3D display is an array of numerous small optical effective elements. These elements are arranged in a certain matrix structure. The computing of the optical modulator matrix is similar to the calculation of the screen image matrix.

Different to the screen image calculations, the result gives the number of the optics element at a position i, j back. This number can be described as a replacement for a part of the optics structure (i.e. an opaque or transparent shape in a parallax barrier). A matrix for an optical modulator could have a much higher resolution as the screen image. The necessary resolution depends on the production technology and the type of optics.

1-Dimensional Matrix

Obviously an optical layer has to be at least 2-dimensional. But as this matrix can be interpreted as instruction for the manufacturing of the optics, the simplest matrix is one-dimensional. An example of an optical layer, where a 1-dimensional matrix is a sufficient description, could be a lenticular screen. Because the lenticulars are arranged in parallel lines, it is possible to describe their positions with only one parameter.

$$W = FractionalPart\left[\frac{k}{n}\right]n \tag{5}$$

2-Dimensional Matrix

A 2-dimensional arrangement can be used for more flexibility and other types of optics. Analog to the screen image, the 2-dimensional form is

$$W = FractionalPart\left[\frac{IntegerPart\left[\frac{k}{q_X}\right]q_A^{'} + IntegerPart\left[\frac{l}{q_Y}\right]q_B^{'}}{n}\right]n \quad (6)$$

Matrix Transformation from Screen Image Matrix

In the 3D display design process, the most important parameters are belonging to the application of a 3D display. These parameters can be used to adjust the display, i.e. for a certain viewing distance, viewing area or depth impression.

Some of these decisions have an influence to the general selection of a certain optics technology; others will affect the number of necessary perspective views, the moiré interferences or the view arrangement and combination matrix.

Therefore the first step in the development process should be the creation of the combination matrix for the major application, while the optics arrangement is a transformation of this underlying rule. In such cases, the whole matrix (or some matrix parameters) of the optical modulator matrix is (are) a function of the screen image matrix.

W=f[V]
with

$$q_A^{'} = f1[q_A]; q_B^{'} = f2[q_B]; q_X^{'} = f3[q_X]; \\ q_Y^{'} = f4[q_Y]; k = f5[i]; l = f6[j]; \quad (7)$$

the value of W is now the function of the parameters of V. See Equation (8).

TYPES OF OPTICAL LAYERS

Integral Photography

In the early 20th century, a technology was introduced to record and reconstruct a real scene by using a number of micro lenses [Lippmann, 1908]. This technology is known as integral photography, the sheet as "fly's eye" lenses array. A main advantage of the integral photography (Figure 7) is the existence of a full parallax in horizontal and vertical direction.

Lenticular Screen

A lenticular (Figure 8) is a lens with refraction in only one direction. This lens type prevents a vertical parallax, but allows a higher resolution in this direction.

Barrier Screen: Transparent / Opaque

Simple in design and easy in production, a black-white barrier might be the ideal candidate for a low cost autostereoscopic display. But the inherent high loss of brightness is reducing the field of applications and prevents those displays from an overall use. As same as in lenses, a barrier screen could be used with a full parallax (pinhole barrier

Box 4.

$$W = FractionalPart\left[\frac{f1[q_A]IntegerPart\left[\frac{fs[x]}{fs[q_X]}\right] + f2[q_B]IntegerPart\left[\frac{f_3[y]}{f_4[q_Y]}\right]}{n}\right]n \quad (8)$$

Figure 7. Integral photography

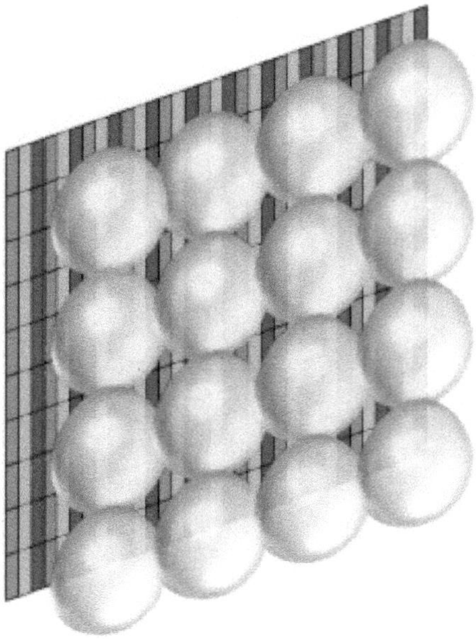

Figure 8. Lenticular screen

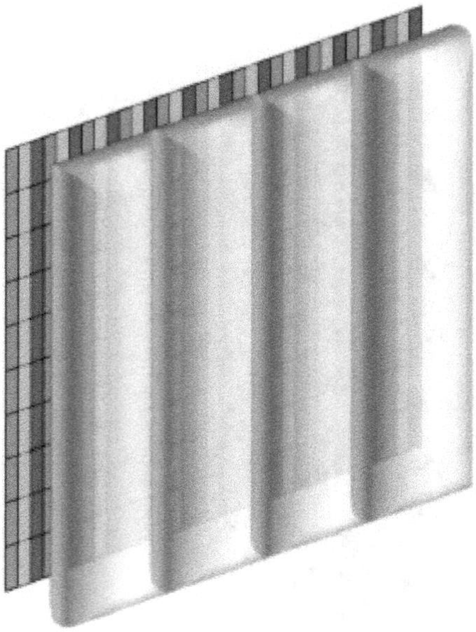

Figure 9. Pinhole barrier

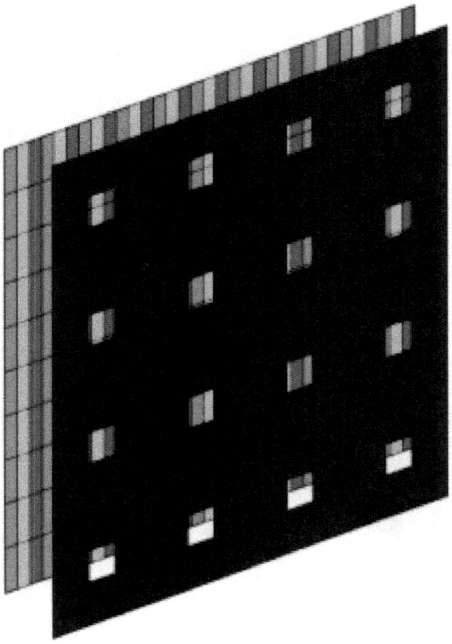

Figure 10. Barrier stripes

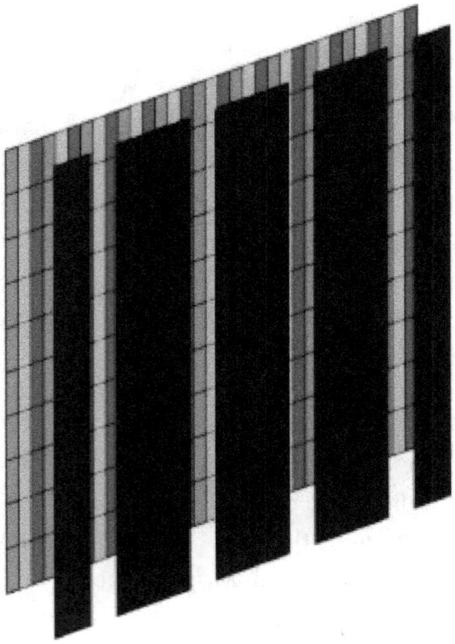

Figure 11. Color barrier

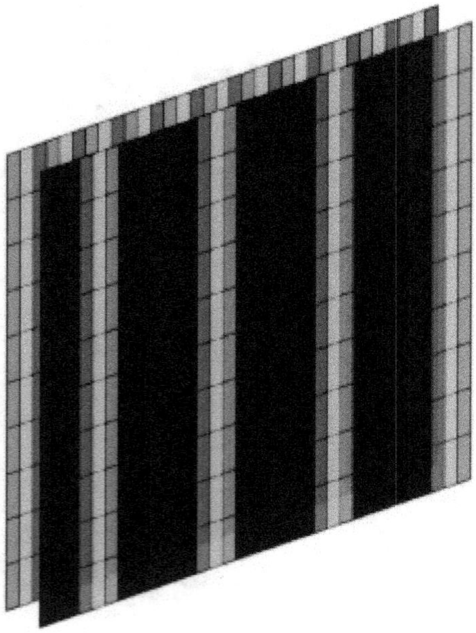

– Figure 9) or only horizontal parallax (barrier stripes – Figure 10).

Color Barrier

To improve the brightness or perceived resolution of a 3D display, an arrangement of color filters can be used. The color filters of the barrier should have near the same wavelength absorption/transmission characteristics as the light of the screen color pixel (Figure 11). Because of their wavelength selective function, such systems are sometimes called as wavelength selective filter array.

Zone Plates

A zone plate (or Fresnel zone plate) is a diffractive optic (Figures 12 and 13). For 3D displays such a zone plate could be used as a simple in-line hologram. If this hologram is reconstructed, it provides a real and a virtual focal point and works like a converging lens and diverging lens in one device.

Figure 12. Circular zone plate

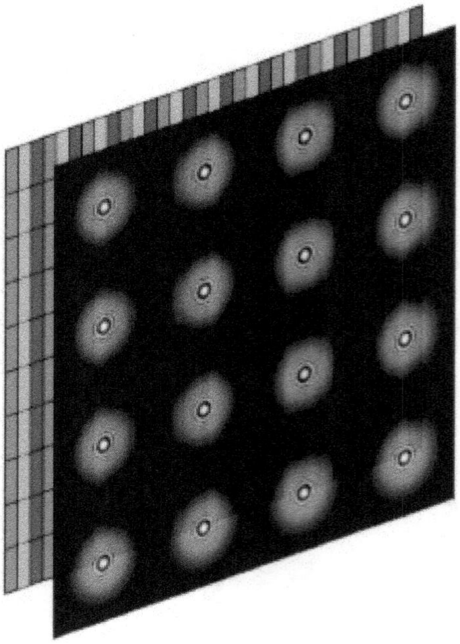

Figure 13. Lenticular zone plate

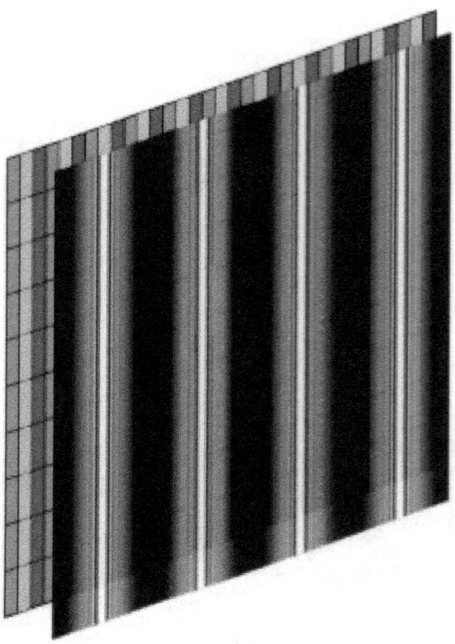

Figure 14. States of a multi layer device

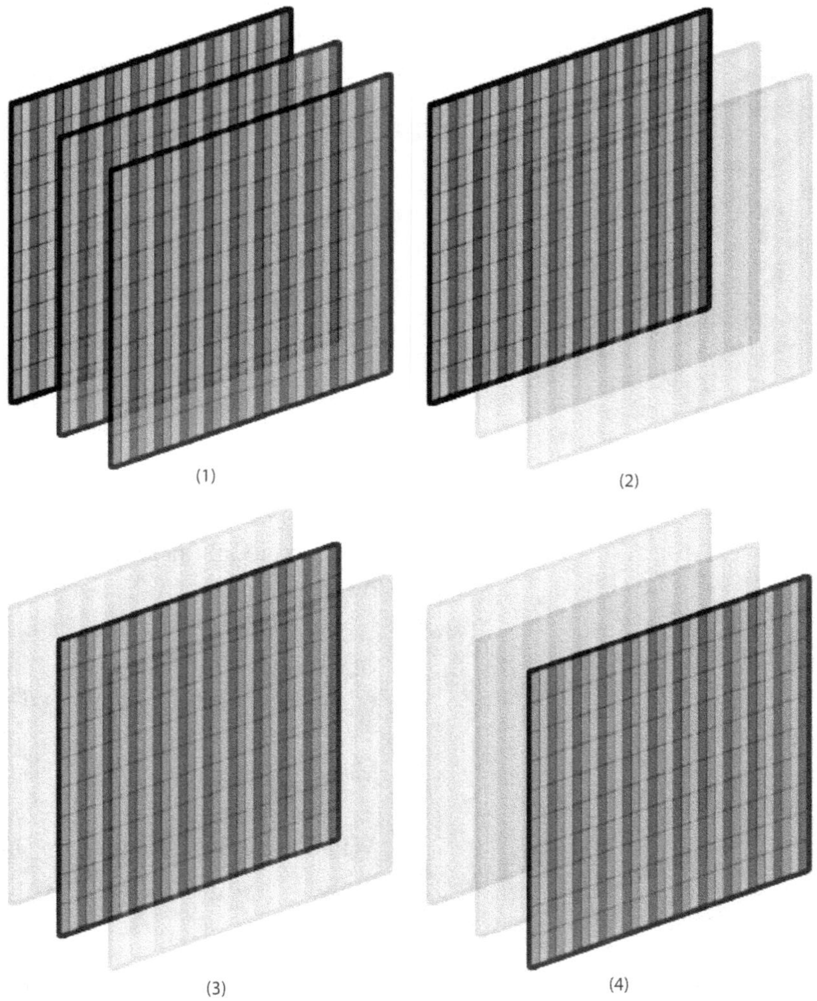

As in other optical layers, a reduction of the vertical parallax is possible (lenticular zone plate).

Multi Layer Devices

A multi layer device reproduces screen images and/or the optical elements on different layers (Figure 14). The layers are static (spatial multiplexing) or switchable (temporal multiplexing). For definition, a multi layer device is only an autostereoscopic display, if a binocular separation is involved.

A 4-dimensional case (spatiotemporal stereoscope) could be possible, but has no practical significance at present. Different to the aforementioned devices, multi layer devices could be optimized for different viewing distances.

Combinations

An autostereoscopic display can be a mix of different devices and optics. In Figure 14, a barrier screen works as a lens aperture for a lens sheet. Other combinations are applicable besides this

Figure 15. Pinhole and fly's eye

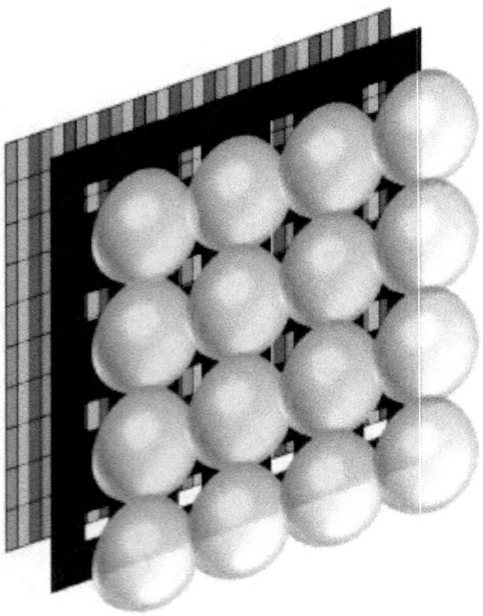

example, but such sophistications are almost based on a benefit-cost analysis.

AUTOSTEREOSCOPIC DISPLAY METRICS

Orthoscopic Viewing Zone

An autostereoscopic display is restricted to a certain viewing area. Only in this "sweet spot" area, the 3D impression is correct (orthoscopic). If the observer is moving outside the orthoscopic viewing zone, the 3D impression becomes blurry. The viewing area in a static autostereoscopic system is mainly defined by two parameters: The optimum viewing distance and the number of views.

Optimum Viewing Distance

Every autostereoscopic 3D display is designed for an optimum viewing distance (a). The 3D optic of the 3D display projects slices of the perspective views in the optimum viewing distance (Figure 16). In this distance, the slices have a certain size (e). The value of e is usually identical with the average eye distance (inter-pupillary distance, ~65mm).

On the left and right side there are more viewing zones. In the transition areas to the next viewing zone, even in the optimum viewing distance, the 3D scene is inverted in depth. This "pseudoscopic" image appears if the total number of views (cycle) has been passed by the observer. The pseudoscopic effect is caused by the inverted image order in the transition zone.

Lateral Freedom of Movement

The illustration in Figure 17 shows the lateral freedom of movement in a 5 view system.

The maximum horizontal dimension (c) in the orthoscopic viewing zone (yellow area) is limited by the product of slice size and number of views (n). If the slice size is constant, a higher number of views results in an enlarged viewing zone.

Viewing Range

Without paying attention to moiré effects, the minimum viewing distance (a_{min}) is defined by:

$$a_{\min} = \frac{b * a}{b + e * n} \tag{9}$$

Under the condition e*n >b, the maximum viewing distance can be calculated with

$$a_{\max} = a + \frac{a * e * n}{b - e * n}$$

But if e*n>=b, the maximum viewing distance would be infinity.

In far viewing distances, the slice size could become wider than the eyes distance. In such cases the impression becomes partly 2-dimensional. Considering the eyes distance (A), the maximum viewing distance is now:

Figure 16. Optimum viewing distance

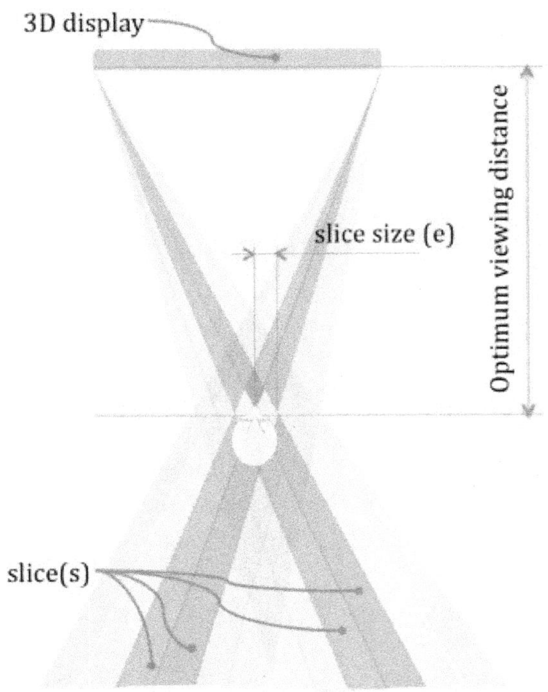

Figure 17. Viewing zone

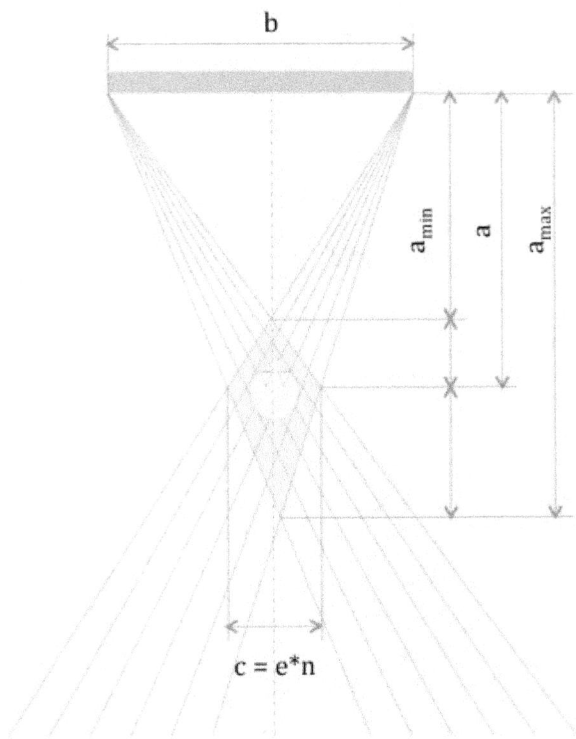

Figure 18. Perceived depth

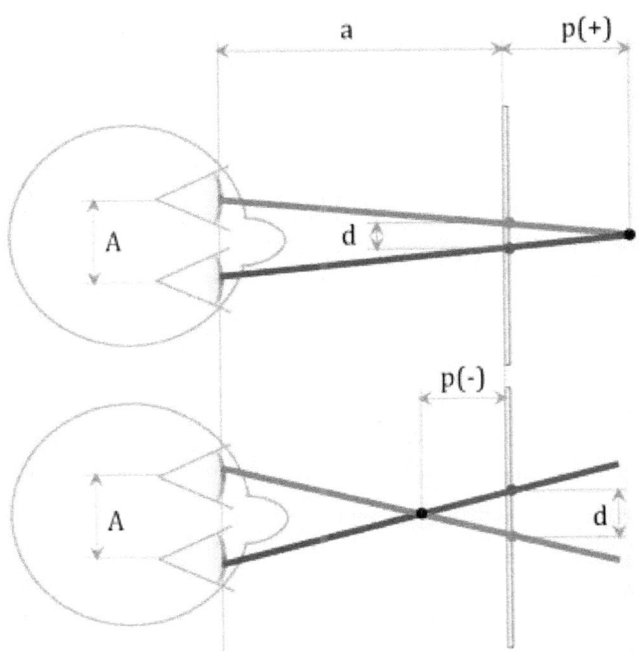

Figure 19. Stereoscopic pyramid, 5 views

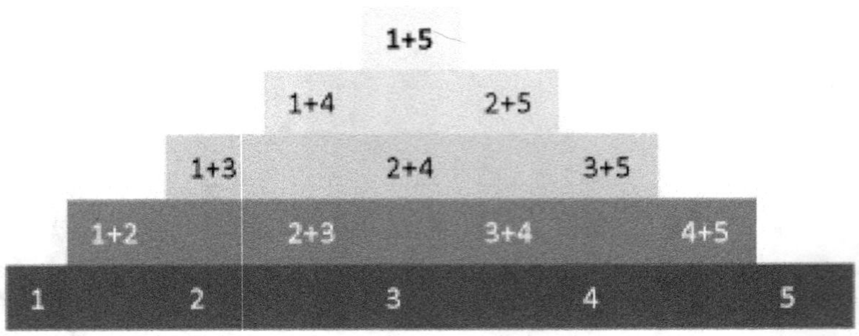

$$a_{\max} = \frac{a * A}{e} \tag{10}$$

Perceived Depth

If an observer is converging to a virtual point, the depth is formed by the screen disparity (d) as shown in Figure 18.

The perceived depth behind the screen (positive parallax) can be calculated with:

$$p(+) = \frac{a}{\left(\frac{A}{d}\right) + 1} \tag{11}$$

And in front of the screen (negative parallax):

$$p(-) = \frac{a}{\left(\frac{A}{d}\right) - 1} \tag{12}$$

Crosstalk vs. Channel Separation

The channel separation is the proportion between the left and right eye view in a 3D device. If the image for one eye can be seen only with this eye (i.e. the left perspective view can be seen only with the left eye, but not with the right eye), the channel separation reached its maximum (100%).

The crosstalk describes the ratio of the visibility for the pseudoscopic image portion divided by the stereoscopic visibility (illuminance at eyes positions E). In an ideal system, the crosstalk is 0. In this chapter, the following relations are used:

$\text{Crosstalk}_{right} = E_{left}/E_{right}$

Channel Separation = 100% -Crosstalk (13)

The real visibility and vision interference of the crosstalk depends on the image parameters (i.e. contrast, textures, edges, colors, shadows….) and parallax value. Even if it is recommended to reduce the crosstalk below 0.1%, it has been shown a 3D impression can be seen even with a crosstalk of more than 10%.

Also the visual system (physiologic and psychologic vision) of the observer effects the 3D perception. People with lower eyesight or visual problems might have in general more difficulties in 3D viewing, while very experienced viewer could accept and enjoy even a massive crosstalk.

It is an obvious truth in stereoscopic cinema that a better separation in between the projections for the left and the right eye will result in a more impressive 3D impression. Also it is well known, a higher crosstalk could cause some discomfort. It is recommended in general reducing the crosstalk for a good 3D vision.

But there is an indication for a higher crosstalk in autostereoscopic systems. As the viewing area is restricted in an autostereoscopic system, a higher crosstalk could enlarge the viewing area and lead to a comfortable viewing experience by having the same 3D impression from almost every point in the viewing area.

View Related

Number of Views

The number of views has not only an influence to the viewing area; also this number defines the quality (density) of correct stereopairs in the viewing area. The orthoscopic stereopairs can be calculated using a pyramid model (Figure 19).

A mathematic equivalent is given with

$$n_{stereo} = n_{view} * \frac{n_{view} - 1}{2} \tag{14}$$

View Flipping

A remarkable flipping of the objects in front of the screen (negative parallax) appears in case of strong depth and short distance if the observer changed the position. If the disparity is lower as 1 pixel for the minimum slice size, the flipping is invisible.

Pseudoscopic Points

On a virtual line, a number of pseudoscopic points can be calculated. A lower number describes a larger orthoscopic area. If the number result is 0, the whole viewing area is orthoscopic.

n_{pseudo} = Pseudoscopic Points =Π*a/c

A pseudoscopic number of 0 could be achieved by tracking mechanism.

FUTURE RESEARCH DIRECTIONS

The main market barriers for autostereoscopic 3D are the limited viewing area and the lack of 3D content. It has been shown that the increase of the number of perspective views will enlarge the viewing area. On the other hand a higher view number decreases the image resolution. Ultra-high resolution panels will support more views in the near future. We will explore if a combination of higher crosstalk and a huge number of views (many thousands) contribute significantly to the viewing area and reduce the conflict of convergence/accommodation. Based on our 4-dimensional formula, a complete virtual test system has to be created, to evaluate the 3D image quality of different matrix combinations (screen and optics).

There are many different multi-view and super-multi-view systems in the market. To drive all these different displays, a universal 3D image and video format supporting a very high number of views will be necessary. This format should allow creating and multiplexing of the required perspective views in real time.

CONCLUSION

Autostereoscopic displays could be described as combination of screen image matrix and optical modulator matrix. As the optical layer is a transformation of the screen image, a complete autostereoscopic system specification requires only the screen image parameters and the dimensions. With these parameters, the display metric and 3D depth impression can be evaluated without a real system. The influence of all display parameters could be simulated, which allows an efficient adaption for different applications.

REFERENCES

DiVerdi, S., Rakkolainen, I., Höllerer, T., & Olwal, A. (2006). A novel walk-through 3D display. In *Proceedings of SPIE 2006* (*Vol. 6055*, pp. 428–437). Electronic Imaging. doi:10.1117/12.643286

Downing, E. A. (1997). *Method and system for three-dimensional display of information based on two-photon upconversion*. (US Patent Application 5,684,621).

Favalora, G. E., et al. (2002). 100 million-voxel volumetric display. In D.G. Hopper (Ed.), *Cockpit Displays IX: Displays for Defense Applications, Proceedings of SPIE: Vol. 4712* (pp. 300-312).

Fuchs, H., & Pizer, St. M. (1986). *Three dimensional display using a varifocal mirror*. (US patent application 4,607,255).

Grasnick, A. (2001). *Three-dimensional representation system*. (US patent application 6,176,582).

Grasnick, A. (2010). Universal 4dimensional multiplexing of layered disparity image sequences for pixel and voxel based display devices. In *Proceedings of SPIE, vol. 7526*.

Jang, J.-S., Oh, Y.-S., & Javidi, B. (2004). Spatiotemporally multiplexed integral imaging projector for large-scale high-resolution three-dimensional display. *Optics Express*, *12*, 557–563. doi:10.1364/OPEX.12.000557

Langhans, K., Bezecny, D., Homann, D., Bahr, D., Vogt, C., Blohm, C., & Scharschmidt, K.-H. (1998). New portable FELIX 3D display. In *Projection Displays IV*. Proceedings of SPIE.

Lippmann, G. M. (1908). *La photographie integrale*. France Academy of the Sciences, Note.

Mach, E. (1905). *Erkenntnis und Irtum*. Leipzig, Germany: Johann Ambrosius Barth.

Spatial Imaging Group @ MIT. (2010). *Holovideo: Mark II.* Retrieved November 18, 2010, from http://www.media.mit. edu/ spi/ HVmark2.htm

Sullivan, A. (2002). *Multi-planar volumetric display system and method of operation using multi planar interlacing*. (US patent application US 6,806,849).

Wolfram Mathematica. (n.d.). *Online documentation*. Retrieved from http://reference. wolfram. com/ mathematica/ guide/ Mathematica.html

ADDITIONAL READING

Dodgson, N. A. (1997) *Autostereo displays: 3D without glasses*. Paper presented at EID: Electronic Information Displays, Esher, UK.

Holliman, N. S. (2006). *Three-Dimensional Display Systems*. In J.P. Dakin and R.G.W. Brown (Ed.), *Handbook of Optoelectronics (Vol II)*. Taylor & Francis, ISBN 0 7503 0646 7.

Kaplan, S. H. (1952). Theory of parallax barriers. *Journal of the SMPTE*, *59*(7).

Mori, Y., Fukushima, N., Yendo, T., Fujii, T., & Tanimoto, M. (2009). View generation with 3D warping using depth information for FTV. *Image Communication*, *24*(1–2), 65–72.

Ruijters, D., & Zinger, S. (2009). IGLANCE: Transmission to Medical High Definition Autostereoscopic Displays. In *3DTV-CONFERENCE 2009: The True Vision - Capture, Transmission and Display of 3D Video* (4 pages). IEEE Press.

KEY TERMS AND DEFINITIONS

Autostereoscopic: This refers to 3D display without the usage of glasses.

Depth Map: This refers to a 2D matrix containing the depth results for every pixel.

Free-Viewpoint Interpolation: This refers to generating intermediate views using interpolation.

Chapter 19
Multi-View Autostereoscopic Visualization using Bandwidth-Limited Channels

Svitlana Zinger
Eindhoven University of Technology, The Netherlands

Yannick Morvan
Philips Healthcare, The Netherlands

Daniel Ruijters
Philips Healthcare, The Netherlands

Luat Do
Eindhoven University of Technology, The Netherlands

Peter H. N. de With
Eindhoven University of Technology, The Netherlands &
Cyclomedia Technology B.V., The Netherlands

ABSTRACT

The increasing popularity of stereoscopic cinema and television paves the way for more advanced stereoscopic technologies, such as high-resolution multi-view autostereoscopic displays. The amount of information conveyed by such displays surpasses, however, the bandwidth capacity of the current broadcasting infrastructure. In this chapter, we will focus on technical solutions to overcome the bandwidth bottleneck that only minimally affect the viewer experience. The presented solutions consist of (1) employing depth-based free-viewpoint interpolation with the aim to reduce the number of views that need to be transmitted, (2) the optimal compression of the depth and texture images while minimizing the resulting image artifacts, and (3) the optimal resolution considerations for a given autostereoscopic display.

DOI: 10.4018/978-1-61350-326-3.ch019

INTRODUCTION

Multi-view autostereoscopic displays add depth impression to the visualized image without requiring the viewer to wear goggles. The presentation of more than two views (which would be sufficient for the stereoscopic effect) allows viewers to move freely within a certain range and still perceive a proper stereoscopic image. A further advantage for the viewer is the ability to slightly look behind objects by a small motion of the viewer.

The transmission of multi-view video to the display is seriously challenged by the bandwidth limitations of the transmission channel. Multiple views of each video frame have to be transmitted (modern multi-view autostereoscopic displays present 8 to 25 views), which increases the required bandwidth considerably. Especially for medical applications, the views are demanded to be of high-resolution and artifacts requirements are very stringent.

In this chapter, several techniques are studied that enable to fit the multi-view video stream in a bandwidth-limited channel. The focus of this chapter is to not only present results for individual steps but to show dependencies and contributions in the complete processing chain.

The chapter begins with a description of the background, which is followed by the three main sections. These sections present solutions for key problems in the overall multi-view video communication. The first main section concerns display and rendering aspects of multi-view presentation. The second section is on multi-view compression, particularly depth compression, as this influences the 3D rendering and the obtained quality. The third main section presents the study on resolution optimization and sampling for professional applications.

BACKGROUND

Multi-view autostereoscopic displays and the problem of video signal transmission for such displays are discussed in this section. We introduce the concept of stereoscopic viewing and discuss the broadcasting options for it.

Multi-View Autostereoscopic Displays

A stereoscopic display presents the viewer with different images for the left and the right eye. Provided that these images contain proper stereoscopic information, the viewer will have the sensation of seeing depth. Principally there are two kinds of stereoscopic displays: the first type requires the viewer to wear goggles or glasses, and the second type, called autostereoscopic display, allows stereoscopic viewing without any external aid. The autostereoscopic effect can be achieved by using lenticular lenses (see Figure 1), or parallax barriers in order to emit different images when viewing under a (slightly) different angle. Modern so-called multi-view autostereoscopic displays provide between 8 and 25 views in order to achieve a smooth transition when the viewer moves his head (van Berkel, 1999; Dodgson, 1997; Maupu et al., 2005; Ruijters, 2009).

Multi-view autostereoscopic displays can be regarded as three-dimensional light field displays (Levoy & Hanrahan, 1996; Isaksen et al., 2000) (or four- dimensional, when also considering time). The dimensions are described by the parameters (x, y, φ), whereby x and y indicate a position on the screen and φ indicates the angle in the horizontal plane in which the light is emitted. The light is further characterized by its intensity and its color.

The multi-view lenticular display device consists of a sheet of cylindrical lenses (lenticulars) placed on top of an LCD in such a way that the

Figure 1. The autostereoscopic lenticular screen. The various subpixels are refracted to different angles by the sheet with the lenticular cylindrical lenses. In this way the left and the right eye are presented with different views.

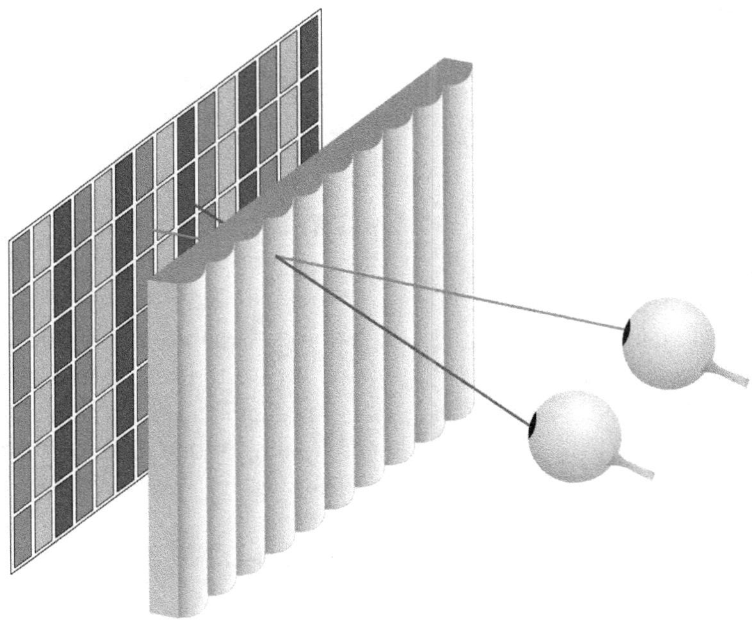

LCD image plane is located at the focal plane of the lenses (van Berkel, 1999). The effect of this arrangement is that LCD pixels located at different positions underneath the lenticulars fill the lenses when viewed from different directions; see Figure 1. Provided that these pixels are loaded with suitable stereo information, a 3D stereo effect is obtained, in which the left and right eyes see different, but matching information.

The fact that the different LCD pixels are assigned to different views (spatial multiplex) leads to a lower resolution per view than the resolution of the LCD grid (Dodgson, 1997). In order to distribute this reduction of resolution over the horizontal and vertical axes, the lenticular cylindrical lenses are not placed vertically and parallel to the LCD column, but slanted at a small angle (van Berkel et al., 1996). The resulting assignment of a set of LCD pixels is specified by the display manufacturer. Note that the red, green, and blue color channels of a single pixel are depicted in different views.

Broadcasting for Multi-View Autostereoscopic Displays

The transmission of a multi-view autostereoscopic video signal encounters several hurdles. These can be capacity limitations; the amount of information that has to be displayed can easily exceed the capacity of the transmission channel. For example, broadcasting 9 uncompressed views of 1280 × 786 pixels with 24 bits per pixel at 20 frames per second requires a channel with a capacity of 9 views × 1280 × 786 pixels × 24 bit × 20 $s^{-1} \approx 4.4$ Gbit/s. Furthermore, there can be a mismatch between the amount of views and their angular interval, as well as other camera parameters at sender and receiver side. This may occur especially when broadcasting to a multitude of heterogeneous receivers. Finally, there can be stringent requirements regarding the image artifacts that are visible, depending on the application of the autostereoscopic visualization (e.g., in medical applications), which can seriously limit the amount of lossy compression that can be applied. In the following sections, we will provide solutions to overcome these hurdles.

FREE-VIEWPOINT INTERPOLATION

Free-viewpoint interpolation allows generation of images in-between camera positions. A common way to obtain such a free-viewpoint interpolated image in accurate manner is to employ a depth image based rendering (DIBR) method (Zinger et al., 2010). Such methods assume the availability of a depth map for each camera image. The depth map encodes the distance to the viewer or camera for the content of each pixel in the camera image (which is called texture image in this context). In this section, we first propose an efficient and accurate DIBR algorithm for multi-view content, and then describe its application to reducing the strain on the transmission channel when broadcasting multi-view autostereoscopic video data.

Image Warping

Image warping is the process of deforming the shape of the image, while interpolating the image content into the new shape. DIBR algorithms are based on warping the image from a camera view to another view (McMillan & Pizer, 1997). Let us specify this in some more detail. Consider a 3D point at homogeneous coordinates $\boldsymbol{P}_w = (X_w, Y_w, Z_w, 1)^T$, captured by two cameras and projected onto the reference and synthetic image plane at pixel positions $\boldsymbol{p}_1$ and $\boldsymbol{p}_2$. The 3D position of the original point $\boldsymbol{P}_w$ in the Euclidean domain can be written as,

$$\boldsymbol{P}_w = (\boldsymbol{K}_1\boldsymbol{R}_1)^{-1} \cdot (\lambda_1\boldsymbol{p}_1 + \boldsymbol{K}_1\boldsymbol{R}_1\boldsymbol{C}_1), \qquad (1)$$

where matrix $\boldsymbol{R}_i$ describes the orientation of the camera i, $\boldsymbol{K}_i$ represents the 3×3 intrinsic parameter matrix of camera i, and $\mathbf{C}_i$ gives the coordinates of the camera center. Assuming that Camera 1 is located at the world-coordinate system origin and looking into the Z direction, i.e. the direction from the origin to $\boldsymbol{P}_w$, we can write the warping equation as,

$$\lambda_2\boldsymbol{p}_2 = \boldsymbol{K}_2\boldsymbol{R}_2\boldsymbol{K}_1^{-1} Z_w\boldsymbol{p}_1 - \boldsymbol{K}_2\boldsymbol{R}_2\boldsymbol{C}_2. \qquad (2)$$

This equation constitutes the 3D image warping equation that enables the synthesis of the virtual view from a reference texture view and a corresponding depth image. This equation specifies the computation for one pixel only, so that it has to be performed for the entire image (see Figure 2).

Proposed DIBR Algorithm

Step 1. Warping Depth Maps and Copying Texture Values to the Corresponding Locations

The depth maps are warped and textures are created for the new viewpoint by copying the texture values to the pixel locations defined by depth map warping. The warping is specified by

$$[D_{warped1}, T_{warped1}] = Warp(HD(D_{ref1})), \qquad (3)$$

$$[\mathrm{D}_{warped2}, \mathrm{T}_{warped2}] = \mathrm{Warp}(\mathrm{HD}(\mathrm{D}_{ref2})),$$

where D_{ref1} and D_{ref2} are depth maps of the first and second reference cameras, respectively, function $HD(.)$ labels the pixels at high discontinuities and $Warp(.)$ is a warping operation, $D_{warped1}$ and $D_{warped2}$ are depth maps, warped from D_{ref1} and D_{ref2}, respectively. Parameters $T_{warped1}$ and $T_{warped2}$ are textures at the new viewpoint. In equation 3, we use the following $HD(.)$ function. The image is warped everywhere, except when the following condition holds (particularly on edges):

Figure 2. Example: surface S' is warped from the reference viewpoint to the virtual viewpoint. The change in the area size of the surface leads to so-called cracks that are processed by a DIBR algorithm. When the depth of an object in a transmitted view is known, the texture image belonging to a virtual camera can be accurately determined.

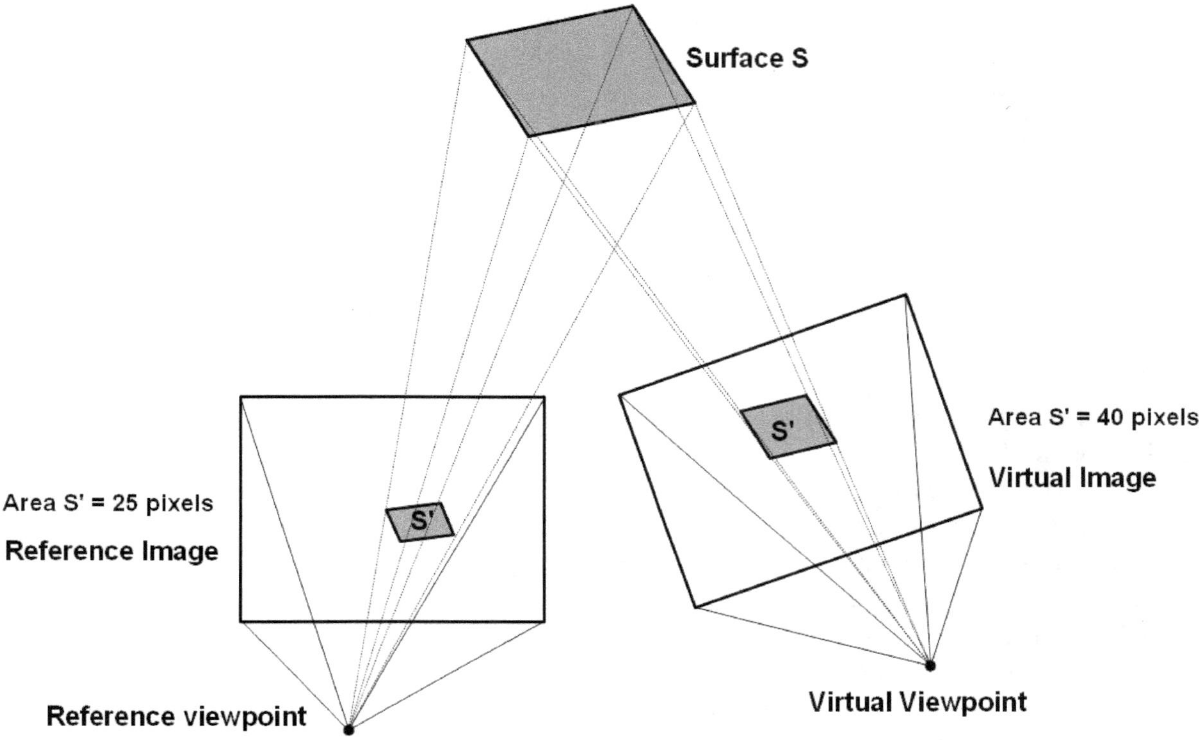

$$\forall_{xy} \in S, \left[\sum_{i=-1}^{1}\sum_{j=-1}^{1} D_{ref}(x+i, y+j)\right] - 9 \cdot D_{ref}(x,y) > T_d. \quad (4)$$

In this condition, S is the image space, D_{ref} denotes the depth map of the reference camera and T_d is a predefined threshold. This function allows to remove only the warped pixels at the edges of high discontinuities, but only restricted to edges at the background side.

Step 2. Median filtering and defining changed pixels

Median filtering is applied to $D_{warped1}$ and $D_{warped2}$ and finds the indexes $Index_{to\_warp1}$ and $Index_{to\_warp2}$ of pixels whose values have changed. This index computation is specified by

$$Index_{to\_warp} = Cracks(Median(D_{warped})). \quad (5)$$

This equation is performed twice for both camera views. Function *Median(.)* is a median filter with a 3x3 window, and *Cracks(.)* detects pixels that have changed during median filtering.

Step 3. Texture Crack Filling by Inverse Warping

The cracks on warped textures are filled in by inverse warping, which is warping from the new view to the reference camera views. This covers the following relation:

$$[D_{warped}, T_{warped}] = Warp^{-1}(Index_{to\_warp}), \quad (6)$$

This is performed also for the two surrounding views with the corresponding input and output label.

Step 4. Create the Texture for the New View

Blending (function *Blend(.)*) the two warped textures and the inpainting, the resulting image gives

$$[D_{new}, T_{new}] = Inpaint(Blend(T_{warped1}, T_{warped2})). \quad (7)$$

After blending the two projected images, disocclusions may still occur. Such areas that cannot be viewed from both reference cameras. The process of filling in those gaps is called inpainting. The disocclusions are not just random regions of an image. They are newly uncovered areas of background without texture information, and certainly not part of foreground objects. When we assume that the disocclusions should be background, we may use the depth information at the edges of the disoccluded region for inpainting with more accurate textures. First, we search for every pixel in the disoccluded region in eight directions for the nearest edge pixel. Then, we only take into account the edge pixels with the lowest depth value (Zinger et al., 2010).

DIBR for Multi-View Data Reduction

In order to reduce the load on the transmission channel, it is possible to transmit fewer views than are displayed on the lenticular screen. The missing views are interpolated after decoding the video stream at the receiver side, see Figure 3.

Let us consider an example of data reduction for transmitting multi-view video. A QuadHD LCD grid consists of 3840×2160 pixels. Assuming that a 9-view QuadHD autostereoscopic display is used, it would make sense to build up a single view in a resolution of 1280×720 pixels (Ruijters and Zinger, 2009). The views that are used for the interpolation algorithm can consist of 32 bits per pixel, where 24 bits are required for the RGB components and 8 bits for depth information. Usage in clinical interventions requires a minimum frame rate of 24 frames per second (fps). When for example 4 views are transmitted at 24 fps, and the others are interpolated, this would require a bandwidth of 4 views × 1280×720 pixels× 24 fps×32 bits = 2.6 Gbit/s (for uncompressed video data) versus 4.4 Gbit/s for 9 views without depth information. Furthermore, the load on the view acquisition or generation side is reduced considerably, since only 4/9 of the data rendered in the naive approach needs to be provided.

COMPRESSION

In this section, we will consider the compression options for multi-view video transmitted to an autostereoscopic display. At first, we discuss the lossy compression of views and introduce the rate-distortion curve for this approach. Then we discuss an innovative algorithm for compressing the depth map, taking into account their specific properties.

Lossy Compression of the Views

After the bandwidth load has been reduced by transmitting a limited number of views, it can be further reduced by lossy compression of the texture and depth images of each view. Specifically, it is possible to further optimize the compression of depth and texture images by jointly encoding them. To illustrate the problem of joint compression of texture and depth, let us consider the following two cases. First, assume that the texture and depth images are compressed at very high and low quality, respectively. In this case, detailed texture is mapped onto a coarse approximation of object

Figure 3. (a) Two configurations for 4 transmitted views, and 9 displayed views. Solid black: transmitted views that can be mapped directly on an output view. Dashed: transmitted views that cannot be mapped on an output view. Light blue: interpolated view. (b) For the white cameras only their parameters (position, field of view, etc.) are transmitted. The missing views are interpolated at the receiver side. Finally, all views are emitted to their respective angle by the lenticular display. Only the images of the gray cameras are rendered and transmitted.

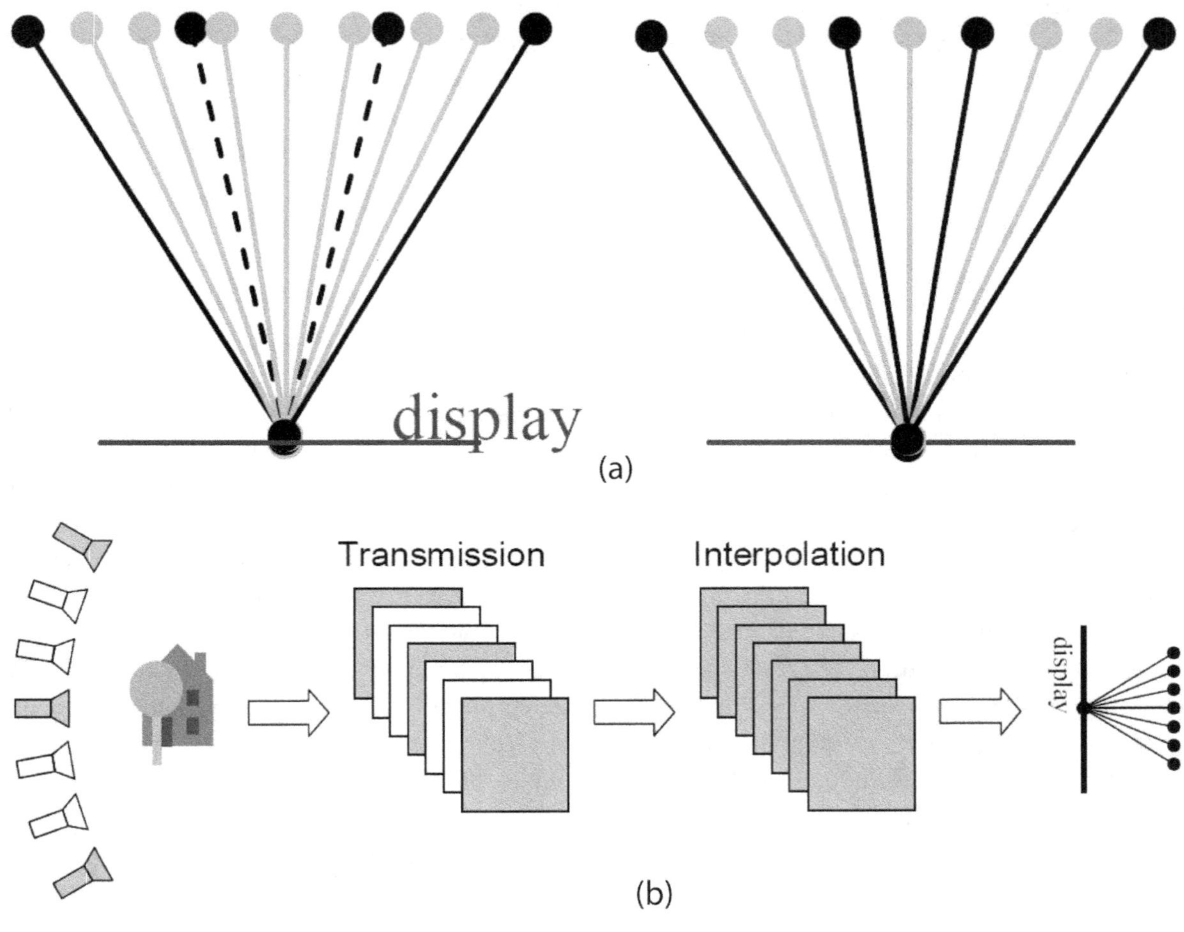

surfaces, which thus yields rendering artifacts. Alternatively, when texture and depth images are compressed at low and high quality, respectively, a high-quality depth image is employed to warp a coarsely quantized texture image, which also yields low-quality rendering. These two simple but extreme cases illustrate that a clear dependence exists between the texture- and depth-quality settings. It goes without saying that this dependency exists in the general case as well. Consequently, the quantization settings for both the depth and texture images should be carefully selected.

To determine the most efficient set of compression ratios for the texture and depth images, the optimal joint texture/depth quantization settings for the encoder has been introduced by Morvan et al. (2007). In order to find the optimal joint quantization settings, the bit-rate control that unifies the texture and depth Rate-Distortion (R-D) functions is created. The algorithm simultaneously combines the depth and texture data into a joint R-D surface model where the rate R is the sum of the depth and texture bit rate and the distortion D corresponds to the rendering quality.

The rendering quality is expressed as a maximal Peak Signal-to-Noise Ratio (PSNR) for every joint bitrate, which is simply the combined depth and texture bit-rate. The R-D surfaces for a "Ballet" video sequence, using the regular settings of H.264 (x264, 2010), can be found in Figure 4. Such a compression optimization was implemented using a slightly extended H.264/MPEG-4 AVC encoder, where the extension involves a joint bit-allocation algorithm. However, note that the proposed extension can be employed as an addition to any encoder, e.g., H.264/MPEG-4 AVC, JPEG-2000. Additionally, this joint encoding model can be readily integrated as a practical sub-system, because it influences the *setting* of the compression system rather than the actual coding *algorithm*. Finally, this optimal setting can be obtained with fast hierarchical search.

The PSNR is calculated in the following way. A virtual viewpoint with the same parameters as the center reference camera is created and to compute the PSNR, a comparison is made between the reference image and the virtual interpolated one. This measurement technique has been described in Mori et al. (2009). The RGB images are first transformed to the YUV color space. Then the PSNR of the Y values is calculated.

Depth Image Compression

The quality of depth images determines the free-viewpoint rendering result and the visual perception of the scene. Therefore, depth image compression is an important issue. Unlike most of the natural images, depth images are normally composed of flat zones separated by sharp edges. This property can be exploited explicitly to define a compression scheme that is better than standard proposals. Morvan et al. (2007) proposed a novel depth image coding algorithm which concentrates on the special characteristics of depth images: smooth regions delineated by sharp edges. The algorithm models these smooth regions using piecewise-linear functions and sharp edges by a straight line. To define the area of support for each modeling function, the image is decomposed into a quadtree that divides the image into blocks of variable size, each block being approximated by one modeling function containing one or two surfaces (see Figure 5). The subdivision of the quadtree and the selection of the type of modeling function are optimized, such that a global rate-distortion trade-off is realized.

In this framework, two classes of modeling functions are used: a class of *piecewise-constant* functions and a class of *piecewise-linear* functions. For example, flat surfaces that show smooth regions in the depth image can be approximated by a piecewise-constant function. Similarly, planar surfaces of the scene like the ground plane and walls, appear as regions of gradually changing grey levels in the depth image. Hence, such a planar region can be approximated by a single linear function. To identify the location of these surfaces in the image, a quadtree decomposition is employed (see Figure 6), which recursively divides the image into variable-size blocks, i.e., nodes of different size, according to the degree of decomposition.

In some cases, the depth image within one block can be approximated with one modeling function. If no suitable approximation can be determined for the block, it is subdivided into four smaller blocks. To prevent that too many small blocks are required along a discontinuity, we divide the block into two regions separated by a straight line. Each of these two regions is coded with an independent function. Consequently, the algorithm chooses between four modeling functions for each leaf as follows.

Figure 4. (a) A frame from the 'Ballet' video sequence; (b) R-D surface for 'Ballet' with H.264 video compression.

(a)

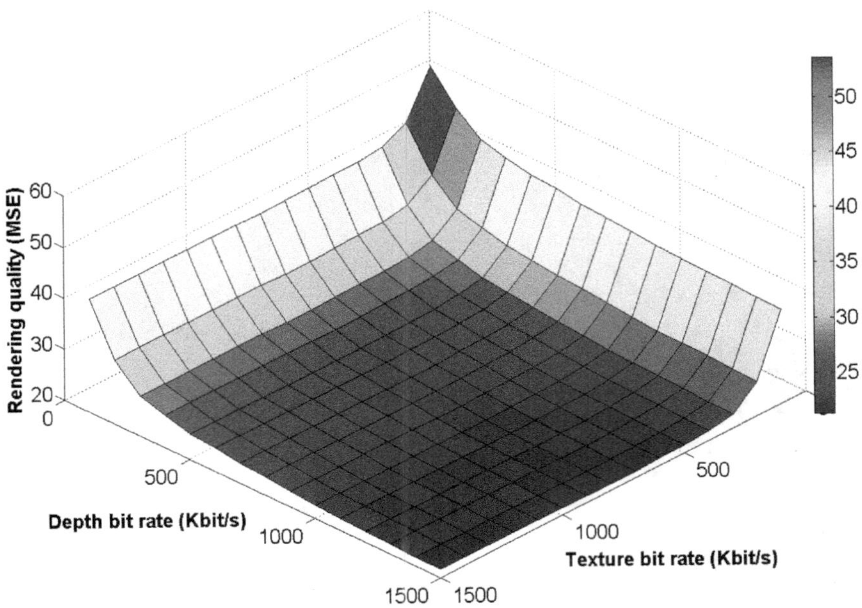

(b)

Figure 5. (a) The original depth image "Teddy"; (b) The corresponding reconstructed depth image using the described algorithm; (c) Superimposed nodes of the quadtree for the picture in (b). (Coding achieved with bit rate=0.12 bit/pixel and PSNR=36.1 dB).

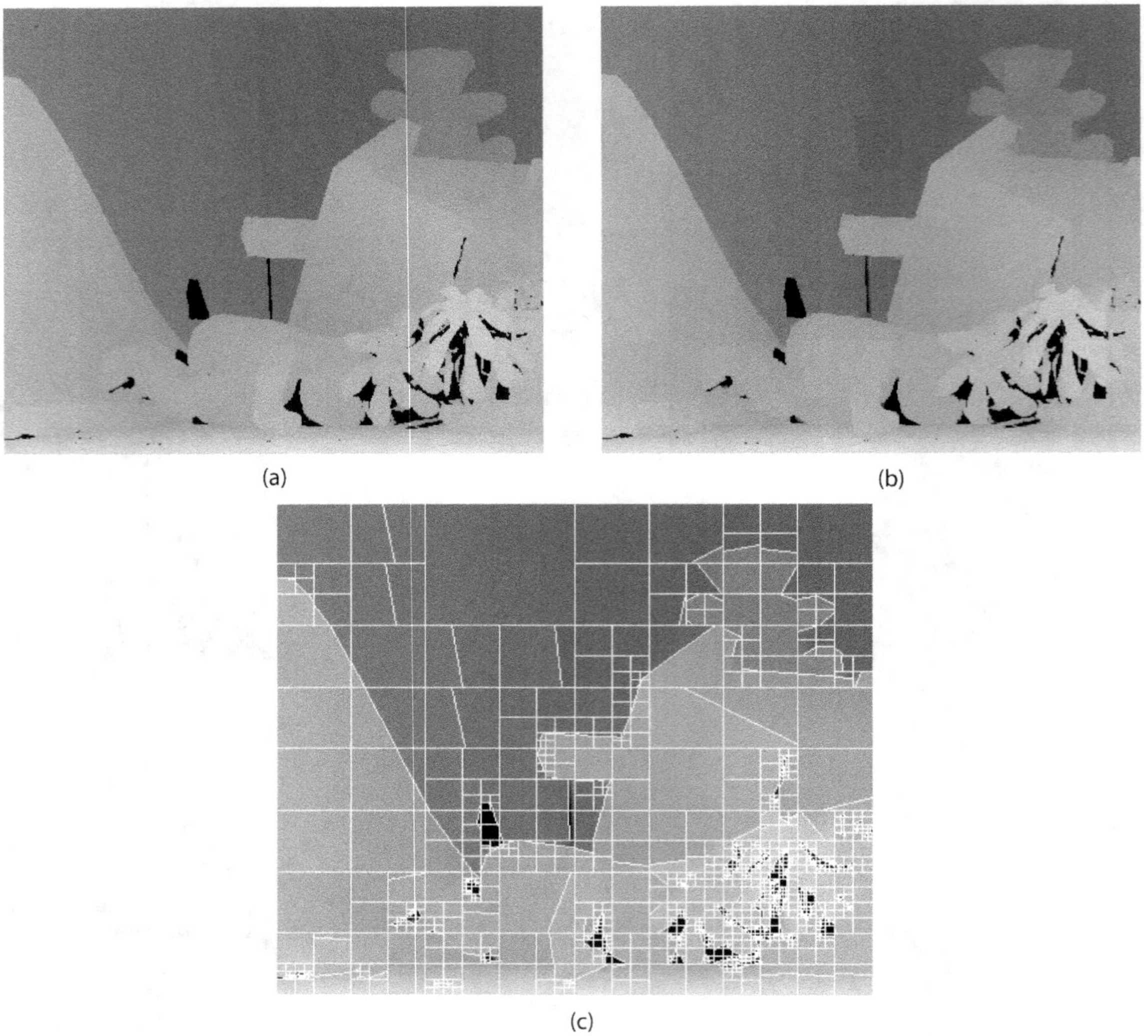

Figure 6. Example of depth image encoding by quadtree decomposition. Each block, i.e., node, of the quadtree is approximated by one linear function.

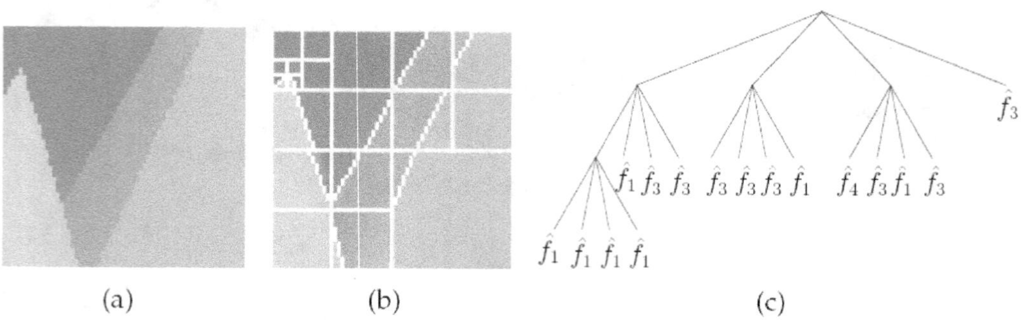

- *Modeling function* $\hat{f}_1$: This function approximates the block content with a constant function.
- *Modeling function* $\hat{f}_2$: This function approximates the block content with a linear function.
- *Modeling function* $\hat{f}_3$: This function subdivides the block into two regions separated by a straight line and approximates each region with a constant function (a wedgelet function);
- *Modeling function* $\hat{f}_4$: This function subdivides the block into two regions separated by a straight line and approximates each region with a linear function (a platelet function);

The selection of a particular modeling function is based on a rate-distortion criterion that trades-off the distortion and rate of the individual functions. More specifically, the algorithm selects for each quadtree block the modeling function that minimizes the Lagrangian R-D cost function according to

$$\tilde{f} = \underset{f_j \in \{\hat{f}_1, \hat{f}_2, \hat{f}_3, \hat{f}_4\}}{\arg\min} \left(D_m\left(\hat{f}_j\right) + \lambda R_m\left(\hat{f}_j\right) \right), \qquad (8)$$

where $D_m\left(\hat{f}_j\right)$ and $R_m\left(\hat{f}_j\right)$ represent the rate and distortion resulting from using a modeling function $\hat{f}_j$. Comparison with JPEG-2000 encoding has shown that this delivers higher compression rates for comparable PSNR, and provides images that lead to fewer artifacts in free-viewpoint interpolation (Morvan et al., 2007).

ADAPTIVE RESOLUTION

Professional applications, such as in the medical domain, require very high quality visualization. This aspect, together with the intrinsic quality of the display and the quality of 3D reconstruction, influences the decision of the surgeon and therefore is of primary importance. When using encoding strategies like H.264 coding, the strain on the bandwidth is the largest when multiple changes occur in the consecutive video frames. In order to cope with the bandwidth constraints in those cases, it is possible to temporarily lower either the temporal resolution (frame rates) or the spatial resolution.

An analysis of the pixel grid of lenticular display is performed in order to determine the optimal spatial resolution (Ruijters, 2009). The maximum information density that can be conveyed by the lenticular display per view, is determined by the way the pixels of the LCD grid are refracted by the lenticular lenses. In modern lenticular displays, the lens array is slanted under a slight angle, which affects the distribution of the set of pixels that are diverted to a particular viewing angle. Though the allocation of the subpixels over the grid is regular, it is not orthogonal. The sampling theory of multi-dimensional signals, described by Dubois (1985), can be used to examine the frequency range that can be transmitted by a certain non-orthogonal grid. Especially the maximum spatial resolution that does not lead to aliasing is of interest. When the resolution is too high, the lenticular display undersamples the transmitted images, and aliasing occurs. Although such images can be low-pass filtered to prevent aliasing, it is preferable to render them immediately at the optimal resolution, in order to keep the bandwidth usage on the transmission channel as low as possible.

The set of subpixels that are refracted to the same angular view can be considered to form a lattice. Let the vectors $\{v_1, v_2, \ldots, v_N\}$ form a basis, not necessarily orthogonal, of $\mathbf{R}^N$. Then lattice $L \subset \mathbf{R}^N$ is defined as a set of discrete points in $\mathbf{R}^N$, formed by all linear combinations of vectors v_1, v_2, ..., v_N with integer coefficients. In order to perform a Fourier transform of a signal, sampled on a lattice, its reciprocal lattice is required. The reciprocal lattice L^* of lattice L is defined as the set of vectors y, such that the dot product between y and x is an integer for all vectors x contained in lattice L. Let V be the matrix, whose columns are the representation of the basis vectors v_1, v_2, ..., v_N in the standard orthonormal basis for $\mathbf{R}^N$. Then matrix W, containing the basis vectors of the reciprocal lattice L^*, is determined by $W^T \cdot V = I$, with I being the $N \times N$ identity matrix. The *Voronoi* cell of a lattice is defined as the set of all points in $\mathbf{R}^N$ closer to the origin than to any other lattice point. The basis V for a given lattice is not unique (i.e., a lattice L can be described by several different basis matrices V). However, any basis for a certain lattice L delivers the same unique Voronoi cell.

Let the Fourier transform of a continuous multi-dimensional signal $u_c(x)$ with x in $\mathbf{R}^N$ be defined as:

$$U_c(f) = \int_{R^N} u_c(x) e^{-j2\pi f \cdot x} dx, \tag{9}$$

where frequency f is in $\mathbf{R}^N$. The Fourier transformation of signal u_c sampled on lattice L is periodical, with lattice L^* as periodicity (Dubois, 1985), leading to,

$$U(f) = \frac{1}{|\det V|} \sum_{r \in L^*} U_c(f + r). \tag{10}$$

Consequently, if a signal that is not bandwidth-limited within the Voronoi cell of lattice L^*, is sampled on lattice L, spectral overlap (i.e., aliasing) occurs.

The transmitted and interpolated views are rendered on an orthogonal grid, and the Voronoi cell of an orthogonal lattice is a simple rectangle. The maximum resolution that can be visualized on the lenticular screen can be examined by fitting this Nyquist frequency rectangle range of the orthogonal grid on the Voronoi cell of the reciprocal lattice of the lenticular sample grid. As long as the rectangle is completely contained within the Voronoi cell, no aliasing occurs. If this is not the case, the spatial resolution is higher than can be visualized by the lenticular screen, and the information loss manifests itself as aliasing artifacts.

As an example for the optimal resolution analysis, we consider a 9-view autostereoscopic display with slightly slanted lenticular lenses. The distribution of the views over the individual subpixels can be found in Figure 7. The composited image can be examined considering only one monochromatic primary color (red, green, or blue), or can be evaluated for all colors together. The basis matrices V of the sample lattice can be established by taking two vectors (nonlinearly dependent) between adjacent lattice points. The LCD pixel distance is used as a metric, which means that two neighboring subpixels (e.g., red and green) have a distance of 1/3 pixel. For example, for the color-independent lattice we take the vectors $v_1 = (5/3, -1)^T$ and $v_2 = (4/3, 1)^T$. This delivers the following basis matrices V with their reciprocals W^T:

$$\begin{aligned} V_{mono} &= \begin{pmatrix} 3 & -1 \\ 0 & -3 \end{pmatrix}, \\ V_{color} &= \begin{pmatrix} 5/3 & 4/3 \\ -1 & 1 \end{pmatrix}, \\ W_{mono} &= \frac{1}{9}\begin{pmatrix} 3 & 0 \\ -1 & -3 \end{pmatrix}, \\ W_{color} &= \frac{1}{9}\begin{pmatrix} 3 & 3 \\ -4 & 5 \end{pmatrix}. \end{aligned} \tag{11}$$

The individual views are rendered on an orthogonal grid, and the Voronoi cell of an orthogonal lattice is a simple rectangle. The maximum resolution that can be visualized on the lenticular screen can be examined by fitting this Nyquist frequency rectangle range of the orthogonal grid on the Voronoi cell of the reciprocal lattice of the lenticular sample grid.

Figure 7. (a) The LCD pixel grid and the view that is associated with each subpixel. The green subpixels that are diverted to view 0 are circled. (b) All subpixels that are diverted to view 0 are circled, independent from their color. (c) The reciprocal lattice of the green subpixels for view 0. The Voronoi cell of the reciprocal lattice is indicated in pink. Dotted rectangle indicated the Nyquist frequency of the 1/3 orthogonal grid is indicated. Since the Voronoi cell does not cover the complete Nyquist frequency range, slight aliasing in the higher frequencies may occur. (d) The reciprocal lattice of the subpixel configuration of view 0, ignoring the color. Since the Nyquist frequency range of the dotted rectangle is fully contained within the Voronoi cell (pink), there is no aliasing in the intensity image.

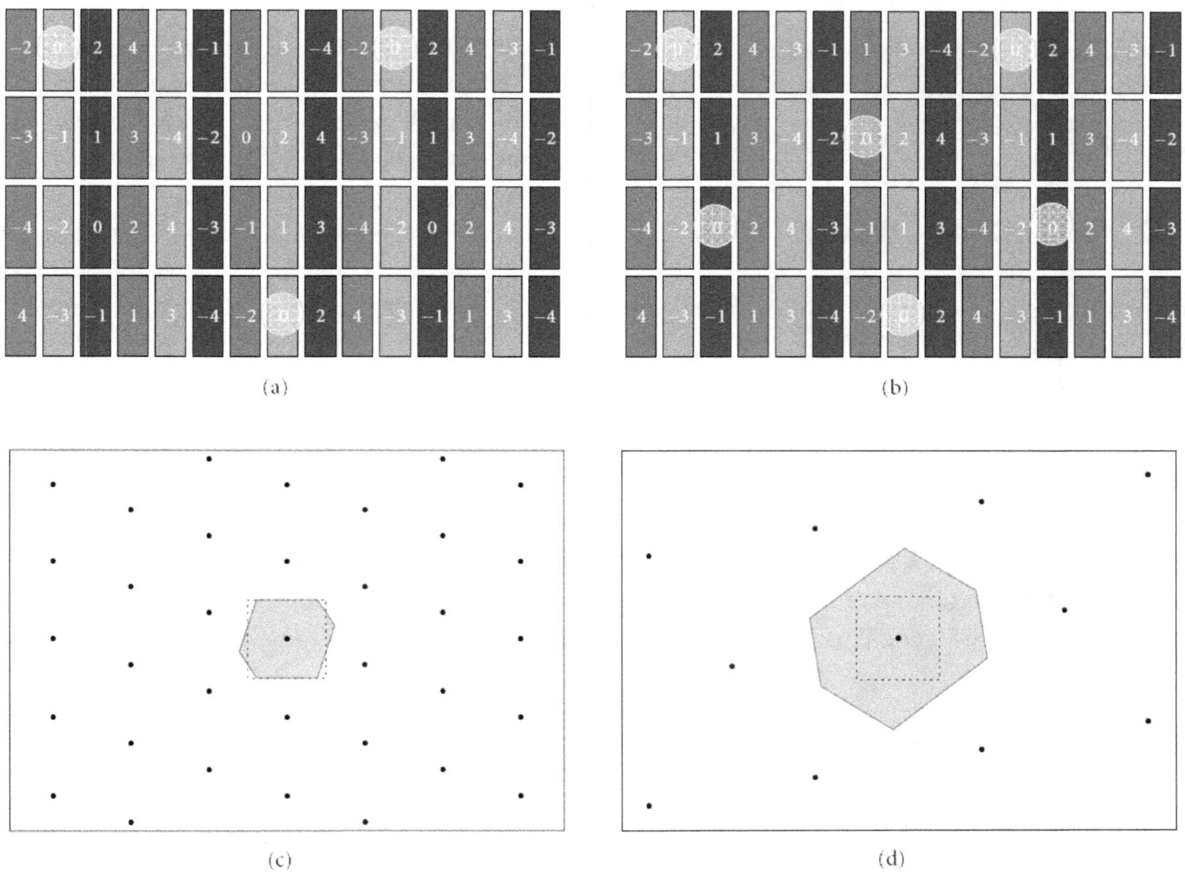

A logical choice for the resolution of the individual views, using a lenticular screen with nine views, seems to be 1/3 of the LCD pixel grid resolution in both directions. After all, this represents the same amount of information: nine views with each $1/3 \cdot 1/3$ of the amount of pixels of the LCD grid. We call this the 1/3 orthogonal grid. The Nyquist frequency rectangle of this resolution has been depicted on top of the Voronoi cell of the reciprocal lattice of the lenticular sample grid in Figure 7. Looking at a single primary color channel (in Figure 7(a) the green subpixels are used, but the lattice is the same for red and blue), it can be noted that the rectangle is not completely encapsulated within the Voronoi cell. This means that for monochromatic red, green, and blue images, there is a slight undersampling in certain directions, and aliasing may occur in the higher frequencies. If the lenticular lattice for a single view is considered, regardless of the colors of the subpixels, then the rectangle is completely contained within the Voronoi cell, such as in Figures 7(b) and 7(d). This implies that for gray-colored images, there is no aliasing when only the intensities are considered, but there may be some aliasing between the colors. In practice, this behavior resembles color dithering for real-world images. High frequent primary-colored structures (such as thin lines) may suffer from slight visible aliasing artifacts, though.

FUTURE RESEARCH DIRECTIONS

The detailed specification in terms of accuracy of the warping function will soon be tested on an experimental real-time platform. Our contribution on free-viewpoint interpolation should be merged with the new techniques emerging from the MPEG 3DAV working group, and benchmarking experiments are ongoing. With respect to video and depth compression, the standardization in the MPEG community tends to focus on fully exploiting the MPEG4 AVC/H.264 coding standard for both signals. In this chapter, we have presented a high-quality alternative for the depth coding, based on dedicated coding of local signal edges in the depth signal. The purpose of this algorithm is to combat the artifacts of MPEG4 AVC compression of the depth signal. However, this alternative has not been standardized. The third discussion regarding the matching of signal resolution to the display has shown that an analysis can reveal when aliasing will occur during the use of autostereoscopic displays. This analysis can be used to adapt the signal resolution such that the quality is optimized. It has been shown that with a lenticular display offering N views, the signal resolution should be $1/N$ times the amount of pixels of the LCD screen. The proposed methodologies are key processing steps for the rapidly developing market of 3D content and displays.

CONCLUSION

This chapter has presented several approaches for 3D multi-view video processing for fitting the limited bandwidth of the transmitting channel. To achieve this, we have considered three solutions: (1) free-viewpoint interpolation based on two surrounding views, (2) video encoding and depth compression, and (3) adaptive resolution processing of the multi-view images for autostereoscopic display. A primary conclusion of this chapter is that when all solutions are jointly applied, the framework allows for the desired bandwidth limitation for the 3D multi-view signals, while maintaining a high quality though with a slight quality loss compared to the unconstrained case. The bandwidth that can be saved by the presented approaches depends very much on the desired image quality and acceptable artifact level, but overall it can be concluded that an average load reduction of 50% (for conservative settings) to more than 95% (for more aggressive compression) can be achieved. Although the individual solutions have been tested and evaluated, the degree to which these solutions should be applied in a system setup, needs to be further evaluated and this also depends on the actual practical system requirements. Next to the techniques described in this chapter it is also possible to reduce the bandwidth bottleneck by increasing the capacity of the broadcasting network (e.g., use solutions like multi-broadcast). However, that is beyond the scope of this chapter. With respect to the viewpoint interpolation, we have found that the processing steps of the algorithm are all useful signal processing functions that solve most of the problems occurring in the view interpolation. A remaining issue is the accuracy of the warping function and the intrinsic quality of the depth map.

REFERENCES

Dodgson, N. A. (1997) *Autostereo displays: 3D without glasses*. Paper presented at EID: Electronic Information Displays, Esher, UK.

Dubois, E. (1985). The sampling and reconstruction of time-varying imagery with application in video systems. *Proceedings of the IEEE, 73*(4), 502–522. doi:10.1109/PROC.1985.13182

Isaksen, A., McMillan, L., & Gortler, S. J. (2000). Dynamically reparameterized light fields. In *Proceedings of the 27th Annual Conference on Computer Graphics and Interactive Techniques (SIGGRAPH '00)*. (pp. 297–306). New Orleans, LA: ACM Press.

Levoy, M., & Hanrahan, P. (1996). Light field rendering. In *Proceedings of the 23rd Annual Conference on Computer Graphics and Interactive Techniques (SIGGRAPH '96)*. (pp. 31–42). New Orleans, LA: ACM Press.

Maupu, D., van Horn, M. H., Weeks, S., & Bullit, E. (2005). 3D stereo interactive medical visualization. *IEEE Computer Graphics and Applications, 25*(5), 67–71. doi:10.1109/MCG.2005.94

McMillan, L., & Pizer, R. S. (1997). *An image based approach to three-dimensional computer graphics*. (Technical Report TR97-013). University of North Carolina at Chapel Hill.

Mori, Y., Fukushima, N., Yendo, T., Fujii, T., & Tanimoto, M. (2009). View generation with 3D warping using depth information for FTV. *Image Communication, 24*(1–2), 65–72.

Morvan, Y., Farin, D., & de With, P. H. N. (2007). Depth-image compression based on an R-D optimized quadtree decomposition for the transmission of multiview images. In *IEEE International Conference on Image Processing*, vol. 5. (pp. V105-V108). San Antonio, TX: IEEE Press.

Morvan, Y., Farin, D., & de With, P. H. N. (2007). Joint depth/texture bit-allocation for multi-view video compression. In *Proceedings of Picture Coding Symposium (PCS)*. Lisboa, Portugal.

Ruijters, D. (2009). Dynamic resolution in GPU-accelerated volume rendering to autostereoscopic multiview lenticular displays. *EURASIP Journal on Advances in Signal Processing, 2009*, Article ID 843753, 8 pages.

Ruijters, D., & Zinger, S. (2009). IGLANCE: Transmission to medical high definition autostereoscopic displays. In *3DTV Conference: The True Vision - Capture, Transmission and Display of 3D Video*. (4 pages). Potsdam, Germany: IEEE Press.

van Berkel, C. (1999). Image preparation for 3D-LCD. In *Proceedings SPIE, Stereoscopic Displays and Virtual Reality Systems VI*, vol. 3639. (pp. 84–91). San Jose, CA: SPIE.

van Berkel, C., Parker, D. W., & Franklin, A. R. (1996). Multiview 3D LCD. In *Proceedings SPIE, Stereoscopic Displays and Virtual Reality Systems III*, vol. 2653. (pp. 32–39). San Jose, CA: SPIE.
x264. (n.d.). *A free h264/avc encoder*. Retrieved July 30, 2010, from http://developers.videolan.org/x264.html

Zinger, S., Do, L., & de With, P. H. N. (2010). Free-viewpoint depth image based rendering. *Journal of Visual Communication and Image Representation, 21*(5-6), 533–541. doi:10.1016/j.jvcir.2010.01.004

ADDITIONAL READING

Do, L., Zinger, S., Morvan, Y., & de With, P. H. N. (2009). Quality improving techniques in DIBR for free-viewpoint video. In *3DTV-CONFERENCE 2009: The True Vision - Capture, Transmission and Display of 3D Video* (4 pages). IEEE Press.

Morvan, Y. (2009). *Acquisition, compression and rendering of depth and texture for multi-view video.* Doctoral dissertation, Technische Universiteit Eindhoven, the Netherlands.

Ruijters, D. (2010). *Multi-modal image fusion during minimally invasive treatment.* Doctoral dissertation, Katholieke Universiteit Leuven, Belgium and Technische Universiteit Eindhoven, the Netherlands.

Ruijters, D., & Zinger, S. (2009). IGLANCE: Transmission to Medical High Definition Autostereoscopic Displays. In *3DTV-CONFERENCE 2009: The True Vision - Capture, Transmission and Display of 3D Video* (4 pages). IEEE Press.

Regarding free-viewpoint interpolation and texture and depth compression, please refer to:Zinger, S., Do, L., & de With, P. H. N. (2010). Free-viewpoint depth image based rendering. *Journal of Visual Communication and Image Representation*, *21*(5-6), 533–541. doi:10.1016/j.jvcir.2010.01.004

Section 4

Robotic Vision

Chapter 20
3D Scene Capture and Analysis for Intelligent Robotics

Ray Jarvis
Monash University, Australia

ABSTRACT

The capability of robots to function effectively in the unstructured real world is dominated by the extent to which supporting sensory and computational resources can capture and analyse the 3D working environments within which they are to carry out their tasks. Many device technologies and computational algorithms have been developed over the last 30 years to enable such capabilities. This chapter chronicles a diverse range of such developments, comparing and contrasting them and indicating their various strengths and weaknesses to support intelligent robotic functionality in various domains of application.

INTRODUCTION

Intelligent Robotics is often defined as the melding of perception, reasoning and actuation or, more succinctly, 'sensor informed purposeful actuation' as embodied in a robotic system. Central to this field is the ability to capture and represent 3D scenes (and environments) and to reason over them to plan and execute useful physical action, whether in the domain of robotic manipulators or mobile robots (or both together).

DOI: 10.4018/978-1-61350-326-3.ch020

The ultimate, but currently impossible, ideal in 3D capture is to imagine a 'magic' powder which can be sprayed uniformly and densely over all surfaces (both external and internal) of objects in the scene and then sucked up into a heap with each particle remembering the location, texture and colour of the surface point it was on. Subsequent analysis would then consist of examining this heap for structures enabling the segmenting out of object surfaces, whole distinct objects, shapes and topology, and also, perhaps, the recognition of components against a pre-developed model data base. Elegant representations of extracted information would support efficient formulation of robotic action plans which may involve the collision-free manipulation of selected objects or efficient obstacle avoiding navigation through the environment to nominated goals or for pursuit/avoidance, hiding, rescue etc. operations. There are many ways in which 3D capture can be carried out using a wide variety of sensor technologies, including ultrasound, passive and active stereopsis, time-of-flight laser systems, silhouette, blur, brightness, attenuation, radar and x-ray. Likewise, there are many methodologies, which can be applied in the scene (environmental) analysis phase to answer such questions as the identity, placement, structural juxtaposition and accessibility of objects in the scene. Unfortunately, the 'magic' powder, as described above, does not exist, but many attempts have been made to approximate its capabilities. Many of these will be described in the chapter.

Once a suitable representation of the robot's working environment can be forged, even if only partially, a variety of motion planning algorithms (for trajectories when considering robotic manipulators and paths when considering mobile robots) supporting robotic hand/eye coordination or mobile robot navigation can be applied. In some circumstances 3D capture and planning are intermingled as in the case of a mobile robot simultaneously exploring and navigating its environment. At the opposite extreme of exploration in an initially unknown environment, it is sometimes possible and appropriate to pre-scan a whole working environment (e.g. shopping mall or city square etc.) in sufficient detail as to allow 'cyber navigation' of that space as a preliminary to replicating the plans with real robots in the real environment. Recently available large scale laser scanning technologies make such an approach very attractive when appropriate, since Virtual Reality like explorations can be used as part of the planning process and specific fixed objects and areas of interest annotated by hand with their functionality properties attached. Also, the localizing task can be reliably and accurately carried out by fusing live, on-board sensor data with the pre-collected model data using various scan matching approaches.

Carrying out a robotic task (once a plan has been devised) can, itself, be a special challenge, depending on the complexity (including dynamics) of the situation, the precision to which the robot can be controlled and its pose (location and orientation) determined, the means of locomotion itself (wheels, tracks, propellers, legs etc.), and the properties of the medium (e.g. underwater, in a vacuum etc.). Often plans need to be revised to accommodate new or changed information to compensate for error or when mission goals are modified. The determination of location and orientation (localization) is often an intensely sensor driven process, especially for mobile robot navigation in natural terrain which may also be initially unknown (not previously mapped or explored).

This chapter will cover all of the above subtopics but concentrate on the 3D capture and scene analysis components, covering many device technologies and analysis methodologies, their strengths and weaknesses and application domains in the context of intelligent robotics, providing many pictorial examples and an extensive bibliography to invite more detailed enquiry.

BACKGROUND

There have been many and varied attempts to provide knowledge of the working environment needed to guide robots in carrying out complex tasks in unstructured environments successfully, reliability and efficiently. Generally, the less structured the environment the more intelligence is required to achieve these goals and the richer must be the environmental knowledge to support this intelligence. In structured factory environments (e.g. an assembly line process) robotic actuation can follow precise pre-trained sequences to carry out the required tasks. Precision and speed are the essential ingredients here. However, when the environment is complex and possibly subject to change, continued sensor vigilance is required to guide adaptation to new situations. Whilst in some such situations (e.g. robot navigation on a plane surface) only 2D representations are necessary to guide path (or trajectory) planning and actuation, more complex, perhaps natural outdoor environments call for 3D representations to guide planning and actuation sequences. Thus, the acquisition and analysis of 3D representations are at the core of intelligent robotics research.

There have been a number of early surveys (Jarvis, 1983; Indyk & Velastin, 1994) describing a range of competing approaches to the acquisition and analysis of 3D data for intelligent robotics and there is insufficient opportunity to cover these in details here. However, some overview of the evolving field of 3D acquisition is appropriate. Four aspects help to reveal the scope of this field, specifically for robotic applications. Firstly, the timelines of live data acquisition is crucial. Whilst some off-line data gathering can be carried out without time constraints, live data is usually required to be acquired in real time during an actual robot mission. Tradeoffs between speed, accuracy and density are often required, these being closely linked to what sensor instruments and computational resources are available. Secondly, whether passive or active methods are acceptable needs resolving. Whilst passive methods require no special energy sources to probe the environment (e.g. ultrasound, light, x-ray, microwave) they tend to be at the mercy of natural variations (night and day, sunlight or cloud, rain etc.). Active methods can cause cross channelling, create hazards (e.g. strong laser beams are eye hazards) and can be detected (obviously inappropriate for stealth in security and military operations) and require energy sources (batteries etc.). The third aspect relates to size, weight range and cost. Clearly a heavy and expensive laser range finder would not be suitable for airborne mapping on a model aircraft. An ultrasound ranging system (Kleeman, 2003), whilst relatively inexpensive and light weight, would not be suitable for large scale outdoor navigation since its ranging domain is restricted to not much beyond 15 metres. The fourth important aspect is what kinds of surfaces can be ranged to. Passive systems based on stereopsis, range from image defocus and optical flow operate only for visually busy surfaces whilst range from shading can operate for visually bland surfaces, but usually the reflective properties of each such surface needs to be known as do some of the lighting properties (e.g. direction of the sun). Ultrasound and laser time-of-flight ranging can cope with visually bland surfaces but these should not be too reflective (e.g. mirrors). Active stereopsis replaces one camera with structured light source to enable bland surface ranging by analysing the projected patterns and knowing the triangulation geometry of the instrument set-up.

Given some form of environmental structure gleaned from sensors or provided through plans etc., there still remains the need to evaluate how best a robot manipulator may pick up and put down selected objects without collision or how a mobile robot might navigate within an obstacle strewn space. Usually shortest path or shortest time solutions are sought but sometimes aspects such as tractability, safety or covertress may also be included in these formulations. In many cases the environmental data is acquired incrementally and the planner has to cope with new and time-varying data. Thus, the efficiency of the path (trajectory) planning methodology is also of concern.

Actuation itself must rely on jointed arms, wheels, tracks, propellers and legs, each modality having its own challenges in the context of the working environment, whether smooth, rugged, under water or in the air (or in space, for that matter). Guiding the actuation process to comply with the path plan is particularly difficult when drift forces, whether from slippage of traction or fluid currents, are imposed.

Figure 1. (a) Triclops passive stereopsis ranging camera; (b) Disparity image.

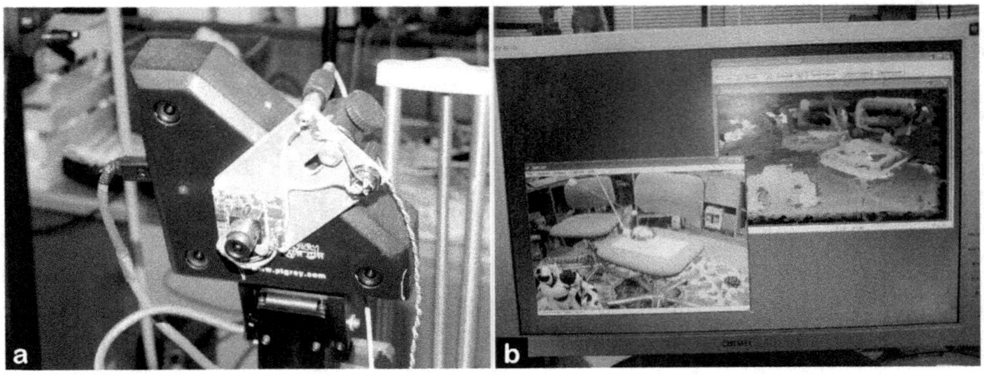

RANGING METHODOLOGY

Rather than attempting to cover every approach to range finding aimed at providing 3D environmental models for intelligent robotic operations a small number of specific examples within the direct experience of the author and for which instrument photographs, technical details and typical results can be supplied have been chosen. These will provide an overview of what principles are involved and what the strengths and weakness are over the range of examples presented.

Passive Stereopsis

Figure 1(a) shows a Point Grey Triclops unit with a three camera passive stereo monochrome camera system. Systems with only two cameras do not allow proper ranging to visual line features aligned to the baseline (parallel to the line between the optical centres of the cameras since the disparity (shift between camera images) cannot be discerned on this direction). By the addition of the third camera this problem is neatly overcome at the cost of some extra instrumentation and computer processing. Range is inversely related to disparity. Figure 1(b) shows an intensity (proportional to disparity) disparity image of a scene with significant visual texture for which passive stereopsis is well suited. Ranging up to several metres is possible with this sensor, given its relatively small camera separation. The larger the base line the more accurately the disparity can be measured but the less is the overlap of images between the cameras for which disparity can be measured at all. Variable base line systems can be used to resolve this trade off for different range dimensions (small versus large scenes). The intensity image from each camera is also available (monochrome). The same vendor also markets two and three camera systems with colour cameras. Poor results are obtained when lighting is insufficient.

Active Stereopsis

To avoid the problems of passive stereopsis where there is insufficient surface texture or visual business in general for which inverse range related disparity cannot be extracted and to introduce high contrast uniform visual structure, one camera of a passive stereopsis pair can be replaced by a projector which is able to cast light pattern over the scene. Stripe patterns running in quadrature to the base line are preferred but, in general, any type of pattern can be used. The single camera views these patterns as distorted from its viewpoint, which is displaced laterally with respect to the projector. The extent of these visual distortions contains disparity and hence range data. Figure 2(a) shows the equipment set-up of a stripe light active stereosis range system. Figure 2(b) shows a reconstructed 3D scene in which striped light range data and colour image data have been fused. Figure 2(c) shows raw 3D data points extracted using this system. One critical problem for light pattern based stereopsis is the difficulty of tracing continuous multiple patterns (e.g. stripes) over range 'jump boundaries' where the identity of an individual stripe can be lost. To resolve this identity issue (which is crucial for the triangulation geometry calculations to extract range) some researchers have used specific patterns for individual stripes, colour and/or intermittent code patterns for each stripe. Our system (Alexander & Ng, 1987) uses a temporal binary coding scheme. As an illustrative example, for a total of 64 stripes (say black on white), one image is recorded with all 64 turned on, another with pairs on and off, a third with groups of four on and off and so on up to 32 on and 32 off. Thus a total of $(\log_2 64)+1$ patterns are used in a temporal sequence. Doubling the resolution needs only one more image. Clearly, the time taken to collect the set of images is $(\log_2 n+1)$ times the acquisition time for one image but this may not be a serious problem if the switching times for patterns are small. In our system a 'light valve' using electronically controlled stripes is used and the seven patterns and corresponding camera images required for our 64-stripe system are dealt with in approximately 0.5 seconds. Had only one stripe at a time been turned on to resolve the ambiguity problem, 64 images would have been needed. In our system each individual stripe can be geometrically identified by the binary number represented in being on and off amongst the 7 images. This is known as a binary coded active stereopsis scheme. Our system can resolve range to less than 1mm for a scene within 50 cm of the apparatus. Unless a very bright light source is used (typically the standard collimated lighting systems of a slide projector is used) there is a limit to the depth of field over which the projected light patterns can be reasonably in focus to provide the high contrast being attempted. Enlarging the lens aperture of the projector increases the brightness of the pattern but lowers the depth of focus. Some trade-off is required. Our system is suitable for 3D scenes within a 50cm on the side cube.

It is relatively easy to fuse colour image pixel values on the 3D range data set and to extract details like surface normal, scene segmentation and planar region extraction is also possible (Jarvis, 1992a; Hoffman, 2000).

Passive Panoramic, Base Line Stereopsis

One of the irritating features of using regular cameras, even wide angled cameras, for robotic vision is that the frame of the image restricts the matching of images which may be taken around the robot. This 'windowing' problem is annoying when trying to match current images with those perhaps acquired earlier in an experiment. Panoramic images acquired by pointing a video camera at a parabolic like mirror with its axis vertical resolves this dilemma very simply, particularly if the image can be unwarped with a simple computation. Furthermore, using two camera/mirror systems displaced along a vertical base line, simple panoramic stereo ranging can be achieved since disparity matches always occur along common radial directions between the cameras. Being able to mechanically separate the cameras along a vertical axis allows for variable base line panoramic stereopsis ranging to cope with a variety of spatial scaling regarding the working environment. Figure 3(f) shows a variable base line panoramic stereopsis ranging system (Lui & Jarvis, 2010) built in the Intelligent Robotics Research Centre at Monash University (of which the author is Director) and Figures 3(a,b) shows a pair of raw panoramic images, Figures 3(c,d) the corresponding unwarped pair of images, Figure 3(e) the disparity image for this scene and Figures 3(g,h) reconstructed range/colour views derived from the disparity calculations and a colour

Figure 2. (a) Striped light active stereopsis equipment; (b) Fused colour and range; (c) 3D data points.

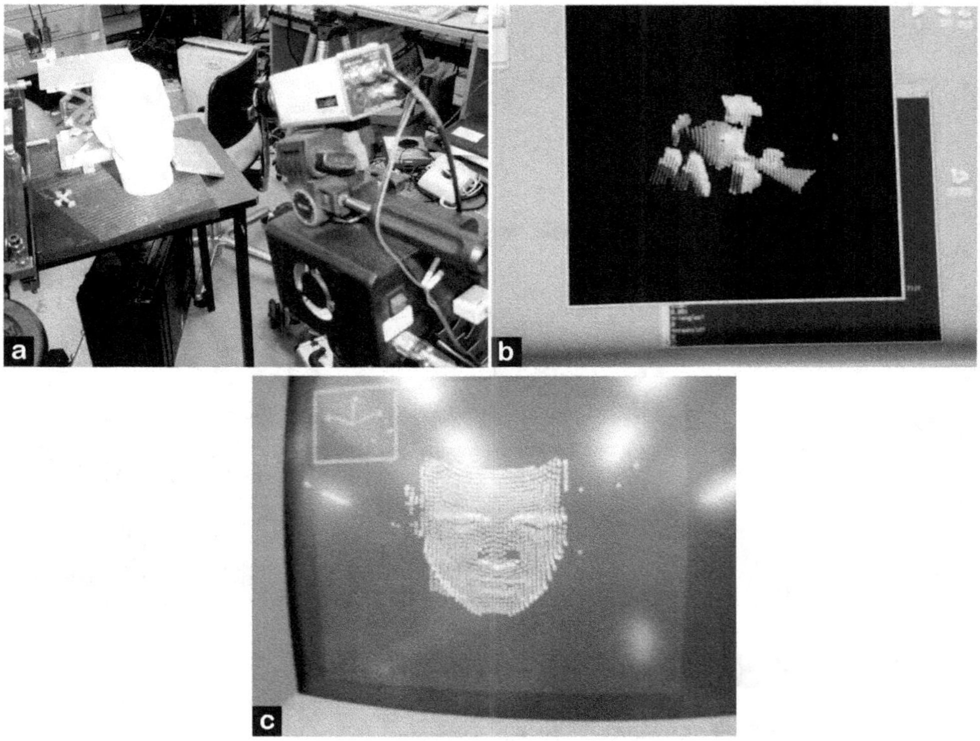

image for this scene. For this system it is possible to choose the optimal base line for a particular scene. The real time performance in calculating the computationally intense disparities through area mask correlation is achieved by utilizing the graphic processor's computational capabilities to the full.

Laser Time-of-Flight Ranging Cameras

Fairly recently, laser cameras which use time-of-flight ranging by means of modulated laser diode light and correlation detection of reflection time have become available on the market but are still quite expensive. These are capable of ranging to 7 metres and can supply range and intensity data at 30 frames per second with low to medium pixel resolution (e.g. PMD-19k). Older models tend to suffer from anomalies near jump boundaries (where range changes abruptly) but new ways of avoiding such effects have emerged. Figure 4(a) shows a typical 3D scene and Figure 4(b) shows the corresponding range image. Whilst the low pixel and range resolution of our system limits the accuracy of complex shape analysis its does provide sufficient information for obstacle detection and human gesture analysis (Li & Jarvis, 2009). The relatively low illumination provided by the laser diodes is no match for the flooding of direct sunlight for our instrument but new models with narrow bandwidths are available for outdoor use. These devices have no problem ranging to visually bland surfaces but specular reflecting surfaces can still be a problem.

Scanning Laser Rangefinders

Whilst a laser range camera, as described above, is a useful instrument for some robotic navigation

Figure 3. (a,b) Pair of raw panoramic images; (c,d) Corresponding unwarped image pair; (e) Disparity image; (f) Variable base-line passive panoramic stereo system; (g,h) Reconstructed range/colour views of scene.

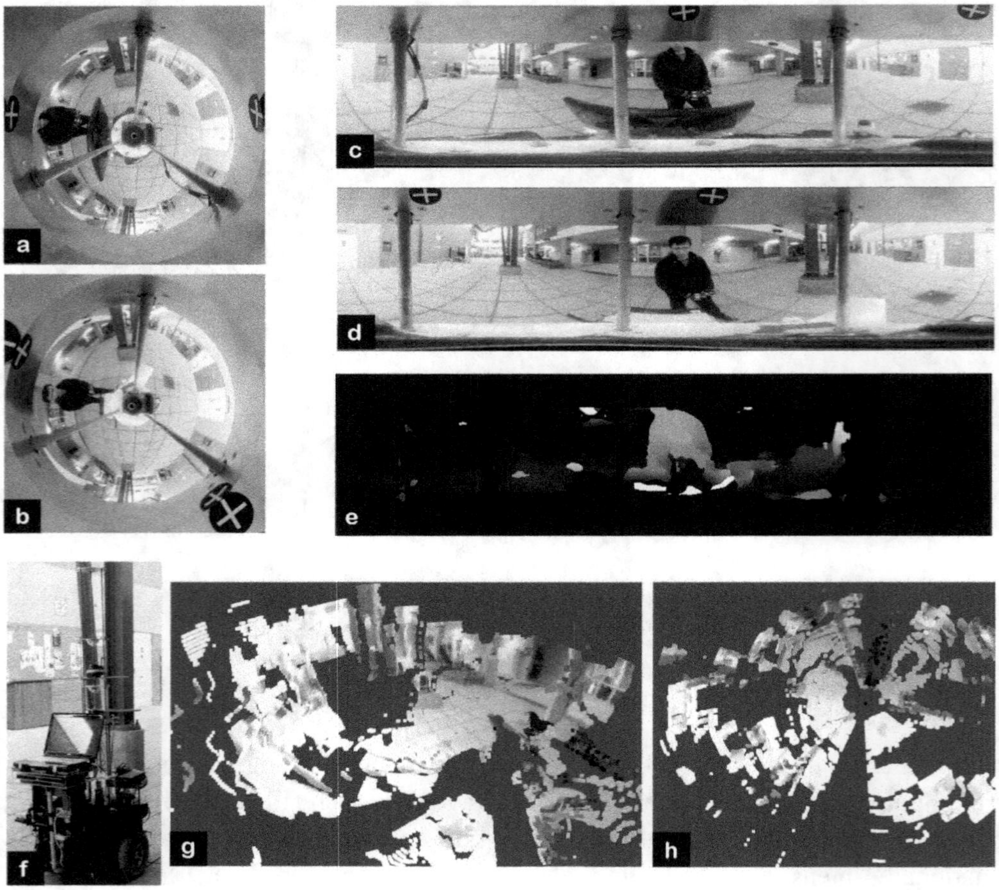

Figure 4. (a) 3D scene; (b) Corresponding range image.

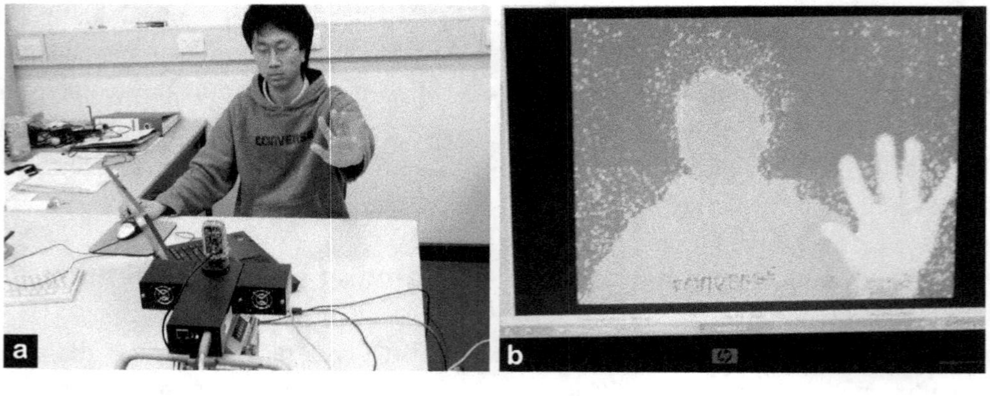

Figure 5. (a) Hokuyo rocking head laser ranging system; (b) 3D objects; (c) Range data points.

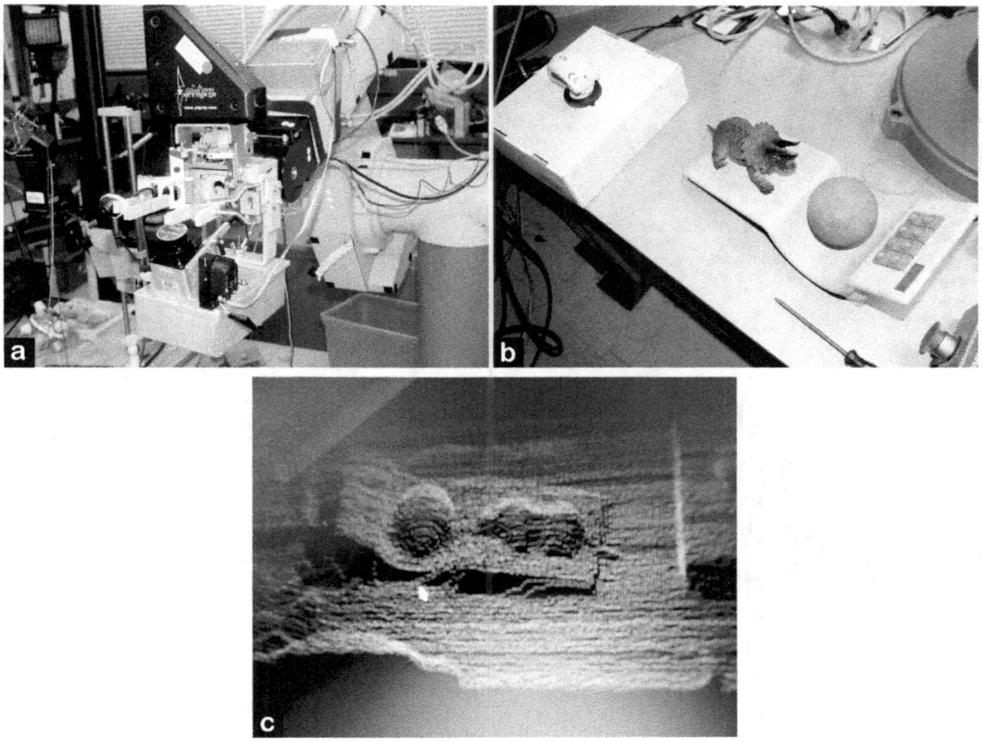

tasks, it is less useful for detailed shape analysis as may be required for robotic manipulation of named objects or the recognition (by shape) of objects in a navigation environment. Scanning laser rangefinders currently available on the market can be used for 3D acquisition by adding dimensions by scanning the collimated laser beam they generate for time-of-flight measurements. Both the Hokuyo and the Erwin Sick laser rangefinders can be used for 3D range acquisition useful for modelling object clusters using a variety of scanning configurations. Both these rangefinders produce a line scan by revolving a mirror which deflects the collimated laser beam in a rotating line scan, the Sick over 180° and the Hokuyo over 240°. These can be used for detecting the existence of obstacles in one specific plane (usually horizontal) for robotic obstacle avoidance. The Hokuyo (URG-04LX-UG01) provides one scan in 0.1 seconds with an angular resolution of 0.36 degrees and range resolution of 1 mm up to 4 metres. The Sick unit scans at 75 Hertz over its 180 degree scan at 0.5° intervals. Figure 5(a) shows a Hokuyo scanning laser range finder in a rocking head configuration using a stepping motor and a crank to nod the head. Whilst the acquisition time for a 3D scan varies according to nod scan resolution, some highly detailed models can be built using this configuration. An example is shown in Figure 5(b). This data is suitable for recognising relatively small objects and supports robot arm/hand manipulation of these objects by guiding the trajectory of the manipulator to detect suitable gripping operations.

Whilst the Sick LMS 200 laser range finder has been used in a number of nodding and rotating configurations (Surmann et al., 2001, May & Surmann, 2007), a particularly interesting set up is shown in Figure 6(a). Here the rangefinder rotates around the axis through the central beam

Figure 6. (a) Erwin sick rotating head laser ranging system; (b) Outdoor scene; (c,e) Indoor scan reconstructions; (e) With colour fusion; (d) 3D scan data for scene of (b).

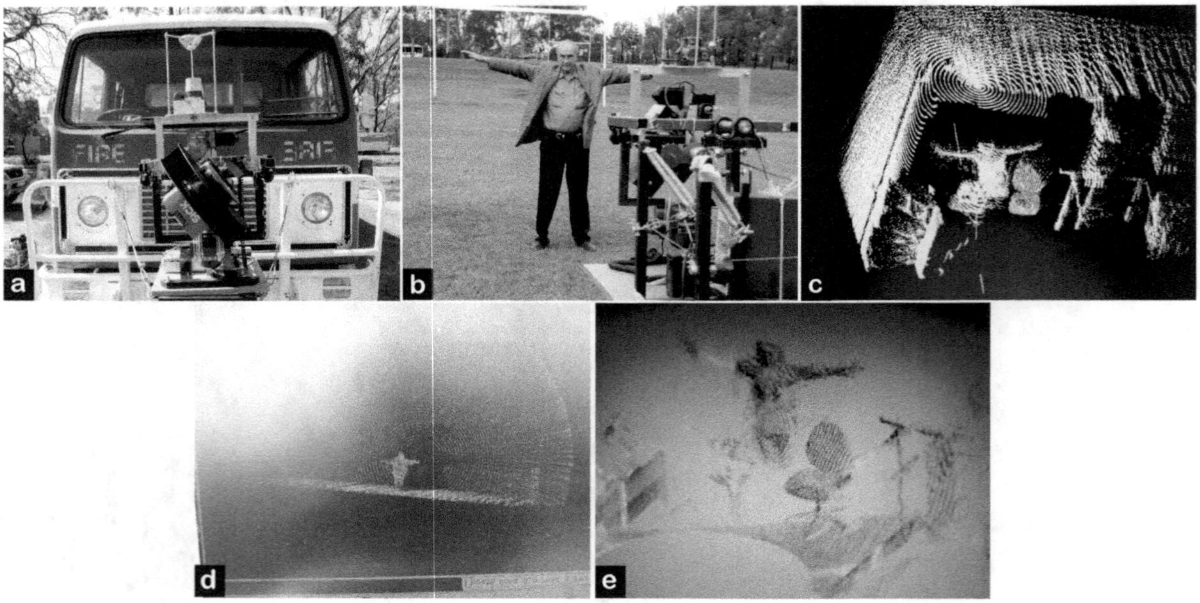

at the 90° position of its 180° scan (Jarvis, 2008). Thus a semicircular planer scan is revolved around that axis. Continuous rotation is achieved by supplying power and data signals via a very low resistance mercury immersion slip ring device so that both high currents and very low noise signals can be transmitted. Solid metal to carbon block contacts as are used in DC motors cannot provide this quality of contact. In our instrument, the range finder completes one rotation (about horizontal axis in our case) in between ≈ 0.5 to 5.0 seconds according to scan density preferences at a time cost. What is elegant about this configuration for obstacle detection, recognition, and avoidance tasks is that the 'central part' of the scan collects spatially denser range measurements than at the periphery, allowing more details to be extracted from the scene components directly in the planned forward path of the mobile robot. A typical scan is shown in Figure 6(d) for the scene of Figure 6(b). This instrument can read up to a maximum range of 30 metres with a 3 mm range resolution. The speed of rotation can be changed continuously at will since an accurate shaft encoder indicates the rotation position without one having to estimate rotation speed. As an extra feature, our scanner has a panoramic mirror based high resolution colour camera system whose axis is fixed in relation to the range scanner frame. Unwarped image data can be registered with range data as seen in Figure 6(e), for which the raw range data is seen in Figure 6(c). The range scanner and panoramic colour vision systems are mounted in a gymbal rig with gyroscopic stabilisation so that the pitch and roll of the robotic vehicle carrying the system can be largely eliminated from the 3D scan. The fused range/colour image data can be constructed on an on-board computer but transmitted to a remote site via radio Ethernet for sensor rich remote control or autonomous navigation overall monitoring and/or supervision.

Large Scale Laser Range Finder Environmental Modelling

Up till now all ranging schemes described were restricted to 'table top' scene scales or limited indoor/outdoor scenes from metres to small tens of metres in extent (e.g. Erwin Sick rotating head laser scanner ranging up to 30 metres). For truly large scale indoor or outdoor environmental modelling the 'big guns' can be used. Two examples will be described here. The first, a Riegl LMS-Z420, is used for off-line gathering of high resolution range and colour data capable of modelling large 3D environments up to 1600 metres across but only by setting up the instrument at strategic locations for minutes to hours at a time and fusing together the various data sets. Obviously, only surfaces visible from a location can be ranged to from that location, so that a complex large environment (e.g. a city square) needs several such data sets to be collected. Figure 7(a) shows the instrument, which is about the size of a large standard fire extinguisher. A laser beam is scanned in a vertical plane using a prism reflector and the whole instrument is rotated slowly about a vertical axis. Our instrument is fitted with a high-resolution colour digital camera, which takes a sufficient number of views around the vertical axis in a separate scan sequence. Figures 7(b,c,d) shows typical 3D reconstructions from indoor 3D data. The fused image/range data can be used for virtual reality cyberspace explorations and is particularly useful for trialling robot navigation strategies before the deployment of the physical robot (Jarvis, 2007). More of this later.

The second large scale laser range finder to be covered here, a Velodyne HDL-54E S2 (see Figure 8(a)), is capable of spinning 64 laser beams, spread evenly in a fan pattern in a plane, about an axis in that plane. It can collect up to 1.8 million samples per second. The rotation speed can be varied from 5 hertz to 15 hertz. We typically use 10 hertz. When the instrument is mounted to scan about a vertical axis on top of a vehicle, whether robotic or not, it can provide range data

Figure 7. (a) Riegl LMS-Z420 laser range scanner; (b,c,d) 3D Indoor colour/range construction examples.

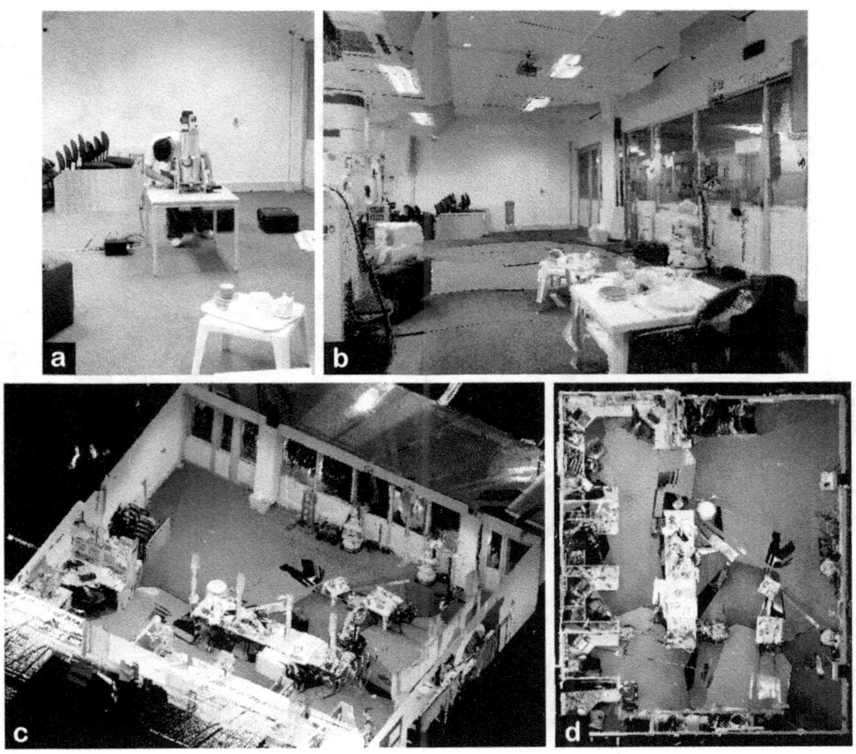

around the vehicle up to 120 metres away at 10 hertz, spanning the elevation range from +2 degrees above the horizontal at –24.8 degrees below the horizontal, thus covering the space where obstacles which could impede the movement of the vehicle reside. A typical outdoor 3D scan is shown in Figure 8(b).

By using a panoramic surveillance camera mounted over the Velodyne it is possible to collect colour image data, which can be fused with the range data. However we are currently not able to do this at 10 hertz. Figure 8(c) shows the Velodyne with a Mobotix panoramic (180 degrees by 360 degrees) camera mounted above it and Figure 8(e) shows a fused colour/range result for an indoor laboratory environment of about 15 metres square.

Cyberspace and Real Robotic Navigation

The robot navigation research community has, over the last ten years or so, expended considerable intellectual energy on the intriguing and important topic of how to construct environment maps, built incrementally from on-board sensor data whilst the robot navigates a previously unknown region and at the same time determines the location and orientation (pose) of the robot within that evolving map. Potentially expanding errors can be carefully managed with clever formulations, usually based on Extended Kalman Filters or Particle Filters, and partially corrected when closure (getting back to a previously measured place) can be reliably identified (Durrant-Whyte & Guivant, 2000).

Figure 8. (a) Velodyne HDL-54E S2 instrument; (b) Typical outdoor velodyne 3D scan; (c) Velodyne/panoramic camera combination on robot, (d) Panoramic colour camera view of laboratory, (e) Range/colour 3D reconstruction of laboratory.

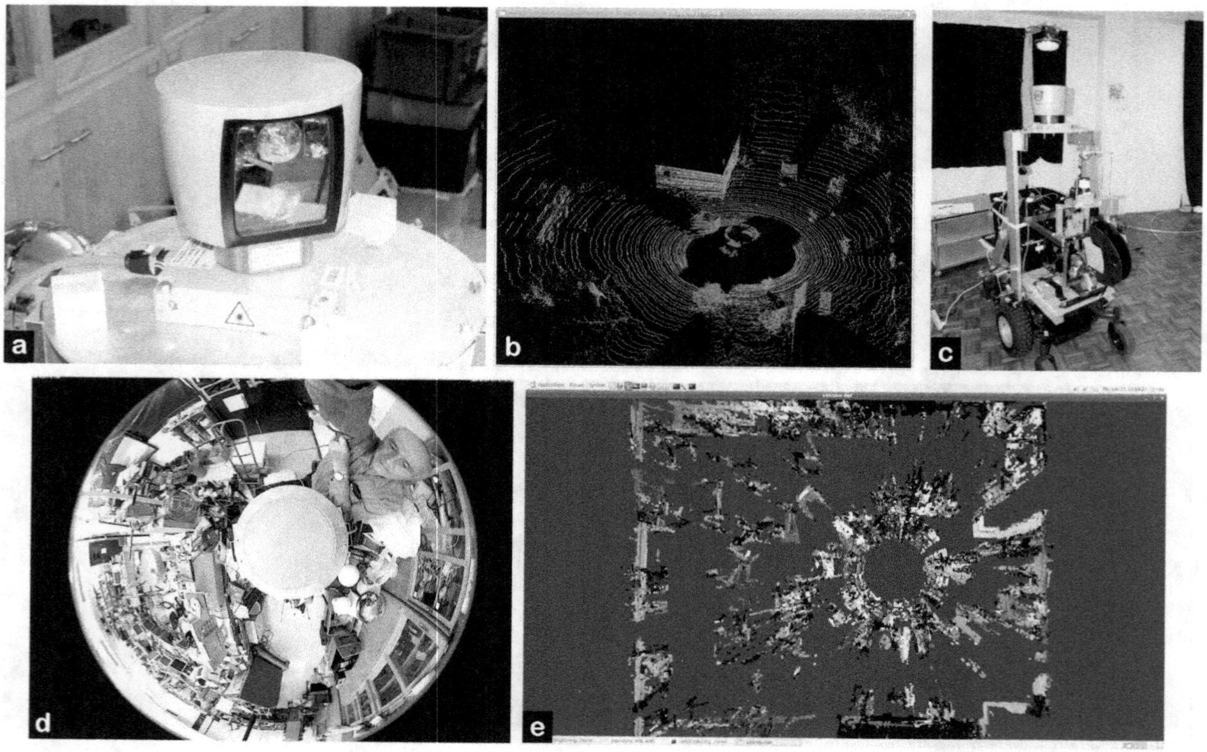

Much discussion has taken place concerning the optimal way of doing this task (Simultaneous Localisation and Mapping – SLAM) and just how reliably one can recognise closure and exploit it.

Whilst this approach is important and intellectually challenging, there are many situations where one-off gathered or previously available map data can be used instead. For example in man-built environments, building plans would be available. In other cases where prior plans are not available or insufficiently complete, using an instrument, such as the Riegl described above, is perfectly justifiable, despite the off-line time and care taken in modelling the environment, if the robot is expected to work in that environment for some time, perhaps on a daily basis.

Using such detailed 3D data avoids the closure problem and also permits the human annotation of important landmarks and functional objects in the environment, which may be relevant to the robot's task. Additionally, complex robot navigational missions can be initially undertaken in the cyberspace, provided by the 3D model data, prior to deploying the real physical robot to actually complete its mission in real space. Thus the gap between simulation and physical actuation can be closed elegantly.

Two robot localisation (determining location and orientation) experiments using Riegl data are described below. Localisation is one crucial requirement for autonomous navigation. Add path planning (covered later) to environmental mapping and localisation and one has the basic components of autonomous robot navigation. In the first experiment the unwarped images from a camera looking up at a panoramic mirror whilst the instrument is moved around an environment previously scanned by a Riegl laser rangefinder are matched against the pre-scanned data. The Haar image compression 'signature' (Liehart & Maydt, 2002) of a live image is matched with similar signatures extracted from the image data fused with the 3D model from a grid of viewpoints using a particle filter (Ho & Jarvis, 2008). The location of the panoramic camera system is determined in real time as it is moved through that environment. The height of the camera can also be extracted. This approach is called 'appearance based' since it relies solely on image matching (here the images used for the database are extracted from the Riegl model data). Figure 9 shows a typical example, which shows part of a localisation trace, the particle filter distribution at the current point in that trace and an insert showing one shot of what is seen by the panoramic camera when a walking frame carrying the panoramic camera system is pushed through the environment by a person. Of course, if the camera system were on a robot its location would be tracked.

The second experiment localises a vehicle with a Velodyne mounted on its roof as it is driven through a bush environment previous scanned by a Riegl range finder (Jarvis & Ho, 2010). In this case range data gathered live from the Velodyne is matched with the 3D data from the Riegl data. Locations of 13cm accuracy were obtained in a large-scale environment of about 150 metres x 100 metres. Figure 10 shows a typical example.

Trajectory/Path Planning

Once an environmental model, even if incomplete, has been acquired and the location of the robot vehicle is determined, optimal collision-free paths to nominated goals can be calculated to complete the requirements of autonomous navigation. A detailed comparison of some well-known path planning methodologies are given in (Jarvis, 2006). Only one simple method will be described here since planning is not the main focus of this chapter. Other approaches are also worth noting (Lozano-Perez, 1983; LaValle & Kuffner, 2000).

In a two dimensional rectangular tessellated floor space with obstacles, each shown as occupied connected cells, it is possible to find optimal collision-free paths from a nominated start point to a nominated goal point using a procedure known as a Distance Transform (Jarvis, 1984, 1994).

Figure 9. Appearance based localization using a Riegl laser range finder and colour image data base

Figure 10. (a) Reconstructed range/colour image outdoor 3D scene; (b) Localisation (Riegl/Velodyne) trace in outdoor environment; (c) Typical distance transform path plan in outdoor environment mapped by Riegll laser range scanner

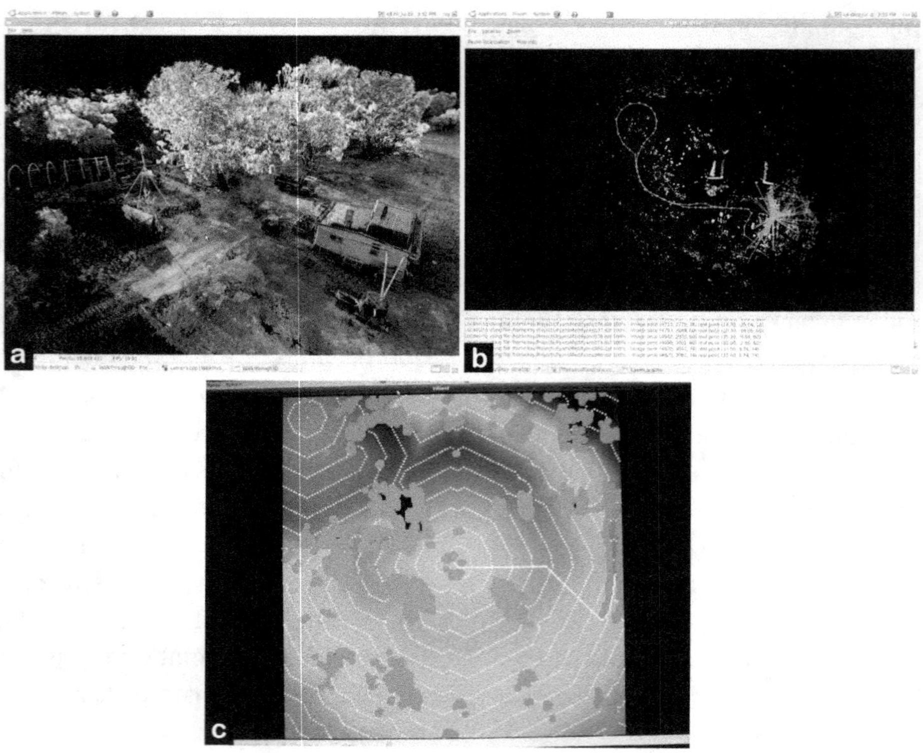

The method permits the allocation of integer distance values to all free (not occupied by obstacle components) cells indicating the number of steps each cell is from the goal. From the given start position following these distance values downhill in a steepest descent sense will lead to the goal in the minimum number of steps. The same approach can be used for 3D paths, spatio-temporal paths and paths in partially known environments. Weighting related to tractability can also be easily accommodated (Jarvis, 2010), as can questions of covertness (Marzouqi & Jarvis, 2003, 2004, 2005). Road paths can also be accommodated (Jarvis, 1992b). Figure 10(c) shows an optimal path from a nominated start point to a nominated goal for an outdoor environment previously scanned by a Riegl range finder.

DISCUSSION AND FUTURE WORK

It is clear that 3D capture plays a vital role for intelligent robotics whether for robotic hand/eye coordination or autonomous mobile robot navigation. When environments become complex and/or large, the representation of the 3D data becomes crucial both in terms of memory storage as well as efficient retrieval. The question of whether a tessellated space or some other representation is best for particular applications keeps coming up. As computation memory and power at reasonable prices keep increasing the debate keeps shifting towards elegance, efficiency and accuracy of algorithms for extracting relevant features to support particular applications. Methods for rendering and segmentation for 3D display require quite different approaches to the support robot manipulation and navigation in unstructured environments.

It would seem reasonable that Environmental Mapping, Robot Navigation and Virtual Reality come together as reinforcing themes, since this kind of approach allows for preliminary robotic simulation trials and human annotation of important feature from a functional point of view.

The debate on whether passive vision (using video cameras) or active devices like laser range finders are better for much intelligent robotics work seems to be favouring the latter as more and more suitable devices come on the market. The questions of detectability and cross-channelling have not yet been fully addressed but have not been dominant themes amongst researches at this time.

Whilst Simultaneous Localisation and Mapping (SLAM) has dominated robotics research literature in recent times, only time will tell if this approach will dominate the commercial side of robots likely to be deployed to do everyday tasks in the presence of humans. The more robust and reliable approach of off-line mapping may take over from the more elegant SLAM solutions just as a matter of practicality and simplicity.

Efforts to find the one complete representation, which support all applications, have so far not been forthcoming. One would hope that in the future such methodologies might emerge and yet one suspects that they will not. Interestingly, in the whole domain of Artificial Intelligence a unifying and universal methodology has likewise eluded discovery.

CONCLUSION

This chapter has outlined a number of instruments and methodologies for capturing and using 3D data to support intelligent robotics, whether for manipulators or mobile vehicles. The improving quality and reducing price of newly emerging 3D sensors will likely change the extent to which real robots will play on an ever increasing role in our everyday lives, accommodating to changeable environment and task complexity and human centric communication modes.

REFERENCES

Alexander, B. F., & Ng, K. C. (1987). 3D shape measurement by active triangulation using an array of coded light stripes. *SPIE: Optics . Illumination and Image Sensing for Machine Vision II, 850*, 199–209.

Durrant-Whyte, H. F., & Guivant, J. (2000). Simultaneous localization and map building using natural features in outdoor environments. *Intelligent Autonomous Systems, 6*(1), 581–588.

Ho, N., & Jarvis, R. A. (2007). *Global localisation in real and cyber worlds using vision.* Australasian Conference on Robotics and Automation 2007, 10th to 12th. Dec., Brisbane, Australia.

Ho, N., & Jarvis, R. A. (2008). *Towards a platform independent real-time panoramic vision based localisation system.* Australasian Conference on Robotics and Automation 08, (ACRA 08), 3rd to 5th Dec., Canberra, Australia.

Hoffman, I. D. (2000). *Three dimensional scene analysis using multiple view range data.* Ph.D. Thesis, Dept. of Electrical and Computer Systems Engineering, Monash University, Victoria, Australia.

Indyk, D., & Velastin, S. A. (1994). Survey of range vision systems. *Mechatronics, 4*(4), 417–449. doi:10.1016/0957-4158(94)90021-3

Jarvis, R., & Ho, N. (2010). Robotic cybernavigation in natural known environments. [Oct., Singapore]. *Accepted for Presentation at Cyberworlds, 2010*, 20–22.

Jarvis, R. A. (1983). A perspective on range finding techniques for computer vision. *IEEE Transactions on Pattern Analysis and Machine Intelligence, 5*(2), 122–139. doi:10.1109/TPAMI.1983.4767365

Jarvis, R. A. (1984). Collision-free trajectory planning using distance transforms. *Journal of the Institution of Engineers, 10*(3), 187–191.

Jarvis, R. A. (1992a). 3D shape and surface colour sensor fusion for robot vision. *Robotica, 10*, 389–396. doi:10.1017/S0263574700010596

Jarvis, R. A. (1992b). Optimal pathways for road vehicle navigation. *Proc. IEEE Tencon*, Nov. 11-13, Melbourne, (pp. 876-880).

Jarvis, R. A. (1994). On distance transform based collision-free path planning for robot navigation in known, unknown and time-varying environments . In Zang, Y. F. (Ed.), *Advanced mobile robots* (pp. 3–31). World Scientific Publishing Co. Pty. Ltd.

Jarvis, R. A. (2006). Robot path planning: Complexity, flexibility and application scope. *Practical Cognitive Agents and Robots*, Nov. 27-28, Perth, Australia, (pp. 3-14).

Jarvis, R. A. (2008). *Sensor rich teleoperation mode robotic bush fire fighting.* International Advanced Robotics Program/EURON WS RISE'2008,International Workshop on Robotics in Risky Interventions and Environmental Surveillance,7th to 8th Jan., Benicassim, Spain.

Jarvis, R. A. (2010). *Terrain-aware path guided robot teleoperation in virtual and real space.* ACHI 2010, St. Maartins, Feb. 10-14.

Jarvis, R. A., Ho, N., & Byrne, J. B. (2007). Autonomous robot navigation in cyber and real worlds. *CyberWorlds 2007*, Hanover, Germany, Oct. 24th to 27th, (pp. 66-73).

Kleeman, L. (2003). Advanced sonar and odometry error modeling for simultaneous localisation and map building. *Proceedings of the IEEE/RSJ International Conference on Intelligent Robots and Systems*, Las Vegas, (pp. 699-704).

LaValle, S. M., & Kuffner, J. J. (2000). Rapidly-exploring random trees: Progress and prospects. In *Proceedings Workshop on the Algorithmic Foundations of Robotics.*

Li, D., & Jarvis, R. (2009). *Real time hand gesture recognition using a range camera.* Australasian Conference on Robotics and Automation (ACRA 2009), Dec.

Lienhart, R., & Maydt, J. (2002). An extended set of Haar-like features for rapid object detection. *Proceedings of 2002 International Conference on Image Processing*, (vol 1, pp. 900-903).

Lozano-Perez, T. (1983). Spatial planning: A configuration space approach. *IEEE Transactions on Computers*, *C-32*(2), 108–120. doi:10.1109/TC.1983.1676196

Lui, W. L. D., & Jarvis, R. (2010). Eye-full tower: A GPU-based variable multibaseline omnidirectional stereovision system with automatic baseline selection for outdoor mobile robot navigation. *Journal of Robotics and Autonomous Systems*, *58*(6), 747–761. doi:10.1016/j.robot.2010.02.007

Marzouqi, M., & Jarvis, J. (2004). Covert robotics: Hiding in known environments. In *Proceedings from 2004 IEEE Conference on Robotics, Automation and Mechatronics*, 1st- 3rd December 2004, Traders Hotel, Singapore (pp. 804-809).

Marzouqi, M. S., & Jarvis, R. A. (2003). Covert path planning for autonomous robot navigation in known environments. *Proc. Australasian Conference on Robotics and Automation*, Brisbane.

Marzouqi, M. S., & Jarvis, R. A. (2005). Covert path planning in unknown environments with known or suspected sentries locations. The IEEE International Conference on Robotics and Automation (ICRA), Spain.

May, S., Pervoelz, K., & Surmann, H. (2007). 3D cameras: 3D computer vision of wide scope . In Obinata, G., & Dutta, A. (Eds.), *Vision systems: Applications*.

Surmann, H., Lingemann, K., Nuchter, A., & Hertzberg, J. (2001). A 3D laser range finder for autonomous mobile robots. Proc. 32nd International Symposium on Robotics, April, 19-22, (pp. 153-158).

ADDITIONAL READING

Arkin, R. C. (1998), Behaviour Based Robotics, The MIT Press Cambridge, Massachusitts, ISBN 0-262-01165-4. 491 pages.

Artac, M, Jogan, M. Leonardis, A., and Bakstein, H. (2005). Panoramic Volumes for Robot Localisation, in Intelligent Robots and Systems,(IROS), Aug., 2668-2674.

Arun, K. S., Huang, T. S., & Blostein, S. D. (1987). Least Square Fitting of Two 3-D Point Sets . *IEEE Transactions on Pattern Analysis and Machine Intelligence*, *9*(5), 698–700. doi:10.1109/TPAMI.1987.4767965

Besl, P. J. and Mc.Kay, N. D. (1992), A Method for registration of 3-D Shapes, IEEE Trans. Pattern Analysis and Machine Intelligence, 1492), 239-256.

Corke, P. I., Strelow, D., & Sing, S. (2004), Omnidirectional Visual Odometry for a Planetary Rover, Proc. International Conference on Intelligent Robots and Systems(IROS).

Fischler, M. A., & Bolles, R. C. (1981). Random Sample Consensus: a Paradigm for Model Fitting with Applications to Image Analysis and Automated Cartography . *Communications of the ACM*, *24*(6), 381–395. doi:10.1145/358669.358692

Fox, D. (2001). KLD-Sampling: Adaptive Particle Filters . In *Advances in Neural Information Processing Systems 14*. MIT Press.

Gibson, J. J. (1996). *The Senses Considered as Perceptual Systems*. Boston: Houghton-Mifflin.

Gregory, R. L. (1970). *The Intelligent Eye*. New York: McGraw-Hill.

Jain, A. K., & Flynn, P. J. (1993). *Three-Dimensional Object recognition Systems*. The Netherlands: Elsevier Science Publishers.

Jarvis, R. A. (1983). Expedient Range Enhanced 3-D Colour Vision . *Robotica*, *1*, 25–31. doi:10.1017/S026357470000103X

Jarvis, R. A. (1984), Robotic Vision Using 3D Space Cube Solid Modelling Derived from Multiple Image Projections, 7th Australian Computer Science Conference, Adelaide,pp 18-1 to 18-11.

Kalman, R. E. (1960), A New approach to Linear Filtering and Prediction Problems, Trans. Of the ASME-Journal of Basic Engineering, 82(Series D0, 35-45. Krotov, E. (1989), Mobile Robot localisation Using a Single Image, Proc. IEEE International conf. on Robotics and Automation, Vol.2, 978-983.

Latombe, J.-C. (1991). *Robot Motion Planning*. The Netherlands: Kluwer Academic Publishers.

Leonard, J. J., Jacob, H., & Feder, S. (1999), A Computationally Efficient Method for large-Scale Concurrent Mapping and Localisation, Proc. 9th international Symposium on Robotics research, Springer Verlag, 169-176.

Lewis, R. A., & Johnston, A. R. (1977), A Scanning Laser Rangefinder for a Robotic Vehicle, Proc. 5th International Joint Conf. on Artificial Intelligence, 762-768.

Marr, D. and Poggio, T.(1976), Cooperative Computation of Stereo Disparity, AI Memo 364, MIT AI Lab., Cambridge, Mass,,June.

Moravec, H. (1980), Obstacle Avoidance and Navigation in the Real World by a Seeing Robot Rover, PhD. Thesis, Stanford University, CA,USA.

Nitzan, D., Bain, A. E., & Duda, R. O. (1977). The Measurement and Use of Registered Reflectance and Range Data in Scene Analysis . *Proceedings of the IEEE*, *65*(2), 206–220. doi:10.1109/PROC.1977.10458

KEY TERMS AND DEFINITIONS

Active Stereopsis: Extraction of range data using the projection of patterns on the scene and one laterally displaced camera.

Appearance Based Vision: Machine understanding of a scene by its visual appearance only.

Computer Vision: Machine understanding of a physical scene through image or range analysis.

Environmental Mapping: The construction and representation of a working environment computational model.

Intelligent Robotics: Sensor informed purposeful actuation of a mechanical device.

Localisation: The determination of the location and pose of a robot in its working environment.

Passive Stereopsis: Extraction of disparity/range information from two or more images using only natural ambient lighting of the 3D scene.

Path or Trajectory Planning: The calculation of a collision-free path for robot navigation or robotic arm manipulation, subject to optimality constraints.

Range Finding: Determining the distances to object surfaces.

Simultaneous Localisation and Mapping (SLAM): The incremental construction of an environmental map as a robotic vehicle navigates through an initially unknown space whilst simultaneously determining its location and pose.

Chapter 21
Stereo Vision Depth Estimation Methods for Robotic Applications

Lazaros Nalpantidis
Royal Institute of Technology (KTH), Sweden

Antonios Gasteratos
Democritus University of Thrace, Greece

ABSTRACT

Vision is undoubtedly the most important sense for humans. Apart from many other low and higher level perception tasks, stereo vision has been proven to provide remarkable results when it comes to depth estimation. As a result, stereo vision is a rather popular and prosperous subject among the computer and machine vision research community. Moreover, the evolution of robotics and the demand for vision-based autonomous behaviors has posed new challenges that need to be tackled. Autonomous operation of robots in real working environments, given limited resources requires effective stereo vision algorithms. This chapter presents suitable depth estimation methods based on stereo vision and discusses potential robotic applications.

INTRODUCTION

Stereo vision is a reliable tool in order to exploit depth data from a scene, apart the pictorial one. The accuracy of the results depends on the choice of the stereo camera system and the stereo correspondence algorithm. Stereo correspondence is a flourishing field, attracting the attention of many researchers (Forsyth & Ponce, 2002; Hartley & Zisserman, 2004). A stereo correspondence algorithm matches pixels of one image (reference) to pixels of the other image (target) and returns

DOI: 10.4018/978-1-61350-326-3.ch021

the corresponding vertical displacement as the reference pixel's disparity, which is proportional to its depth. Thus, stereo vision is able to retrieve the third dimension of a scenery and, therefore, its importance is obvious in issues such as traversability estimation, robot navigation, simultaneous localization and mapping (SLAM), as well as in many other aspects of production, security, defense, exploration and entertainment.

Stereo correspondence algorithms can be grouped into those producing sparse and those giving dense output. Feature based methods stem from human vision studies and are based on matching segments or edges between two images, thus resulting in a sparse output. This disadvantage is counterbalanced by the accuracy and the speed of calculations. However, robotic applications demand more and more dense output. This is the reason why most of the relevant literature is focused on stereo correspondence algorithms that produce dense output. In order to categorize and evaluate them a context has been proposed (Scharstein & Szeliski, 2002). According to this, dense matching algorithms are classified in local and global ones. Local methods (area-based) trade accuracy for speed. They are also referred to as window-based methods because disparity computation at a given point depends only on intensity values within a finite support window. Global methods (energy-based) on the other hand are more time consuming but very accurate. Their goal is to minimize a global cost function, which combines data and smoothness terms, taking into account the whole image. Of course, there are many other methods that are not strictly included in either of these two broad classes. A detailed taxonomy and presentation of dense stereo correspondence algorithms can be found in (Scharstein & Szeliski, 2002). Additionally, the recent advances in the field as well as the aspect of hardware implementable stereo algorithms are covered in (Nalpantidis, Sirakoulis, & Gasteratos, 2008b).

ISSUES OF ROBOTICS-ORIENTED STEREO VISION

While a heavily investigated problem, stereo correspondence is far from being solved. Furthermore, the recent advances in robotics and related technologies have placed more challenges and stricter requirements to the issue. However, common problems related to outdoor exploration, such as possible decalibration of the stereo system and tolerance to non-perfect lighting conditions, have been barely addressed. Robotic applications demand stereo correspondence algorithms to be able to cope with not ideally captured images of the working environments of the robots (see Figure 1) and at the same time to be able to provide accurate results operating in real-time frame rates. Some of the open issues of robotics-oriented stereo vision methods are the handling of non-ideal lighting conditions, the requirement for simple calculation schemes, the use of multi-view stereo systems, the handling of miscalibrated image sensors, and the introduction of new biologically inspired methods to robotic vision.

Non-Ideal Lighting Conditions

The correctness of stereo correspondence algorithms' depth estimations is based on the assumption that the same feature in the two stereo images should have ideally the same intensity. However, this assumption is often not valid. Even in the case that the gains of the two cameras are perfectly tuned, so as to result in the same intensity for the same features in both images, the fact that the two cameras shoot from a different pose, might result in different intensities for the same point, due to shading reasons. In general, stereo image pairs captured in real life environments often suffer from differentiations in illumination, as those shown in Figure 2. Moreover, in real environments, which is the case for robotic applications, the illumination is far from being ideal (Klancar, Kristan, & Karba, 2004; Hogue, German, & Jenkin, 2007).

Figure 1. Robots equipped with stereo cameras in a real environments

The issue is usually treated by using robust pixel dissimilarity measures, which are able to compensate for lightness differentiations. Such a measure is the Zero mean Normalized Cross-Correlation (ZNCC) (Binaghi, Gallo, Marino, & Raspanti, 2004; Corke, 2005). However, this measure's computation is rather demanding and indirect approaches have been employed (Sun & Peleg, 2003). On the other hand, Ogale and Aloimonos in (Ogale & Aloimonos, 2005a, 2005b, 2007) propose and use a compositional approach to unify many early visual modules such as segmentation, shape and depth estimation, occlusion detection and local signal processing. As a result this method can process images with contrast, among others, mismatches. The first-stage dissimilarity measure used in this method is the phase differences from various frequency channels. Apart from these dissimilarity measures, a luminosity-compensated dissimilarity measure (LCDM) has been proposed in (Nalpantidis & Gasteratos, 2010b) and will be discussed in detail in a following section of this chapter.

Simplicity of Computations

Autonomous robots rely on their own decision-making algorithms (De Cubber, Doroftei,

Figure 2. A stereo image pair suffering from illumination differentiations

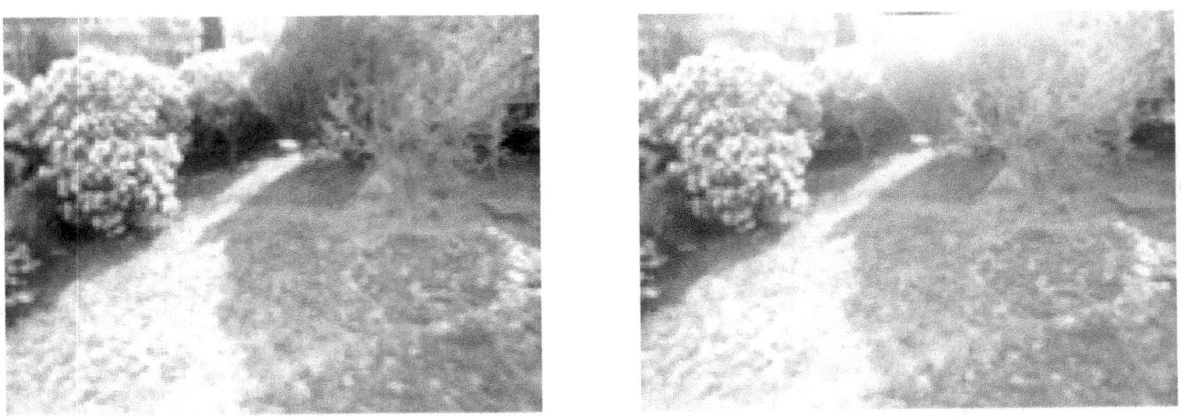

Nalpantidis, Sirakoulis, & Gasteratos, 2009). In the case of stereo vision-based navigation, the accuracy and the refresh rate of the computed disparity maps are the cornerstone of its success (Schreer, 1998). The most urgent constraint in autonomous robotics is the real-time operation and, consequently, such applications usually utilize local algorithms (Labayrade, Aubert, & Tarel, 2002; Soquet, Aubert, & Hautiere, 2007; Kelly & Stentz, 1998; Zhao, Katupitiya, & Ward, 2007; Konolige et al., 2006; Agrawal, Konolige, & Bolles, 2007). The hardware implementation of already proposed algorithms found in literature is not always straightforward (Nalpantidis et al., 2008b). Nevertheless, the hardware implementation of efficient and robust stereo algorithms able to provide real-time frame rates, especially in the case of moving robots, is very appealing. The allure of hardware implementations is that they easily outperform the algorithms executed on a general-purpose computer and, thus, the achieved frame-rates are generally higher. Furthermore, the power consumed by a dedicated hardware platform, e.g. ASIC or FPGA, is considerably lower than that of a common microprocessor and the computational power of the robot's onboard available computers is left intact.

Multi-View Stereo Vision

Early previous work focused on developing stereo algorithms mostly for binocular camera configurations. However, redundancy can lead to more accurate and reliable depth estimations. More recently, due to significant boost of the available computational power, vision systems using multiple cameras are becoming increasingly feasible and practical. The transition from binocular to multi-ocular systems has the advantage of potentially increasing the stability and accuracy of depth calculations. The continuous price-reduction of vision sensors allowed the development of multiple camera arrays ready for use in many applications. For instance, Yang et al. (Ruigang, Welch, & Bishop, 2002) used a five-camera system for real-time rendering using modern graphics hardware, while Schirmacher et al. (Schirmacher, Li, & Seidel, 2001) increased the number of cameras and built up a six-camera system for on-the-fly processing of generalized Lumigraphs. Moreover, developers of camera arrays have expanded their systems so as to use tens of cameras, such as the MIT distributed light field camera (Yang, Everett, Buehler, & Mcmillan, 2002) and the Stanford multi-camera array (Wilburn, Smulski, Lee, & Horowitz, 2002). These systems are using 64, and 128 cameras respectively. Most of the aforementioned camera arrays are utilized for real-time image rendering. On the other hand, a research area that could also be benefited by the use of multiple camera arrays is the so-called cooperative stereo vision; i.e., multiple stereo pairs being considered to improve the overall depth estimation results. To this end, Zitnick (Zitnick & Kanade, 2000) presented an algorithm for binocular occlusion detection and Mingxiang (Mingxiang & Yunde, 2006) expanded it to trinocular stereo.

The system proposed in (Nalpantidis, Chrysostomou, & Gasteratos, 2009) is a combination of quad-camera sensor hardware and a custom-tailored software algorithm. The sensory configuration of the presented system consists of four identical cameras. The four cameras are placed so as their optical axes to have parallel orientation and their principal points to be co-planar, residing on the corners of the same square, as shown in Figure 3(a). The images captured by the upper-left camera are considered as the reference images of each tetrad. Each one of the other three cameras produces images to be corresponded to the reference images. Thus, for each tetrad of images three, differently oriented, stereo pairs result, i.e. an horizontal, a vertical and a diagonal one. The concept, as well as the result of such a group of cameras is presented in Figure 3(b).

The hardware configuration, i.e. the four cameras' formation, produces three stereo image pairs.

Figure 3. (a) The quad-camera configuration and (b) the results (up-left) and scene capturing (right) using a quad-camera configuration

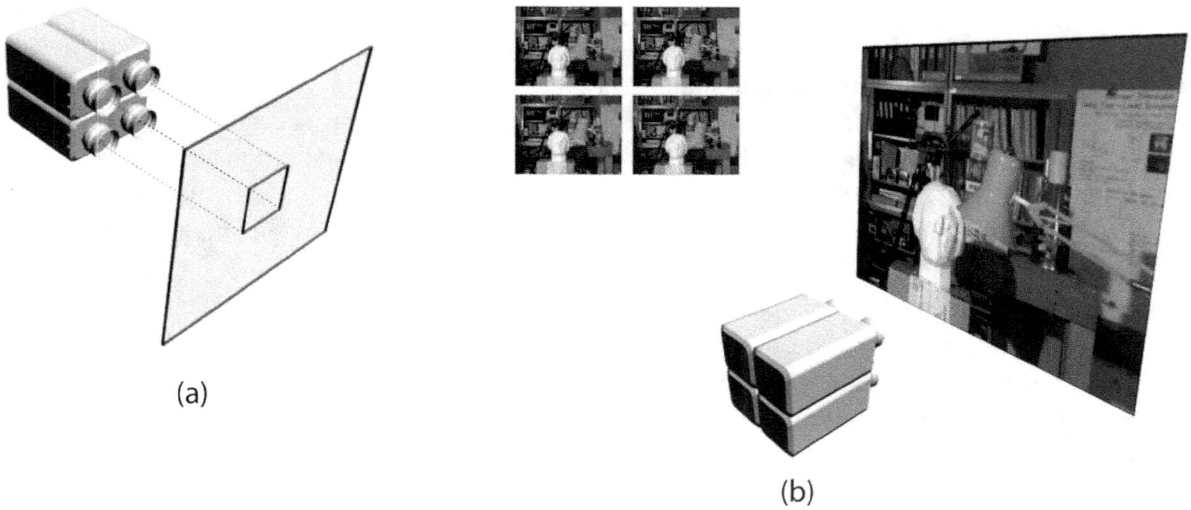

Each pair is submitted to a simple and rapid stereo correspondence algorithm, resulting, thus, in a disparity map. For each disparity map a certainty map is calculated, indicating each pixel's reliability. Finally, the three disparity maps are fused, according to their certainties for each pixel. The outcome is a single disparity map that incorporates the best parts of its producing disparity maps. The percentage of pixels whose absolute disparity error is greater than 1 in the non-occluded, all, and near discontinuities and occluded regions using the aforementioned method are 10.8%, 12.6%, and 31.5% respectively. These results are significantly better than the results that would have been obtained if only a pair of images were to be used. The combined hardware and software system is able to produce accurate dense depth maps in frame rate suitable for autonomous robotic applications.

Uncalibrated Stereo Images

The issue of processing uncalibrated images is common to applications where the sensory system is not explicitly specified. The plethora of computations most commonly require the massive parallelization found in custom tailored hardware implementations. Moreover, the contemporary powerful graphics machines are able to achieve enhanced results in terms of processing time and data volume.

A hierarchical disparity estimation algorithm implemented on programmable 3D graphics processing unit is reported in (Zach, Karner, & Bischof, 2004). This method can process either rectified or non-rectified image pairs. Bidirectional matching is utilized in conjunction with a locally aggregated sum of absolute intensity differences (SAD). This implementation, on an ATI Radeon 9700 Pro, can achieve up to 50 fps for 256×256 pixel input images. The FPGA implementation presented in (Jeong & Park, 2004) uses the dynamic programming search method on a Trellis solution space. It copes with the vergent cameras case, i.e. cameras with optical axes that intersect arbitrarily, producing non-rectified stereo pairs. The image pairs received from the cameras are initially rectified using linear interpolation and then, during a second step, the disparity is calculated. The architecture has the form of a linear systolic array using simple processing elements. The design is canonical and simple to be implemented in parallel. The resulting system can process 1280×1000 pixel images with up to 208 disparity levels at 15 fps. An extension of the previous method is presented in (Park & Jeong, 2007). The main difference is that information from previously processed lines are incorporated so as to enforce better inter-scanline consistency. The running speed is 30 fps for 320×240 pixel images with 128 disparity levels. The number of utilized processing elements is 128. The percentage of pixels with disparity error larger than 1 in the non-occluded areas is 2.63, 0.91, 3.44, and 1.88 for the Tsukuba, Map, Venus and Sawtooth image sets, respectively. Finally, (Masrani & MacLean, 2006) proposes the utilization of a local weighted phase-correlation method. The platform used is the Transmogrifer-4 system containing four Altera Stratix S80 FPGAs. The system performs rectification and left-right consistency check to improve the accuracy of the results. The speed for 640×480 pixel images with 128 disparity levels is 30 fps.

The stereo vision algorithm presented in (Nalpantidis, Amanatiadis, Sirakoulis, & Gasteratos, in press) is inspired by motion estimation techniques. It is based on a fast-executed SAD core for correspondence search in both directions of the input images. The results of this core are enhanced using sophisticated computational techniques; Gaussian weighted aggregation and 3D cellular automata (CA) rules are used. The hierarchical iteration of the basic stereo algorithm is achieved using a fuzzy scaling technique. The aforementioned characteristics provide results of improved quality, being at the same time easy to be hardware implemented. Consequently, the presented algorithm is able to cope with uncalibrated input images. The presented scheme is block search-based and does not perform scanline pixel matching. As a result, it does require neither camera calibration nor image rectification. However, it is clear that block search approaches require more computational resources since the number of pixels to be considered is greatly increased. In order to address this problem, the presented algorithm employs a variation of a motion estimation algorithm (Yin, Tourapis, Tourapis, & Boyce, 2003), which is used for JVT/H.264 video coding (Wiegand, Sullivan, Bjntegaard, & Luthra, 2003). The adaptation of compression motion estimation algorithms into disparity estimation schemes can be effective both in accuracy and complexity terms, since compression algorithms also attempt to achieve complexity reduction while maintaining coding efficiency. On the other hand, CA have been employed as a intelligent and efficient way to refine and enhance the stereo algorithm's intermediate results. Let the maximum expected horizontal disparity for a stereo image pair be D. The dimensions of the stereo pixel matching search block are $D \times D$. For each search block, the disparity value is determined by the horizontal distance of the (single pixel sampling) best match in terms of minimum SAD, as shown in Figure 4.

In the first stage, the disparity algorithm finds the best match on the quadruple sample grid (circles). Then, the algorithm searches the double pixel positions next to this best match (squares) to assess whether the match can be improved and if so, the single pixel positions next to the best double pixel position (triangles) are then explored. The general scheme of the presented hierarchical matching disparity algorithm between a stereo image pair is shown in Figure 5.

Each of the intermediate disparity maps of the first two steps is used as initial conditions for the succeeding, refining correspondence searches. In order to perform the hierarchical disparity search three different versions of the input images are employed and the stereo correspondence algorithm is applied to each of these three pairs. The quadruple search step is performed as a normal pixel-by-pixel search, on a quarter-size version of the input images. That is, each of the initial images has been down-sampled to 25% of their initial dimensions. The quadruple search is performed by applying the stereo correspondence algorithm in $(D/4) \times (D/4)$ search regions, on the down-sized image pair (D being the maximum

Figure 4. Quadruple, double and single pixel sample matching algorithm

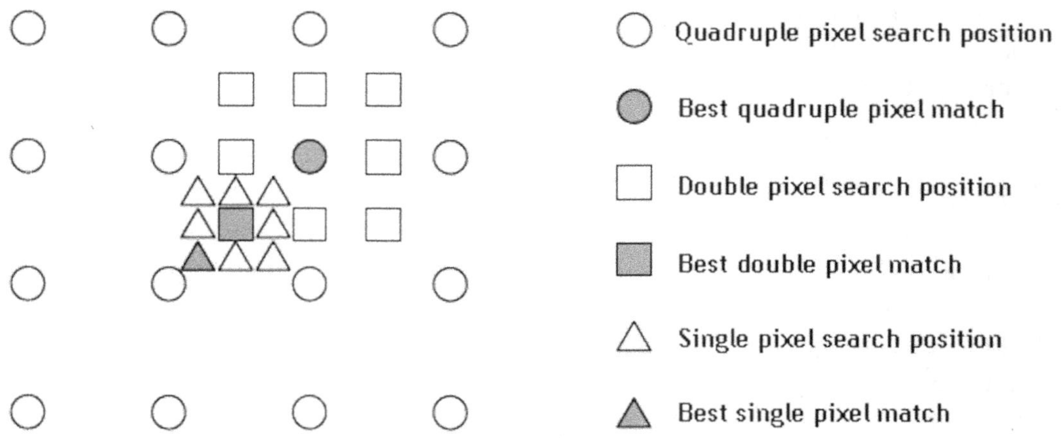

expected horizontal disparity value in the original image pair). The choice of the maximum searched disparity *D/4* is reasonable as the search is performed on a 1/4 version of the original images. This method provides good depth estimation with limited calculations, even for not calibrated input. The experimentally calculated Normalized Mean Square Error (NMSE) for radially distorted (10%) images of the Tsukuba, Venus, Teddy, and Cones data sets is 0.0712, 0.0491, 0.1098, and 0.0500 respectively. These results were in all cases less than 0.112% different from the results obtained for the original non-distorted image pairs and show that the discussed algorithm is robust against input's miscalibrations and distortions.

Biologically Inspired Methods

The success of the human visual system (HVS) in obtaining depth information from two 2D images still remains a goal to be accomplished by machine vision. Incorporating procedures and features from HVS in artificial stereo-equipped systems, could improve their performance. The key concept behind this transfer of know-how from nature to engineering is identifying, understand-

Figure 5. General scheme of the presented hierarchical matching disparity algorithm. The search block is enlarged for viewing purposes

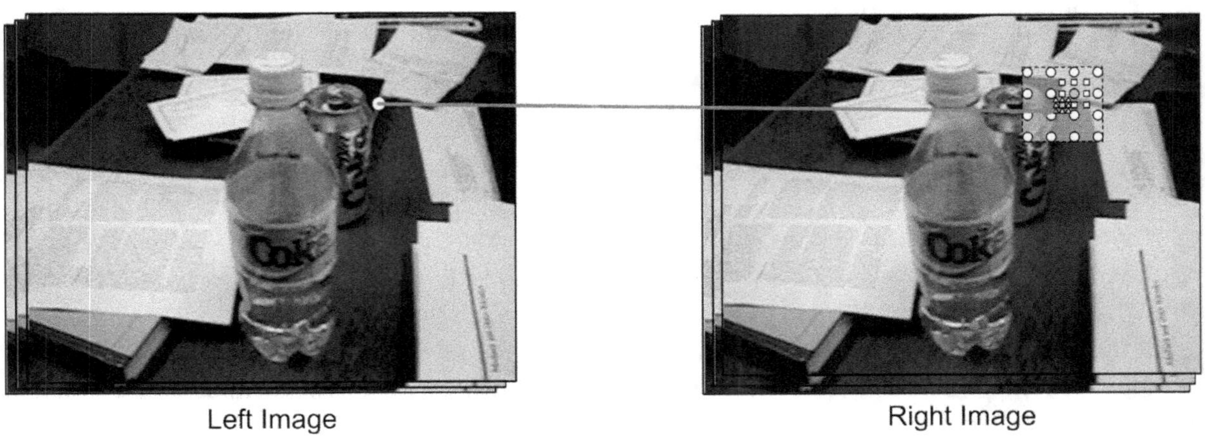

ing and expressing the basic principles of natural stereoscopic vision, aiming to improve the state-of-the-art in machine vision. These principles are mainly involved in the aggregation step that most existing algorithms employ.

HVS has been studied by many branches of the scientific community. Physics have expressed color information through color spaces, while biology has investigated the response of the eyes to it and the physiology of the eye. Psychophysics has studied the relationship between individual stimuli's changes and the perceived intensity, which is applicable to vision as well as all the other modalities. On the other hand, the gestalt school of psychology suggested grouping as the key for interpreting human vision.

Gestalt is a movement of psychology that deals with perceptual organization. Gestalt psychology examines the relationships that bond individual elements so as to form a group (Forsyth & Ponce, 2002). As a consequence, a pattern emerges instead of separate parts. This pattern has generally completely different characteristics to its parts. Some of the gestalt rules by which elements tend to be associated together and interpreted as a group are the following:

- **Proximity:** elements that are close to each other.
- **Similarity:** elements similar in an attribute.
- **Continuity:** elements that could belong to a smooth larger feature.
- **Common fate:** elements that exhibit similar behavior.
- **Closure:** elements that could provide closed curves.
- **Parallelism:** elements that seem to be parallel.
- **Symmetry:** elements that exhibit a larger symmetry.

Gestalt laws have proven themselves to be precious tools in interpreting the way the human perceives his environment through vision. While all the laws are valuable in order to understand the context of an image, basic image processing tasks could be restricted to using the most basic ones. In order to express an image processing task through the prism of the gestalt theory, pixels should be considered as the elements. The correlation degree between them should be treated as the bonding relationship of the elements. The basic but at the same time important gestalt laws of proximity, similarity and continuity can then be applied in order to perform the given task.

As far as machine stereo vision is concerned, biological and psychological findings can be incorporated in the expression of proper correlation functions. Real life is the ultimate resource for finding right solutions in many fields of robotics, computer science and electronics (Mead, 1990; Shimonomura, Kushima, & Yagi, 2008; Berthouze & Metta, 2005). The natural selection process is a strict judge that favors the more effective solutions for each problem. Applying ideas borrowed from other sciences in technological problems can lead to very effective results. Consequently, further blending of biological and psychological findings with computer vision indicates a promising direction towards simple and accurate computer vision algorithms.

DEPTH MAPS COMPUTATION

The majority of stereo correspondence algorithms can be described using more or less the same structural set (Scharstein & Szeliski, 2002; Nalpantidis et al., 2008b). The basic building blocks are:

1. Computation of a matching cost function for every pixel in both the input images.
2. Aggregation of the computed matching cost inside a support region for every pixel in each image.
3. Finding the optimum disparity value for every pixel of one picture.
4. Refinement of the resulted disparity map.

Every stereo correspondence algorithm makes use of a matching cost function in order to establish correspondence between two pixels. The results of the matching cost computation comprise the disparity space image (DSI). DSI is a 3D matrix containing the computed matching costs for every pixel and for all its potential disparity values (Muhlmann, Maier, Hesser, & Manner, 2002). Usually, the matching costs are aggregated over support regions. These regions could be 2D or even 3D (Zitnick & Kanade, 2000; Brockers, Hund, & Mertsching, 2005) ones within the DSI cube. Due to the aggregation step it is not single pixels that will be matched, but image regions. Aggregation of the matching cost values is a common and essential technique in order to suppress the effect of noise that usually leads to false matching. The selection of the optimum disparity value for each pixel is performed afterwards. It can be a simple winner-takes-all (WTA) process or a more sophisticated one. In many cases it is an iterative process as depicted in Figure 6. An additional disparity refinement step is frequently adopted. It is usually intended to interpolate the calculated disparity values, giving sub-pixel accuracy or assign values to not calculated pixels. The general structure of the majority of stereo correspondence algorithms is shown in Figure 6.

Given that iterative methodologies are generally not suitable for robotic applications, due to computation time restrictions, the main differentiations among the robotics-oriented covered algorithms have to do with the dissimilarity measure and the dissimilarity measure's aggregation scheme that they employ.

Dissimilarity Measures

Detecting conjugate pairs in stereo images is a challenging research problem known as the correspondence problem, i.e., to find for each point in the left image, the corresponding point in the right one (Barnard & Thompson, 1980). To determine these two points from a conjugate pair, it is necessary to measure the (dis-)similarity of the points. The point to be matched without any ambiguity should be distinctly different from its surrounding pixels. Several algorithms have been proposed in order to address this problem. However, every algorithm makes use of a matching cost function so as to establish correspondence between two pixels. The matching cost function is a measure that quantitatively expresses how much dissimilar (or equivalently similar) two image pixels are. There is a number of such measures that have been used in robotic vision algorithms, e.g. the absolute intensity differences (AD), the squared intensity differences (SD), the zero normalized cross correlation (ZNCC), phase-based measures, and the luminosity-compensated dissimilarity measure (LCDM). Each one of them has its merits and disadvantages regarding computational complexity and lighting-differentiations tolerance. An evaluation of various matching costs can be found in (Scharstein & Szeliski, 2002; Mayoral, Lera, & Perez-Ilzarbe, 2006; Hirschmuller & Scharstein, 2007).

Figure 6. General structure of stereo correspondence algorithm

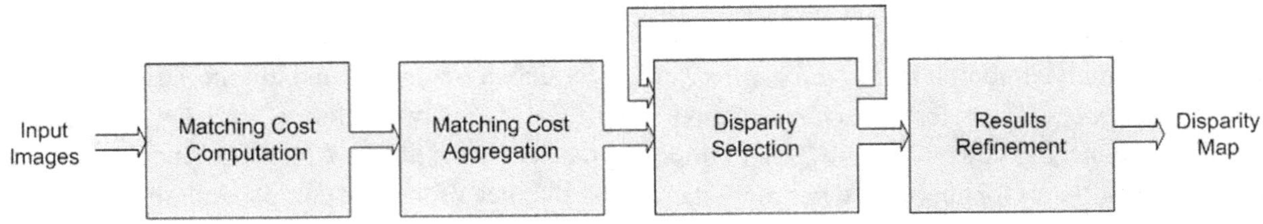

AD is the simplest measure of all. It involves simple subtractions and calculations of absolute values. As a result, it is the most commonly used measure found in literature. The mathematical formulation of AD is:

$$AD(x,y,d) = \left|I_l(x,y) - I_r(x,y-d)\right| \quad (1)$$

where I_l, I_r are the intensity values in left and right image, (x, y) are the pixels coordinates and d is the disparity value under consideration.

SD is somewhat more accurate in expressing the dissimilarity of two pixels. However, the higher computational cost of calculating the square of the intensities' difference is not usually justified by the accuracy gain. It can be calculated as:

$$SD(x,y,d) = \left(I_l(x,y) - I_r(x,y-d)\right)^2 \quad (2)$$

The normalized cross correlation calculates the dissimilarity of image regions instead of single pixels. It produces very robust results, on the cost of computational load. Its mathematical expression is:

$$NCC(x,y,d) = \frac{\sum_{x,y\in W} I_l(x,y)\cdot I_r(x,y-d)}{\sqrt{\sum_{x,y\in W} I_l^2(x,y)\cdot \sum_{x,y\in W} I_r^2(x,y-d)}} \quad (3)$$

where W is the image region under consideration.

The LCDM, as introduced in (Nalpantidis & Gasteratos, 2010b), provides stereo algorithms with tolerance against difficult lighting conditions. The images are initially transformed from the RGB to the HSL colorspace. The transition from the RGB colorspace, which is the usual output of contemporary cameras, to the HSL is straightforward and does not involve any complicated mathematical computations (Gonzalez & Woods, 1992). The HSL colorspace representation is a double cone, as shown in Figure 7(a). In this colorspace, H stands for hue and it determines the human impression about which color (red, green, blue, etc) is depicted. Each color is represented by an angular value ranging between 0 and 360 degrees (0 being red, 120 green and 240 blue). S stands for saturation and determines how vivid or gray the particular color seems. Its value ranges from 0 for gray to 1 for fully saturated (pure) colors. The L channel of the HSL colorspace stands for the Luminosity and it determines the intensity of a specific color. It ranges from 0 for completely dark colors (black) to 1 for fully illuminated colors (white).

Consequently, the HSL colorspace inherently expresses the lightness of a color and demarcates it from its qualitative characteristics. That is, an object will result in the same values of H and S regardless the environment's illumination conditions. According to this assumption, the proposed dissimilarity measure disregards the values of the L channel in order to calculate the dissimilarity of two colors. The omission of the vertical (L) axis from the colorspace representation leads to 2D circular disk, defined only by H and S, as show in Figure 7(b).

The transition from the 3D colorspace representation to the 2D one, can be conceived as a floor plan projection of the double cone, when observed along the vertical (L) axis. Thus, any color can be described as a planar vector with its initial point being the disc's center. As a consequence, each color P_k can be described as a polar vector or equivalently as a complex number with modulus equal to S_k and argument equal to H_k. That is, a color in the luminosity indifferent colorspace representation can be described as:

$$P_k = S_k e^{iH_k} \quad (4)$$

As a result, the difference, or equivalently the luminosity-compensated dissimilarity measure (LCDM), of two colors P_1 and P_2, shown with dashed line in Figure 7(b) can be calculated as the difference of the two complex numbers:

Figure 7. Views of the HSL color space representation. (a) The double cone representation; and (b) the horizontal slice at L=0

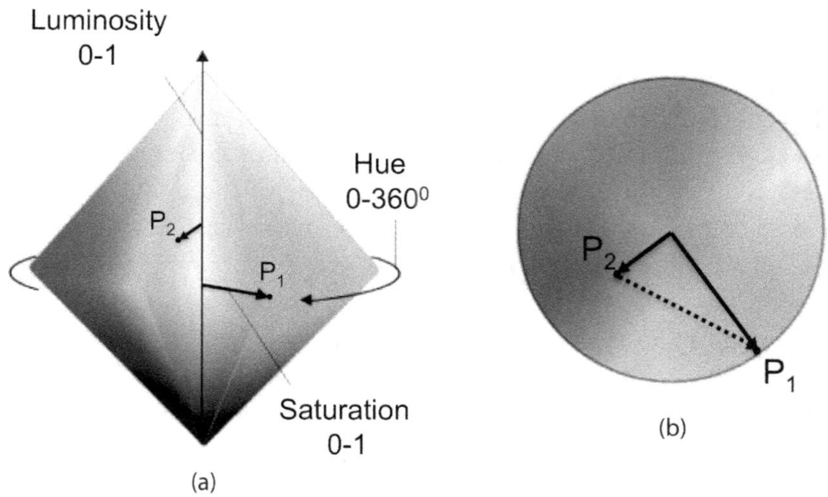

$$LCDM(P_1,P_2) = \left|\vec{P}_1 - \vec{P}_2\right| = \sqrt{S_1^2 + S_2^2 - 2S_1S_2\cos(H_1 - H_2)} \quad (5)$$

This equation is the mathematical formulation of the proposed LCDM dissimilarity measure. It takes into consideration any chromatic information available, except the luminosity. Thus, it can tolerate and compensate for any difference and non-uniformity of the lighting conditions. The proposed measure ignores some information (*L*), in contrast to the typical AD or SD. As a result, these latter measures are expected to perform somewhat better for totally ideal lighting conditions, which is the case for synthetic and carefully captured test images. On the other hand, any deviation from ideal lighting conditions is supposed to leave the proposed LCDM unaffected, while AD or SD will result in more and more false-matches.

Aggregation Schemes

The dissimilarity values for all the considered disparity values calculated in the first step of a stereo correspondence algorithm comprise the DSI. These results can be aggregated inside fix-sized square windows for constant value of disparity. The width of the window plays an important role on the final result. Small windows generally preserve details but suffer from noise, whereas big windows have the inverse behavior. The window's actual dimensions are chosen so as to keep a balance between the loss of detail and the emergence of noise, given the algorithm's details and the operating situations. The simplest scenario of aggregation is the constant support weight aggregation (CSW), i.e. that of simply summing the values of pixels within each support window.

The summation of the dissimilarity values can also be a weighted one. For instance, in the aggregation scheme used in (Nalpantidis, Sirakoulis, & Gasteratos, 2008a) each pixel is assigned a weight w(i,j,d), the value of which results from the 2D Gaussian function of the pixels Euclidean distance from the central pixel. The center of the function coincides with the central pixel and has a standard deviation equal to the one third of the distance from the central pixel to the nearest window-border. The Gaussian weight function remains the same for fixed width of the support window. Thus, it can be considered as a fixed mask that can be computed once, and then applied to all the windows.

However, the accuracy of local algorithms that employ CSW aggregation is generally considered low. Indeed, methods that use fixed support regions or even adaptively variable in size and/or shape support regions for aggregation of the computed dissimilarity values have been proven to produce results of inferior quality during the past.

Methods based on extended local methodologies sacrifice some of their computational simplicity in order to obtain more accurate results (Mordohai & Medioni, 2006). Adaptive support weights (ASW) based methods (Yoon & Kweon, 2006a; Gu, Su, Liu, & Zhang, 2008) achieve this, by using fix-sized support windows, whose pixels contribution in the aggregation stage varies depending on their degree of correlation to the windows' central pixel. Despite the acceptance that these methods have enjoyed, the determination of a correlation function is still an active topic.

An ASW-based method for correspondence search is presented in (Yoon & Kweon, 2006a). The support-weights of the pixels in a given support window are adjusted based on color similarity and geometric proximity to reduce the image ambiguity. The difference between pixel colors is measured in the CIELab color space, as this color space is based on measurements on the typical observer and, therefore, the distance of two points in this space is proportional to the stimulus perceived by the human eye. The running time for the Tsukuba image pair with a 35x35 pixels support window is about 0.016 fps on an AMD 2700 processor. The error ratio is 1.29%, 0.97%, 0.99%, and 1.13% for the Tsukuba, Sawtooth, Venus and Map image sets, respectively. These figures can be further improved through a left-right consistency check. The same authors propose a pre-processing step for correspondence search in the presence of specular highlights in (Yoon & Kweon, 2006b). For given input images, specular-free two-band images are generated. The similarity between pixels of these input-image representations can be measured using various correspondence search methods such as the simple SAD-based method, the adaptive support-weights method (Yoon & Kweon, 2006c) and the dynamic programming (DP) method (Lei, Selzer, & Yang, 2006).

Another ASW-based approach is presented in (Nalpantidis & Gasteratos, 2010a). Assigning the right significance weights to each pixel during aggregation has been achieved using the ideas of gestalt theory. The three basic gestalt laws get the following meaning:

- **Proximity (or equivalently Distance):** The closer two pixels are the more correlated to each other they are.
- **Intensity similarity (or equivalently Intensity dissimilarity):** The more similar the colors of two pixels are the more correlated they are.
- **Continuity (or equivalently Discontinuity):** The more similar is the depth of two pixels the more probable it is that they belong to the same larger feature and thus the more correlated they are.

Thus, gestalt theory can be used in order to determine to which degree two pixels are correlated.

The remaining question is exactly how much a correlated pixel to another should contribute to it during the aggregation process. In other words, it is necessary to establish an appropriate mapping between correlation degree and contribution. It is well known, since the 19th century, that HVS interprets physical stimuli in a psychological, non linear rather than in an absolute, linear manner. This psychophysical relationship has been investigated in depth and many explaining theories have been expressed (Pinoli & Debayle, 2007). The Weber-Fechner law is one of those theories and is widely acceptable. It indicates a logarithmic correlation between the subjective perceived intensity and the objective stimulus intensity.

The mathematical expression of this psychophysical law can be derived considering that the change of perception is proportional to the relative change of the causing stimulus:

$$p = -k \cdot \ln \frac{S}{S_0} \tag{6}$$

where p is the perceived stimulus intensity, S is the actual stimulus intensity, and k is a positive constant determined by the nature of the stimulus. The algorithm presented in (Nalpantidis & Gasteratos, 2010a) calculates the correlation degree by proposing and using a mathematical expression of three gestalt laws, namely the laws of proximity, similarity and continuity. While all the gestalt laws are significant for image understanding applications, these three can be considered to be the essential ones in image processing. Trying to express the gestalt laws in the form of mathematical equations is a difficult task and requires a lot of consideration, as there is no all-satisfying solution. Albeit gestalt psychology theory describes the qualitative characteristics of perceptual organization, a quantitative description, although desired, is not always available. A mathematical expression for the above mentioned gestalt laws has been proposed. The normalized contribution of each of them is subject to the psychophysical law of Weber-Fechner. The mathematical expression is the following:

Let x, y be the coordinates of the central pixel, x', y' the coordinates of a pixel lying inside its support region and d the disparity value currently being considered. Proximity of the two pixels is taken into consideration using their Euclidean distance on the image plain. The distance of the pixel (x', y') from the pixel (x, y) is calculated as:

$$distance(x', y')_{(x,y)} = \sqrt{(x - x')^2 - (y - y')^2} \tag{7}$$

The color dissimilarity of the two pixels can be estimated by the AD of their color intensities. This metric should not be confused with the AD calculated in the first step of the algorithm, since the former ones are calculated for pixels of the same image. Thus, the dissimilarity between the pixels (x', y') and (x, y) is calculated as:

$$dissimilarity(x', y')_{(x,y)} = \frac{1}{3} \sum_{C \in R,G,B} \left| I_C(x, y) - I_C(x', y') \right| \tag{8}$$

However, the AD calculated in the first step of this algorithm can be used to estimate the continuity of the pixels (x, y) and (x', y'). The continuity of two pixels can be described by the possibility that they both have the same depth, i.e. to share the same disparity value. The normalization of the AD calculated in the first step, results in an expression of the possibility that the true disparity value for the pixel (x', y') is not d. This possibility measure express the complement of continuity, i.e. the discontinuity. The less likely it is for a pixel (x', y') to have a disparity value d, the less it should bias the central pixel (x, y) in favor of the same disparity value d. The discontinuity between the pixels (x', y') and (x, y) is calculated as:

$$discontinuity(x', y', d)_{(x,y,d)} = \frac{AD(x', y', d)}{\max(AD)} \tag{9}$$

The last three equations quantify the gestalt theory. On the other hand the exact impact of those expression on the final result, is obtained by applying the Weber-Fechner. The values for distance, dissimilarity and discontinuity used hereafter are normalized towards its respective maximum value. Consequently, the factor S_o of the Weber-Fechner law for this case is equal to one and can be neglected. Thus, the weighting factor due to each gestalt law can be calculated:

$$w_{dist}(x', y', d)_{(x,y,d)} = -k_1 \cdot \ln\left(distance(x', y', d)_{(x,y,d)}\right) \tag{10}$$

$$w_{dissim}(x',y',d)_{(x,y,d)} = -k_2 \cdot \ln\left(dissimilarity(x',y',d)_{(x,y,d)}\right) \tag{11}$$

$$w_{discon}(x',y',d)_{(x,y,d)} = -k_3 \cdot \ln\left(discontinuity(x',y',d)_{(x,y,d)}\right) \tag{12}$$

These three weights are combined into one by multiplication, providing a general total weight:

$$w_{tot} = w_{dist} \cdot w_{dissim} \cdot w_{discon} \tag{13}$$

The total weight is calculated for both the left and the right input images, obtaining $w_{tot,l}$ and $w_{tot,r}$, respectively. However, distance and discontinuity are the same for both images considering the same pixel. Consequently, only dissimilarity has to be separately calculated for each image. Finally, the ASW aggregation, taking into consideration the weighting factor for each pixel is performed and results into the aggregated DSI:

$$DSI(x,y,d) = \frac{\sum\left(w_{tot,l} \cdot w_{tot,r} \cdot AD(x,y,d)\right)}{\sum\left(w_{tot,l} \cdot w_{tot,r}\right)} \tag{14}$$

TRAVERSABILITY ESTIMATION

Autonomous robots' behavior greatly depends on the accuracy of their decision-making algorithms. Reliable depth estimation is commonly needed in numerous autonomous behaviors. Autonomous navigation (Hariyama, Takeuchi, & Kameyama, 2000), obstacle avoidance (Nalpantidis, Kostavelis, & Gasteratos, 2009), localization and mapping, and traversability estimation are just a few of them (Murray & Little, 2000; Sim & Little, 2009). Vision-based solutions are becoming more and more attractive due to their decreasing cost as well as their inherent coherence with human imposed mechanisms. In the case of stereo vision-based navigation, the accuracy and the refresh rate of the computed disparity maps are the cornerstone of its success (Iocchi & Konolige, 1998; Schreer, 1998). However, robotic applications place strict requirements on the demanded speed and accuracy of vision depth-computing algorithms. Depth estimation using stereo vision, comprises the stereo correspondence problem. Stereo correspondence is known to be very computational demanding. The computation of dense and accurate depth images, i.e. disparity maps, in frame rates suitable for robotic applications is an open problem for the scientific community. Most of the attempts to confront the demand for accuracy focus on the development of sophisticated stereo correspondence algorithms, which usually increase the computational load exponentially. On the other hand, the need for real-time frame rates, inevitably, imposes compromises concerning the quality of the results. However, results' reliability is of crucial importance for autonomous robotic applications.

A wide range of sensors and various methods have been proposed in the relevant literature, as far as traversability estimation techniques are concerned. Some interesting details about the developed sensor systems and proposed detection and avoidance algorithms can be found in (Borenstein & Koren, 1990) and (Ohya, Kosaka, & Kak, 1998). Movarec has proposed the Certainty Grid method in (Moravec, 1987) and Borenstein (Borenstein & Koren, 1991) has proposed the Virtual Force Field method for robot obstacle avoidance. Then the Elastic Strips method was proposed in (Khatib, 1996, 1999) treating the trajectory of the robot as an elastic material to avoid obstacles. Moreover, (Kyung Hyun, Minh Ngoc, & M. Asif Ali, 2008) present a modified Elastic Strip method for mobile robots operating in uncertain environments. Review of popular obstacle avoidance algorithms covering them in more detail can be found in (Manz, Liscano, & Green, 1993) and (Kunchev, Jain, Ivancevic, & Finn, 2006).

The traversability estimation systems found in literature involve the use of one or a combination of ultrasonic, Laser. infrared (IR) or vision senors (Siegwart & Nourbakhsh, 2004). The use of ultrasonic, Laser and IR sensors is well-studied and the depth measurements are quite accurate and easily available. However, such sensors suffer either from achieving only low refresh rates (Vandorpe, Van Brussel, & Xu, 1996) or being extremely expensive. On the other hand vision sensors, either monocular, stereo or multicamera ones, can combine high frame rates and appealing prices.

Stereo vision is often used in vision-based methods, instead of monocular sensors, due to the simpler calculations involved in the depth estimation. Regarding stereo vision systems, one of the most popular methods for obstacle avoidance is the initial estimation of the so called v-disparity image (De Cubber et al., 2009). This method requires complex calculations and is applied in order to confront the noise in low quality disparity images (Labayrade et al., 2002; Zhao et al., 2007; Soquet et al., 2007). However, if detailed and noise-free disparity maps were available, less complicated methods could have been used.

Such a method is found in (Nalpantidis, Kostavelis, & Gasteratos, 2009). The disparity map obtained by a stereo correspondence algorithm is used to extract useful information about the navigation of a robot. Contrary to many implementations that involve complex calculations upon the disparity map, the proposed decision making algorithm involves only simple summations and checks. This is feasible due to the absence of significant noise in the produced disparity map. The goal of the developed algorithm is to detect any existing obstacles in front of the robot and to safely avoid it, by steering the robot left, right or to moving it forward. In order to achieve that, the developed method divides the disparity map into three windows, as in Figure 8.

In the central window, the pixels p whose disparity value $D(p)$ is greater than a defined threshold value T are enumerated. Then, the enumeration result is examined. If it is smaller than a predefined rate r of all the central windows pixels, this means that there are no obstacles detected exactly in front of the robot and in close distance, and thus the robot can move forward. On the other hand, if this enumeration's result exceeds the predefined rate, the algorithm examines the other two windows and chooses the one with the smaller average disparity value. In this way the window with the fewer obstacles will be selected. The values of the parameters T and r play an important role to the algorithm's behavior. Small values of T in conjunction with small values of r favor the hesitancy in moving forward, ensuring obstacle avoidance but at the same time being susceptible to false alarms due to noise. On the other hand, the opposite scenario is less susceptible to false alarms but may be proven risky.

Traversability estimation is also a significant part of visual mapping applications. A 2D map can be computed from stereo image pairs. Using the disparity map obtained form a stereo correspondence algorithm a reliable v-disparity image can be computed (Labayrade et al., 2002; Zhao et al., 2007). The terrain in the v-disparity image is modeled by a linear equation. The parameters of this linear equation can be found using Hough transform (De Cubber et al., 2009), if the camera-

Figure 8. Depth map's division in three windows

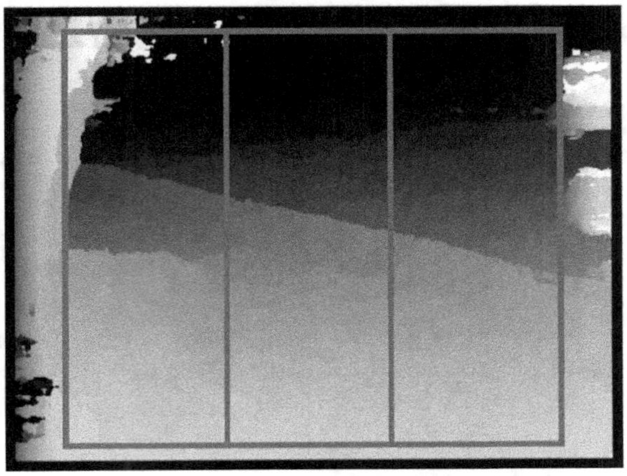

environment system's geometry is unknown. However, if the geometry of the system is constant and known (which is the case for a camera firmly mounted on a robot exploring a flat, e.g. indoor, environment) the two parameters can be easily computed beforehand and used in all the image pairs during the exploration. A tolerance region on either side of the terrain's linear segment is considered and any point outside this region is considered as an "obstacle". The linear segments denoting the terrain and the tolerance region overlaid on the v-disparity image are shown in Figure 9(a). For each pixel corresponding to an "obstacle" the local coordinates are computed. The local map, e.g. the one shown in Figure 9(b), is an occupancy grid of the environment consisting of all the points corresponding to "obstacles".

CONCLUSION

Stereo vision is a tested, useful and popular tool for inferring the depth of a scene with only passive optical sensors. Robotics, on the other hand, evolves rapidly and demand methods that can serve autonomous behaviors. Within this context, stereo correspondence algorithms need to provide accurate depth maps, in real-time frame-rates, confronting, at the same time, any difficulties imposed by the robots' environments.

In this chapter, the most interesting research issues of the robotics-oriented stereo vision field have been covered and solutions and possibilities have been presented. Such issues involve the handling of non-ideal lighting conditions, the requirement for simple calculation schemes, the use of multi-view stereo systems, the handling of miscalibrated image sensors, and the introduction of new biologically inspired methods to robotic vision. Various stereo correspondence algorithms that have non-iterative computational structure and are able to cope with real life images have been discussed. The dissimilarity measures, as well as the aggregation schemes that they employ have been examined.

Since many stereo vision-based robotic applications demand such characteristics, the presented stereo correspondence algorithms comprise effective solutions, which can be used as the cornerstone of more advanced autonomous robotic behaviors. Last, such applications of stereo vision within the domain of mobile robotic applications are covered. More specifically, the use of the obtained depth maps by algorithms that analyze the traversability

Figure 9. (a) V-disparity images for the image and (b) the corresponding local map

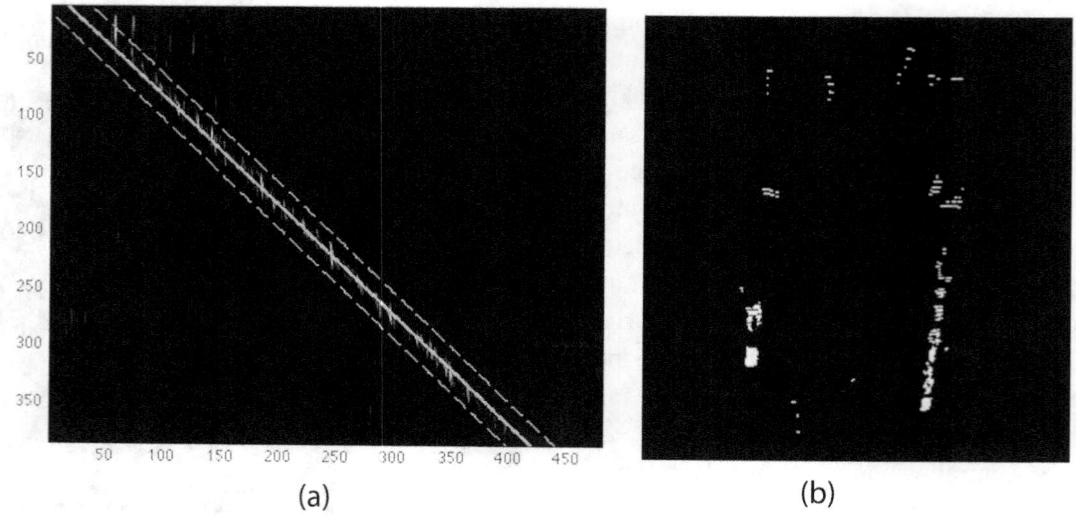

of the field in order the robot to avoid possible obstacles has been examined.

REFERENCES

Agrawal, M., Konolige, K., & Bolles, R. (2007). *Localization and mapping for autonomous navigation in outdoor terrains: A stereo vision approach.* In IEEE Workshop on Applications of Computer Vision. Austin, Texas, USA.

Barnard, S. T., & Thompson, W. B. (1980). Disparity analysis of images. *IEEE Transactions on Pattern Analysis and Machine Intelligence*, *2*(4), 333–340. doi:10.1109/TPAMI.1980.4767032

Berthouze, L., & Metta, G. (2005). Epigenetic robotics: Modelling cognitive development in robotic systems. *Cognitive Systems Research*, *6*(3), 189–192. doi:10.1016/j.cogsys.2004.11.002

Binaghi, E., Gallo, I., Marino, G., & Raspanti, M. (2004). Neural adaptive stereo matching. *Pattern Recognition Letters*, *25*(15), 1743–1758. doi:10.1016/j.patrec.2004.07.001

Borenstein, J., & Koren, Y. (1990). Real-time obstacle avoidance for fast mobile robots in cluttered environments. *IEEE Transactions on Systems, Man, and Cybernetics*, *19*(5), 1179–1187. doi:10.1109/21.44033

Borenstein, J., & Koren, Y. (1991). The vector field histogram-fast obstacle avoidance for mobile robot. *IEEE Transactions on Robotics and Automation*, *7*(3), 278–288. doi:10.1109/70.88137

Brockers, R., Hund, M., & Mertsching, B. (2005). Stereo vision using cost-relaxation with 3D support regions. In *Image and Vision Computing New Zealand* (pp. 96-101).

Corke, P. (2005, November). Machine vision toolbox. *IEEE Robotics & Automation Magazine*, *12*(4), 16–25. doi:10.1109/MRA.2005.1577021

De Cubber, G., Doroftei, D., Nalpantidis, L., Sirakoulis, G. C., & Gasteratos, A. (2009). *Stereobased terrain traversability analysis for robot navigation.* In IARP/EURON workshop on robotics for risky interventions and environmental surveillance. Brussels, Belgium.

Forsyth, D. A., & Ponce, J. (2002). *Computer vision: A modern approach.* Upper Saddle River, NJ: Prentice Hall.

Gonzalez, R. C., & Woods, R. E. (1992). *Digital image processing.* Boston, MA: AddisonWesley Longman Publishing Co., Inc.

Gu, Z., Su, X., Liu, Y., & Zhang, Q. (2008). Local stereo matching with adaptive support-weight, rank transform and disparity calibration. *Pattern Recognition Letters*, *29*(9), 1230–1235. doi:10.1016/j.patrec.2008.01.032

Hariyama, M., Takeuchi, T., & Kameyama, M. (2000). Reliable stereo matching for highly-safe intelligent vehicles and its VLSI implementation. In *IEEE Intelligent Vehicles Symposium* (p. 128133).

Hartley, R., & Zisserman, A. (2004). *Multiple view geometry in computer vision* (2nd ed.). Cambridge University Press. doi:10.1017/CBO9780511811685

Hirschmuller, H., & Scharstein, D. (2007, June). *Evaluation of cost functions for stereo matching.* In IEEE Computer Society Conference on Computer Vision and Pattern Recognition. Minneapolis, Minnesota, USA.

Hogue, A., German, A., & Jenkin, M. (2007). Underwater environment reconstruction using stereo and inertial data. In *IEEE International Conference on Systems, Man and Cybernetics* (p. 2372-2377). Montreal, Canada.

Iocchi, L., & Konolige, K. (1998). *A multiresolution stereo vision system for mobile robots*. In Italian AI Association Workshop on New Trends in Robotics Research.

Jeong, H., & Park, S. (2004). Generalized trellis stereo matching with systolic array. In *International Symposium on Parallel and Distributed Processing and Applications* (vol. 3358, p. 263-267). Springer Verlag.

Kelly, A., & Stentz, A. (1998, May). *Stereo vision enhancements for low-cost outdoor autonomous vehicles*. In International Conference on Robotics and Automation, Workshop Ws-7, Navigation of Outdoor Autonomous Vehicles.

Khatib, O. (1996). Motion coordination and reactive control of autonomous multi-manipulator system. *Journal of Robotic Systems*, *15*(4), 300–319.

Khatib, O. (1999). Robot in human environments: Basic autonomous capabilities. *The International Journal of Robotics Research*, *18*(7), 684–696. doi:10.1177/02783649922066501

Klancar, G., Kristan, M., & Karba, R. (2004). Wide-angle camera distortions and non-uniform illumination in mobile robot tracking. *Journal of Robotics and Autonomous Systems*, *46*, 125–133. doi:10.1016/j.robot.2003.11.001

Konolige, K., Agrawal, M., Bolles, R. C., Cowan, C., Fischler, M., & Gerkey, B. P. (2006). Outdoor mapping and navigation using stereo vision. In *International Symposium on Experimental Robotics* (vol. 39, pp. 179-190). Brazil: Springer.

Kunchev, V., Jain, L., Ivancevic, V., & Finn, A. (2006). Path planning and obstacle avoidance for autonomous mobile robots: A review. In International Conference on Knowledge-Based and Intelligent Information and Engineering Systems (vol. 4252, pp. 537-544). Springer-Verlag.

Kyung Hyun, C., Minh Ngoc, N., & Asif Ali, R. (2008). A real time collision avoidance algorithm for mobile robot based on elastic force. *International Journal of Mechanical . Industrial and Aerospace Engineering*, *2*(4), 230–233.

Labayrade, R., Aubert, D., & Tarel, J.-P. (2002). Real time obstacle detection in stereovision on non flat road geometry through V-disparity representation. In *IEEE Intelligent Vehicle Symposium* (vol. 2, pp. 646-651). Versailles, France.

Lei, C., Selzer, J., & Yang, Y.-H. (2006). Region-tree based stereo using dynamic programming optimization. *IEEE Conference on Computer Vision and Pattern Recognition, 2,* 2378-2385.

Manz, A., Liscano, R., & Green, D. (1993). A comparison of realtime obstacle avoidance methods for mobile robots . In *Experimental Robotics ii* (pp. 299–316). Springer-Verlag. doi:10.1007/BFb0036147

Masrani, D. K., & MacLean, W. J. (2006). A real-time large disparity range stereo-system using FPGAS. In *IEEE International Conference on Computer Vision Systems* (vol. 3852, p. 13-20).

Mayoral, R., Lera, G., & Perez-Ilzarbe, M. J. (2006). Evaluation of correspondence errors for stereo. *Image and Vision Computing*, *24*(12), 1288–1300. doi:10.1016/j.imavis.2006.04.006

Mead, C. (1990). Neuromorphic electronic systems. *Proceedings of the IEEE*, *78*(10), 1629–1636. doi:10.1109/5.58356

Mingxiang, L., & Yunde, J. (2006, Dec). Trinocular cooperative stereo vision and occlusion detection. IEEE International Conference on Robotics and Biomimetics, 1129-1133.

Moravec, P. (1987). Certainty grids for mobile robots. In *NASA/JPL Space Telerobotics Workshop* (Vol. 3, pp. 307-312).

Mordohai, P., & Medioni, G. G. (2006). Stereo using monocular cues within the tensor voting framework. *IEEE Transactions on Pattern Analysis and Machine Intelligence*, *28*(6), 968982. doi:10.1109/TPAMI.2006.129

Muhlmann, K., Maier, D., Hesser, J., & Manner, R. (2002). Calculating dense disparity maps from color stereo images, an efficient implementation. *International Journal of Computer Vision*, *47*(1-3), 79–88. doi:10.1023/A:1014581421794

Murray, D., & Little, J. J. (2000). Using real-time stereo vision for mobile robot navigation. *Autonomous Robots*, *8*(2), 161–171. doi:10.1023/A:1008987612352

Nalpantidis, L., Amanatiadis, A., Sirakoulis, G. C., & Gasteratos, A. (in press). An efficient hierarchical matching algorithm for processing uncalibrated stereo vision images and its hardware architecture. *IET Image Processing.*

Nalpantidis, L., Chrysostomou, D., & Gasteratos, A. (2009, December). Obtaining reliable depth maps for robotic applications with a quad-camera system. In International Conference *on Intelligent Robotics and Applications* (vol. 5928, p. 906-916). Singapore: Springer-Verlag.

Nalpantidis, L., & Gasteratos, A. (2010a). Biologically and psychophysically inspired adaptive support weights algorithm for stereo correspondence. *Robotics and Autonomous Systems*, *58*, 457–464. doi:10.1016/j.robot.2010.02.002

Nalpantidis, L., & Gasteratos, A. (2010b). Stereo vision for robotic applications in the presence of non-ideal lighting conditions. *Image and Vision Computing*, *28*, 940–951. doi:10.1016/j.imavis.2009.11.011

Nalpantidis, L., Kostavelis, I., & Gasteratos, A. (2009). Stereovision-based algorithm for obstacle avoidance. In *International Conference on Intelligent Robotics and Applications* (vol. 5928, pp. 195-204). Singapore: Springer-Verlag.

Nalpantidis, L., Sirakoulis, G. C., & Gasteratos, A. (2008a). A dense stereo correspondence algorithm for hardware implementation with enhanced disparity selection. In *5th Hellenic Conference on Artificial Intelligence* (vol. 5138, pp. 365-370). Syros, Greece: Springer-Verlag.

Nalpantidis, L., Sirakoulis, G. C., & Gasteratos, A. (2008b). Review of stereo vision algorithms: From software to hardware. *International Journal of Optomechatronics*, *2*(4), 435–462. doi:10.1080/15599610802438680

Ogale, A. S., & Aloimonos, Y. (2005a, April). Robust contrast invariant stereo correspondence. In *IEEE International Conference on Robotics and Automation* (pp. 819-824).

Ogale, A. S., & Aloimonos, Y. (2005b). Shape and the stereo correspondence problem. *International Journal of Computer Vision*, *65*(3), 147–162. doi:10.1007/s11263-005-3672-3

Ogale, A. S., & Aloimonos, Y. (2007). 04). A roadmap to the integration of early visual modules. *International Journal of Computer Vision*, *72*(1), 9–25. doi:10.1007/s11263-006-8890-9

Ohya, A., Kosaka, A., & Kak, A. (1998). Vision-based navigation of mobile robot with obstacle avoidance by single camera vision and ultrasonic sensing. *IEEE Transactions on Robotics and Automation*, *14*(6), 969–978. doi:10.1109/70.736780

Park, S., & Jeong, H. (2007). Real-time stereo vision FPGA chip with low error rate. In *International Conference on Multimedia and Ubiquitous Engineering* (pp. 751-756).

Pinoli, J. C., & Debayle, J. (2007). Logarithmic adaptive neighborhood image processing (LANIP): Introduction, connections to human brightness perception, and application issues. *EURASIP Journal on Advances in Signal Processing*, (1): 114–135.

Ruigang, Y., Welch, G., & Bishop, G. (2002). Real-time consensus-based scene reconstruction using commodity graphics hardware. *10th Pacific Conference on Computer Graphics and Applications*, (pp. 225-234).

Scharstein, D., & Szeliski, R. (2002). A taxonomy and evaluation of dense two-frame stereo correspondence algorithms. *International Journal of Computer Vision*, *47*(1-3), 7–42. doi:10.1023/A:1014573219977

Schirmacher, H., Li, M., & Seidel, H.-P. (2001). *On-the-fly processing of generalized lumigraphs* (pp. 165–173). Eurographics.

Schreer, O. (1998). Stereo vision-based navigation in unknown indoor environment. In *5th European Conference on Computer Vision* (vol. 1, pp. 203-217).

Shimonomura, K., Kushima, T., & Yagi, T. (2008). Binocular robot vision emulating disparity computation in the primary visual cortex. *Neural Networks*, *21*(2-3), 331–340. doi:10.1016/j.neunet.2007.12.033

Siegwart, R., & Nourbakhsh, I. R. (2004). *Introduction to autonomous mobile robots*. Massachusetts: MIT Press.

Sim, R., & Little, J. J. (2009). Autonomous vision-based robotic exploration and mapping using hybrid maps and particle filters. *Image and Vision Computing*, *27*(1-2), 167-177. (Canadian Robotic Vision 2005 and 2006)

Soquet, N., Aubert, D., & Hautiere, N. (2007). Road segmentation supervised by an extended V-disparity algorithm for autonomous navigation. In *IEEE Intelligent Vehicles Symposium* (pp. 160-165). Istanbul, Turkey.

Sun, C., & Peleg, S. (2003). Fast panoramic stereo matching using cylindrical maximum surfaces. IEEE Trans. *SMC Part B*, *34*, 760–765.

Vandorpe, J., Van Brussel, H., & Xu, H. (1996). Exact dynamic map building for a mobile robot using geometrical primitives produced by a 2d range finder. In IEEE International Conference on Robotics and Automation (pp. 901-908). Minneapolis, USA.

Wiegand, T., Sullivan, G., Bjntegaard, G., & Luthra, A. (2003). Overview of the H.264/AVC video coding standard. *IEEE Transactions on Circuits and Systems for Video Technology*, *13*(7), 560–576. doi:10.1109/TCSVT.2003.815165

Wilburn, B., Smulski, M., Lee, K., & Horowitz, M. A. (2002). The light field video camera. In *Media Processors* (p. 29-36).

Yang, J. C., Everett, M., Buehler, C., & Mcmillan, L. (2002). A real-time distributed light field camera. In *Eurographics Workshop on Rendering* (pp. 77-86).

Yin, P., Tourapis, H., Tourapis, A., & Boyce, J. (2003). Fast mode decision and motion estimation for JVT/H.264. In *IEEE International Conference on Image Processing* (vol. 3, pp. 853–856).

Yoon, K.-J., & Kweon, I. S. (2006a). Adaptive support-weight approach for correspondence search. *IEEE Transactions on Pattern Analysis and Machine Intelligence*, *28*(4), 650–656. doi:10.1109/TPAMI.2006.70

Yoon, K.-J., & Kweon, I. S. (2006b). Correspondence search in the presence of specular highlights using specular-free two-band images. In *7th Asian Conference on Computer Vision* (vol. 3852, pp. 761-770). Hyderabad, India: Springer.

Yoon, K.-J., & Kweon, I. S. (2006c). Stereo matching with symmetric cost functions. In *IEEE Computer Society Conference on Computer Vision and Pattern Recognition* (vol. 2, pp. 2371-2377).

Zach, C., Karner, K., & Bischof, H. (2004). Hierarchical disparity estimation with programmable 3D hardware. In *International Conference in Central Europe on Computer Graphics, Visualization and Computer Vision* (pp. 275-282).

Zhao, J., Katupitiya, J., & Ward, J. (2007). Global correlation based ground plane estimation using V-disparity image. In *IEEE International Conference on Robotics and Automation* (pp. 529-534). Rome, Italy.

Zitnick, C. L., & Kanade, T. (2000). A cooperative algorithm for stereo matching and occlusion detection. *IEEE Transactions on Pattern Analysis and Machine Intelligence*, *22*(7), 675–684. doi:10.1109/34.865184

KEY TERMS AND DEFINITIONS

Dense Stereo Correspondence Algorithm: A stereo correspondence algorithm that estimates disparity values for all the image pixels.

Disparity Map: An image constituted by the disparity values of each pixel, being thus equivalent to a depth map.

Disparity: The difference of an observed point's image coordinates when viewed under different viewpoints.

Dissimilarity Measure: A function that quantitatively expresses how much dissimilar two image pixels are.

Sparse Stereo Correspondence Algorithm: A stereo correspondence algorithm that estimates disparity values for some of the image pixels.

Stereo Correspondence: The procedure of matching pixels between two images that derive from the same scene.

Traversability Estimation: The procedure of determining whether there are obstacles or not in a field.

Chapter 22
Stereo-Vision-Based Fire Detection and Suppression Robot for Buildings

Chao-Ching Ho
National Yunlin University of Science and Technology, Taiwan

ABSTRACT

A stereo-vision-based fire detection and suppression robot with an intelligent processing algorithm for use in large spaces is proposed in this chapter. The successive processing steps of our real-time algorithm use the motion segmentation algorithm to register the possible position of a fire flame in a video; the real-time algorithm then analyzes the spectral, spatial, and motion orientation characteristics of the fire flame regions from the image sequences of the video. The characterization of a fire flame was carried out by using a heuristic method to determine the potential fire flame candidate region. The fire-fighting robot uses stereo vision generated by means of two calibrated cameras to acquire images of the fire flame and applies the continuously adaptive mean shift (CAMSHIFT) vision-tracking algorithm to provide feedback on the real-time position of the fire flame with a high frame rate. Experimental results showed that the stereo-vision-based mobile robot was able to successfully complete a fire-extinguishing task.

INTRODUCTION

Fire incidents can cause loss of lives and damage to property. Damage due to fire has always been a major area of concern for museums, warehouses, and residential buildings. Conventional fire detection sensors (e.g., ionization and photoelectric detectors) and fire sprinkler systems monitor only particular points in space. In most cases, conventional point-type detectors are installed on walls or on a ceiling. The delays in the activation

DOI: 10.4018/978-1-61350-326-3.ch022

of fire detection sensors and sprinklers in large spaces are a major problem. Hence, the monitoring capabilities of point-type sensor devices are limited to a certain distance, and they are ineffective for monitoring large areas. These devices are not sufficiently flexible to detect fire incidents, and many fire-detection sensors and sprinklers are required to be installed very close to the monitoring areas. Comparatively, the video camera is a volume sensor, and potentially monitors a larger area and has a much higher probability of successful early detection of fire flames. Video surveillance technology is suitable for early detection of fires due to its low detection delay, good resolution, and high localization accuracy. Early detection of fires can certainly expedite fire-fighting efforts, and consequently, fires can be extinguished before they spread to other areas. To monitor large spaces, the use of a mobile fire-fighting robot is a more flexible alternative than installing a large number of detectors and sprinklers. When a fire is detected, the fire-fighting robot can move to the position of the fire flame and safely evacuate an object from the fire area. Stereo vision systems can provide the robot with precise depth information about a target. Hence, the use of two cameras instead of one increases the suppression efficiency and adaptability of the robot while detecting and evacuating a burning object.

BACKGROUND

In recent times, research on the detection of fire flames using surveillance cameras with machine vision has gained momentum. The image processing approach involves the extraction of the fire flame pixels from a background by using frame difference technologies. Healey et al. (1993) presented a fire detection algorithm using a color video input with a pre-partitioning scheme under some restricted conditions, without rejecting the similar fire-like alias. Phillips et al. (2002) and Celik et al. (2007) conducted studies on computer vision by using spectral analysis and the flickering property of fire flame pixels to recognize the existence of fires at a scene. Hue and saturation are adopted as feature vectors to extract the fire pixels from the visual images (Chen, 2003). Fire flame features based on the HSI(hue, saturation, intensity) color model are extracted, and regions with fire-like colors are roughly separated from the image by the color separation method (Horng, 2005). Then, the image difference method based on chromatics is used to remove spurious fire-like regions such as objects with similar fire colors or areas reflected from fire flames. A fuzzy-based dominant flame color lookup table is created, and fire regions are automatically selected (Wang, 2006). However, either the solution does not consider the temporal variation of flames or the approach is too complicated to process in real time.

Fire suppression systems usually use water to extinguish fires due to its good ability to suppress fire. Chen et al. (2004) developed a water-spraying-based fire suppression system. The fire searching method is realized based on computer vision theory using one CCD camera that is installed at the end of a fire monitor chamber. However, it is necessary to calculate the changes in the space coordinates of the fire with displacement and the pivot angle of the CCD camera in the fire searching process. Ho (2009) proposes a fire-tracking scheme based on CAMSHIFT. The CAMSHIFT algorithm is applied to track the trajectory and compute the 2D positions of the specified moving fire-fighting robot in real time with the aid of a vision system. Yuan (2010) adopted the computer vision techniques to extract color and motion characteristics for real-time fire detection. However, the system was designed to move a water gun along a fixed path using computer-based control. Hence, the monitor ranges of the scene are limited and not sufficiently flexible.

FIRE DETECTION AND SUPPRESSION ROBOT

Issues, Controversies, Problems

The basic problems associated with conventional fire detection sensors are that they are not sufficiently reliable due to the time delay between the start of the fire and nonzero input in the detection sensor (Podržaj, 2008). A fire-fighting robot, equipped with vision-based fire detection technology, is capable of eliminating the conventional sensor delay factor and providing fire detection coverage for a larger area with minimized cost. The overall objective of the mobile robot system is to develop an autonomous fire detection technology that is capable of identifying fires at early stages, alerting fire staff, and extinguishing the fire.

To enable fire detection by the robot, the main objective here is to develop a color vision-based fire detection system that is accurate and performs efficiently. The image processing approach involves the extraction of the fire flame pixels from a background by using frame difference technologies. To achieve the segmentation of fire features, color processing is a better alternative than gray-scale processing. Color processing can avoid the generation of false alarms due to the variations in lighting conditions (e.g., natural background illumination) better than gray-scale processing. However, vision-based fire flame detection still has great technical challenges. Fire flames are non-rigid objects and do not exhibit primitive image features and variability in density, lighting, etc. (Ho, 2009).

For robot navigation and fire extinguishing, robots with machine vision have been attracting substantial attention recently. A visual feedback mechanism commonly used in a robot system is the machine vision measuring system capable of non-contacting measurement; this type of measurement is advantageous because the actual environment does not always allow contact with the surface at which the measurement is performed. One common problem with the acquired images is the accompanying noises, namely, disturbances caused by irrelevant objects in the background or foreground or by substandard illumination. As a visual tracking system must be capable of recording image features in real time, another problem with most computer-based vision systems is their limited performance due to inadequate computing power for motion tracking after processing resources, including the memory, have been allocated to the tasks of feature extraction and template pattern matching, which accounts for the handling of the disturbances mentioned above (Sumi, 1995; Di Stefano, 2003). Due to the high cost of image capturing equipment and image processing components, the need for substantial computing resources by visual servoing algorithms is not often taken into consideration. Despite expected improvements on the performance of visual servo with image processing and control components operating on separate processors, the additional cost involved with the extra components presents yet another problem.

SOLUTIONS AND RECOMMENDATIONS

Image Capturing

Image capturing and processing are the two major challenges in the construction of a stereo visual tracking system. The main goals of these respective processes are to provide visual feedback and identify the objects one wants to track (in this case a burning target), while simultaneously determining their 3D positions. In the proposed stereo tracking system, the video-signal-capturing process is conducted by stereo sensors, which are two low-cost CMOS cameras with pixel resolutions of 320 x 240. The captured synchronized concurrent video frames are then transmitted via USB to a PC, the image processor, and then buffered in the PC's system memory. The frames are eventually displayed in the capture windows using the PC's windows driver model (WDM) functions. The windows driver model can reach a high video frame rate, while the two CMOS cameras can also achieve a frame rate of 30 images per second; therefore, real-time tracking can be achieved.

Calibration of Image-Based System

It might be expected that after the raw images are obtained, the next step would be to process them. However, camera calibration and robot hand-eye transformation are two fundamental steps that need to be taken into consideration before any reliable image processing or even further visual servoing can be performed. To obtain the 3D position of an object portrayed in 2D images, the relationship between the coordinates of the points in a 3D frame, in this paper the tank, and the coordinates of their corresponding points projected onto the 2D imagery plains must be found. The first step toward this goal is recognizing the basic characteristics of the cameras used to retrieve the images. These characteristics, generally grouped into intrinsic and extrinsic parameters, are shown in Figure 1.

Camera calibration can be used to obtain the intrinsic and extrinsic parameters of the camera. The intrinsic parameters, which are independent of a camera's position in the physical environment, describe the camera's focal length (f_x, f_y), principal point (C_x, C_y), and distortion coefficients. On the other hand, the extrinsic parameters offer information on the transformation between the coordinate systems of the camera and the concrete world, including a 3D translation vector t that provides the translational components and a rotation matrix R that provides the rotational components. For stereo camera systems, it is also necessary to obtain the relative extrinsic parameters between the two cameras. The combination of a camera's, or an array of cameras', intrinsic and extrinsic parameters presents the full set of data needed to locate, in the outside world, the corresponding position of a point in images taken by the camera(s). M.-C. Villa-Uriol (2004) and

Figure 1. The robot is calibrated to recognize the intrinsic and extrinsic parameters of the camera

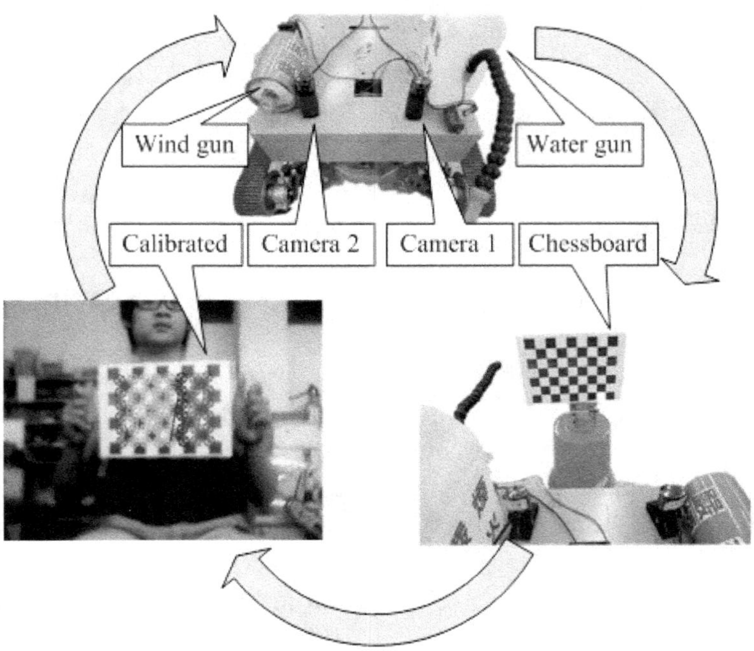

Hutchinson (2006) provided a complete review on the fundamentals and techniques of camera calibration.

In this work, static pattern calibration is applied to attain the cameras' two sets of parameters and empower automatic edge detection. The key to this calibration process is the use of a flat checkerboard with a known geometric pattern. The dimensions of each square on the board are predetermined. Points on the board model plane and their projections in digital images taken by the to-be-calibrated cameras are passed as parameters to the intrinsic calibration routine.

The extrinsic calibration methods used here also employ digital checker-patterned board images. Assume that for every point M in the physical world reference frame we have $m = RM + t$, where m is the 3D coordinates of the point's counterpart in the camera reference frame, while R and t are the rotational and translational matrices, respectively, for the coordinate transformation between the physical world and camera reference frames. The extrinsic parameters of the to-be-calibrated cameras can be calculated linearly with the help of one array that stores the coordinates of chosen reference points in the digital checkerboard images and another array that contains the corresponding points in the physical world. The collection of reference points is selected on the checker-patterned model plane for the two cameras used in this chapter.

Vision-Based Fire Detection

The real-time fire-fighting robot is guided by a vision-based CAMSHIFT tracking surveillance system and is equipped with a water gun to extinguish the burning targets. The vision-based fire flame detection algorithm consists of five steps: (1) moving pixels or regions in the current frame of a video are determined with the motion history image (MHI); (2) the HSI colors of moving pixels are checked; (3) if the histogram of moving pixels is correlated with the fire flame color histogram, then the disordered measurement and temporal analysis are performed to determine if fire flame colored pixels flicker or not; and (4) Back-Projection and CAMSHIFT are applied to track the fire flame region; (5) the distance between the robot and the burning target is calculated through binocular stereo.

Moving Motion Segmentation with Motion History Image (MHI)

The MHI is a scalar-valued image where intensity is a function of the recency of motion (Davis, 1999; Bobick, 2001; Bradski, 2002). This moving history representation can be used to determine the current movement of the object and to segment and measure the motions induced by the object (e.g., fire flame) in a video scene. MHI representations have the following advantages: a range of times from frame to frame to several seconds may be encoded in a single image, direct recognition of the motion itself is possible, motion recognition is not computationally taxing and real-time implementation is possible, and the motion within the detecting scene can be monitored. An MHI is used to represent how the fire flame is moving, since the outward boundaries of the fire flame are less prone to misdetection than the source regions of fire flame. In an MHI, the pixel intensity is a function of the motion history at that location; in the MHI, brighter values correspond to a more recent motion. It should be noted that the final motion locations appear brighter in the MHI.

Correlation of Spectral Characteristics

The first step in detecting possible fire flame pixel candidates is to transform the color space into HSI color space and then carry out analysis. The HSI color system projects the standard red-green-blue (RGB color model) color space along its principle diagonal in terms of white to black shades to avoid the influence of lighting changes (Castleman, 1996). Hue is the dominant color (red, green, and blue) of an area, and saturation is the colorfulness of an area in proportion to its brightness. Intensity is related to the color luminance, e.g., human skin occupies a small portion of the H and S spaces. The advantages of the HSI space are the intuitiveness of the components and the explicit discrimination between luminance and chrominance. The hue, saturation, and intensity components of the HSI model are normalized into the following ranges: $0° \leq$ hue $\leq 360°$, $0 \leq$ saturation ≤ 255, and $0 \leq$ intensity ≤ 255. The computed fire flame spectral histogram correlation coefficient is measured by the compare correlation analysis. The template of fire flame spectral histogram, which is based on empirical analysis results for the fire flames with colors from red to yellow, was created to detect the flame-colored pixels (Horng, 2005). Hue is an attribute of the pure color of the image scene, and it was demonstrated that it can be used in assessing the prospect of numerical labelling of the flame colors (Huang, 2008). Hence, the detection of flame pixels is carried out using the hue channel histogram correlation analysis with the fire flame template, which maps the hue value of general flames to be distributed from 0° to 60°.

Chaotic Spatial Structure Analysis

The moving object regions with disordered ratios of perimeter to area for the extracted fire flame region Ω are defined as: P/A, where P represents the perimeter of the region and A represents the area of the region. As the complexity of a shape increases (i.e., the perimeter increases with respect to the area), the value associated with the disordered ratio Ω increases. The chaotic and turbulent nature of a region can be detected by relating the extracted spatial features to the fire flame likelihood region and the smoke likelihood region (Chen, 2004). The likelihood that a flame-like region is a flame region is highly correlated with the parameter Ω.

Temporal Analysis

It is not always sufficient to detect fire flame correctly based on color information. There are many objects, with similar color properties as the fire flame spectrum. The key to distinguishing between the flame and flame-colored objects is the nature of their motion. The flames in a fire dance around, so any particular pixel will only see fire for a fraction of the time. This kind of temporal periodicity is commonly known as flickering. The flicker of fire flame causes the spectral values in the fire flame region to fluctuate in time. The flicker in fire is also used as additional information. The candidate regions are checked to see whether they continuously appear and disappear over time. The level crossing rate LCR is utilized for validating these extracted fire flickering regions. Temporal variation for each pixel is computed by finding the level crossing rate of the most likely fire flame candidate region above the heuristic threshold value among consecutive frames. The heuristic threshold is determined based on the fire flame models in recorded video sequences.

Back-Projection

If a visual servoing system is applied to natural backgrounds, color data usually provides more reliable and flexible information than monochrome data (Kim, 1996; LeGrand, 1996). The CAMSHIFT tracking engine is based on the histogram projection algorithm (Swain, 1990), which is a useful technique for color object recognition, especially for object identification in complex background surroundings. Histogram back-projection is a primitive operation that finds and identifies the association between pixel values in a grabbed image and the values in a particular histogram bin. Histogram and back-projection performed on any consecutive frame would generate a probability image on which the value of each pixel represents the probability of the exact same pixel from the input belonging to the target histogram that was used. Given that m histogram bins are used, we can define n image pixel locations. Thus, we have histograms $\{\hat{y}_u\}$, $u=1,\ldots, m$ and pixel locations $\{x_i\}$, $i=1,\ldots, n$. Let us also define a function $c{:}R^2 \rightarrow \{1,\ldots, m\}$ that associates a pixel at location x_i^* with a histogram bin index $c(x_i^*)$. Then, the histograms can be computed with the equation $\hat{y}_u = \sum_{i=1}^{n} \delta\left[c\left(x_i^*\right) - u\right]$. In all cases, the values in the histogram bins are rescaled to fit within the discrete pixel range of the possible output 2D probability distribution image with the function $\left\{\widehat{p}_u = \min\left(\frac{UPPER}{\max\left(\hat{y}\right)}\hat{y}_u, UPPER\right)\right\}_{u=1\ldots m}$. That is, the values in the histogram bins, which originally lie in the range $[0, max(\hat{y}_u)]$, now lie in the new range $[0, UPPER]$. In the end, the input pixels with the highest probability of being in the sample histogram will be mapped onto a 2D histogram back-projection image with the highest visible intensities.

CAMSHIFT Tracking

The CAMSHIFT algorithm is a non-parametric technique that can track a specified target's 2D position efficiently across a series of images. When tracking the 2D position of a colored object, in our case a fire flame colored region, the CAMSHIFT operates on a color probability distribution image derived from color histograms. The center and size of the targeted object region is computed and used as settings for the search window on the following frame of the video sequence. Figure 2 shows the images of the tracked object in the digital pictures that have been recognized and processed, and the green bounding ellipse presents the fire flame region tracked by the CAMSHIFT algorithm. The calculation of the color probability distribution is not performed on the entire image, but only on limited regions surrounding the current CAMSHIFT window, which includes images of the specified object that are transformed into a discrete probability image. This tends to result in a large reduction in the computational costs. The CAMSHIFT algorithm can be summarized by the following steps:

- **Step 1:** A region of interest (ROI) window is selected to be the sample image for future color probability distribution computation. In tracking procedures, this window is placed over the targeted object.
- **Step 2:** A mean shift search window is initially centered at the first frame's data point position.
- **Step 3:** The color distribution of the region centered at the mean shift search window is calculated, producing a discrete probability image. The mean location (the centroid) of the discrete probability image can be found within the search window by first obtaining moment values. Given that $I(x, y)$ is the pixel value function for the intensity of the discrete probability image at point (x, y) in the search window, one can compute the zeroth moment for that point.
- **Step 4:** The mean shift algorithm (step 3) is iterated, replacing (x, y) with the corresponding (x_c, y_c), until the centroid of the search window region's generated probability image converges to a constant point. This point should be at the center of the tracked target. The zeroth moment (distribution area) and the mean location (the centroid) are stored.
- **Step 5:** The size of the search windows is set as a function of the zeroth moment found in Step 4 to match the size of the tracked object, and the center of the search window is placed on the following frame at the mean location found in Step 4. The process is then repeated, beginning at Step 3.

Figure 2. The image at the 169th frame is detected as a flame by searching the candidate regions, which are above the LCR threshold. The CAMSHIFT tracking algorithm is employed to track the movement of flame pixels, and the flame region is bounded by the green ellipse.

Depth Calculated through Binocular Stereo

The purpose of using two cameras instead of one to track a burning target provides data on the burning target's position in a third dimension. The calculation of depth via binocular stereo is a common way to extend one's knowledge of a scene from 2D to 3D (Jain, 1995). First, feature points are grouped into teams of two—one each in the two images obtained by each of the two stereo cameras—to create a set of stereo pairs. Then, for every pair, each point is envisioned as a ray. The two rays would intersect in the actual 3D world. Now suppose the 3D coordinates of the two points in a stereo pair are P_1 and P_2; the rotation matrices and translation vectors for the transformation between the two camera coordinate systems and the stereo coordinate system are R_1, R_2 and t_1, t_2. Therefore, given a point M in the stereo coordinate system, we have $P_1 = R_1M + t_1$ and $P_2 = R_2M + t_2$. Finally, the point where the two rays intersect is found to be the point M that produces the P_1 and P_2 closest to the rays (represented by straight lines) in their respective camera coordinate systems. This approach transforms the coordinate systems of the two cameras into absolute coordinates. In this work, the real-time 3D depths of the burning target, tracked with the presently discussed 3D point measuring method involving the intersecting rays, are rendered using Open Graphics Library (Wright, 2000). The specified corner positions of the chessboard are processed and analyzed with the calibrated intrinsic and extrinsic parameters and the absolute coordinates are calculated via binocular stereo.

a. LCR probability distribution image of flame.
b. The moving flame pixels are separated.
c. The CAMSHIFT tracking algorithm is employed.

STEREO-VISION-BASED FIRE EXTINGUISHING MOBILE ROBOT

The real-time stereo-vision-based wheeled mobile robot (WMR) is composed of a stereo-vision-based CAMSHIFT tracking algorithm, two CMOS cameras (with stereo vision) for capturing images, and a water gun for extinguishing the burning target. To equip a mobile robot with 3-dimensional range detection capability, the calibration of the binocular vision system should be carried out. The inside geometry information of and spatial relationship between the two cameras used can be acquired by calibrating their intrinsic and extrinsic parameters. The binocular image processing system first finds the same feature points of objects, then compares them to see if they are the same based on the epipolar geometry constraint. The next step is to calculate the 3-dimensional coordinate of the object using the triangular perspective theory, and to compute the distance between the robot and the object.

A fuzzy controller is used to control the robot's rotating direction to track the target on the front side. The robot follows the commands of the fuzzy reasoning module to manipulate the mobile rotation. As soon as the tracked target is followed without colliding with obstacles, the fuzzy reasoning visual system orders the mobile robot to cease rotating. In this work, the two main input variables for the fuzzy controller are the sum of the target's horizontal offset position from the stereo image pairs $X_s = X_l + X_r$, as shown in Figure 3, and the followed target's depth position relative to the robot T_d. The values of X_s and T_d are rescaled to fit the range [–1, 1]. The value of X_s consists of five fuzzy regions: left far (LF), left near (LN), zero (ZE), right near (RN), and right far (RF). For simplicity, these five standard triangular membership functions for the fuzzy region variables {LF, LN, ZE, RN, and RF} are used. The value of T_d consists of three fuzzy regions: near (N), middle (M), and far (F), which represent the target's depth distance relative to the mobile robot.

The rotating signal θ_r at the fuzzy controller output consists of five singletons: left large (LL), left small (LS), zero (ZE), right small (RS), and right large (RL). With the values of the four singletons in hand, the controller output u can then be calculated using the defuzzifier formula. The fuzzy rules are listed in Table 1, which also represent the fuzzy associative matrix. The lines and columns correspond to the target's relative depth distances and horizontal offset position values, respectively (inputs to the fuzzy reasoning system), while the values of the matrix correspond to a robot-rotating signal (output of the fuzzy reasoning system). Overall, the fuzzy reasoning system is governed by the min/max inference technique and the center of gravity for the defuzzification step. The two cameras of the proposed visual servoing system track the target independently of each other with their CAMSHIFT tracking engine.

EXPERIMENTAL RESULTS

The main purpose of this work is to design a wheeled mobile robot with stereo machine vision

Figure 3. Two main input variables for the fuzzy controller are Xs (Xl + Xr) and Td

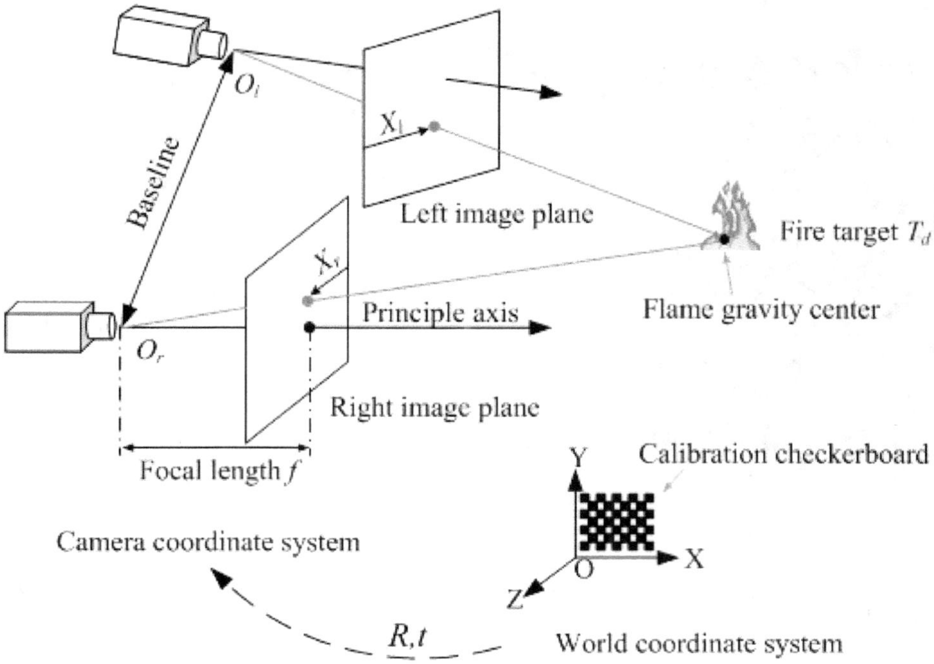

to extinguish a burning target. The stereo visual system is used to analyze the images to get the 3D coordinate position of the burning target on the ground, mark the region and its gravity center, and then the mobile robot is driven to extinguish the burning object with a water gun. The position and orientation of the stereo cameras with respect to the real world's coordinate system could be obtained through extrinsic calibration routines. Both cameras are mounted at the front part of the mobile robot for the best effective view of the targeted object. As shown in Figure 4, the robot follows the commands of the fuzzy reasoning module to track the burning target with the fire detection function enabled and hence, the robot rotates to approach the fire flame on the front side. Demo video clips of the experiment are available on the Web site (http://www.youtube.com/watch?v=dgTSsX2ezMU and http://www.youtube.com/watch?v=JkgC-Q7ogE8).

a. Frame 100, first camera
b. Frame 100, second camera

Table 1. Fuzzy associative matrix for mobile robot steering

θ_r		T_d		
		N	*M*	*F*
X_s	***LF***	***LS***	***LS***	***LL***
	LN	***LS***	***LS***	***LS***
	ZE	***ZE***	***ZE***	***ZE***
	RN	***RS***	***RS***	***RS***
	RF	***RS***	***RS***	***RL***

Figure 4. Position deviations between the robot and the fire and viewpoint from the robot's first and second camera after applying the visual servoing rules

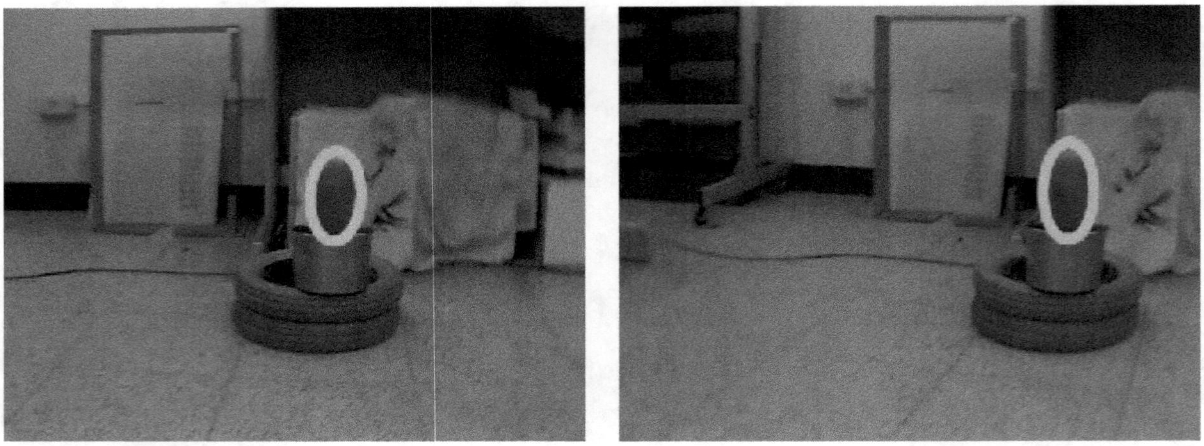

(a) (b)

FUTURE RESEARCH DIRECTIONS

Designing a visual tracking system to avoid the in-path obstacles is a complex task because a large amount of video data must be transmitted and processed in real time. The main task for the target tracking without hitting the obstacles is obstacle detection, which is essential for a safe autonomous mobile robot. Detecting obstacles requires an active perception of the surroundings. Laser scanners have the great advantage of providing accurate depth information that has to be computed from calibrated stereo images if cameras are used for the same task (Ho, 2009). The real-time object tracking and collision avoidance method for mobile robot navigation in indoor and outdoor environments using stereo vision fused with laser sensors is an emerging trend.

Multi-sensor fusion is necessary to cut down on the number of false alarms, since it can reduce the effects of errors in measurements. A variety of multi-sensors were fused together in sensor packages and evaluated based on a set of cost and performance criteria. Vision-based real-time detection for early fire flame detection can be fused with the multiple sensors in order to have a more robust video-based fire detection system. The development of a more sophisticated algorithm, versus the simple threshold rule, for multi-sensor detectors is currently under investigation.

In addition, multiple fires may break out simultaneously, and hence the conducting research on tracking multiple fire regions concurrently and extinguishing dynamically by the behavior of the flame is great technical challenges.

Significant progress has been made in visual servoing during the last few years. Several robust tracking algorithms have been developed, which can track objects in real time in simple scenarios. The proposed framework can find further applications in versatile fields like automated surveillance, human computer interaction, video retrieval, traffic monitoring, and vehicle navigation. Further, motion estimation is a very active area of research in which new solutions are continuously being developed. One challenge in tracking is to develop robust algorithms suitable for tracking objects via hardware logic IP (system-on-chip technology). The combination of different sensors employed in visual servoing (e.g., audio and force sensors) is also a new direction for further development. Such hybrid sensor-based servoing provides additional information that can be used in conjunction with a video-based tracker to solve problems like severe occlusion or estimating tracking more robustly.

CONCLUSION

This chapter proposes a stereo-vision-based wheeled mobile fire detecting and fighting system that has been successfully implemented and shown to work in non-ideal real-world residential buildings. Spectral, spatial, and temporal motion features and a heuristic-based classifier are employed to extract real fire flame data and are adopted for helping the validation of that fire flame. Stereo vision tracking can be achieved by applying the CAMSHIFT algorithm and using two low-cost, calibrated USB cameras, which enable high-speed image capturing. A computationally efficient and robust implementation of the visual measurement and servo mobile robot can be used to obtain reliable real-time online 3D positioning of a particular burning object. Moreover, the fire-fighting system provides more safety in fire fighting and is very economical when incorporated with other fire alarm systems for use in large spaces. Intelligent and automatic control of the fire-fighting robot improves its detection efficiency and suppression adaptability. Experimental results show that real-time fire flame detection and suppression is achieved even under non-ideal lighting conditions.

ACKNOWLEDGMENT

This research was partially supported by the National Science Council in Taiwan under Grant NSC 98-2218-E-224-003, NSC 100-2221-E-224-016 and Industrial Technology Research Institute. Special thanks are given to anonymous reviewers for their helpful suggestions.

REFERENCES

Bobick, A., & Davis, J. W. (2001). The representation and recognition of action using temporal templates. *IEEE Transactions on Pattern Analysis and Machine Intelligence*, *23*(3), 257–267. doi:10.1109/34.910878

Bradski, G. R., & Davis, J. W. (2002). Motion segmentation and pose recognition with motion history gradients. *Machine Vision and Applications*, *13*(3), 174–184. doi:10.1007/s001380100064

Castleman, K. R. (1996). *Digital image processing*. Upper Saddle River, NJ: Prentice Hall Press.

Celik, T., Özkaramanlı, H., & Demirel, H. (2007). Fire and smoke detection without sensors: Image processing based approach. *Proceedings of 15th European Signal Processing Conference* (pp. 1794-1798).

Chen, T., Yuan, H., Su, G., & Fan, W. (2004). An automatic fire searching and suppression system for large spaces. *Fire Safety Journal*, *39*(4), 297–307. doi:10.1016/j.firesaf.2003.11.007

Chen, T. H., Kao, C. L., & Chang, S. M. (2003). *An intelligent real-time fire-detection method based on video processing*. Paper presented at the IEEE 37th Annual 2003 International Conference on Carnahan.

Chen, T. H., Wu, P. H., & Chiou, Y. C. (2004). *An early fire-detection method based on image processing*. Paper presented at the International Conference on Image Processing.

Davis, J. (1999). *Recognizing movement using motion histograms*.

Di Stefano, L., Mattoccia, S., & Mola, M. (2003). *An efficient algorithm for exhaustive template matching based on normalized cross correlation*. Paper presented at the 12th International Conference on Image Analysis and Processing.

Ho, C.-C. (2009, 16-19, August). *Machine vision based 3D scanning system.* Paper presented at the International Conference on Electronic Measurement & Instruments, Beijing.

Ho, C.-C. (2009). Machine vision-based real-time early flame and smoke detection. *Measurement Science and Technology, 20*(4), 045502(045513pp).

Huang, H. W., & Zhang, Y. (2008). Flame colour characterization in the visible and infrared spectrum using a digital camera and image processing. *Measurement Science & Technology, 19*(8), 085406. doi:10.1088/0957-0233/19/8/085406

Hutchinson, T. C., Kuester, F., Doerr, K. U., & Lim, D. (2006). Optimal hardware and software design of an image-based system for capturing dynamic movements. *IEEE Transactions on Instrumentation and Measurement, 55*(1), 164–175. doi:10.1109/TIM.2005.860872

Jain, R., Kasturi, R., & Schunck, B. G. (1995). *Machine vision*. McGraw-Hill.

Kim, K. I., Oh, S. Y., Kim, S. W., Jeong, H., Lee, C. N., Kim, B. S., et al. (1996). *An autonomous land vehicle PRV III.* Paper presented at the IEEE Intelligent Vehicles Symposium.

LeGrand, R., & Luo, R. C. (1996, 22-28 April). *Position estimation of selected targets.* Paper presented at the International Conference on Robotics and Automation, Minneapolis, MN

Podr aj, P., & Hashimoto, H. (2008). Intelligent space as a framework for fire detection and evacuation. *Fire Technology, 44*(1), 65-76.

Sumi, K., Hashimoto, M., & Okuda, H. (1995). *Three-level broad-edge matching based real-time robot vision.* Paper presented at the IEEE International Conference on Robotics and Automation.

Swain, M. J., & Ballard, D. H. (1990). *Indexing via color histograms.* Paper presented at the Third International Conference on Computer Vision.

Villa-Uriol, M. C., Chaudhary, G., Kuester, F., Hutchinson, T., & Bagherzadeh, N. (2004). *Extracting 3D from 2D: Selection basis for camera calibration.* Paper presented at the 7th International Conference on Computer Graphics and Imaging (CGIM).

Wang, S. J., Tsai, M. T., Ho, Y. K., & Chiang, C. C. (2006). Video-based early flame detection for vessels by using the fuzzy color clustering algorithm. *Proceedings of International Computer Symposium, 3*, 1179–1184.

Wright, R. S. Jr, & Sweet, M. R. (2000). *OpenGL superBible* (2nd ed.). Waite Group Press.

Yuan, F. (2010). An integrated fire detection and suppression system based on widely available video surveillance. *Machine Vision and Applications, 21*(6), 941–948. doi:10.1007/s00138-010-0276-x

ADDITIONAL READING

Bradski, G. R. (1998). Computer vision face tracking for use in a perceptual user interface. *Intel Technology Journal, 2*(2), 12–21.

Bradski, G. R., & Davis, J. W. (2002). Motion segmentation and pose recognition with motion history gradients. *Machine Vision and Applications, 13*(3), 174–184. doi:10.1007/s001380100064

Chen, T. H., Yin, Y. H., Huang, S. F., & Ye, Y. T. (2006, Dec. 2006). *The smoke detection for early fire-alarming system base on video processing.* Paper presented at the International Conference on Intelligent Information Hiding and Multimedia Signal Processing.

Davis, J. W., & Bobick, A. (1997, June). *The representation and recognition of action using temporal templates.* Paper presented at the Proceedings of IEEE Conference on Computer Vision and Pattern Recognition, Puerto Rico.

Fang, J., Jie, J., Hong-Yong, Y., & Yong-Ming, Z. (2006). Early fire smoke movements and detection in high large volume spaces. *Building and Environment*, *41*(11), 1482–1493. doi:10.1016/j.buildenv.2005.05.036

Ho, C.-C. (2010). Image processing algorithm for tracking and picking an object for an open platform mobile robot. *Key Engineering Materials*, *419*, 569–572. doi:10.4028/www.scientific.net/KEM.419-420.569

Ho, C.-C. (2010). A stereo vision based target tracking and obstacle avoidance. *Key Engineering Materials*, *419*, 565–568. doi:10.4028/www.scientific.net/KEM.419-420.565

Ho, C.-C., Chen, M.-C., & Lien, C.-H. (2011). Machine vision-based intelligent fire fighting robot. *Key Engineering Materials*, *450*, 312–315. doi:10.4028/www.scientific.net/KEM.450.312

Ho, C.-C., & Kuo, T.-H. (2009). *Real-time video-based fire smoke detection system.* Paper presented at the IEEE/ASME Conference on Advanced Intelligent Mechatronics.

Ho, C.-C., & Shih, C.-L. (2007). *Real-time tracking and stereo vision-based control of a goldfish-catching system.* Paper presented at the Tthe 35th International MATADOR Conference.

Ho, C.-C., & Shih, C.-L. (2008). Machine vision based tracking control of a ball-beam system. *Key Engineering Materials*, *381-382*, 301–304. doi:10.4028/www.scientific.net/KEM.381-382.301

Ho, C.-C., & Shih, C.-L. (2008). A real-time visual servo architecture for a golden fish catching system. *Measurement and Control*, *41*(5), 151–154.

Ho, C.-C., & Shih, C.-L. (2009). A real-time fuzzy reasoning based control system for catching a moving goldfish. *International Journal of Control. Automation and Systems*, *7*(5), 755–763. doi:10.1007/s12555-009-0508-x

Horng, W.-b., Peng, J.-w., & Chen, C.-y. (2005, March 19-22). *A new image-based real-time flame detection method using color analysis.* Paper presented at the International Conference on Networking, Sensing and Control, Tucson, Arizona, USA.

Kang, K. (2007). A smoke model and its application for smoke management in an underground mass transit station. *Fire Safety Journal*, *42*(3), 218–231. doi:10.1016/j.firesaf.2006.10.003

Ko, B. C., Cheong, K. H., & Nam, J. Y. (2008). Fire detection based on vision sensor and support vector machines. *Fire Safety Journal.*

Kopilovic, I., Vagvolgyi, B., & Sziranyi, T. (2000, 09/03 - 09/07). *Application of panoramic annular lens for motion analysis tasks: surveillance and smoke detection.* Paper presented at the 15th International Conference on Pattern Recognition, Barcelona, Spain.

Lai, C. L., Yang, J. C., & Chen, Y. H. (2007, May). *A real time video processing based surveillance system for early fire and flood detection.* Paper presented at the The 24th IEEE Instrumentation and Measurement Technology Conference.

Liu, C.-B., & Ahuja, N. (2004). *Vision based fire detection.* Paper presented at the The 17th International Conference on Pattern Recognition, Urbana.

Phillips, W. III, Shah, M., & da Vitoria Lobo, N. (2002). Flame recognition in video. *Pattern Recognition Letters*, *23*(1-3), 319–327. doi:10.1016/S0167-8655(01)00135-0

Schroeder, D. (2004). Evaluation of three wildfire smoke detection systems. [FERIC]. *Advantage-Forest Engineering Research Institute of Canada*, *5*(24), 8.

The references to materials directly cited in the chapter have been included in the reference section. However, there are many additional readings relevant to stereo vision, some of which are listed in this section.

Toreyin, B. U., Dedeoglu, Y., & Cetin, A. E. (2005). *Flame detection in video using hidden markov models.* Paper presented at the International Conference on Image Processing.

Toreyin, B. U., Dedeoglu, Y., & Cetin, A. E. (2005). *Wavelet based real-time smoke detection in video.* Paper presented at the The 13th European Signal Processing Conference, Antalya, Turkey.

Toreyin, B. U., Dedeoglu, Y., & Cetin, A. E. (2006). *Contour based smoke detection in video using wavelets.* Paper presented at the The 14th European Signal Processing Conference, Florance, Italy.

Toreyin, B. U., Dedeogˇlu, Y., Gudukbay, U., & Cetin, A. E. (2006). Computer vision based method for real-time fire and flame detection. *Pattern Recognition Letters, 27*(1), 49–58. doi:10.1016/j.patrec.2005.06.015

Wieser, D., & Brupbacher, T. (2001, March 25-28). *Smoke detection in tunnels using video images.* Paper presented at the 12th international conference on automatic fire detection, Gaithersburg, USA.

Xiong, Z., Caballero, R., Wang, H., Finn, A. M., Lelic, M. A., & Peng, P.-Y. (2007). *Video-based smoke detection: possibilities, techniques, and challenges.* Paper presented at the Fire Suppression and Detection Research and Applications, Orlando, FL.

Yuan, F. (2008). A fast accumulative motion orientation model based on integral image for video smoke detection. *Pattern Recognition Letters, 29*(7), 925–932. doi:10.1016/j.patrec.2008.01.013

KEY TERMS AND DEFINITIONS

3D Tracking: Position based tracking algorithm to provide feedback on the real-time position of the specified target with a high frame rate

Fire Surveillance System: Video based fire surveillance technology is suitable for early detection of fires due to its low detection delay, good resolution, and high localization accuracy.

Fire-Fighting Robot: The mobile robot is able to successfully complete a fire-extinguishing task.

Stereo Vision: 3D position can be generated by means of two calibrated cameras.

Video Based Fire Flame Detection: Analyze the spectral, spatial, and motion orientation characteristics of the fire flame regions from the image sequences of the video to detect fire.

Video Based Smoke Pattern Recognition: Analyze the spectral, spatial, and motion orientation characteristics of the fire smoke regions from the image sequences of the video to detect fire smoke.

Visual Servoing: A visual tracking system to track an object and video data must be transmitted and processed in real time in order to feedback to the system.

Section 5
3D Imaging Applications

Chapter 23
3D DMB Player and Its Reliable 3D Services in T-DMB Systems

Cheolkon Jung
Xidian University, China

Licheng Jiao
Xidian University, China

ABSTRACT

This chapter introduces a 3D DMB player which can provide realistic 3D services to consumers in terrestrial-digital multimedia broadcasting (T-DMB) systems. This chapter also provides a parameter approximation method which can create auto-stereoscopic images reliably in the 3D DMB player. Since the bit-budget for the transmission of additional data stream is strictly limited in current T-DMB systems, depth-image-based rendering (DIBR) techniques have been studied to provide 3D services in mobile devices. In order to create the auto-stereoscopic images reliably in the 3D DMB player, exact parameters such as convergence distance, scale factor, and far/near clipping plane should be given in contents. However, some contents contain unknown or inappropriate parameter values in a real environment. This makes it extremely difficult to create auto-stereoscopic images and provide consumers with reliable 3D services. Therefore, we explain how to approximate the rendering parameters by taking mobile display size into consideration. Experimental results show that the parameter approximation method can create auto-stereoscopic images reliably in the 3D DMB player.

INTRODUCTION

At present, three dimensional television (3D TV) is being considered as one of the next generation broadcasting technologies because it can provide consumers with more realistic and life-like visual home entertainment experiences. Up to now, many researchers have paid much attention to the development of 3D TV broadcasting technologies. Consequently, auto-stereoscopic 3D displays based on different perspective views have been developed and their related 3D services have been provided. Above all, 3D services over

DOI: 10.4018/978-1-61350-326-3.ch023

Figure 1. The general process of depth-image-based rendering (DIBR). (a) Reference image. (b) Depth image. (c) Virtual two views. (d) Auto-stereoscopic image.

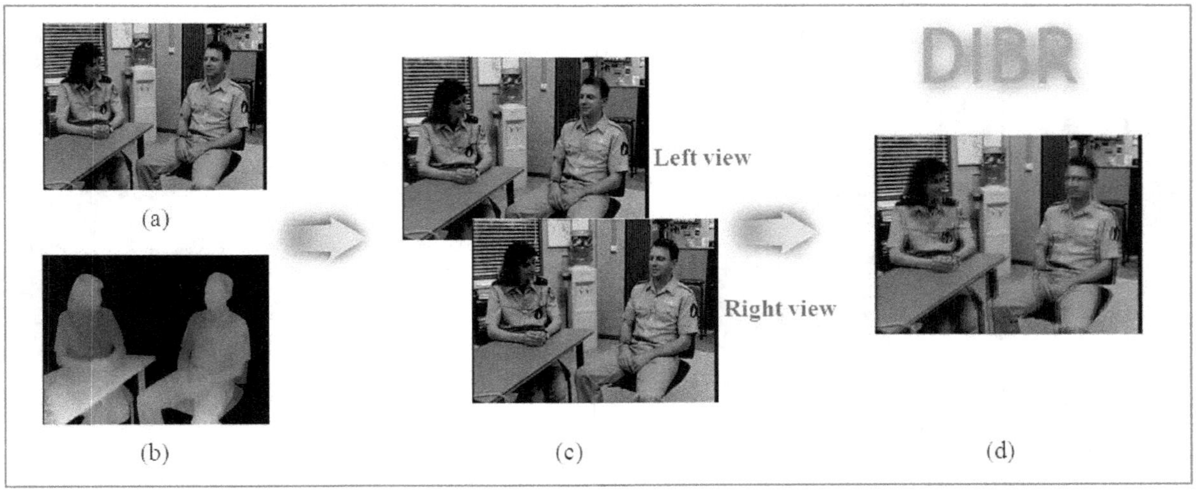

T-DMB are very attractive because the single user environment of T-DMB is suitable for glassless 3D viewing with mobile displays. However, the T-DMB system has limitation of the bit budget for the transmission of additional video streams because of its limited bandwidth (Park et al., 2009; Schreer et al., 2005; Jung et al., 2008; Yun et al., 2009; Jung et al., 2010c). Actually, the bit-budget for the transmission of additional data stream is strictly limited in the current state of T-DMB. The T-DMB system supports about 2Mbps of useful data rate in the 1.536MHz channel. The additional data stream should be served in bitrates below about 64Kbps, which are insufficient to compress the additional color video stream efficiently. Therefore, the ATTEST project, which started in March 2002 as part of the European Information Society Technologies (IST), has proposed a DIBR technique because depth information as the additional data can be compressed efficiently below 64Kbps (Park et al., 2009; Fehn, 2004).

In DIBR, left and right virtual views, which form auto-stereoscopic image, are rendered by reference image and its corresponding depth image in auto-stereoscopic displays as shown in Figure 1. To maintain the backward compatibility with traditional 2D broadcasting, regular 2D color video in digital TV format is used in the reference image (Figure 1(a)). Its corresponding depth image, which stores depth information of 8-bit gray values with 0 at the furthest place and 255 at the nearest place, is just added with the same spatiotemporal resolution (Figure 1(b)) (Lee et al., 2007; Zhang & Tam, 2005; Hur et al., 2005; Lee et al., 2009). The two virtual views are shown in Figure 1(c); and their auto-stereoscopic image is shown in Figure 1(d). The auto-stereoscopic image is produced by interleaving the two virtual views as shown in Figure 2(a), and the glassless 3D services from the auto-stereoscopic images can be provided to consumers in the displays by the parallax barrier as shown in Figure 2(b).

Many researchers have studied on the DIBR techniques for 3D data services over T-DMB (Park et al., 2009; Fehn, 2004; Zhang & Tam, 2005; Choi et al., 2009; Oh et al., 2009). Fehn (2004) provided the detailed descriptions of the 3D TV system introduced by the ATTEST project including compression and transmission. Also, the high-quality DIBR technique using the shift-sensor camera setup was introduced in Fehn's work (2004). Zhang & Tam (2005) proposed the

Figure 2. Formation of the auto-stereoscopic image and glassless 3D services by parallax barrier. (a) Interleaving. (b) Parallax barrier.

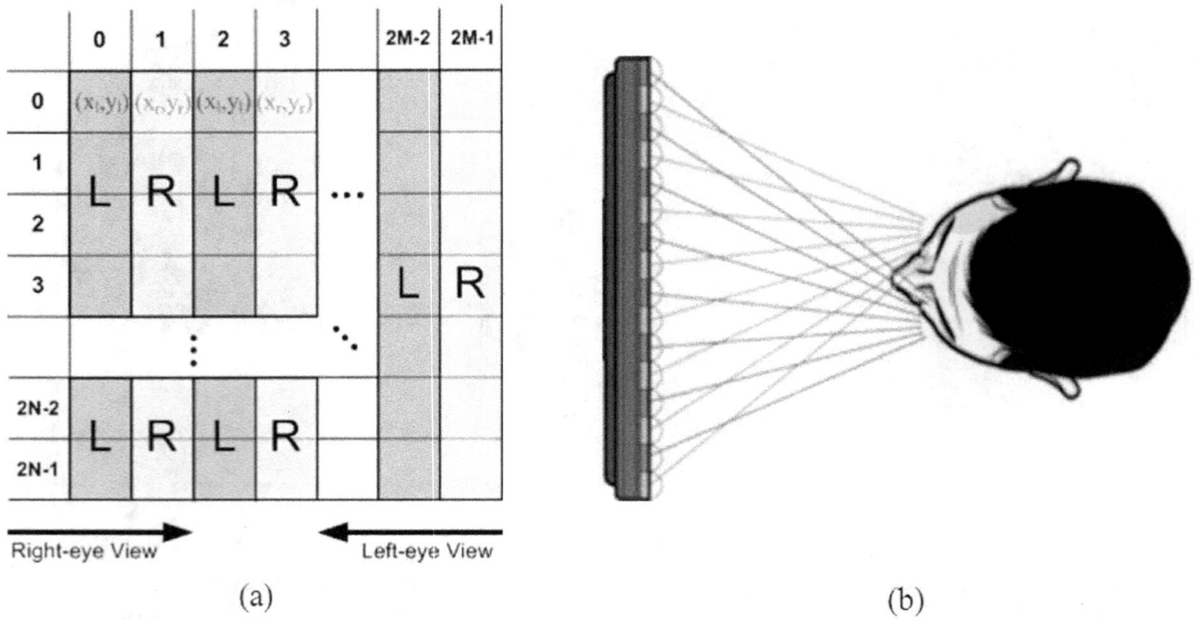

depth pre-processing algorithm using an asymmetric filter to reduce the disocclusion areas. One of the inherent problems in DIBR is the disocclusion that scene area, which is occluded in the reference image, become visible in any of the virtual left and right views. The asymmetric filter was able to reduce the disocclusion areas efficiently and maintain good depth quality successfully while creating the auto-stereoscopic images. Above all, the service architecture for the DIBR based 3D services was proposed in Park et al.'s work (2009). Figure 3 shows the block diagram of 3D services based on DIBR in Park et al.'s work (2009). As can be seen, it mainly consists of two parts: transmitter and receiver parts. In the transmitter part, a depth sequence is preprocessed for taking advantages of both reduction of disocclusion area (hole) and distortion minimization when virtual views are created. The preprocessed depth sequence and the reference sequence are coded by the H.264/AVC baseline encoder and transmitted through T-DMB channel. In the figure, D_o and I_o are original depth and reference images, respectively; and D and I are decoded depth and reference images, respectively. T_1 and T_2 are two thresholds of adaptive smoothing filters for depth preprocessing. In the receiver part, once both the reference stream and its corresponding depth stream are received, the auto-stereoscopic image sequence is created by applying 3D warping, hole filling, and interleaving simultaneously. The created auto-stereoscopic image sequence is displayed through 3D mobile displays with parallax barriers. In the transmitter part, the depth preprocessing is based on adaptive smoothing techniques which are explained in Park et al.'s work (2009). It is applied before encoding and this is due to the fact that T-DMB receivers should guarantee the real-time rendering with the limited computation power. Therefore, simple and fast 3D warping, hole-filling, and interleaving algorithms are used in the receiver part.

In this chapter, we introduce the 3D DMB system and the 3D DMB player which can provide

Figure 3. Block diagram of the DIBR-based 3D service over T-DMB (Park, Jung, Oh, Lee, Kim, Lee, Lee, Yun, Hur & Kim, 2009).

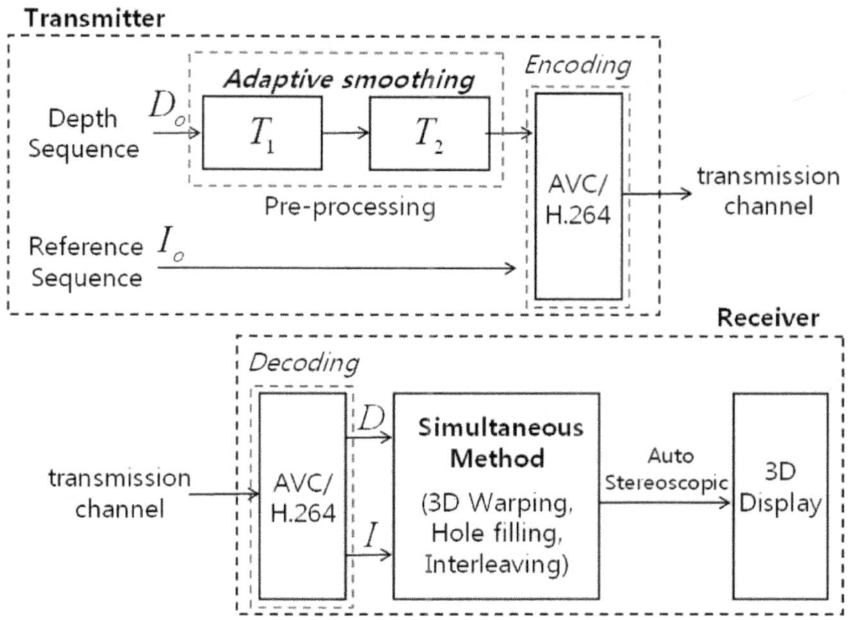

3D services to consumers. Here, the 3D DMB player implements realistic 3D services based on the DIBR technique. For providing reliable 3D services, exact parameters such as convergence distance, scale factor, and far/near clipping plane should be given for DIBR according to 3D contents and mobile displays. However, in a real environment, some parameters contain unknown values of actual contents when we create auto-stereoscopic images because of the content provider's mistakes or transmission errors. Moreover, inappropriate values are given in some parameters, which make it extremely difficult to create auto-stereoscopic images in mobile displays. Therefore, this chapter provides a novel method to approximate rendering parameters which can create auto-stereoscopic images rapidly and provide consumers with reliable 3D services of various contents. By the parameter approximation method, we have created auto-stereoscopic images rapidly and implemented reliable 3D services of various contents.

OVERVIEW OF THE 3D DMB SYSTEM

Since 3D services are provided to consumers over T-DMB, it is impossible to understand the 3D DMB system without knowledge of the current T-DMB system. T-DMB is a digital radio system for sending multimedia to mobile devices such as mobile phones, portable media player (PMP), and personal digital assistant (PDA).

The key points of T-DMB are to provide personality, mobility, and interactivity to consumers. To be more concrete, T-DMB is based on Eureka-147 digital audio broadcasting (DAB) system by extending multimedia protocol stacks to provide mobile TV services (Lee et al., 2008a; Lee et al., 2008b). It incorporates the latest media coding technologies such as MPEG-4 part 3 BSAC (bit sliced arithmetic coding), HEAAC v2 (high efficiency advanced audio coding) and part 10 H. 264 /AVC (advanced video coding) to achieve high performances. Moreover, T-DMB supports interactive data services by utilizing MPEG-4

BIFS (binary format for scenes). Therefore, it can provide audio-associated data service as well as downloadable data services. Representative services of T-DMB are internet service, on-line shopping, and pay-per-view (PPV). In addition, the T-DMB system is able to provide new 3D video and data services to consumers because auto-stereoscopic mobile displays are quite feasible due to the advancement of 3D LCD technologies. The disparity is relatively small in the auto-stereoscopic mobile displays because of their small sizes. Therefore, eye strain and visual fatigue can be reduced (Lee et al., 2009). As shown in Figure 4, the 3D DMB system can be classified into two main groups: broadcasting server and receiver (Lee et al., 2010; Kim, 2008; Oh et al., 2007). In the broadcasting server, broadcasting 3D contents are created from stereoscopic camera and multichannel microphone. The 3D contents are coded by 3D DMB encoder and transmitted through 2D T-DMB channel. If the 3D DMB services are based on DIBR, reference and preprocessed depth sequences are coded by the H.264/AVC baseline encoder and transmitted through conventional 2D T-DMB channel. Here, G.704 is an ITU-T standard for synchronous frame structures and gives functional characteristics of interfaces associated with network nodes. In the case of the receiver part, glassless 3D DMB portable receivers are employed to play 3D video data. Recently, implementation of 3D DMB receiver for 3D data service is presented in Lee et al.'s work (Lee et al., 2010). The presented receiver is based on MPEG-4 BIFS technology, and implemented to perform the functionalities for 3D data service such as parsing of scene description, decoding and rendering of stereoscopic image pairs according to the MPEG-4 BIFS technology. The 3D T-DMB receivers should have equipments to display stereoscopic images including parallax barrier strip displays and lenticular screens. In addition, it is necessary to create stereoscopic images reliably in the receivers (Lee et al., 2008a; Yun et al., 2008).

Although many studies have been made on auto-stereoscopic displays, 3D video coding, and DIBR based 3D services, there is little results on how to set rendering parameters accurately such as convergence distance, scale factor, far/near clipping plane in DIBR (Park et al., 2009; Jung et al., 2008; Fehn, 2004; Lee et al., 2009; Cho et al., 2007). However, if the exact values are not assigned to the parameters, it is impossible to create auto-stereoscopic images and provide consumers with reliable 3D services. In this chapter, we explain a novel method to set the rendering parameters accurately by taking mobile display size into consideration. We also introduce our 3D DMB player which can provide realistic and reliable 3D services to consumers.

REALISTIC AND RELIABLE 3D SERVICES ON THE 3D DMB PLAYER

Depth Preprocessing in the Transmitter Part

As mentioned above, one of the inherent problems in DIBR is disocclusion which is commonly referred as 'hole'. As shown in Figure 5, the holes inevitably occur because the scene area, which is occluded in the reference image, become visible in any of the virtual left and right views. The holes are caused by the disoccluded regions of Figure 5(a) and appear in white regions of Figure 5(b). Since there is no information to fill the holes in both the reference image and its corresponding depth image, it is not easy to handle the holes. Thus, to minimize the holes and preserve the depth information, two different smoothing filters are sequentially applied to original depth images in the transmitter part (Park et al., 2009). The main idea of the depth preprocessing is to apply two different adaptive smoothing filters sequentially to original depth images. The first filter is a discontinuity-preserving smoothing filter which removes noise and preserves original depth information.

Figure 4. System configuration of the 3D DMB system (Kim, 2008). G.704 is an ITU-T standard for synchronous frame structures and gives functional characteristics of interfaces associated with network nodes.

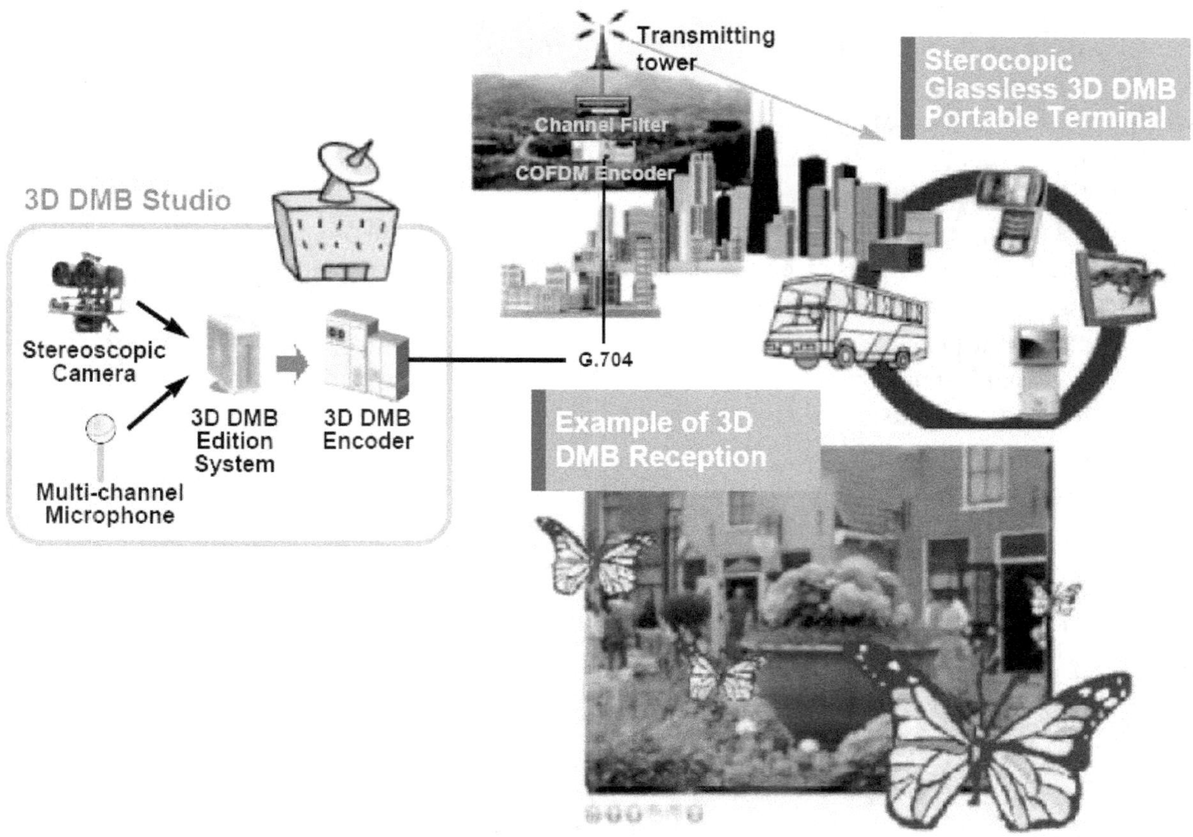

Figure 5. Holes caused by disoccluded regions. (a) Cause of disoccluded regions. (b) Virtual left view of 'Ballet' sequence (white pixels are the disoccluded regions).

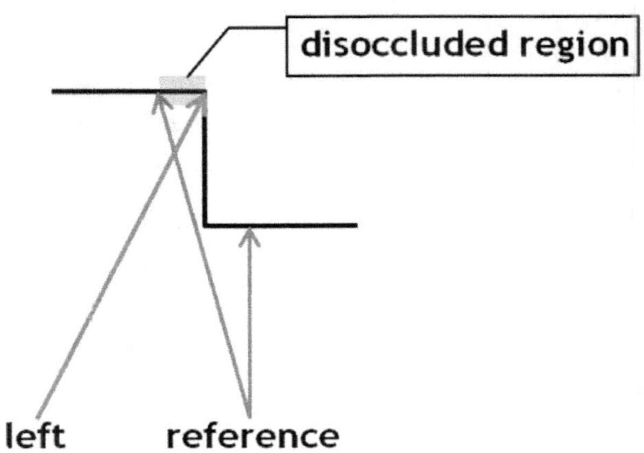

(a)

(b)

The second filter is a gradient-based smoothing filter which smoothes original depth images in the horizontal direction and reduces holes.

The two adaptive smoothing filters iteratively convolve the input depth images to be smoothed with a 3x3 mask whose coefficients reflect the specific measurements at each point (Park et al., 2009; Park et al., 2008; Jung & Jiao, 2010a; Jung et al., 2010d). Notice that the depth preprocessing procedure is employed before encoding and this is due to the fact that T-DMB receivers should guarantee the real-time rendering with the limited computational power. Then, the reference and preprocessed depth sequences are coded by the H.264/AVC baseline encoder and transmitted over T-DMB. Figure 6 shows the depth preprocessing procedure using the adaptive smoothing filters. In this figure, left and right windows show original and preprocessed depth images, respectively.

3D DMB Player and Reliable DIBR in the Receiver Part

3D DMB Player

The target platform of the DIBR based 3D services is the portable player with the auto-stereoscopic 3D display. The auto-stereoscopic display is made by direction-based techniques such as parallax barrier strip displays and lenticular screens. Accordingly, each eye of the viewer can see only the corresponding view by directing the light emitted by pixels of distinct two (left and right) perspective views exclusively to the appropriated eye. In our 3D DMB player, the special parallax barrier has been affixed to the display of the SONY VAIO

Figure 6. Depth preprocessing procedure using adaptive smoothing techniques (left window: original depth image, right window: preprocessed depth image)

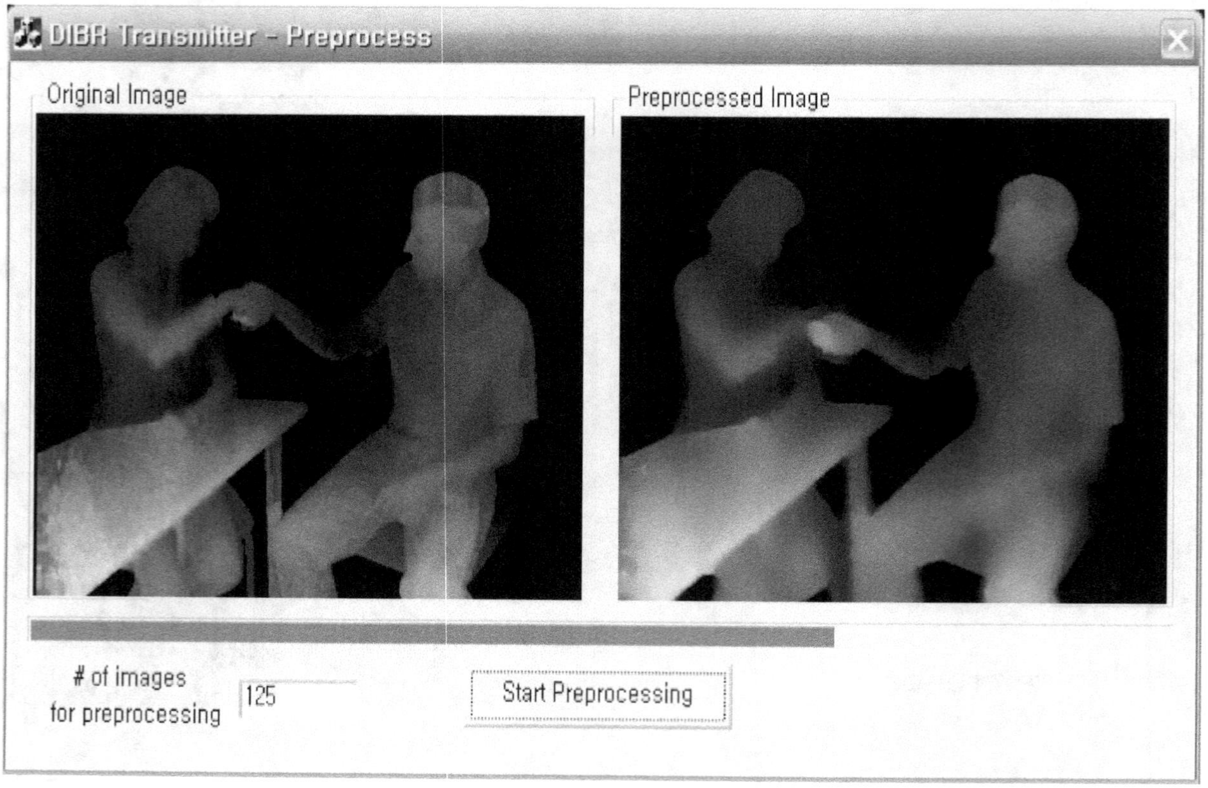

VGN-UX17LP(CPU: Intel Core™ Solo Processor U1400) for viewing 3D auto-stereoscopic images as shown in Figure 7. The DIBR technique of the receiver consists of three main steps: 3D image warping, hole filling, and interleaving. Because of the limited computational power and memory of mobile devices, a look-up-table (LUT) based simultaneous method has been used for real-time rendering (Park et al., 2009; Choi et al., 2009). The main idea of the LUT based simultaneous method is to conduct three steps of warping, hole filling, and interleaving simultaneously by directly representing the value of each pixel in the two virtual views on the auto-stereoscopic image plane. In the 3D image warping step, the distance of pixel movement for depth values is calculated by the pre-constructed LUT. Then, the holes mainly appear around the boundary of objects, and are filled simply by linear interpolation of neighborhood pixels. Finally, auto-stereoscopic images for 3D displays are created by interleaving the two virtual views on the 3D DMB player.

Reliable DIBR Technique

As mentioned earlier, the DIBR technique in the receivers consists of 3D image warping, hole filling, and interleaving. In the 3D image warping, a depth distance, $Z(v)$, is computed by a depth value of depth images, v, using z_{far} (i.e., far clipping plane) and z_{near} (i.e., near clipping plane) as follows (Park et al., 2009; Jung et al., 2008).

$$Z(v) = \frac{1}{\frac{1}{z_{near}}\left(\frac{v}{255}\right) + \frac{1}{z_{far}}\left(1 - \frac{v}{255}\right)}, \qquad v \in [0, \cdots, 255] \tag{1}$$

As can be seen in equation (1), two parameters, z_{far} and z_{near}, are requirements for getting the depth distance, $Z(v)$. However, most of actual contents do not provide these two parameter values. In this case, receivers should request transmitters to send them additionally, which can cause delays in rendering the contents. Moreover, if transmitters do not find and give the parameters, it is very difficult to create auto-stereoscopic images in

Figure 7. Our 3D DMB player: the special parallax barrier is affixed to the display of the SONY VAIO VGN-UX17LP for viewing 3D auto-stereoscopic images

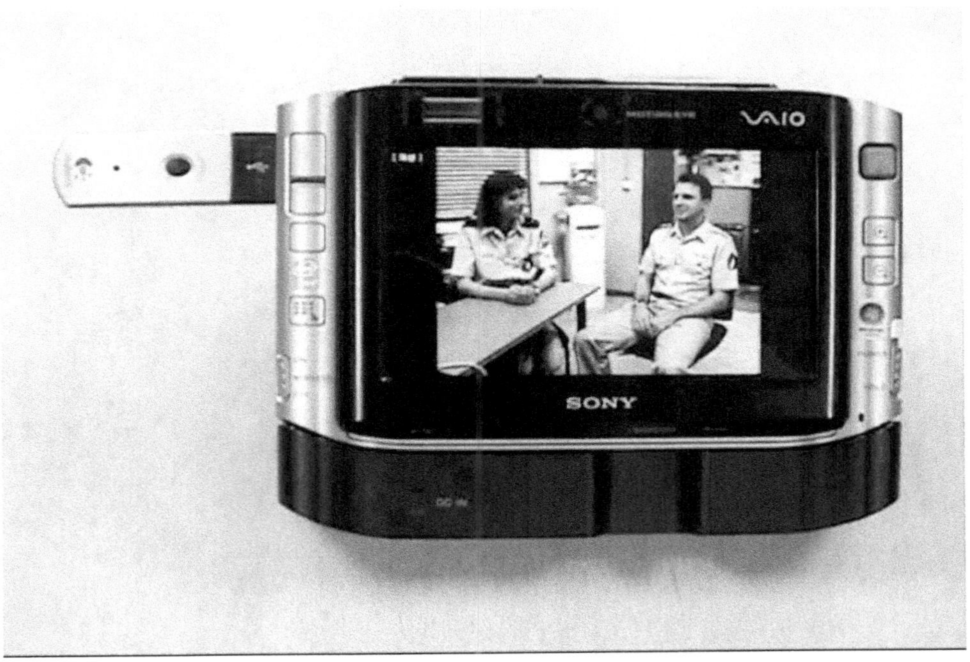

the receivers. In order to solve the problem and be able to provide reliable 3D services, equation (1) is approximated as follows.

If we assume that z_{far} is 255 and z_{near} is 1, equation (1) is calculated as follows (Jung & Jiao, 2010b).

$$Z(v) = \frac{255}{\frac{254}{255}v + 1}, \qquad v \in [0, \cdots, 255] \tag{2}$$

Since 254/255 can be approximated to 1, we can express equation (2) as the following form.

$$Z(v) = \frac{255}{v+1}, \qquad v \in [0, \cdots, 255] \tag{3}$$

Depth values can be converted into depth distances using equation (3) on contents which have unknown values of the two parameters z_{far} and z_{near}. In addition, we can make virtual left and right views which generate auto-stereoscopic images implementing equation (3). Figure 8 shows the geometry of the virtual camera setup for generating virtual views. The parameters f and t_c denote the focal length and baseline distance between virtual cameras C_l and C_r, respectively. From the geometry, the pixel positions (x_c, y), (x_l, y), and (x_r, y) of the reference view, and the two virtual views corresponding to each point P with the depth distance Z have the following relationships (Park et al., 2009; Jung et al., 2008):

$$x_l = x_c + \frac{\alpha_\mu t_c}{2Z(x_c, y)} - \frac{\alpha_\mu t_c}{2Z_c} \tag{4}$$

Figure 8. The geometry of the camera setup for generating virtual views in DIBR

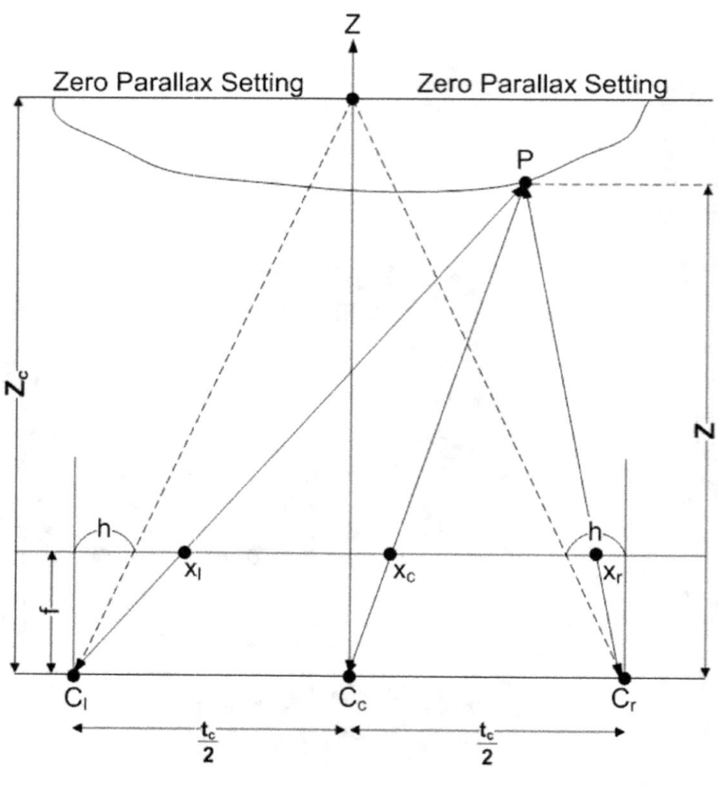

$$x_r = x_c - \frac{\alpha_\mu t_c}{2Z(x_c, y)} + \frac{\alpha_\mu t_c}{2Z_c} \tag{5}$$

where Z_c and α_u are the convergence distance and scale factor between the two virtual cameras, respectively. Here, the scale factor α_u is obtained by dividing the focal length f into pixel size. Moreover, it has been reported that users may feel much eyestrain if differences between x_l and x_r are more than 3% of the image width (Fehn, 2004). Accordingly, the maximum differences are adjusted to meet 3% of the image width. Based on the aforementioned assumptions, α_u can be calculated as follows.

$$x_l - x_r = \frac{\alpha_u t_c}{Z(x_c, y)} - \frac{\alpha_u t_c}{Z_c} < 0.03 \times w \tag{6}$$

where w is the image width. We assume that t_c =60mm because general distances between left and right eyes are about 60mm. To maximize the differences between x_l and x_r, we set $Z(x_c,y)$=1 and Z_c=255. Then, α_u is computed by equation (7).

$$\alpha_u = 0.0005 \times w \tag{7}$$

Left and right virtual views are created reliably using equations (4)-(7), even if contents contain unknown or inappropriate parameter values. By equations (4) and (5), disocclusion area (hole) is inevitably produced. The holes are filled using linear interpolation algorithm in the two virtual views. Then, auto-stereoscopic images are generated by interleaving the two virtual views. As a result, it is possible to provide consumers with reliable 3D services even if contents contain unknown or inappropriate parameter values. Moreover, the computational cost of the rendering procedure is reduced because the depth distances are computed by equation (3) instead of equation (1). It is very effective for mobile devices with the limited computational power and memory.

EXPERIMENTAL RESULTS

To evaluate the effectiveness of the parameter approximation method, five typical sequences, each consisting of the reference sequence and the corresponding depth sequence, were used for the tests. All experiments were performed on the 3D DMB player. As shown in Figure 9, they are 'Interview' and 'Orbi' from Heinrich-Hertz-Institute (HHI), 'Ballet' and 'Breakdancer' from Microsoft Research, and 'Etri_CG' from Electronics and Telecommunications Research Institute (ETRI), respectively. All test sequences for the experiments are adjusted to the size of 320x240 pixels. Since the image width w is 320, the scale factor α_u is 0.16 by equation (7). In addition, a depth distance $Z(v)$ corresponding to a depth value v is computed by equation (3). $Z(v)$ corresponding to v is distributed as shown in Figure 10. The distribution is nearly equivalent to that by equation (1). By equation (6), the maximum differences between x_l and x_r are approximately 10, and this means that consumers would feel much eyestrain if disparity between x_l and x_r is more than 10 pixels.

Figure 11 shows two auto-stereoscopic images of the 'Interview' sequence each created by equations (1) and (3). It can be observed that the parameter approximation method can create nearly identical auto-stereoscopic images obtained by equation (1) on the 3D DMB player. Figure 12 shows the success rate of creating auto-stereoscopic images from the test sequences. In the table, P means that the test sequence is successfully played on the 3D DMB player. The success rate evaluates how many sequences are successfully played from the 5 test sequences by the proposed method. It can be observed that our method is able to create auto-stereoscopic images of various contents reliably on the 3D DMB player. That is, even if contents do not contain some of parameter values such as near/far clipping plane, our DMB player can create auto-stereoscopic images reliably. Therefore, reliable and realistic 3D services can be provided to consum-

Figure 9. Examples of five test sequences (left: reference images, right: depth images). (a) Interview. (b) Orbi. (c) Ballet. (d) Breakdancer. (e) Etri_CG.

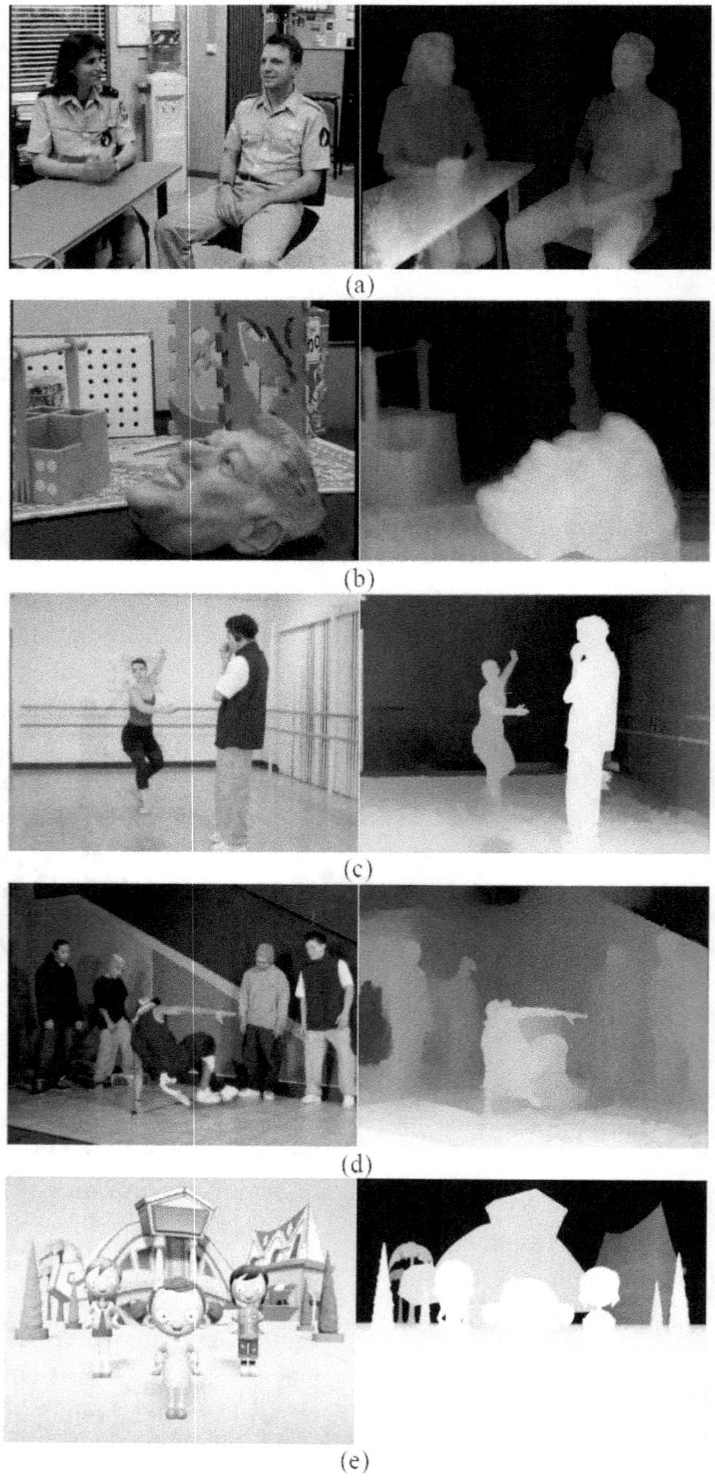

Figure 10. A depth distance Z(v) versus a depth value v

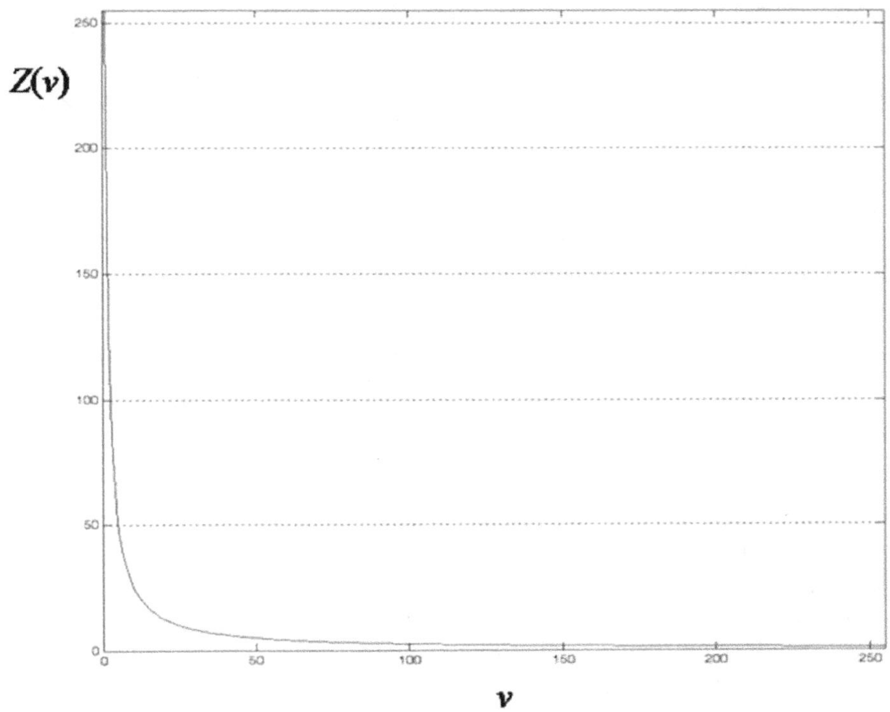

ers by the parameter approximation method. Moreover, the parameter approximation method has the merit of reducing computational cost because of its simplified computation. To verify the effectiveness of the computational cost, we measured time it took to create the auto-stereoscopic images using separately equations (1) and (3). As shown in Figure 13, the average process-

Figure 11. Auto-stereoscopic images of the 'Interview' sequence (a) created by equation (1) and (b) created by equation (3)

(a) (b)

Figure 12. The success rate of creating auto-stereoscopic images from the test sequences (the success rate evaluates how many sequences are successfully played from the 5 test sequences and P means that the test sequence is successfully played on the 3D DMB player)

Sequences	Result
Interview	*P*
Orbi	*P*
Ballet	*P*
Breakdancer	*P*
Etri_CG	*P*
Success Rate	1.0

ing time of the parameter approximation method is 0.053 sec per image (s/image). Such results indicate that our method reduces 26.4% (0.019 s/image) of the processing time as compared to equation (1). This means that our method is very effective in terms of computational cost for creating auto-stereoscopic images as well. Additionally, the 3D DMB player can play both 2D and 3D contents. If consumers want to view 3D contents on the player, they have only to switch on the parallax barrier button located at the top of the player.

FUTURE RESEARCH DIRECTIONS

The glassless 3D services can be implemented in the DMB player by the parallax barrier. However, the viewer's eye should be located at a specific sweet position because each of the auto-stereoscopic images is pointing to a specific eye position as shown in Figure 7. It is inevitable that viewers keep their eyes firmly fixed on the sweet position to consume 3D contents in the DMB player. Therefore, camera eye tracking is required to solve this problem and improve the

Figure 13. Comparison of average processing times for creating auto-stereoscopic images by equations (1) and (3) (Experiments were performed on the 3D DMB player (CPU: Intel Core™ Solo Processor U1400). The unit of this test is sec per image (s/image).)

Sequences	(1)	(3)
Interview	0.073	0.053
Orbi	0.071	0.053
Ballet	0.072	0.054
Breakdancer	0.072	0.053
Etri_CG	0.074	0.053
Average	0.072	0.053

3D viewing environment for the parallax barrier auto-stereoscopic displays (Kim & Kim, 2008; Strandvall, 2009; Zhang & Zhang, 2010; Zhu & Ji, 2005). Further studies on the real-time robust eye tracking under variable lighting conditions and face orientations are needed. In addition, it has been reported that the depth perception rate is undesirably decreased by coding (Park et al., 2009). This is due to the fact that high frequency components of depth images are filtered by lossy compression techniques such as H.264 and thus coding artifacts occur. Consequently, further research should be carried out to improve the depth perception rate in the future.

CONCLUSION

In this chapter, we have introduced a 3D DMB player which provides consumers with realistic 3D services. We have also provided an efficient DIBR technique for reliable 3D services on the 3D DMB player. For the reliable 3D services, we have explained a novel method to approximate rendering parameters by taking mobile display size into consideration. By the parameter approximation method, we can create auto-stereoscopic images reliably even if contents contain unknown or inappropriate parameter values. Our method also reduces computational cost because of its simplified computation. Experimental results show that our method reduces 26.4% of the processing time as compared to conventional methods. Therefore, it is demonstrated that our method can create auto-stereoscopic images of various contents reliably and rapidly on the 3D DMB player. Consequently, our method can provide consumers with realistic and reliable 3D services over T-DMB as well. We believe that our method will contribute to the popularization of 3D services based on DIBR over T-DMB.

ACKNOWLEDGMENT

The partial work reported in this manuscript was done while the first author was with the Sungkyunkwan University in Republic of Korea. The authors would like to thank the anonymous reviewers for their useful comments and Dr. Gwangsoon Lee and Prof. Joongkyu Kim for their valuable discussions. This work was supported by the National High Technology Research and Development Program (863 Program) of China (Nos. 2008AA01Z125 and 2009AA12Z210), the National Science and Technology Ministry of China (Nos. XADZ2008159 and 51307040103), the National Natural Science Foundation of China (Nos. 61050110144, 60702062, 60803097, 60972148, 60971128, and 60970066), the National Research Foundation for the Doctoral Program of Higher Education of China (No. 200807010003), the Key Project of Ministry of Education of China (No. 108115), the Fund for Foreign Scholars in University Research and Teaching Programs (the 111 Project) (No. B07048), and the Fundamental Research Funds for the Central Universities (Nos. JY10000902001, JY10000902038, and JY10000902032).

REFERENCES

Cho, S., Kwon, H., Hur, N., Kim, J., & Lee, S. I. (2007). Stereoscopic video codec for 3D video service over T-DMB. In *Proceedings of IEEE International Conference on Consumer Electronics* (pp. 1-2).

Choi, S., Jung, C., Lee, S., Kim, J. K., Jung, K., Lee, G., et al. (2009). 3D DMB player and its realistic 3D services over T-DMB. In *Proceedings of IEEE International Symposium on Multimedia*, (pp. 440-441).

Fehn, C. (2004). Depth-image-based rendering (DIBR), compression and transmission for a new approach on 3D-TV. In *Proceedings of SPIE Conf. Stereoscopic Displays and Virtual Reality Systems* (pp. 93-104).

Hur, N., Tam, W. J., Speranza, F., Ahn, C., & Lee, S. I. (2005). Depth-image-based stereoscopic image rendering considering IDCT and anisotropic diffusion. In *Proceedings of IEEE International Conference on Consumer Electronics* (pp. 381–382).

Jung, C., & Jiao, L. (2010a). Novel Bayesian deringing method in image interpolation and compression using a SGLI prior. *Optics Express*, *18*(7), 7138–7149. doi:10.1364/OE.18.007138

Jung, C., & Jiao, L. (2010b). Reliable depth-image-based rendering using parameter approximation in mobile devices. *IEICE Electronics Express*, *7*(10), 666–671. doi:10.1587/elex.7.666

Jung, C., Jiao, L., Kim, H., & Kim, J. K. (2010d). Spatial-gradient-local- inhomogeniety: An efficient image denoising prior. *Journal of Electronic Imaging*, *19*(3), 033005. doi:10.1117/1.3466800

Jung, C., Jiao, L., Oh, Y., & Kim, J. K. (2010c). Depth-preserving DIBR based on disparity map over T-DMB. *Electronics Letters*, *46*(9), 628–629. doi:10.1049/el.2010.3457

Jung, K., Park, Y. K., Kim, J. K., Lee, H., Yun, K., Hur, N., & Kim, J. (2008). Depth image based rendering for 3D data service over T-DMB. In *Proceedings of 3DTV Conference* (pp. 237-240).

Kim, H. J., & Kim, W. Y. (2008). Eye detection in facial images using zernike moments with SVM. *ETRI Journal*, *30*(2), 335–337. doi:10.4218/etrij.08.0207.0150

Kim, J. (2008). *3DTV and mobile 3D AV services*. Korea-EU ICT Forum. Retrieved June 16, 2008, from http://www.eurosouth korea-ict.org/ documents/ forum_presentations/ Jinwoong% 20Kim_3DTV% 20and% 20Mobile% 203D% 20AV% 20Services.pdf

Lee, B. H., Yun, K., Hur, N., Kim, J., & Lee, S. I. (2009). Stereoscopic contents authoring system for 3D DMB data service. In *Proceeding of SPIE Conference on Electronic Imaging* 7237 (72311D).

Lee, B. H., Yun, K., Park, M., Hur, N., & Kim, J. (2008a). A study on the trend of mobile 3D services. *Ettrends*, *23*, 99–111.

Lee, G., Lee, H., Yun, K., Lee, B., Hur, N., Kim, J. W., & Lee, S. I. (2010). Implementation of 3D T-DMB receiver for three-dimensional data service. In *Proceeding of IEEE International Conference on Consumer Electronics* (pp. 93-94).

Lee, H., Cho, S., Yun, K., Hur, N., & Kim, J. (2008b). *A backward-compatible, mobile, personalized 3DTV broadcasting system based on T-DMB. Three-Dimensional Television, Signals and Communication Technology* (pp. 11–28). Springer.

Lee, H., Yun, K., Hur, N., Kim, J., Min, B. C., & Kim, J. K. (2007). A structure for 2D/3D mixed service based on terrestrial DMB system. In *Proceedings of 3DTV Conference* (pp. 1-4).

Lee, S., Oh, Y., Lee, S., Jung, C., Kim, J. K., Lee, G., & Hur, N. (2009). An efficient parameter setting for 3D service based on DIBR over T-DMB. In *Proceedings of KIPS Spring Conference* (pp. 75-77).

Oh, K. J., Kim, M., Yoon, J. S., Kim, J., Park, I., & Lee, S. … Ho, Y. S. (2007). Multi-view video and multi-channel audio broadcasting system. In *Proceedings of 3DTV Conference* (pp. 1-4).

Oh, K. J., Yea, S., & Ho, Y. S. (2009). Hole filling method using depth based in-painting for view synthesis in free viewpoint television and 3-D video. In *Proceeding of Picture Coding Symposium* (pp. 233-236).

Park, Y. K., Jung, K., Oh, Y., Lee, S., Kim, J. K., & Lee, G. (2009). Depth-image-based rendering for 3DTV Service over T-DMB. *Signal Processing Image Communication*, *24*, 122–136. doi:10.1016/j.image.2008.10.008

Park, Y. K., Park, S. L., & Kim, J. K. (2008). Retinex method based on adaptive smoothing for illumination invariant face recognition. *Signal Processing*, *88*(8), 1929–1945. doi:10.1016/j.sigpro.2008.01.028

Schreer, O., Kauff, P., & Sikora, T. (2005). *3D video communication: Algorithms, concepts and real-time systems in human centred communication*. USA: Wiley Press. doi:10.1002/0470022736

Strandvall, T. (2009). Eye tracking in human-computer interaction and usability research . *Lecture Notes in Computer Science*, *5727*, 936–937. doi:10.1007/978-3-642-03658-3_119

Yun, K., Lee, B., Lee, G., Lee, H., Jung, K., Hur, N., & Kim, J. (2009). A study on the trend of 3DTV broadcasting technology standardization and service. *Ettrends*, *24*, 143–151.

Yun, K., Lee, H., Hur, N., & Kim, J. (2008). Development of 3D video and 3D data services for T-DMB. In *Proceeding of SPIE Conference on Electronic Imaging,* 6803 (68030Z).

Zhang, L., & Tam, W. J. (2005). Stereoscopic image generation based on depth images for 3DTV. *IEEE Transactions on Broadcasting*, *51*, 191–199. doi:10.1109/TBC.2005.846190

Zhang, Z., & Zhang, J. (2010). A new real-time eye tracking based on nonlinear unscented Kalman filter for monitoring driver fatigue. *Journal of Control Theory and Applications*, *8*(2), 181–188. doi:10.1007/s11768-010-8043-0

Zhu, Z., & Ji, Q. (2005). Robust real-time eye detection and tracking under variable lighting conditions and various face orientations. *Computer Vision and Image Understanding*, *98*(1), 124–154. doi:10.1016/j.cviu.2004.07.012

ADDITIONAL READING

ETSI EN 300 401 v1.3.3 (2001). *Radio Broadcasting Systems; Digital Audio Broadcasting (DAB) to mobile, portable and fixed receivers*. European Telecommunications Standards Institute (ETSI).

ETSI EN 300 401 v1.4.1 (2006). *Radio Broadcasting Systems; Digital Audio Broadcasting (DAB) to mobile, portable and fixed receivers*. European Telecommunications Standards Institute (ETSI).

TTAS KO-07.0024 (2005). *Radio Broadcasting Systems, Specification of the Video Services for VHF Digital Multimedia Broadcasting (DMB) to Mobile, Portable and Fixed Receivers.* Telecommunications Technology Association in Korea (TTA).

KEY TERMS AND DEFINITIONS

ATTEST: The advanced three-dimensional television system technologies (ATTEST) project aims to prove a novel concept for a 3D-TV broadcast chain, and their essential requirements are the backwards compatibility with existing 2D broadcast and flexibility to support a wide range of different 2D and3D displays; It is part of IST.

BIFS: The binary format for scenes (BIFS) is a binary format for two or three dimensional audiovisual content, and is based on VRML and part 11 of the MPEG-4 standard; MPEG-4 BIFS is used in DMB.

BSAC: The bit sliced arithmetic coding (BSAC) is an MPEG-4 standard (ISO/IEC 14496-3 subpart 4) for scalable audio coding, and its support for scalability allows for nearly transparent sound quality at 64 kbps and graceful degradation at lower bit rates.

DIBR: The depth-image-based rendering (DIBR) is defined as the process of creating two (left and right) virtual views of a real-world scene from mono-scopic color video and its corresponding per-pixel depth information.

DMB: The digital multimedia broadcasting (DMB) is a digital radio transmission technology developed by South Korea as part of the national IT project for sending multimedia such as TV, radio, and data-casting to mobile devices such as mobile phones.

Eureka-147 DAB: The Eureka-147 digital audio broadcasting (DAB) is the most commonly used and is coordinated by the World DMB Forum, which represents more than 30 countries; it was defined in the late 1980s, and is now being introduced in many countries.

HEAAC: The high efficiency advanced audio coding (HEAAC) is an extension of advanced audio coding (AAC) using spectral band replication (SBR) and parametric stereo (PS), and designed to increase coding efficiency at low bitrates by using partial parametric representation of audio.

IST: Information Society Technologies (IST) is one of the thematic priorities in the European Union Sixth Framework Program for research and technological development set during the period of 2002-2006.

Chapter 24
3D Scanner, State of the Art

Francesco Bellocchio
Università degli Studi di Milano, Italy

Stefano Ferrari
Università degli Studi di Milano, Italy

ABSTRACT

The digital models of real objects are used today in many fields: medicine, archeology, and entertainment are some examples of areas in which these models are applied. Generally, the first step of the creation of a real object's 3D model consists in capturing the geometrical information of the physical object. Real objects can be small as coins or big as buildings: the different requirements have brought to the development of a very variegated set of techniques for the acquisition of geometrical information of the object. The aim of this chapter is to present and explain the techniques the 3D scanners are based on and compare them in terms of accuracy, speed, and applicability, in order to understand advantages and disadvantages of the different approaches.

INTRODUCTION

A digital tridimensional model is a numerical representation of the visual features of the object. From the digital model, it is possible to compute a realistic representation of the object in a bidimensional image. This image through the use of some techniques as prospective and shading, can emulate the human eye perception, giving a realistic representation of the object tridimensionality. A 3D visualization system, generally, is composed of two elements: the scene, a mathematical representation of the tridimensional objects, and the render, the technique to compute the 2D images of the scene.

The applications based on the tridimensional model processing are today very diffuse thanks to the increasing availability of tridimensional graphic devices and the decreasing trend of the cost of computational power. These applications are used in many fields such as design, archeology, medicine and entertainment. The chance to use digital 3D model can have many advantages. It is possible to use the model for digital simulation or

DOI: 10.4018/978-1-61350-326-3.ch024

to create a modified digital version of the object. In the entertainment field, the 3D modeling allows to use real objects or people for the creation of characters or environments in the digital animation. Furthermore, for the study of large objects like buildings or geographical regions it can be very useful handling scalable digital models.

The digital 3D model can come from two different ways: Computer Aided Design (CAD) and physical object measurement. In a CAD environment, simple objects can be represented through simple equations: for instance, the equation $x^2+y^2+z^2 = r^2$ can be used for representing a sphere with radius r. Although these simple equations can seem limitative, the set of representable objects can be extended through a technique called Constructive Solid Geometry (CSG); this technique is based on the combination of simple solid objects (e.g., cube, cone, sphere) in order to create more complex objects through operations as union, intersection, difference (e.g., a tube can be seen as a difference of two cylinders with different radius). Anyway, this method is not suitable to describe a large class of real objects and then is not commonly used. Nowadays, the CAD software allows the creation of very complex models that are, generally, based on Non-Uniform Rational B-Spline (NURBS) (Piegel, 1997), a mathematical model that allows the generation of curves and surfaces with great flexibility and precision. The NURBS is suitable for handling both analytic and freeform shapes.

On the other hand, the digitization through physical object measurement is a process that allows for obtaining the 3D model in a semi-automatic way. It is based on the measurement of geometric features of the object and on its visual features as the color and texture. With respect to the CAD, the digitization is characterized by a generally faster creation process and a higher (or, at least, measurable) level of accuracy. Furthermore, the digitization, being substantially a measurement process, does not require artistic abilities for the operator.

The applications that make use of digitization form a huge class. For instance:

- The archeology and the arts are characterized by two divergent necessities: it is very important to preserve an artwork, but it is also desirable that many people can appreciate it. The virtual museum allows a larger public access than a real museum without risk for the exposed objects and, at the same time, it can be a way to attract visitors to the real museum. The user interested to a single artwork has the possibility of explore it directly and in a deep way. In fact, if an artwork is placed in a theca, the field of view can be strongly limited, while, as the 3D model can be observed from different points of view and at different scales, every details of the object can be appreciated from its realistic virtual copy. Furthermore the 3D modeling can improve both the study of an artwork and the accuracy of the cataloging.
- There are several applications in which 3D model of human parts are used. In virtual fashion a model of each customer is acquired, allowing the computation of the cloth size. Then, the model can be dressed with different clothes in order to drive the customer through the shopping. In medicine, the 3D models of organs can help the physician in the diagnosis; for instance, the 3D ultrasonography is used to check the fetal morphology.
- The representation of an object through the quantification of its features allows for performing an efficient comparison of different objects belonging to the same class. This concept is applied in different contexts. For example in manufacturing it is applied for the quality control, while in security it is applied for identity identification through anthropometric measurement (biometrics). Both the applica-

tions can take advantages from the use of 3D digitization.

- The virtual environments are very important for the training for critical tasks or dangerous procedures where the human error have to be limited at the minimum. An example is the virtual surgery, where the doctor can practice on a virtual patient for gaining experience before performing the surgery. Furthermore, surgeons can practice operations multiple times without the use of limited resources as cadavers or animals. One of the first applications of virtual training was flight simulator. It was used both for the training and the evaluation of pilots. Often, in these applications the 3D models are combined with haptic devices, in order to enrich the virtual reality experience with tactile sensations.
- The design and reverse engineering find a great help in the use of 3D models. Some designers, especially in the architecture field, prefer to create physical prototypes first (e.g., using clay) and then digitize them. The 3D models can then be included in simulation or presentation. Furthermore, in some cases legacy objects whose digital version is not available have to be reengineered or included in new projects; for these cases, as well as for reverse engineering, a 3D digitization is usually a less expensive and a more accurate solution than the manual modeling.
- In the last decade, the number of movies with 3D digital characters is strongly increased. Besides 3D animation movies, the use of a 3D model of the actor allows to avoid laborious makeup sessions for adding particular physical features or the use of stuntmen for dangerous scenes. Also, many video games are inspired to real persons, as sport games, for example. Since the realism of the game is directly related to the fidelity of the avatars, the advantages of the use of an accurate 3D scanner is evident.

Different applications can have very different requirements. For example, for the reconstruction in virtual archeology a good accuracy and a low invasivity are required, but generally the acquisition time is not an important constraint. Instead, in videoconference applications, real-time processing is needed, while the quality of the modeling is secondary. Furthermore, as the information is generally about the facial mimic, the acquisition and the reconstruction techniques can be based on a face model. In the industrial quality control is important to have a fast reconstruction at a low cost because the same operations are repeated for many objects of the same type. Hence, the a priori knowledge on the object can be used to achieve a more robust and fast reconstruction.

The different requirements for 3D digitization applications have brought to different systems for realizing the 3D reconstruction. The typical device to collect the geometric information of the real object is the 3D scanner. There are many types of 3D scanner and they are very variegated. Generally these devices realize a measure of the 3D coordinates of points sampled on the surface of the target object. The idea is to collect a cloud of points in order to allow the computation of the model of the real object surface. Some kinds of these devices are able to capture the texture information of the scanned object too; this information can be very useful for the creation of more realistic tridimensional model graphical representation.

The physical principle exploited for computing the 3D coordinates of the points divides the scanners in different categories, as sketched in Figure 1. In the following sections, the devices that belong to each category will be described, their working principles will be explained, and their advantages and disadvantages will be consequently discussed.

Figure 1. 3D scanner taxonomy

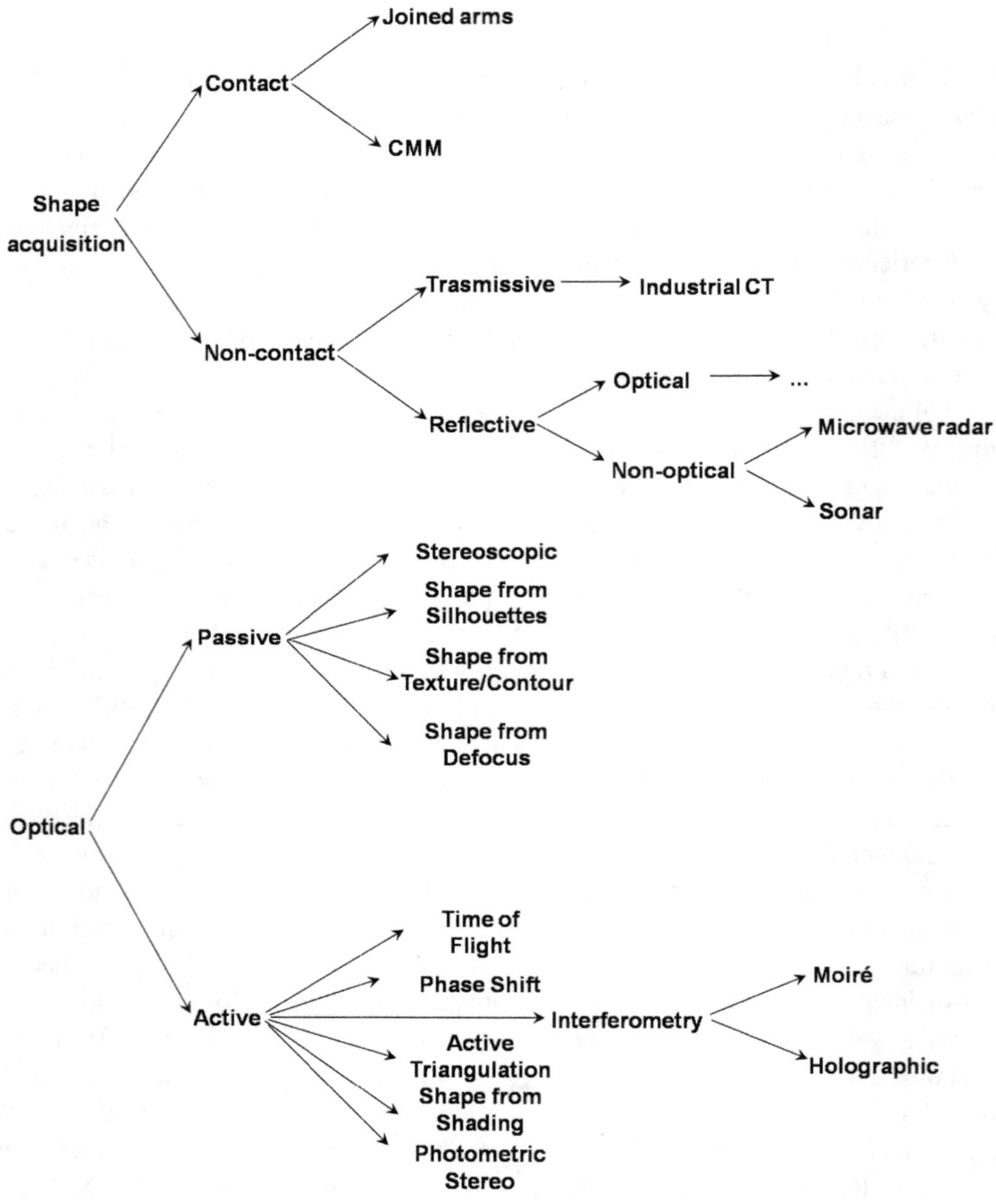

3D SCANNERS

Two different approaches are used by the 3D scanner systems for measuring the geometrical features of an object. The first one is based on the interaction between a sensor and the surface of the object; the systems using this approach are called contact 3D scanner. The second one is based on the interaction between a radiation (electromagnetic or sound) and the surface of the object, in this case the systems are called non-contact 3D scanners. In the following these two categories will be illustrated in depth.

Contact 3D Scanners

In contact 3D scanners the surface of the object is probed through the physical touch. There are mainly two types of these systems: Coordinate Measuring Machine (CMM) and Joined Arm. The first is composed of a tactile probe attached to a vertical arm, which can be moved along the horizontal plane (Figure 2). The movement of the probe is allowed by the three orthogonal axes in a typical three dimensional coordinate system. The probed coordinates result directly from the displacement of the actuators along each axis. The object is placed on a reference plane where the probe can explore it. The movement of the probe can be performed both automatic and manually operated. Generally, these systems enjoy a good accuracy (Helmel Checkmaster 112-102 allows for an accuracy of 9 μm ("Checkmaster Manual CMMs," 2010)) and they are used mostly in manufacturing. The disadvantages are that the working volume is bounded by the structure and that the acquisition direction is only vertical.

The Joined Arm is composed of a chain of articulated links with a probe as end effector. The 3D coordinates of the probe result as the composition of the rototraslations operated by each link. Since, as for the CMM, the point coordinates are computed through the position of the mechanical components, these systems are generally sensitive to temperature and humidity variations. For this reason, in order to provide good performance, they require a high mechanical technology to be realized.

The main differences with respect to the CMM are in the mechanical structure. Since generally the arms have a greater degree of freedom, they can be used for a larger class of objects. The arms are typically manually operated, while the CMM can be more easily automated. Both these devices can be very precise (Cam2 Quantum Arm has an accuracy of 0.018mm ("The FaroArm Family, "2010)), but they are relatively slow compared to the non-contact scanner systems. Furthermore, these methods are invasive and so they are not suitable for delicate object (e.g., archaeological artifacts). Another disadvantage is the price, as these systems are not generally cheap. It should be noticed that the tactile probe of both these system can be substituted with another kind of sensor, for realizing a non-contact measurement. In this case the systems are no longer belonging to the class of contact 3D scanners.

Non-Contact 3D Scanners

In non-contact systems, the sampling of the surface is performed by the interaction between some kind of radiation and the object surface itself. Depending if the radiation is supposed to pass through the object or if it is reflected by the object surface, these systems can be divided in two sub-categories: transmissive and reflective.

Transmissive Systems: Industrial Computed Tomography

In transmissive systems the object has to be positioned between the emitter (which irradiate the object) and receiver (which collect the radiation attenuated by the object). The main representative of this category is the Industrial Computed Tomography. The radiation, a beam of high energy photons generated by an X-ray tube, penetrates the target object and is captured by a 2D detector as a digital radiograph image. The 3D models are reconstructed from a set of 2D X-ray images of the object taken from different views. The views are obtained rotating the object, which, to this aim, is positioned on a turn table that can rotate with a high precision (0.25 to 1 degree steps are commonly adopted). From this series of 2D radiographs through, generally, the back-projection algorithm (Feldkamp, 1984) it is possible to compute a 3D voxel model. The three-dimensional resolution of the obtained model ranges from a few micrometers to hundreds of micrometers, and depends on the pixel size of the X-ray detector.

Figure 2. Coordinate measuring machine

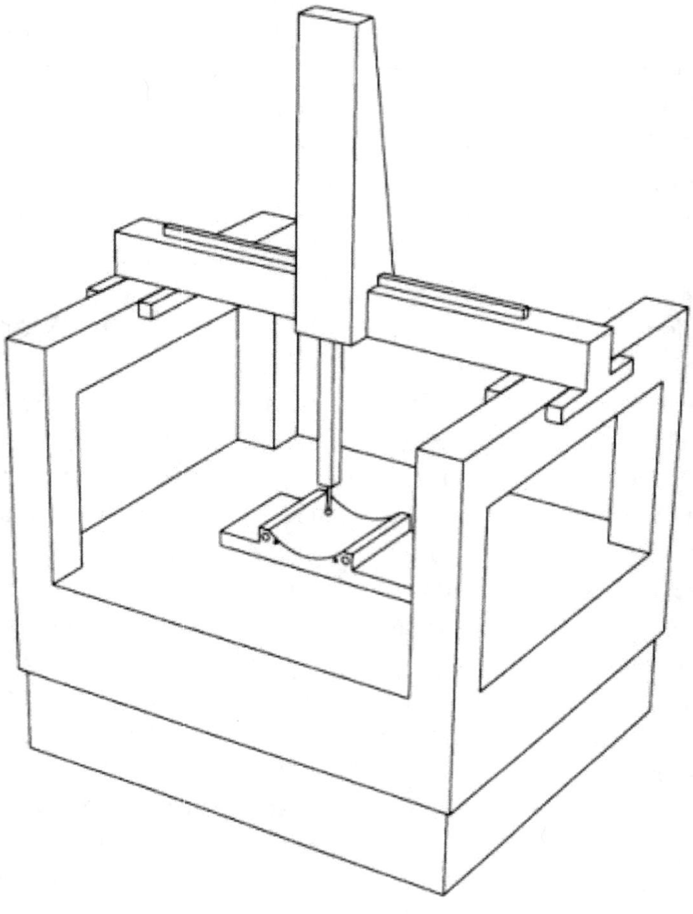

This kind of system allows the reconstruction of both external and internal surfaces and the method is unaffected by certain visual object properties (dark, reflective or transparent surfaces). The structure of the hardware makes the system suitable for relatively small objects.

It should be noted that the density and the thickness of the object affect the energy collected by the X-rays detector. Furthermore the reconstruction of the model from the 2D radiograph images is computationally intensive. An example of this system is Nikon XT H 22 5 ("Nikon Metrology - XT H 225," 2010), it has an accuracy of 0.001 mm but the scanning volume is limited to 30 cm × 30 cm × 30 cm (Table 1).

Reflective Systems

The reflective systems exploit the radiation reflected by the object surface for estimating the position of the points of the surface. They can be classified from the type of radiation they use. In particular, optical systems use optical radiation (wavelength between 100 nm and 300 μm), while non-optical systems use sound or non-optical electromagnetic radiation to make the measurements. Since optical systems form the main category of 3D scanners, they will be considered in deep in the next section.

The class of non-optical systems is composed by devices based on radar and sonar systems. Al-

Table 1. Some examples of industrial 3D scanners

Model	Measuring Method	Scan Range (depth of field)	Accuracy	Acquisition Speed
Leica ScanStation C10	Tof, pulsed range finder	300 m	2 mm (at 50 m)	50,000 points/s
Leica HDS6200	Tof, phase shift	80 m	5 mm (at 25 m)	1,000,000 points/s
Metris MCA II	Physical contact	3.6 m	0.1 mm	
3D Digital Corp. e-scan	Laser triangulation	300 – 650 mm	0.135 mm (at 300 mm)	700 points/s
Nikon XT H 225	Computed Tomography	300 mm	0.001 mm	
Faro Laser ScanArm	Physical contact/ Laser triangulation	3.7 m	0.016 mm	

though the radiations exploited are very different (the radar uses electromagnetic microwaves, the sonar uses sound or ultrasound waves), both of them are based on the principle of measuring the time-of-flight of the emitted radiation: from the time required for the wave to reach the object and return to the system, knowing the speed of the utilized radiation, it is possible to estimate the distance covered by the radiation, which can be considered equal to the double of the distance of the object from the scanning device. As this principle is used also for a class of optical scanners, it will better explained in the next section.

Due to the use of microwave radiations, radar systems have a very large depth of field, up to 400 km, and can perform ground penetrating reconstructions. A typical application is for air defense. These systems are quite expensive and generally have low accuracy.

When a sonic wave is used, as in sonar systems, the measurement is insensitive to the optical properties of the object and can be applied for the reconstruction in environments where the optical radiation would be distorted or too much mitigated, as the underwater setting. These systems are characterized by a low accuracy due to low signal to noise ratio.

OPTICAL 3D SCANNERS

The ability of reconstructing an object without physically touching it has important advantages: it is applicable to delicate objects (e.g., archeology artifacts), and the use of radiation allows, generally, high speed of acquisition and wide reconstruction (e.g., landscape reconstruction). Furthermore, due to the availability of inexpensive optic sensors, very low cost systems can be realized. Depending on the source of the radiation (device emitted or environmental), these optical 3D scanners can be divided in two sub-categories: passive and active systems.

Passive Systems

The passive systems do not emit any kind of radiation themselves; they usually use the reflected ambient radiation. Generally, they are based on the use of Charge-Coupled Devices (CCDs), the classical sensors that are embedded in the commercial digital cameras. The sensors collect images of the scene, eventually from different points of view or with different optical setup. Then, the images are analyzed in order to compute the 3D coordinates of some points in the scene.

The passive scanner can be very cheap; normally, they do not need particular hardware but typically do not yield dense and highly accurate digitization. Often, with these scanners the 3D points' computation is not easy and a heavy computational effort can be required. Examples of these systems are based on stereoscopy, shape-from-silhouettes, shape-from-texture (or contour) and defocus (Wholer, 2009).

Stereoscopic Systems

The stereoscopic systems are based on the analysis of two (or more) images of the same scene, seen from different points of view. The 3D points of the scene are captured as their 2D projection in the taken images. If the corresponding 2D points are found on couples of images, their projected rays can be estimated and the 3D coordinates of the point are recovered as the intersection of the projection rays (Figure 3). This reconstruction method is known as triangulation. It should be noticed that this method requires the complete knowledge of the camera parameters: their (relative) position and orientation, but also their internal parameters: focal length, optical centre, CCD size, and distortion parameters.

The camera parameters are determined during a phase called calibration. Generally, this phase is performed before the scanning session, using a particular scene, such as a chessboards or simple objects, where the correspondence problem (i.e., the matching between the projections of the same points in the 3D space on the acquired images) can be easily solved. It is also possible to compute an estimation of the calibration parameters directly from the images of the object (Mckinley, 2001).

The real problem of this kind of system is the computation of the correspondence pairs of the 3D points. For this reason the stereoscopic technique is generally used for the reconstruction of particular objects in which the correspondence problem can be solved easily. Since using standard image processing techniques it is relatively simple to extract peculiar points (such as the corners of an object) from an image, these methods are applied for the reconstruction of building or, in general, of objects in which the edges are evident.

A possible approach for reducing the computational complexity of the correspondence problem consists in capturing many images in which the point of view slightly changes. Since the position of a point on an image will be slightly different from that on the next image, the search for the correspondence for each point can be performed only in a small portion of each image. However, it should be considered that in this case the complexity of the estimation of the calibration parameters can increase.

Figure 3. Stereoscopic system

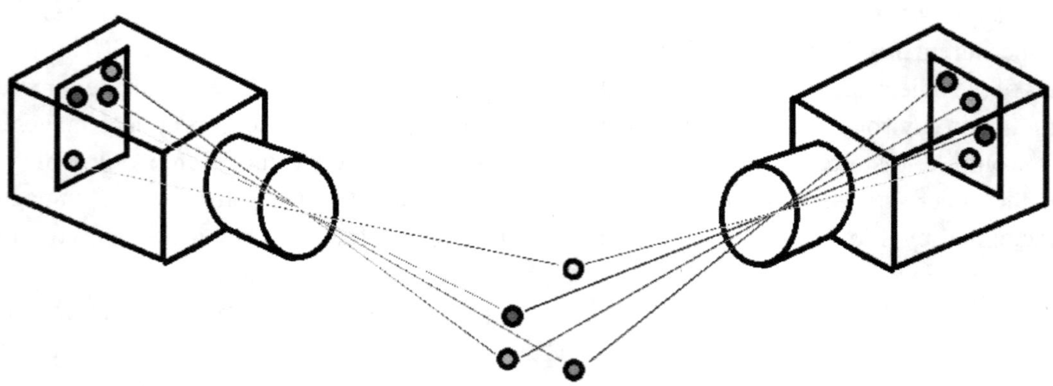

The main advantages of these techniques are the potential low cost of the hardware needed and the non-invasivity of the method. The generally low accuracy and the sensitivity to the calibration phase limit the diffusion of these systems in real applications.

Shape-from-Silhouettes

The silhouette systems (Potmesil, 1987; Vaillant, 1992) compute the model as composition of the contours of the object taken from different points of view. To this aim, the typical scanner of this category is composed by a turn table (where the object is placed on), a flat background (which simplify the contour extraction procedure), and a single camera.

While the object rotates the camera capture an image from which the contour is extracted. Each contour can be seen as a cone of projected rays that contains the object. The intersection of these cones determines the approximate shape of the object. This system has the benefit that is realizable easily and with low cost hardware, but has the strong limitation that only convex object can be accurately reconstructed. In fact, the cavities of an object are not visible in the projected silhouettes and then they cannot be reconstructed, which limit the use of these systems in real applications.

Shape-from-Texture & Shape-from-Contour

Techniques that extract information about the object's shape from the its texture or contour provides useful clues for 3D digitization and are interesting results of the computer vision theory, but are rarely implemented for real 3D scanners. In fact, these techniques are not able to compute the 3D coordinate of object points, but only the surface curvature (up to a scale parameter) or its orientation.

Shape-from-texture is grounded on the hypothesis that the surface of the object is covered by a texture characterized by a pattern that is repeated with regularity. By means of the analysis of the texture distortion, it is possible to compute the curvature of the surface. The surface normals are estimated from the analysis of the local inhomogeneities (Aloimonos, 1986). Furthermore, a diffuse illumination of the scene is required, as the shading can influence the texture analysis.

A similar technique is called shape-from-contour. In this case the surface orientation is computed through the analysis of the distortion of a planar object. For example, if the object contour is known to be a circle (e.g., a coin), while the contour of the acquired object is elliptical, it is possible to estimate the surface orientation that realizes this distortion.

Shape-from-Defocus

In the shape-from-defocus systems (Levin, 2007), the defocus produced by a lens is driven to allow the extraction of depth information. In these scanners, a conventional camera captures several images of the same scene using different focal lengths. Generally, this method can make use of a single camera that records all the images for the different focus set-ups. The frequency content of the same region in different images is used for identifying in which image the considered region is on focus. Since from the focal length the distance of the plane of focus from the optical centre is determined, then knowing the region on focus for a given focal length gives the distance of that region from the camera too. Typically, these systems are not able to make very precise reconstruction, as the accuracy depends on the set-up (depth field). Besides, this technique can be applied only on texturized objects. However, these systems can be realized using low cost hardware, and, being optical passive, they are non-invasive.

Active Systems

The active systems emit some kind of radiation and the interaction between the object and the radiation is captured by a sensor. From the analysis of the captured data, knowing the features of the emitted radiation, the coordinates of the points can be obtained. As a matter of fact, they are the most common scanner systems. Among the several kinds of scanners that belong to this category, the most exploited principles are: time of flight (ToF), phase shift and active triangulation. However, interferometry scanners found application for specific problems, such as the digitization of very small objects, while illuminant-based techniques have theoretic interest especially for applications where the color of the object have to be captured.

Time-of-Flight

Time-of-flight (ToF) systems measure the distance from scanner to surface points through the measure of the time employed by the radiation to reach the object and come back to the scanner. Knowing the speed of the radiation and the roundtrip time, it is possible to compute the distance and, knowing the direction of the emitted radiation, the 3D points' coordinates (Gambino, 2005). Hence, changing the direction of the emission, the system can cover the entire field of view.

Depending on the type of waves used, such devices are classified as optical radar (optical waves), radar (electromagnetic waves of low frequency) and sonar (acoustic waves). The optical signal based systems are the most used type. Such systems are sometimes referred to as LIDAR (LIght Detection And Ranging) or LADAR (LAser Detection And Ranging). These systems are characterized by a relatively high speed acquisition (10,000–100,000 points per second) and their depth of view can reach some kilometers.

Generally, the optical ToFs accuracy is limited because the high speed of radiation that is used. In fact, for measuring the distance with 1 mm accuracy, it is necessary to be able to measure a time range in the order of picoseconds. Hence, these systems are generally applied in long-range 3D measurement of large object, such as building and geographic features. The optical properties and the orientation of the surface with respect to the emitted ray affect the energy collected by the photo detector and can cause loss of accuracy.

As said above, these systems are often used to geographic reconstruction; the aerial laser scanning is probably the most advanced and efficient technique to survey a wide natural or urban territory. These systems, mounted on an airplane or on a helicopter, work emitting/receiving up to 100,000 laser beams per second. The laser sensor is often coupled with a GPS satellite receiver that allows recovering the scanner position for each acquired point. Hence, each point can be referred to the same reference system and the acquired points (which can form a dense cloud of points) can be related to a cartographic reference frame, for an extremely detailed description of the covered surface (Visintini, 2007). ToF scanners are often used in environment digitization. A relatively recent application is the digital crime scene reconstruction; through the digital model the police are helped in the scene analysis task. For this aim, the typical scanner model is composed by a rotating head which permits a wide field of view; for example the model Leica ScanStation C10 ("Leica ScanSystem C 10", 2010) has a field of view of 360° horizontal and 270° vertical.

Another kind of ToF system is the Zcam, produced by 3DVSystems (Yahav, 2007), which provide in real-time the depth information of the observed scene. The scene is illuminated by the Zcam which emits pulses of infra-red light. Then it senses the reflected light from scene pixel-wise. Depending on the sensed distance the pixels are arranged in layer. The distance information is output as a grey level image, where the grey value correlates to the relative distance.

Figure 4. Phase shift system

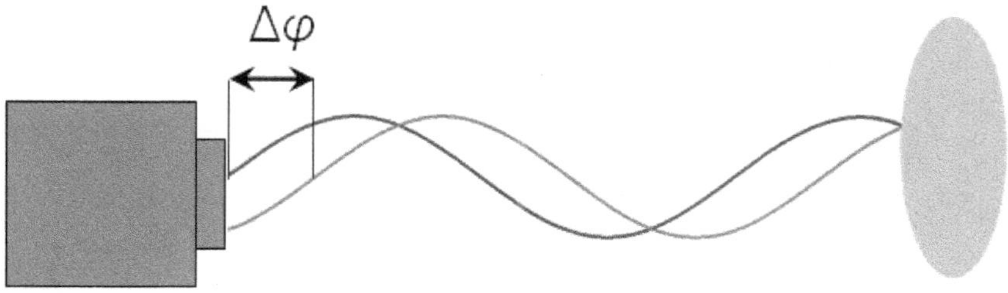

Phase Shift

Phase shift systems use a laser beam whose power is sinusoidally modulated over the time (Figure 4). From the phase difference between the emitted and reflected signal, it is possible to compute the roundtrip distance. In fact, the phase difference between the emission and reflection signal is proportional to the travelled distance, as: $d = c\ \Delta\phi / 4\pi f$, where d is the object distance, $\Delta\phi$ is the phase shift, c is the light speed, and f is the light frequency. Since the phase can be distinguished only in the same period, the periodicity of the signal generates ambiguity. To resolve this ambiguity, multiple frequency signals are used. This method has performances quite similar to the ToF method, but can reach a higher acquisition speed. An example of this system is Leica HDS6200 ("Leica HDS6200," 2010), it has an accuracy of 2 mm at 25 m and an acquisition speed of 1,000,000 points per second (Table 1).

Active Triangulation

In active triangulation systems the scene is illuminated by a coherent light source from one direction and viewed from another. These systems primarily differ by the light structure used (a single spot or a laser sheet beam or coded light) and the scanning method (moving the object or moving the scanning mechanism).

If the source is a low-divergence laser beam, the interaction of this radiation with the surface object will produce a spot, which can be easily detected by a sensor (typically a CCD). The orientation and the position of the source and the sensor are typically known. From the spot location on the sensor, the line between the sensed spot and the camera centre point can be computed. As the laser line is known, the 3D point will results as the intersection point between the camera line and the laser line ("EScan Specifications," 2010). Hence, the point 3D coordinates can be calculated by triangulation (Figure 5). If the laser orientation and position are not known, it is possible to calculate the coordinates using two or more cameras as in stereoscopic method. In this way, the system acquires one point per frame, while using a different light source more points per frame can be captured. In fact, when a laser sheet or a matrix spot is used, it is possible to reconstruct more points at a time for a single frame.

As the laser sheet illuminates a plane in the space, the camera captures the contour resulting from the intersection of this plane and the object surface. Then, for any image pixel on the contour, the corresponding 3D point on the object surface is found by intersecting the ray passing through the pixel and the laser 3D plane equation. The use of a matrix spot allows to sample a region instead of a line (as done by the laser sheet), and it can be potentially the faster solution for surface acquisition. However, the problem of matching every beam with its projected point acquired by the camera is more complex than in the single beam case.

Figure 5. Active triangulation system

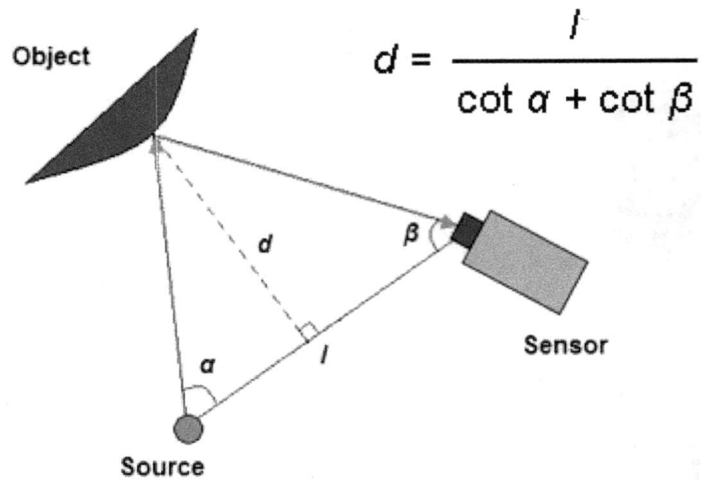

Whenever a moving object have to be acquired, the use of a matrix spot would be required, with strong constraints on the speed of acquisition. However, generally, in such an applications, this technique is not used because the problem of the points correspondence. The typical approach used in this case is instead the projection of a structured light pattern.

There are many different techniques based on the projection of structured pattern, and generally they make use of a calibrated camera-projector pair (Salvi, 2004). The aim of these techniques is to characterize each point by projecting a different light pattern for a different direction. Hence, the illumination is used like a code, allowing the correct identification of each direction. The encoding is realized using different strategies as colored stripes (Wust, 1991) or time-coded stripes (Rusinkiewicz, 2002). The colored stripes encoding presents an important problem: both the surface color of the object and the ambient light influence the color of the reflected light. For this reason, to reconstruct a colored (or textured) object other kinds of coding are preferred.

In (Rusinkiewicz, 2002), a structured-light rangefinder scanner using temporal stripe coding is proposed. Using a projector and a camera synchronized at 60 Hz, four successive frames are exploited to acquire a 115×77 matrix points. For each frame, a set of black/white stripes is projected. Observing as a pixel change its color (from white to black and from black to white) in different frames, it is possible to compute which stripe is illuminating the pixel and then, through the triangulation, the 3D position of the point. Actually, the entities that carry the code are not the stripes, but the stripes boundaries: in this way, a more efficient coding is possible. In fact, a single stripe can carry a bit (the stripe can be black or white), while a boundary can carry two bit (it can have a stripe on the left white and on the right black, and so on).

In (Huang, 2006), it is proposed another very efficient scanner system. This system uses three phase-shifted sinusoidal grayscale fringe patterns, to provide pixel-level resolution. The projector and a camera are synchronized at 120 Hz with a resolution of 532×500 points per frame; the system accuracy is 0.05 mm. For each pixel, the phase from the three pattern intensities is calculated. This information determines the correspondence between the image field and the projection field. The phase map calculated from the three images camera can be converted to the depth map by a phase-to-height conversion algorithm based on triangulation.

With this system it is possible to realize a real-time reconstruction. For example, the system is able to measure human faces, capturing 3D dynamic facial changes. In order to provide a high definition real-time reconstruction, a Graphics Processing Unit (GPU) is employed to compute the 3D coordinates points. These devices have a highly parallel structure that makes them more effective than typical CPUs for a range of complex algorithms (Zhang, 2006). GPUs are very useful in the reconstruction problems because typically these problems are characterized by parallelizable computations.

All the active triangulation system, generally, are characterized by a good accuracy and are relatively fast. The strong limitation of these systems is the size of scanning field, since the depth of field is proportional to the sensor/emitter displacement and the emitter power. Then these systems are not usable for digitization of large objects. Furthermore object's color and ambient illumination may interfere with the measurement.

Shape-From-Shading and Photometric Stereo

The shape-from-shading problem consists in the estimation of the three-dimensional shape of a surface from the brightness of an image of that surface. The first formulation of this problem was proposed in the 70's (Horn, 1975). The work showed that the problem implies the solution of a nonlinear first-order differential equation called the brightness equation.

Today, the shape-from-shading problem is known to be an ill-posed problem, which does not have a unique solution (Brooks, 1983). What makes difficult to find a solution for this problem is often illustrated by the concave/convex ambiguity that is the fact that the same shading can be obtained both for a surface and its inverted surface, for a different direction of the illuminant. Moreover, this kind of ambiguity can be widely generalized. In (Belhumeur, 1999), it is showed that, given the illuminant direction and the Lambertian reflectance (the fraction of light that is reflected, aka albedo) of the surface, the same image can be obtained by a continuous family of surfaces, which depends linearly by three parameters. In other words, neither shading nor shadowing of an object observed from a single viewpoint can provide the exact 3D structure of the surface.

However the problem can be solved under simplified conditions. The first one is the use of directional lighting with known direction and intensity. But, again, this simplification is not enough and, in order to solve the problem, knowledge about reflection properties of the surface of the object is also required. In particular, the surface should be Lambertian, namely the apparent brightness of the surface has to be the same when the observer change the angle of view, and the albedo should be known. As the method implies the use of a known radiation, it can be considered as belonging to the active systems class. Under these conditions the angle between the surface normals and the incident light can be computed. However in this way the surface normals are derived as cones around the light direction. Hence, the surface normal in a given point is not unique and it is derived considering also the values of the normals in a neighborhood of the considered point and making the assumption that the surface is smooth.

When a photometric stereo technique is used, the problem is simplified by illuminating the scene from different positions (Higo, 2009; Hernández, 2008). With this technique, introduced in (Woodham, 1980), it is possible to estimate the local surface orientation by using several images of the same surface taken from the same viewpoint, but under illumination that comes from different directions. The light sources are ideally point sources, which position is known with respect the reference system, oriented in different directions. The lights are activated one at a time, for each captured frame, so that in each image there is a well-defined light source direction from which to measure the surface orientation. Analyzing the sequence of intensity changes of a region, a unique value for the surface normal can be derived. In general, for a Lambertian surface, three different light directions are enough to solve uncertainties and compute the normals.

This approach is more robust with respect to shape-from-shading, but the use of synchronized light sources implies a more complicated 3D system, which can strongly limit the acquisition volume. On the other hand, the availability of images taken with different lighting conditions allows a more robust estimation of the color of the surface.

Moiré Interferometry

The Moiré interferometry is a technique used for detecting and measuring deformations in a quasi-planar surface. The method utilizes the interference effect between some form of specimen grating and reference grating (Idesawa, 1977).

The principle of the method is that projecting parallel equispaced planes or fringes on the surface of the object and observing the scene from a different direction, the observed fringes will appear as distorted by the surface shape. By comparing the observed fringes with a reference fringes (by means the interference), the measurement of displacement from the plane can be obtained. In more detail, measuring the fringe distances obtained from the superimposition of the grating projected with the grating observed and knowing the projection and observation angles, the z coordinate can be determined. This technique allows a high accuracy (on the order of micrometers), but for a very small field of view. In fact, the grating projected have to be very dense (e.g., with 1000–2000 lines/mm). This characteristic limits the method to microscopic reconstruction.

Holographic Interferometry

A hologram is the recording of the interference pattern formed by a reference laser beam and the same beam reflected by the target object. It can be obtained by splitting a laser beam in two parts: one is projected onto the object and the other one goes directly to the camera.

The holographic interferometry is a technique for measuring vertical displacement by comparing the holograms of the same object at different states. In particular, vertical displacements can be estimated comparing images taken while the object is moved along the vertical axis. The images are analyzed to detect the peak of the interference pattern for each pixel, which allows for computing the height of the considered pixel. The systems based on this technique are quite expensive, but allows sub-nanometer measurement. Since the field of view is very small, generally, the method is applied for objects of size of few millimeters.

Hybrid Techniques

In the previous sections, an overview of many techniques characterized by complementary strength and weakness has been presented. For exploiting the advantages of each approach many real systems can implement more than one technique. For instance CAM2 Laser ScanArm V3 ("3D Measurement," 2010) is a Joined Arm where the probe is an active triangulation laser scanner, combining the precision and the speed of the active system with the mobility of the Joined Arm.

Another example is a system that combines photometric techniques with structured light (Lu, 2010). The reconstruction is performed using a multi-resolution scheme where the structured light method is used to acquire the low resolution geometry of the surface and the photometric stereo is used to capture the fine surface normals. The result is a high resolution model. A further example is represented by the optical 3D scanner system that can be enriched with a couple of emitter and receiver that exploit another kind of radiation (such as sonic, microwave radiation) for the objects or environments in which the optical radiation cannot be applied.

The combination of multiple techniques generally allows a more robust systems and an improvement of the accuracy. The price to be paid is the complexity and then the cost of the system. Sometime, in the digitization of a single object it can be useful to employ several scanners (Levoy, 2000). For instance, it happens for large objects scanning, where an accurate 3D measurement cannot be realized in a single session. In this case, the use of scanner with a large field of view (FoV) can be used to capture the main shape of the object, while the details can be captured in several scanning sessions using an 3D scanner with an higher resolution (but a lower FoV). As, generally, each system provides a points cloud, it is possible to combine the different data to realize a single tridimensional model. The operation of transforming the different clouds of points such that they refer to same coordinate system is called registration. This operation consists in the computation of a rototraslation matrix for each points cloud by identifying the overlapping subsets of points. Among the algorithms for addressing this task, probably the most famous is Iterative Closest Points (ICP) (Besl, 1992).

FUTURE RESEARCH DIRECTIONS

There are many aspects of scanners systems that should be considered for future research. The first aspect concerns the strategies which can be applied to face the situations in which a single system is not able to obtain a good measure. For example, the properties of the object material (e.g., transparency or reflectance) can degrade the measure obtained using reflective systems (Tongbo, 2007; Hullin, 2008). In (Hullin, 2008) it is proposed a technique based on the immersion of the transparent physical object in a fluorescent liquid, which highlights where the laser sheet impacts the object surface allowing the reconstruction.

A second aspect regards the diffusion of the acquisition systems. The 3D scanners present on the market are generally high cost devices. There are some studies about how obtain an acquisition system using *off-the-shelf* hardware (e.g., webcam, projectors, etc.) (Ho, 2009; Drenik, 2008; Reznicek, 2008). These methods generally allow to obtaining low cost systems, but they are characterized by a low accuracy. An important research direction regards the improvement of the accuracy of 3D scanner based on common use devices. The trend of the improvement of cameras and projectors resolution is a further incentive for the development of low cost and high accuracy acquisition systems.

Besides the costs, other features can be improved. For instance, the field of view can be a strategic feature for application such as environmental scanning, where sensors with large FoV are required (De Ruvo, 2010). In (Hu, 2009) a laser scanner that allows a large scale reconstruction is presented. Besides featuring a large FoV (360° horizontal and 330° vertical), it integrates the structured light measurements with stereo photogrammetry for accurately locating the edges of the objects of the scene. The power and computational costs of a system or its physical features, such as the weight and the size, can be obstacles in its use for mobile robots or portable devices. The implementation on low demanding hardware and the integration of different techniques can allow for obtaining relatively low cost system for unmanned vehicles (Nagai, 2009; Ryde, 2009).

Another aspect regards the use of several techniques in order to exploit the different features of each of them. For example, for the detailed reconstruction of big object can be useful using a kind of scanner for a global low accurate reconstruction and adding the details using another kind of scanner that operates with high accuracy but in limited regions of the object. The merge of the two techniques allows a high accurate reconstruction of a big object (Zheng, 2008). In general, using a combination of several approaches guarantees a more robust and more accurate reconstruction. Hence this field of research can determine important improvement for the scanning problem.

CONCLUSION

In this chapter, an overview of the techniques used to implement the 3D scanning devices has been provided. The systems have been classified in a taxonomy that privileges, as criterion for the classification, the physical principle exploited to extract the 3D information. However, other properties of the scanning systems can be used as classification key. For instance, among others, the accuracy, the resolution, or the speed of acquisition can characterize a scanner system, but these properties are more related to an actual implementation of the systems than to a class of scanners and hence are not suited for a structured treatment of the subject. On the other hand, these properties have to be considered when a scanning system has to be chosen and are often critical for the choice. Obviously, there is no way for indicating a scanner system as the best one, because each model has been designed for a specific field of application.

In (Catalan, 2007) a method for evaluating the 3D scanners is suggested. It considers some important features (e.g., field of view, accuracy, physical weight, scanning time) and for each feature it associates a weight. By giving a score for each feature of each scanners considered, a single final score can be computed as the sum of each score. Anyway, probably the three principal aspects that should be considered in order to choose a 3D scanner are the properties of the objects to acquire (size and material features), the accuracy required, and the budget, under auxiliary constraints such as the speed of acquisition required and the environmental conditions.

Some attention should be paid also to the human aspects: some models require a deep knowledge of the principles exploited by the scanners and can be used only by trained people. An important aspect to note about every 3D scanner is the calibration procedure. Generally the 3D scanners have different setup and the points cloud reconstruction is possible only if the set-up parameters are known. The aim of calibration phase is the estimation of setup parameters. This phase is critical for many types of scanner and the time spent for it can vary from some minutes to hours with respect an acquisition time of just some seconds or less. Furthermore, the precision of the system is, typically, strongly connected to the quality of calibration executed.

However, as the technological advances improves the computational power and the performance of the devices, more attention is paid by the scanner designers for making the systems more users friendly. In fact, in the last decade many research works are related to the estimation of the calibration parameter without a proper calibration stage. In this track, an interesting approach, mainly oriented to the stereoscopic techniques, is the passive 3D reconstruction, which allows the estimation of the calibration parameters after the acquisition session.

Since devices for the fruition of 3D contents are becoming widely available to the consumer market, compact and easy-to-use devices for producing 3D contents are likely to be proposed. Hence, it can be envisioned that the miniaturization of components such as CCD sensors or pico-projectors will be exploited for implementing small, point-and-click optical devices.

REFERENCES

Aloimonos, Y. (1986). Detection of surface orientation from texture I: The case of plane. *IEEE Conf. on Computer Vision and Pattern Recognition* (pp. 584–593).

Batlle, J., Mouaddib, E., & Salvi, J. (1998). Recent progress in coded structured light as a technique to solve the correspondence problem: A survey. *Pattern Recognition, 31*(7), 963–982. doi:10.1016/S0031-3203(97)00074-5

Belhumeur, P. N., Kriegman, D. J., & Yuille, A. L. (1999). *The bas-relief ambiguity* (pp. 33–44). IJCV.

Besl, P., & McKay, N. (1992). A method for registration of 3D shapes. *IEEE Trans. on PAMI, 17*(8).

Brooks, M. (1983). Two results concerning ambiguity in shape from shading. *AAAI-83* (pp. 36–39).

Catalan, R. B., Perez, E. I., & Perez, B. Z. (2007). Evaluation of 3D scanners to develop virtual reality applications. *Fourth Congress of Electronics, Robotics and Automotive Mechanics* (pp. 551–556).

De Ruvo, P., De Ruvo, G., Distante, A., Nitti, M., Stella, E., & Marino, F. (2010). An environmental 3-D scanner with wide fov geometric parameters set up. *2010 IEEE Int. Conf. on Imaging Systems and Techniques (IST)* (pp. 111–114).

Drenik, M., & Kampel, M. (2008). An evaluation of low cost scanning versus industrial 3D scanning devices. *Image and Signal Processing, CISP, 08*, 756–760.

EScan. (n.d.). *Specifications*. Retrieved October 12, 2010, from http://escan3d.com/ ?page_id=11

FARO. (2010). *3D measurement-Portable measurement devices-Portable CMMs-FARO*. Retrieved July 15, 2010, from http://www.faro.com/ content. aspx? ct=en&content= pro&item= 1&subitem= 58

Feldkamp, L. A., Davis, L. C., & Kress, J. (1984). Practical conebeam algorithm. *Journal of the Optical Society of America. A, Optics and Image Science, 1*, 612–619. doi:10.1364/JOSAA.1.000612

Gambino, M. C., Fontana, R., Gianfrate, G., Greco, M., Marras, L., Materazzi, M., et al. Pezzati, L. (2005). *A 3D scanning device for architectural relieves based on time-of-flight technology*. Berlin, Germany: Springer.

Helmel. (n.d.). *Checkmaster manual: CMMs*. Retrieved October 12, 2010, from http://http:// www.helmel.com/ Checkmaster.htm

Hernández, C., Vogiatzis, G., & Cipolla, R. (2008). Multiview photometric stereo. *IEEE Trans. on PAMI, 30*(3), 548–554. doi:10.1109/TPAMI.2007.70820

Higo, T., Matsushita, Y., Joshi, N., & Ikeuchi, K. (2009). *A hand-held photometric stereo camera for 3D modeling*. Proc in Computer Vision and Pattern Recognition.

Ho, C. (2009). Machine vision based 3D scanning system. *Electronic Measurement & Instruments, 2009. ICEMI '09* (pp. 4-445–4-449).

Hoppe, H. (1994). *Surface reconstruction from unorganized points*. Unpublished doctoral dissertation, Dept. of Computer Science and Engineering, University of Washington.

Horn, B. (1975). Obtaining shape from shading information . In Winston, P. H. (Ed.), *The Psychology of Computer Vision*.

Hu, S., & Zhang, A. (2009). 3D laser omnimapping for 3D reconstruction of large-scale scenes. *2009 Joint Urban Remote Sensing Event* (pp. 1–5).

Huang, P., & Zhang, S. (2006). Fast three-step phase shifting algorithm. *Applied Optics, 45*(21), 5086–5091. doi:10.1364/AO.45.005086

Hullin, M. B., Fuchs, M., Ihrke, I., Seidel, H. P., & Lensch Hendrik P. A. (2008). Fluorescent immersion range scanning. *ACM Transactions on Graphics (Proceedings of SIGGRAPH 2008), 27*(3), 87:1–87:10.

Idesawa, M., Yatagai, T., & Soma, T. (1977). Scanning moiré method and automatic measurement of 3D shapes. *Applied Optics*, *16*(8), 2152–2162. doi:10.1364/AO.16.002152

Industrial Computed Tomography Systems. Retrieved October 12, 2010, from http://www.xviewct.com/

Leica HDS6200 - The latest ultra high-speed. Retrieved October 12, 2010, from http://hds. leica-geosystems.com/ en/ Leica-HDS6200_64228.htm

Leica ScanSystem C 10. Retrieved October 12, 2010, from http://www.leica- geosystems.com/ en/ 79411.htm

Levin, A., Fergus, R., Durand, F., & Freeman, W. T. (2007). Image and depth from a conventional camera with a coded aperture. *ACM Transactions on Graphics, 26*(3), 70–1–70–9.

Levoy, M., Rusinkiewicz, S., Ginzton, M., Ginsberg, J., Pulli, K., & Koller, D. … Fulk, D. (2000). The digital Michelangelo project: 3D scanning of large statues. *Proc. ACM SIGGRAPH 2000.*

Lu, Z., Tai, Y. W., Ben-Ezra, M., & Brown, M. S. (2010). *A framework for ultra high resolution 3D imaging* (pp. 1205–1212). Proc in Computer Vision and Pattern Recognition.

Mckinley, T. J., McWaters, M., & Jain, V. K. (2001). 3D Reconstruction from a stereo pair without the knowledge of intrinsic or extrinsic parameters. *DCV '01 Proceedings of the Second International Workshop on Digital and Computational Video* (pp. 148–155).

Nagai, M., Tianen Chen, Shibasaki, R., Kumagai, H., & Ahmed, A. (2009). UAV-borne 3-d mapping system by multisensor integration. *IEEE Trans. on Geoscience and Remote Sensing, 47*(3), 701–708.

Nikon Metrology - XT H 225. Retrieved October 12, 2010, from http:// http://www.nikonmetrology.com/ products/ x-ray_ and_ ct_ inspection/ industrial_ x-ray_ and_ct/ xt_h_225_ industrial_ ct_scanning/

Piegel, L., & Tiller,, W. (1997). *The NURBS book*. Springler-Verlag.

Potmesil, M. (1987). Generating octree models of 3d objects from their silhouettes in a sequence of images. *CVGIP*, *40*, 1–29.

Reznicek, J., & Pavelka, K. (2008). New low cost 3D scanning techniques for cultural heritage documentation. *The International Archives of the Photogrammetry, Remote Sensing and Spatial Information Sciences, Beijing, 2008*, 37.

Rusinkiewicz, S., Hall-Holt, O., & Levoy, M. (2002). Real-time 3D model acquisition. In *Proc. of the 29th Conf. on Comp. Graph. and Int. Tech* (pp. 438–446). ACM Press.

Ryde, J. (2009). An inexpensive 3D scanner for indoor mobile robots. *Proc. of the IEEE/RSJ Int. Conf, on Intelligent Robots and Systems, 2009* (pp. 5185–5190).

Salvi, J., Pagès, J., & Batlle, J. (2004). Pattern codification strategies in structured light systems. *Pattern Recognition, 37*, 827–849. doi:10.1016/j.patcog.2003.10.002

The FaroArm Family. (2010). Retrieved October 12, 2010, from http://www.faro.com/ FaroArm/ Home.htm

Tongbo, C., Lensch, H. P. A., Fuchs, C., & Seidel, H.-P. (2007). *Polarization and phase-shifting for 3D scanning of translucent objects* (pp. 1–8). Computer Vision and Pattern Recognition.

Vaillant, R., & Faugeras, O. (1992). Using extremal boundaries for 3D object modeling. *IEEE Transactions on Pattern Analysis and Machine Intelligence*, *2*(14), 157–173. doi:10.1109/34.121787

Visintini, D., Spangher, A., & Fico, B. (2007). The VRML model of Victoria Square in Gorizia (Italy) from laser scanning and photogrammetric 3d surveys. *Proc in Web3D 2007* (pp. 165–169).

Wohler, C. (2009). *3D computer vision: Efficient methods and applications*. Springer, 2009.

Woodham, R. J. (1980). Photometric method for determining surface orientation from multiple images. *Optical Engineering (Redondo Beach, Calif.)*, *19*(1), 139–144.

Wust, C., & Capson, D. W. (1991). Surface profile measurement using color fringe projection. *Machine Vision and Applications*, *4*, 193–203. doi:10.1007/BF01230201

Yahav, G., Iddan, G. J., & Mandelboum, D. (2007). 3D imaging camera for gaming application. *Consumer Electronics, 2007. ICCE*, *2007*, 1–2.

Zhang, S., Royer, D., & Yau, S. (2006). GPU-assisted high-resolution, real-time 3d shape measurement. *Optics Express*, *14*(20), 9120–9129. doi:10.1364/OE.14.009120

Zheng, H., Saupe, D., Roth, M., Bohler, A., & Opuchlik, P. (2008). Efficient 3D shape acquisition and registration using hybrid scanning data. *Proceedings of 3DPVT'08 - the Fourth International Symposium on 3D Data Processing, Visualization and Transmission* (pp. 251–258).

ADDITIONAL READING

Bernardini, F., & Rushmeier, H. (2002). The 3D Model Acquisition Pipeline. *Computer Graphics Forum*, *21*(2), 149–172. doi:10.1111/1467-8659.00574

Blais, F. (2004). Review of 20 Years of Range Sensor Development. *Journal of Electronic Imaging*, *13*(1), 231–243. doi:10.1117/1.1631921

Boulanger, P. (1999). *Knowledge representation and analysis of range data*. Tutorial Notes, Proceedings of the 2nd International Conference on 3D Digital Imaging and Modeling.

El-Hakim, S. F., Beraldin, J. A., Picard, M., & Godin, G. (2004). Detailed 3D reconstruction of large-scale heritage sites with integrated techniques. *IEEE Computer Graphics and Applications*, *24*(3), 21–29. doi:10.1109/MCG.2004.1318815

Ihrke, I., Kutulakos, K. N., Lensch, H. P. A., Magnor, M., & Heidrich, W. (2008). State of the Art in Transparent and Specular Object Reconstruction. *In STAR Proceedings of Eurographics* (pp. 87–108).

Kazunori, O., Toyokazu, K., & Satoshi, T. (2009). Development of 3D laser scanner for measuring uniform and dense 3D shapes of static objects in dynamic environment. *Proc. in IEEE ROBIO, 2008*, 2161–2167.

Kien, D. T. (2005). *A Review of 3D Reconstruction from Video Sequences. Technical report*. The Netherlands: University of Amsterdam.

Koceski, S., Koceska, N., Zobel, P. B., & Durante, F. (2009). Characterization and modeling of a 3D scanner for mobile robot navigation. *Proc in Control and Automation, MED*, *09*, 79–84.

Morimoto, M., & Fujii, K. (2005). A portable 3D scanner based on structured light and stereo camera. *IEEE International Symposium on Communications and Information Technology, ISCIT 2005*, 1, (pp. 569–572).

Shi, B., Matsushita, Y., Wei, Y., Xu, C., & Tan, P. (2010). *Self-calibrating photometric. Proc. in stereo* (pp. 1118–1125). Computer Vision and Pattern Recognition.

Wohler, C. (2009). *3D Computer Vision: Efficient Methods and Applications*. Springer.

KEY TERMS AND DEFINITIONS

3D Measurement: Procedure for obtaining the geometrical properties of a real object.

3D Model: Numerical representation of the geometrical (and sometime also the visual) properties of an object.

3D Scanner: Measuring device for obtaining the 3D model of a real object.

Active 3D Scanner: Family of 3D scanners which make use of an emitted radiation for performing the 3D measurement.

Cloud of Points: Representation of the geometry of an object by means of a set of 3D points sampled on the surface of the object.

Object Digitization: Procedure for obtaining the 3D model of an object.

Passive 3D Scanner: Family of 3D scanners which make use only of the reflected environmental light for performing the 3D measurement.

Triangulation: Technique for estimating the 3D coordinates of a point from its 2D projections.

Chapter 25
3D Imaging for Mapping and Inspection Applications in Outdoor Environments

Sreenivas R. Sukumar
The University of Tennessee, USA

Andreas F. Koschan
The University of Tennessee, USA

Mongi A. Abidi
The University of Tennessee, USA

ABSTRACT

This chapter is aimed at introducing the fundamentals of three-dimensional (3D) imaging to scientists, students, and practitioners while also documenting recent developments in the ability to rapidly digitize real-world environments. We begin with a survey of popular 3D sensing options and list factors that challenge 3D imaging in outdoor environments. The survey guides the reader towards the choice of a 3D sensor for his or her application of interest. Then, we describe 3D data acquisition strategies and integration methodologies for multi-view range data from laser scanners, multi-view image data from cameras mounted on a mobile platform and multi-sensor localization based 3D mapping. We explain the steps involved in creating 3D models from raw sensor data for each of these data acquisition strategies. Finally, we document research results obtained in the Imaging, Robotics and Intelligent Systems Laboratory at the University of Tennessee, Knoxville from 3D imaging prototypes developed for automated pavement runway inspection and urban mapping.

DOI: 10.4018/978-1-61350-326-3.ch025

INTRODUCTION

Over the last decade, many research efforts have focused on the development of 3D imaging solutions for photo-realistic 3D scene building, 3D scene description, and 3D data visualization. The efforts broadly fall into three categories:

1. building ready-to-deploy system prototypes with easy-to-use acquisition interfaces;
2. formulating and implementing algorithms for processing and integration of acquired datasets;
3. demonstrating the advantages of 3D sensing-based innovations over existing methodologies.

As researchers following and contributing to the literature in the area of 3D sensing for mapping and inspection applications, this chapter draws upon our experience from several projects including 3D under-vehicle inspection (Sukumar et al., 2007), reverse-engineering of automotive components (Page et al., 2009), road surface inspection (Yu et al., 2007), terrain modeling (Sukumar et al., 2006) and mapping of hazardous environments (Grinstead et al., 2006). The breadth of applications that we addressed in the last decade helped us realize that 3D imaging system design, especially in outdoor environments, can be a challenging problem with a steep learning curve. We hope to document some of the lessons learned in operating 3D sensors outside of controlled laboratory environments in large-scale outdoor environments. Our anticipation is that with augmented reality concepts and three-dimensional television making the foray into the consumer markets, we will soon be witnessing an increased demand for generating photo-realistic 3D immersive environments for these new devices. This chapter, we hope, will act as a knowledge dissemination source for entrepreneurs and early researchers who wish to learn the challenge and solution space with 3D sensing in outdoor environments.

With this motivation, we begin by introducing the fundamentals of 3D sensing and mapping in the background section. We will address questions like - What sensing methods are available? What accuracy can one expect from these sensors? Is a particular sensing methodology too slow? Is a sensing technique illumination sensitive? We present a brief overview of the underlying principles of active and passive 3D sensing methods and explain why some sensors are better suited than others to outdoor mapping applications. This section briefly explains concepts of 3D shape extraction using principles of stereoscopy, triangulation, time-of-flight etc. and provides appropriate reference links for detailed descriptions. We then expand upon the particular challenges and requirements of 3D sensing in outdoor environments. Based on these requirements, we argue that very few sensing methods are suitable for real-world deployment challenges. We pick laser-scanner (both triangulation-based and time-of-flight) based systems and image-based 3D reconstruction systems as potential sensors for outdoor mapping applications and discuss them in greater detail.

In the section following the background, we describe data integration strategies for multi-view range data, multi-view image data and mobile scanning using line profile scanners. We explain the range data integration approach as that of imaging a scene of interest from different viewpoints and registering the multi-view scans into one common co-ordinate system. Such methods have already been used in applications for site verification (Sequiera et al., 2007), building information models for energy efficiency simulations (Okorn et al., 2010) etc. We then contrast the range data integration approach with mobile scanning, where the concept is to mount 3D sensors on a manned/unmanned or remotely operated mobile platform equipped with a suite of different localization sensors. The line-profiles from the 3D sensors, after further processing and alignment based on localization information, deliver geometrically accurate, geographically meaningful photo-realistic

Figure 1. Classification of popular 3D imaging methods based on the physics of range sensing. Of these methods, we find the time-of-flight approach and the triangulation approach meeting accuracy, resolution, and real-time acquisition requirements for our large-scale mapping applications. The passive image-based pose and 3D structure recovery, particularly shape from motion, shows immense potential as a relatively inexpensive solution (please see the additional reading section for descriptive references for the 3D sensing methods).

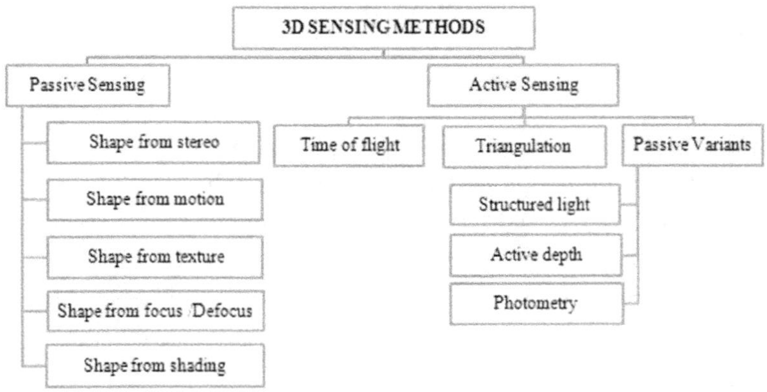

3D models of the scenes of interest. This approach has been exploited by the mobile robotics research community for 3D mine mapping (Thrun, 2003) and urban mapping (Früh & Zakhor, 2001) etc. We explain design ideas and mobile scanning methodologies that use multiple laser line profile scanners and also present a multi-sensor approach using localization hardware for 3D mapping in detail.

We present research results obtained in the Imaging, Robotics and Intelligent Systems Laboratory (IRIS) at the University of Tennessee, Knoxville (UTK) while implementing and testing our localization-based 3D sensing system in a separate Section. The specific focus is on the following applications listed below:

- Automated Runway/Road Pavement Inspection – addressing the need for illumination-independent automated detection and analysis of airport runway cracks, in contrast to camera-based systems that are still under development.
- Urban mapping – catering to the gaming and simulation community by providing augmented photo-realistic immersive environments.
- Mapping of hazardous environments – arising from the need for mapping and archiving large scale radioactive or hazardous environments for future cleanup and maintenance.

With real-world deployment experience, we demonstrate 3D imaging prototypes as significant improvements over existing camera-based vision systems. In the conclusions section, we identify issues and challenges ahead of us in new sensor development and design, the need for better localization sensors and the ability to integrate uncertain multi-sensor data. Our hope is that the conclusions drawn in this section will inspire innovation in each of these fertile areas for future research.

BACKGROUND: 3D SENSING METHODS

Overview of 3D Sensing Techniques

In this subsection, we discuss 3D sensing methods and weigh their suitability for mapping in outdoor environments. We begin by presenting a classification of different techniques in Figure 1. The classification is based on a study similar to the review of Blais (2004) on 3D range sensing in which different 3D sensing methods are categorized as passive or active. The major difference between active and passive techniques is that active sensors cast an external source of illumination (as a structured pattern, lasers etc.) to infer depth while passive methods are primarily image-based.

We begin our discussion with passive techniques for 3D imaging. The popular stereo approach is similar to the way humans perceive depth and involves two cameras taking a picture of the same scene from two different locations at the same time. Just like our eyes, image-based passive 3D reconstruction methods take 2D pictures as projective inputs of the 3D world and recover depth using computational substitutes of human perception. One computational approach is to estimate depth information by matching correspondences in the images from the two cameras and applying epipolar geometry. The alternate approach avoids matching individual pixels and instead models disparity between stereo pairs into a regularized global energy function that is iteratively optimized for a tradeoff between intensity disparity and smoothness support from neighboring pixels. With additional knowledge of camera parameters and focal length the disparity at each pixel estimated using the energy function is converted to range measurements. An extension of passive stereo that uses only a single camera is the shape from motion method. Shape from motion algorithms are also based on epipolar geometry, the difference from stereo being that frames in a video are considered as data of the same scene taken from different viewpoints. Passive triangulation algorithms, both shape from stereo and shape from motion, are challenged by the ill-posed problem of correspondence in stereo matching. We will be revisiting shape from motion principles in detail in the next Section.

Another scheme to extract 3D shape is via the principle of focusing and defocusing. The methods infer range from two or more images of the same scene, acquired under varying focus settings. By continuously varying the focus of a motorized lens and estimating the amount of blur for each focus value, the best focused image is determined. A model linking focus values and distance is then used to approximate distance. The decision model makes use of the law of thin lenses and computes range based on the focal length of the camera and the image plane distance from center of the lens. However, this method has its limitation in the fact that blur estimation influences the focal length computation and the derived range. The system required for the imaging process is best suited for microscopy applications, but not as well-suited for wide-area mapping.

While shape from stereo, shape from motion, shape from focus/defocus infer 3D geometry from two or more images, there exist methods for shape recovery from a single image. Shape from texture and shape from shading techniques fall in this category. Shape from shading uses the patterns of light and shading for establishing a fundamental equation from a single image relating the image intensity and 3D surface slope. The fundamental equation, the idea of the reflectance map, and a Lambertian assumption about the surface helps approximate the underlying shape by solving a set of differential equations (Trucco & Verri, 1999). In real-world environments, the physics and the mathematics required to solve for structure becomes complicated. The Lambertian model assumption for shading in outdoor conditions can be grossly inaccurate leading to error and discontinuity in the recovered 3D structure. Shape from texture

(Witkin, 1981) is also a single image technique that leverages the distortion observed in texture created during the imaging process when a 3D point in space is projected into a 2D plane. The method selects a representation scheme adequate for the texture cues in the image, computes the chosen distortion parameters in a representation scheme, and combines the distortion with texture gradients to estimate local orientation of the surface at each pixel.

Recently, Saxena et al. (2008), Hoeim et al. (2005) and Criminisi et al. (2000) propose methods for inferring 3D structure from a single image. Criminisi et al. (2000) recover 3D structure by finding the vanishing point and vanishing line using line segments in the 2D image. Hoeim et al. (2005) use spatial features to define superpixels to classify and associate pixels in the image to different 3D planes. Saxena et al. (2008) implement a machine learning approach to estimating 3D structure by supervised learning of monocular depth cues using ground truth range data. These methods are very good tools for the extraction of low-level 3D information from a 2D scene such as the distance between two object features in an image, identifying planar regions in the image, or classifying the ground, horizon and sky. These methods are not best suited for reconstructing a scene at a desired accuracy. In other words, techniques recovering 3D structure from a single image are very useful methods when further image acquisition in the scene is not possible and the only evidence of the scene-of-interest is a photograph.

Our interest in this chapter is the ability to build systems that can quickly image a scene of interest in outdoor environments. Most of the passive techniques discussed thus far fall short of the desired depth accuracy of the recovered 3D structure. The passive variants for shape recovery listed under the active sensing category in Figure 1 improve upon the passive methods discussed thus far by introducing an additional source of illumination. For example, the structured lighting approach projects a pre-designed pattern of pixels, usually in the form of grids and bars, and observes the deformation of the pattern on the surface of the object to learn about the 3D shape. Photometric methods use additional hardware in the form of an optical receiver that includes a photo sensor configured to detect spatio-temporal modulated optical signals directed at the scene from a set of spatially separated optical transmitters. The receiver also converts the optical signals from each of the optical transmitters to a corresponding electronic signal that is further analyzed to determine geometric properties of the scene using principles of interferometry. Active depth estimation using holography is another idea that uses a special interference pattern created in a photosensitive medium like photographic film. The third dimension of depth is inferred from the combined beams of the interference pattern projected and reflected off the surface of interest. Spatial interferometry based sensors provide high accuracy for applications requiring short range (on the order of a few meters), but can have issues with dynamically changing scenes or when the scene is imaged using a mobile platform.

We are left with two methods from the classification chart, namely the active triangulation and time-of flight systems. Both these systems are laser-based. With the active triangulation scheme, a laser in the visible spectrum (usually a line laser) illuminates the scene. The laser line traces the surface shape of the scene as a curve which is imaged using a high speed camera. The camera samples the curve traced by the laser into points representing the scene. By using a special calibration procedure to estimate depth, the surface profiles can be mapped into a metric 3D structure. The idea is that by moving the camera and laser arrangement relative to the scene of interest while simultaneously accumulating and sampling the curve traced by the laser profile we can build a 3D model of the entire scene. This approach can be configured to a high degree of accuracy and readily lends to applications where the scene is static and a mobile platform can be used to reconstruct the scene. But, being camera-based, such a system will have the same field-of-view restrictions as the passive methods. On the other hand, the time-of-flight systems are based on physical principles of estimating distance from a scene by shooting out a laser and sensing the reflection. With the knowledge of the speed of the laser, the observed time taken for the laser to travel, reflect and return is then used to compute the distance from the laser source. The time-of-flight approach does not provide high accuracy as the laser triangulation methods but usually spans a larger field of view and range.

Each acquisition method has its own advantages and disadvantages. Based on the application of interest and the application requirements a practitioner has to consider several factors before building a 3D imaging system. We list some of the factors in the following paragraphs.

Factors to Consider in Choosing a 3D Sensor

We discussed popular 3D sensing options available to us. However, if we were to pick a sensor or design a new one for a new application of interest, we have to evaluate sensors across several limiting factors. We list a few factors that we researchers typically use before conceptualizing a 3D system configuration.

- **Depth accuracy and spatial resolution:** The most critical aspect in the choice of a sensor is the depth accuracy and the spatial resolution. For inspection applications, being able to detect very fine features that are only a few millimeters deep, long, and wide is significant whereas with mapping applications, centimeter level accuracy may be sufficient.
- **Field-of-view and range of operation:** Each 3D sensing method discussed thus far operates at a fixed range - a minimum and maximum distance between the sensor and the scene. The accuracy, resolution and the precision of the sensor is specific to this range. Also, for a fixed range both laser scanners and camera based systems have limited field of view. Both field-of-view and the range of operation are as important as depth accuracy and spatial resolution while choosing a 3D sensor. As an example, while designing a robot for 3D under-vehicle inspection applications (Sukumar et al., 2006), we expected that the range of operation and the field of view will be limited by the ground clearance of an automobile. Our choice of the sensor for the robotic imaging system had to be based on the need to accommodate the variance in the ground clearance of a variety of cars and automobiles from different manufacturers (typically varying from a minimum of 10 centimeters in compact cars to 90 centimeters in large trucks). In an application where a robot carries a 3D sensor to map the under-carriage of an automobile factors like size and weight of the sensor also become important.
- **Speed of acquisition:** The speed of acquisition determines how much area we can image per unit time using a sensor. The acquisition rate indirectly dictates the density of sampled points in the final 3D model. Today, 3D area imagers require a few seconds to capture a square-meter of a scene while 3D line profilers are the high speed acquisition devices capable of capturing several thousand profiles a second.
- **Sensor cost:** We had noted earlier that accuracy and precision of active sensing methods is much more reliable than image-based passive 3D sensing methods. But, laser scanners can be expensive - in the range of a few thousand United States dollars. Vision-based 3D sensing on the

other hand, using shape from stereo or shape from motion techniques, only cost as much as the camera and the integration software; this is potentially orders of magnitude less than a high resolution 3D laser range scanner.

- **Photo-realism:** Most mapping applications require a photo-realistic reconstruction of the scene - an accurate spatial and spectral construction with geometry and color information. This requirement can be a limiting factor in the choice of the sensor. Today, some of the laser-range scanners are packaged with a high-resolution camera that is calibrated to register the 2D color information onto the 3D point cloud from its sensing electronics. However, if the laser scanners are operated as line-profile scanners instead of area-scanners, color integration has to be considered as an additional task during integration.
- **Power requirements:** One often ignored aspect of scanning in outdoor environments is the need for transportable power. While most camera-based systems can run on off-the-shelf batteries, laser scanners and other hardware for 3D acquisition are power-hungry devices. The voltage and the wattage requirements have to be considered during the system design. Most 3D sensors today are designed to operate out of line supply. For mobile scanning applications in outdoor environments, access to a line supply may not always be possible. Applications such as runway inspection, requiring several laser range scanners to operate over a period of several hours, need deep-cycle batteries with heavy-duty inverters as a source of power. Low-power line-profile scanners can be operated using cheap off-the-shelf thermal batteries for several minutes.
- **Availability of software development toolkits:** Most commercial off-the-shelf sensors do come with acquisition software. Even if not packaged along with a sensor purchase, the software may be available for an additional cost. Sensors still in the prototype phase may not have software support. Even when acquisition software is available, we have realized that software development toolkits (SDKs) instead of pre-compiled packaged software are more useful for the design of a custom 3D imaging system. SDKs enable programmable control, acquisition and integration of the sensor data considerably reducing the time between acquisition and model integration while also leveraging innovative system-design using the sensor.
- **Ambience and illumination assumptions:** Some sensors require specific illumination configurations to operate and may not be readily amenable to outdoor environments. Some sensors can also be very sensitive to illumination changes. For example, photometric stereo-based techniques can be thrown off by sunlight while most other camera-based passive techniques are illumination sensitive. The reliance of sensor accuracy on ambient conditions, specular nature of the material, etc. is not a desired characteristic for an outdoor mapping system. Even an active system such as a structured light system may not be effective in an outdoor environment. The intensity of active and structured illumination can be overwhelmed by the intensity of sunlight. The workaround for illumination sensitivities that a practitioner might be able to employ is to conduct the 3D data acquisition in the dark at night without sensor limitations.

Figure 2. Data acquisition and integration strategies for 3D modeling in outdoor environments. We also illustrate the multi-view approach where several snapshots of the same scene are acquired in contrast to the continuous acquisition of line profiles aligned based on position and orientation information from localization hardware.

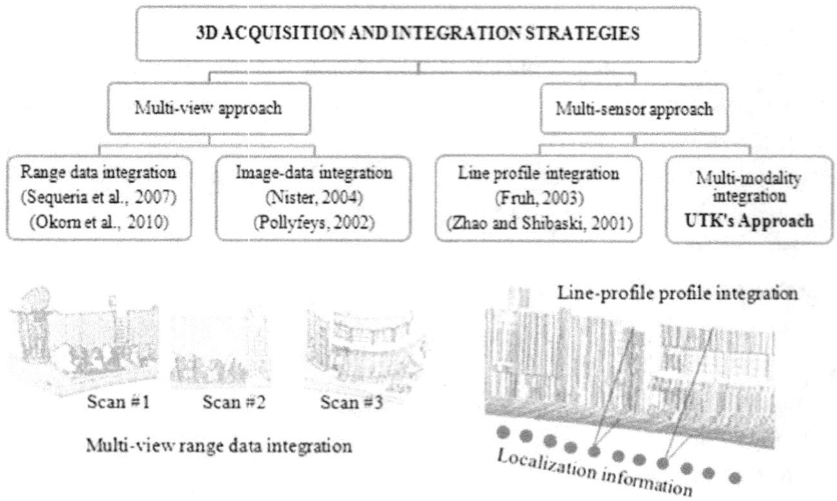

On evaluating sensing techniques along the aforementioned factors for mapping and inspection applications we are able to narrow down our options. In outdoor environments where centimeter accuracy is sufficient, time-of-flight-based scanners appear to be the ideal choice. Vision-based systems, especially those using shape from motion algorithms, sound very promising for centimeter level accuracy also. However, for crack inspection and detection millimeter accuracy is desired and only laser triangulation systems are able to digitize high accuracy high fidelity 3D geometry at the rate of a few thousand profiles in one second. But, we already know that triangulation systems have a limited field of view and sunlight can overwhelm the structured light (even if the illuminant is a high power laser). In such situations when large areas need to be digitized, our recommendation is to use an array of laser triangulation sensors with each sensor mounted with an optical filter tuned at the wavelength of the illuminating laser.

DATA ACQUISITION AND INTEGRATION STRATEGIES FOR 3D MODELING IN OUTDOOR ENVIRONMENTS

After choosing a sensing mechanism, we still need to understand different data formats from sensors. The data format influences how we collect and integrate data into 3D models. We broadly categorize data acquisition strategies into the multi-view approach and the multi-sensor approach in Figure 2 and discuss ideas to integrate 3D structure recovered from range data, image data and range-profile data in this Section. We begin with the multi-view range data integration, and later present 3D image data integration followed by multi-modality integration techniques implemented at UTK.

Multi-View Range-Data Integration

A typical off-the-shelf time-of-flight laser range scanner is designed to operate both in continuous mode (delivering the 3D structure of its entire field-of-view) or in line-profile mode (sampling one line at a time) as illustrated in Figure 2. The continuous mode operation output is usually a range image or a scattered point cloud. The output while operating as a line scanner is a profile that needs localization information to be aligned. The multi-view approach applies to most 'area' imagers like a digital camera or a range scanner set to digitize its entire field of view. The strategy is that the imaging system acquires a single snapshot of the outdoor scene at a time. The imaging system is then transported to different vantage points from where we scan the scene again. We illustrated this in the bottom left inset in Figure 2. We showed the multi-view scans in red, green and blue point clouds. We can register these point clouds to a common co-ordinate system and build the 3D model of the environment. This can be done manually using software programs like MeshLab, Rapidform, or GeoMagic, or using implementations of the Gaussian Fields framework (Bougherbel, 2005). With the multi-sensor approach, the strategy is to use line profile scanner(s) along with localization hardware like global positioning systems and inertial measurement units to align profiles sampling the 3D structure of the scene.

Image-Based Multi-View 3D Reconstruction

For each image pair in the sequence, discrete features are detected, sifted to find the corresponding matches between the successive image frames, and then used to determine the motion estimate of the camera platform between the views. We use a calibration approach where the intrinsic parameters of the camera used in the system are estimated by collecting images of planar grid patterns at different orientations and feeding it to Zhang's calibration method (Zhang, 2000). We prefer the calibrated approach to uncalibrated approaches, such as (Pollefeys, 1999), because of the dependable accuracy with the 3D structure and the reduced computations aiding real-time localization. A block diagram for image-based motion and structure estimation following (Pollefeys et al., 1999) is outlined in Figure 3.

After offline calibration, the online motion and structure estimation from images begins with feature detection. There are a number of feature detectors available for this task, and the standard Harris corner detector (Harris & Stephens, 1988) appears to be the most common in the literature due to its robustness to noise, stability and performance (Schmid et al., 1998). In Figure 4, we show the Harris corners as red and green markers on two successive frames on one of our experi-

Figure 3. Block diagram for image-based motion and structure estimation following (Pollefeys et al., 1999)

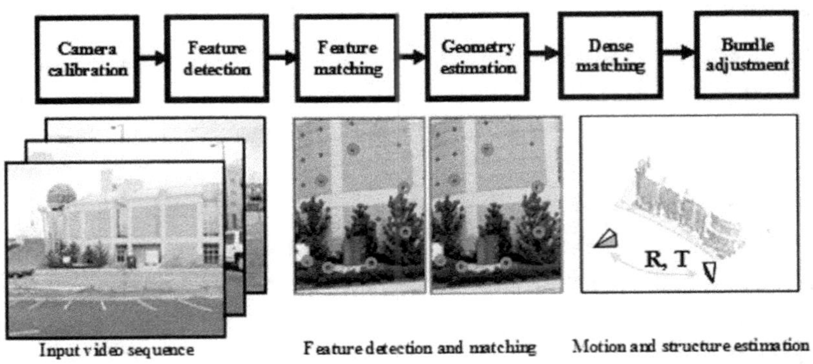

Figure 4. Pictorial description of the structure and motion estimation algorithm on video frames. The top images represent two successive frames collected while experimenting in the downtown area of Knoxville, Tennessee. The bottom left image shows the motion vectors estimated from the image data and the bottom right image shows the result of outlier rejection. The inliers of the motion matches are then used to compute the 3D motion of the camera that generated these images.

mental datasets in the downtown area of Knoxville, Tennessee. The Harris features are used as the starting locations for a window-based intensity correlation matching. This matching process is typically an $O(N^2)$ operation which can be accelerated by reducing the search space. We do this by restricting the search range to those features in the second image that lie within a distance of R pixels of the same feature in the first image. This radius is determined based on the assumed range of velocities of the mobile platform, as compared to the acquisition rate of the camera. The resulting correspondences are then filtered using an algorithm based on dominant mode of the correct matches.

The bottom images in Figure 4 show the image frames with the motion tracks of observed features superimposed. Notice that while the majority of the feature tracks indicate motion along the same direction, some of them exhibit anomalous behavior. Such anomalous motion tracks are called outliers which result mainly from noise – either through the estimated motion of features in the scene or by false matches from the correlation stage. Removal of these noisy feature tracks increases the video localization system's robustness to noise, and provides a more accurate estimate of the platform's pose.

The most common method to remove such outliers and make pose estimation robust to noise is through a "random sample consensus algorithm" (RANSAC). The RANSAC approach is a probabilistic solution, introduced by Fischler and Bolles (1981). A small subset of feature correspondences are randomly selected from the set of all feature tracks estimated after correlation matching. This subset of correspondences defines a fundamental matrix F for the image pair. F is a rank 2, 3 by 3 matrix. For simple projection models, the minimum cardinality of the feature track subset is three. Several modifications are available based on motion assumptions and the projection model such as the 5-point (Nister, 2004), 6-point (Hartley & Zisserman, 2000) and 8-point (Hartley, 1997) matching algorithms that will recover the matrix F as a linear system relating correspondences between successive image frames. Next, the epipolar error e_i is computed for all feature tracks, measuring the distance of each feature from its corresponding epipolar line, defined by the computed elements F_i. Since, the subset of correspondences can contain errors, we will have to evaluate a sufficiently large number of such feature subsets. Each evaluation will provide a hypothesis about the state of the camera system and the structure. The RANSAC procedure iterates through all these hypotheses to choose a set of feature tracks that have maximal support. The cost function in RANSAC is usually the mean epipolar error. If the mean epipolar error for a subset is less than that from previous iterations, the current fundamental matrix and its associated mean epipolar error become the best estimate for this two-frame motion. The process is then iterated until convergence within a threshold. We illustrate this procedure in Figure 5 in a simplified line fitting example.

The green markers in Figure 5 represent the inliers and the red the outliers. The problem of estimating F then in this simplified line-fitting example is to randomly select two points from the data of matches and then seek the support from other matches iteratively. The linear model F that relates the image features in successive frames is evaluated based on the distance between each feature track and the linear motion model. Through the iterative procedure, several hypotheses are evaluated and the one that converges to maximal support is chosen. The number of minimal subsets M_h (equation 1) to evaluate depends on the feature detector. In equation (1), p refers to the probability that a pixel in the image is a feature, ε is the error associated with the location of features detected by the feature detector and s is the choice of the n-point matching algorithm. If using the 5-point algorithm, $s = 5$. When the threshold for support search is set appropriately in RANSAC, the algorithm has been proved to be robust in rejecting outliers as shown in Figure 4.

$$M_h = \log(1-p) \,/\, \log(1-(1-\varepsilon)^s) \qquad (1)$$

When the iterative procedure is complete, the best estimate of F is computed by removing all features that were considered outliers with an epipolar error greater than two pixels and recomputing F using all the inlying feature matches.

Figure 5. Inlier classification using RANSAC. This n-point matching algorithm generates different hypotheses by randomly sampling the motion matches and fits a model-based on the minimal subset. Competing hypotheses are iteratively scored based on a threshold to choose the one with maximal support.

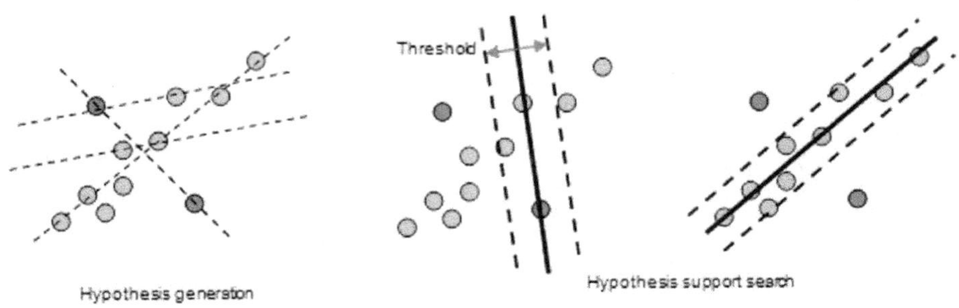

The pre-computed camera calibration matrix K is used to calculate the *essential matrix E* via

$$E = K^T F K \tag{2}$$

The next step of motion calculation is to extract the translation and rotation parameters from E. It can be shown that the translation vector T_s is the solution to $min\|E^T T_s\|$ - the unit eigenvector with the smallest eigenvalue of the matrix EE^T. The sign of the translation vector can be determined by using the constraint that the imaged scene must be in front of the camera. Determining the solution to the rotation matrix R involves solving

$$\min \left\| (R^T [-T_s]_x - E^T) \right\|, \tag{3}$$

which can be efficiently solved using a quaternion form. The output of this motion estimation system is a 5 degree of freedom (DOF) motion state with an unknown scale factor γ. We use an absolute distance measurement from the onboard laser range scanner to provide the scale factor γ.

So far, we described the state estimation from images. We now describe the structure estimation process. Though several methods exist for 3D reconstruction from images (Ma et al., 2003), the fast factorization approach for projective reconstruction appears to be the most suited for our application. Note, that the structure estimation algorithm that we use is not a two-frame method but a multi-frame method to counter the effect of vibrations in the robotic platform. For the sake of simplicity, the geometry estimation was explained based on the two-frame method in earlier paragraphs and is easily extendable to multiple frames.

Let us now consider recovering the projective structure from matched features in a video frame. Suppose the jth point in the ith frame, $\mathbf{x}_{ij}$ is projected from the scene point X_j by $\lambda_{ij}\mathbf{x}_{ij} = P_i X_j$, where λ_{ij} and P_i denote the projective depths and projection matrices, respectively. Given N_p matched points in N_f frames we have:

$$\begin{bmatrix} \lambda_{11}\mathbf{x}_{11} & \lambda_{12}\mathbf{x}_{12} & \cdots & \lambda_{1N_p}\mathbf{x}_{1N_p} \\ \lambda_{21}\mathbf{x}_{21} & \lambda_{22}\mathbf{x}_{22} & \cdots & \lambda_{2N_p}\mathbf{x}_{2N_p} \\ \vdots & \vdots & \vdots & \vdots \\ \lambda_{N_f1}\mathbf{x}_{N_f1} & \lambda_{N_f2}\mathbf{x}_{N_f2} & \cdots & \lambda_{N_fN_p}\mathbf{x}_{N_fN_p} \end{bmatrix} = \begin{bmatrix} P_1 \\ P_2 \\ \vdots \\ P_{N_f} \end{bmatrix} \begin{bmatrix} X_1 & X_2 & \cdots & X_{N_p} \end{bmatrix}, \tag{4}$$

where the matrix on the left hand side is the measurement matrix.

The initial depth values can be set to unity or be obtained with Sturm and Triggs' method (Sturm and Triggs, 1996). After the depth values are normalized, we find the nearest rank-4 approximation of the measurement matrix using SVD (Singular Value Decomposition), based on which the camera matrices and 3D reconstructed points are derived. These reconstructed points are re-projected into each image to obtain new estimates of the depths. The process is repeated until the variations in the projective depths are negligible. The step-by-step iterative projective reconstruction algorithm can be summarized as shown below.

1) Normalize the image data using isotropic scaling.
2) Start with an initial estimate of the projective depth values.
3) Normalize the depth values
   (3.1) Rescale each column of the measurement matrix so that $\sum_{i=1}^{N_f} \lambda_{ij}^2 \mathbf{x}_{ij}^T \mathbf{x}_{ij} = 1$.
   (3.2) Rescale each triplet of rows of the measurement matrix so that $\sum_{j=1}^{N_p} \lambda_{ij}^2 \mathbf{x}_{ij}^T \mathbf{x}_{ij} = 1$.
   (3.3) Repeat until there is no significant change in the measurement matrix.
4) Form the measurement matrix, find its nearest rank-4 approximation using SVD and

decompose to find the camera matrices and 3D points.

5) Re-project the points into each image to obtain new estimates of the depths.
6) If the variations in the projective depths are small enough, stop. Otherwise repeat (3)-(6). The initial depths can be set to ones or obtained using Sturm and Triggs' method or a method from (Mahmud and Hebert, 2001).

The output of the procedure is a projective reconstruction that we need to transform into a metric reconstruction. Towards that goal, we need to find a projective transformation matrix H and update the projective reconstruction by HX_j. Using the dual absolute quadric Ω^* we have $\omega_i^* \sim P_i\Omega_1^*P_i^T$ where $\omega_i^* = K_iK_i^T$ with K_i as the camera's intrinsic matrices (Sturm & Triggs, 1996). A linear solution of Ω^* can be obtained by imposing additional constraints on the camera's intrinsic parameters, such as zero skew, unit aspect ratio, and zero principal point, and the rank-3 property is applied for improved accuracy. Note that we already computed the K matrix from apriori calibration. The projective transformation matrix is obtained by forcing $H\Omega_1^*H^T = diag(1,1,1,0)$ and projective reconstruction is elevated to metric reconstruction by $P_{E,i} = P_iH^{-1}$ and $X_{E,j} = HX_j$. Finally, bundle adjustment is carried out to minimize the projection errors over several frames by computing the $\min\sum_{i,j}\left\|\mathbf{x}_{ij} - P_{E,i}X_{E,j}\right\|^2$. Once the sparse points are reconstructed, dense matching is carried out to transform each pixel in the 2D image into a 3D point. Based on the sampling density requirement on the final 3D model, interpolation is also carried out.

Our experience with implementing image-based motion estimation encourages the implementation of vision-based navigation in structured environments with buildings where GPS sensors can fail. However, we also realize that motion estimation from video can be perturbed by several factors. Illumination change, wind, weather, type of motion (as in rotation only or translation only), moving objects in the scene, multiple layers of objects in the camera's field of view can affect pose recovery. At the feature detection stage, there is error about the pixels that are mistakenly classified as features, and in environments with a significant amount of vegetation several features in the image could appear as potential matches in the correlation matching phase creating confusion with several hypothesis with support in the RANSAC stage. The structure estimation is completely dependent on the initial sparse reconstruction and the estimation of the fundamental matrix. We believe that uncertainty analysis has to be performed at the geometric estimation stage to improve the localization accuracy.

THE UNIVERSITY OF TENNESSEE APPROACH

We have included three different 3D acquisition methods (triangulation-based, time-of-flight and structured lighting) into our architecture. The reason behind including these methods was through initial experiments, where we concluded that the triangulation-based 3D sensing matched our requirements for high speed and high accuracy crack inspection (though we realized we might have to use an array of sensors for larger fields of view); for urban terrain like buildings the RIEGL time-of-flight-based scanner was the better option, and for mapping terrain for simulators, the SICK scanner's accuracy and resolution seemed to be sufficient. Hence, our software and hardware design accommodates all three types of scanners. In this section, our focus is more on the image acquisition system than the processing. We had argued that the map of any environment can only be as good as the 3D sensing method and the localization hardware. The modular architecture was hence formulated to include different types of sensors making the 3D mapping process less dependent on the application in hand. Different versions of the 3D mapping system that we have developed are shown in Figure 6.

The sheet-of-light triangulation-based 3D sensor (IVP RANGER SC386) that we use is capable of acquiring 2000 profiles/second that corresponds to 6 mm separation between successive profiles driving at 30 miles/hour. In terms of accuracy, our system that was placed 70 cm above the road surface and configured for a baseline of 70 cm and a triangulation angle of 45 degrees gives 1 mm accurate depth information. The price that we pay, however, in using such a system is the field-of-view. We are able to scan a 0.6 m wide section of the road using a single sensor. We believe using an array of such sensors is a possible solution to large area micro-scale data collection. For other applications such as terrain and urban mapping, we list the specifications of the sensors we recommend in Figure 7.

To assign global references to our integrated maps, we collect physical location information by setting up a GPS base station and placing a receiver on the mobile platform. The GPS data is accurate up to 3 cm in the motion direction and gives us 10 samples of position information in one second. The GPS can be thought of as sam-

Figure 6. The sensor architecture lends to different levels of modularity with large-scale mapping. After testing a development prototype on a push cart (middle row), we have improved towards application-specific professional packaging (bottom row).

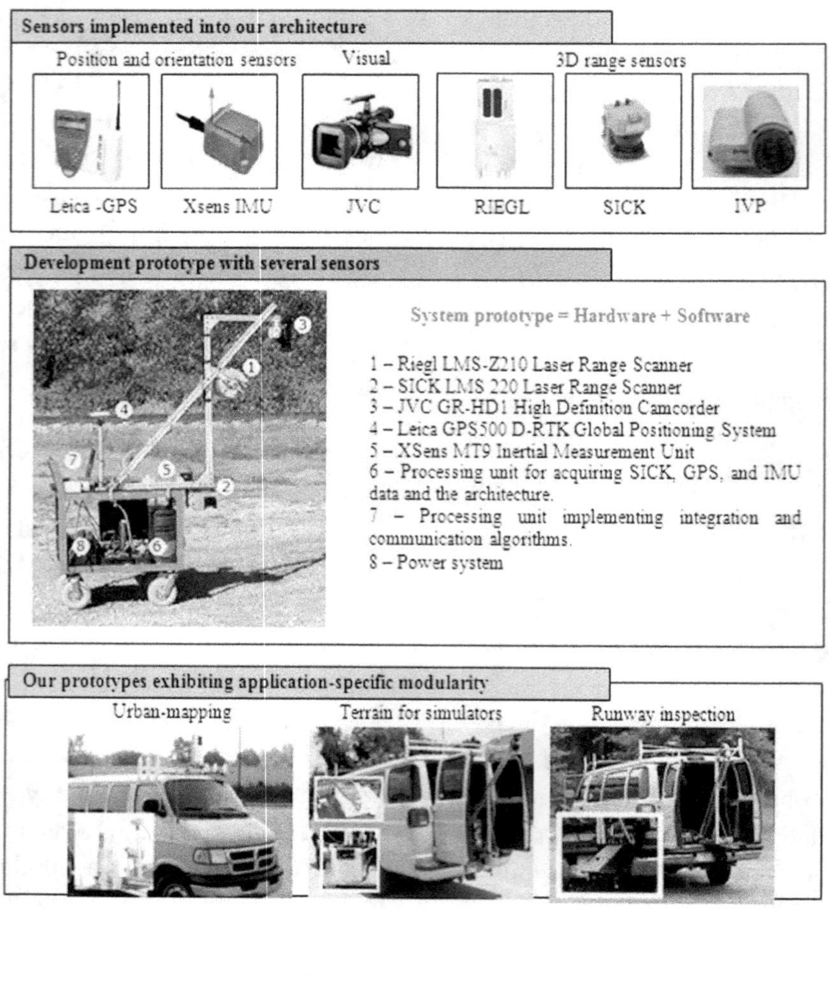

Figure 7. Specification of the components in our modular approach along with some design notes towards reproducing our system. We have included the size and weight factors to emphasize the portability and robustness. We have also provided the sensor characteristics and their expected accuracy that will later be used as a bound in the noise model for the sensors.

RIEGL

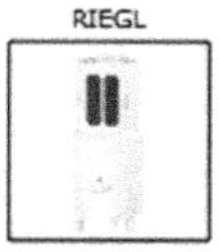

The time-of-flight range scanner is a large scale mapping instrument with an accuracy of 5 cm that can digitize 3m – 300 m far objects at 21 profiles/sec. This scanner is used for mapping large scale urban environments centimeter accuracy is sufficient.

Our differential GPS system consists a base station and the rover. The base station is placed in an area with no obstructions and the rover with one GPS radio receiver and antenna is mounted on the mobile platform (with a common axis and in line with 3D sensor) to communicate with the base station. With clear weather and satellite availability positional accuracy of up to 2 cm is achievable.

The triangulation-based laser scanner can provide depth accuracy of 1 mm configured for a baseline of 70 cm and stand off 70 cm and a triangulation angle of 45 degrees gives 1m width of view at 2000 profiles/sec. We recommend the use of this system for applications like crack inspection that require detail and accuracy.

Is placed as close as possible to the 3D range sensor and measures orientation angles accurate up to a tenth of a degree. Inertial measurements are generally noisy and have to be characterized before we can use the data for 3D profile alignment.

The range accuracy of the high-speed, time-of-flight range scanner is up to 5 mm in mapping 16 m wide profiles of the scene at 30 profiles/sec. This scanner can be noisy and might require prior noise characterization. We use this scanner in mapping road terrain where cm accuracy is sufficient.

Placed on a rod for increased field-of-view orthogonal to the road surface and also pre-registered with the range profile. This high resolution video camera can provide a 1024 x 720 image frames at a frequency of 30 Hz.

Sensor Performance	RIEGL LMS-Z210	SICK LMS200	Leica -GPS GS500	Xsens IMU MT9	Video camera JVC-HD1
Measurement Range	300 m	8 m	worldwide	300 deg/sec	visible
Acquisition Rate	10K points/sec	35 Hz	10 Hz	100 Hz	30 Hz
Resolution	1 cm	1 cm	1 cm	0.5 deg	1280 x 720i
Accuracy	5 cm	1.5 cm	2 cm	< 1 deg	n/a
Computer Interface	RS232/Parallel	RS422	RS232	RS232	IEEE 1394
Operating Voltage	12 VDC	24 VDC	10.5-32 VDC	3.4-12 VDC	7.2 VDC
Dimensions (W x L x H)	21x21x44 cm	16x19x16 cm	13x13x9 cm	58x58x22 mm	12x27x10 cm
Weight	13 kg	4.5 kg	600 g	50 g	1.3 kg

pling the 3D motion of the mobile platform that houses the sensors. In the prototypes shown in Figure 6, we see a video camera mounted on a rod, whose image axis is orthogonal to the surface of interest. We prefer the orthogonal field-of-view for generating texture because it makes the registration of range and intensity profiles easy and considerably improves integration time without having to consider CCD calibration and rectification of images. In addition to using video for texture, we use additional cameras to help estimate the motion of the sensor platform as back up to GPS satellite signals that may intermittently be unavailable during certain time intervals of the day. As a backup localization system, and for compensating for the vibrations and the resulting oscillations on the mobile platform caused by the suspension system in unstructured terrain, we used an inertial measurement unit (IMU) for measuring the orientation Euler angles (roll, pitch and yaw) of the sensor mount during data collection.

We have built our system by choosing sensors based on the application requirements. These sensors fall into two categories as pose-recovery sensors (GPS, IMU, cameras) and structure-recovery sensors (cameras and laser scanners) with the potential of also using some structure recovery sensors to infer pose. Though the idea of mapping appears trivial once the GPS provides global location and IMU provides relative orientation information to align the 3D profiles from the laser scanners into a global co-ordinate system, we have to discuss several issues before we actually deliver an integrated 3D model. In this section, we describe the procedure for integrating the data collected from multiple sensors into one complete single multi-modal 3D dataset. The processing steps that we implement are shown in a block diagram in Figure 8. The task of spatial alignment

Figure 8. Block diagram for integrating multi-sensor data into a 3D model. The sensors provide localization and structure information which is fused and aligned into a 3D model. We have included an uncertainty analysis step before the alignment to handle dynamic situations in the real world.

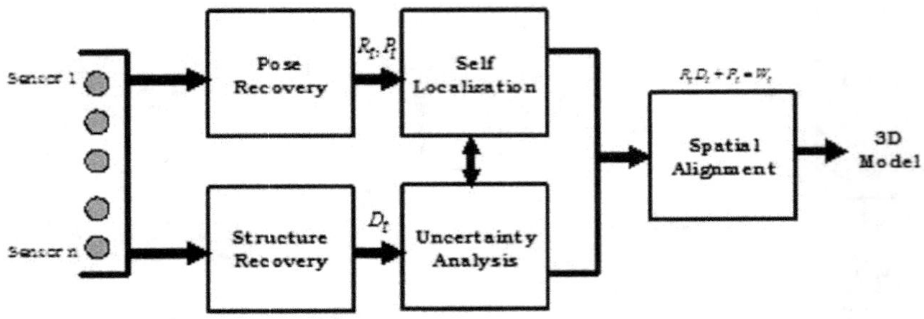

is not trivial because each of the measurement systems has its own reference coordinate system differently oriented in free space. As a first step toward integration and fusion of the data, we use the GPS coordinate frame as our reference frame and transform the range and intensity profiles to that frame without losing geographic location information of the scene.

We need to deal with another important issue before transforming the data to the real-world coordinates. We attribute this issue to different acquisition rates from different sensors. The GPS supplies data at a frequency of 10 Hz, the video camera at 30 Hz, the IMU at 100 Hz, while the range profiles are acquired at nearly 2000 Hz. We have two ways of resolving this issue: (1) Discard the range data and use the profiles that are time synchronized with the GPS data or (2) use all the points of the range data and align the profiles based on an interpolated GPS path at the time instants when we have acquired the range data. We lose more information in discarding acquired data by choosing the former solution. We hence suggest cubic spline interpolation of the GPS path as a 3D curve at time stamps recorded by the range sensor. The IMU orientation data also needs to be interpolated. Having characterized our IMU sensor, we apply moving average smoothing techniques to reduce the noise in its measurements before interpolation. The uncertainty analysis block takes care of the belief propagation on sensor data before spatial alignment in such situations.

We denote the Euler angles of roll, pitch and yaw from the IMU by $(\omega_t, \varphi_t, \kappa_t)$ and the 3D range measurements at a particular time t by $D_t = (x^t_r, y^t_r, z^t_r)$. We assume that we have already interpolated the localization sensor data to synchronize in time with the range profiles. Let the GPS measurements be (x_g, y_g, z_g) considering the moment arm distance along each dimension of the range sensor from the GPS receiver. Now the mapping to the real-world co-ordinate system W_t of the profile acquired for that instantaneous time t can be computed using

$$R_tD_t+P_t = W_t \tag{5}$$

Where

$$R_t = \begin{bmatrix} \cos\varphi\cos\kappa & \sin\omega\sin\varphi\cos\kappa+\cos\omega\sin\kappa & -\cos\omega\sin\varphi\cos\kappa+\sin\omega\sin\kappa \\ -\cos\varphi\sin\kappa & -\sin\omega\sin\varphi\sin\kappa+\cos\omega\cos\kappa & \cos\omega\sin\varphi\sin\kappa+\sin\omega\cos\kappa \\ \sin\varphi & -\sin\omega\cos\varphi & \cos\omega\cos\varphi \end{bmatrix} \tag{6}$$

$D_t = [x^t{}_r, y^t{}_r, z^t{}_r]^T$ is the measurement from the 3D range sensor and

$P_t = [x^t{}_g, y^t{}_g, z^t{}_g]^T$ is the position of the range sensor through GPS measurements.

The transformation and alignment based on multi-sensor data collected over a time period

Figure 9. Spatial integration of multi-sensor data requires a global reference frame and interpolation to consider different sampling rates of sensors. The range profiles are in a local co-ordinate frame that is transformed into the GPS co-ordinate frame based on the self-localization data and integrated as a textured 3D model.

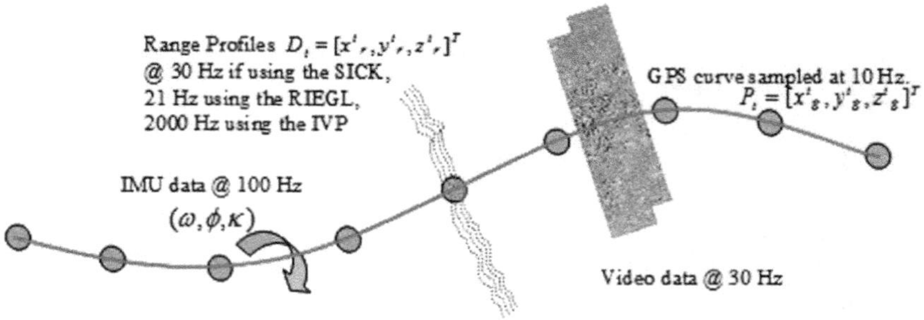

gives us an unorganized point cloud of data that, for visualization purposes, we triangulate using the method described by Hoppe in (Hoppe et al., 1992). Our experience indicates that triangulation should be performed on smaller patches as the data is acquired and later merged into a large 3D dataset. The dense point cloud is converted into a mesh that can be textured using the color images from the video. By the design of our setup and initial hardware registration step, we can map the color pixels in the CCD to the range profile as a quick method for multi-modal visualization. The process of digitizing a real world scene by (i) sampling the geometry as points and profiles D_t, (ii) sampling color using cameras and (iii) aligning geometry and color in a global co-ordinate frame can be better understood from illustration in Figure 9.

APPLICATIONS

Airport Runway/Road Pavement Inspection

The main goal of road surface inspection is being able to identify crack patterns, rut depths and the roughness of the cracks. The depth information is of particular significance in airfields because the rating scheme for the runway surface (Walker, 2004) is not dependent on the length and width of the cracks alone, as is the case with pavement distress applications, but also on the depth. Crack depths on the order of a few millimeters require high precision distance measurements. Therefore, the design requirements for a comprehensive airfield data collection system should address accuracy and precision in three dimensions of measurement, speed of acquisition, time required for post processing, and ease of visualization and evaluation.

With current data collection methods confirming the necessity to integrate several heterogeneous technologies, we further identify the scope for improvements in system design by addressing the time of acquisition and processing and list the important characteristics of a real-time deployable system. An ideal road data collection system must operate in real-time data acquisition and real-time post processing speeds. The duration required for data analysis should not overwhelm the time required for acquisition. A single pass data collection should be sufficient for cost-effective distress identification and localization in roads and runways. The critical aspects in the design are the accuracy and robustness of the system and its extendibility to arbitrary terrain, which would represent an improvement over current

Figure 10. Summary of technologies demonstrated for road surface inspection

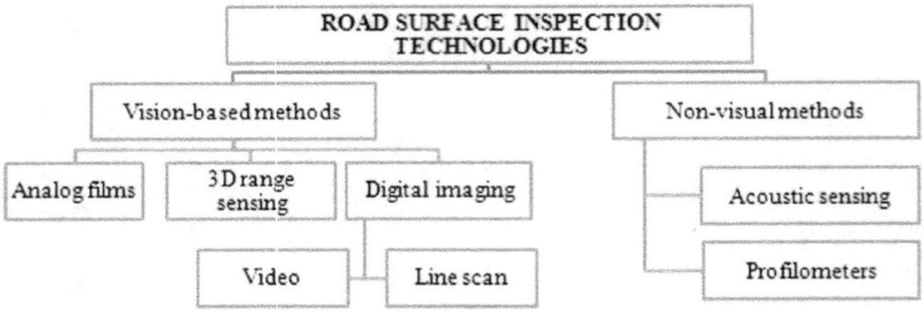

state-of-the-art methods assuming relatively planar surfaces in their design. We also note that in addition to detecting and classifying cracks, depth information is also important for the detection of object debris and other anomalies like vegetation that should not exist on an asphalt or concrete runway. An advanced system is needed that is also able to operate independently from illumination requirements.

State of the Art

Related work towards pavement distress, especially on airport runways and army maintained highways, dates back to the early 1980's. The pavement management system (PMS) concept was proposed by the U.S Army (TM 5-623, 1982) and has since then undergone metamorphosis, keeping pace with improving imaging technology. However, transportation departments met with limited real-time success using digital imaging techniques towards automatic crack detection and filling (McGhee, 2004) until the late nineties. Non-visual sensors and several improvements on image-based methods were proposed during this period. We summarize these methods in Figure 10 and discuss the advantages and disadvantages of the different types of sensing methodologies.

Today, analog films have been completely replaced by digital cameras. Among digital systems, video cameras are preferred to line scan methods for the ease of use without special illumination requirements, though line scan methods offer very high resolution data. Such video-based vision systems have two major drawbacks in extension to pavement inspection. They do not provide sufficient depth information and also have ambient illumination requirements. Range sensors that directly give depth measurements have limited field of view while profilometers and acoustic sensors, though inexpensive, can only provide low resolution and low dynamic range.

In 1987 Mendelsohn listed several of these methods including acoustic sensors and profilometers, and suggested that the imaging modality was a promising approach (Mendelsohn, 1987). At that time, the processing and image acquisition speeds challenged the feasibility of a fast and efficient inspection system. Several surveys were conducted to make an assessment of the feasibility of incorporating image acquisition and processing methods for both development and implementation of automated road surface inspection (Howe & Clemena, 1998; Wang, 2000). The conclusions of the surveys, encouraged by improving hardware and processing equipment, have led to most of the commercial video-based systems available today; these primarily consist of an array of high speed imaging sensors supported with illumination equipment. The video data from such systems (Meignen, 1997) promises to be sufficient for distress detection, but requires additional spatial information for crack filling after detection and maintenance. A potential solution AMPIS (Chung

Figure 11. Multi-modal integrated 3D data of an area of interest with three small zoomed in sections of areas with different roughness and depth of cracks. The zoomed in sectional views show the color and the color-coded range data side-by-side. Our system is calibrated for high accuracy (order of millimeters) to even sense depth variations caused by the asphalt chips on the surface.

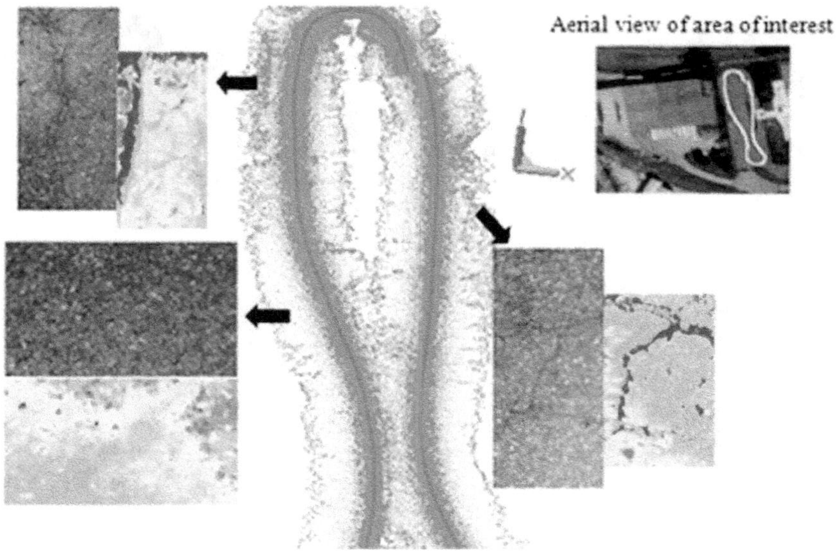

et al., 2003) was proposed that combined GPS information with video to create GIS-like databases of road surfaces. AMPIS claims improved road network identification, pavement inspection for better maintenance and data management over the base framework of PMS.

Hass et al. (1992) proposed a system that incorporated a laser range sensor for depth measurements to overcome the shortcomings of the video-based system. They concluded that combining laser range data and video image data can provide overall better accuracy and speed of crack detection, although due to the time consuming aspect of laser range sensing in 1992, they demonstrated range imaging for crack verification after the detection using the video-based system. Several 3D approaches have been demonstrated since then. Laurent et al. (1997) proposed a synchronized laser scanning mechanism to capture high precision 3D range and texture profiles. Bursanescu and Blais (1997) reiterated a 3D optical sensor as the answer to high resolution and high accuracy acquisition and redesign a sensor to meet the specific requirements of the pavement inspection application. They demonstrated six such sensors mounted on a mobile platform acquiring data at normal highway speeds. We were able to implement design ideas from several of these papers into our approach.

Results with the UTK Approach

To demonstrate proof of concept, we tested our system on several pavements with different types of cracks and present here some of those results. One such area of interest is shown as an inset along with the GPS path on a satellite map overlaid on Google Maps and the multi-modal integrated dataset in Figure 11. The discontinuity in the GPS path shown on the inset image is because we did not return precisely to the starting point. To draw attention to the resolution at which we have imaged we show some magnified images of cracks and rough asphalt surfaces in the same figure.

We have color-coded the depth to emphasize the cracks. The small cracks on the right inset are

Figure 12. Large swaths of digitized areas at very high resolution. (a) Alligator cracks detected on the road surface. (b) Detection of foreign object debris based on 3D information. (c) Textured visualization of asphalt pavements.

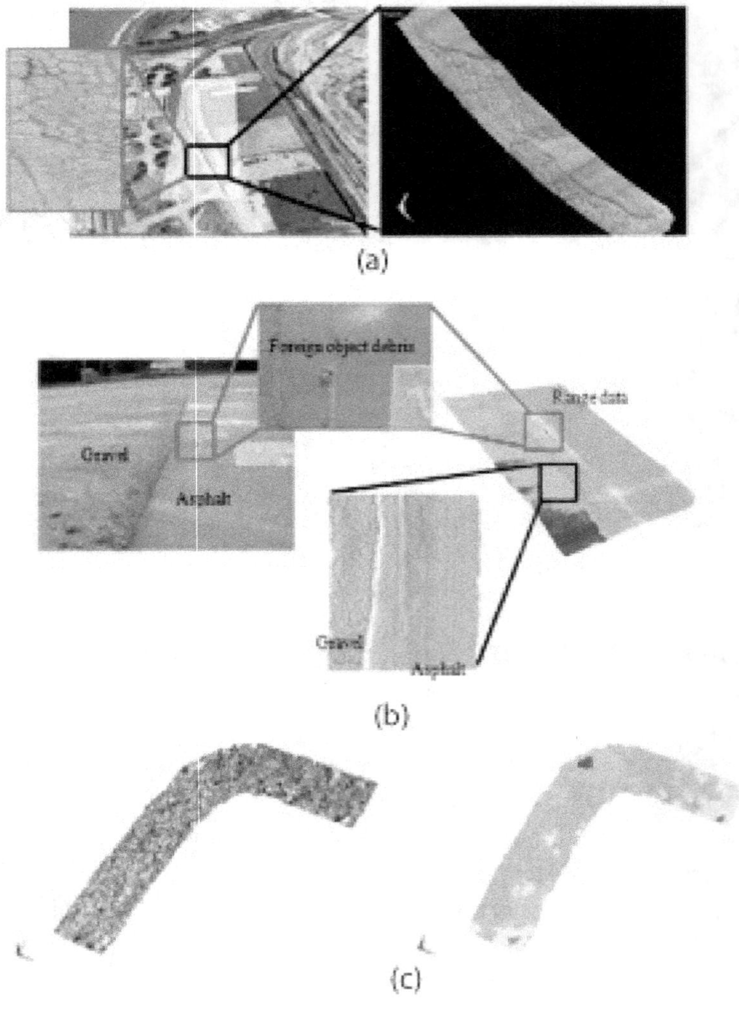

about 2 cm wide and 1 cm deep while the longitudinal crack in the top-left inset is 3 cm wide and 3 cm deep. We have not shown the entire path (75m) at that high resolution considering the size of the data and memory resources required to render the model. We illustrate the ability to digitize large swaths of data where even large-area alligator cracks can be detected in Figure 12. Figure 12 (a) zooms in on a small distressed section of a road digitized using our system. The deep alligator cracks and a longitudinal crack are visible in the 3D model. Figure 12 (b) demonstrates our ability to detect foreign object debris (the red section of the color-coded image) with relative ease compared to commercial video-based systems. With the gray shaded inset, we also note that the dataset in Figure 12 (b) shows perceivable geometric details that can differentiate gravel and asphalt surfaces. Figure 12 (c) is a texture mapped 3D model of an area inside a parking lot. The video sensor in our system is used to generate the texture and is registered with the 3D range profiles.

Improvements Over the State-of-the-Art Systems

With the 3D models that we have integrated, crack detection has become easy with simple threshold-based algorithms giving us fast and accurate results. We have overcome the illumination requirements of the contemporary video-based systems and are able to scan while driving at 30 miles/hour at 3 mm depth accuracy on the cracks and 6mm distance between profiles. With data samples from four sensors supplying data at different rates, we have integrated accurate photo-realistic 3D models for surface condition archival and convenient visualization. Our modular design enables the replacement of sensors to map larger areas, albeit with slightly reduced accuracy based on the requirements of the application in hand. Our datasets encapsulate color and geometric information and allow application of existing color-based and geometry-based crack detection/classification algorithms. Our output of real world terrain as triangle mesh datasets can easily be used as input to finite element analysis based vehicle-terrain simulators.

URBAN MAPPING

The motivation for large-scale terrain mapping is improved strategic planning in security situations. Unmanned vehicles have been deployed in several defense and security applications to provide a priori information about unknown unstructured environments with minimal risk to human life (Gage, 1995). These vehicles are armed with sensors and are capable of avoiding obstacles to navigate in an unknown environment as well as reporting concerns in different scenarios such as a battlefield (Freiburger et al., 2003), civilian security (Courtright, 1991), disaster management (Murphy, 2004), or in a patrol/surveillance mission (Klarquist, 1999). In such missions, the 3D environment map of the surveyed area of interest is useful feedback for organizing future action and deployment of resources in a much more efficient manner. For this reason, we require a modular multi-sensor system and processing package that can be mounted on unmanned vehicles/mobility platforms to generate photo-realistic, geometrically accurate geo-referenced 3D models of the area of interest. Such a system should be able to generate 3D models without making any assumptions about the vehicle trajectory and ambient illumination and should also consider the uncertainties involved in a dynamic unstructured environment. Real-time data collection and processing is also desired.

We are focused on digitizing real world environments without having to worry about the failure of the GPS or the inconsistencies in vision-based recovery. Hence, an independent modular system with the processing interface dedicated for mapping can expedite the map building process and improve mapping accuracy. Based on the level of detail that we desire in the environment, such a system should be modular and flexible in the system design, making the data collection and processing less cumbersome. Also, the map building process using the unmanned vehicles that are usually operated in stealth mode should be independent of ambient illumination capable of acquiring visual results both during the day and in the night. The hope and promise is that such a system would be faster and more realistic than computer graphics based design.

State-of-the-Art Methods

In the early attempts at terrain modeling, large swaths of coarse terrain data were acquired using airborne video systems (Baillard & Maître, 1999). Moving away from air-borne systems to easily accessible ground vehicles, an inexpensive approach of recovering 3D structure of buildings and cityscapes from video (Pollefeys et al., 2000) was demonstrated on cases where the shape could be recovered using stereo principles from successive image frames. Zhao and Shibaski (1997) demonstrated that using range sensors and a line CCD as extra data for registration and integration to create textured 3D models of urban environments was a faster and efficient approach to urban scene modeling compared to the aerial survey that was the state-of-the art at that time.

The MIT City scanning project (Antone & Teller, 2000) that inferred structure using spherical nodules was another effort in that direction. Inspired by Zhao and Shibaski (2001), Christian Früh (2001) came up with the idea for urban mapping using two laser range profilers in an orthogonal arrangement along with digital cameras. He demonstrated the system mounted on a truck and driving at normal highway speeds to collect data that was processed offline. With his orthogonal arrangement, he was able to compute centimeter level accuracy by matching successive laser scans against each other and between the two sensors. The horizontal laser scans were used to approximate a component of the acquisition vehicle's motion. With the vertical scanner providing the façade of the urban structure, he proposed two different approaches in using information from aerial maps to minimize global localization error using laser scans alone. One of those methods was to use cross correlation and the other a Markov-Monte Carlo technique to acquire 3D models in a matter of few minutes subject to traffic conditions. The two major drawbacks of this approach are the availability of the aerial map and the magnitude of global error that accumulated over 100m of data.

Zhao and Shibaski (2001) further improved on Konno et al. (2000) who proposed three single-row laser range scanners and six line cameras mounted on a measure vehicle (GeoMaster), with a system equipped with a GPS/INS/Odometer-based navigation system. Their sensor mount outputs three kinds of data sources: laser range points, line images, and navigation data. Either the laser range points or the line images are in the sensor's local coordinate system at the time of measurement. They are synchronized with the navigation data using the sensor's local clock and integrated into 3D models offline. The motivation behind these urban scanning projects described so far are more on digitization than accuracy of digitization with expected errors on the order of a few centimeters. These methods did not address the uncertainty in the measurement process and the dynamic environment towards map building.

Results with the UTK Approach

We tested our system acquiring several miles of data in and around Knoxville, Tennessee and present some of the results using a RIEGL scanner in Figure 13. There are three examples depicted in the figure. The first one in Figure 13(a) shows a shopping center digitized and textured by driving our imaging prototype along the road in the parking lot in front of the shopping center.

The second one is the Women's Basketball hall of fame building near the University of Tennessee campus (Figure 13 (b)). We also show the path of our acquisition platform on a satellite image as an inset in the same figure. These models are accurate up to a few centimeters and extremely dense with each model consisting of no more than 100,000 points. Mapping the Hall of fame building was a challenge. The building is along a curve in the road and mapping using image-based techniques was non-trivial. We also had reservations about the availability of GPS signals, as the building was very close to the urban canyon in the downtown area. Our instrumentation and integration method successfully handled the situation, resulting in the accurate and photo-realistic model.

We present another model integrated using our system in Figure 13(c). We mapped a 1 kilometer long path around the mall area on Chapman Highway, Knoxville without any prior knowledge about the area. We have shown magnified sections of the Goody's store to indicate the sampling density achievable using our system without having to

Figure 13. Large areas of urban environments digitized at very high geometric resolution with high fidelity texture. (a) BI-LO shopping complex in Knoxville. (b) The women's Basketball Hall of Fame building. (c) The 3D rendering of a shopping mall with the zoomed inset of the Goody's store on Chapman Highway in Knoxville. The sampling density of digitization and the photo-realistic rendering are key enhancements with our systems. Our output models are triangle meshes that are easy to embed in immersive virtual environments.

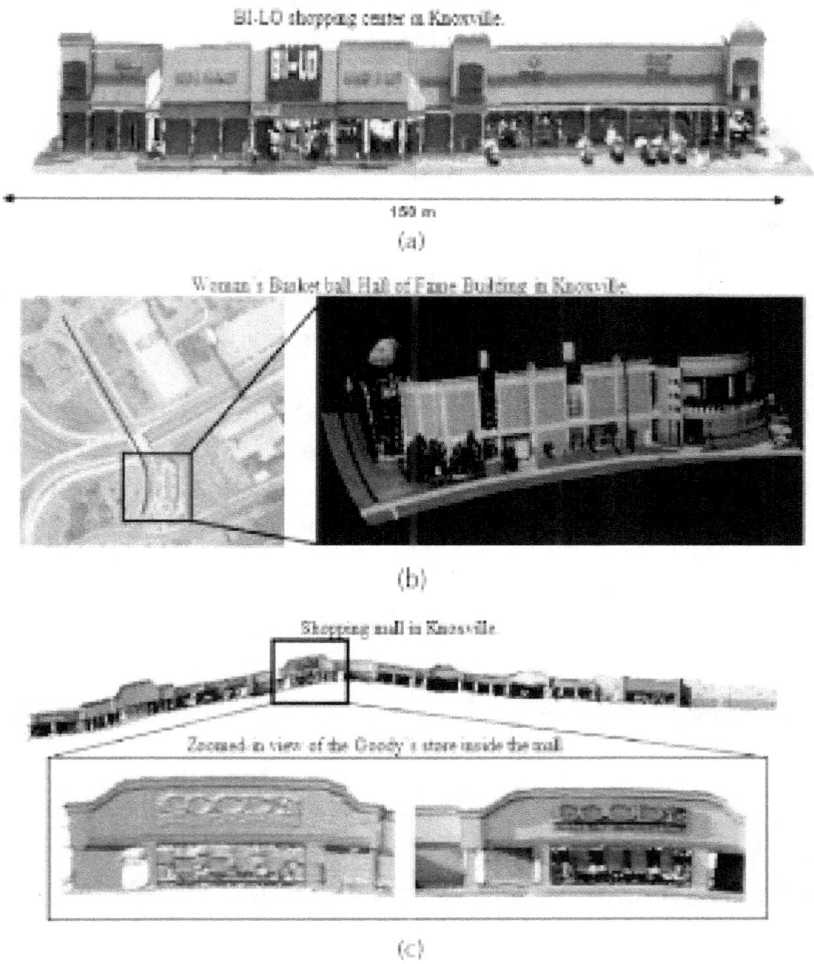

compromise on the texture quality. Our acquisition took about 10 minutes, further emphasizing our ability to quickly produce 3D models of urban environments. For large datasets spanning several miles, we process datasets offline. A mile of data usually takes approximately one hour of processing on off-the-shelf desktop computers.

Improvements Over the State-of-the-Art Systems

In essence, we have documented mobile mapping prototypes consisting of four main components: hardware for 3D geometry and texture acquisition; hardware for positioning and orientation (pose and trajectory) measurement; a mobile platform which moves the sensing package past the environment to be digitized; and software to perform the necessary information fusion to combine the data from different sensing modalities and to process the resulting model to fit the application at hand. While other researchers have developed 3D terrain acquisition systems, these tend to be fixed in regards to the hardware and the fusion methods used. In contrast, our system treats the components independently with the following improvements in accuracy, resolution and photorealism. Our system promises mm- to cm-level accuracy as required. Our contributions over the state-of-the-art are particularly with respect to the accuracy at which we are able to image and simplicity in integration towards efficient processing and realistic visualization. The modularity inherent in our design allows the system to be as robust to real world environments as the individual components, at the same time being independent of application-specific hardware modifications. In other words, our design is capable of easy integration when mounted on an aerial vehicle or a ground vehicle based on operational need without requiring excessive reconfiguration. The modular design enables us to treat accuracy and resolution as parameters of the system to suit the application in hand.

CONCLUSION

In spite of the tremendous advances in 3D sensor design, 3D sensors are sensitive to several factors. Up to now, no design exists for a single 3D camera/device, analogous to the digital camera design for 2D pictures, that can adapt to work with the same quality and reliability both indoors and outdoors. 3D sensing appears to be affected by a multitude of factors in addition to the known issues with color cameras. Moreover, handling occlusions has to be addressed in 3D sensing. When using a mobile system or a range scanner, objects in the scene can occlude other objects within the field-of-view especially in outdoor environments. Filling up missing data caused by view-occlusions can be a time consuming task requiring several acquisitions.

The fusion approach we presented tries to compensate for some of the issues with 3D sensing. However, it has its limitations. If the sensors are all functional and perfect, there would be essentially no error in the integrated 3D map after the spatial alignment. However, the sensors are noisy and can fail. The noise manifests in the localization measurements and also in the 3D structure measurements. Uncertainty in the state recovered during self localization propagates as uncertainty into the integrated map. Hence, before we claim robustness to noise and a bound on the error in the integrated 3D map, we have to handle uncertainty from both of these sources. Our modular design of including different sensors minimizes the measurement uncertainty in the geometric samples of the scene, but we still have to deal with the uncertainty in localization. Also, if the mapping has to be performed autonomously, localization appears to be much more significant and a more challenging problem requiring models for predicting expected uncertainty from the sensors. There is a significant need for less noisy localization/pose estimation sensors and better uncertainty handling methods.

ACKNOWLEDGMENT

The authors would like to thank Brad Grinstead, Si-Jie Yu, and Yi Yao for their contributions to the work presented in this chapter. This work was supported in part by the University Research Program in Robotics under grant DOE-DE-FG52-2004NA25589, by the DOD/RDECOM/NAC/ARC Program under grant R01-1344-18, and by the US Air Force under grant FA8650-10-1-5902.

REFERENCES

Antone, M. E., & Teller, S. (2000). Automatic recovery of relative camera positions in urban scenes. *Computer Vision and Pattern Recognition, 2*, 282–289.

Baillard, C., & Maître, H. (1999). 3-D reconstruction of urban scenes from aerial stereo imagery: A focusing strategy. *Computer Vision and Image Understanding, 76*(3), 244–258. doi:10.1006/cviu.1999.0793

Blais, F. (2004). Review of 20 years of range sensor development. *Journal of Electronic Imaging, 13*(1), 231–240. doi:10.1117/1.1631921

Boughorbel, F., Koschan, A., & Abidi, M. (2005) Automatic registration of 3D datasets using Gaussian fields. In *Proc. IEEE International Conference on Image Processing ICIP2005*, vol. III, (pp. 804-807).

Bursanescu, L., & Blais, F. (1997). Automated pavement distress data collection and analysis: A 3D approach. In *Proc. of. Conf. on Recent Advances in 3-D Digital Imaging and Modeling*, (pp. 311-317).

Chung, H. C., Girardello, R., Soeller, T., & Shinozuka, M. (2003). Automated management for pavement inspection system. In *Proc. of the SPIE Smart Structures and Materials Symposium, 5057*, (pp. 634-644).

Courtright, M. L. (1991). Unmanned vehicles go to war. *Machine Design, 63*(25), 60–64.

Criminisi, A., Reid, I., & Zisserman, A. (2000). Single view metrology. *International Journal of Computer Vision, 40*(2), 123–148. doi:10.1023/A:1026598000963

Fischler, M. A., & Bolles, R. C. (1981). Random sample consensus: A paradigm for model fitting with applications to image analysis and automated cartography. *Communications of the ACM, 24*(6), 381–395. doi:10.1145/358669.358692

Freiburger, L. A., Smuda, W., Karlsen, R. E., Lakshmanan, S., & Ma, B. (2003). ODIS the under-vehicle inspection robot: Development status update. In *Proc. of SPIE Unmanned Ground Vehicle Technology V*, vol. 5083, (pp. 322-335).

Früh, C., & Zakhor, A. (2001). Fast 3D model generation in urban environments. In *Proc. of the International Conference on Multi-sensor Fusion and Integration for Intelligent Systems*, (pp. 165-170).

Gage, D. (1995). UGV history 101: A brief history of Unmanned Ground Vehicle (UGV) development efforts. *Unmanned Systems Magazine, 13*(3), 9–16.

Grinstead, B., Sukumar, S. R., Page, D. L., Koschan, A. F., Abidi, M. A., & Gorsich, D. (2006). Mobile scanning system for the fast digitization of existing roadways and structures. *Sensor Review Journal, 26*(4), 283–289. doi:10.1108/02602280610691999

Haas, C. (1992). *Investigation of a pavement crack-filling robot*. Pittsburgh, PA: Report to the Strategic Highway Research Program.

Harris, C., & Stephens, M. J. (1988). A combined corner and edge detector. In *Proc. of the Alvey Vision Conference*, (pp. 147-152).

Hartley, R. I. (1997). In defense of the eight-point algorithm. *IEEE Transactions on Pattern Analysis and Machine Intelligence, 19*(6), 580–593. doi:10.1109/34.601246

Hartley, R. I., & Zisserman, A. (2000). *Multiple view geometry in computer vision*. Cambridge University Press.

Hoiem, D., Efros, A., & Hebert, M. (2005). Automatic photo pop-up. In *Proc. of ACM SIGGRPAPH*, vol. 24, (pp. 577–584).

Hoppe, H., DeRose, T., Duchamp, T., McDonald, J., & Stuetzle, W. (1992). Surface reconstruction from unorganized points. *ACM SIGGRAPH Computer Graphics*, *26*(2), 71–78. doi:10.1145/142920.134011

Howe, R., & Clemena, G. G. (1998). *An assessment of the feasibility of developing and implementing an automated pavement distress survey system incorporating digital image processing. Rep. No. VTRC 98-R1*. Virginia Transportation Research Council.

Klarquist, W. N., Bonner, K. G., & Gothard, B. M. (1999). Demo III: Reconnaissance, surveillance, and target acquisition (RSTA) preliminary design. In *Proc. of SPIE Mobile Robots XIII and Intelligent Transportation Systems*, vol. 3525, (pp. 232-242).

Konno, T. (2000). A new approach to mobile mapping for automated reconstruction of urban 3D model. In *Proc. of the International Workshop on Urban Multi-Media/3D Mapping*, (CDROM).

Laurent, J., Talbot, M., & Doucent, M. (1997). Road surface inspection using laser scanners adapted for the high precision measurements of large flat surfaces. In *Proceedings of the International Conference on Recent Advances in 3-D Digital Imaging and Modeling*.

Mahamud, S., Hebert, M., Omori, Y., & Ponce, J. (2001). Provably-convergent iterative methods for projective structure from motion. In *Proc. of the Conference on Computer Vision and Pattern Recognition*, (pp. 1018-1025).

McGhee, K. H. (2004). *NCHRP Report 334: Automated pavement distress collection techniques. Transportation Research Board*. Washington, D.C.: National Research Council.

Meignen, D., Bernadet, M., & Briand, H. (1997). One application of neural networks for defects using video data bases: Identification of road distresses. In *Proc. DEXA Workshop*, (pp. 459-464).

Mendelsohn, D. H. (1987). Automated pavement crack detection: An assessment of leading technologies. In *Proc. 2nd North American Conference on Managing Pavements, 3*, (pp. 297-314).

Murphy, R. R. (2004). Activities of the rescue robots at the World Trade Center from 11-21 September 2001. *IEEE Robotics & Automation Magazine*, *11*(3). doi:10.1109/MRA.2004.1337826

Nister, D. (2004). An efficient solution to the five-point relative pose problem. *IEEE Transactions on Pattern Analysis and Machine Intelligence*, *26*(6), 756–777. doi:10.1109/TPAMI.2004.17

Okorn, B. E., Xiong, X., Akinci, B., & Huber, D. (2010). Toward automated modeling of floor plans. In *Proc. of the Symposium on 3D Data Processing, Visualization and Transmission*.

Page, D., Koschan, A., & Abidi, M. (2007). Methodologies and techniques for reverse engineering - The potential for automation with 3D laser scanners . In Raja, V., & Fernandes, K. (Eds.), *Reverse engineering - An industrial perspective* (pp. 11–32). London, UK: Springer.

Pollefeys, M., Gool, L. V., Vergauwen, M., Verbiest, F., Cornelis, K., & Tops, J. (2002). Video-to-3D. In *Proceedings of the Symposium on Photogrammetric Computer Vision, A*, (pp. 252-257).

Pollefeys, M., Koch, R., & Gool, L. V. (1999). Self-calibration and metric reconstruction in spite of varying and unknown internal camera parameters. *International Journal of Computer Vision*, *32*, 7–25. doi:10.1023/A:1008109111715

Pollefeys, M., Koch, R., Vergauwen, M., & Gool, L. V. (2000). Automated reconstruction of 3D scenes from sequences of images. *ISPRS Journal of Photogrammetry and Remote Sensing*, *55*(4), 251–267. doi:10.1016/S0924-2716(00)00023-X

Saxena, A., Chung, S., & Ng, A. (2008). 3-D depth reconstruction from a single still image. *International Journal of Computer Vision, 76*(1), 53–69. doi:10.1007/s11263-007-0071-y

Schmid, C., Mohr, R., & Bauckhage, C. (1998). Comparing and evaluating interest points. In *Proc. of the International Conference on Computer Vision*, (pp. 230-235).

Sequeira, V., Boström, G., & Gonçalves, J. G. M. (2007). 3D Site modelling and verification - Usage of 3D laser techniques for verification of plant design for nuclear security applications. In A. Koschan, M. Pollefeys, & M. Abidi (Eds.), *3D imaging for safety and security*, (pp. 249-278). Dordrecht, The Netherlands: Springer.

Sturm, P., & Triggs, B. (1996). A factorization based algorithm for multi-image projective structure and motion. In *Proc. European Conf. on Computer Vision*, Cambridge, (pp. 709-720).

Sukumar, S. R., Page, D. L., Gribok, A., Koschan, A. F., Abidi, M. A., Gorsich, D. J., & Gerhart, G. R. (2006). Robotic 3D imaging system for under vehicle inspection. *Journal of Electronic Imaging*, 15(3), (033008).

Sukumar, S. R., Page, D. L., Koschan, A., & Abidi, M. A. (2007). Under vehicle inspection with 3D imaging. In A. Koschan, M. Pollefeys, & M. Abidi (Eds.), *3D imaging for safety and security*, (pp. 249-278). Dordrecht, The Netherlands: Springer.

Sukumar, S. R., Yu, S.-J., Page, D. L., Koschan, A. F., & Abidi, M. A. (2006). Multi-sensor integration for unmanned terrain modeling. In *Proc. SPIE Unmanned Systems Technology VIII*, vol. 6230, (pp. 65-74).

Thrun, S., Haehnel, D., Ferguson, D., Montemerlo, M., Triebel, R., & Burgard, W. … Whittaker, W. (2003). A system for volumetric robotic mapping of abandoned mines. In *Proceedings of the Intl. Conference on Robotics and Automation*, vol. 3, (pp. 4270-4275).

Trucco, E., & Verri, A. (1998). *Introductory techniques for 3D computer vision*. Englewood Cliffs, NJ: Prentice-Hall.

Walker, D. (2004). *Asphalt airfield pavements, pavement surface evaluation and rating. University of Wisconsin*. Madison: Transportation Information Center.

Wang, R., & Spelke, E. (2002). Human spatial representation: Insights from animals. *Trends in Cognitive Sciences, 6*(9). doi:10.1016/S1364-6613(02)01961-7

Witkin, A. P. (1981). Recovering surface shape and orientation from texture. *Artificial Intelligence, 17*, 17–45. doi:10.1016/0004-3702(81)90019-9

Yu, S.-J., Sukumar, S. R., Koschan, A., Page, D. L., & Abidi, M. A. (2007). 3D reconstruction of road surface using an integrated multi-sensory approach. *Optics and Lasers in Engineering, 45*(7), 808–818. doi:10.1016/j.optlaseng.2006.12.007

Zhang, Z. (2000). A flexible new technique for camera calibration. *IEEE Transactions on Pattern Analysis and Machine Intelligence, 22*(11), 1330–1334. doi:10.1109/34.888718

Zhao, H., & Shibasaki, R. (1997). Automated registration of ground-based laser range image for reconstructing urban 3D object. *International Archive on Photogrammetry and Remote Sensing, 32*(3-4W2).

Zhao, H., & Shibasaki, R. (2001). High accurate positioning and mapping in urban area using laser range scanner. In *Proc. of the IEEE Intelligent Vehicles Symposium*.

ADDITIONAL READING

Besl, P. J. (1988). Active, optical range imaging sensors . *Machine Vision and Applications, 1*(2), 127–152. doi:10.1007/BF01212277

Chen, F., Brown, G. M., & Song, M. (2000). Overview of three-dimensional shape measurement using optical methods. *Optical Engineering (Redondo Beach, Calif.)*, *39*(10).

Crossley, S., Seed, N. L., Thacker, N. A., & Ivey, P. A. (2004). Improving accuracy, robustness and computational efficiency in 3D computer vision. *Image and Vision Computing*, *22*(5), 399–412. doi:10.1016/j.imavis.2003.12.006

Cyganek, B., & Siebert, J. P. (2009). *An Introduction to 3D Computer Vision Techniques and Algorithms*. Hoboken, NJ: Wiley. doi:10.1002/9780470699720

Faugeras, O. (1993), *Three-Dimensional Computer Vision: A Geometric Viewpoint*, MIT Press, Cambridge, Ma.

Forsyth, D., & Ponce, J. (2002). *Computer Vision: A Modern Approach*. Upper Saddle River, N.J.: Prentice Hall.

Girod, B., Greiner, G., & Niemann, H. (2000). *Principles of 3D Image Analysis and Synthesis*. Dordrecht, The Netherlands: Springer.

Grimson, E. L. (1981), *From Images to Surfaces*, MIT Press, Cambridge, Ma.

Gruen, A., & Huang, T. S. (2001). *Calibration and Orientation of Cameras in Computer Vision*. Berlin: Springer.

Hartley, R., & Zisserman, A. (2003). *Multiple View Geometry in Computer Vision* (2nd ed.). Cambridge, England: Cambridge University Press.

Klette, R., Schlüns, K., & Koschan, A. (1998). *Computer Vision - Three-Dimensional Data from Images*. Singapore: Springer.

Koschan, A., Pollefeys, M., & Abidi, A. (2007). *3D Imaging for Safety and Security*. Dordrecht, The Netherlands: Springer. doi:10.1007/978-1-4020-6182-0

Levoy, M., Pulli, K., Curless, B., Rusinkiewicz, S., Koller, D., Pereira, L., et al. (2000). The Digital Michelangelo Project: 3D Scanning of Large Statues. In *Proc. of the ACM SIGGRAPH*, (pp. 131-144).

Ma, Y., Huang, K., Vidal, R., Košecká, J., & Sastry, S. (2004). Rank Conditions on the Multiple-View Matrix. *International Journal of Computer Vision*, *59*(2), 115–137. doi:10.1023/B:VISI.0000022286.53224.3d

Ma, Y., Soatto, S., Kosecka, J., & Sastry, S. (2004). *An Invitation to 3-D Vision: From Images to Geometric Models*. New York: Springer.

Marr, D. (1982), *Vision: A Computational Investigation into the Human Representation and Processing of Visual Information*, MIT Press, Cambridge, Ma.

Nakamura, K., Saito, H., & Ozawa, S. (2000), 3D Reconstruction of Book Surface Taken from Image Sequence with Handy Camera, In *Proc. 15th International Conference on Pattern Recognition*, 4, (pp.4575).

Pollefeys, M., Koch, R., Vergauwen, M., & Van Gool, L. (2000). Automated reconstruction of 3D scenes from sequences of images. *ISPRS Journal of Photogrammetry and Remote Sensing*, *55*(4), 251–267. doi:10.1016/S0924-2716(00)00023-X

Shirley, L. G., & Hallerman, G. R. (1996). *Application of tunable lasers to laser radar and 3D imaging, Technical Report #1025*. Lexington: MIT Lincoln Lab.

Sjodahl, M., & Synnergren, P. (1999). Measurement of shape by using projected random patterns and temporal digital speckle photography. *Applied Optics*, *38*(10), 1990–1997. doi:10.1364/AO.38.001990

Sukumar, S. R. (2008). *Uncertainty Minimization in Robotic 3D Mapping Systems Operating in Dynamic Large-Scale Environments*, Doctoral Dissertation, The University of Tennessee, Knoxville.

Thacker, N. A., Clark, A. F., Barron, J., Beveridge, L., Courtney, P., & Crum, W. R. (2008). Performance characterization in computer vision: A guide to best practices . *Computer Vision and Image Understanding, 109*(3), 305–334. doi:10.1016/j.cviu.2007.04.006

Trolinger, J. D. (1996). Ultrahigh resolution interferometry . *Proceedings of the Society for Photo-Instrumentation Engineers, 2861*, 114–123.

Tyler, C. W. (2010), *Computer Vision: From Surfaces to 3D Objects*, CRC Press, Boca Raton, Fl.

Wöhler, C. (2009). *3D Computer Vision - Efficient Methods and Applications*. Heidelberg: Springer.

KEY TERMS AND DEFINITIONS

Bundle Adjustment: The bundle adjustment procedure simultaneously refines the 3D coordinates describing the 3D scene geometry along with the parameters of relative motion and intrinsic optical characteristics of the camera used to acquire the images that are used in the 3D reconstruction. Most feature-based 3D reconstruction algorithms use a bundle adjustment procedure as an attempt to minimize the reprojection error of image points in a 3D world.

Camera Calibration: The term camera calibration can refer to (a) geometric calibration or (b) photometric calibration. Geometric calibration is the process of finding/estimating the parameters of the camera (usually using projection models allowing for some distortion) that transforms the 3D real world scene into a gridded 2D image. Photometric calibration on the other hand is more relevant to photometric and stereoscopic methods, refers to the mapping of colors in one image to another. Both photometric calibration and geometric calibration is typically performed before the data acquisition.

Triangulation: Based on trigonometry and geometry (especially the properties of triangles), triangulation is the process of determining the location of a point by measuring angles to it from two end points of a fixed baseline. Every real-world scene point can be considered as the vertex of a triangle consisting of the scene point and the two end points of a fixed baseline. With angles at the base line known and the length of one of the sides of the triangle (baseline) known, triangulation refers to the trigonometric formulation to estimate the distance of the scene point from the fixed baseline acquisition setup.

Chapter 26
3D Laser Scanner Techniques:
A Novel Application for the Morphological Study of Meteorite Impact Rocks

Mercedes Farjas
Universidad Politécnica de Madrid, Spain

Jesús Martinez-Frias
NASA Astrobiology Institute, Spain

Jose María Hierro
Universidad Politécnica de Madrid, Spain

ABSTRACT

The use of 3D scanning systems for acquiring and analyzing the external shape features of arbitrary objects has different applications in different cultural, scientific, and technological fields. In this work, 3D laser scanning techniques are used, for the first time, to our knowledge, as a novel and non-destructive application for the morphological study of meteorite impact rocks. The subject of the study was a rock displaying impact textures and associated with the Karikkoselkä impact crater (Finland) (Lehtinen et al. 1996). This methodology permitted: (1) a computerized three-dimensional modelling to be carried out on the bulk impact-related rock; (2) other more specific characterizations to be performed, such as detailed topographic studies of its surface impact features; (3) some physical properties of the rock to be determined (volume); (4) the shatter cone impact texture to be completed with a realistic estimation of its convergence angle; and (5) a broad demonstration of the significance and effectiveness of 3D laser scanning techniques as a complementary tool for the study of this type of meteoritic impact-related rocks.

DOI: 10.4018/978-1-61350-326-3.ch026

INTRODUCTION: IMPACT ROCK (KARIKKOSELKA IMPACT CRATER)

Numerous studies on terrestrial rocks have shown they have different types of geological features which can be used as micro or macro-markers of large impact events (Koeberl & Martinez-Ruiz, 2003). Certain shock effects have been shown to be uniquely and unequivocally associated with meteorite impact craters. These include, among others, multiple sets of microscopic planar deformation features, mainly in quartz and feldspar grains, high-pressure mineral phases (e.g. stishovite, coesite) and shatter cones (impact textures which form when the shockwave from a meteorite impact event passes through and modifies the target rocks). The study and detailed characterization of these impact-related signatures in terrestrial environments encompasses the use of classical mineralogical and geochemical techniques and the development of theoretical, numerical and experimental models.

The impact rock used in the work (Figure 1) was sampled by one of the authors (JMF) in the Karikkoselka area of Finland and forms part of the lithotheque at the Spanish Centro de Astrobiologia. The well known northern European Fennoscandian Shield has at least 32 impact structures, of which 10 are located in Finland (see Earth Impact Database, Spray, 2009). This shield is extremely important for impact cratering research, since it is well exposed, easily accessible and has been mapped in detail (Plado & Pesonen, 2002). The Karikkoselka structure (62°13' N; Long. 25°15' E) was proposed as a meteorite impact crater by Lehtinen et al. (1996). The crater is the smallest of those so far identified in Finland, with a diameter of 1.5 km and a depth of 150 m. The geological setting corresponds to the Central Finnish Granite Complex dating from the Paleoproterozoic Age. The target rock is of porphyritic granite from a site where many shatter cones have been found. In general terms, shatter cones can be defined as unusual, striated, horse-tailed conical fractures, measuring from millimetres to meters in length aand produced in different types of rocks by the passage of a shock wave (Sagy et al. 2002), although the mechanism by which they are formed is not well understood (Dawson, 2009). The directional striated surfaces of shatter cones are positive/negative features. An extremely interesting (and useful) feature of shatter cones is that the tips point toward the origin of the shockwave. This means that they can be used to reconstruct the location, size and shape of prehistoric impact craters that have subsequently been modified by later processes. Shatter cones are usually formed at pressures between 2 and 6 GPa, although rocks have been found that had been subjected to pressures around 25 GPa (Milton 1977). Further information on impact textures can be found, for example, in Dietz 1947, Amstutz, 1965, Milton, 1977, Roach et al. 1993, Gibson & Spray, 1998, Baratoux & Melosh, 2003 Sagy et al. 2002, 2004, Lugli et al. 2005, Dawson, 2009, Ferrière L. & Osinski, 2010).

The impact rock from the Karikkoselka impact crater (Figure 1) is a red-coloured granitic specimen, measuring 131.10 mm x 82.30 mm x 92.80 mm, with a rough surface marked with incomplete but well-defined shatter cone textural features.

3D LASER SCANNER TECHNIQUES

We began our research projects, by questioning site and artefact documentation methodologies when participating in periodical archaeological campaigns. We first became acquainted with laser scanners in 2003 through the Leica company, which put one at our disposal for a data acquisition test. The modelling subject chosen was the statue of Cibeles in Madrid and the project was duly carried out after obtaining the necessary licences. The experience was briefly analysed in the journal of the Colegio Oficial de Ingenieros Técnicos y Topografía (Farjas & Sardiña, 2003). The images thus obtained were quite surprising

Figure 1. Meteorite-impact related rock which was used in the project. It comes from the impact crater area of Karikkoselka (Finland). Note its surface textures, which are indicative of an impactogenic origin. The detailed analysis by using 3D laser techniques is extremely important to both determine and characterize the morphological features caused by the shock condition.

and, impressed by the device's capacity for data capture, to further analyse its possibilities for data handling we decided to carry out a series of projects. One of the first of these was the 3D modelling of a replica of the statue of a Xian warrior. The model was obtained to an accuracy of within 1 cm. Using our knowledge of photogrammetry we carried out several tests and had to adapt the shapes and colours to the new data acquisition system.

We then applied laser scanner technology to objects belonging to the national heritage and compared the results with those obtained by photogrammetry, publishing the results of the first analyses (Farjas & Bravo 2007). A partial summary of these projects was presented in Farjas, M. & García Lázaro F.J. 2008.

Later on we undertook the modelling of the Abrigo de Buendía prehistoric site, and then used the same methodology for the laser scanning survey of Los Zarpazos in the Atapuerca archaeological site (Burgos, Spain). Other projects was used to test the capacity of laser scanning to reproduce natural spaces in three dimensions, with the later addition of data from a short range scanner (Farjas et al. 2009).

In the laser scanner projects the general procedure consisted of:

- Pre-editing of the data capture. If the scan was too dense, re-sampling or segmentation was carried out.
- Registration of each point cloud in the chosen project reference system, generally local or global.

- Elimination of unwanted or erroneous points.
- Three-dimensional modelling.

The final results consisted of 3D models of the site, orthophotos, cartographies and videos. After the processing the data acquired by laser scanner, orthophotos can be obtained from the 3D point clouds. As these already contain the radiometric information, the process is a direct orthogonal projection of the point clouds onto a reference system defined by the user. The system does not necessarily have to be parallel to the scans, but can be whichever is best adapted to the zone geometry. The orthophoto can then be exported to a CAD document for cartographic editing.

We experimented with different methodologies for the registration of artefacts (Farjas 2007). In Atapuerca a short-range Konica Minolta laser scanner was used to include scratch marks made by tigers in the model. The work carried out so far is focussed on a methodological comparison of new systems of data acquisition with traditional photogrammetry and the latest versions with correlation options presently available. In general, we would recommend that before undertaking a project with laser scanner systems the objectives should be evaluated so as to form a basis for a decision on the best data acquisition method that meets the needs of the project.

Carrying out a study of the techniques involved in 3D laser scanner modelling of archaeological sites and artefacts, in previous works we examine experiences in the use of laser scanner technologies. Laser technology has been shown to have a wide range of uses in the field of three-dimensional object registration. It is used in architecture and engineering to survey buildings, domes and bridges as well as to inspect components during factory production processes. At the present we apply these new technologies in new science fields.

In geology, a subject involved in the present study, the graphic registration of rocks has traditionally been accomplished by techniques such as photography or hand-sketches. Our aim is to analyse the possibilities of laser scanning for the storage, diffusion and metric analysis of rocks of morphometric interest.

For the study, we used a 3-D *NextEngine* scanner (Figure 2), which captures objects in full colour by low-cost precision multi-laser systems applying 3D optical triangulation techniques.

The scanner has its own data-processing software (*ScanStudio HD),* which can explore, align, merge and clean the scanned images. It can export data to different types of files (STL, OBJ, VRML, U3D among others), produce results in the form of 3D models compatible with design programs such as *SolidWorks, 3DS Max, ZBrush, Rhino, Modo, Matemática,* and print them in ZCrop, Stratasys and other 3D printers.

The scanner dimensions are: 224mm long, 91mm wide and 277mm high. It contains laser optics, cameras and processing equipment and uses four classes of 1M 10 mW matrices (650nm wavelength), solid state lasers and twin CMOS RGB 3.0 megapíxel image sensors to capture geometry and colour textures. Studio lighting includes white light illuminators with triphosphorus for the whole colour range. Acquisition speed is 50,000 points per second. There are no pre-

Figure 2. NextEngine scanner

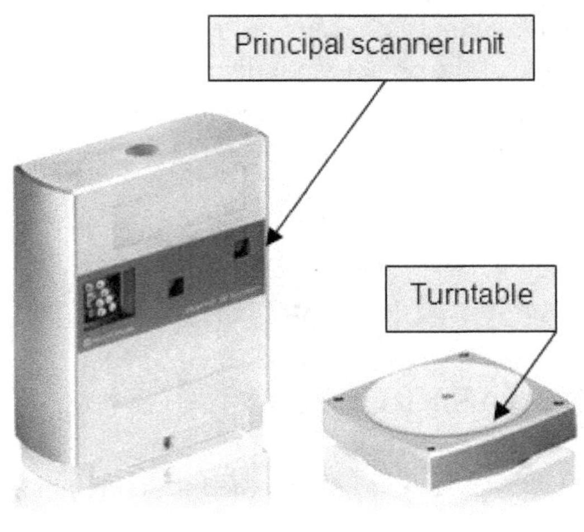

established object limits and there are two types of scan: *wide* and *macro*, according to object size and resolution of the output files. In *macro* mode, the visible area is 130 x 97mm for object-scanner distances around 178mm. In wide mode, visible area is 343×256mm for a range of around 406mm. Resolution, colour texture and precision are all different in the two modes. *Macro* uses a resolution of 200 dpi and 400 dpi point density on the surface and achieves a precision of ± 0.127 mm. In wide mode, resolution is up to 75 dpi, density 150 dpi and precision ± 0.381mm.

The auxiliary equipment includes a turntable controlled by *NextEngine* (*ScanStudio HD*) software, which can stand weights of up to 9 kilos, which is both stable and useful for 360° scanning. Rotation intervals can be set so that the object turns through a certain angle for each scan in the sector. Various 3D sections of the object can be acquired and later the sections can be merged into a single model using the same program.

The methodological possibilities of the device were explored in a pre-study phase, using different objects and textures, in order to define the final scanning procedure and the different methods of processing the data acquired (Figure 3).

The results of the pre-study phase were used to calculate the scanning times required for modelling rocks with impactogenic textures, also the best configurations for different surface features and the methods that could be used to merge the different unmarked scans and without altering the surface of the object (an indispensable condition of the project was that it should be *non-contact*). The trials were thus used as a basis for defining optimal project methodology.

3D MODELLING OF IMPACTOGENIC ROCK FROM THE KARIKKOSELKÄ LAKE

The work was divided into the following phases:

- Data acquisition.
- Data treatment and processing.
- Visualisation of results.

Before data acquisition could begin, the following parameters had to be defined:

- Scanning system.

As mentioned above, the *NextEngine* scanner is equipped with two complete scanning systems, with two cameras and two laser sets with the corresponding optical systems to obtain high-precision results at all possible distances. There were two data acquisition options:

- *Macro*: used for scanning small objects (e.g. mobile phones) with good resolution. In this scanning mode, the ideal scanner-object distance is 6.5" (16.5cm) with a maximum precision of ±0.381mm.
- *Wide*: used for larger objects (about the size of a shoe box). Recommended scanner-object distance is 17" (43cm) and maximum precision ±0.381mm.

It should be mentioned here that for dark, glossy or transparent objects it may be necessary to apply a thin coating of *PowderPen* (talcum powder) to reduce reflectance, as we discovered with the first pyramid-shaped object in the trials. For objects that have no natural marks on their surface to facilitate alignment, these can either be given small artificial marks or they can be scanned together with a second object with clearly defined features ((e.g. a ruler) to help in merging the different scans.

- Positioning of the object and choosing location and number of scans.

Before scanning, it is advisable to make a detailed visual study of the object in order to choose the right position on the turntable. The object's

Figure 3. Pre-study trials carried out with the chosen laser scanner equipment

Photographs	Scans

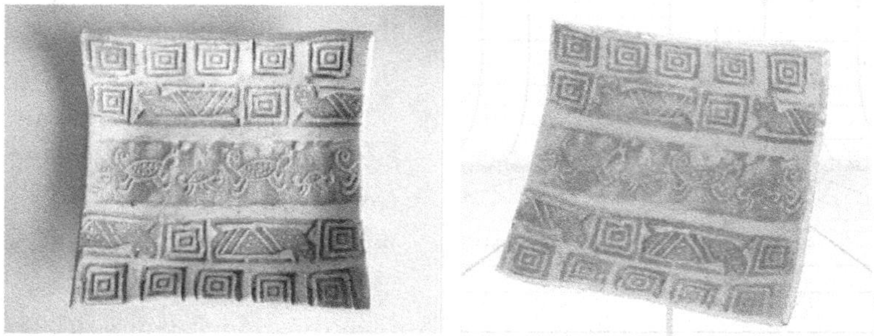

First trial: *ceramic object*
Objective: *to check the results obtained for modelling surface textures and reproducing fine graphic details.*

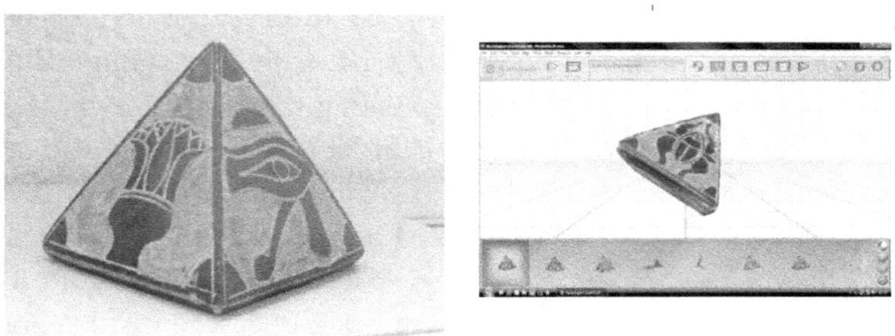

Second trial: *object in the shape of a pyramid*
Objecive: *to analyse dark colours and glossy surfaces. A thin layer of PowderPen was applied to reduce reflectance.*

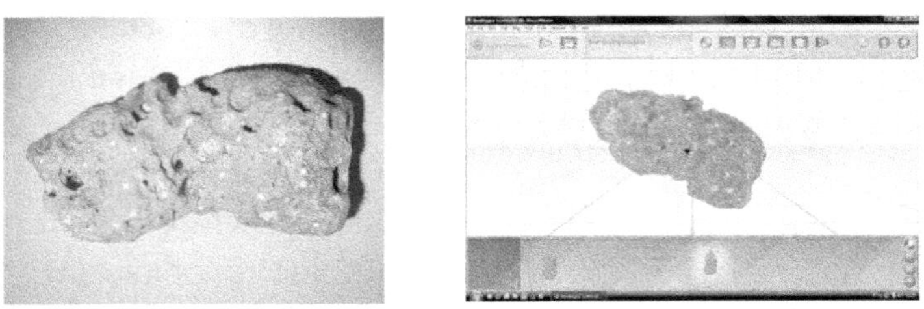

Third trial: *stone with multiple round surface cavities*
Objective: *to analyse surfaces with hidden areas.*

morphology needs to be assessed so that it can be placed in whatever position will reduce the number of required scans to the minimum. The aim should be to scan the largest possible surface area with the minimum possible number of scans. In this way we not only improve precision but also reduce the volume of data acquired and thus make data handling easier.

After data acquisition comes the data handling phase. Most laser scanners are equipped with their own specific software applications to deal with the large number of points acquired in each scan. It should be remembered that in this stage a traditional CAD system would not be able to deal with this information without the help of the laser scanner software. The general procedure consists of the following phases:

- Design or project of the capture positions.
- Data acquisition and scanning of object.
- Alignment of scans from different positions of the rock on the turntable (the set of images by sectors for each position of the rock is known as a *family*).
- Elimination of noise and cleaning of information.
- Alignment of the models obtained from each position of the rock on the turntable (*scan families*).
- Simplification of the model.
- Merging of the different scan families.
- Manual cleaning and refinement of model (optional).
- Preparing the model for treatment by CAD programs.
- Exporting the results.

Most of these processes are done interactively. Visualisation of the results can be performed before or after data handling. The quality of the results will depend to a large extent on the decisions taken on correctly ordering the procedures. The following results can be obtained from the 3D scan files:

- Point clouds.
- Triangular mesh surfaces.
- A solid object without photorealistic surface textures.
- A solid object with photorealistic surface textures.
- Parallel aligned sections of the object.

In the phase previous to the study of the Karikkoselkä rock, specific features and handling restrictions were analysed, the total number of scans required for complete 3D imaging were defined and the ideal scanning distance was calculated.

As our aim was to achieve maximum scanning precision and the sample was small in size, the *macro* scanning option was chosen, since the visible area of this mode (130 x 97mm) covered the rock dimensions and offered a precision of 0.127mm.

We also bore in mind the fact that the rock surface contained a series of small dark minerals (biotite, amphiboles) that could be used as common points for different scans and would also be useful in the alignment phase. To guarantee the identification of common points in different scans, a ruler graduated in centimetres was placed on the rock over the scanned space.

After a study of its morphology, it was decided to place the rock in three different positions (providing three scan *families*) to capture the entire surface. A total of twelve scans were obtained: six in Position 1 (Family A – first scan), three in Position 2 (Family B – second scan) and three in Position 3 (Family C – third scan).

The laser scanner's integrated camera was used to capture images and select the sweep zone in each position. The 3D laser scanner is connected to a laptop computer when operating which provides control of all actions carried out. It also captures, saves and processes the information sent to it from the scanner (Figure 4).

The aim of the first scan was to obtain a digital register of the maximum surface of the rock in order to optimise the total number of subsequent scans. The 360° of a complete revolution of the turntable were divided into six 60° segments, each of which registered the surface exposed in the segment.

When the rock has been placed on the turntable and the program has been started, the configuration of the scanned parameters is indicated on the computer screen and data capture can begin (Figure 5).

As mentioned above, the *macro* maximum precision option was chosen combined with the slowest standard speed "SD" (95s) to obtain the

Figure 4. Data capture process. The 3D laser scanner is connected to a laptop computer when operating which provides control of all actions carried out. It also captures saves and processes the information sent to it from the scanner.

Figure 5. View of the parameter configuration screen of first scan

highest degree of detail from the rock surface and the highest quality in the final results.

The sliding object-colour control on the device was set at 50% and finish was set to *shiny* as the rock surface was considered to have a glossy texture. The smallest triangle size was selected to improve the precision of the results. Smoothness was set at the lowest setting of 1 (from 1 to 5) as the surface was somewhat rough due to shatter cones and other surface markings.

Finally, the autoalign option was deactivated and the manual option was chosen for aligning scans and scan families, in order to have a more complete control of the process and improve precision. After setting all the variable elements in the menu and centring the rock in the viewer, the *scan* command was given.

In the second and third scanning positions, the aim was to cover the shaded zones, which could not be previously registered. These consisted of the lower zone on which the rock had rested on the turntable and the upper zone on which the ruler had been placed.

Scanning parameter configuration was the same in both cases as that used in the first position, the only difference being the zone registered in each position. In the second position the bottom zone was registered and the upper zone was registered in the third position.

DATA TREATMENT METHODOLOGY

Aligning Scans from Each Rock Position (Align Family)

After finishing the scanning program, the scans from each family must be aligned. To do this, two scans from each family are selected on each of which at least three identical positions have been identified (families A, B and C) (Figure 6). After this procedure, the complete set of scans can be automatically aligned.

To align the six scans belonging to Family A, work began with scans A1 and A6 and the Align command was given. The degree of precision achieved in the alignment was 0.381mm. Families B and C were treated in the same way, with a precision in B of 0.381mm and the highest was obtained in Family C at 0.0254mm.

Eliminating Noise (Trim)

When all the families have been correctly aligned, the next step is *trimming*, which consists of eliminating noise and any irrelevant elements that may have been registered during the rock scans. This is a laborious process that requires a certain ability for spatial visualisation and also mastery of the

Figure 6. Results obtained in the alignment of scans from each position. To align the six scans belonging to Family A, work began with scans A1 and A6 and the Align command was given. The degree of precision achieved in the alignment was 0.381mm. Families B and C were treated in the same way, with a precision in B of 0.381mm and the highest was obtained in Family C at 0.0254mm.

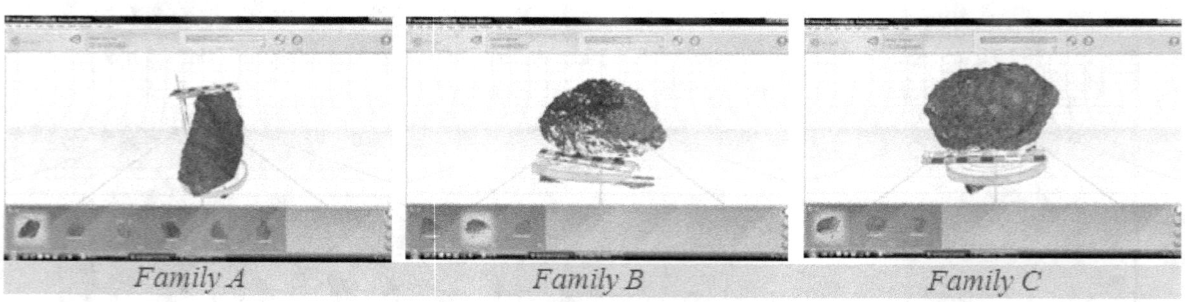

Figure 7. Removing noise from the model in Position 1(Family A)

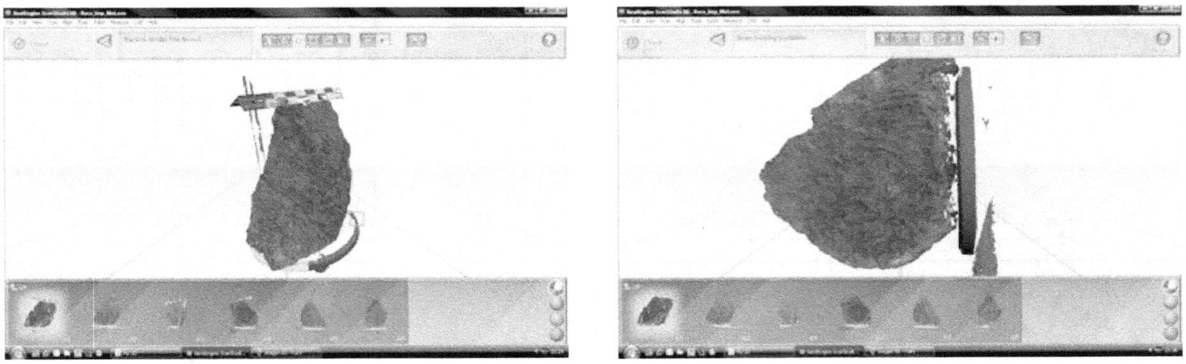

rotate, drag and zoom commands activated by the mouse buttons.

The data is cleaned up by eliminating the auxiliary elements used in the rock scans, including the ruler used for alignment, the turntable and the vertical support bar. Some noise also had to be removed from around the rock (Figure 7).

All the cutting tools were used, the polygonal most of all, being the best adapted to the shapes of the elements that had to be eliminated. Even though it was also the slowest option, it offered the best guarantee against the involuntary removal of valuable elements. The final results in each of the three positions can be seen in Figure 8. All the elements registered in the scanning phase that did not belong to the rock have been eliminated.

Align Families Option

At this point we now have three 3-dimensional images, corresponding to the three positions of the rock (scan families), which have been aligned and cleaned of extraneous matter but are independent of each other. In other words, we have the rock divided into three separate blocks. In this stage of the treatment a single model is obtained by means of the common points in the blocks. The alignment process is exactly the same as that described above for the alignment of scan families in which at least three common points must be identified between two families.

After studying the results of the noise elimination process, the next procedure was to align the scans. The A and C families were selected for this stage, since they had more clearly identifiable shared common points than the other possible

Figure 8. Results obtained in the trimming process

Family A *Family B* *Family C*

Figure 9. Results of scan alignment

Alignment of Families A and C *Alignment of Families A and B*

combinations (Figure 9). Since there were now no auxiliary elements left in the model to help with alignment (the ruler had been removed in the previous stage) this work had to be done with reference to the surface features of the sample and

Figure 10. Information on the number of points and triangles in the model

thus required extreme care so that the process was slowed down considerably.

The precision achieved in aligning Families A and C and Families A and B was 0.015 in (0.381mm).

Fuse Option

After aligning all the scans, we now have a model composed of different overlapping meshes. With the merging tool, the aim is to simplify the data into a single mesh of the scanned object without overlaps and to eliminate any gaps in the model.

The merging process was long and complex and several unsuccessful attempts were made. A problem arose when all the merging options had been configured and the *fuse* command was given. At this point the program suddenly closed down without warning, so it was decided to reduce the number of points and triangles. The simplification process is described below. Information on the number of points and triangles was obtained from the main menu (*Model Information* option) (Figure 10).

According to the *NextEngine* instructions, between one and one and a half million points were needed to obtain results, so the *RE-Generate Scan(s)* command was given (*Fuse* menu). The number of points and triangles was reduced to about half the original number, although merging could have been carried out with higher numbers than those recommended (Figure 11).

The tolerance of the final merging was 0.0050" (0.127mm) and the finished model was then stored (Figure 12).

Manual Refinement and Polishing of the Model (Polish)

As stated above, manual model refinement is an optional process. We decided to leave out this stage so as not to alter the rock's virtual morphology.

Figure 11. Information on number of points and triangles after the number had been reduced

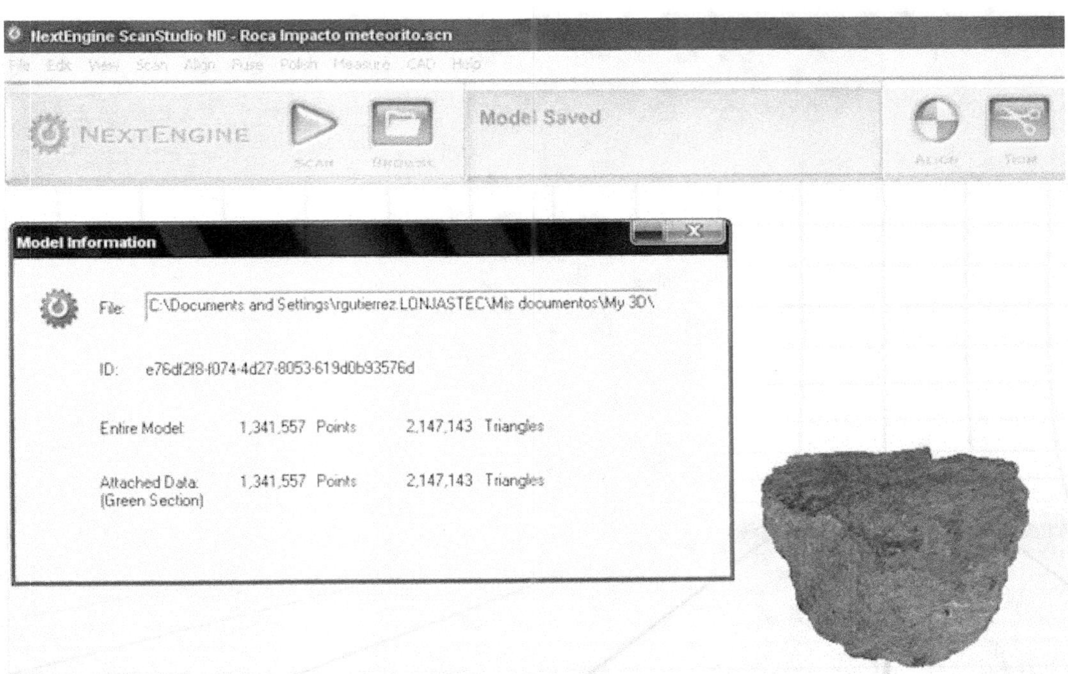

Table 1. Results of the merging of scans

MERGING OF SCANS	Standard deviation O (mm)
FAMILY A = 1+2+3+4+5+6	0.3810
FAMILY B = 1+2+3	0.3810
FAMILY C = 1+2+3	0.0254
A+C	0.3810
A+C+B	0.3810

Saving and Exporting

By means of the *File* menu and the *Save* and *Save as* commands, all the changes made to the model and the information can be saved in a file in any of the directories. The file containing the model was exported in different formats (PLY, OBJ, STL, VRML, XYZ, U3D, IGES and STEP) so that it could later be studied with different programs and converted, via bridge programs, to AutoCad or other 3D systems for geometrical analysis.

After completing all the data capture and treatment processes, the final results can be visualised in four different modes: the realistic model (Figure 13), the colourless solid model (Figure 14), model with triangles (Figure 15) and the points model (Figure 16). This section will explain some of the possible options and analyses that can be performed.

One of the possibilities is to produce a cartographic document, an example of which is shown in Figure 17.

MODEL PRECISION ANALYSIS

After obtaining the three-dimensional model of the rock with impactogenic textures, a study was carried out to determine the degree of precision achieved, which involved an analysis of each of the factors that influenced the model generation process. The first variable to be considered was the uncertainty of the data acquisition process, due

Figure 12. Results of merging

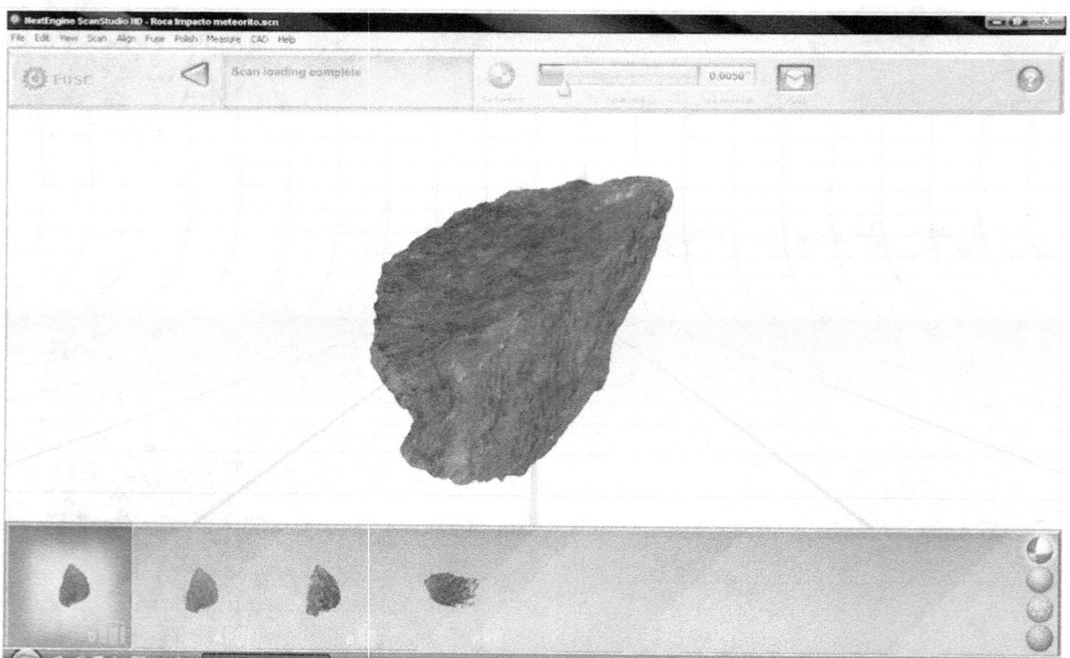

Figure 13. Realistic model

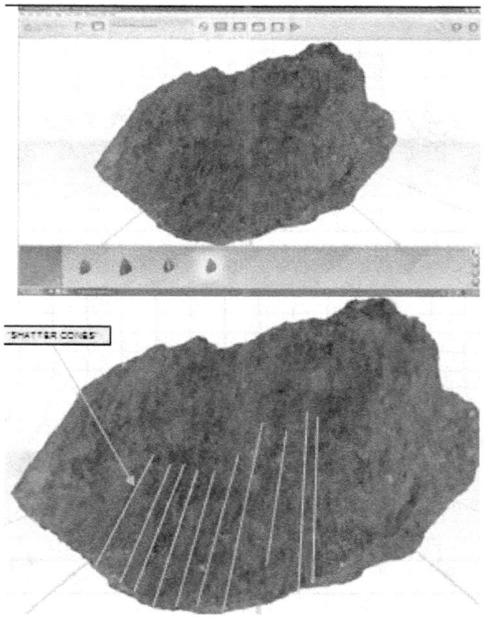

Figure 14. Solid model

Figure 15. Mesh model

to the technical characteristics of the measuring device. According to the manufacturer's technical specifications, the precision of the absolute position of each of the measured points is 0.127mm at a distance of 16.5cm. This uncertainty is known as *instrument error* and can be expressed as e_i:

e_i = 0.1270 mm

The second factor that influences the final precision is the accuracy with which the scans are merged, which depends on the standard deviation of the transformation calculations, for which the *ScanStudio HD* program was used. The results are shown in Table 1.

This parameter will be given in the complete model by the quadratic component of the errors made in joining pairs and is expressed as:

$$e_u = \sqrt{0,38102 + 0,38102 + 0,02542 + 0,38102 + 0,38102} = 0,7620\ mm \quad (1)$$

After obtaining all of the errors that affect the model, the total uncertainty can be expressed as the quadratic component of the values that form these variables:

$$e_T = \sqrt{e_i^2 + e_u^2} = 0,7725\ mm \quad (2)$$

To verify this result, measurements were taken of three characteristic distances on the rock itself by means of a caliper with a precision of 0.01mm: these distances were: maximum length, width and height, as shown in Figure 18.

The results obtained with the caliper were as follows:

Figure 16. Points model

Figure 17. Solid model

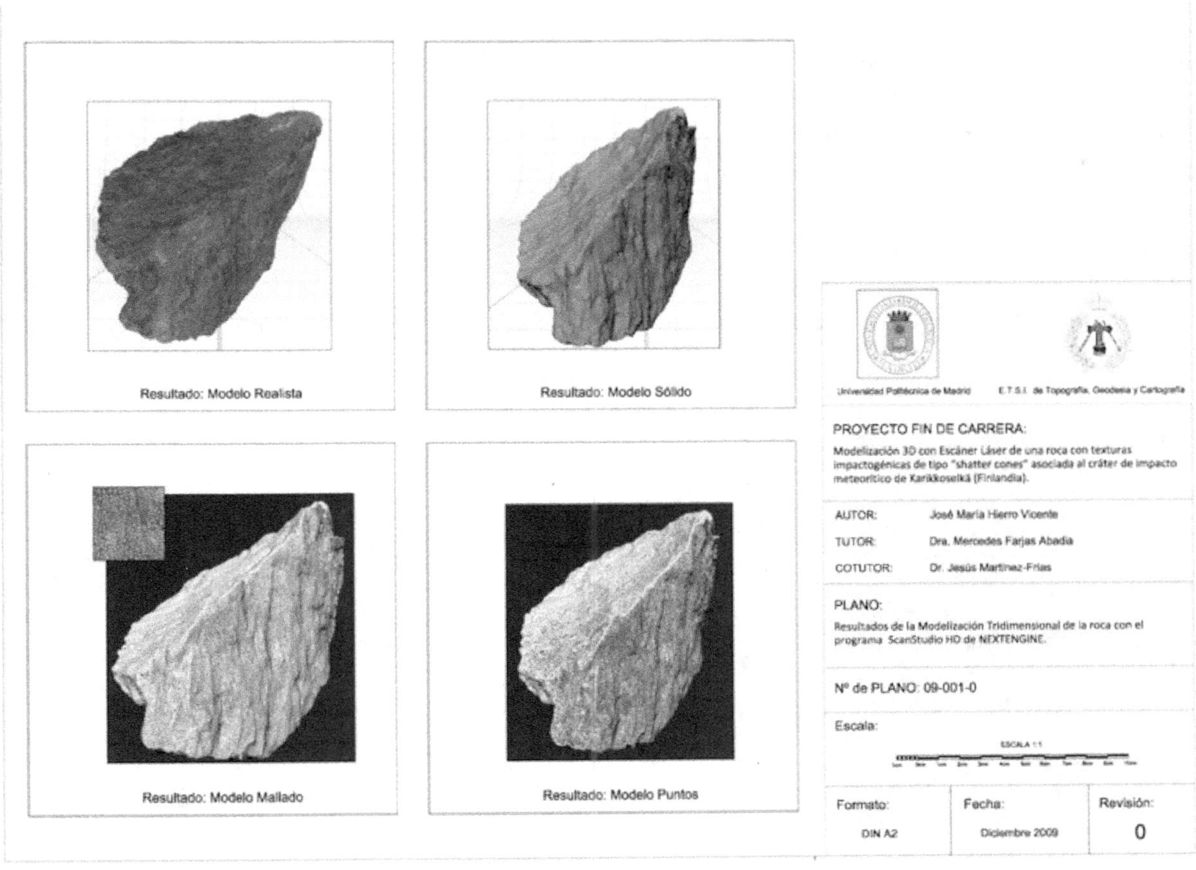

Table 2. L1 error calculation table

l1(mm) MiniMagics	L1(mm) Caliper	ABSOLUTE ERROR L1-l1(MM)	RELATIVE ERROR	MAXIMUM ERROR IN LENGTH	ARITH. ERROR	STANDARD DEVIATION
131.823	131.100	0.723	0.55%	0.62%	0.42%	0.15%
131.301	131.100	0.201	0.15%			
131.811	131.100	0.781	0.60%			
131.853	131.100	0.753	0.57%			
131.043	131.100	0.057	0.04%			
131.666	131.100	0.566	0.43%			
131.213	131.100	0.113	0.09%			
131.915	131.100	0.815	**0.62%**			
131.894	131.100	0.794	0.61%			
131.825	131.100	0.725	0.55%			

- Maximum length (L_1): 131.10mm.
- Maximum width (L_2): 82.30 mm
- Maximum height (L_3): 92.80 mm

Since the traditional method of measuring by caliper is more common than generating point clouds by laser scanners, the measurements obtained in this comparison were taken as the "real" values and the differences found as the "absolute errors". The ratio between the absolute error and the real value determines the relative error. The arithmetic mean of the relative errors will be the arithmetic error, and if we calculate the square root of the square of the sum of the absolute errors divided by N (number of readings) the standard deviation is obtained.

The same distances were analysed ten times each with *MiniMagics* software. To obtain the measurements of the three series, the 3D model was rotated several times and the zoom was used to select the required points (Figure 19).

In order to determine the precision of these dimensional measurements, the uncertainty calculations were carried out as shown in Tables 2, 3 and 4.

If we consider the relative values obtained and select the least favourable, i.e. 0.63% and apply it to the greatest rock length magnitude, i.e.

Figure 18. Solid model: Measuring maximum length, width and height with a caliper

Measuring maximum length with a caliper. *Measuring maximum width with a caliper.* *Measuring maximum height with a caliper.*

Figure 19. Measured by the MiniMagics program

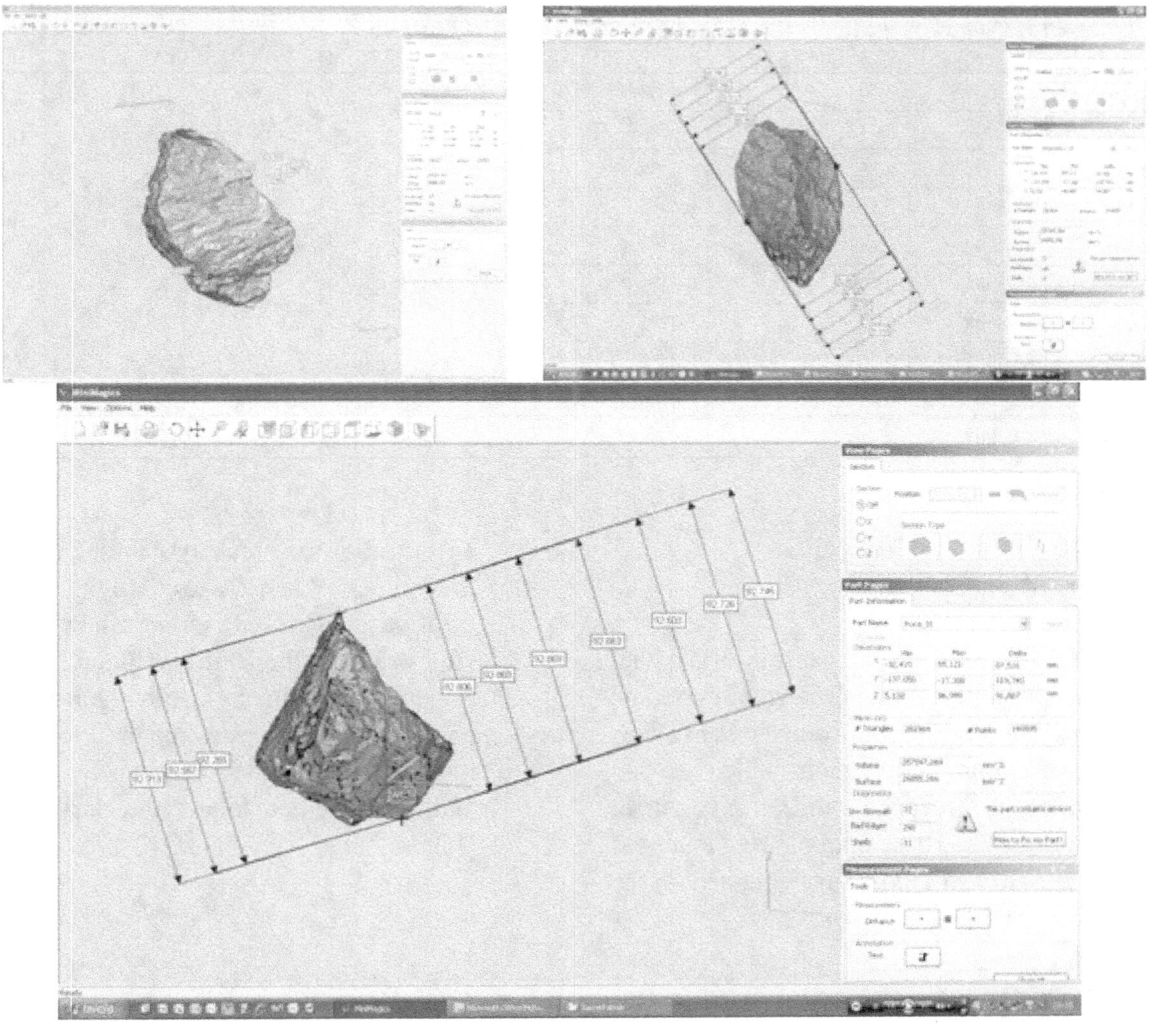

L_1=131.100mm, we can estimate that the maximum error made when obtaining a distance between two points on the model is 0.826mm. However, if we consider maximum arithmetic error, i.e. the mean value of the relative errors, and again apply it to the length of greatest magnitude, a mean error of 0.65mm is obtained. This could be considered as an estimator of the possible mean error when determining distances with the program.

If we take the arithmetic mean of the ten lengths of each of the series and compare it with the means obtained from the caliper, we will obtain a value for the precision with which the work has been carried out (Tables 5, 6 and 7).

From this data we can conclude that the precision obtained in scanning the rock is 0.05%, i.e. 0.651mm. This value is somewhat lower than the precision that can be attained with *NextEngine* for a scanning distance of 0.77mm.

Table 3. L2 error calculation table

l2(mm)	L2(mm) Caliper	ABSOLUTE ERROR l2-L2 (mm)	RELATIVE ERROR	MAXIMUM ERROR IN LENGTH	ARITH. ERROR	STANDARD DEVIATION
82.706	82.300	0.406	0.49%	0.63%	0.32%	0.12%
82.258	82.300	0.042	0.05%			
82.512	82.300	0.212	0.26%			
82.819	82.300	0,519	**0.63%**			
82.553	82.300	0.253	0.31%			
82.309	82.300	0.009	0.01%			
82.080	82.300	0.220	0.27%			
82.237	82.300	0.063	0.08%			
81.917	82.300	0.383	0.47%			
81.814	82.300	0.486	0.59%			

DIMENSIONAL ANALYSIS OF THE MODEL

The *ScanStudio HD* version that we worked with includes a *Demo Option* with a series of optional *CAD tools*. Our version could only work in *Demo* with the *Orient* and *Spline* tools, which showed the results on the screen but could not export them to other programs. These options were analysed and the *CAD Tools Demo* results obtained for the model were as follows:

- *Orient*: allows the rock to be turned through the three X, Y and Z axes until the desired position is achieved, acting either on the cube or on the model itself (Figure 20).
- *Spline*: allows the rock to be split into as many parallel planes as required to obtain the intersection lines between the planes and the model, creating a new data family (Figure 21).

Table 4. L3 error calculation table

l3(mm)	L2(mm) Caliper	ABSOLUTE ERROR l2-L2 (mm)	RELATIVE ERROR	MAXIMUM ERROR IN LENGTH	ARITH. ERROR	STANDARD DEVIATION
92.806	92.800	0.006	0.01%	0.58%	0.16%	0.07%
92.869	92.800	0.069	0.07%			
92.869	92.800	0.069	0,07%			
92.863	92.800	0.063	0.07%			
92,503	92.800	0.297	0.32%			
92.726	92.800	0.074	0.08%			
92.745	92.800	0.055	0.06%			
92.265	92.800	0.535	**0.58%**			
92.567	92.800	0.233	0.25%			
92.913	92.800	0.113	0.12%			

Table 5. Calculation of the mean values of distances obtained with MiniMagics

L1(mm)	Mean value L1(mm)	L2(mm)	Mean value L2 (mm)	L3 (mm)	Mean value L3 (mm)
131.923	131.751	82.706	82.341	92.806	92.713
131.301		82.258		92.869	
131.811		82,512		92,869	
131.853		83.119		92.863	
132.043		82.953		92.503	
131,666		82.309		92,726	
131.213		82,080		92.745	
131.915		82.237		92.265	
131.894		81.917		92.567	
131.825		81.314		92.913	

Table 6. Calculation of the mean values of distances obtained with MiniMagics

MEASURE	REAL (mm)	MEASURE (mm)	ABSOLUTE ERROR (mm)	RELATIVE ERROR
L1	131.10	131.75	0.6514	0.50%
L2	92.30	82.34	0.0405	0.05%
L3	92.80	92.66	0.136	0.15%

In order to carry out a deeper metric analysis of the model, we started by trying to use the AutoCad program. To pass from one .obj file to another with AutoCad.dwg format, we used *OBJ Import for AutoCad* (SYCODE), a company that develops software for computer-assisted design (CAD) systems. These solutions come in the form of independent applications or plug-ins that work within the principal CAD systems: *AutoCAD, Inventor, 3D Studio Max, Maya, Pro/ENGINEER, Kubotek, SolidWorks, Solid Edge, SpaceClaim, Alibre Design, Rhinoceros, IronCAD, INOVATE, IntelliCAD, Bricscad, Acrobat* and *SketchUp*, etc.

The results obtained were in the form of a file of 72MB, a size which excessively complicated the working of AutoCad 2008, as this software has not enough capacity to manage files of these characteristics. AutoCad showed the rock on the screen in the form of a mesh (Figure 22) and when we tried to transform this into a solid view of the rock the program was inclined either to shut down without warning or, after a long wait, give up the task as impossible.

We therefore abandoned this option and continued the search for a program better adapted to the characteristics of the file that would allow us to carry out the dimensional analysis that was the objective of our study. After a lengthy search, it was decided to work with the free 3D *MiniMagics* visor from the *Materialise Group,* which can execute 3D files with .STL extension. This com-

Table 7. Results of estimators

Arithmetic error	0.23%
Standar deviation	0.17%
Maximum error in length	0.50%

Figure 20. Orienting

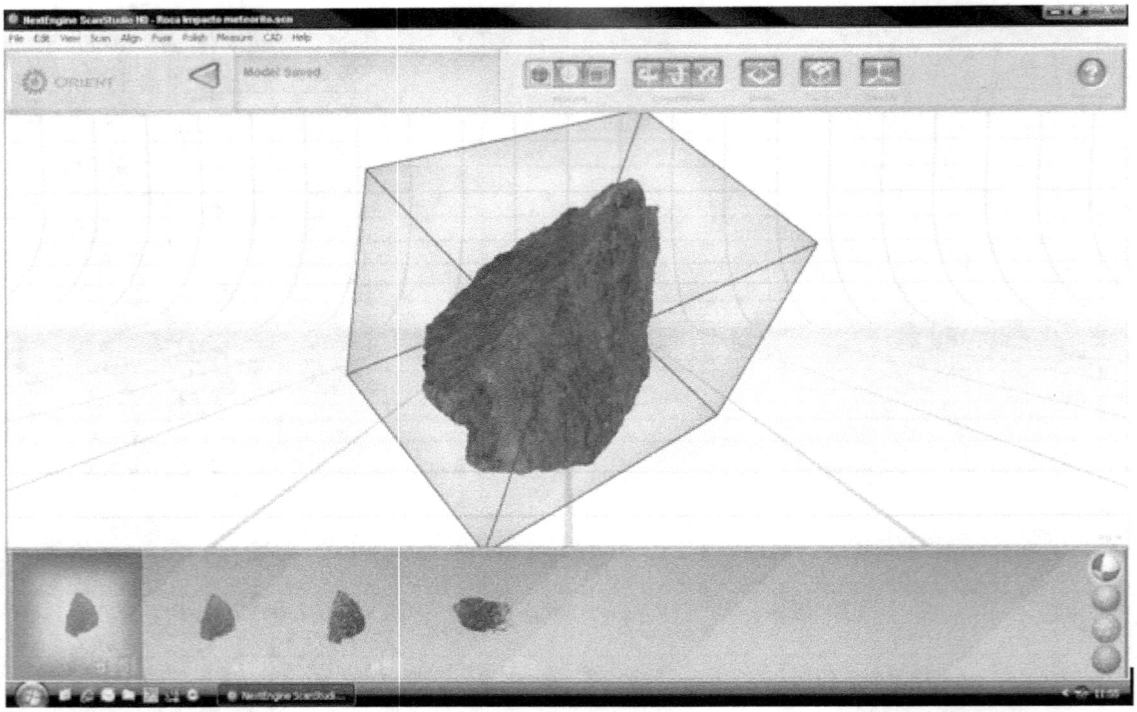

pany, from Leuven in Belgium, is well-known for the development of industrial and medical prototypes. Besides possessing the largest capacity for rapid prototyping equipment in Europe, it has a worldwide reputation for providing innovative software solutions. It is a leader in 3D digital printing and software and plays a leading role in dental image-processing and surgery simulation.

Figure 21. Spline options

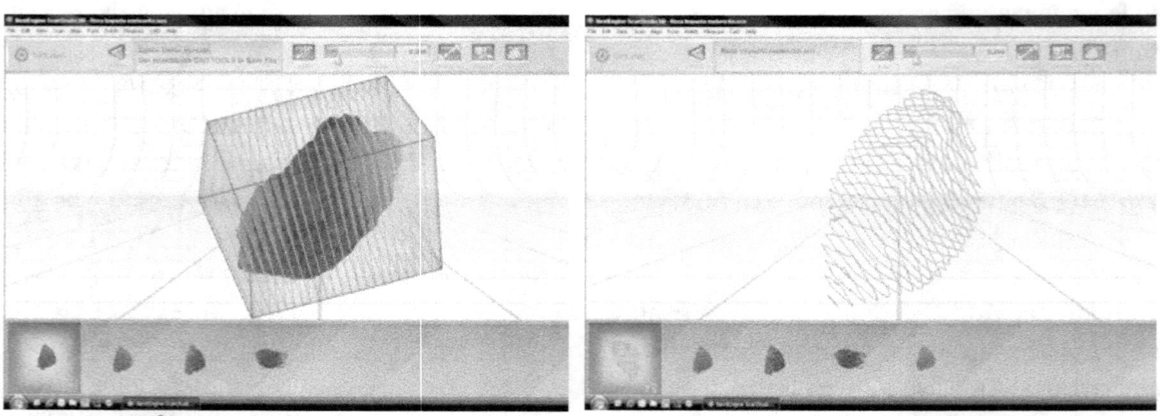

Selecting Spline *Spline results*

Figure 22. 3D view of the rock in AutoCad

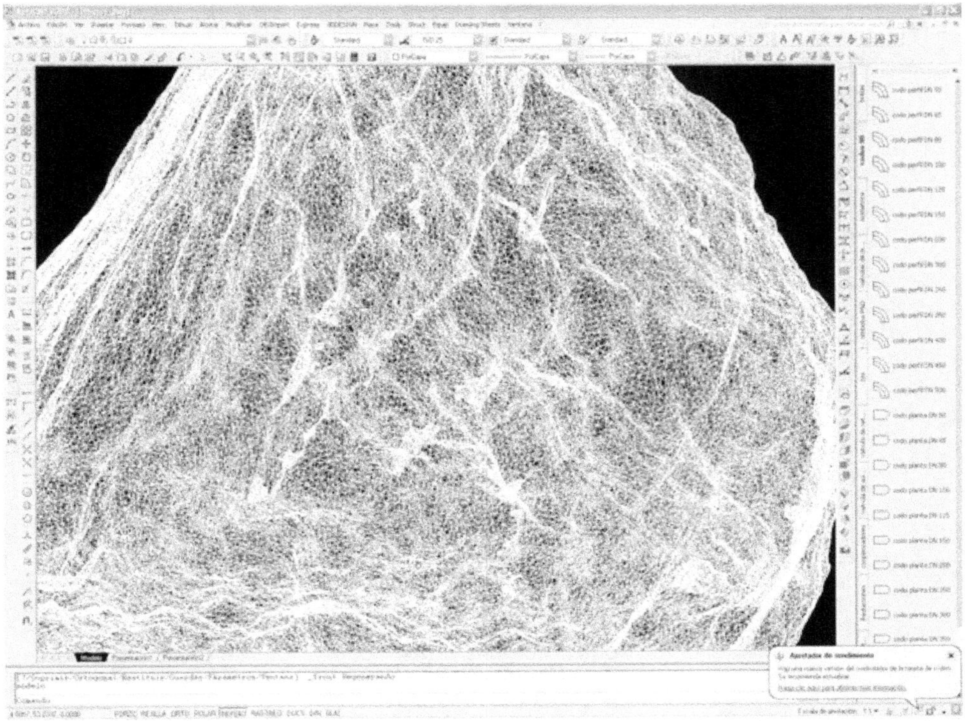

We now had two software systems by which to obtain the results: the treatment of the scanned 3D model by the *ScanStudio HD* model, or the model imported from the *MiniMagics* program.

A dimensional analysis was performed with MiniMagics. In this, selected points could be chosen from the 3D model and the basic geometrical analysis could be carried out on screen. The surface area and the volume of the rock were thus calculated in this way (Figure 23).

RESULTS AND CONCLUSION

This paper describes for the first time a novel application of 3D laser scanner techniques for the morphological study of meteorite impact rocks. The results obtained from the preliminary stage of the research, which consisted of testing the analytical procedures on a wide variety of objects and surface textures, including rocks, were crucial for establishing the most favourable modelling conditions under which the study of the real impact

Figure 23. Volume and surface of the rock in MiniMagics

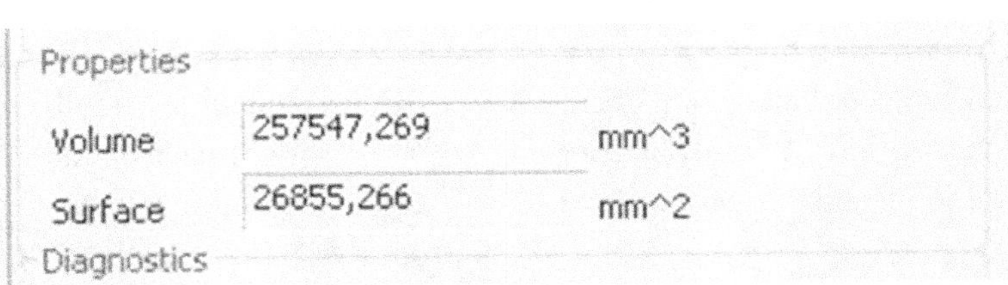

Figure 24. Measurements of the convergence angle in MiniMagics program

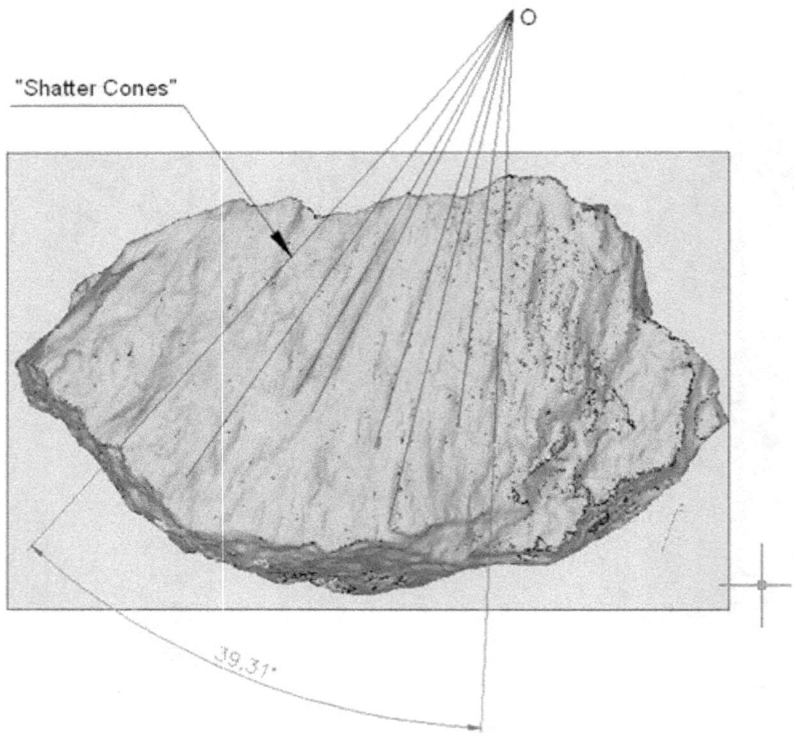

material from the Karikkoselka meteorite impact crater was carried out. The three-dimensional modelling of the rock was successfully achieved. The finished model can be visualized in four different ways: (1) a realistic model; (2) a colourless model; (3) a network model using triangles, and (4) a network model using dots. This modelling, together with the possibility of orienting and splining the virtual image obtained, facilitated the dimensional analysis of the rock (volume: 257547.269 mm^3; surface: 26855.266 mm^2, with a scanning precision of 0.50% = 0.651 mm), as well as determining other characteristics on its surface, including investigation of the shatter cones. For this, in addition to assembling a detailed graphic representation of the superficial roughness of the impact rock (maximum depth of shatter cone striae: 2.47 mm and maximum width: 14.03mm) (Figure 25), it was also possible to make a geometrical estimation of the convergence angle (Figure 24) of the incomplete shatter cone: 39.31°

This work confirms the importance and efficiency of 3D laser scanning techniques as a complementary tool for the scientific study of this type of rocks impacted by meteorites, and opens a new line of research in the context of meteorites and planetary geology. It can be used in both the field and laboratory, as well as for scientific and museological purposes (e.g. geological heritage, remote accessing, non-presential teaching).

FUTURE RESEARCH DIRECTIONS

The results obtained confirm that the application of 3D laser techniques for the morphological study of meteorite impact rocks yields extraordinarily useful information (which can be quantified and processed), regarding the effects caused by the

Figure 25. Measurements of the maximum depth of shatter cone striae and of the maximum width in the MiniMagics program

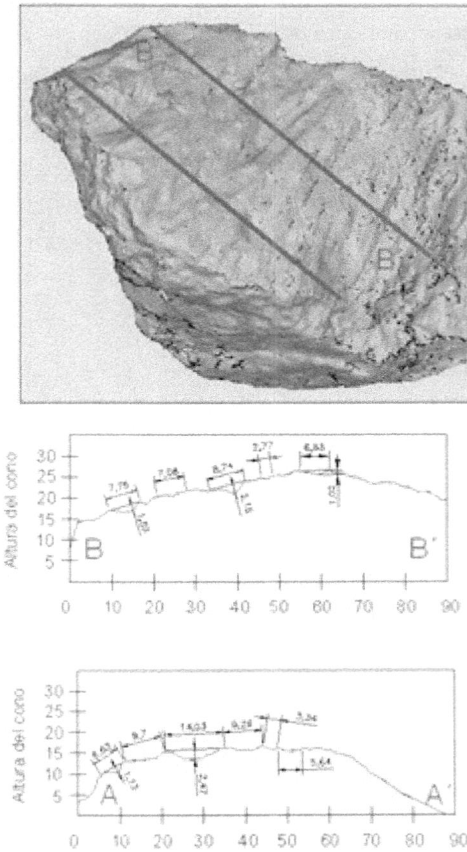

meteoritic impacts on terrestrial target rocks. This research study opens a new field of work in relation with meteorites and planetary geology studies, not only for a more complete characterization of the geological features associated to the impact-related materials, but also because it could complement the classical mineralogical and cosmogeochemical studies. In the future, we attempt to carry out this type of computerized analysis directly on some selected meteorites, showing morphological and textural features (e.g. oriented shape, friction striae, fusion crust, etc) which are reflecting the complex ablation processes which they undergone during their atmospheric entry.

ACKNOWLEDGMENT

The authors are grateful for the institutional support received from TIERRA (Thematic Network of Earth Sciences) of the Spanish *RedIris* (advanced communications network for the scientific community) through the website: http://tierra.rediris.es/merge/3Dmeteorites, which is specifically devoted to the ETSITGC (UPM) - CAB (CSIC/INTA) joint studies on 3D modelling, laser techniques and meteorites, planetary geology and astrobiology.

REFERENCES

Amstutz, G.C. (1965) A morphological comparison of diagenetic cone-in-cone structures and shatter cones. *Annals of the New Cork Academy of Sciences, 123-A2,* 1050-1057.

Baratoux, D., & Melosh, H. J. (2003). The formation of shatter cones by shock wave interference during impacting. *Earth and Planetary Science Letters, 216*(1-2), 43–54. doi:10.1016/S0012-821X(03)00474-6

Dawson, E. (2009). Meteorite impact shatter cones - Adiabatic shear bands? *9th International Conference on Mechanical and Physical Behaviour of Materials under Dynamic Loading*, Royal Mil. Acad., Brussels, Belgium, (vol. 2, 1471-1477).

Dietz, R. S. (1947). Meteorite impact suggested by the orientation of shatter cones at the Kentland, Indiana, Disturbance. *Science, 105*(2715), 42–43. doi:10.1126/science.105.2715.42

Farjas, M. (2007a). *El registro en los objetos arqueológicos: Métrica y Divulgación.* Madrid: Reyferr.

Farjas, M., et al. (2009). *Automatic point-cloud surveys in prehistoric sites documentation and modelling.* Paper presented at the 37th CAA 2009 Conference, Computer applications in Archaeology: Making history interactive, Williamsburg, Virginia, USA. http://www.caa2009.org/ articles/ Farjas_ Contribution163_ a.pdf

Farjas, M., & Bravo, A. (2007b). *Tecnologías de representación 3D en los procesos de documentación del patrimonio pétreo. Ciencia, Tecnología y Sociedad para una Conservación Sostenible del Patrimonio Pétreo* (pp. 47–57). Madrid: Restauradores Sin Fronteras.

Farjas, M., & Sardiña, C. (2003). Novedades Técnicas: Presentación del equipo Cyrax 2500 de Leica Geosystem. *Topografía y Cartografía, 116*, 70–71.

Ferrière, L., & Osinski, G. R. (2010). *Shatter cones and associated shock-induced microdeformations in minerals –New investigations and implications for their formation.* Paper presented at the 41st Lunar and Planetary Science Conference.

Gibson, H. M., & Spray, J. G. (1998). Shock-induced melting and vaporization of shatter cone surfaces: Evidence from the Sudbury impact structure. *Meteoritics & Planetary Science, 33*(2), 329–336. doi:10.1111/j.1945-5100.1998.tb01637.x

Koeberl, C., & Martinez-Ruiz, F. (Eds.). (2003). *Impact markers in the stratigraphic record.* New York, NY: Springer-Verlag, Impact Studies Series.

Lehtinen, M., Pesonen, L. J., Puranen, R., & Deutsch, A. (1996). Karikkoselka-A new impact structure in Finland. *Lunar and Planetary Science, 27*, 739.

Lugli, S., Reimold, W. U., & Koeberl, C. (2005). Silicified cone-in-cone structures from Erfoud (Morocco): A comparison with impact-generated shatter cones. 8th International Meeting on Response of the Earth System to Impact Processes (IMPACT), Mora, Sweden. *Impact Tectonics*, Impact Studies Series, (pp. 81-110).

Milton, D. J. (1977). Shatter cones - An outstanding problem in shock mechanics. In: Impact and explosion cratering: Planetary and terrestrial implications; *Proceedings of the Symposium on Planetary Cratering Mechanics,* Flagstaff, Ariz., (pp. 703-714). New York, NY: Pergamon Press, Inc.

Plado, J., & Pesonen, L. J. (Eds.). (2002). *Impacts in Precambrian shields*. Berlin, Germany: Springer Verlag.

Roach, D. E., Fowler, A. D., & Fyson, W. K. (1993). Fractal fingerprinting of joint and shatter-cone surfaces. *Geology, 21*(8), 759–762. doi:10.1130/0091-7613(1993)021<0759:FFOJAS>2.3.CO;2

Sagy, A., Fineberg, J., & Reches, Z. (2004). Shatter cones: Branched, rapid fractures formed by shock impact. *Journal of Geophysical Research. Solid Earth, 109*(B10), B10209. doi:10.1029/2004JB003016

Sagy, A., Reches, Z., & Fineberg, J. (2002). Dynamic fracture by large extraterrestrial impacts as the origin of shatter cones. *Nature, 418*(6895), 310–313. doi:10.1038/nature00903

ADDITIONAL READING

Adams, L. (1992). *Programación gráfica. Técnicas avanzadas de modelado, acabado y animación 3D*. Madrid, Spain: ANAYA Multimedia SA.

ArcTron Ltd. (2005). *Documentación sobre sistemas 3D*. Retrieved September, 2010 from http://www.arctron.com/.

Atin Sinha. New Frontiers in Manufacturing Education: Rapid Prototyping, 3D Scanning and Reverse Engineering. *American Society for Engineering Education*. Retrieved October 24, 2009, from http://155.225.14.146/ asee-se/ proceedings/ ASEE2009/ papers/ PR2009075SIN.PDF)

Barber, D., Mills, J., & Bryan, P. (2004). Towards A Standard Specification For Terrestrial Laser Scaning . In *Cultural Heritage. Presented at International Society for Photogrammetry and Remote Sensing*. Antalya: Istambul.

Borg, C. E., & Margin, M. (2003). Escáneres 3D de largo alcance: ¿Avanzando hacia una herramienta híbrida o hacia una metodología híbrida? *Datum XXI, 1*(5), 42–46.

Bracci, S., Falletti, F., & Scopigno, M. M. R. (2004). *Explorando David: diagnóstico y estado de la conservación*. Italia: Giunti Press.

Breuckmann, B., et al. (2009). *Surface Scanning-New Perspectives for Archaeological Data Management and Methodology.* Presented at the 37th International CAA Conference. Williamsburg, Virginia, USA.

Callieri, M., Cignoni, P., Dellepiane, M., & Scopigno, R. (2009) *Pushing Time-of-Flight scanners to the limit* Presented at the 10th International Symposium on Virtual Reality, Archaeology and Cultural Heritage VAST (2009), Viena, Austria, page 85-92 - 200

Callieri, M., Cignoni, P., Ganovelli, F., Montani, C., Pingi, P., & Scopigno, R. (2008). *Integrazione di dati tridimensionali e campionamento del colore a fini di documentazione e visualizzazione interattiva*. Firenze, Italy: EDIFIR, Edizioni Firenze.

Demir, N., Bayram, B., Alkış, Z., Helvaci, C., Çetin, I., & Vögtle, T. (2004). *Laser scanning for terrestial photogrammetry, alternativa system or combined with traditional system?* Presented at International Society for Photogrammetry and Remote Sensing, Istambul.

Escarpa, F. J. (2002). Introducción a los sistemas de medición tridimensional con láser. *Mapping-Interactivo, Artículo 198*. Retrieved October 4, 2008, from http://www.mapping interactivo.com/ plantilla- egeo.asp? id_ articulo=198

Farjas, M. (2003). Las Ciencias cartográficas en la arqueología: la búsqueda de la métrica en los modelos de divulgación científica. *DATUM, XXI*, 4–12.

Farjas, M. (2006). *Aulaweb Topografía II*. Madrid, Spain: Escuela Técnica Superior de Ingenieros en Topografía, Geodesia y Cartografía. Universidad Politécnica de Madrid.

Farjas, M., et al. (2009). *3D Scanner Virtual Modeling Versus Arcchaeological Tracings (First Part)*. Presented at the 15th International Conference on Virtual Systems and Multimedia, VSMM 2009. Vienna, Austria.

Farjas, M., & García-Lázaro, F. J. (Coordinadores) (2008). *Modelización tridimensional y sistemas láser escáner 3D aplicados al Patrimonio Histórico.* Madrid, Spain: La Ergástula.

Geosystems, L. (2005). Documentación Técnica y Manuales de Referencia sobre sistemas láser Retrieved November 7, 2007, from from http://www.leica-geosystems. com/ es/ index.htm.

Guidi, G., Remondino, F., Morlando, G., Del Mastio, A., Uccheddu, F., & Pelagotti, A. Performances Evaluation of a Low Cost Active Sensor for Cultural Heritage Documentation. Retrieved January 17, 2009, from http://www.photogrammetry. ethz.ch/ general/ persons/ fabio/ guidi_ etal_ O3D07.pdf

Hong, E., Lee, I., & Lee, J. Measurement of Rock Joint Roughness by 3D Scanner. Retrieved May, 2006, from http://www.korea.ac.kr/ ~semlab/

Lerma García, J. L., Van Genechten, B., Heine, E., & Santana Quintero, M. (2008). *Theory and practice on terrestrial laser scanning. Training Material Based on Practical Applications* Valencia, Spain: Universidad Politécnica de Valencia. ISBN: 978-84-8363-312-0. Ref. UPV: 2008-220

Levoy, M., Rusinkiewicz, S., Ginzton, M., & Ginsberg, J. (2000). *The digital Michelangelo project: 3d scanning of large statues.* Departament of Computer Science and Engineering. University of Washington. Retrieved October 6,2004 from http://graphics. stanford.edu/ papers/ dmich-sig00/ dmich-sig00.html

Lodeiro, J. M. (1995). *Aplicaciones de la topografía en la documentación arquitectónica y monumental.* Madrid, España: Colegio Oficial de Ingenieros Técnicos en Topografía.

López, F. E., Márquez, I., Franco, D., Ramírez, F. Elaboración de Bustos por medio de Digitalización 3D y Prototipos Rápidos" Internet, web de *Universidad Autónoma de Nuevo León Facultad de Ingeniería Mecánica y Eléctrica.* Retrieved September 8, 2008, from http://mecanica-uanl. com/ pdfs/ A3_211.pdf.

Martínez-Frías, J., & Örmo, J. Impactos Meteoríticos. *Centro de Astrobiologia.* Retrieved April 16, 2005, from http://cab.inta.es/ and *Red Temática TIERRA,* retrieved June 1, 2005, from http://tierra.rediris.es

Mensi, S. A. (2005). Documentación Técnica y Manuales de Referencia sobre sistemas Láser Retrieved July, 2006, from http://www.mensi. com/ Website2002/ index2.html.

MiniMagics. Retrieved February 8, 2008, from http://www.materialise.com/ materialise/ view/ en/ 2562719- MiniMagics.html

nub3d S.L. (2005). Documentación Técnica y Manuales de Referencia sobre sistemas Láser Retrieved September, 2007, from (http://www. nub3d.com/ Spanish/).

Scopigno, R.; Cignoni, P.; Montani. C. (2007). High quality digital acquisition and virtual presentation of three-dimensional models. *Archeologia e Calcolatori,* Volume 18, page 163-179 - Dicembre (Supplemento 1)

Shulz, T., & Ingensand, H. (2004). *Terrestrial Laser Scanning: Investigations and Applications for High Precision Scanning.* Presented at the meeting FIG Working Week of Athens. Retrieved October 13, 2004

Spray, J. (2009). *Earth Impact Database.* Planetary and Space Science Centre.

Strait, S. G., Smith, N. E., & Penkrot, T. The Promise of Low Cost 3D Laser Scanners. *Marshall University, The University of Texas at Austin.* Retrieved September 19, 2007, from http:// paleoview3d. marshall.edu/ laser2.php.

Todd, B. E., & William, K. Hemphill, Steven C. Wallace. 3D Scanning Fossils for Archiving and Animation: A New Frontier for Digital Media. *American Society for Engineering Education.* Retrieved June 3, 2009, from http://edgd.asee. org/ conferences/ proceedings/ 63rdMid/ papers/ emma_ 3D_ Scanning_ Fossils_ poster.pdf

University of New Brunswick Fredericton. *New Brunswick, Canada.* Retrieved March 10, 2009, from http://www.unb.ca/ passc/ ImpactDatabase/

Vozikus, G. (2004). *Laser Scanning: New method for recording and documentation in Archaeology.* Presented at the meeting FIG Working Week of Athens. Retrieved October 13, 2004 from http:// www.fig.net/ pub/ athens/ papers/ wsa1/ WSA1_ 4_ Vozikis_ et_ al.pdf

Walton, E. L., Herd, C. D. K., & Duke, M. J. M. (2009). Mineralogy, petrology and cosmogenic radionuclide chemistry of the Buzzard Coulee H4 chondrite." *Dept. of Earth & Atmospheric Sciences, University of Alberta, Edmonton, Canada.* Retrieved October, 2009, from ftp://ftp.lpi.usra. edu/ pub/ outgoing/ lpsc2009/ full308.pdf

KEY TERMS AND DEFINITIONS

3D Laser Scanning: Data acquisition system of a given surface or object in a systematic, automated manner, within a coordinate system.

Impact Rocks: Stones associated with meteorite impact craters.

Meteorite: Stony, stony-iron or metallic natural object (normally from the asteroids, the Moon or Mars) that are the remains of a meteoroid that has reached the earth's surface. Recently meteorites have been also found on Mars.

Morphological Modelling: 3D representation of a surface.

Shatter Cones: Small, cone-shaped fractures formed by the shock of a meteorite impact.

Chapter 27
3D Camera Tracking for Mixed Reality using Multi-Sensors Technology

Fakhreddine Ababsa
University of Evry Val d'Essonne, France

Iman Maissa Zendjebil
University of Evry Val d'Essonne, France

Jean-Yves Didier
University of Evry Val d'Essonne, France

ABSTRACT

The concept of Mixed Reality (MR) aims at completing our perception of the real world, by adding fictitious elements that are not perceptible naturally such as: computer generated images, virtual objects, texts, symbols, graphics, sounds, smells, et cetera. One of the major challenges for efficient Mixed Reality system is to ensure the spatiotemporal coherence of the augmented scene between the virtual and the real objects. The quality of the Real/Virtual registration depends mainly on the accuracy of the 3D camera pose estimation. The goal of this chapter is to provide an overview on the recent multi-sensor fusion approaches used in Mixed Reality systems for the 3D camera tracking. We describe the main sensors used in those approaches and we detail the issues surrounding their use (calibration process, fusion strategies, etc.). We include the description of some Mixed Reality techniques developed these last years and which use multi-sensor technology. Finally, we highlight new directions and open problems in this research field.

DOI: 10.4018/978-1-61350-326-3.ch027

INTRODUCTION

In MR applications the vision-based approaches are often used to achieve the camera tracking. Vision-based techniques estimate the camera pose using only the visual information extracted from the acquired images. In most MR applications, camera tracking remains a difficult task which must be accurate and stable. It is known that a non-robust tracking or not enough accurate can generate a "jitter" effect on the Real/Virtual registration, and often leads to tracking failure. In order to deal with this problem, some MR systems use artificial markers, called also fiducials. The main idea consists in placing in the environment several markers among which the content, the size, the position and the orientation are known by the system. By using image processing methods, the MR systems can then extract and identify the markers and thus localize the camera. However, theses methods suffer generally from a lack of accuracy when the markers are occluded or in the case of blurring effect generating by abrupt motion of the camera. Other MR systems use Markerless tracking approaches in order to estimate the camera pose. The principle consists in using salient geometric features (points, edges, silhouettes) existing naturally in the scene. In this case, the registration between the Real and Virtual worlds is realized thanks to the alignment of the 2D information extracted from the images with the 3D model of the scene. These methods usually give a more precise solution than marker-based techniques. However, their main disadvantage lies in the reliability of the 2D-3D matching process. Indeed, an erroneous matching would engender false camera pose estimation. Furthermore, vision-based approaches remain very sensitive to working conditions. Their performances decrease significantly when they are used in uncontrolled environments where situations such as change in brightness, occlusions and sudden motion arise rather often. Multi-sensors techniques which combine various technologies and methods seem to open a new way to resolve the lack of robustness of vision-based methods. They generally fuse a vision-based tracking approach with measurements obtained from localization sensors (inertial, GPS, etc.) to compensate for the shortcomings of each technology when used alone.

The objective of this chapter is to present some original solutions which use multi-sensors technology in order to estimate the camera localization.

STATE OF THE ART

The idea of combining several kinds of sensors is not recent. The first multi-sensors system appeared with robotic applications where, for example, Vieville et al. (1993) proposed to combine a camera with an inertial sensor to automatically correct the path of an autonomous mobile robot. This idea has been exploited these last years by the community of Mixed Reality. Several works proposed to fuse vision and inertial data sensors, using a Kalman filter (You et al., 1999) (Ribo et al., 2002) (Hol et al., 2006) (Reitmayr & Drummond, 2006) (Bleser & Stricker, 2008) or a particular filter (Ababsa et al., 2003) (Ababsa & Mallem, 2007). The strategy consists in merging all data from all sensors to localize the camera following a prediction/correction model. The data provided by inertial sensors (gyroscopes, magnetometers, etc.) are generally used to predict the 3D motion of the camera which is then adjusted and refined using the vision-based techniques. The Kalman filter is generally implemented to perform the data fusion. Kalman filter is a recursive filter that estimates the state of a linear dynamic system from a series of noisy measurements. Recursive estimation means that only the estimated state from the previous time step and the current measurement are needed to compute the estimate for the current state. So, no history of observations and/or estimates is required.

You et al. (1999) developed a hybrid sensor combining a vision system with three gyroscopes to estimate the orientation of the camera in an outdoor environment. Their visual tracking allows refining the obtained estimation. The system described by Drummond and Reitmayr (2006) combines an edge-based tracking with inertial measurements (angular velocity, linear acceleration, magnetic fields). The visual tracking is used for accurate 3D localization while the inertial sensor compensates errors due to sudden motion and occlusion. The measurements of gravity and magnetic field are used to limit the drift problem. The gyroscope is employed to automatically reset the tracking process. Data provided by the two sensors are combined with an extended Kalman filter using a constant velocity model. More recently, Reitmayr and Drummond (2007) proposed to use the GPS positions to re-initialize visual tracking when it fails. Thus, initialization of the visual tracking is obtained by defining a search area represented by an ellipse centred on the GPS position.

Recently, Bleser and Stricker (2008) proposed to combine a texture-based tracking with an inertial sensor. The camera pose is predicted from data provided by the accelerometers using an Extended Kalman filter (EKF). In order to estimate the pose, the EKF fuse the 2D/3D correspondences obtained from the image analysis and the inertial measurements acquired from the inertial sensor. A rendering of CAD model (textured patches) is made using the predicted poses. This allows aligning iteratively the textured patches in the current image to estimate the 2D motion and to update the estimate given by the filter. Natural feature points are tracked by a KLT (Kanade Lucas Tomasi) tracker. The motion model assumes constant acceleration and constant angular velocity. This approach needed offline preparation for generating a textured CAD model of the environment.

Hu et al. (2004) proposed to combine a camera, a GPS and an inertial gyroscope sensor. The fusion approach is based on PPM (Parameterized model matching algorithm). The road shape model is derived from the digital map with respect to GPS position, and matches with road features extracted from the real images. The fusion is based on a predictor-corrector control theory. After checking data integrity, GPS data will start a new loop and reset gyro's integrated. Gyro's prediction will be feedback into the gyro integration module as a dynamical correction factor. When the image feature tracking is failed, gyro's prediction data is used for the camera pose estimation.

Ababsa and Mallem (2007) proposed a particle filter instead of the Kalman filter. Particle filters (PF), also known as methods of Monte-Carlo sequential, are sophisticated techniques for estimating models based on simulation. PFs are generally used to estimate Bayesian models. They represent an alternative to extended Kalman filter, their advantage is that they approach the optimal Bayesian estimation using enough samples. Ababsa et al. merged data from fiducial-based method with inertial data (gyros and accelerometers). Their fusion algorithm is based on a particle filter with sampling importance resampling (SIR). As the two sensors have different sampling frequency, the authors implemented two complementary filters. Thus, if there is no data of vision (e.g. occlusion), the system uses only data from the inertial sensor and vice versa.

Aron et al. (2007) used the inertial sensor to estimate the orientation of the camera only when the visual tracking fails. The orientation allows tracking the visual primitives by defining a search area in the image to perform the features matching. A homography is estimated from this set of matched features to estimate the camera pose. The errors of the inertial sensor are taken into account to optimize the search area. Unlike the approach proposed by Aron et al. (2007) which only estimates the camera orientation, Maidi et al. (2009) used an inertial sensor to estimate both the position and the orientation. Their multimodal system allows tracking fiducials and handling occlusions by combining several sensors and techniques depending on the existing conditions in the environment. When the target is partially

Figure 1. Mixed reality system for outdoor application

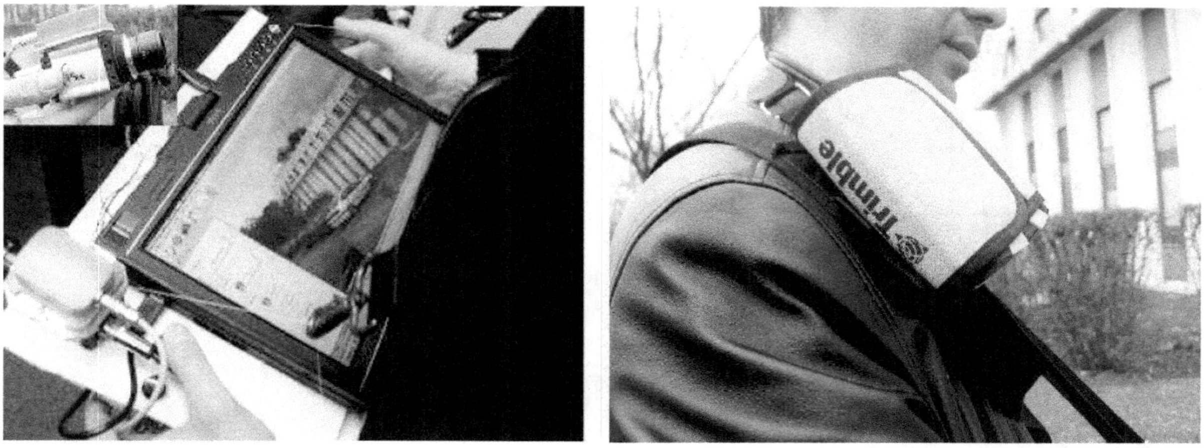

occluded, the system uses a point-based tracking. In presence of a total occlusion of the fiducials, inertial sensor helps to overcome the vision failure. However, the estimation of position from acceleration produces drift over time resulting in a tracking failure.

The idea of combining sensors following the assistance scheme seems more interesting than the data fusion. Indeed, assistance approach makes the system more intelligent so that it can adapt itself to different situations and uses at each time only the data provided by the available sensors. In next sections we discuss issues and problems dealing with conceiving such systems and we give in details of some original solutions.

MULTI-SENSORS SYSTEM USING ASSISTANCE SCHEME

MR systems used in outdoors environments must satisfy several criteria in order to be accepted by the end users. Existing projects in this field aim at developing systems encompassing accurate multi-sensors based 3D localization, a realistic visualization via mobile devices and interaction techniques according to the mobility aspect and the needs of end users. According to these criteria, such systems are generally composed of tablet PC which consists of a handheld display device and the processing unit. This device is connected to the 3D localization system, usually composed of three sensors (see Figure 1): a GPS receiver worn by the user and an inertial sensor attached rigidly to a camera. The GPS returns a global positioning. The inertial sensor estimates 3D orientations, accelerations, and angular velocity and 3D magnetic fields. The camera is used for both the visual feedback and recovering the camera poses. The objective for this section is to carry out a generic solution for the 3D localization adaptable to different types of outdoor environments.

Using the assistance scheme implies that the system must be subdivided in two subsystems: a main subsystem and an auxiliary one. The main subsystem corresponds to the visual tracking because it is more accurate. The auxiliary subsystem is used only when the visual tracking fails; it is composed of the GPS and the inertial sensors. Figure 2 provides a flow chart to describe the 3D localization process using our assistance scheme.

Vision Subsystem

The camera pose is computed using its intrinsic parameters and the knowledge of the position of

Figure 2. The system data flow scheme

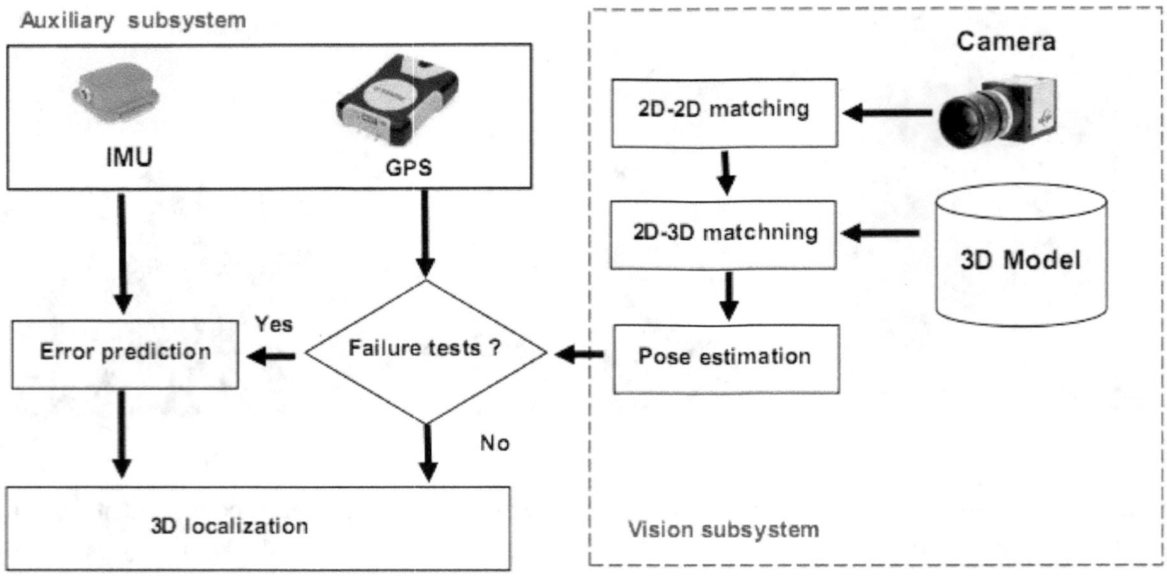

3D reference points (3D model of the scene) and their 2D projections into the image (2D/3D matching). Indeed, according to the pinhole model, the camera pose is formalized as an error minimization between 2D points and the projection of 3D points using the camera pose parameters. Several algorithms can be used to perform this non-linear minimization problem such as Newton method or orthogonal iteration (OI) algorithm (Lu et al., 2000). In this case, the pose estimation is formulated as a minimization of metric error based on the collinearity in the scene space. In addition, the vision subsystem often needs an initialization setup. This step is delicate; it consists in matching the 3D visible points of the model with their 2D projections in the initial view in order to estimate the initial localization of the camera. The obtained 2D/3D matching must be maintained from one image to another in order to update the pose estimation. For this, the vision subsystem uses a point-based visual tracking where the tracked points correspond to the 3D feature extracted from the 3D model. The initial matching is updated with 2D/2D visual tracking. Moreover, the estimated pose must verify some coherence criteria to check neither it is plausible or not. A failure makes the system mostly rely on the Auxiliary subsystem.

Auxiliary Subsystem

This subsystem (Zendjebil et al, 2008), composed of GPS and inertial sensor, replaces the vision subsystem when this one fails. The position and orientation given by the vision subsystem are substituted by the absolute position provided by the GPS receiver and the orientation given by the inertial sensor. The use of the Auxiliary subsystem is not limited only to replace the vision subsystem. The Auxiliary subsystem is also used to initialize the vision subsystem. Moreover, from the position and orientation given by this subsystem, we can measure the accuracy of the 3D localization estimated by the vision subsystem by defining some confidence intervals. The Auxiliary subsystem is composed of two modules: prediction and correction. The prediction module is used to predict accuracy errors of the localization system. It is based on online training of the error between the two subsystems. Once the localization system switches to the Auxiliary subsystem, the error is predicted following a Gaussian model and used to improve the position and the orientation provided by the GPS and the inertial sensor. The two parts composing the system interact continuously with each other. Also, the use of GPS for position estimation solves the problem of inertial sensor's drift, which is used only for orientation estimation.

System Calibration

Certain prerequisites are essential for the proper functioning of such system. In fact, each sensor provides data in its own reference frame. The inertial sensor computes the orientation between a body reference frame attached to itself and a local level reference frame. Also, the GPS position is expressed in an earth reference frame defined by WGS84 (World Geodetic System) standard. For registration, we need to estimate continually the camera pose which relates the world reference frame to the camera reference frame. Thus, the 3D localization provided by the Auxiliary system must be aligned with the camera reference frame. The several sensors must be aligned in a unified reference frame in order to have the same position and orientation of the point of view. So, the hybrid sensor must be calibrated to determine the relationships between the several sensors and thus to unify the measurements. The accuracy of the Auxiliary subsystem depends on the accuracy of the calibration processes. In this case two calibration processes are performed offline. The first one consists in estimating the relationship between inertial sensor and camera (Inertial/Camera calibration). The second one estimates the transformation which maps the GPS position to the camera position (GPS/Camera transformation).

Initialization: 2D/3D Matching

The initialization process is an important issue for the vision-based localization approaches. It represents the process that matches 3D visible points with their 2D projections in the initial view. A bad matching affects the 3D localization estimation. However, there are not reliable and accurate automatic methods. We can find some approaches that are based on objects recognition (Zollner et al., 2008) or rendering patches (Bleser & Stricker, 2008). These approaches require a substantial database (respectively objects images and patches). The main idea is to avoid a full manually points matching done by user. One solution consists in making a rendering of a wire frame model with a fixed point of view or using the position and orientation given by the Auxiliary subsystem. Then, the user manually registers the projected model over the real view by moving around the camera. Once the registration is validated, the second step consists in identifying the 2D correspondences. For this, the process detects the corners close to the projections of the 3D points using Harris detector (Harris, 1993). Then, the initialization setup performs 2D-2D matching. To improve the 2D-2D matching, a descriptor-based method is used. So, a SURF descriptor (Bay et al., 2008) is associated to each 3D point. SURF (Speeded Up Robust Features) is a scale and rotation invariant detector and descriptor. The use of this descriptor allows obtaining a robust and efficient matching procedure. Indeed, around the 2D projection of the 3D points, we detected Harris corners. Once the descriptor of the detected points is computed, the process looks for the most similar point which has the shortest distance between its descriptor and the descriptor of the 3D points. A RANSAC (Fischler & Bolles, 1981) algorithm is used to discard outliers.

Visual Tracking

Once the vision system is initialized the visual tracking can start. To estimate the camera pose, we must keep the 2D/3D matching for each current view. This can be achieved by using a frame-to-frame 2D points tracking. Tracking consists in following features from one frame t-1 to another frame t. Several approaches can be used such as correlation matching methods; however they are very expensive in computing time. To track 2D features in real time, the chosen method must be fast and accurate. For that Tomasi and Kanade (1991) Tracker can be adopted. This algorithm used an optical flow computation to track features points or a set of predefined points from the previous image I_{t-1} to the current image I_t. Therefore, this algorithm tracks a set of 2D points associated to visible 3D points. Briefly, 2D points are searched in the neighborhood of its position in view t-1 based on the minimization of brightness difference. To minimize the time computation, the KLT tracker uses a pyramid of images for the current view. Therefore, tracking is done at the coarsest level and then propagate to the finest. This allows following the features over a long distance with great precision. The approach is fast and accurate, but it requires that the tracked points are always visible. So the approach does not handle occlusions.

Failure Tests

The pose estimated by vision can be wrong. So, we need to handle errors in order to switch to the Auxiliary localization subsystem. The errors are due to several factors mainly occlusions, sudden motion and the change of brightness. These conditions affect the visual tracking. Therefore, some criteria are defined to quantify the quality of the estimated pose. If one of these criteria is not verified, the pose is rejected and the system switches to the Auxiliary subsystem.

Number of Tracked Points

The number of 2D/3D matching points affects the accuracy of the minimization process used to estimate the camera pose. Indeed, the more we have a large set of 2D/3D matched points, the more the estimated pose is accurate and vice versa. For this, we define a minimum number of matching. Below this threshold, it is considered impossible to estimate the pose with the vision subsystem.

Projection Error

The number of matched points is not sufficient to ensure the accuracy of the pose estimation; the projection error criterion can also be used. This error represents the average square of the difference between the projection of 3D points using estimated pose and the 2D points. If the error is large, greater than an empirical threshold, the pose is considered wrong.

Confidence Intervals

The data provided by the Auxiliary subsystem can also be used as an indicator of the pose validation. In fact, from the position and orientation given by the Auxiliary subsystem, confidence intervals are defined. They are represented by an ellipsoid centered by the orientation provided by the inertial sensor and an ellipse which center is determined by the 2D position given by GPS. The axes of the ellipse or the ellipsoid can be defined $3*\sigma$ (standard deviation of the offset between the camera pose and Auxiliary estimation) or empirically. If the pose computed by the vision subsystem is included in these confidence intervals (position in the ellipse and the orientation in ellipsoid), the pose is considered correct.

Error Prediction

The estimation of the 3D localization provided by the combination of the GPS and the inertial sensor is less accurate then the vision-based estimation. The computation of the produced error is important in the localization process. Indeed, it allows quantifying the quality of measurements in order to improve the 3D localization estimation provided by the Auxiliary subsystem. The error represents the offset between the camera

Figure 3. The state machine scheme of 3D Localization system's operation

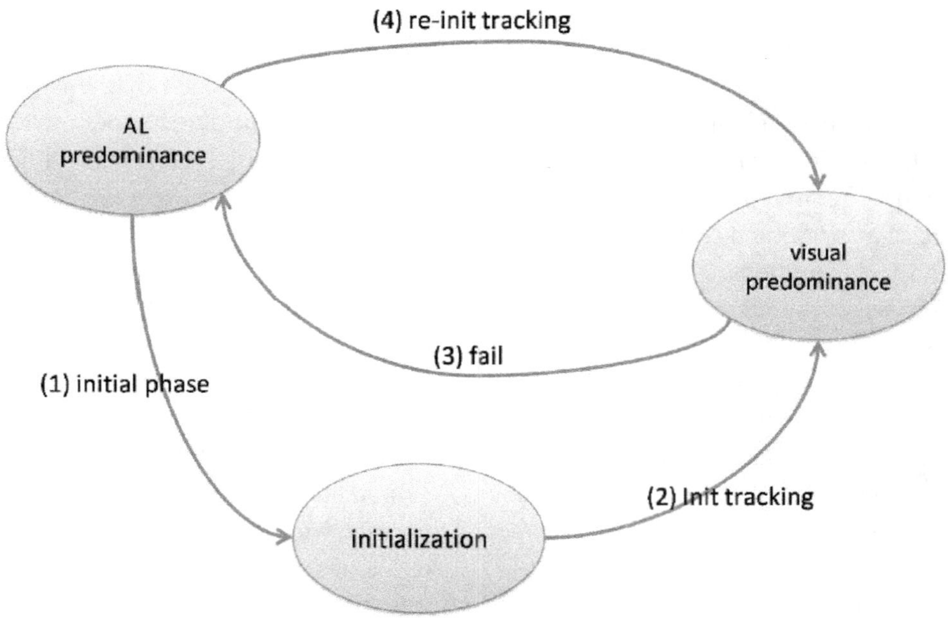

pose and the position and orientation deduced from GPS and inertial sensor. When the vision fails, this error must be predicted. For that, the error is modeled as a regression with a Gaussian process (Williams, 1997). The Gaussian process is a stochastic process which generates samples and can be used as a prior probability distribution over functions in Bayesian inference. During visual tracking, the offset between the Auxiliary subsystem and the vision subsystem is recorded for the online training step. When the visual tracking fails, the Gaussian process predicts the offset made by GPS and the inertial sensor. This offset which is represented by the mean error is used to correct the estimation of the 3D localization.

System Operation

The localization system operates using a finite state machine scheme (see Figure 3). A finite state machine is an abstract model composed of a finite number of states, transitions between those states, and actions. This formalism is mainly used in the theory of computability and formal languages.

We identify three states: the Auxiliary predominance state, the initialization state and the visual predominance state. The transitions between different states are as follows: At the initialization state, the Auxiliary subsystem provides an estimation of the pose (1). This estimation is refined with vision subsystem (2). When the visual tracking fails, the Auxiliary subsystem takes over to estimate the 3D localization (3). Since the Auxiliary subsystem is less accurate than the vision subsystem, the estimation is corrected taking into account the predicted error. Thereafter, the estimation is used to re-initialize the visual tracking (4).

System Behavior in Real Conditions of Use

The proposed system is developed using ARCS (Didier et al., 2009) (Augmented Reality System Component), a component-programming system. ARCS allows to prototype rapidly Augmented and Mixed Reality applications and facilitates interfacing multiple heterogeneous technologies. On the one hand, ARCS uses a programming paradigm of classical components specially designed to meet the constraints imposed by the MR applications (especially real-time constraint). On the other hand, ARCS is based on a finite state machine which allows switching from one state to another state called sheets. This feature facilitates the implementation of our hybrid system. We tested this 3D localization system on real data acquired in outdoor and real conditions.

The camera was calibrated offline using the Faugeras and Toscani (1987) algorithm in order to compute its intrinsic parameters. The hybrid sensor was calibrated using a set of reference data (GPS positions and images for GPS/Camera calibration and inertial sensor orientations and images for Inertial/Camera calibration). Several experiments have been achieved to study the behavior of the proposed system when used in outdoor environments. The first experiment considers a straight line as a truth data. The origin of this line is defined in front of the origin of the world reference frame. This line is sampled, and for each sample we take a set of data acquisitions, namely images and GPS positions. The sensors are mounted on a tripod to ensure more stability. The reference measurements are taken with a telemeter which accuracy is about 0.15m. For each acquired image, we calculated the position and the orientation of the camera.

From GPS data and the transformation estimated during the calibration step, we deduce the absolute position with respect to the world reference frame associated to the real scene. By comparing the different estimated positions to the reference positions, we find a mean offset about (1.8374m; 1.4810m). The same GPS positions compared to the camera's positions give a mean error equal to (1.7321m; 1.4702m) with a standard deviation (1.8314m; 1.0116m). The second experiment focused on the relative position between two successive fixed positions. In average the offset between the reference position and that obtained with the GPS is about 0.7817m with a standard deviation equal to 1.06m. Similar values are given by the vision subsystem, i.e. an offset mean about 0.8743m with a standard deviation of 0.9524m. Therefore, these results demonstrate that the movement provided by the two subsystems is consistent. The third experiment performed several continuous recordings of GPS/camera positions. The two sensors are time-stamped in order to synchronize them and to retrieve the set of data acquired at the same time. The positions given by the vision and the GPS without correction are compared and the obtained errors are about 0,9235m in the x-axis (with a standard deviation of 0.6669m) and 0.8170m in the y-axis (with a standard deviation of 0.6755m).

In addition, in order to study the error prediction approach we first used a set of 76 data acquired in continuous manner to perform the error training. Then, the Gaussian process is used with the last 30 data to predict errors. The mean offset between the predicted error and the real one is about (μ_x = 0.2742m; σ_x = 0.4799) and (μ_y = 0.5757m; σ_y = 0.5097m). The positions provided by the GPS receiver are then corrected using this predicted error. This allows improving the 3D localization provided by the Auxiliary subsystem. To assess the accuracy of the inertial sensor, we compared the orientations produced from the gyroscope to those computed by the vision pose estimation algorithm. For that, a video with several orientations in an outdoor environment has performed. Both orientations have the same behavior. However, in some cases, we found that external factors can affect the inertial measurements, particularly in defining the local reference frame where the x axis is in the direction of the local magnetic north. This causes errors in the orientation estimation.

To solve this problem the rotation between the local reference frame associated to inertial sensor and the world reference frame is re-estimated continuously. The behavior of the whole system is also tested. The initialization process allows having the matching of the 3D visible points from the 3D model with their projections in the first view. From this 2D/3D matching, the set of 2D points

Figure 4. Registration of the 3D model using the poses obtained with our hybrid system

(a) #0686 (b) #1053 (c) #1054 (d) #1055

are defined and tracked frame to frame. For each frame, the wire frame model is registered using the positions and orientations obtained from the hybrid localization system. In Figure 4, the green color projection is obtained from the positions and orientations provided by the vision subsystem. The wire frame model is well superimposed on the real view which demonstrates the accuracy of the camera pose estimation. In magenta, the projected model is obtained with the positions and orientations provided by the Auxiliary subsystem. Figure 4 show that when vision fails, the localization system switches to the Auxiliary subsystem to provide 3D localization. The localization is corrected with the predicted error which contributes to improve the estimation.

Figure 5 show that during the occlusion of the tracked points, the Auxiliary subsystem provides always an estimation of the position and orientation of the camera. Therefore, even when a total occlusion occurred, the system can provide a rough estimation of the 3D localization. This would not be the case if we used individually the camera.

CONCLUSION AND FUTURE WORKS

In this chapter, we presented a generic solution for 3D camera localization using multi-sensors technology. The system combines a camera, a GPS and an inertial sensor; it is designed to work in outdoor environments. Instead to fusion all data, the proposed system is based on an assistance scheme. It is composed of two parts which work in a complementary manner and controlled by a finite state machine allowing continuous 3D localization. The vision subsystem, representing the main part, uses a point-based visual tracking. Once the vision fails, the system switches to Auxiliary subsystem which is composed of the GPS/inertial sensors. The Auxiliary subsystem is less accurate then the vision subsystem, especially the GPS positioning. Hence, a prediction stage is performed to improve the accuracy of the Auxiliary

Figure 5. Registration of the 3D model using the auxiliary subsystem: Occlusion case

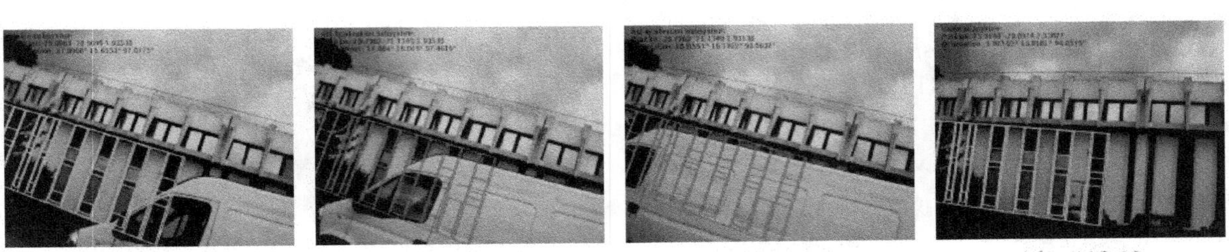

(a) #1236 (b) #1239 (c) #1245 (d) #1269

subsystem. Furthermore, the Auxiliary subsystem is used to define confidence intervals to validate visual tracking. The 3D localization provided by the two subsystems is used to learn, on-line, the errors made by the Auxiliary subsystem. The two subsystems interact continuously to each other. The obtained results are quite satisfactory with respect to the purpose of MR systems. They have shown that the proposed system has quite good accuracy compared to other approaches.

The system was tested in outdoor environment and has demonstrated its capacity to adapt itself to the several conditions occurred in such environments. For example, when a total occlusion of the scene model is occurred, the Auxiliary system takes over the 3D localization estimation until the vision becomes operational. However to increase the robustness and the efficiency of the whole system, improvements must be made in several parts. Actually, within the implemented vision-based method, the tracked points must be always visible. So, one challenge is to develop a tracking method which can handle visual occlusions and update automatically the set of tracked points by adding, in real time, new visible points. In addition, other markerless tracking approaches can be combined with the point tracker such as edge-based methods (Ababsa & Mallem, 2006) to improve the accuracy of the vision-based pose estimation. Also, the fusion process can be optimized if we consider the motion dynamic of the camera given by the IMU sensor. On the other hand, the experiments have shown that the GPS signal can be obstructed when the user is quite near the buildings. So, when the system switches to the Auxiliary subsystem, the position could not be estimated. This problem can be solved by adding other kinds of positioning sensors which can replace the GPS (RFID, WIFI, etc.). The main idea is to develop a ubiquitous tracking system composed of a network of complementary sensors which can be solicited separately and in real time in terms of the situations occurred in the environments.

ACKNOWLEDGMENT

This work is supported by the RAXENV project funded by the French National Research Agency (ANR).

REFERENCES

Ababsa, F., Didier, J. Y., Mallem, M., & Roussel, D. (2003). Head motion prediction in augmented reality systems using Monte Carlo particle filters. In the 13th International Conference on Artificial Reality and Telexixtance (ICAT'03). Tokyo, Japan, (pp. 83-88).

Ababsa, F., & Mallem, M. (2006). Robust line tracking using a particle filter for camera pose estimation. ACM Symposium on Virtual Reality Software and Technology (VRST'06). Limassol, Cyprus, (pp. 207-211).

Ababsa, F., & Mallem, M., (2007). Hybrid 3D camera pose estimation using particle filter sensor fusion. *Advanced Robotics, the International Journal of the Robotics Society of Japan (RSJ), 21,* 165–181

Aron, M., Simon, G., & Berger, M. (2007). Use of inertial sensors to support video tracking. *Computer Animation Virtual Worlds, 18*(1), 57–68. doi:10.1002/cav.161

Bay, H., Ess, A., Tuytelaars, T., & Van Goo, L. (2008). Surf: Speeded up robust features. [CVIU]. *Computer Vision and Image Understanding, 110*(3), 346–359. doi:10.1016/j.cviu.2007.09.014

Bleser, G., & Stricker, D. (2008). Advanced tracking through efficient image processing and visual-inertial sensor fusion. In *IEEE International Conference on Virtual Reality*, (pp. 137–144).

Didier, J. Y., Otmane, S., & Mallem, M. (2009). *Arcs: Une architecture logicielle reconfigurable pour la conception des applications de réalité augmentée. Technique et Science Informatiques (TSI)*. Innovations en Réalité Virtuelle et Réalité Augmentée.

Faugeras, O. D., & Toscani, G. (1987). Camera calibration for 3D computer vision. *In International Workshop on Industrial Applications of Machine Vision and Machine Intelligence*, (pp. 240–247).

Fischler, M. A., & Bolles, R. C. (1981). Random sample consensus: a paradigm for model fitting with applications to image analysis and automated cartography. *Communications of the ACM*, *24*(6), 381–395. doi:10.1145/358669.358692

Harris, C. (1993). Tracking with rigid models . In Blake, A. (Ed.), *Active vision* (pp. 59–73).

Hol, J. D., Schon, T. B., Gustafsson, F., & Slycke, P. J. (2006). Sensor fusion for augmented reality. In *IEEE International Conference on Information Fusion*, (pp. 1–6). Florence, Italy.

Hu, Z., Keiichi, U., Lu, H., & Lamosa, F. (2004). Fusion of vision, 3D gyro and GPS for camera dynamic registration. In *International Conference on Pattern Recognition* (ICPR'04), Vol. 3, Washington, DC, USA, (pp. 351–354).

Lu, C. P., Hager, G. D., & Mjolsness, E. (2000). Fast and globally convergent pose estimation from video images. *IEEE Transactions on Pattern Analysis and Machine Intelligence*, *22*(6), 610–622. doi:10.1109/34.862199

Maidi, M., Ababsa, F., & Mallem, M. (2009). Vision-inertial tracking system for robust fiducials registration in augmented reality. *In Symposium on Computational Intelligence for Multimedia Signal and Vision Processing*, Nashville (USA).

Reitmayr, G., & Drummond, T. (2006). Going out: Robust model-based tracking for outdoor augmented reality. In *ACM/IEEE International Symposium on Mixed and Augmented Reality*, Santa Barbara, California, USA

Reitmayr, G., & Drummond, T. (2007). *Initialisation for visual tracking in urban environments*. In ACM/IEEE International Symposium on Mixed and Augmented Reality, Nara, Japan

Ribo, M., Lang, P., Ganster, H., Brandner, M., Stock, C., & Pinz, A. (2002). Hybrid tracking for outdoor augmented reality applications. *IEEE Computer Graphics and Applications*, *22*(6), 54–63. doi:10.1109/MCG.2002.1046629

Tomasi, C., & Kanade, T. (1991). *Detection and tracking of point features*. Carnegie Mellon University Technical report CMU-CS-91-132, April 1991.

Viéville, T., Romann, F., Hotz, B., Mathieu, H., Buffa, M., & Robert, L. ... Audren, J. T. (1993). *Autonomous navigation of a mobile robot using inertial and visual cues*. In International Conference on Intelligent Robots and Systems.

Williams, C. (1997). *Prediction with Gaussian processes: From linear regression to linear prediction and beyond. Technical report*. Neural Computing Research Group.

You, S., Neumann, U., & Azuma, R. (1999). Orientation tracking for outdoor augmented reality registration. *IEEE Computer Graphics and Applications*, *19*(6), 36–42. doi:10.1109/38.799738

Zendjebil, I. M., Ababsa, F., Didier, J. Y., & Mallem, M. (2008). On the hybrid aid-localization for outdoor augmented reality applications. In *ACM Symposium on Virtual Reality Software and Technology*, (pp. 249–250). Bordeaux, France.

Zollner, M., Pagani, A., Pastarmov, P., Wuest, H., & Stricker, D. (2008). Reality filtering: A visual time machine in augmented reality . In *VAST European Association for Computer Graphics* (pp. 71–77). Eurographics.

Chapter 28
Recovering 3D Human Body Postures from Depth Maps and Its Application in Human Activity Recognition

Nguyen Duc Thang
Kyung Hee University, Korea

Md. Zia Uddin
Kyung Hee University, Korea

Young-Koo Lee
Kyung Hee University, Korea

Sungyoung Lee
Kyung Hee University, Korea

Tae-Seong Kim
Kyung Hee University, Korea

ABSTRACT

We present an approach of how to recover 3D human body postures from depth maps captured by a stereo camera and an application of this approach to recognize human activities with the joint angles derived from the recovered body postures. With a pair of images captured with a stereo camera, first a depth map is computed to get the 3D information (i.e., 3D data) of a human subject. Separately the human body is modeled in 3D with a set of connected ellipsoids and their joints: the joint is parameterized with the kinematic angles. Then the 3D body model and 3D data are co-registered with our devised algorithm that works in two steps: the first step assigns the labels of body parts to each point of the 3D data; the second step computes the kinematic angles to fit the 3D human model to the labeled 3D data. The co-registration algorithm is iterated until it converges to a stable 3D body model that matches the 3D human posture reflected in the 3D data. We present our demonstrative results of recovering body postures in full 3D from continuous video frames of various activities with an error of about 6^0-14^0 in the estimated kinematic angles. Our technique requires neither markers attached to the human subject nor multiple cameras: it only requires a single stereo camera. As an application of our body posture recovery technique in 3D, we present how various human activities can be recognized with the body joint angles derived from the recovered body postures. The features of body joints angles are utilized over the conventional binary body silhouettes and Hidden Markov Models are utilized to model and recognize various human activities. Our experimental results show the presented techniques outperform the conventional human activity recognition techniques.

DOI: 10.4018/978-1-61350-326-3.ch028

INTRODUCTION

Through several million years of human evolution, stereopsis is one of the unique functions in the human vision system, allowing depth perception: it is a process of combining two images projected to two human eyes to create the visual perception of depth. Learned from the human stereoscopic system, a stereo camera was invented to synchronously capture two images of a scene with a slight difference in the view angle from which depth information of the scene can be derived. The depth information is generally reflected in a 2-D image called a depth map in which the depth information is encoded in a range of grayscale pixel values. Since its first commercial product in 1950s, *Stereo Realist*, introduced by the David White Company, there have been continuous developments of a stereo camera until now with the latest products such as a digital stereo camera, Fujifilm FinePix Real 3D W1 and a stereo webcam, Minoru 3D. Lately, 3D movies, in which depth information is added to RGB images, have received a lot of attention with the latest success of a film, *Avatar* released in 2009. Watching 3D movies and 3D TVs with the special viewing glasses is becoming a part of our lives these days.

Another area where the depth information could be valuable is the field of human computer interaction (HCI). In this area, 3D motion information of a user is utilized to better control external devices such as computers and games. In the conventional ways, capturing 3D human motion or movement (i.e., a sequence of human postures) is typically done using optical markers or motion sensors. Such systems are capable of producing some kinematic parameters of human motion with high accuracy and speed using wearable optical markers or sensors. However, it is inconvenient to a user who needs to wear specially designed optical markers or sensor-suits when running these systems. This disadvantage combined with the high cost equipment makes the systems impractical in daily use applications. In the case of using motion sensors, a user has to hand-hold controllers equipped with accelerometers or gyroscopes. One good example is the Wii controller of Nintendo which uses optical sensors and accelerometers to recognize the hand motion of the user to control the games. Lately, some efforts are being made to capture the whole body movement without the markers or motion sensors. Using a stereo camera and its derived depth map is one of options, since depth maps may provide sufficient 3D information to derive human body motions in 3D. Although this approach should open a new possibility for various novel applications in HCI such as games and u-lifecare, obtaining human body postures in 3D directly from depth maps is not very straightforward.

There have been some attempts to develop marker-less systems to estimate human motion from a sequence of monocular images or RGB images, only reflecting 2-D information. Because the 3D information of the subject is lost, the efforts to reconstruct the 3D motion of the subject from only monocular images face difficulties with ambiguity and occlusion that lead to inaccurate results (Yang & Lee, 2007). Therefore, most marker-less systems use multiple cameras to capture 3D human motion. Through such systems, the 3D information of the observed human subject is captured from different directional views, thereby providing better results of recovered human motion in 3D (Knossow et al., 2008; Gupata et al., 2008). However, it is usually complicated to setup such a system, because it requires enough space where the cameras can be installed. Also it requires synchronization of the cameras. Thus, there are always some tradeoffs between the flexibility of using a single camera and the ability to get the 3D information using multiple cameras.

Another way of recovering a series of human postures or motion in full 3D is to utilize the information in depth maps. However, there has been little effort to recover 3D human body postures using this approach. Some conventional works to estimate human body postures from depth maps can be classified in the following two approaches: namely the *matching-based* approach and the *model-based* approach. In the *matching-based* approach (Yang & Lee, 2007), one tries to match a depth map with a set of generated human body postures to find the most compatible human body posture in the depth map. In the *model-based* approach (Urtasun et al., 2006), one creates a human body model and fits the model to the given depth map to estimate its corresponding human body posture. In this chapter, we present an approach of recovering human body postures from depth maps based on the framework of the *model-based* approach. However, in our approach we have added a novel step of detecting human body parts and incorporated it into our co-registration algorithm such that human body postures can be estimated in a more efficient and generalized framework.

The chapter begins with a survey of the conventional approaches including the use of optical markers and multiple cameras to capture 3D human body postures. We discuss their advantages and disadvantages in comparison with our approach of recovering 3D human body postures directly from depth maps without using optical markers or multiple cameras. In the following sections, we present technical details of our method with examples and demonstrations. Subsequently, as an application of our technique in human activity recognition (HAR), we present a section of how various human activities can be recognized with the derived body joint angles from the recovered body postures. We conclude the chapter with future research directions.

BACKGROUND

In general, there are two main frames of human motion (or a time-series of postures) capture systems. One is the optical system (i.e., video sensor based), which uses video cameras to obtain images and applies image processing techniques to reconstruct human motion from the acquired images. The other is the non-optical (i.e., motion sensor based) system, which uses gyroscopes (to measure angular velocity), accelerometers (to measure acceleration), or magnetic sensors (to measure the position and orientation of magnetic markers) to capture human motion. Here, we mainly focus on the systems using optical devices.

Most conventional optical systems to acquire human motion commonly use markers. Basically, the users are required to wear optical markers, so that the cameras can locate the position of the human body parts where the markers are attached. To avoid the effects of occlusion, additional cameras are installed at different locations. The number of the cameras might be up to several hundreds to make sure the full coverage around the human subject. In this method, the kinematic parameters are estimated using the relative locations of the detected markers. For instance, the kinematic angles at the knee joint are estimated based on the 3D coordinates of the detected markers at the ankle, knee, and crotch. The main advantages of the method are fast processing speed and high accuracy. For example, capturing human body postures via VICON exhibits a recording frame rate up to 240 frames-per-second that is enough to capture human activities with fast movements. However the devices for this approach are very expensive.

Nowadays, there are increasing research efforts to develop a marker-less system to recover human body postures in 3D from video. Obviously, the video is conveniently recorded with a normal camera to provide a sequence of monocular images. The articulated human body model was reconstructed from some detected regions of the human body in monocular images using the inverse kinematics (Taylor, 2000). In other approaches, a probabilistic model was designed to establish the relationship between the human postures and the cues from images like color, contours, and silhouettes. Machine learning techniques such as the sampling by the Monte-Carlo method (Lee & Cohen, 2006) were applied to find the human body posture most probabilistically compatible with the information given in the images. However, as the depth information is lost (i.e., the 3D object is projected into a 2-D image), there will be an ambiguity of reconstructing a 3D human posture from a monocular image. The appearance of a human subject in an image might also correspond to many possible configurations of the human posture in 3D. Due to this limitation, most previous researches based on a monocular image concentrate only on detecting the human body parts (Hua et al., 2005; Ramanan et al., 2007; Roberts et al., 2007).

Rather than processing on a single image, a lot of attempts have been proposed to utilize monocular images acquired with multiple cameras to get more accurate results of recovering human body postures. For instance, a setup with multiple cameras described in (Horaud et al., 2009; Knossow et al., 2007) was composed of six cameras installed at different locations to estimate motion of a tracked subject. Typically, the information in monocular images with different directional views is combined to reconstruct the 3D data of a human subject. The 3D data might be presented by 3D voxels or by a cloud of 3D points. Thus, with each presentation of 3D data, there are different ways to reconstruct human body postures. In (Sundaresan & Chellapa, 2008), the authors presented a method to segment the 3D voxels into different body parts and registered each part by one quadric surface to reconstruct the articulated human model. To segment the 3D voxels, they mapped the voxels' coordinates into a new domain using the Laplacian Eigenmaps where they could discover the skeleton structure (1-D manifolds) of the 3D data. Based on this skeleton structure, they could assign the 3D data to corresponding human body parts using probabilistic registration. Some other methods like ISOMAP (Chu et al., 2003; Tenenbaum et al., 2000), Locally Linear Embedding (Roweis & Saul, 2000), or Multidimensional Scaling (Cox & Cox, 2001) are also available to recover the human skeleton structure of the 3D voxels. Meanwhile, with another form of representation of 3D data, a cloud of 3D points, in (Plankers & Fua 2003), the authors modeled the human body with an isosurface, called the *soft object*. The shape of the *soft object* was controlled by the kinematic parameters of the human model. The least-square estimator was used to minimize the differences between the *soft object* and the cloud of 3D points, consequently finding the human body posture most fitted with the 3D data. Rather than using a single surface like the *soft object*, in (Horaud et al., 2009), they used a set of surfaces with ellipsoids to present the human body. In order to perform the registration of the ellipsoids to the 3D data, each 3D point was cast into one ellipsoid using the datum distance and the least-square estimator was utilized to draw the ellipsoids close the 3D data.

Although the marker-less systems using multiple cameras to recover human body postures can overcome the disadvantages of the system using a single camera with the ambiguities and occlusions of the 3D data when presented in a monocular image, there are still some remaining limitations in the multiple camera-based approaches. For instance, there is a need for extra software and hardware to support the transfer of large video data from multiple cameras over a network. Also, the data acquired with more than one camera must be calibrated to compute the 3D coordinate of each pixel of the recorded images within the same coordinate system. Moreover, the multiple cameras require a complicated installation. Therefore, using a single stereo camera should be more flexible and practical in the recovery of human body postures. As mentioned, there are two types of approaches of recovering human body postures from depth maps.

The first is the *matching-based* approach in which a set of human body postures is generated and compared with a depth map derived from a stereo camera to find the best matching posture. In (Yang & Lee, 2007), about 100,000 human postures, presenting most appearances of the human body in 3D, were created and stored in an exemplar database. However, with a large number of human body postures, the authors had to develop an efficient algorithm to organize and retrieve the human body posture stored in the database. To avoid generating all possible human postures, in (Olivier et al., 2009), only a limited number of human postures at the time index *t* that are close to the human body posture estimated at the time index *t-1* were generated. This method evaluated the discrepancies between the created human postures and the 3D information of the new depth map given at the time index *t* to find the human posture best compatible with the depth map. The drawback of this method is that with the limited number of generated postures, the accuracy of estimating human body postures tends to be low. In the opposite case, with the increased number of generated postures, the time needed to search for an appropriate human posture gets prolonged.

Apart from the *matching-based* approach, the *model-based* approach (Urtasun et al., 2006) estimates human body postures directly from depth maps without using a set of temporary postures for matching. This approach models an articulated human body in 3D and formulates an estimation problem to minimize the difference between the human model and the information in a depth map to recover a human posture. Our technique of recovering human body postures presented in this chapter is based on the framework of this *model-based* approach. However, we have extended and generalized the approach by developing a co-registration algorithm with an additional step of detecting human body parts in 3D before fitting the human body model to 3D data (Thang et al., 2010a).

HOW TO RECOVER 3D HUMAN BODY POSTURES FROM DEPTH MAPS

The overall steps of our method of recovering human body postures from depth maps are presented in Figure 1. First, we preprocess a pair of stereo images to obtain a depth map and calculate the 3D information (i.e., 3D data) from the depth map. Separately, we create our articulated human body model using a set of ellipsoids and parameterize the model with kinematic joint angles. Finally, we co-register the body model to the 3D data of the depth map to estimate the joint angles. Our co-registration involves the following two main steps:

- **Labeling:** The labeling step assigns a label of each human body part (i.e., an ellipsoid) to each point of the 3D data using the information and cues from RGB images.
- **Model Fitting:** after the body part labeling, the model fitting step fits each point to its corresponding ellipsoid by minimizing the distance between them.

This two-step co-registration process is iterated to minimize the differences between the 3D human body model and the observed 3D data. Finally, the algorithm finds the best human pos-

Figure 1. Essential steps of our methodology of recovering 3D human body postures from depth maps

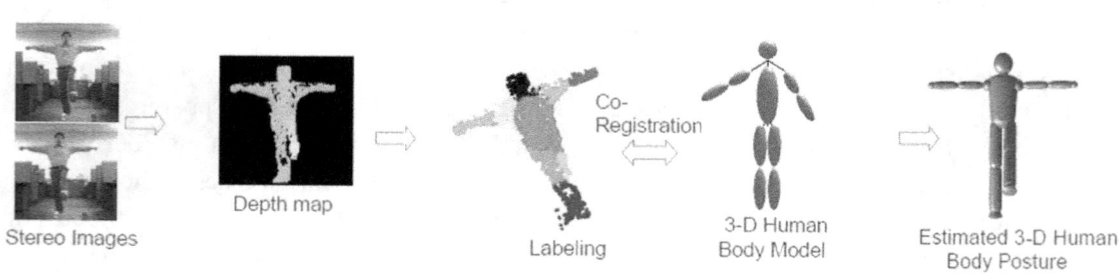

ture on a frame-by-frame basis. In the following sub-sections, more details of each process are presented.

Preprocessing of Stereo Images

As mentioned, a stereo camera is used to capture a pair of images in a time sequence containing human motion. For each pair of images, we apply the stereo matching algorithm (Cech & Sara, 2007) to compute the pixel disparities between them, generating a depth map that decodes the 3D information of the scene: the pixel with higher disparity value is closer to the camera than other pixels. Continuously, we perform the background modeling and subtraction (Wang et al., 2003) in a RGB image to get the binary silhouette of a human subject and use the binary silhouette to extract the region of interest in the depth map containing only the 3D information of the human subject. Then, for each pixel belonging to the human body region in the depth map, we calculate its coordinate in the 3D space in order to estimate the kinematic joint angles of the human posture correctly. The depth value Z_w of a pixel in the 3D coordinate system is computed by

$$Z_w = \frac{f_c b}{d} \tag{1}$$

where f_c is the focus length, b the base-line, and d a disparity value of the pixel. The two remained coordinate X_w and Y_w are computed by

$$X_w = \frac{uZ_w}{f_c}, \tag{2}$$

$$Y_w = \frac{vZ_w}{f_c}, \tag{3}$$

where u and v are the column and row index of the pixel in the depth map.

3D Human Body Modeling

We create the articulated human body with a set of ellipsoids where each ellipsoid represents one human body part as shown in Figure 2(a). For the convenience of transformation computations, we formulate the equation of each ellipsoid in the 4-D projective space as,

$$q(X) = X^T Q_\theta^T S^T D S Q_\theta X - 2 = 0 \tag{4}$$

where the constant matrix $D = diag[a^{-2}, a^{-2}, b^{-2}, 1]$, $b \geq a$ determines the size of the ellipsoid. The constant matrix **S** locates the center of the ellipsoid in the local coordinate attached to the ellipsoid. Q_θ is the skeleton-induced transformation matrix. $X = [x,y,z,1]^T$, indicating the coordinate of a 3D point in the 4-D projective space. Each segment of the human body model is controlled by a series of transformations specified by the kinematic parameters at each body joint, therefore Q_θ is a matrix function of $\theta = (\theta_1, \theta_2, \ldots, \theta_n)$, where $\theta_1, \theta_2, \ldots, \theta_n$ are n kinematic parameters. We separate Q_θ into a series of matrices where each matrix is computed based on a single parameter,

$$Q_\theta = Q_n(\theta_n) Q_{n-1}(\theta_{n-1}) \ldots Q_1(\theta_1) \tag{5}$$

where $Q_1(\theta_1), Q_2(\theta_2), \ldots, Q_6(\theta_6)$ are of six degrees of freedom (DOF) (i.e., three translations and three rotations) that determine the transformation from the global coordinate system to the local coordinate system attached at the body hip. The other matrix element, $Q_i(\theta_i) = Tr_i R(\theta_i)$ with $i > 6$ is the transformation matrix from the local coordinate system attached to the body segment i to the local coordinate system attached to the body segment $i+1$, where Tr_i is the constant translation matrix dependent on a skeleton structure and $R(\theta_i)$ is the rotation matrix at each body joint around the x-, y-, or z-axis. We can assign the value of the matrix Tr_i by an identity

Figure 2. Two examples of running E-steps to detect the body part labels. (a) Initial models. (b) The label assignments found by the first iteration of E-step. (c) The label assignments found by the last iteration of E-step.

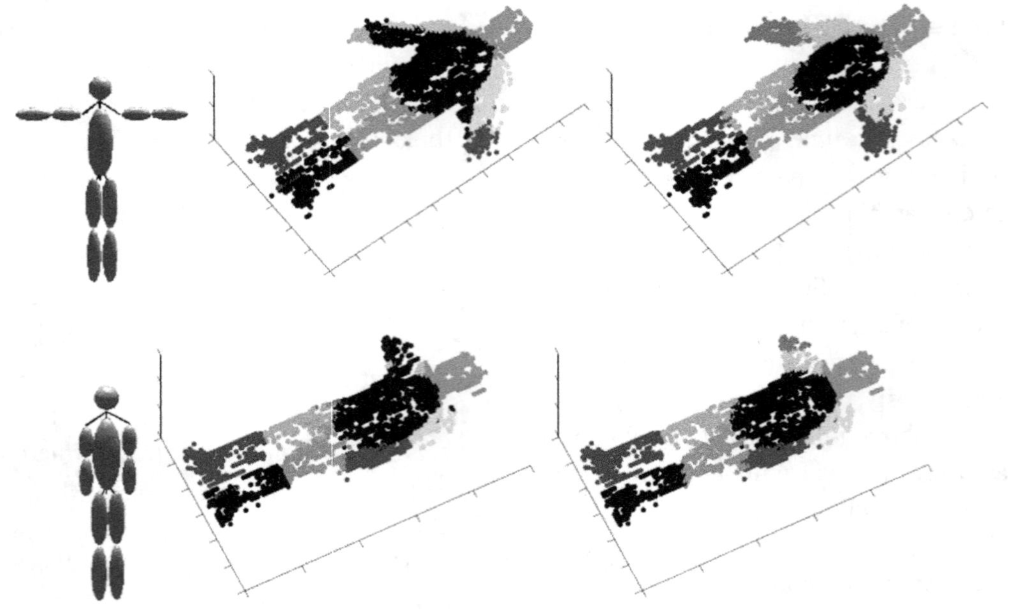

matrix if we want to add more than one DOF to a body joint.

Our defined human model is composed of 14 body segments, nine joints (i.e., two knees, two hips, two elbows, two shoulders, and one neck), and 24 DOF (i.e., two DOF at each joint and six DOF for the transformation from the global coordinate system to the local coordinate system at the body hip). In addition, another human body model using the super quadric can be created for better display of the results as in Figures 3 and 4. The formulation of the super-quadric surface without any transformation (rotation or translation) is derived as

$$\left(\frac{x}{a}\right)^2 + \left(\frac{y}{b}\right)^2 = \left(1 + \frac{sz}{c}\right)\left(1 - \left(1 - \frac{2z}{c}\right)^d\right) \quad (6)$$

where a, b, and c determine the size of the super-quadric along the x-, y-, and z-axis, respectively.

Mathematical Relationship Between Human Body Model and Depth Information

In this section, we introduce a probabilistic distribution that represents the relationship between the human body posture specified by the kinematic parameters and the information in the corresponding depth map and RGB image. Let $D = (X_1, X_2, \ldots, X_N)$ denote N points of the 3D data computed from a depth map and I denote a RGB image. The supplementary variable $V = (v_1, v_2, \ldots, v_N)$ is used to label the body part where each point should belong to. The posterior probability between the label V and the kinematic parameter θ

Figure 3. Experimental results with (a) elbow movements in the horizontal direction, (b) elbow movements in the vertical direction, (c) knee movements, and (d) shoulder movements. From the left column to the right: RGB images, depth maps, and recovered human postures in the front view and +45^0 view.

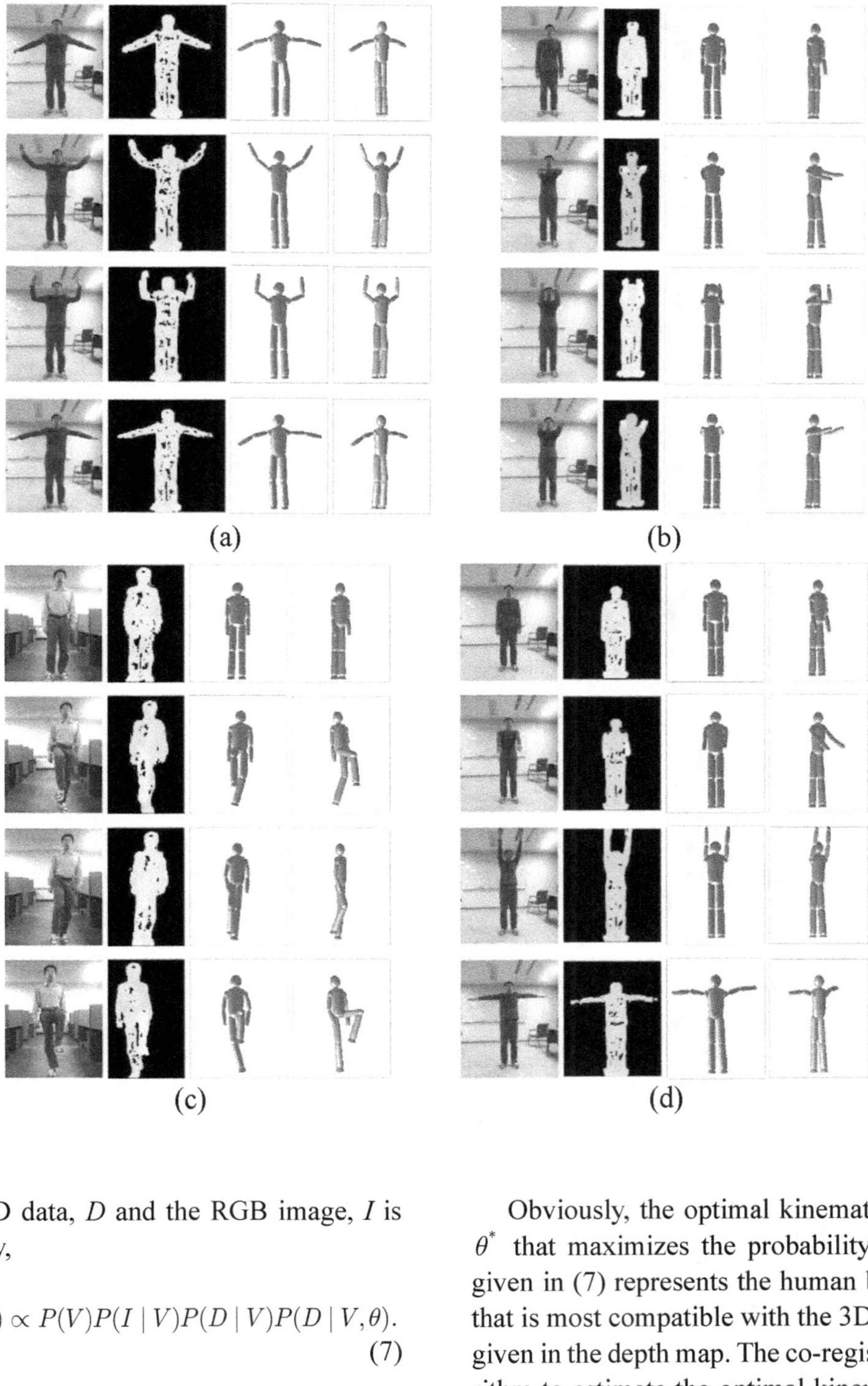

given the 3D data, D and the RGB image, I is expressed by,

$$P(V,\theta \mid I,D) \propto P(V)P(I \mid V)P(D \mid V)P(D \mid V,\theta). \quad (7)$$

Obviously, the optimal kinematic parameter θ^* that maximizes the probability distribution given in (7) represents the human body posture that is most compatible with the 3D information given in the depth map. The co-registration algorithm to estimate the optimal kinematic parameter θ^*, recovering the correct human body

Figure 4. Experimental results with (a) a walking sequence (top). Recovered human postures are depicted in the front view (middle) and -45⁰ view (bottom) and (b) an arbitrary activity sequence (top). Recovered human postures are depicted in the front view (middle) and -45⁰ view (bottom).

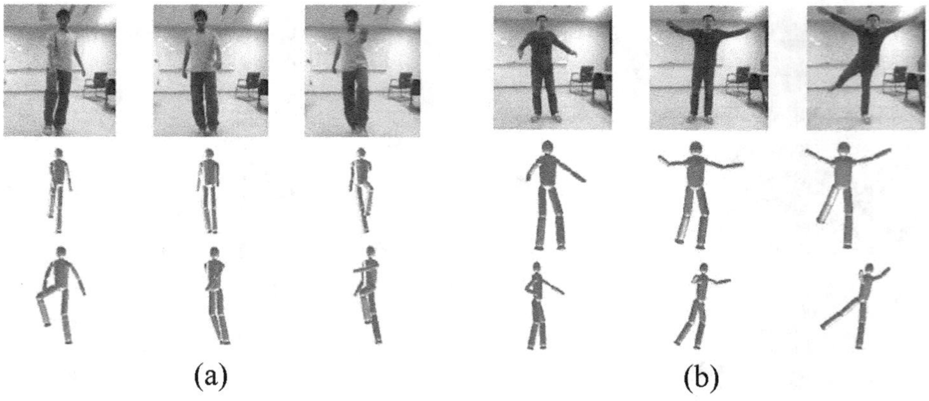

posture from the given depth map is presented in the next section.

a) Smoothness Prior

The smoothness prior found from the Potts model (Boykov et al., 2001) is given by

$$P(V) = \prod_{i=1}^{N} \prod_{j \subset N_i} P(v_i, v_j) \quad (8)$$

where N_i is a set of neighbors of the point i and $P(v_i, v_j)$ is defined by

$$P(v_i, v_j) = \begin{cases} 1 & if\ v_i \neq v_j \\ e^{\gamma} & if\ v_i = v_j \end{cases} \quad (9)$$

where γ is a positive constant. The smoothness prior $P(v_i, v_j)$ is used to derive the label of each point toward the same label of its neighbors that makes the labeling outcomes smooth and removes outliers. Here, the neighbors of one pixel in a depth map lie inside a circle with the center at the pixel's location with its radius *d=3*.

b) Image Likelihood

The RGB image containing the information of a human subject in a color space can be used to detect some human body parts, providing extra information for assigning the labels of 3D data. The detection results are integrated into equation (7) by the likelihood term $P(I|V)$,

$$P(I \mid V) = \prod_{i=1}^{N} \varphi(I \mid v_i). \quad (10)$$

In our work, we perform the face and torso detection to calculate the probability of one point inside the detected regions getting a label *'head'* or *'torso'*. The face areas are located by detecting the skin color in the HSV color space (Conaire et al., 2007).

$$\varphi(I \mid v_i = head) = \begin{cases} e^{c} & \text{the pixel } i \text{ is marked as 'face'} \\ 1 & \text{other wise} \end{cases} \quad (11)$$

where c is a positive constant.

The likelihood of a pixel labeled as *'torso'* is computed based on the function $f(r_i)$,

$$f(r_i) = \kappa e^{-d(r_i)} \quad (12)$$

where $d(r_i)$ is the algebraic distance from a point r_i in a RGB image with a coordinate $[x^r, y^r, 1]^T$ in the 3D prospective space to the center of the body O_{body}: O_{body} lies in a middle between the center of the face and the center of a binary silhouette. K is a positive constant. The algebraic distance $d(r_i)$ is computed by,

$$d(r_i) = r_i^T Q_e^T D_e Q_e r_i - 1 \tag{13}$$

where $\mathbf{D}_e$ and $\mathbf{Q}_e$ are the 3×3 matrices that configure the size and shape of the ellipse representing the torso. The likelihood to assign a point as a '*torso*' is given by,

$$\varphi(I \mid v_i = torso) = \begin{cases} f(r_i) & d(r_i) \leq 1 \\ 1 & \text{other wise.} \end{cases} \tag{14}$$

c) Pairwise Geodesic Relationship among 3D Points

The geodesic distance is measured by the length of the shortest path between two points on a curved surface. During the movement and deformation of a non-rigid object like the human body, the geodesic distance between any two points on the boundary surface of the object is preserved. Therefore, we utilize this property of the geodesic distance to derive the geodesic constraints between any two points of the 3D data representing the human body.

Since there are a large number of 3D points, we need a large number of computations to estimate the geodesic distance among all pairs of the 3D points. In order to reduce the number of computations, we assign a set of close points into a group, called a *cell*. All 3D points belonging to the same cell receive the same geodesic constraint. Computing the geodesic distance by the shortest path distance in graph using the Dijkstra's algorithm (Dijkstra, 1959), we express $P(D|V)$ by

$$P(D \mid V) = \prod_{i=1}^{N} \prod_{j_c=1}^{N_c} P_{geo}(D \mid v_i, v_{j_c}) \tag{15}$$

$$P_{geo}(D \mid v_i, v_{j_c}) = \begin{cases} e^{-\alpha} & d(v_{i_c}, v_{j_c}) < d_{\min}(v_{i_c}, v_{j_c}) \\ e^{-\beta} & d(v_{i_c}, v_{j_c}) > d_{\max}(v_{i_c}, v_{j_c}) \end{cases} \tag{16}$$

where i_c is the cell that holds the point i, $d(v_{i_c}, v_{j_c})$ the geodesic distance between the cell i_c and j_c, N_c the number of cells, and (α, β) two positive constants. Two values, $d_{\min}(v_{i_c}, v_{j_c})$ and $d_{\max}(v_{i_c}, v_{j_c})$ define the lower and upper bound for the geodesic distance between a pair of labels. The two related labels assigned to two 3D points that are too far or too close are penalized to reduce the belief in these assignments.

d) Reconstruction Error

The discrepancies between the human model created by a set of connected ellipsoids and the cloud of 3D points are measured by the total Euclidean distances from each 3D point to the ellipsoid corresponding to the label of this point. Thus, the Euclidean distance is considered as another factor to assign the label of each point during the registration process. $P(D \mid V, \theta)$ is defined by

$$P(D \mid V, \theta) = \prod_{i=1}^{N} e^{-\frac{d^2(X_i, \theta, v_i)}{2\sigma^2}} \tag{17}$$

where $d(X_i, \theta, v_i)$ is the Euclidean distance from the point X_i to the ellipsoid v_i and the constant σ is variance. The Euclidean distance is calculated by the distance from one point to the nearest point lying on the ellipsoid surface. In general, to compute the Euclidean distance $d(X_i, \theta, v_i)$, we need to solve a sixth-degree polynomial equation (Heckbert, 1994). However, with the symmetric ellipsoid defined in our articulated human model, a sixth-degree polynomial equation is simplified to a fourth-degree polynomial that has an analytical solution allowing us to compute its roots.

Co-Registration of 3D Human Body Model and 3D Depth Information

A human body posture that best matches the observed 3D data is subject to the kinematic parameter θ^* that maximizes the posterior probability given in (7),

$$\theta^* = \arg\max_\theta \sum_V P(V,\theta \mid I,D). \tag{18}$$

To solve this optimization problem, the EM algorithm is a suitable choice with the incorporation of the latent variable, *V*. Let *Q(V)* be the probability distribution of the label *V*. Our algorithm to estimate a human body posture from a given depth map is formulated in an EM framework with the following two key steps:

- **E-step:** Assuming that the current value of the kinematic parameter θ is θ_{old}, E-step estimates the label assignments by computing the probability distribution $Q(V) = P(V \mid \theta_{old}, I, D)$ of the label given the information of the RGB image and the 3D data of the depth map.
- M-step: With the label assignment $Q_{old}(V)$ found by E-step, M-step maximizes $E_{Q_{old}(V)}[\log(P(V,\theta \mid I,D))]$ or equivalently minimizes the reconstruction error between the model and the cloud of 3D points to estimate a new optimal value of the kinematic parameter θ.

The two-step co-registration process is iterated to minimize the differences between the 3D model and the observed data and finally the correct human posture is found. More details of those two steps are presented as follows.

a) E-step: Labeling

It is intractable to calculate the exact distribution *Q(V)* of the label *V*. Therefore, we approximate the distribution *Q(V)* by using the mean field approach (Toyoda & Hasegawa, 2008). The logarithm of *Q(V)* is given by

$$\log Q(V) \propto \sum_{i=1}^{N} g_i(v_i) + \sum_{i=1}^{N}\sum_{j \subset N_i} g_{ij}(v_i, v_j) + \sum_{i=1}^{N}\sum_{j_c=1}^{Nc} h(v_i, v_{j_c}) \tag{19}$$

where $g_i(v_i)$ is the sum of the logarithms of the image likelihood in (10) and the reconstruction error in (17), $g_{ij}(v_i, v_j)$ the logarithm of the smooth prior in (8), and $h(v_i, v_{j_c})$ the logarithm of the geodesic constraints in (15),

$$h(v_i, v_{j_c}) = \log P_{geo}(D \mid v_i, v_{j_c}). \tag{20}$$

The probability of a pixel *i* having a label v_i, $q_i(v_i) = P(v_i \mid \theta, I, D)$ is iteratively updated until it approaches to a stable value by an equation

$$q_{i_{step+1}}(v_i) = \frac{1}{Z_{i_{step}}(v_i)} \exp\left\{ g_i(v_i) + \sum_{j \subset N_i}\sum_{v_j} q_{j_{step}}(v_j) g_{ij}(v_i, v_j) + \sum_{j_c=1}^{N_C}\sum_{v_{j_c}} q_{step}^{j_c}(v_{j_c}) h(v_i, v_{j_c}) \right\} \tag{21}$$

where $Z_{i_{step}}(v_i) = \sum_{v_i} q_{i_{step}}(v_i)$ is a normalization factor and $q_{step}^{j_c}(v_{j_c}) = E[q_{j_{step}}(v_{j_c})]$ an average probability of all pixels j belonging to the cell j_c. We use $\frac{1}{Z_{i_0}(v_i)} \exp\{g_i(v_i)\}$ as an initial value of $q_{i_0}(v_i)$. For simplification, we set $q_{step}^{j_c}(v_{j_c} = \varepsilon) = 1$ when the probability of the cell j_c belonging to the ellipsoid ε is largest and $q_{step}^{j_c}(v_{j_c}) = 0$ for $v_{j_c} \neq \varepsilon$. In Figure 2, we show two examples of running E-step to detect the body part labels from the 3D data.

b) M-step: Model Fitting

After the probability distribution of the label variables is estimated from E-step, M-step computes a new value of the kinematic parameter θ as the solution of the optimization problem

$$\arg\max_\theta E_{Q(V)}[\log P(D \mid \theta, V)] \tag{22}$$

Here, we remove the terms in Equation (7) independent of θ. Equation (22) can be rewritten as

$$-\arg\max_\theta \sum_{\varepsilon=1}^{N_\varepsilon} \sum_{i=1}^{N} q_i(v_i = \varepsilon) d^2(X_i, \theta, v_i = \varepsilon) \tag{23}$$

$$\text{or } \arg\min_\theta \sum_{\varepsilon=1}^{N_\varepsilon} \sum_{i=1}^{N} q_i(v_i = \varepsilon) \left\| X_i - Z_i(\theta)^\varepsilon \right\|^2$$

where N_ε is the number of ellipsoids and $Z_i(\theta)^\varepsilon$ the nearest point of X_i lying on the surface of the ellipsoid ε. To reduce the number of computations, we set $q_i(v_i = \varepsilon) = 1$ for ε satisfying $q_i(v_i = \varepsilon) \geq q_i(v_i \neq \varepsilon)$ and $q_i(v_i) = 0$ for $v_i \neq \varepsilon$. We solved the non-linear optimization problem in (23) by the Levenberg-Marquardt method (Murray et al., 1994; Sundaresan et al., 2004).

To summarize, we describe the presented algorithm in Table 1.

Results of Recovering Human Body Postures from Depth Maps

In our experiments, we used a stereo camera, Bumblebee 2.0 of Point Grey Research, to capture stereo image pairs with their resolution at 640 × 480. We asked our subjects to perform various motions in front of the stereo camera as depicted in Figure 3. Note that a sequence of frames in a video stream is shown from top-to-bottom in a column. In Figures 3(a) and 3(b), the movements of the elbows in the horizontal and vertical directions were evaluated in our experiments. The subjects raised their hands up to create an angle about 90^0 between the upper hand and lower hand, then brought their hands down. In the next experiment shown in Figure 3(c), the subject in video performed an activity at their knee joints. The subject lifted his right leg up to a 90^0 between the upper leg and lower leg then he did the same motion with the other leg. In addition, we considered the body movements created by the combination of the two kinematic angles at the shoulders as in Figure 3(d). To evaluate the reconstruction error, we generated the ground-truth of the estimated kinematic angles by using the hand-label method (Gupta et al., 2008; Lee & Cohen, 2006). Some points were hand-labeled to determine the position of the body joints in the RGB images such as hand, elbow, shoulder, etc. Using the 3D information estimated from the depth maps, we computed the coordinate of these labeled points in 3D and then calculated the ground-truth angles. Then, we compared the kinematic angles of the recovered human body postures against the ground-truth angles and obtained the mean error of about 6^0~14^0 in the estimated kinematic angles.

In order to track the movements of the whole human body, the subjects were asked to perform complicated activities with all arms and legs. Figure 4 shows two video sequences and the recovered human body postures reflected in those sequences in two view angles. The average distance between the 3D points and the ellipsoids of the human model were used to evaluate the error measurements of the reconstructed postures. The average distance D_t of the frame t was computed by

$$D_t = \sum_{i=1}^{N} d_t(i) / N \tag{24}$$

Table 1. The co-registration algorithm used to estimate human body postures from depth maps

1. At the time index t, initialize the value of the kinematic parameter θ_t with the value of the kinematic parameter θ_{t-1} estimated at the time index t-1
2. **E- step:** Compute $g_i(v_i)$ from the sum of the logarithms of the image likelihood in (10) and the reconstruction error in (17) and use $\exp\{g_i(v_i)\} / Z_{i_0}(v_i)$ as an initial value of $q_{i_0}(v_i)$
3. Compute $g_{ij}(v_i, v_j)$ from the logarithm of the smooth prior in (8) and $h(v_i, v_{j_c})$ from the logarithm of the geodesic constraints in (15)
4. Update

$$q_{i_{step+1}}(v_i) = \frac{1}{Z_{i_{step}}(v_i)} \exp\left\{ g_i(v_i) + \sum_{j \subset N_i} \sum_{v_j} q_{j_{step}}(v_j) g_{ij}(v_i, v_j) + \sum_{j_c=1}^{N_C} \sum_{v_{j_c}} q_{step}^{j_c}(v_{j_c}) h(v_i, v_{j_c}) \right\}$$

5. If $q_{i_{step+1}}(v_i)$ has not converged, go back to step 3
6. **M-step:** Estimate new values of the kinematic parameter

$$\theta_t = \arg\min_\theta \sum_{\varepsilon=1}^{N_\varepsilon} \sum_{i=1}^{N} q_i(v_i = \varepsilon) \left\| X_i - Z_i(\theta)^\varepsilon \right\|^2$$

7. If θ_t has not converged, go back to step 2

where $d_t(i)$ is the Euclidean distance between the point i and the nearest ellipsoid of the human model and N is the number of points. The mean error distance D_t for the walking and arbitrary sequences depicted in Figure 4 came out to be 0.06m and 0.04m respectively.

HUMAN ACTIVITY RECOGNITION USING BODY JOINT ANGLES

Human Activity Recognition (HAR) is defined as recognizing various human activities utilizing external sensors such as acceleration, motion, or video sensors. In recent years, HAR from video has evoked considerable interests among researchers in computer vision and image processing communities (Robertson & Reid, 2006). A key reason for this is its potential usefulness of the outcomes of such recognition in practical applications such as human computer interaction, automated surveillance, smart home, and human healthcare applications. A general method for video-based HAR starts with the extraction of key features from images and comparing them against the features of various activities. Thus, activity feature extraction, modeling, and recognition techniques become essential elements in this regard.

In general, 2-D binary silhouettes of human body shapes are the most common representations of human activity that have been applied for video-based HAR (Yamato et al., 1992; Carlsson & Sullivan, 2002; Niu & Abdel-Mottaleb, 2004; Niu & Abdel-Mottaleb, 2005; Uddin et al., 2008; Uddin et al., 2009). For instance, in (Yamato et al., 1992), a binary silhouette-based HAR system was proposed to transform the time sequential silhouettes into a feature vector sequence through the binary pixel-based mesh feature extraction from every image. Then, the features were utilized to recognize several tennis actions with Hidden Markov Models (HMMs). In (Carlsson & Sullivan, 2002), a silhouette matching key frame-based approach was applied to recognize forehand and backhand strokes from tennis videos. Regarding binary silhouette-based features, Principal Com-

Figure 5. Processes involved in the binary silhouette and 3D body joint angle-based HAR

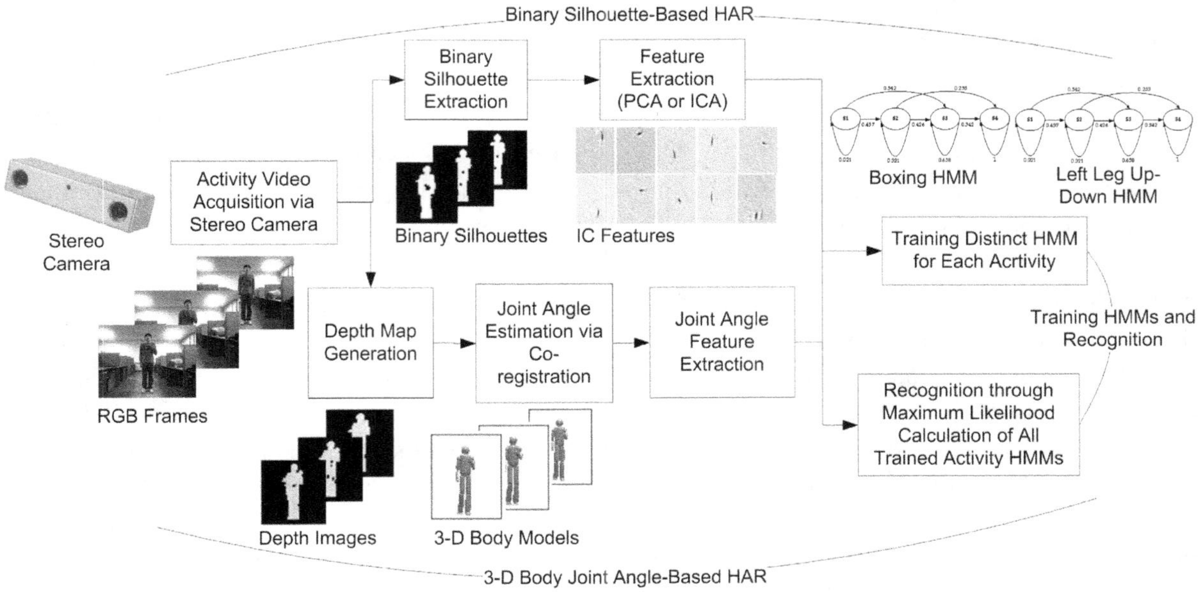

ponent Analysis (PCA), a feature extractor based on the second-order statistics, is most commonly applied (Niu & Abdel-Mottaleb, 2004; Niu & Abdel-Mottaleb, 2005; Uddin et al., 2009). After applying PCA, some top PCs (i.e., eigenvectors) are chosen to produce global features representing most frequently moving parts of the human body in various activities. In (Niu & Abdel-Mottaleb, 2004; Niu & Abdel-Mottaleb, 2005), the authors utilized PC features from binary silhouettes and optical flow-based motion features in combination with HMM to recognize different view-invariant activities. The top flow of Figure 5 shows the typical processing components of the binary silhouette-based HAR. Once the binary silhouettes are obtained from RGB images, some prominent features, obtained through the feature extraction process, are then applied to a recognition technique to train and recognize various human activities.

Recently, more advanced HAR techniques have been introduced in terms of new features and more powerful feature extraction techniques. Although binary silhouettes are commonly employed to represent a wide variety of body configurations, they also produce ambiguities by representing the same silhouette for different postures from different activities: especially for those activities that are performed toward the video camera. Thus, the binary silhouettes do not seem to be a good choice to represent human body postures in different activities. In this regard, depth silhouettes for human body representations can be a solution. In the case of depth-based silhouette representation, the pixel values are set on the basis of the distance to the camera and hence it can provide better activity information than the binary silhouettes. In (Uddin et al., 2008; Uddin et al., 2009), the authors proposed to use a new feature extraction technique called Independent Component Analysis (ICA) to produce prominent local features from time-sequential depth silhouettes to be used with HMMs and obtained superior HAR performance than the binary silhouette-based approaches.

However, depth silhouettes do not convey truly 3D information of the human body postures and hence generates the similar problems as binary silhouettes: they represent the human body in different activities from one angle view of depth. As the human body consists of limbs connected with joints, if one is able to obtain their 3D joint angle information, one can form much stronger features than conventional silhouette features that will lead to significantly improved HAR. In this section, we present an application of HAR based on our estimated 3D body joint angle features and HMM. From the time-sequential activity video frames, the joint angles are first estimated by co-registering a 3D human body model to the stereo information and then mapped into codewords to generate a sequence of discrete symbols for an HMM of each activity. With these symbols, each activity HMM is trained and used for activity recognition. The bottom of Figure 5 shows the basic processes regarding 3D body joint angle-based HAR. It indicates that after obtaining the depth images, joint angles are estimated via co-registration and represented as features to feed into the HMMs to train and recognize different human activities. Some more details of the essential processing steps are given below.

3D Joint Angle Features in Human Activities

Once we obtain the joint angles of the 3D human body for each video frame as discussed earlier, we can utilize these to represent various human activities effectively. The estimated joint angles from a video frame of a particular activity form a feature vector: thus, each activity video clip is represented in a sequence of joint angle feature vectors as $(F_1,F_2,\ldots,F_T)$, where T is the length of the activity video. Therefore, the 3D joint angle features from video can really contribute in distinguishing an activity from another: especially those activities that are not discernible with the conventional binary or depth silhouette-based approaches.

Training and Recognition via HMM

HMM has been applied extensively to solve a large number of spatiotemporal pattern recognition problems including human activity recognition because of its capability of handling sequential information in space and time with its probabilistic learning capability for recognition (Lawrence & Rabiner, 1989; Niu & Abdel-Mottaleb, 2004; Niu & Abdel-Mottaleb, 2005; Uddin et al., 2008; Uddin et al., 2009). Basically, HMM is a stochastic process where an underlying process is usually unobservable but it can be observed through another set of stochastic processes that produces observation symbols. To learn a video-based human activity in a HMM, the symbol sequences obtained from the training image sequences of distinct activities are used to optimize the corresponding HMM. Finally, the trained HMMs are used to calculate the maximum likelihood for recognition.

Technically, HMM is a collection of finite states connected by transitions. Every state is characterized by transition and symbol observation probabilities. A generic HMM is expressed as $H = \{S,\pi,A,B\}$ where S denotes possible states, π the initial probability of the states, A the transition probability matrix between the hidden states and B the observation probability from every state. If the number of activities is N then there will be a dictionary $(H_1,H_2,\ldots,H_N)$ of N trained models. To estimate HMM parameters, one could use the Baum-Welch algorithm (Lawrence & Rabiner, 1989).

We choose a four-state and left-to-right HMM in this study to model sequential events of each human activity. To recognize each test activity, the obtained observation symbol sequence $O=\{O_1,O_2,\ldots,O_T\}$ through the vector quantization process is used to determine the proper activity HMM from all the trained activity HMMs by means of the highest likelihood as

$$decision = \arg\max_{i=1,2,\ldots,M}\{P(O \mid H_i)\} \tag{25}$$

where H_i indicates i^{th} HMM and M number of activities. More details on regarding training and testing of HMMs for human activity recognition are available in our previous work (Uddin et al., 2008; Uddin et al., 2009).

Results of Recognizing Various Human Activities

We had built a database of six different activities (namely, left hand up-down, right hand up-down, both hands up-down, boxing, left leg up-down, and right leg up-down) to be trained and recognized via our 3D joint angle and HMM-based approach. A total of 15 and 40 image sequences of each activity were prepared to be used for training and recognition respectively.

We started our experiments with the traditional binary silhouette-based HAR. Table 2 shows the experimental results of HMM-based HAR utilizing the IC features of binary silhouettes and joint angle features of 3D body model respectively. As ICA is superior to PCA by extracting the local binary silhouette features (Uddin et al., 2009), it was utilized for HAR where 150 features were considered in the feature space. Binary silhouettes were not appropriate to recognize the activities used in our experiments, yielding a much lower mean recognition rate of 58.33%. On the contrary, utilizing the 3D body joint angle features, we obtained a mean recognition rate of 92.50%, which is far better than that of the binary silhouette-based HAR. The experimental results show that the 3D joint angle features are remarkably superior to the conventionally used silhouette features. The body joint angle features seem to be much more sensitive toward complex activities that are not discernable with the body silhouettes.

FUTURE RESEARCH DIRECTIONS

As presented, our human motion capturing system using a stereo camera is potentially applicable to various biomedical and HCI areas. However, due to the existing errors of recovered kinematic angles, our system might face difficulty with practical applications requiring high accurate results of estimating motion. For instance, in biomechanics

Table 2. Experimental results of video-based HAR using binary silhouettes vs. joint angles

Approach	Activity	Recognition Rate	Mean	Standard Deviation
Binary Silhouette-Based HAR	Left hand up-down	47.50%	58.33	16.78
	Right hand up-down	60		
	Both hands up-down	67.50		
	Boxing	30		
	Left leg up-down	72.50		
	Right leg up-down	72.50		
Joint Angle-Based HAR	Left hand up-down	87.50	92.50	4.18
	Right hand up-down	97.5		
	Both hands up-down	87.50		
	Boxing	95		
	Left leg up-down	92.50		
	Right leg up-down	95		

measurements, some systems need small errors of recovered kinematic angles in order to analyze the detailed motion of a tracked subject. In health care areas, a human motion capturing system can be used to help a handicap person to learn how to walk, run, etc. However, the system with large errors of estimated kinematic angles might cause adverse effects to the treatment of the patient. The other difficulty of our method relates to estimating human motion from tricker movements or rapid changes of trackers' locations. In this situation, there are large variations of the human postures between two consecutive frames. A part of information used to assign the label of 3D data might get inaccurate, causing a missing calculation of some body parts. For such reasons, we plan our future work to improve the reliability of our presented techniques and its robustness to handle the rapid and complex changes of human postures in a video sequence. The concerns are addressed by developing better labeling method with investigating more information to detect human body parts from RGB images as exampled in (Ninh et al., 2009). Also in the model fitting part of our algorithm, a large number of 3D points processed in the algorithm slow down the co-registration process and take into account outliers in computations that affect the recovering results. To mitigate this problem, we recently suggested a way of utilizing clusters of 3D points being assigned the same label of a body part and computing the kinematic parameters with a small number of clusters (Thang et al., 2010b). This greatly reduced the computational time, eliminated the presence of outliers, and made the presented techniques more practical.

As a practical application, we presented our work of HAR using the derived feature of joint angles, which proved its superior performance over the conventional feature of body silhouettes. We believe that our presented work in this chapter should be able to find its use in other applications such as advanced HCI, video games, smart homes, smart hospitals, etc.

CONCLUSION

In this chapter, we have presented our marker-less system to recover human body postures in 3D from a sequence of depth maps acquired by a single stereo camera. We have described our methodology including how to estimate the 3D data of a depth map, how to create a human body model, and how to co-register the human body model to the 3D data. Our experimental results with real video data have shown that our method successfully recovers human body postures from depth maps: our validation indicates an error range of about 6^0-14^0 in the estimated joint angles. In addition, as an application of our technique, we have presented a HAR work using the derived body joint angles. Again our experimental results with real video data show that our HAR system produces significantly better recognition rates than the conventional approaches in which binary silhouettes are utilized to recognize human activities.

ACKNOWLEDGMENT

This research was supported by the MKE (The Ministry of Knowledge Economy), Korea, under the ITRC (Information Technology Research Center) support program supervised by the NIPA (National IT Industry Promotion Agency) (NIPA-2010-(C1090-1021-0003)).

REFERENCES

Boykov, Y., Veksler, O., & Zabih, R. (2001). Fast approximate energy minimization via graph cuts. *IEEE Transactions on Pattern Analysis and Machine Intelligence*, *23*(11), 1222–1239. doi:10.1109/34.969114

Carlsson, S., & Sullivan, J. (2002). Action recognition by shape matching to key frames, *IEEE Computer Society Workshop on Models versus Exemplars in Computer Vision*, (pp. 263-270).

Cech, J., & Sara, R. (2007). Efficient sampling of disparity space for fast and accurate matching, *Proceedings of IEEE International Conference on Computer Vision and Pattern Recognition*, (pp. 1-8).

Chu, C.-W., Jenkins, O. C., & Mataric, M. J. (2003). Markerless kinematic model and motion capture from volume sequences. *Proceedings of IEEE International Conference on Computer Vision and Pattern Recognition*, (pp. 475-482).

Conaire, C. O., O'Connor, N. E., & Smeaton, A. F. (2007). Detector adaption by maximizing agreement between independent data sources. *Proceedings of IEEE International Conference on Computer Vision and Pattern Recognition*, (pp. 1-6).

Cox, T. F., & Cox, M. A. A. (2001). *Multidimensional scaling*. Boca Raton, FL: Chapman and Hall.

Dijkstra, E. W. (1959). A note on two problems in connexion with graphs. *Numer Math, 1*, 269–271. doi:10.1007/BF01386390

Gupta, A., Mittal, A., & Davis, L. S. (2008). Constraint integration for efficient multiview pose estimation with self-occlusions. *IEEE Transactions on Pattern Analysis and Machine Intelligence, 30*(3), 493–506. doi:10.1109/TPAMI.2007.1173

Heckbert, P. S. (1994). *Graphics germs IV*. San Diego, CA: Academic Press.

Horaud, R., Niskanen, M., Dewaele, G., & Boyer, E. (2009). Human motion tracking by registering an articulated surface to 3D points and normals. *IEEE Transactions on Pattern Analysis and Machine Intelligence, 31*(1), 158–163. doi:10.1109/TPAMI.2008.108

Hua, G., Yang, M., & Wu, Y. (2005). Learning to estimate human pose with data driven belief propagation. *Proceedings of IEEE International Conference on Computer Vision and Pattern Recognition*, (pp. 747-754).

Knossow, D., Ronfard, R., & Horaud, R. (2008). Human motion tracking with a kinematic parameterization of extremal contours. *International Journal of Computer Vision, 79*(3), 247–269. doi:10.1007/s11263-007-0116-2

Lawrence, R., & Rabiner, A. (1989). Tutorial on hidden Markov models and selected applications in speech recognition . *Proceedings of the IEEE, 77*(2), 257–286. doi:10.1109/5.18626

Lee, M. W., & Cohen, I. (2006). A model-based approach for estimating human 3D poses in static images. *IEEE Transactions on Pattern Analysis and Machine Intelligence, 28*(6), 905–916. doi:10.1109/TPAMI.2006.110

Murray, R. M., Li, Z., & Sastry, S. S. (1994). *A mathematical introduction to robotic manipulation*. Boca Raton, FL: CRC Press.

Ninh, H., Han, T. X., Walther, D. B., Liu, M., & Huang, S. (2009). Hierarchical space-time model enabling efficient search for human actions. *IEEE Transactions on Circuits and Systems for Video Technology, 19*(6), 808–820. doi:10.1109/TCSVT.2009.2017399

Niu, F., & Abdel-Mottaleb, M. (2004). View-invariant human activity recognition based on shape and motion features. *Proceedings of the IEEE Sixth International Symposium on Multimedia Software Engineering* (pp. 546-556).

Niu, F., & Abdel-Mottaleb, M. (2005). HMM-based segmentation and recognition of human activities from video sequences. *Proceedings of IEEE International Conference on Multimedia & Expo* (pp. 804-807).

Olivier, B., Pascal, C. C., & Arnaud, B. (2009). Fast nonparametric belief propagation for real-time stereo articulated body tracking. *Computer Vision and Image Understanding*, *113*(1), 29–47. doi:10.1016/j.cviu.2008.07.001

Plankers, R., & Fua, P. (2003). Articulated soft objects for multiview shape and motion capture. *IEEE Transactions on Pattern Analysis and Machine Intelligence*, *25*(9), 1182–1187. doi:10.1109/TPAMI.2003.1227995

Ramanan, D., Forsyth, D. A., & Zisserman, A. (2007). Tracking people by learning their appearance. *IEEE Transactions on Pattern Analysis and Machine Intelligence*, *29*(1), 65–81. doi:10.1109/TPAMI.2007.250600

Roberts, T. J., McKenna, S. J., & Ricketts, I. W. (2007). Human pose estimation using partial configurations and probabilistic regions. *International Journal of Computer Vision*, *73*(3), 285–306. doi:10.1007/s11263-006-9781-9

Robertson, N., & Reid, I. (2006). A general method for human activity recognition in video . *Computer Vision and Image Understanding*, *104*(2), 232–248. doi:10.1016/j.cviu.2006.07.006

Roweis, S. T., & Saul, L. K. (2000). Nonlinear dimensionality reduction by locally linear embedding. *Science*, *290*(5500), 2323–2326. doi:10.1126/science.290.5500.2323

Sundaresan, A., & Chellapa, R. RoyChowdhury, R. (2004). Multiple view tracking of humans modeled by kinematic chains. *Proceedings of IEEE International Conference on Image Processing*, (pp. 1009-1012).

Sundaresan, A., & Chellapa, R. (2008). Model driven segmentation of articulating humans in Laplacian Eigenspace. *IEEE Transactions on Pattern Analysis and Machine Intelligence*, *30*(10), 1771–1785. doi:10.1109/TPAMI.2007.70823

Taylor, C. J. (2000). Reconstruction of articulated objects from point correspondences in a single uncalibrated image. *Computer Vision and Image Understanding*, *80*(3), 349–363. doi:10.1006/cviu.2000.0878

Tenenbaum, J. B., De Silva, V., & Langford, J. C. (2000). A global geometric framework for nonlinear dimensionality reduction. *Science*, *290*(5500), 2319–2323. doi:10.1126/science.290.5500.2319

Thang, N. D., Kim, T.-S., Lee, Y.-K., & Lee, S. (2010a). Estimation of 3D human body posture via co-registration of 3D human model and sequential stereo information. *Applied Intelligence*. doi:.doi:10.1007/s10489-009-0209-4

Thang, N. D., Kim, T.-S., Lee, Y.-K., & Lee, S. (2010b). Fast 3D human motion capturing from stereo data using Gaussian clusters. *International Conference on Control, Automation and Systems*, (pp. 1428-1431).

Toyoda, T., & Hasegawa, O. (2008). Random field model for integration of local information and global information. *IEEE Transactions on Pattern Analysis and Machine Intelligence*, *30*(8), 1483–1489. doi:10.1109/TPAMI.2008.105

Uddin, M. Z., Lee, J. J., & Kim, T.-S. (2009). Independent shape component-based human activity recognition via hidden Markov model. *Applied Intelligence*, *33*(2). doi:.doi:10.1007/s10489-008-0159-2

Uddin, M. Z., Truc, P. T. H., Lee, J. J., & Kim, T.-S. (2008). Human activity recognition using independent component features from depth images, *Proceedings of the 5th International Conference on Ubiquitous Healthcare*, (pp. 181-183).

Urtasun, R., Fleet, D., & Fua, P. (2006). Temporal motion models for monocular and multiview 3D human body tracking. *Computer Vision and Image Understanding*, *104*(2), 157–177. doi:10.1016/j.cviu.2006.08.006

Wang, L., Tan, T., Ninh, H., & Hu, W. (2003). Silhouette analysis-based gait recognition for human identification. *IEEE Transactions on Pattern Analysis and Machine Intelligence, 25*(12), 1505–1518. doi:10.1109/TPAMI.2003.1251144

Yamato, J., Ohya, J., & Ishii, K. (1992). Recognizing human action in time-sequential images using hidden Markov model, *Proceedings of IEEE International Conference on Computer Vision and Pattern Recognition*, (pp. 379-385).

Yang, H. D., & Lee, S. W. (2007). Reconstruction of 3D human body pose from stereo image sequences based on top-down learning. *Journal of Pattern Recognition, 40*(11), 3120–3131. doi:10.1016/j.patcog.2007.01.033

ADDITIONAL READING

Besl, P., & McKay, N. (1992). A method for registration of 3D shapes. *IEEE Transactions on Pattern Analysis and Machine Intelligence, 14*(2), 239–256. doi:10.1109/34.121791

Bregler, C., Malik, J., & Pullen, K. (2004). Twist based acquisition and tracking of animal and human kinematics. *International Journal of Computer Vision, 56*(3), 179–194. doi:10.1023/B:VISI.0000011203.00237.9b

Cedras, C., & Shah, M. (1995). Motion-based recognition: A survey. *Image and Vision Computing, 13*(2), 129–155. doi:10.1016/0262-8856(95)93154-K

Chang, I., & Lin, S.-Y. (2010). 3D human motion tracking based on a progressive particle filter. *Journal of Pattern Recognition, 43*(10), 3621–3635. doi:10.1016/j.patcog.2010.05.003

Cheung, K., Baker, S., & Kanade, T. (2005). Shape-from-silhouette across time part I: Theory and algorithm. *International Journal of Computer Vision, 62*(3), 221–247. doi:10.1007/s11263-005-4881-5

Corazza, S., Mündermann, L., Gambaretto, E., Ferrigno, G., & Andriacchi, T. P. (2010). Markerless motion capture through visual hull, articulated ICP and subject specific model generation. *International Journal of Computer Vision, 87*(1-2). doi:10.1007/s11263-009-0284-3

Darby, J., Li, B., & Costen, N. (2010). Tracking human pose with multiple activity models. *Journal of Pattern Recognition, 43*(9), 3042–3058. doi:10.1016/j.patcog.2010.03.018

Dimitrijevic, M., Lepetit, V., & Fua, P. (2006). Human body pose detection using Bayesian spatio-temporal templates . *Computer Vision and Image Understanding, 104*(2-3), 127–139. doi:10.1016/j.cviu.2006.07.007

Felzenswalb, P., & Huttenlocher, D. (2005). Pictorial structures for object recognition. *International Journal of Computer Vision, 61*(1), 55–79. doi:10.1023/B:VISI.0000042934.15159.49

Forsyth, D. A., & Ponce, J. (2003). *Computer vision - a modern approach.* Upper Saddle River, New Jersey: Prentice Hall.

Fossati, A., Dimitrijevic, M., Lepetit, V., & Fua, P. (2010). From canonical poses to 3D motion capture using a single camera. *IEEE Transactions on Pattern Analysis and Machine Intelligence, 32*(7), 1165–1181. doi:10.1109/TPAMI.2009.108

Kakadiaris, I. A., & Metaxas, D. (1998). Three-dimensional human body model acquisition from multiple views. *International Journal of Computer Vision, 30*(3), 191–218. doi:10.1023/A:1008071332753

Lee, H. J., & Chen, Z. (1985). Determination of 3D human body posture from a single view. *Computer Vision Graphics and Image Processing*, *30*(2), 148–168. doi:10.1016/0734-189X(85)90094-5

Lee, M. W., & Nevatia, R. (2009). Human Pose Tracking in Monocular Sequence Using Multilevel Structured Models. *IEEE Transactions on Pattern Analysis and Machine Intelligence*, *31*(1), 27–38. doi:10.1109/TPAMI.2008.35

Martin, F., & Horaud, R. (2002). Multiple camera tracking of rigid objects. *The International Journal of Robotics Research*, *21*(2), 97–113. doi:10.1177/027836402760475324

Mikic, I., Trivedi, M. M., Hunter, E., & Cosman, P. C. (2003). Human body model acquisition and tracking using voxel data. *International Journal of Computer Vision*, *53*(3), 199–223. doi:10.1023/A:1023012723347

Moeslund, T. B., Hilton, A., & Krüger, V. (2006). A survey of advances in vision-based human motion capture and analysis. *Computer Vision and Image Understanding*, *104*(2), 90–126. doi:10.1016/j.cviu.2006.08.002

Mori, G., & Malik, J. (2006). Recovering 3D human body configurations using shape contexts. *IEEE Transactions on Pattern Analysis and Machine Intelligence*, *28*(7), 1052–1062. doi:10.1109/TPAMI.2006.149

O'Rourke, J., & Badler, N. I. (1980). Model-based image analysis of human motion using constraint propagation. *IEEE Transactions on Pattern Analysis and Machine Intelligence*, *2*(6), 522–536.

Ong, E.-J., Micilotta, A. S., Bowden, R., & Hilton, A. (2006). Viewpoint invariant exemplar-based 3D human tracking. *Computer Vision and Image Understanding*, *104*(2), 178–189. doi:10.1016/j.cviu.2006.08.004

Poppe, R. (2007). Vision-based human motion analysis: An overview. *Computer Vision and Image Understanding*, *108*(1-2), 4–18. doi:10.1016/j.cviu.2006.10.016

Roberts, T. J., McKenna, S. J., & Ricketts, I. W. (2007). Human pose estimation using partial configurations and probabilistic regions. *International Journal of Computer Vision*, *73*(3), 285–306. doi:10.1007/s11263-006-9781-9

Scharstein, D., & Szeliski, R. (2002). A taxonomy and evaluation of dense two-frame stereo correspondence algorithms. *International Journal of Computer Vision*, *47*(1-3), 7–42. doi:10.1023/A:1014573219977

Shankar, R. R., Allen, Y. Y., Shankar, S., & Yi, M. (2010). Robust algebraic segmentation of mixed rigid-body and planar motions from two views. *International Journal of Computer Vision*, *88*(3), 425–446. doi:10.1007/s11263-009-0314-1

Sidenbladh, H., & Black, M. J. (2003). Learning the statistics of people in images and video. *International Journal of Computer Vision*, *54*(1-3), 181–207. doi:10.1023/A:1023765619733

Signal, R., Balan, A., & Black, M. J. (2010). HumanEva: Synchronised video and motion capture dataset and baseline algorithm for evaluation of articulated human motion. *International Journal of Computer Vision*, *87*(1-2), 4–27. doi:10.1007/s11263-009-0273-6

Sminchisescu, C., Kanaujia, A., & Metaxas, D. (2006). Conditional models for contextual human motion recognition. *Computer Vision and Image Understanding*, *104*(2-3), 210–220. doi:10.1016/j.cviu.2006.07.014

KEY TERMS AND DEFINITIONS

Depth Map: A 2-D image representing the depth information of a scene using gray-scaled colors.

Human Computer Interaction (HCI): A research of interaction between users and computers.

Maker-Based Human Motion Capture: An approach of capturing human motion by attaching markers to the human body. The trajectories of the markers detected in 3D space provide the motion information of the tracked subject.

Markerless-Based Human Motion Capture: An approach of capturing human motion without using markers.

Stereo Camera: A type of camera composed of two or more lenses to allow taking some pictures of a scene in alternate view angles to estimate the information of depth.

Stereo Matching Algorithm: An algorithm used to generate the depth map from a pair of images captured by a stereo camera.

Stereopsis: A process of combining two images received from two human eyes to create a 3D sensation about viewed objects.

Chapter 29
3D Face Recognition using an Adaptive Non-Uniform Face Mesh

Wei Jen Chew
The University of Nottingham, Malaysia

Kah Phooi Seng
The University of Nottingham, Malaysia

Li-Minn Ang
The University of Nottingham, Malaysia

ABSTRACT

Face recognition using 3D faces has become widely popular in the last few years due to its ability to overcome recognition problems encountered by 2D images. An important aspect to a 3D face recognition system is how to represent the 3D face image. In this chapter, it is proposed that the 3D face image be represented using adaptive non-uniform meshes which conform to the original range image. Basically, the range image is converted to meshes using the plane fitting method. Instead of using a mesh with uniform sized triangles, an adaptive non-uniform mesh was used instead to reduce the amount of points needed to represent the face. This is because some parts of the face have more contours than others, hence requires a finer mesh. The mesh created is then used for face recognition purposes, using Principal Component Analysis (PCA) and Linear Discriminant Analysis (LDA). Simulation results show that an adaptive non-uniform mesh is able to produce almost similar recognition rates compared to uniform meshes but with significant reduction in number of vertices.

INTRODUCTION

With the advancement of technology, face recognition using 3D images is slowly becoming a feasible option. This is because the ability to capture 3D images of faces accurately and efficiently is now becoming a reality. In the past, face recognition systems used 2D images to perform identification (Turk & Pentland, 1991; Belhumeur, Hespanha & Kriegman, 1997). Although 2D images can produce good results, they still suffer from a few problems, like illumination and pose changes (Zhao, Chellappa, Rosenfeld & Phillips, 2000).

DOI: 10.4018/978-1-61350-326-3.ch029

Since 3D images are able to overcome problems usually faced when using 2D images for face recognition, like illumination and pose changes, hence active research should be performed on face recognition using 3D images so that when the technology to capture 3D images efficiently finally arrives, the software to perform good recognition would also be ready.

An area that is worth investigating with regard to 3D face recognition is the representation of the 3D face image. At the moment, a common method to capture a 3D face image is to use a 3D scanner. Basically, the scanner will use either laser or light and shine onto the face, thereby obtaining a range image that consists of points with x, y and z coordinates. The x and y coordinate is the width and height of the point while the z coordinate represents the depth of that face point from the scanner. Therefore, the 3D image of the face from one direction is obtained.

Range image is able to represent the face accurately and has been used for 3D face recognition. They are able to show the exact contours of the face when the point matrix is dense enough. However, the image size can be huge due to the amount of data stored which takes up storage space or causes long transmission time. Besides that, some area of the face may not need a large amount of points to represent it, causing some points to be redundant.

Therefore, face meshes have been proposed in the past to represent a 3D face image for face recognition since they require fewer points. A mesh consists of many small polygons that make up the face. The advantage of using a face mesh is that each polygon represents a small area of a face. If the polygon used is a triangle, then a small area that may be represented by many data points in a range image can be represented by only 3 points in a mesh image. The points in a mesh are known as vertices while the lines joining the vertices are known as edges.

In this chapter, the aim is to create a mesh that is able to represent a face sufficiently for 3D face recognition while using minimum number of vertices. Hence, it is proposed that an adaptive non-uniform face mesh be built to represent the original face range image. This mesh will then be used for face recognition purposes to determine if the proposed face mesh is a feasible alternative.

BACKGROUND

A typical mesh consists of many uniform little triangles that cover the whole face, as shown in Figure 1. (Xu et al., 2004) method converts a point cloud face into a mesh, first by using a coarse mesh and then subsequently refining it to a finer dense mesh to represent a face. After that, recognition is done using the face meshes. Although recognition usually concentrates on the face area around the eyes, nose and mouth, the whole face was converted into a finer dense mesh for recognition.

(Ansari et al., 2007) used a general mesh model of a face and deformed this model according to the range image of the face, estimating the depth of the triangles using plane fitting. To obtain a smoother mesh, they subdivide each triangle into 4 smaller triangles and then deform the mesh again using plane fitting to get a more accurate mesh for the face. After that, recognition is performed using a voting-based classifier. However, the criteria to determine whether the mesh is accurate enough for the face were not discussed.

(Tanaka et al., 1993) also performed subdivision on their triangles. However, they only divided their mesh triangle into 2 triangles instead of 4. The meshes were subdivided according to their surface curvature and will only stop the subdivision at a certain predetermined threshold.

(Wu et al., 2001) proposed a geometric mesh simplification scheme for constructing multi-resolution meshes. Using the Face Constriction Process (FCP), they introduce a mesh simplification scheme that was also effective in preserving the face features. This was achieved by using the

Figure 1. Uniform face mesh

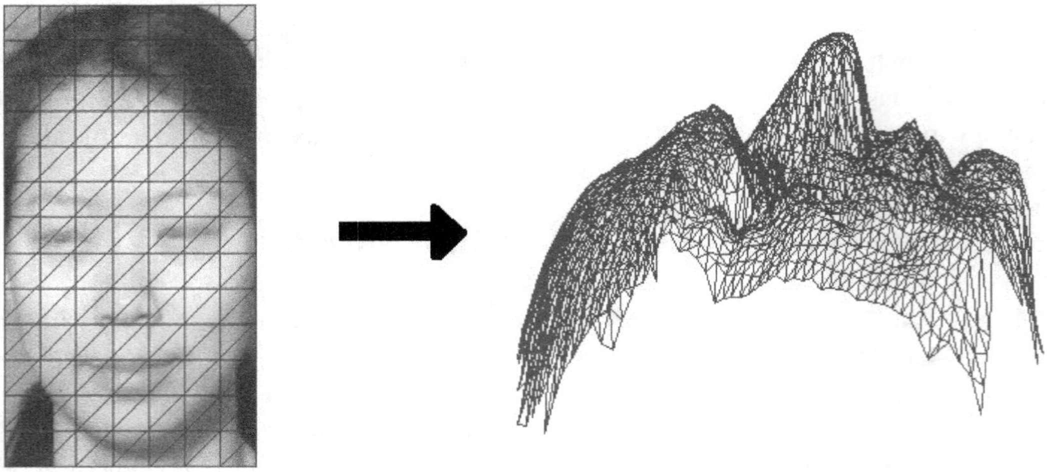

FCP validity checking and weight ordering equation. It was claimed that their scheme has simple computation, saves time and is easy to implement.

(Lee et al., 2005) simplifies their meshes through vertex elimination and integration. This is achieved by removing vertices that meet the specified values in minimum distance and curvature as well as reducing the number of meshes that display identical shapes. They claim that their proposed method reduces the face model interpolation time and is able to produce 3D face models without distorting the original model's shape, size and figuration.

(Fahn et al., 2002) performs mesh simplification by using the quadratic error metric. Using a series of edge contractions guided by the quadratic error metric to produce a progressive mesh, high fidelity approximations were created. It is claimed that this method is simple, fast and memory efficient.

However, even meshes can be modified to further reduce the number of points representing a face. (Fang et al., 2004) simplified their triangle mesh using a triangle list. Basically, they record out their mesh vertices in a list and then arrange them in a certain order. To reduce the amount of triangles, 2 vertices are then collapsed, hence substituting one vertex with an existing one. This is done directly using the list since it has been rearranged in a certain order. The aim of their paper was to be able to transmit their mesh progressively.

FACE MESH MANIPULATION

Based on what was discussed earlier, most research creates meshes that are uniform in size throughout the whole face. Finer meshes would be needed to represent areas of the face that have many contours like the nose, eyes and mouth. Hence, to obtain good recognition rates, a face would need to be represented by a fine mesh.

However, not every area of the face needs to be represented by fine meshes since areas like the forehead can sufficiently be represented by coarser meshes. Hence, a non-uniform face mesh that can adapt the type of mesh with the face area is proposed. This adaptive non-uniform mesh would be able to produce good recognition rates with lower amount of vertices.

Commonly, 3D images are captured and stored in range image format. Therefore, the first step is to convert the original range image format into a mesh. This can be performed using the plane fitting method. An adaptive non-uniform mesh can then be obtained by either building a non-uniform mesh directly or building a uniform mesh first and then manipulating the mesh using methods like edge collapse or Face Constriction Process.

Range Image

Range image is an image with height, width and depth values for each point or pixel. The height and width values would be similar to those in a 2D plane while the depth value would be the distance of the point from the 3D scanner. The depth value would be from the direction of capture. Since the depth value should be constant in any type of illumination, they would be able to provide more consistent values for different environments compared to a 2D image. Having the height, width and depth values also enables the image to be rotated around, hence able to handle different pose position. However, to have an accurate representation of a face, many individual points packed closely together would be needed, therefore causing a range image file size to be large. This can cause storage and transmission problems.

Plane Equation

A method to reduce the amounts of points used to represent a 3D face image is to convert the range image originally captured to a mesh. This is because an area in a mesh formed by 3 points is able to represent a group of range image points. To convert from range image to mesh, each triangle in the mesh would be superimposed on top of a group of range image points. Then, to obtain the depth, or z coordinate, of each point on the triangle, the plane equation for the group of range image points is calculated using the equation (1) (Bourke P., 1989).

$$Ax + By + Cz + D = 0 \quad (1)$$

where the normal to the plane is the vector (A,B,C).

Therefore, to estimate the depth value of each point on the mesh triangle, the 3 pairs of x and y coordinates are inserted into the equation to obtain each z values. This results in a mesh that has different depth at different points, corresponding to the range image.

Edge Collapse and Vertex Split

To manipulate a face mesh to change the sizes of the triangles inside the mesh, a method commonly used is edge collapse. Basically, an edge will be removed, causing two vertices to combine into one vertex, hence reducing the amount of points used for the mesh. The opposite of an edge collapse is the vertex split, which means splitting one vertex into two. These two methods will change the sizes of the triangles surrounding the collapse or split. An edge collapse creates lesser but larger triangles while a vertex split creates more but smaller triangles. Therefore, the collapse should be performed at parts of the face with less contours while the split should be performed at parts of the face with many contours. Figure 2 shows an example of an edge collapse and vertex split.

Face Constriction Process (FCP)

In this method, instead of removing an edge or vertices, the whole triangle is removed instead. There are 2 different methods to rebuild the hole in the mesh. (Gieng et al., 1997) replaces the removed triangle with a vertex and the 3 adjacent triangles surrounding the removed triangle are converted to edges. As for (Hamann, 1994) method, the investigated triangle as well as the surrounding triangles were removed and then the hole will be retriangulated. The method changes the topology of the mesh more drastically. Figure 3 shows an example of FCP by (Gieng et al. 1997).

Although the mesh manipulation methods are able to split the mesh, however each manipulation method would require extra computation. For the

Figure 2. Example of edge collapse and vertex split

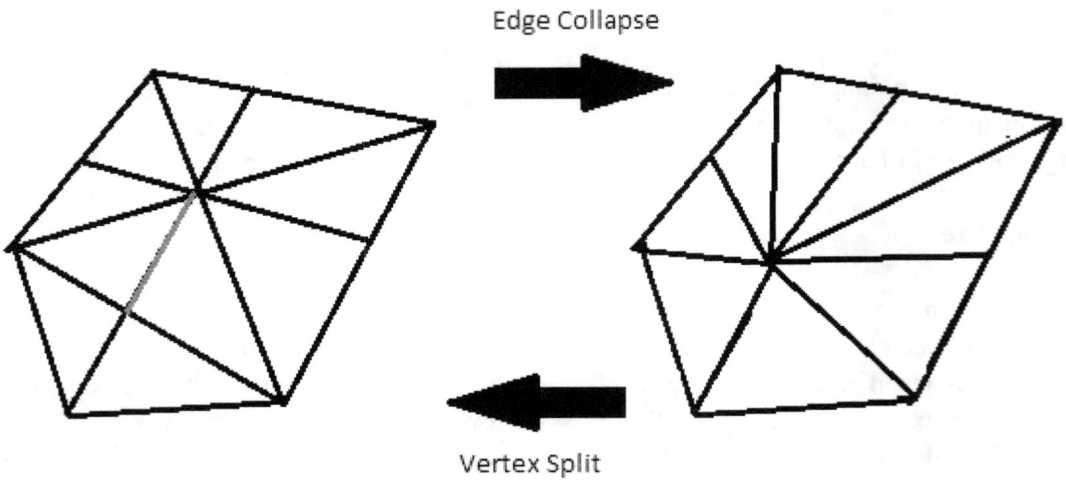

edge collapse and vertex split method, which vertex to split and which edge to collapse have to be determined using a set of rules. For the splitting triangle method, which point to split at as well as which edge the point will be placed would also need to be determined. As for the FCP method, after removing the triangles, the point to join back all the points to cover the hole left would need to be computed. Hence, it is proposed that the adaptive non-uniform mesh is built directly instead of having a uniform mesh and then manipulating it using the methods discussed.

FACE RECOGNITION METHOD

For the recognition section, it is proposed that PCA (Turk & Pentland, 1991) and LDA (Belhumeur et al., 1997) are performed and then to determine the identity of the unknown probe face when compared with the faces in the database, Nearest Neighbour Euclidean Distance would be used. The reason for using both PCA and LDA is because although PCA is good for dimensionality reduction, it lacks discrimination ability (Mandal et al., 2007). Therefore, LDA is performed after PCA to optimize classification.

Figure 3. Example of FCP

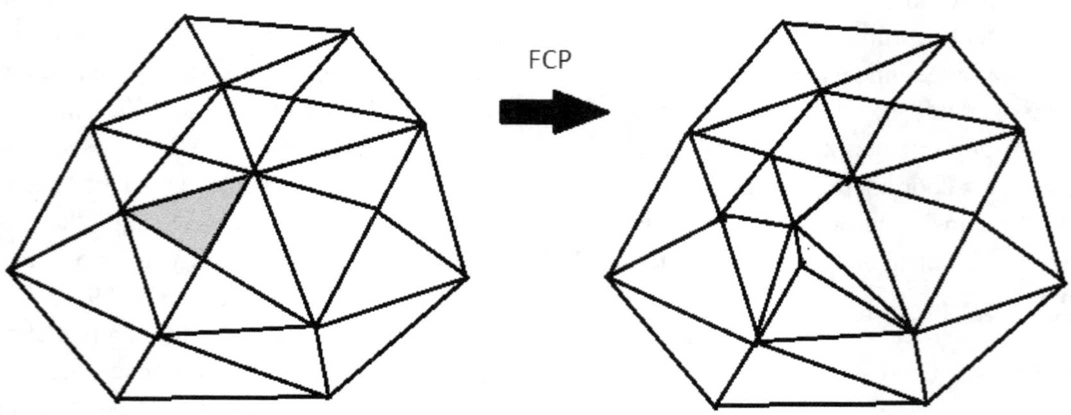

Principal Component Analysis (PCA)

This type of analysis is a statistical algorithm that is used to approximate the original data with lower dimensional feature vectors (Turk & Pentland, 1991). To use PCA for face recognition, the first step is to convert each range image in the database into 1D vector by concatenating the rows or columns into a long vector. Then, the mean is calculated using equation (3) by summing the entire database 1D vector together and then dividing by the amount of faces in the database. Each face is then centered by subtracting the mean image from each face image using equation (2) (Turk & Pentland, 1991).

$$\overline{x}^{i} = x^{i} - m \tag{2}$$

where

$$m = \frac{1}{p}\sum_{i=1}^{p} x^{i} \tag{3}$$

Next, the data matrix is created by combining the centered database image side-by-side to create a data matrix. The covariance matrix is then be calculated by multiplying the data matrix with its transpose, as in equation (4) (Turk & Pentland, 1991).

$$\Omega = \overline{X}\overline{X}^{T} \tag{4}$$

This is followed by the calculation of the eigenvalues and eigenvectors for the covariance matrix using equation (5) (Turk & Pentland, 1991).

$$\Omega V = \lambda V \tag{5}$$

where V is the eigenvectors set and λ is the corresponding eigenvalues

An eigenspace is created by the sorted eigenvectors matrix. Finally, the centered training images are projected into the eigenspace created. The projection is the dot product of the centered training image with each of the ordered eigenvectors calculated.

Linear Discriminant Analysis (LDA)

For LDA, the first step is to calculate the within class scatter matrix which shows the amount of scatter between training images in the same class. The scatter matrix is calculated using equation (6) where S_i is the scatter matrix and m_i is the mean of the training images (Belhumeur et al., 1997).

$$S_i = \sum_{x \in X_i} (x - m_i)(x - m_i)^{T} \tag{6}$$

The within class scatter matrix, which is the sum of all the scatter matrices, is calculated using equation (7) where S_w is the within class scatter matrix and C is the number of classes (Belhumeur et al., 1997).

$$S_w = \sum_{i=1}^{C} S_i \tag{7}$$

Next, the between class scatter matrix is calculated using equation (8) where S_B is the between class scatter matrix, n_i is the number of images in the ith class and m is the total mean of all training images (Belhumeur et al., 1997).

$$S_B = \sum_{i=1}^{C} n_i (m_i - m)(m_i - m)^{T} \tag{8}$$

The generalized eigenvectors (V) and eigenvalues (λ) of the within class and between class matrices are calculate using equation (9) (Belhumeur et al., 1997).

$$S_B V = \lambda S_W V \tag{9}$$

Sorting the non-zero eigenvectors from high to low according to the corresponding eigenvalues, the Fisher basis vector is formed. Calculating the dot product of the training images with each of the Fisher basis vectors, the training images will be projected onto the Fisher basis vectors.

To identify the test image, it is projected onto the Fisher basis vector and the Euclidean distances between the test and training images is calculated. The class that the test image belongs to is indicated by the shortest Euclidean distance.

PROPOSED ADAPTIVE NON-UNIFORM FACE MESH CONSTRUCTION METHOD

In this chapter, the aim is to build a face mesh from range image so that it contains minimal amount of vertices yet is still able to provide reasonable recognition rate. It is proposed that an adaptive non-uniform face mesh would be a feasible option since different parts of the face would require different sizes of mesh. This is because part of the face with more contour changes like the eyes, nose and mouth area would require finer meshes compared to the forehead area.

Although the mesh manipulation methods discussed earlier shows potential in changing a uniform mesh to non-uniform, extra computation would be needed to determine the factors like the splitting and collapsing points. Therefore, it is decided that instead of building a uniform face mesh first and then refining it through those mesh manipulation methods, it was decided that the adaptive non-uniform mesh would be directly built from the range image, using only the plane fitting method.

For the proposed method, first, a coarse mesh is placed over the face range image and then plane fitting is performed to obtain the depth of each of the triangles in the mesh. However, before that triangle is accepted as part of the non-uniform face mesh built, the distance of all the range image points within that triangle will be calculated using equation (10).

Point Distance =

$$\sum_{i=1}^{n} \frac{A \times x(i) + B \times y(i) + C \times z(i) + D}{\sqrt{A^2 + B^2 + C^2}} \quad (10)$$

where n is the total number of image points within the triangle

If the average distance is above a certain threshold set, then it means that the estimated plane is unable to accurately represent that group of range image points. Therefore, that triangle will be removed from the mesh and the range image points will remain. However, if the average distance is below or equal to the threshold, then the triangle will be accepted as part of the non-uniform face mesh built and that group of range image points will be removed from the range image. Once the whole image is tested with the coarse mesh, a finer mesh would be introduced and whole process will be repeated again with the remaining range image points.

Using this method, the non-uniform mesh is able to be adapted to the range image to accurately represent the face. Instead of using fine and coarse meshes at predetermined face areas, this method enables the mesh to decide whether a certain area of the face requires coarse or finer meshes. Hence, no vertices would be wasted on areas that do not require a finer mesh.

However, due to the depth value of each triangle in the mesh is calculated separately, therefore the mesh is still disjointed. To smoothen out the mesh, all the z-coordinates surrounding a pixel location are recorded and the z-coordinate of the investigated location will be the median of all the surrounding values. To eliminate outliers in the face mesh, all the z-coordinates must be more than Q1-(1.5xIQR) and less than Q3+(1.5xIQR) where Q1, Q3 and IQR are the 1st quartile, 3rd quartile and inter-quartile range respectively. The final adaptive face mesh obtained is shown in Figure 4.

Figure 4. Adaptive face mesh

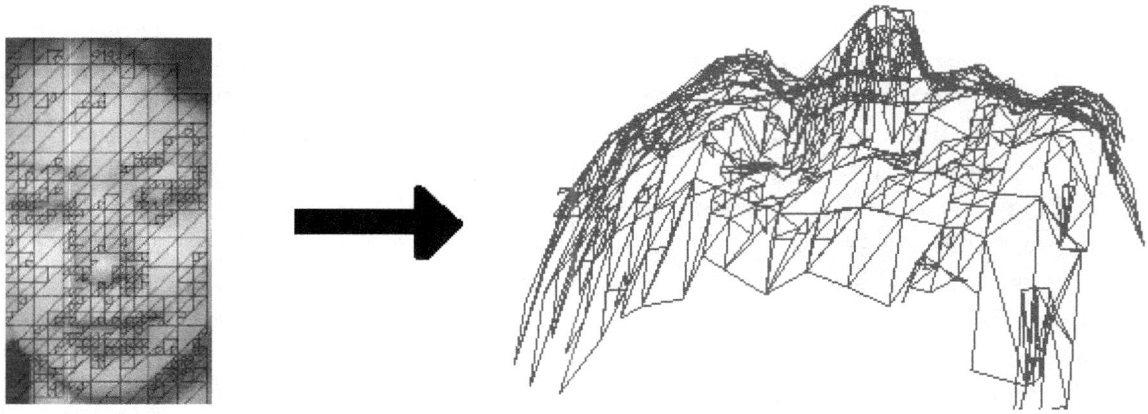

It can be observed from Figure 4 that the finer meshes are concentrated at the eyes, nose and mouth areas while the forehead and cheek areas uses coarser meshes.

Next, to prove that the proposed adaptive non-uniform face mesh is able to produce good recognition rates, the proposed mesh recognition rates are compared to other uniform face meshes. In this chapter, face recognition was performed using Principal Component Analysis (PCA) (Turk & Pentland, 1991) followed by Linear Discriminant Analysis (LDA) (Belhumeur et al., 1997). This is because although PCA is good for dimensionality reduction, it lacks discrimination ability. Therefore, LDA is performed after PCA to optimize classification (Mandal et al., 2007).

To obtain the training and probe data to perform PCA and LDA, it was decided that the z-axis depth data be used. The face area needed for this is 100 pixels above the nose tip, 50 pixels below the nose tip and 50 pixels to the left and right of the nose tip. A set area was used to produce a consistent training and probe set.

Since the mesh does not have points at every location of the image, the training and probe set points were found by first determining which triangle on the mesh does the wanted point falls into. Once the triangle is found, the triangle plane equation is calculated using the (11), (12), (13) and (14).

$$A = y_1(z_2 - z_3) + y_2(z_3 - z_1) + y_3(z_1 - z_2) \tag{11}$$

$$B = z_1(x_2 - x_3) + z_2(x_3 - x_1) + z_3(x_1 - x_2) \tag{12}$$

$$C = x_1(y_2 - y_3) + x_2(y_3 - y_1) + x_3(y_1 - y_2) \tag{13}$$

$$D = -(x_1(y_2z_3 - y_3z_2) + x_2(y_3z_1 - y_1z_3) + x_3(y_1z_2 - y_2z_1)) \tag{14}$$

where (x_1,y_1), (x_2,y_2) and (x_3,y_3) are the 3 corner points of the triangle.

Once the above information is calculated, the plane equation of the triangle is known. Then, by inserting the x and y coordinate of the investigated point into the plane equation, the z coordinate of the mesh at that location is obtained.

After that, PCA followed by LDA can be performed on both training and probe sets. The training image that has the shortest Euclidean distance to the probe image in LDA eigenspace will be determined as the identity of the unknown probe person.

RESULTS & DISCUSSION

For this chapter, the UND 3D face database was used (Flynn et al., 2003; Chang et al., 2003). Each subject in this database comprises of 2 files, which are a .ppm image file and a .abs data file. The data file contains the number of rows information, number of columns information and values for 4 matrices, which are the flag, x-coordinates, y-coordinates and z-coordinates. The flag matrix uses a 1 to indicate that the position has a valid value and 0 to indicate otherwise. The training set created for this experiment consists of 50 different people and there are 3 meshes for each person. This means that there were a total of 150 training meshes used. As for the probe set, it consists of 30 different people.

Face recognition was performed using PCA followed by LDA on the 4 different types of face meshes. The 4 meshes are a fully coarse mesh, a fully fine mesh, a fixed non-uniform mesh and the proposed adaptive non-uniform mesh. A fixed non-uniform mesh contains fine mesh between the eye level to the mouth level while the rest of the face is represented by a coarse mesh. Table 1 shows the recognition rate for the four different meshes.

From Table 1, it is observed that a fully fine mesh provided the highest recognition rate, while a fully coarse mesh gave the lowest recognition rate. This is expected since a fine mesh is able to represent the contours of a face more accurately compared to a coarse mesh. Both non-uniform meshes gave slightly lower recognition rate compared to a fully fine mesh. Hence, this shows that these non-uniform meshes could be a feasible alternative to represent the faces since they have the advantage of being constructed by fewer vertices as shown in Table 2.

From Table 2, it is observed that a fully coarse mesh contains the least amount of vertices while a fully fine mesh contains the most amount of vertices. However, with such a large difference in their recognition rate, therefore it would not be practical to use a fully coarse mesh to represent a face for recognition purposes even though they contain fewer vertices.

However, for non-uniform meshes, their recognition rate is only slightly lower than a fully fine mesh recognition rate yet only contain about less than half the amount of vertices used for a fully fine mesh. Hence, a non-uniform mesh could be considered as a file size saving alternative to represent a face.

Comparing both the non-uniform meshes, it can be observed that they have similar recognition rates but the proposed adaptive non-uniform mesh contains an average of about 1000 vertices less than the fixed non-uniform mesh. This is because the adaptive mesh is built according to the contours of each individual face, using finer meshes only at parts of the face that needs them. Comparatively, the fixed mesh just assumes that the area between the eyes and mouth would need finer meshes, thereby some places within this area that does not need finer meshes would also be included.

Table 1. Recognition rate for different meshes

Mesh Types	Recognition Rate
Fully Coarse Mesh	73%
Fully Fine Mesh	87%
Fixed Non-Uniform Mesh	83%
Adaptive Non-Uniform Mesh	83%

Table 2. Average number of vertices

Mesh Types	Average Number of Vertices
Fully Coarse Mesh	922
Fully Fine Mesh	8814
Fixed Non-Uniform Mesh	3832
Adaptive Non-Uniform Mesh	2711

FUTURE RESEARCH DIRECTIONS

Currently, a basic recognition method is used to compare the recognition ability of various different face meshes. Hence, the next step is to improve the recognition rate that can be obtained by an adaptive non-uniform face mesh. This will further encourage the use of this type of meshes.

CONCLUSION

In this chapter, the objective was to build a face mesh that is able to produce good recognition rates while having minimal number of vertices. It was proposed that an adaptive non-uniform face mesh be built to achieve this target, instead of obtaining an initial face mesh and then trying to simplify it using mesh manipulation methods. The mesh would be able to adapt different mesh size to different parts of the face. The proposed adaptive face mesh was built from the range image using plane fitting. The depth of each triangle in the mesh was estimated using this method and then the average distance of all the range values within the triangle was calculated. If the average distance was within a certain threshold, then the triangle depth is accepted, if not, a smaller triangle will be used in that area to create the mesh. The face recognition rates obtained in simulations show that the proposed adaptive non-uniform face mesh was able to produce good recognition rates while using less than half the amount of vertices needed to build a fully fine face mesh. Hence, this type of face mesh should be further researched since they have the advantage of containing less vertices which causes the file size to be smaller, therefore having the ability to reduce storage space and transmission time.

REFERENCES

Ansari, A., Abdel-Mottaleb, M., & Mahoor, M. H. (2007). *3D face mesh modeling from range images for 3D face recognition.* Accepted in IEEE International Conference on Image Processing, ICIP.

Belhumeur, P. N., Hespanha, J. P., & Kriegman, D. J. (1997). Eigenfaces vs. Fisherfaces: Recognition using class specific linear projection. *IEEE Transactions on Pattern Analysis and Machine Intelligence, 19*(7), 711–729. doi:10.1109/34.598228

Bourke, P. (1989). *Equation of a plane*. Retrieved January 5, 2009, from http://local.wasp. uwa.edu. au/ ~pbourke/ geometry/ planeeq/

Chang, K., Bowyer, K. W., & Flynn, P. J. (2003). Face recognition using 2D and 3D facial data. *ACM Workshop on Multimodal User Authentication*, (pp. 25-32).

Chin-Shyurng, F., Hung-Kuang, C., & Yi-Haur, S. (2002). Polygonal mesh simplification with face color and boundary edge preservation using quadric error metric. *ISMSE, 2002*, 174–181.

Fang, T.-Z., Hu, Z.-G., & Jin, W.-K. (2004). Progressive transmission of single-resolution mesh image. *Electronics Letters, 40*(16). doi:10.1049/el:20040598

Flynn, P. J., Bowyer, K. W., & Phillips, P. J. (2003). *Assessment of time dependency in face recognition: An initial study* (pp. 44–51). Audio and Video-Based Biometric Person Authentication.

Gieng, T. S., Hamann, B., & Joy, K. I. (1997). Smooth hierarchical surface triangulations. *Proceedings of IEEE Visualization, 97*, 379–386.

Hamann, B. (1994). A data reduction scheme for triangulated surfaces. *Computer Aided Geometric Design, 11*, 197–214. doi:10.1016/0167-8396(94)90032-9

Hyun Cheol, L., Eun Seok, K., & Gi Tack, H. (2005). Fast 3D face modeling using mesh optimization and vertex integration. *Proceedings of 7th International Workshop on Enterprise networking and Computing in Healthcare Industry.* HEALTHCOM 2005.

Mandal, B., Jiang, X. D., & Kot, A. (2007). *Dimensionality reduction in subspace face recognition.* IEEE International Conference on Information, Communications and Signal Processing, Singapore.

Tanaka, H. T., & Kishino, F. (1993). Adaptive mesh generation for surface reconstruction: Parallel hierarchical triangulation without cracks. *Proc. IEEE 10th International Conference on Pattern Recognition*, (pp. 88-94).

Turk, M., & Pentland, A. (1991). Eigenfaces for recognition. *Journal of Cognitive Neuroscience*, *3*(1), 71–86. doi:10.1162/jocn.1991.3.1.71

Wu, J. H., Hu, S. M., Tai, C. L., & Sun, J. G. (2001). An effective feature-preserving mesh simplification scheme based on face constriction. [Tokyo, Japan: IEEE Computer Society Press.]. *Proceedings of Pacific Graphics*, *2001*, 12–21.

Xu, C., Wang, Y., Tan, T., & Quan, L. (2004). *Face recognition based on 3D mesh model.* Accepted by SPIE Defense & Security Symposium. Retrieved from http://www.nlpr. ia.ac.cn/ english/ irds/ publications.htm

Zhao, W., Chellappa, R., Rosenfeld, A., & Phillips, P. J. (2000). *Face recognition: A literature survey.* UMD CfAR Technical Report CAR-TR-948.

ADDITIONAL READING

Achermann, B., & Bunke, H. (2000). *Classifying range images of human faces with Hausdorff distance.* in *Proc. ICPR*, pp.809- 813.

Achermann, B., Jiang, X., & Bunke, H. (1997). *Face recognition using range images.* in *Proc. Int. Conf. on Virtual Systems and MultiMedia*, pp.129-136.

Ajmal, S. Mian, M. Bennamoun and R. Owens (2006). *Automatic 3D Face Detection, Normalization and Recognition.* Third International Symposium on 3D Data Processing, Visualization and Transmission *(3DPVT).*

Besl, P., & McKay, N. (1992). A method for registration of 3-D shapes. *IEEE Transactions on Pattern Analysis and Machine Intelligence*, *14*(2), 239–256. doi:10.1109/34.121791

Braje, W. L., Kersten, D., Tarr, M. J., & Troje, N. F. (1998). Illumination effects in face recognition. *Psychobiology*, *26*(4), 371–380.

Chen, Y., & Medioni, G. (1992). Object modeling by registration of multiple range images. *Image and Vision Computing*, *10*(3), 145–155. doi:10.1016/0262-8856(92)90066-C

Colbry, D. (2006). *Human Face Verification by Robust 3D Surface Alignment.* Doctoral dissertation, Department of Computer Science, Michigan State University, Michigan, United States of America.

Fabry, T., Vandermeulen, D., & Suetens, P. (2008). *3D Face Recognition using Point Cloud Kernel Correlation.* 2nd IEEE International Conference on Biometrics: Theory, Applications and Systems. BTAS 2008.

Garcia, E., Dugelay, J.-L., & Delingette, H. (2001). *Low Cost 3D Face Acquisition and Modeling. itcc*, p. 0657, International Conference on Information Technology: Coding and Computing (ITCC '01).

Georghiades, A. S., Belhumeur, P. N., & Kriegman, D. J. (2001). From few to many: Illumination cone models for face recognition under variable lighting and pose. *IEEE Transactions on Pattern Analysis and Machine Intelligence*, *23*(6), 643–660. doi:10.1109/34.927464

Gordon, G. (1991). Face recognition based on depth and curvature features . In *SPIE Proc* (*Vol. 1570*, pp. 234–247). Geometric Methods in Computer Vision. doi:10.1117/12.48428

Hesher, C., Srivastava, A., & Erlebacher, G. (2003). *A novel technique for face recognition using range imaging.* in *Proc. 7th Int. Symposium on Signal Processing and Its Applications*, pp.201-204.

Irfano˘glu, M. O., G¨okberk, B., & Akarun, L. (2004). *3D Shape-Based Face Recognition Using Automatically Registered Facial Surfaces.* in *Proc. ICPR,* vol.4, pp.183-186.

Lao, S., Sumi, Y., Kawade, M., & Tomita, F. (2000). *3D template matching for pose invariant face recognition using 3D facial model built with iso-luminance line based stereo vision.* in *Proc. ICPR*, vol.2, pp.911-916.

Lu, X. (2006). *3D Face Recognition Across Pose and Expression.* Doctoral dissertation, Department of Computer Science & Engineering, Michigan State University, Michigan, United States of America.

Lu, X., Colbry, D., & Jain, A. K. (2004). *Three-Dimensional Model Based Face Recognition.* in *Proc. ICPR.*

Lu, X., & Jain, A. K. (2005). *Integrating range and texture information for 3d face recognition.* In: Proc. 7th IEEE Workshop on Applications of Computer Vision (WACV'05), Breckenridge, CO.

Lu, X., & Jain, A. K. (2005). *Deformation Analysis for 3D Face Matching.* wacv-motion, pp. 99-104, Seventh IEEE Workshops on Application of Computer Vision (WACV/MOTION'05) - Volume 1.

Moreno, A. B. A´. Sa'nchez, J.F. Ve'lez, F.J. D'ıaz (2003). *Face recognition using 3D surface-extracted descriptors.* in *Proc. IMVIPC.*

Nagamine, T., Uemura, T., & Masuda, I. (1992). *3D facial image analysis for human identification.* in *Proc. ICPR*, pp.324-327.

Song, Yu. Wenhong Wang, Yanyan Chen (2009). *Research on 3D Face Recognition Algorithm.* First International Workshop on Education Technology and Computer Science. ETCS '09.

Tanaka, H., Ikeda, M., & Chiaki, H. (1998). *Curvature-based face surface recognition using spherical correlation.* in *Proc. ICFG,* pp.372-377.

Tsalakanidou, F., Tzovaras, D., & Strintzis, M. G. (2003). Use of depth and colour eigenfaces for face recognition. *Pattern Recognition Letters, 24*(9-10), 1427–1435. doi:10.1016/S0167-8655(02)00383-5

Wang, Y., Chua, C.-S., & Ho, Y.-K. (2002). Facial feature detection and face recognition from 2d and 3d images. *Pattern Recognition Letters, 23*(10), 1191–1202. doi:10.1016/S0167-8655(02)00066-1

Zhao, W., & Chellappa, R. (2000). *Sfs based view synthesis for robust face recognition.* in Proc. Int'l Conf Automatic Face and Gesture Recognition, (Los Alamitos, CA, USA), p. 285, IEEE Computer Society.

KEY TERMS AND DEFINITIONS

3D Face Recognition: Face recognition using 3D images.

Adaptive Mesh: Mesh that adapts to the surface of each different face.

Linear Discriminant Analysis (LDA): Feature extractor.

Nearest Neighbour: Classifier.

Plane Fitting: Method used to obtain mesh from range image.

Principal Component Analysis (PCA): Feature extractor.

Range Image: Image with depth values.

Chapter 30
Subject Independent Facial Expression Recognition from 3D Face Models using Deformation Modeling

Ruchir Srivastava
National University of Singapore, Singapore

Shuicheng Yan
National University of Singapore, Singapore

Terence Sim
National University of Singapore, Singapore

Surendra Ranganath
Indian Institute of Technology, Gandhinagar, India

ABSTRACT

Most of the works on Facial Expression Recognition (FER) have worked on 2D images or videos. However, researchers are now increasingly utilizing 3D information for FER. As a contribution, this chapter zooms in on 3D based approaches while introducing FER. Prominent works are reviewed briefly, and some of the issues involved in 3D FER are discussed along with the future research directions. In most of the FER approaches, there is a need for having a neutral (expressionless) face of the subject which might not always be practical. This chapter also presents a novel technique of feature extraction which does not require any neutral face of the test subject. A proposition has been verified experimentally that motion of a set of landmark points on the face, in exhibiting a particular facial expression, is similar in different persons. The presented approach shows promising results using Support Vector Machine (SVM) as the classifier.

DOI: 10.4018/978-1-61350-326-3.ch030

INTRODUCTION

Computers have now become a part of our day to day lives as they are being widely used for running industries, account keeping, entertainment, health, shopping, communication and so on. It is believed that in the near future the environment around us will embed intelligent devices which will assist us in our day to day activities. E.g. when you come back tired from the office work, your music player will play your favorite musical program and if you are drowsy and want to take a nap, lights will dim automatically.

To make the above scenario a reality, computing devices need to become socially intelligent as well, apart from their computational intelligence (Vinciarelli et. al., 2009). Computers need to understand day to day needs of humans, which may be physical, mental, emotional and so on. This is motivating the researchers to delve into the field of Human Computer Interaction (HCI).

An important avenue in HCI is human emotion recognition which aims at automatically understanding human emotions and present desired responses. Emotion is much of an internal state and sometimes even human beings find it difficult to understand internal feelings of a person. This makes human emotion recognition even much more difficult for computers. From the perspective of computers, there are different modalities reflecting emotions of a person such as facial expression, voice, spoken words, hand and body gestures etc. Out of these modalities, it has been found that facial expressions of a speaker account for about 55 percent of the effect conveyed in human communication, while 38 percent of the rest is conveyed by voice intonation and 7 percent by spoken words (Pantic and Rothkrantz, 2000). Considering the importance of facial expressions in conveying emotions, automatic Facial Expression Recognition (FER) is developing as an important area of research.

Applications of Facial Expression Recognition

FER enables automation of services that require a good appreciation of the emotional state of the user. For example, if we understand the emotion of customers, a system can recommend products that they may be interested in. Similarly, there are many other areas where FER finds useful applications.

Facial expressions have been widely used in clinical research to study schizophrenia, which is a neuropsychiatric disorder in which patients have difficulty in recognizing and expressing emotions. Techniques of facial expression recognition have been used to analyze such abnormalities. Research has also been performed to recognize the facial expressions of epileptic patients while they undergo seizures. This helps in understanding the cerebral organization when seizures take place.

In academia, FER is used for understanding receptivity of students towards an automated tutoring system. Facial expressions of the students can reflect whether he finds the lecture interesting or not. Based on this feedback, the tutoring system adjusts the speed of instruction; slowing down if the student is bored and speeding up if student is grasping well.

In the field of advertising, FER is used for understanding the emotional responses of consumers towards television advertisements or towards different consumer products while they are shopping on the internet. Recent application of FER has come up in the form of development of technologies such as targeted advertisement where the advertisement on the billboards adapts to the facial expressions of the viewer.

FER techniques are helpful in making robots more social. Recently robots are being designed in such that they are able to interact socially with each other and also with humans. The capability of FER is very helpful in facilitating robots to communicate with humans because for humans, the face to face communication is a real-time process operating in times of the order of 40 milliseconds which is very fast for robots (Bartlett et al., 2003)

As far as surveillance is concerned, it has been performed mainly using face recognition rather than facial 'expression' recognition. Generally face recognition algorithms are effective only as long as the subject doesn't exhibit any expression. This is not easy to achieve especially in camera surveillance systems where people are unaware of being tracked. FER is very useful in these situations. FER systems have been developed for such real-time situations. Another security application is to hide the identity of an internet user during a chat without hiding his expressions. In such a case, the person's expressions are communicated using an animated avatar.

Objectives of the Chapter

Computers can recognize facial expressions from different kinds of data most common of which are videos and static images due to the ease of their being easy to capture. With the development of 3D imaging technologies, it is possible to go beyond 2D data and get 3D models of faces just as we have images in 2D. With the availability of 3D data, more works are coming up to recognize facial expressions using 3D data. The present chapter addresses the area of 3D FER with the following objectives:

- To introduce the area of Facial Expression Recognition (FER) with a review of the prominent works. Special attention is given to works using 3D model based approaches.
- To provide a discussion on 3D model based approaches for FER, challenges and possibility of future research in this area.
- To present a novel feature extraction methodology for FER using 3D facial models.

The main contribution of the chapter is in the form of a discussion on FER with special emphasis on methods using 3D face models. Challenges and future research directions in this field have been discussed. To the best of our knowledge other works have discussed FER only in more general terms, not in the context of using 3D imaging technologies. To complement the discussion, a novel approach for FER using 3D face models is presented (Srivastava et al., 2010) which is subject independent; means it can recognize expressions of people which the computer has not seen before. It is proposed and validated experimentally that the motion of a set of landmark points on the face in exhibiting a particular facial expression, is similar in different persons. This motion is used to model the deformation of the face when a certain expression is exhibited. These deformation models are used for the task of FER.

BACKGROUND

Representing Emotions

The actual aim of facial expression recognition is to recognize the underlying emotions. In order to facilitate gauging emotions from the facial movements, psychologists have used different ways to represent human emotions. The representations can be categorized into message judgment and sign judgment approaches (Cohn, 2006). Using message judgment representation, there is a particular facial expression attributed to an emotion. Thus, the way to recognize emotion is to just recognize the facial expression. On the other hand sign judgment approach does not directly recognize emotion but it describes certain external facial movements. It tries to code all possible perceptible changes occurring on a face due to expressions. This approach stops at this stage without going into the mental state of the person. Further analysis is needed to recognize emotions.

One of the popular representations of emotions under the message judgment approach is in terms of six basic universal emotions as proposed

Figure 1. A typical framework for facial expression recognition algorithms

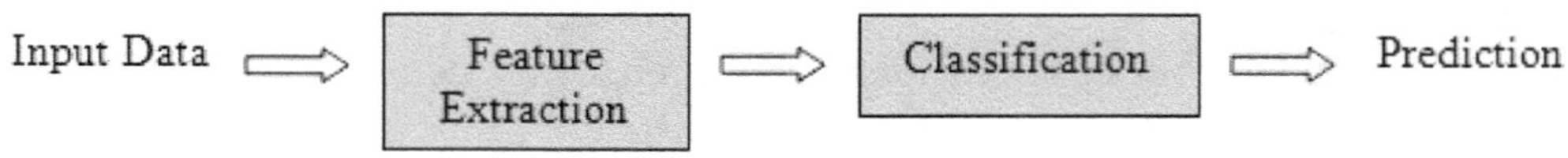

by Ekman & Friesen (1971). The six basic emotions are anger, disgust, fear, happiness, sadness and surprise. Most of the existing works on FER have recognized facial expressions corresponding to these six basic emotions. This description of emotions can be easily labeled and has an intuitive understanding for humans. But a disadvantage of this representation is that many facial expressions that we encounter in our day to day life cannot be categorized into these six expressions.

Another way to represent emotions under message judgment approach is using the dimensional approach where emotions are characterized by two dimensions viz. evaluation and activation. Evaluation determines whether the emotion is positive or negative while activation determines the intensity of the emotion. A wide range of emotions can be represented using this system. However, in this system, expressions for fear and anger cannot be distinguished. Also, trained labelers are required for labeling emotions using the dimensional approach.

Using sign judgment approach, emotions can be defined in terms of facial actions such as raising eyebrows, pulling lips apart and so on. A facial expression can be represented as a set of facial actions as defined in the Facial Action Coding System (FACS) introduced by Ekman and Friesen (1978). Facial actions are called Action Units (AUs). An advantage of Action Units is that they can be combined to define a wide range of facial expressions corresponding to emotions beyond the six basic emotions. However due to interpersonal variations in display of emotions, it is hard to associate a specific set of AUs with a particular emotion consequently most of the work just recognize AUs without recognizing emotions.

A Typical FER Algorithm

Given the input data, an FER algorithm can be broken down into two main stages, viz. feature extraction and classification, as shown in Figure 1. The techniques for feature extraction and classification depend on the type of the input data.

Types of Input Data

FER algorithms are meant for a wide range of applications. Depending on the application, there can be different types of data to be analyzed for FER. The different types on input data can be:

1. **Single image:** Only one image is available which shows the facial expression to be recognized. It is usually accompanied by an expressionless image for each person in the database.
2. **Video:** An image sequence depicting temporal evolution of the facial expression.
3. **Multiple images:** Multiple images means that the images are taken of the same person over a significant period of time say with a gap of a few months between two images.
4. **Group images or videos:** The images or videos contain groups of persons rather than one person only.
5. **3D static models:** This type of data contains the 3D models of faces which are used for FER. Only a single face model is used which bears a facial expression. In many cases a 3D face model of the neutral expression is also required.
6. **3D dynamic models:** Just as videos are sequences of images, recently sequences of

3D face models are also being used as data for FER.

7. **Both images and 3D models:** Information from both images and 3D models can be combined for getting relevant features for FER.

The different types of data listed above can be captured either in a controlled environment or in a real world environment. A controlled environment means that the face is facing the camera, has minimal rigid motion, sufficient lighting is there and the face is unoccluded. Apart from these environmental conditions, the facial expressions are usually posed i.e. they are acted out. A real world environment may have variations in facial pose and illumination. The face may get occluded sometimes and the facial expressions may be spontaneously exhibited.

Different types of data listed above can be classified into 2D or 3D data based on the dimensionality of the space in which face is visible. 2D data refers to the images or videos captured by a camera. Such data does not have any depth information. When depth information is also extracted from the scene, the resulting data is in 3D. Another major classification of the facial data can be made on the basis of inclusion of temporal information in them. A static data refers to just one image (or a face model in 3D) depicting a particular facial expression. While a dynamic data contains a sequence of such images (or 3D models) each depicting the progress of the expression with time. The types of data can be 2D static, 2D dynamic, 3D static and 3D dynamic based on the above two bases of classification. It is important to choose a suitable type of data for evaluating FER algorithms so that real life situations can be dealt with.

As far as choosing between static and dynamic data is concerned, recently, it has been proposed that temporal dynamics are very important to distinguish between posed and natural facial expressions. Experiments were also conducted to show the importance of motion in identifying subtle facial expressions (Ambadar et al., 2005). Through the experiments it was established that the inherent dynamic property in motion is beneficial for reducing the ambiguity in recognizing facial expressions. Temporal dynamics can be captured using dynamic data.

In choosing between 3D and 2D, researchers had to use 2D data only due to lack of technology for capturing 3D facial data. However, 2D image (or video) based approaches encountered various limitations. The efficacy of the approach was considerably affected when pose or illumination changes were involved. Also, in exhibiting an expression, the facial muscles move and the change is reflected in the facial skin. The skin motion is in 3D which cannot be captured accurately with 2D images of the face. With the improvement of 3D imaging technologies, researchers tried to address these shortcomings of 2D FER by utilizing the full 3D information about the face.

Feature Extraction

Feature extraction refers to the processing of raw data to extract characteristics distinguishing different classes of expressions. Usually the raw data is in the form of an image or a video (image sequence) from which face is detected, using standard face detection techniques. Details on face detection techniques can be found in Yang et al. (2002). Once the faces are detected, they are normalized to remove variations in them mainly due to size and illumination.

Feature extraction techniques can be based on geometrical displacement of facial features (geometry based approach) or the change in facial appearance (appearance based approach). In the geometry based approach, prominent landmark points such as eye and mouth corners are identified on the face. Motion of these points is modeled for each facial expression and these models provide features for further classification. On the other hand appearance based approaches apply image filters such as the Gabor filter or Haar like filters on the extracted facial region. Filter coefficients serve as features. Appearance based approaches extract information from the whole face and the dimensionality of feature vectors is usually much higher as compared to that in geometry based approaches. However, on of the major limitation of geometry based is the need for manual intervention to locate the facial landmark points.

It is still a debatable issue whether appearance based approaches are better or geometry based. However, researchers are also looking into hybrid techniques utilizing both geometry and appearance for feature extraction.

Classification

Classification is the stage where the test facial expression is classified into one of the output expression classes based on the extracted features. Output categories are mostly either the six basic expressions or a certain number of Action Units. However, it is being argued that apart from the basic emotions, there are also many other emotions which are experienced in our daily experience. These emotions include fatigue, pain, thinking, embarrassment etc. and can be referred to as subtle emotions.

Among the various classification techniques, one of the most popular techniques is Support Vector Machines (SVMs). Given an image, SVMs can predict the facial expression in it and in the case of image sequence, SVMs can be applied on a frame by frame basis i.e. on each image frame a prediction result can be obtained and the final result can be a combination of results for each frame. However, in the case of image sequences, extra temporal information is present. Emotions in adjacent frames are related to each other and a frame by frame prediction does not utilize this relationship. In the case of image sequences, the spatio-temporal classification using Hidden Markov Models (HMMs) is used more often as compared to SVM since HMMs utilize the relationship between emotions in adjacent frames.

Classification result can either be definitive or probabilistic. E.g. In a test video, a probabilistic prediction will tell the probability of the video containing each of the output emotions while definitive prediction assigns one of the output emotions to the video.

Prominent Works on FER

Attempts to automatically analyze facial expressions date back to 1978 with the work of Suwa et al. (1978). Most of the researchers have used either geometry based or appearance based approaches which have been introduced in the previous section on feature extraction. Among the geometry based approaches, Yacoob and Davis (1996) used optical flow in the regions of mouth, eyebrows and eyes for modeling non-rigid facial motions in image sequences. A planar face model was used for modeling rigid facial movements. Extracted motion parameters were used in a rule based framework to predict 6 basic facial expressions and the neutral face. Pantic and Rothkrantz (2004) extracted facial landmark points on static images depicting both frontal (person facing the camera) and profile views (side-views) of faces. Displacement of these points from neutral face to expressive face was used to recognize 22 AUs from frontal images and 24 AUs from profile images. When information from both views was combined, 32 different AUs were recognized. Yeasin et al. (2006) proposed a spatio-temporal approach for recognizing six basic expressions from video and also for computing the levels of interest i.e. the intensity of the expression. Optical flow vectors projected onto a lower dimension using PCA were used as basic features. Experiments were conducted on videos depicting both posed and spontaneous expressions. Kotsia and Pitas (2007) used a Candide (Ahlberg, J., 2001) grid to track facial movements in a video till the frame corresponding to the highest expression intensity.

The displacements of nodes of the grid from the first to the maximum expression intensity frame were used to recognize six basic expressions or a set of AUs.

Different appearance based approaches were compared by Donato et al. (1999) and using both Gabor filters and Independent Component Analysis, 96% accuracy was achieved in recognizing 12 AUs. One of the major works using appearance based approach was by Bartlett et al. (2005) who applied Gabor filters on the frames of input video and used the output magnitudes for recognizing 17 AUs. Different recognition engines such as Adaboost, SVMs and Linear Discriminant Analysis (LDA) were compared. Feature selection was explored using Adaboost so as to reduce the dimensionality of the feature vectors before feeding to SVM or LDA classifiers. Best results were obtained using Adaboost followed by SVM. Apart from Gabor filters, Haar filters were used by Whitehill and Omlin (2006) in conjunction with Adaboost for AU recognition and showed better performance as compared to Gabor filter based approach. Valstar et al. (2004) used the concept of multilevel motion history images (MMHI). MMHI representation is an extension of temporal templates which are 2D images showing motion history i.e. where and when motion occurred in an image sequence. 21 AUs were recognized comparing two classification schemes: (i) a two-stage classifier combining a kNN-based and a rule-based classifier, and (ii) a SNoW classifier.

Considering individual limitations of appearance based and geometry based approaches, researchers have also combined the two approaches for FER. Tian et al. (2001) recognized 16 AUs and the neutral face by analyzing both permanent facial features (related to eyes, mouth, nose etc.) and transient facial features such as wrinkles, furrows etc in videos. They proposed multistate face and facial component models for this task. Ashraf et al. (2007) used Active Appearance Models (AAMs) to automatically recognize pain from video. AAMs were used to track the facial motions defined by both shape and appearance parameters. Different representations from AAM were used and classification was performed using SVMs. Wang et al. (2007) used 2D and 3D geometric features and appearance features extracted from 2D images to quantify the difference in the way schizophrenic patients and healthy persons exhibit facial expressions. Zhou et al. (2010) have recognized facial events in video using an unsupervised method, Aligned Cluster Analysis (ACA), and a multi-subject correspondence algorithm. Faces have been tracked across the video using AAMs. Geometric features are in the form of certain facial distances while SIFT descriptors computed at points around the outer outline of the mouth and on the eyebrows serve as appearance based features.

State of the Art in 3D FER

For evaluating 3D FER algorithms, there is hardly any publicly available facial expression database. To the best of our knowledge, only publicly available databases are BU-3DFE database (Yin et al., 2006) and Bosphorus database (Savran et al., 2008) containing static 3D face models and BU-4DFE database (Yin et al., 2008) containing dynamic 3D face models.

The approaches for 3D FER can be classified based on whether they utilize dynamics or not. As far as the dynamic approach is concerned, one of the earliest attempts was by Gokturk et al. (2002). Major facial actions of opening and closing the mouth and eyebrow raising are identified apart from neutral and smile expressions. The face was modeled using 19 landmark points on the face. It was assumed that the deformation of the face from neutral can be expressed as a linear combination of a small number of known basis vectors. Basis vectors were computed using Singular Value Decomposition (SVD) of the 3D shape trajectory matrix. Coefficients of the linear combination were used as features for classification.

Sun and Yin (2008) proposed a spatio-temporal approach using 3D dynamic geometric facial model sequences, to tackle FER problems. The approach integrated a 3D facial surface descriptor and Hidden Markov Models (HMM) to recognize facial expressions. Extensive experiments were performed to explore three types of HMMs viz. temporal 1D-HMM, pseudo 2D-HMM (a combination of a spatial HMM and a temporal HMM), and real 2D-HMM. Other prominent works using 3D dynamics are those of Dornaika and Davoine (2008) and Chang et al. (2005).

Among the earliest works on static 3D face data, Yabui et al. (2003) used range images for the task of FER. A variant of the eigenspace method called the 'Eigenspace Method based on Class features (EMC)' was used to get an eigenspace for classification. Srivastava and Roy (2009) used 3D residues for FER. Residues are the displacement of facial feature points from neutral when an expression is exhibited. Residues have been only used in 2D but its application in 3D proved effective than using other approaches as shown in their work. Wang et al. (2006), Soyel and Demirel (2007) and Tang and Huang (2008) have also performed FER using static 3D data.

Research in 3D FER is still in early stages and there is a lot of possibility of research in 3D FER as discussed later in the chapter.

3D FER: ISSUES OF CONCERN

Acquiring 3D Data

Considering the wide applicability of FER, it is important that 3D face models are easily acquired in such real life applications. Usually 3D face models are either captured using 3D scanners or reconstructed from video. 3D scanners yield a dense face mesh with the number of vertices large enough to sample the entire surface information of the face. Having a dense sampling is an advantage with the 3D scanners. With advances in 3D imaging, now it is possible to get 3D videos in real time, at a capture rate as high as 500fps (Dimensional Imaging). Such data is often referred to as 4D data due to an added temporal dimension to 3D objects. However, even when fast scanners are available, they are not very suitable for day to day applications considering their cost. A complete 3D scanning system's cost can range from USD 40K-150K or even $400K (3D Scan Company). Using lasers for scanning is also harmful for the eyes.

3D face models can also be reconstructed from videos of the person whose expressions are to be recognized if the person moves his face to show different views of his face. This technology is known as face reconstruction. The basic methodology of face reconstruction uses stereo algorithms which use the information about the relative orientation of camera and the face. Using this information for different views, the 3D location of some prominent points on the face is obtained with reference to a fixed origin. 3D location of a large number of points on the face approximates a 3D model of the whole face. Once the face is reconstructed, it can be tracked using a generic 3D face model.

One of the major challenges in face reconstruction is to reduce the computational complexity involved. Computational complexity prevents a real time application involving face reconstruction. Another challenge is that reconstruction is accurate only for small non-rigid motions. But when facial expressions are involved, reconstruction can be erroneous. Facial hair also deteriorates quality of reconstruction. There are a few other disadvantages of this approach such as frame selection, sequence segmentation, structure fusion and bundle adjustment (See Kien, 2005 for details).

An alternative to getting the full 3D face model is to fit a generic 3D face model with a fewer vertices to a face in a video. Fitting is usually done based on the locations of a few control points on the face. Once the face model is fitted in the first frame of the video, the face can be tracked

Figure 2. Ways to represent 3D face models. a) 3D surface + texture, b) 3D surface + triangular mesh, c) 3D surface (Stylianou and Lanitis, 2009)

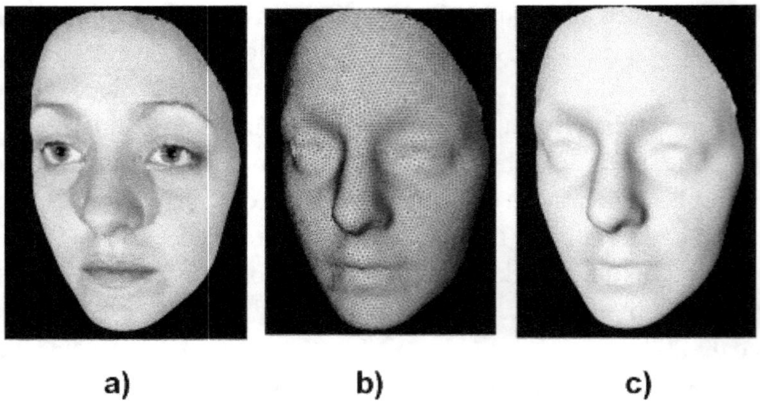

with the help of this 3D model. The process of fitting the face model and 3D face tracking can be achieved in an automated manner in real-time (Zhu and Ji, 2004). This approach is very much suitable for real life applications.

Once the 3D face model is acquired, it can be represented either as a shaded 3D surface, 3D surface overlaid with a triangular mesh or 3D surface overlaid with texture (See Figure 2). However, for computational purposes, triangular mesh representation is used. A triangular mesh is a graph, $G = (V, T)$ with V denoting the set of vertices, $V = \{\overline{p} = (x, y, z)\}$ and T denoting a set of triangles $T = \{(\overline{p}_i, \overline{p}_j, \overline{p}_k)\}$, where $\overline{p}_i, \overline{p}_j, \overline{p}_k \in V$ (Stylianou and Lanitis, 2009).

Feature Extraction

A suitable feature extraction methodology needs to be developed for FER using 3D face models. In this step there are certain challenges faced by researchers. Some of the challenges are for FER in general and not just limited to 3D FER. Prominent challenges in feature extraction are:

- **Automation**: In many works there is a manual intervention needed to detect facial landmark points in 3D face reconstruction (Stylianou and Lanitis, 2009). This is not desirous for real life applications. 3D reconstruction can be combined with techniques for automatic facial landmark detection from 2D images. However, most of such techniques can only detect few facial landmarks and that too only in a frontal or near-frontal face.
- **Dealing with subtle expressions:** Apart from the six basic facial expressions, many times humans display mixed emotions which cannot be explicitly categorized into one of these six expressions. In order to characterize such expressions, an algorithm which gives the probability of each expression will be more effective. Also, the expressions many times do not convey the emotions of a person. It is still a challenge to get the emotion of a person.
- **Analysis of intensity of expressions**: Expressions can be displayed with varying intensities. Most of the research deals only with the high intensities of expressions. But in practical applications, especially in the case of analyzing temporal evolution of facial expression, it is necessary to analyze lower intensities of facial expressions as well. In daily intercourses, humans usually display lower intensity expressions.

- **Environmental changes:** Real life situations involve a lot of change in the environment, especially in outdoor applications. This change can be of lightning conditions, background clutter, occlusion or confusing background patterns. Such changes can have serious affects on the feature extraction process.

Subject Independence

Apart from the above mentioned challenges, another challenge is to obtain subject independency in a close to natural environment. Most of the FER approaches using 3D face models require a neutral 3D model of the face i.e. the facial model when the person exhibits no expression. However, having the neutral face of a person is only possible when the person is known and his neutral face has been previously captured. There are other difficulties as well when a neutral model is needed. E.g. Even if a person is monitored using a video camera; it is difficult to ascertain when the face is actually neutral. Requirement of a neutral 3D model limits the application of FER methods. The method proposed in the following section overcomes this limitation as it does not need any neutral 3D model of the test subject. Soyel and Demirel (2007) have also proposed a method no requiring neutral face model. They use 5 normalized facial distances for FER. However, interpersonal variations may affect the accuracy of their method. E.g. using facial distances, a large horizontal distance between the two lip corners (mouth opening) may indicate lip stretching. But a person can have a wider mouth as compared to others and even in the neutral state the value of mouth opening may be large which may wrongly indicate lip stretching. These interpersonal differences will create more ambiguity in low intensity expressions where facial distances do not change much from the neutral. This limitation in their approach has also been analytically shown by Srivastava and Roy (2009).

A NOVEL APPROACH FOR 3D FER

3D Dataset Used

One of the most important factors for FER from 3D face models is availability of a suitable database. Acquisition of 3D models is more difficult as compared to acquiring images. Because of this difficulty there was lack of a widely recognized database for Facial Expression Recognition. Yin et al. (2006) at Binghamton University constructed a 3D facial expression database (BU-3DFE database) for facial behavior research which has been used for evaluation of the presented algorithm.

The BU-3DFE database contains triangulated 3D mesh models and 2D facial textures for 100 subjects. Each subject has 3D models for 4 intensities of 6 expressions and a neutral, making a total of 2500 3D models. Intensity of an expression refers to different stages of development of an expression. A low intensity level is closer to a neutral face and intensity increases as the expression progresses in time towards the peak. The spatial coordinates (x, y, z) and color (R, G, B) values are provided for all vertices in each facial model. Apart from this, the database also provides the spatial positions and color for 83 corresponding facial points on each facial model. These 83 points correspond to prominent facial features such as corners and contours. Some of the sample images for 3D models in the database are given in Figure 3. Figure 4 shows the 83 points marked on the face.

This algorithm uses only the spatial locations of the 83 landmark points. This is beneficial in dealing with variations in image appearance due to changing illumination, occlusions etc. Feature extraction will be more robust to such disturbances because color information is not used. This works starts with the assumption that positions of these facial points are provided although while implementation these points need to be extracted automatically.

Figure 3. Sample images of 3D models from the BU-3DFE database. 4 levels of intensities are displayed by the same person for each expression. Expressions displayed from left to right are Anger, Disgust (Top row), Fear, Happiness (Middle row) and Sadness, Surprise (Bottom row).

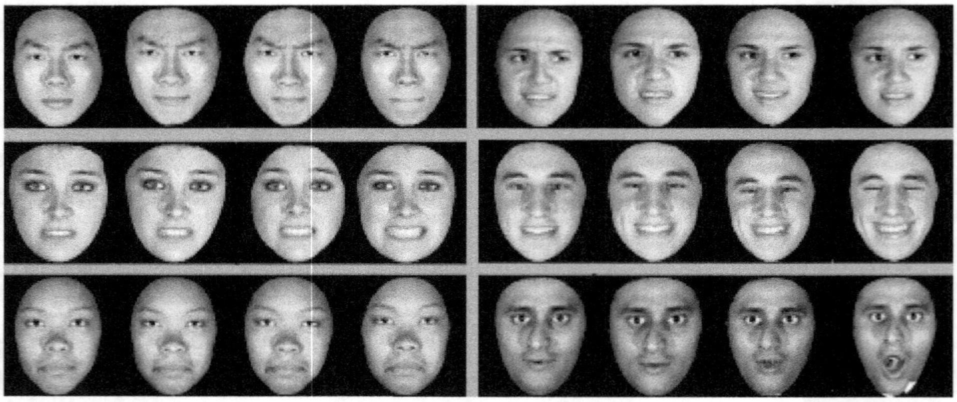

Figure 4. 83 landmark points used for analysis in the presented work

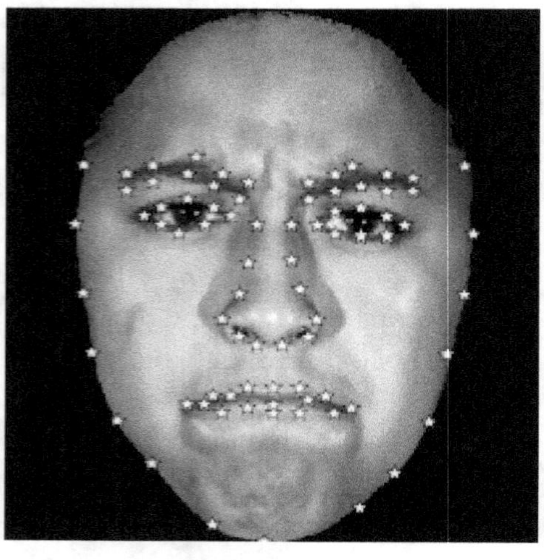

Figure 5. a. Movement of right lip corner while exhibiting happiness. We propose that this motion will be similar in different persons. b. Classification scheme for the presented algorithm. F_i represents the relevant feature set for the ith expression while D_i represents the decision value estimates of the query expression for the ith expression.

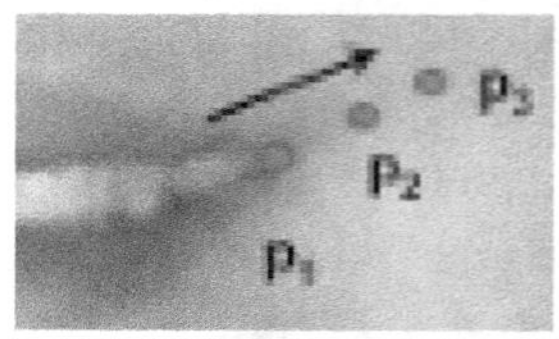

An advantage of using only positions of landmark points for our analysis is that we need not know the spatial locations of all the vertices of the 3D face model. This allows us to fit a generic 3D face model to the first frame of the video and then track the face. Advantages of getting the 3D data using this approach have already been discussed before as compared to other approaches.

The presented approach models the facial deformations when expressions are exhibited. Facial deformation is more prominent in some areas as compared to others. We can assume that deformation is indicated by the movement of a few prominent facial landmark points. To understand how an expression is modeled using motion of landmark points, consider figure 5a., where the right lip corner position moves from point p_1 to p_3 when 'happiness' expression is exhibited. We propose that direction of movement of this point will be similar in different persons when they exhibit 'happiness'. Modeling deformation at this point is a way to model this similarity.

Let $f_p^i(n)$ represent the pth landmark point on the ith person at a temporal instant n. Here $i = 1$ to 100 (corresponding to 100 persons in the database), $p = 1$ to 83 (corresponding to 83 facial landmark points); and $n = 1$ to N_s where N_s is the number of temporal samples found out by cubic spline interpolation. With these representations, pseudocode of the presented algorithm for feature extraction followed by classification is as follows:

Pseudocode of the Presented Algorithm

1. Inputs:
   - 3D models of the test face, with Cartesian (*x*, *y*, *z*) coordinates given at 83 landmark points corresponding to four levels of expression intensity.
   - 3D models X_i^e for N training examples, where $i = 1,2,3\ldots N$ and e refers to the expression. Each example has four gradations of the expression e.
2. For each X_i^e, find 3D position at each facial point in following steps:
   - Set origin of the coordinate system at the nose tip.
   - Using Cubic spline interpolation find the Cartesian coordinates of the point at intensities between gradations j and $j + 1$, $j \in [1, 3]$. Let these coordinates be $f_p^i(n)$ with $n = 1$ to N_s denoting the samples.
   - Transform the coordinates of $f_p^i(n)$ from Cartesian (*x*, *y*, *z*) to spherical coordinate system (r, θ, ϕ).
   - Using equation 1, find $\vec{\theta}_{p0}^{i0}$ and $\vec{\phi}_{p0}^{i0}$, the parameters giving the 3D position of the p_0^{th} landmark point of i_0^{th} person.
3. Using equation 2, find the feature vector for the ith person.
4. Perform feature selection using the significance ratio test.
5. Classify the test data X_{tes} using one vs. all scheme of SVM.

Details of the algorithm are given in the next section.

EXPERIMENTS AND RESULTS

Cubic Spline Interpolation

In the database, there are four gradations of expressions available corresponding to only four temporal samples from neutral to peak of the expression. 4 temporal samples are insufficient to estimate motion direction reliably. Cubic spline interpolation is used to find more samples. After interpolation at a landmark point; say the left eye corner; we have N_s (= 61 in our case) temporal samples for each subject. After interpolation, the spatial positions of the points are transformed from Cartesian to the spherical (r, θ, ϕ) coordinate system. This is because just two parameters θ and ϕ directly give the directional information about the 3D position of a point with respect to the origin while Cartesian system needs three parameters, *x*, *y* and *z* for this purpose. For further analysis, let the θ and ϕ values for $f_p^i(n)$ be represented by $\theta_p^i(n)$ and $\phi_p^i(n)$, respectively.

Deformation Modeling

To model the motion direction, the $\theta_p^i(n)$ and $\phi_p^i(n)$ values are used for a fixed value of p (landmark point) say, p_0. During this modeling, it is required to find out if there is any pattern in the $\theta_{p0}^i(n)$ and $\phi_{p0}^i(n)$ values for different persons. To find this pattern, out of the 60 subjects in the training dataset, a subset I composing of 30 subjects was randomly selected. 2D histograms of $\theta_{p0}^i(n)$ and $\phi_{p0}^i(n)$ values were plotted for all the subjects in the selected subset. This procedure was repeated

Figure 6. a. Histograms for the distribution of $\theta^{i}_{p0}(n)$ and $\phi^{i}_{p0}(n)$ over 5 runs. b. Separate histogram for $\theta^{i}_{p0}(n)$ c. Separate histogram for $\phi^{i}_{p0}(n)$ corresponding to the 2D histogram in run1 and anger expression.

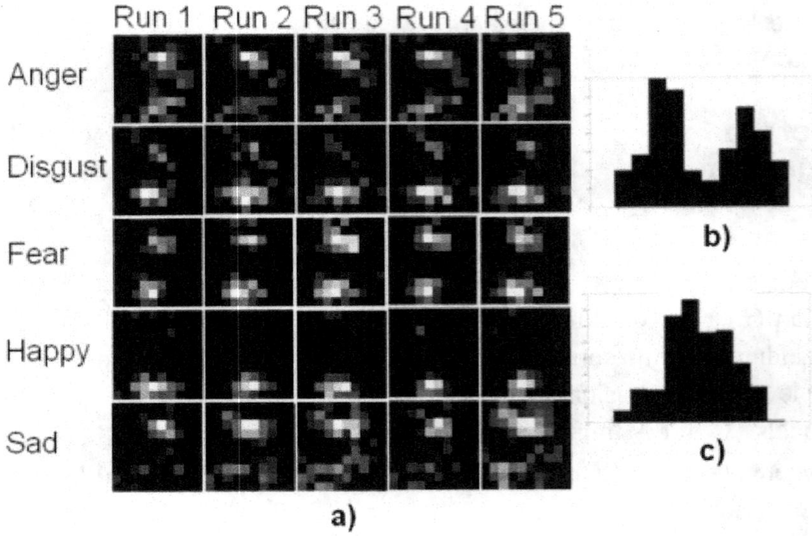

5 times to check the consistency of the patterns. The 2D histograms, are shown as images in Figure 6a. The figure shows that the histograms are consistent over all the 5 runs. This consistency was found for other landmark points as well. This shows that the temporal samples at the landmark point for different subjects belong to a specific nature of distribution and thus the motion of landmark point is similar in different persons. Also, the difference in between the histograms of different expressions clearly shows that the motion directions are discriminative and can be used as features for FER.

The 2D histograms are projected on each of the θ and ϕ axes, to obtain two 1D histograms showing the distribution of $\theta^{i}_{p0}(n)$ and $\phi^{i}_{p0}(n)$, separately (Figure 6b. and c.). Deformations of the landmark point p_0 for each person are modeled as these 1D distributions. In order to represent these distributions, the parameters of these distributions are used. Few typical parameters for any probability distribution of a real-valued random variable are mean, variance, skewness and kurtosis. Using these parameters the features for p_0 are given by $\vec{\theta}^{i0}_{p0}$ and $\vec{\phi}^{i0}_{p0}$ which are defined as follows:

$$\begin{aligned} \vec{\theta}^{i0}_{p0} &= \left[\theta^{i0}_{(1)p0} \quad \theta^{i0}_{(2)p0} \quad \theta^{i0}_{(3)p0} \quad \theta^{i0}_{(4)p0}\right] \\ \vec{\phi}^{i0}_{p0} &= \left[\phi^{i0}_{(1)p0} \quad \phi^{i0}_{(2)p0} \quad \phi^{i0}_{(3)p0} \quad \phi^{i0}_{(4)p0}\right] \end{aligned} \tag{1}$$

where, $\theta^{i0}_{(1)p0}$, $\theta^{i0}_{(2)p0}$, $\theta^{i0}_{(3)p0}$ and $\theta^{i0}_{(4)p0}$ represent the mean, variance, skewness and kurtosis of $\theta^{i0}_{p0}(n)$ with n varying from 1 to N_s. Similar notations are used for ϕ as well.

For the i_0^{th} person, the feature vector is given by

$$\vec{x}_{i0} = \left[\vec{\theta}^{i0}_{1} \quad \vec{\theta}^{i0}_{2} \quad \ldots \quad \vec{\theta}^{i0}_{p} \quad \ldots \quad \vec{\theta}^{i0}_{83} \quad \vec{\phi}^{i0}_{1} \quad \vec{\phi}^{i0}_{2} \quad \ldots \quad \vec{\phi}^{i0}_{p} \quad \ldots \quad \vec{\phi}^{i0}_{83}\right] \tag{2}$$

Classification

There are total 83 landmark points used for feature extraction. These points are distributed all over the face as shown in Figure 4. However, many of these points will be irrelevant for a particular E.g. when a person is happy, he will; in general; stretch his lips wide apart. So, the motion of the lip corners is relevant. While one doesn't generally frown when happy. Consequently, the motion of the eyebrows will not be that much relevant. We need to perform feature selection to select relevant features. However since there are different features which might be relevant for distinguishing each expression from the others, features are selected considering the problem of discriminating one expression at a time from the others. Therefore, six individual classifiers are implemented using one vs. all scheme of a Support Vector Machine (SVM) classifier. Feature selection is performed using the significance ratio test (Weiss & Indurkhya, 1998).

After feature selection on the test data, classification is performed as per the scheme given in figure 5b. For a test expression, classifiers for each expression give a decision value estimate indicating how much the probability of the test expression belonging to each expression is. Predicted expression corresponds to that expression for which classifier gives the maximum decision value.

Results

For a two class classification a reliable performance measure is the Receiver Operating Characteristic (ROC) curves. The performance of a binary classification is indicated by the area under the ROC curve. The closer the AUC is to 1 the better the performance of classification is. Results of individual classification are given in the form of the ROC curves along with areas under the curves in Figure 7. It can be seen that the Areas Under the Curve (AUCs) are very close to 1 for all the individual binary classifiers except in the case of fear vs. other expressions. This shows that less discrimination between fear and other expressions. Final classification results are presented in table 2 a in the form of a confusion matrix. In the confusion matrix an element (i, j) (i.e. row corresponding to expression i and column to expression j) shows the number of test samples which were predicted to belong to expression j when the sample actually belongs to expression i. This way the diagonal elements in the confusion matrix show the correct recognition rate for a particular expression.

Table 1. Areas under the ROC curves for individual classifiers for the six expressions. A higher value of AUC indicates better separation.

Expression	Area Under the Curve (AUC)
Anger	0.9588
Disgust	0.9543
Fear	0.7081
Happiness	0.9642
Sadness	0.9261
Surprise	0.9887

We see from the classification results, that the highest recognition rate of 98.8% has been achieved for 'disgust'. This can be attributed to the fact that the most common facial motion for disgust is wrinkling of nose. Since our proposition was based on similarity of motion of landmark points, the more similar is this motion among different persons for an expression, the better the recognition will be for that expression. The average recognition rate is 88.1% if we ignore the results for the Fear expression. Even when we include results of Fear expression, the average recognition rate is 80.3%. As was evident by the ROC curve, there was less discrimination in between Fear and the other expressions. This is also substantiated by the results.

The presented method identifies the six basic expressions; however, it was not designed to

Figure 7. ROC curves for individual classifiers for the six expressions: a. Anger, b. Disgust, c. Fear, d. Sadness, and e. Surprise

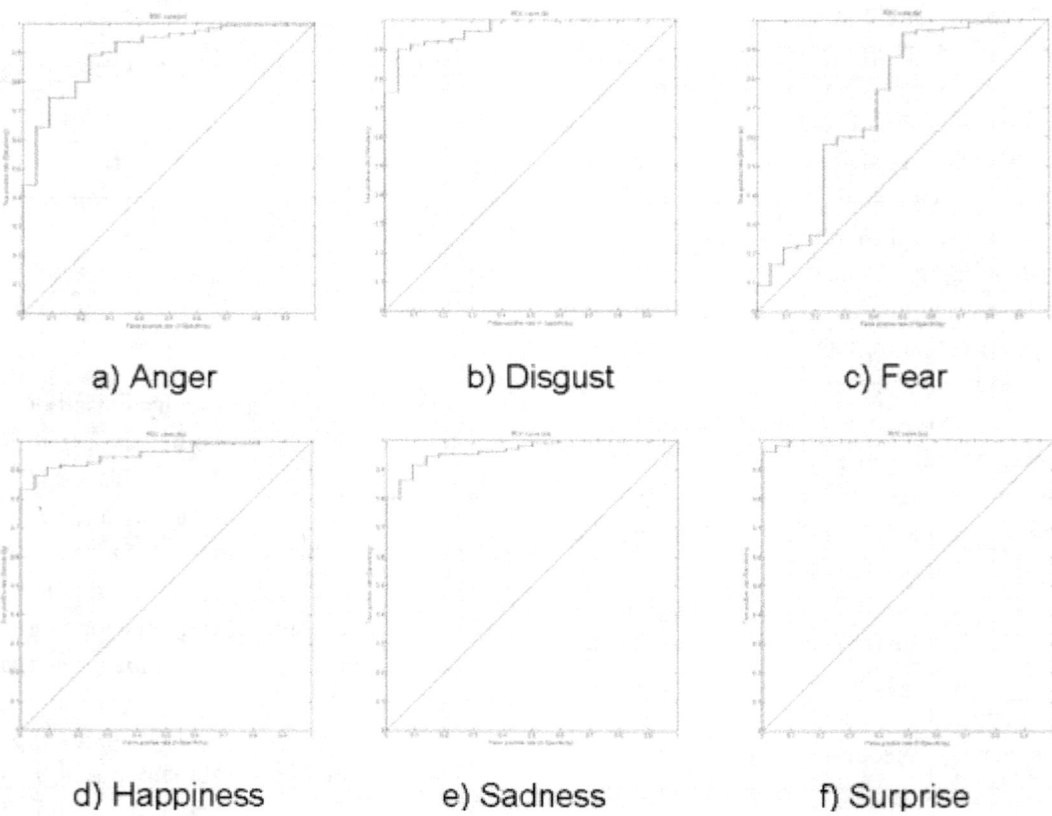

Table 2. Confusion matrices of the classification results using the presented method. Training: 60 subjects, Testing: 22 subjects.(ARR=80.4%).

	Anger	**Disgust**	**Fear**	**Happiness**	**Sadness**	**Surprise**
Anger	**79.6**	6.8	1.1	1.1	8.0	3.4
Disgust	0.0	**98.8**	0.0	1.2	0.0	0.0
Fear	4.5	9.1	**41.5**	35.8	9.1	0.0
Happiness	0.0	3.9	0.0	**96.1**	0.0	0.0
Sadness	9.1	4.5	9.6	0.0	**72.3**	4.5
Surprise	0.0	0.0	6.1	0.0	0.0	**93.9**

recognize the expressionless face. This is because the algorithm is based on deformation modeling and we assume that expressionless face means deformation is negligible and thus modeling deformation is meaningless.

Comparison with a Related Work

We have presented a novel method for FER which does not require any 3D model for neutral face. We compare our approach with work by Soyel and Demirel (2007) using facial distances, which also does not require neutral 3D face model. The limitation of their work has been already discussed in this chapter under the topic of subject independence. Experiments were conducted for both methods on the same set of training and test data for 10 runs and the Average confusion matrix across those runs is displayed in Tables 2 and 3. It is to be noted that we use all 4 intensity of facial expression in the experiments and that might be the reason for a low accuracy for the other method since as mentioned before, using facial distances can be less robust for lower intensity facial expressions. Soyel and Demirel (2007) do not specifically mention the intensity of expressions that they have used in their experiments. It can be observed that the presented algorithm (ARR=80.4%) performs better than the related algorithm (ARR=77.7%). However the other algorithm performs much better for Fear.

FUTURE RESEARCH DIRECTIONS

The presented algorithm was evaluated using 3D static face models for FER. As landmark points are used, a dense 3D model is not necessary for implementation of the algorithm. The algorithm can be implemented using 3D data generated from a video sequence using a generic mesh with a much fewer vertices as compared to that in a full 3D model. A video sequence is easily obtainable in a real life situation. Considering 3D FER, there is a wide possibility of future research in the following areas:

Formation of a Representative Database

For development of algorithms for 3D FER, availability of a 3D facial expression database is a must. Databases presently available have been captured under controlled conditions mainly because available 3D scanners cannot operate satisfactorily in outdoor real life situations. However, for applying the FER algorithms in daily life applications it becomes necessary to evaluate them under natural conditions instead of a laboratory setup.

Another issue in facing a natural environment is about the nature of expressions in real life. Paul Ekman, one of the pioneers in research on emotions emphasizes that there are many facial expression that do not correspond to emotions (Ekman, 1978 and Ekman, 1979). These subtle facial expressions; such as agreement, interested, flirting etc; can also act as social signals or constituents of social behavior (Vinciarelli et al., 2009). The analysis of these expressions is one of the parts of Social Signal Processing (SSP) and the efforts in this direction are still in infancy. Despite of being in infancy SSP is now attracting attention of the research community. In fact, the MIT Technology Review magazine has presented 'reality mining' as one of the 10 emerging technologies that are most likely to 'change the way we live'. Reality

Table 3. Confusion matrices of the classification results using the method proposed by Soyel and Demirel (2007). Training and test data same as in Table 2.(ARR=77.7%).

	Anger	Disgust	Fear	Happiness	Sadness	Surprise
Anger	**73.7**	8.1	4.5	0.0	13.6	0.0
Disgust	0.0	**80.1**	2.9	2.2	14.8	0.0
Fear	4.5	4.1	**75.5**	1.8	14.1	0.0
Happiness	0.0	9.5	3.2	**85.5**	1.8	0.0
Sadness	9.1	4.5	14.5	0.0	**67.5**	4.4
Surprise	2.0	0.0	9.1	5.0	0.0	**83.9**

mining refers to automatic analysis of everyday social interactions in groups of several tens of individuals (Vinciarelli et al., 2009). It is one of the main applications of SSP and involves analysis of facial expressions as well.

This opens up possibilities of development of a 3D facial expression database with faces captured in outdoor environment and exhibiting natural expressions.

Automatic Facial Landmark Point Detection

In the present work, we assumed that the location of the facial landmark points will be provided. But this is a non-trivial assumption. Considering the state-of-the-art techniques for landmark detection, points can be detected only either for frontal or for profile view of the face but these techniques fail with variations in facial pose. Pose variations play a major role in real life.

Dealing with Computational and Storage Complexities

Yin et al. (2008) report facing problems in processing and storage of their 3D dynamic models. Because of these limitations, presently it is possible to record only short duration 3D videos where the person begins from neutral and deliberately shows expression and then comes back to neutral. Natural expressions are prolonged and so longer 3D videos need to be recorded. Also, the BU-4DFE database in its present form takes around 500GB of storage space. Techniques need to be devised for compact storage of the 3D models especially the dynamic models.

CONCLUSION

With the increasing involvement of computers in our day to day lives, Human Computer Interaction (HCI) is being researched a lot nowadays. Emotion recognition is a vital part of HCI and since emotions are mostly conveyed by facial expressions, Facial Expression Recognition (FER) is a research field which finds application in many avenues. FER has been performed since late 1970s; however most of the work was done using 2D images or videos. In 1990s, researchers started exploring the use of depth information as well for FER. Due to the additional depth information and other advantages of 3D data such as pose and illumination invariance, 3D FER is gaining pace in recent years. Researches have shown that dynamics of expression are crucial for analysis of facial expressions. This makes FER from 3D dynamic data more effective than using 3D static data.

3D dynamic faces can be constructed by using laser scanners or by using 3D reconstruction techniques. However, from the point of view of application of 3D FER techniques in real world, these techniques are not very feasible due to their computational complexity or inaccuracy. A better approach to get 3D data can be to fit a generic 3D model to faces and then track them.

Most of the current works on 3D FER require a 3D face model for the neutral face. This paper presented a novel feature extraction technique for facial expression recognition using 3D models, in which there is no need to have a 3D model corresponding to the neutral (expressionless) face of the person whose expression is being analyzed. This makes the method very relevant in the real life situations where it might not be always possible to have the neutral facial model corresponding to a test expression. One vs. all scheme of classification was implemented using Support Vector Machines (SVM). For each individual classifier, a feature selection was performed using the significance ratio test. Promising classification results support the presented feature extraction technique.

In spite of the promise in the current 3D FER techniques, there are a few challenges that need to be tackled in order to make these techniques easy to implement in real life situations. This chapter highlighted these challenges and also proposed future research directions.

REFERENCES

Ahlberg, J. (2001). *CANDIDE-3—An updated parameterized face. (Report No. LiTH-ISY-R-2326)*. Sweden: Department of Electrical Engineering, Linkoping University.

Ambadar, Z., Schooler, J. W., & Cohn, J. F. (2005). Deciphering the enigmatic face: The importance of facial dynamics to interpreting subtle facial expressions. *Psychological Science, 16*(5), 403–410. doi:10.1111/j.0956-7976.2005.01548.x

Ashraf, A. B., Lucey, S., Cohn, J. F., Chen, T., Ambadar, Z., & Prkachin, K. … Theobald, B. J. (2007). The painful face: Pain expression recognition using active appearance models. In *International Conference on Multimodal Interfaces* (pp. 9-14).

Bartlett, M. S., Littlewort, G., Fasel, I., & Movellan, J. (2003). *Real time face detection and facial expression recognition: Development and applications to human computer interaction*. In Computer Vision and Pattern Recognition Workshop on Computer Vision and Pattern Recognition for Human-Computer Interaction.

Bartlett, M. S., Littlewort, G., Frank, M., Lainscsek, C., Fasel, I., & Movellan, J. (2005). *Recognizing facial expression: Machine learning and application to spontaneous behavior*. In IEEE Computer Society Conference on Computer Vision and Pattern Recognition.

Chang, Y., Vieira, M., Turk, M., & Velho, L. (2005). Automatic 3D facial expression analysis in videos. In *Analysis and Modeling of Faces and Gestures: Second International Workshop* (pp. 293-307). Berlin, Germany: Springer.

Cohn, J. F. (2006). Foundations of human computing: Facial expression and emotion. In *Proceedings of the 8th International Conference on Multimodal Interfaces*. Association for Computing Machinery.

Dimensional Imaging. Retrieved July 28, 2010, from http://www.di3d.com/ index.php

Donato, G., Bartlett, M. S., Hager, J. C., Ekman, P., & Sejnowski, T. J. (1999). Classifying facial actions. *IEEE Transactions on Pattern Analysis and Machine Intelligence, 21*(10), 974–989. doi:10.1109/34.799905

Dornaika, F., & Davoine, F. (2008). Simultaneous facial action tracking and expression recognition in the presence of head motion. *International Journal of Computer Vision, 76*(3), 257–281. doi:10.1007/s11263-007-0059-7

Ekman, P. (1978). Facial signs: Facts, fantasies, and possibilities . In Sebeok, T. (Ed.), *Sight, sound, and sense* (pp. 124–156). Bloomington, IN: Indiana University Press.

Ekman, P. (1979). About brows: Emotional and conversational signals . In von Cranach, M., Foppa, K., Lepenies, W., & Ploog, D. (Eds.), *Human ethology: Claims and limits of a new discipline: Contributions to the colloquium* (pp. 169–248). Cambridge, UK: Cambridge University Press.

Ekman, P., & Friesen, W. V. (1971). Constants across cultures in the face and emotion. *Journal of Personality and Social Psychology, 17*(2), 124–129. doi:10.1037/h0030377

Ekman, P., & Friesen, W. V. (1978). *Facial action coding system: A technique for the measurement of facial movement*. Palo Alto, CA: Consulting Psychologists Press.

Fasel, B., & Luettin, J. (2003). Automatic facial expression analysis: A survey. *Pattern Recognition, 36*(1), 259–275. doi:10.1016/S0031-3203(02)00052-3

Gokturk, S. B., Bouguet, J. Y., Tomasi, C., & Girod, B. (2002). Model-based face tracking for view-independent facial expression recognition. In *Proceedings of the Fifth IEEE International Conference on Automatic Face and Gesture Recognition*, (pp. 287-293). IEEE Computer Society.

Kien, D. T. (2005). *A review of 3D reconstruction from video sequences. Relation, 10(1.107), 2244*. Citeseer.

Kotsia, I., & Pitas, I. (2007). Facial expression recognition in image sequences using geometric deformation features and support vector machine. *IEEE Transactions on Image Processing*, *16*(1), 172–187. doi:10.1109/TIP.2006.884954

Pantic, M., & Rothkrantz, L. (2000). Automatic analysis of facial expressions: The state of the art. *IEEE Transactions on Pattern Analysis and Machine Intelligence*, *22*(12), 1424–1445. doi:10.1109/34.895976

Pantic, M., & Rothkrantz, L. (2004). Facial action recognition for facial expression analysis from static face images. *IEEE Transactions on Systems, Man and Cybernetics . Part B*, *34*(3), 1449–1461.

Savran, A., Alyuz, N., Dibeklioglu, H., Celiktutan, O., Gokberk, B., Sankur, B., & Akarun, L. (2008). Bosphorus database for 3D face analysis . In *The First COST 2101*. Workhop on Biometrics and Identity Management.

3D*Scan Company*. Retrieved July 28, 2010, from http://www.3dscanco. com/ about/ 3d-scanning/ faq.cfm

Soyel, H., & Demirel, H. (2007). Facial expression recognition using 3d facial feature distances. [Berlin, Germany: Springer.]. *Lecture Notes in Computer Science*, *4633*, 831–838. doi:10.1007/978-3-540-74260-9_74

Srivastava, R., & Roy, S. (2009). *3D facial expression recognition using residues*. In IEEE Region 10 Conference, TENCON.

Srivastava, R., Sim, T., Yan, S., & Ranganath, S. (2010). *Feature selection for facial expression recognition using deformation modeling*. In International Conference on Digital Image Processing.

Stylianou, G., & Lanitis, A. (2009). Image based 3D face reconstruction: A survey. *International Journal of Image and Graphics*, *9*(2), 217–250. doi:10.1142/S0219467809003411

Sun, Y., & Yin, L. (2008). Facial expression recognition based on 3D dynamic range model sequences. In *Proceedings of the 10th European Conference on Computer Vision: Part II* (pp. 58-71). Berlin, Germany: Springer-Verlag.

Suwa, M., Sugie, N., & Fujimora, K. (1978). A preliminary note on pattern recognition of human emotional expression. *International Joint Conference on Pattern Recognition* (pp. 408-410).

Tang, H., & Huang, T. (2008). 3D facial expression recognition based on automatically selected features. In *IEEE Computer Society Conference on Computer Vision and Pattern Recognition Workshops* (pp.1-8)

Tian, Y. L., Kanade, T., & Cohn, J. F. (2001). Recognizing action units for facial expression analysis. *IEEE Transactions on Pattern Analysis and Machine Intelligence*, *23*(2), 97–115. doi:10.1109/34.908962

Valstar, M., Pantic, M., & Patras, I. (2004). Motion history for facial action detection from face video. In *IEEE International Conference on Systems, Man and Cybernetics*, vol.1, (pp. 635-640).

Vinciarelli, A., Pantic, M., & Bourlard, H. (2009). Social signal processing: Survey of an emerging domain. *Image and Vision Computing*, *27*(12), 1743–1759. doi:10.1016/j.imavis.2008.11.007

Wang, J., Yin, L., Wei, X., & Sun, Y. (2006). *3D facial expression recognition based on primitive surface feature distribution*. In IEEE International Conference on Computer Vision and Pattern Recognition.

Wang, P., Kohler, C., Barret, F., Gur, R., & Verma, R. (2007). *Quantifying facial expression abnormality in schizophrenia by combining 2D and 3D features*. In IEEE International Conference on Computer Vision and Pattern Recognition.

Weiss, S. M., & Indurkhya, N. (1998). *Predictive data mining: A practical guide*. Morgan Kaufmann Publishers.

Whitehill, J., & Omlin, C. W. (2006). Haar features for FACS AU recognition. In *International Conference on Automatic Face and Gesture Recognition* (pp. 217-222).

Yabui, T., Kenmochi, Y., & Kotani, K. (2003). Facial expression analysis from 3D range images: Comparison with the analysis from 2D images and their integration. In *International Conference on Image Processing* (pp. 879-882).

Yacoob, Y., & Davis, L. S. (1996). Recognizing human facial expression from long image sequences using optical flow . *IEEE Transactions on Pattern Analysis and Machine Intelligence*, *18*(6), 636–642. doi:10.1109/34.506414

Yang, M. H., Kriegman, D. J., & Ahuja, N. (2002). Detecting faces in images: A survey. *IEEE Transactions on Pattern Analysis and Machine Intelligence*, *24*(1), 34–58. doi:10.1109/34.982883

Yeasin, M., Bullot, B., & Sharma, R. (2006). Recognition of facial expressions and measurement of levels of interest from video. *IEEE Transactions on Multimedia*, *8*(3), 500–507. doi:10.1109/TMM.2006.870737

Yin, L., Chen, X., Sun, Y., Worm, T., & Reale, M. (2008). *A high-resolution 3D dynamic facial expression database*. In IEEE International Conference on Face and Gesture Recognition.

Yin, L., Wei, X., Sun, Y., Wang, J., & Rosato, M. J. (2006). A 3D facial expression database for facial behavior research. In *Proceedings of the 7th International Conference on Automatic Face and Gesture Recognition* (pp. 211–216). IEEE Computer Society.

Zhou, F., De la Torre, F., & Cohn, J. F. (2010). *Unsupervised discovery of facial events*. In IEEE International Conference on Computer Vision and Pattern Recognition.

Zhu, Z., & Ji, Q. (2004). 3D face pose tracking from an uncalibrated monocular camera. In *Proceedings of the 17th International Conference on Pattern Recognition,* (pp. 400-403). IEEE Computer Society.

ADDITIONAL READING

Basso, C., Paysan, P., & Vetter, T. (2006). Registration of expressions data using a 3D morphable model. In *Proceedings of the 7th IEEE International Conference Automatic Face and Gesture Recognition*.

Benedikt, L., Cosker, D., Rosin, P. L., & Marshall, D. (2008). 3D Facial Gestures in Biometrics: from Feasibility Study to Application. In *2nd IEEE International Conference on Biometrics: Theory, Applications and Systems* (pp 1-6).

Blanz, V., Scherbaum, K., & Seidel, H. P. (2007). Fitting a Morphable Model to 3D Scans of Faces. In *Proceedings of International Conference on Computer Vision* (pp. 1-8).

Blanz, V., & Vetter, T. (1999). A morphable model for the synthesis of 3D faces. In *Proceedings of the 26th annual conference on Computer graphics and interactive techniques* (pp. 187-194). Association For Computing Machinery Press/Addison-Wesley Publishing Co.

Bowyer, K. W., Chang, K., & Flynn, P. (2006). A survey of approaches and challenges in 3D and multi-model 3d+2d face recognition. *Computer Vision and Image Understanding, 101*(1), 1–15. doi:10.1016/j.cviu.2005.05.005

Chibelushi, C.C. & Bourel, F. (2003). Facial expression recognition: A brief tutorial overview. *CVonline: On-Line Compendium of Computer Vision.*

Cootes, T. F., Edwards, G. J., & Taylor, C. J. (2001). Active appearance models. *IEEE Transactions on Pattern Analysis and Machine Intelligence, PAMI, 23*(6), 681–685. doi:10.1109/34.927467

Darwin, C., Ekman, P., & Prodger, P. (2002). *The expression of the emotions in man and animals.* United States: Oxford University Press.

Dornaika, F., & Raducanu, B. (2009). Facial expression recognition for HCI applications . In *Encyclopedia of Artficial Intelligence* (pp. 625–631). IGI Global Publishers.

Ekman, P., & Rosenberg, E. L. (Eds.). (1997). *What the face reveals.* United States: Oxford University Press.

Essa, I. A., & Pentland, A. P. (1997). Coding, analysis, interpretation, and recognition of facial expressions. *IEEE Transactions on Pattern Analysis and Machine Intelligence, 19*(7), 757–763. doi:10.1109/34.598232

Hahnel, M., Wiratanaya, A., & Kraiss, K. F. (2006). Facial Expression Modelling from Still Images Using a Single Generic 3D Head Model. *Lecture Notes in Computer Science, 4174,* 324–333. doi:10.1007/11861898_33

Kittler, J., Hilton, A., Hamouz, M., & Illingworth, J. (2005). 3D assisted face recognition: A survey of 3D imaging, modeling and recognition approaches. In *Proceedings of the IEEE Workshop on Advanced 3D Imaging for Safety and Security.*

Kobayashi, H., & Hara, F. (1997). Facial interaction between animated 3D face robot and human beings. In *IEEE International Conference on Systems, Man, and Cybernetics* (pp. 3732-3737).

Kuo, C. J., Huang, R. S., & Lin, T. G. (2002). 3-D facial model estimation from single front-view facial image. *IEEE Transactions on Circuits and Systems for Video Technology, 12*(3), 183–192. doi:10.1109/76.993439

Lin, I. C., & Ouhyoung, M. (2005). Mirror MoCap: Automatic and efficient capture of dense 3D facial motion parameters from video. *The Visual Computer, 21*(6), 355–372. doi:10.1007/s00371-005-0291-5

Picard, R. W. (2000). *Affective Computing.* MIT Press.

Romdhani, S., Blanz, V., Basso, C., & Vetter, T. (2005). Morphable Models of Faces . In Li, S., & Jain, A. (Eds.), *Handbook of Face Recognition* (pp. 217–245). doi:10.1007/0-387-27257-7_11

Scheenstra, A., Ruifrok, A., & Veltkamp, R. C. (2005). A survey of 3D face recognition methods. In *Proceedings of the 5th International Conference of Audio- and Video-Based Biometric Person Authentication.*

Wen, Z., & Huang, T. S. (2004). *3D Face Processing.* Kluwer Academic Publishers.

Zhao, W., & Chellapa, R. (2006). *Face processing-advanced modeling and methods.* Academic Press.

KEY TERMS AND DEFINITIONS

3D Face Model: A listing of the (*x, y, z*) coordinates of a fixed number of points lying on the facial surface.This model may also be accompanied by a texture image containing color (RGB) values at these vertices.

Facial Expression Recognition: A research area in which computers are made to recognize the facial expressions of humans.

Feature Extraction: For a classification problem, feature extraction is a method to extract characteristics of the data that can be used to discriminate between different classes of the data.

Human Computer Interaction (HCI): It is a study on how human beings and computers interact with each other.

Landmark Points: Prominent points on an object (Face in our case). Usually these points are easily discernible such as corners.

Subject Independence: A subject independent algorithm can work on any person even if not seen before.

Support Vector Machines (SVMs): A supervised learning tool which is used for classification and regression problems. It finds a separating hyperplane in a higher dimension.

Chapter 31
3D Thumbnails for 3D Videos with Depth

Yeliz Yigit
Bilkent University, Turkey

S. Fatih Isler
Bilkent University, Turkey

Tolga Capin
Bilkent University, Turkey

ABSTRACT

In this chapter, we present a new thumbnail format for 3D videos with depth, 3D thumbnail, which helps users to understand the content by preserving the recognizable features and qualities of 3D videos. The current thumbnail solutions do not give the general idea of the content and are not illustrative. In spite of the existence of 3D media content databases, there is no thumbnail representation for 3D contents. Thus, we propose a framework that generates 3D thumbnails from layered depth video (LDV) and video plus depth (V+D) by using two different methodologies on importance maps: saliency-depth and layer based approaches. Finally, several experiments are presented that indicate 3D thumbnails are illustrative.

INTRODUCTION

Today, the popularity of 3D media usage in computerized environment and the research on 3D content generation is increasing. 3D contents are frequently used in various applications such as computer games, movies and even in home environmental systems and this reputation leads visualization of 3D contents such as 3D videos and 3D images becoming more significant. For visualizing the 3D contents, thumbnail representation is used in order to provide a quick overview of multimedia files in order to allow a quick scanning over a large number of data. By using the traditional methods, the thumbnail generally shows the first frame of the video and for images, visual representation is generated by using shrinking, manual cropping or uniform scaling. However, these approaches do not preserve the important parts of the multimedia files and resulting thumbnails do not give the general idea of the content. Furthermore, in spite of the existence of

DOI: 10.4018/978-1-61350-326-3.ch031

3D content databases, there is no standardization on the thumbnail representation for 3D contents while their usage area is widespread. Thus, the thumbnail representation is very crucial to get a quick overview of the content rather than downloading from the database and processing it.

Therefore, we propose a thumbnail generation system that creates meaningful, illustrative visual representations of 3D video with depth contents without losing perceivable elements in the selected video frame by using saliency-depth and layer based methodologies. Moreover, in order to represent the 3D contents realistically and enhance depth perception, the resulting thumbnail should be in 3D. Thus, the framework constructs geometries of important objects as polygon meshes and adds 3D effects such as shadow and parallax mapping. Figure 1 illustrates a layout that holds resultant 3D thumbnails for 3D videos with depth.

While creating 3D thumbnails, it is required to select suitable 3D video formats since compression and coding algorithms of 3D videos show diversity according to the varieties of 3D displays: classical two-view stereo video (CSV), video plus depth (V+D), layered depth video (LDV) and multi-view video plus depth (MDV). Some of these formats and coding algorithms are standardized by MPEG, since standard formats and efficient compression are crucial for the success of 3D video applications (F.Institute, 2008). For our framework, V+D and LDV formats are eligible because of simplicity and the depth information they provide. V+D format provides a color video and an associated depth map that stands for geometry-enhanced information of the 3D scene. The color video is original video itself and the depth map is a monochromatic, luminance-only video. Besides, LDV is an extension of V+D format. It contains all information that V+D satisfies with an extra layer called background layer which includes foreground objects and the associated depth map of the background layer. By using the properties of V+D and LDV videos, we develop two different thumbnail generation methods based on the information they present. These proposed methodologies create meaningful thumbnails without losing perceivable visual elements in the selected original video frame.

Figure 1. 3D thumbnails on a 3D grid layout

In this chapter, the previous work on 3D video formats and thumbnail generation methods, the proposed framework that generates 3D thumbnails from video plus depth (V+D) and layered depth video (LDV), two 3D thumbnail generation methodologies based on 3D meshes and parallax mapping, and several experiments showing effectiveness and recognizability of 3D thumbnails, are presented.

BACKGROUND

We discuss 3D video formats and thumbnail generation approaches under two different subsections since our approach combines them.

3D Video Formats

Recently, several numbers of researches on 3D imaging and video formats are rapidly progressing. 3D video formats are roughly divided into two classes: *N*-view video formats and geometry-enhanced formats. The first class represents the multi-view video with *N* views. Conventional stereo video (CSV) is the least complex and most popular format of *N*-view video for stereoscopic applications.

Otherwise, geometry-enhanced information is provided for 3D video formats in the second class. Multi-view video + depth video (MDV), layered-depth video (LDV) and video plus depth (V+D) are examples of geometry-enhanced formatted videos. As it is referred from its name, MDV has more than one view and associated depth maps for each view. This depth data is used to synthesize a number of arbitrary dense intermediate views for multi-view displays (Gundogdu, 2010). LDV is one variant of MDV, which further reduces the color and depth data by representing the common information in all input views by one central view and difference information in residual views (Müller, 2008). Besides, foreground objects are stored on the background layer in the LDV format with associated depth information. Since geometry-enhanced formats are complex and more data is stored, the disadvantage of MDV and LDV is the requirement of the intermediate view synthesis. In addition to this, high-quality depth map generation is required beforehand and errors in depth data may cause considerable degradation in quality of intermediate views. On the other hand, the special case, V+D codes one color video and associated depth map and the second view is generated after decoding.

V+D and LDV formats are appropriate for creating effective thumbnails for 3D videos with depth in order to try the efficiency of our framework for both simple and complex 3D formats. Furthermore, our thumbnail generation system uses depth information that V+D and LDV formats satisfy for generating 3D thumbnails.

In addition to this, a video frame should be selected in order to create illustrative thumbnails. There are considerable number of researches on video summarization and frame selection that are based on clustering-based (Farin, 2002), keyframe-based (Mundur, 2006), rule-based (Lienhart, 1997) and mathematically-oriented (Gong, 2002) methods. However, for our work, video summarization and frame selection issues are out of scope. Thus, we apply a saliency-based frame selection. In this case, for each frame of the 3D input video, the saliency is computed and the frame that has the highest saliency value is selected.

Thumbnail Generation

Our goal is to create thumbnails from 3D videos with depth without losing perceivable elements on the selected original frame. Thus, it is essential to preserve the perceivable visual elements in an image for increasing the recognizable features of the thumbnail. Computation of important elements and performing non-uniform scaling to image are involved in the proposed thumbnail representation. This problem is similar to proposed methodologies for image retargeting (Setlur, 2005).

Manually by standard tools such as (Adobe, 2010) and (Gimp, 2101), image retargeting can be achieved by standard image editing algorithms such as uniform scaling and cropping. Nevertheless, important regions of the image cannot be conserved with uniform scaling and cropping. Moreover, when the input image contains more than one important object, it leads contextual information lost and quality of the image degrades.

In addition to this, automatic cropping techniques based on visual attention have been proposed (Suh, 2003) which can be processed by saliency maps (Itti, 1998) and face detection (Bregler, 1998; Yow, 1998). Nevertheless, the main disadvantage of this technique is that it only performs for a single object and this leads loss of multiple features.

Another way to generate thumbnails is by using epitomes. Epitome is the miniature and the condensed version of the input image which contains the most important elements of the original image (Jojic, 2003). Despite the conservation of important elements, this method is suitable when the image contains repetitive unit patterns.

The main work behind our approach is Setlur (2005) image retargeting algorithm since it works for multiple objects by preserving recognizable features and maximizes the salient content. This method segments the input image into regions by using mean-shift algorithm, identifies important regions by a saliency based approach, extracts them, fills the resulting gaps, resizes the filled background into a desired size and pastes important objects onto it by using computed aspect ratios according to importance values of objects.

MAIN FOCUS OF THE CHAPTER

The objective behind this work is to create helpful and demonstrative 3D thumbnails for various types of 3D video formats. Since the proposed methods for generating thumbnails do not preserve the important features and do not give the idea of the content, we suggest a new thumbnail format, 3D thumbnail.

Moreover, today 3D is popular in computer games, movies and home environment applications. Besides, everything included user interfaces will be in 3D soon. Thus we have generated the resulting thumbnail in 3D because by using 3D layouts, more objects can be illustrated on the thumbnail with a realistic look. 3D contents which are represented by multiple thumbnails can be epitomized with a single thumbnail by preserving the important objects in a 3D layout. Lastly, in spite of the popularity and widespread usage of 3D content databases, there is no standardized thumbnail representation for 3D contents such as 3D videos.

In 3D content databases, the 3D contents are signified in 5 or 6 images in order to help users identify the 3D content. Instead of using several numbers of 2D thumbnails for giving information about the content, it is sufficient to use single 3D thumbnail that satisfies geometry-enhanced information.

The inputs of our system are V+D and LDV formatted 3D videos. V+D and LDV formats are suitable for our system since the associated depth maps are essential for generating 3D thumbnails. On the other hand, with the purpose of trying different thumbnail generation methods based on saliency-depth and layer information, and the efficiency of our framework over both simple and complex 3D formats, V+D and LDV formats are appropriate.

In order to create 3D thumbnails for V+D formatted videos, the first step is to segment the selected frame of the input color video into regions with the aim of finding the important regions. Then, by a saliency-depth based approach, importance map is obtained and important objects are extracted from the original frame. The resulting frame with gaps are filled by reconstructing the blanks with the same texture as the given input color frame by successively adding pixels and the frame is resized to a standard thumbnail size. Next, with the intention of generating the saliency-depth based retargeted image, the aspect ratios and positions of the important objects are determined by a constraint-based algorithm and scaled important objects are pasted on to the resized background. Finally, 3D mesh that represents the 3D thumbnail is created by using the retargeted color frame and the associated depth map.

On the other hand, the thumbnail generation algorithm for LDV is similar to the proposed approach for V+D except some steps. Firstly, as well as the input color video, the associated background layer is also segmented into regions. Secondly, instead of finding salient regions and classifying them as important objects, foreground objects on the background layer are assumed to be important objects. Apart from these steps, the remaining procedure is same as the one for V+D.

The 3D thumbnail generation methodology for V+D and LDV are explained in detail in the next section.

METHODOLOGY

System Overview

The input of our framework is the specific video frame of a 3D video with depth (Either LDV or V+D) as RGB color map and the associated depth map. For LDV, besides the input color video, the associated background layer is additional essential input for importance map extraction.

Firstly, the input color map is segmented into regions. Then, the importance map is extracted by using a saliency-depth approach for V+D formats. This step is different when the type of the content is LDV since the background layer is utilized with the purpose of importance map generation. Thus, while stating salient and foremost objects as important for V+D formats, the foreground objects on the background layer of LDV formats are important. After mapping the importance values, important objects are extracted, later to be exaggerated and the resulting gaps are filled with same texture as the given input color frame. Afterwards, the background is resized to the standard thumbnail size which has 192x192 resolutions. Then, important objects are pasted onto the resized background by a constraint-based algorithm. Finally, we apply two different methods for the resultant 3D thumbnail: 3D mesh-based and parallax-mapped techniques.

Image Segmentation

In order to find the objects on the color map and assign their importance values, it is necessary to segment the color map into regions. There are three proposed image segmentation methods: mean-shift (Comaniciu, 2002), graph-based (Felzenszwalb, 2004) and hybrid segmentation (Pantofaru, 2005). These three approaches are evaluated in the work of Pantofaru (2005) by considering correctness and stability of the algorithms. According to the results, both the mean-shift and hybrid segmentation methods create more realistic segmentations than the graph-based approach with a variety of parameters and both of the methods are stable. Since the hybrid segmentation algorithm is the combination of mean-shift and graph-based segmentation, it is more computationally expensive. Thus, we have preferred the mean-shift algorithm for its power and flexibility of modeling.

In Computer Vision, the mean-shift segmentation has a widespread usage. This algorithm takes spatial radius h_s, color radius h_r and the minimum region area M as parameters with the input color map. In this algorithm, the first step is to convert RGB color map into $L\alpha\beta$ color space since the method uses CIE-Luv color space which has Gaussian smoothed blue-yellow, red-green and luminance planes. The next step is to determine and label the clusters by neighboring pixels within a spatial radius h_s, and color radius h_r. As the parameters are set by users, we set h_s as 6, h_r as 5 and M as 50 after some trials for our system.

Note that this step is also applied to the background layer of LDV formats in order to identify the foreground objects. Thus, the color map and background layer are segmented into regions by using mean-shift algorithm for LDV.

Importance Map Extraction

The importance map extraction approach works differently for V+D and LDV. Saliency-depth based method is applied to V+D, while a layer based importance map extraction is used for LDV.

Saliency-Depth Based Importance Map

For V+D videos, importance map extraction is based on the saliency and depth information. After the segmentation of the color map, three steps are achieved for generating the importance map: computation of saliency based on color map, computation of saliency based on depth map and computation of overall saliency map.

Most of the physiological experiments verify that human vision system is only aware of some parts of the incoming information in full detail. In order to locate the points of interest, the saliency concept is proposed. The graph-based visual saliency image attention model is used for saliency computation (Harel, 2007). It is a bottom-up visual saliency model that is constructed in two steps: Constructing activation maps on certain feature channels and normalization.

Graph-based visual saliency method contains three steps: *Feature extraction*, *activation* and *normalization of the activation map*. In the *feature extraction* step, the features such as color, orientation, texture, intensity are extracted from the color map through linear filtering and the calculation of center-surround differences for each feature type is completed. In the *activation step,* single or multiple activation maps are extracted by using feature vectors and subtracting feature maps at different scales such as henceforth, center, surround. Finally, the *normalization of the activation map* is performed. The goal of this step is to concentrate mass on activation maps by normalizing the effect of feature maps and summing them into the final saliency value of the pixel based on the color map.

Figure 2 shows several numbers of results that are based on graph-based visual saliency image attention model. The detailed explanation of the algorithm can be found in the work of Harel (2007).

After calculating saliency for each pixel on the color map, the depth saliency is computed. It is observed that depth is another factor to decide whether an object is important or should be ignored. In other words, closer objects should be more essential than the ones that are distant. Thus, we add the depth saliency for each pixel on the color map by using the associated depth map. A simple equation that is adapted from the work of Longurst (2006) is used in order to calculate the depth importance. The equation uses a model of exponential decay to get a typical linear model of very close objects.

Figure 2. Graph-based visual saliency image attention model. (a) Original Image; (b) Salient parts of the image (red – most salient); (c) Resulting saliency maps.

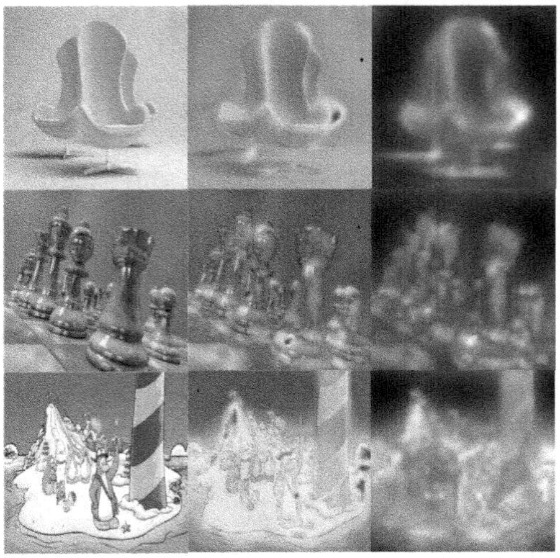

The last step is to compute the overall saliency. For each region that is segmented by mean-shift algorithm, the calculation of the overall saliency of the region is processed by averaging the sum of the color-based and depth-based saliency of pixels belonging to the region.

Layer Based Importance Map

For LDV videos, we follow a layer based approach with the aim of importance map extraction since foreground objects on the background layer are assumed to be important. In other words, the closer objects should be more salient than the distant ones. Thus, the segmented regions on the background layer are extracted from the color map at the end of this step. This approach is simpler than the saliency-depth based approach which is applied for V+D, because we use the features of LDV as our basis.

Background Resynthesis

After extracting important objects from the original color map, the background resynthesis step takes place. In this case, resynthesis refers to filling gaps of the extracted area with information from the surrounding area. This step is based on Harrison (2002)'s inpainting method. The algorithm reconstructs the gaps with the same texture as the given input color map by successively adding pixels that are selected. The procedure has two stages: pixel analysis and filling. In the first stage, relationships between pixels on the color map are analyzed and the value of each pixel that can be obtained by neighboring pixels is established. In the second stage, until all blank locations on the color map are filled, pixels are added by using the results of the pixel analysis stage. The procedure is capable of reproducing large features from the color map, even though it only examines interactions between pixels that are close to neighbors. Then, the color map is resized to 192x192 (standard thumbnail size).

Pasting of Important Objects

The next step is to paste important objects onto the new background. The constraint-based algorithm is utilized with the aim of pasting each object according to their importance values from the most important to least (Setlur, 2005). The goal is to preserve the relative positions of the important regions in order to keep the resized images layout similar to the original color map. For this algorithm, there are four constraints: positions of the important objects must stay the same, aspect ratios of the important objects must be maintained, the important objects must not overlap in the retargeted background if they are not overlapping in the original color map, and the background color of the important objects must not change.

From the most important object to least, this step reduces the change in position and the size of the important objects and the algorithm seeks whether the four conditions are satisfied or not. The aspect ratio and the position of the important objects are calculated according to the original and the retargeted color map.

3D Thumbnail Generation

In the 3D thumbnail generation stage, we apply two different approaches to get the 3D visual effect on the retargeted color map: 3D mesh generation and parallax mapping techniques.

3D Mesh Generation Technique

With the purpose of the generation of a 3D mesh from the retargeted color map by using the associated depth map, we follow a simple algorithm as illustrated in Figure 3.

The inputs of the 3D mesh generation algorithm are the retargeted color map and the corresponding depth map. In order to create the geometry of the resulting 3D mesh, the vertices that describe points and corner locations of the mesh in 3D space should be extracted. Thus, positions of vertices on the x-y coordinate are obtained from the retargeted color map and the depth values of

Figure 3. The flow order of the 3D mesh generation technique

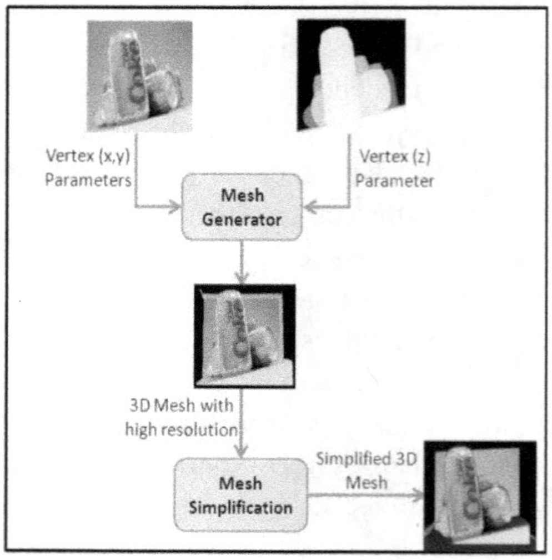

the corresponding vertices are acquired from the depth map. After obtaining all vertices, the construction of triangular faces between vertices to form the actual 3D mesh is achieved. Next step is to compute the texture data and face normals. Since the thumbnails should be in a simple format and a several numbers of thumbnails should be displayed in applications, it is necessary to consider the performance. Therefore, the constructed 3D mesh should be simplified because for a retargeted image which has 192x192 resolutions, there exist 36864 vertices and 73728 faces without simplification and this makes simultaneous rendering of multiple thumbnails impossible. For achieving the simplification, we use an edge collapse algorithm based on the quadric metric approach (Garland, 1998). This method produces high quality approximations of polygon models rapidly. In order to process simplification, iterative contractions of vertex pairs are used and the surface error approximations are maintained by using quadratic matrices. In addition to this, the algorithm joins unconnected regions by reducing arbitrary vertex pairs. Thus, after simplification it is guaranteed to have meshes that contains up to 4000 faces.

Parallax Mapping Technique

Parallax mapping is a shading technique and the enhancement of bump mapping or normal mapping which is proposed by Tomomichi (2001). It is applied to textures in 3D rendering applications such as 3D games, virtual environment applications etc. and also known as offset mapping or virtual displacement mapping.

Parallax mapping is a simple method to give motion parallax effects on a polygon. In other words, 2D textures have more apparent depth when this approach is applied. The combination of traditional normal and height mapping creates 3D effect without the use of additional vertices. By adding depth to 2D textures, the final render appears to have a much higher polygon count than it actually has. Finally, it is a per-pixel shape representation and can be accomplished using the current generation of 3D hardware.

USER EVALUATION

In order to evaluate the performance of the resultant 3D thumbnails from 3D videos with depth and the proposed 3D thumbnail generation methods (3D mesh vs. parallax-mapped based), we have performed two experimental studies.

Subjects

15 voluntary subjects participated: 13 males and 2 females with a mean age of 25.17. Two of the subjects were novice and others were experienced users with a computer science background.

Equipment

All experiments were performed on the Sharp Actius RD3D with 15-inch XGA (1024-by-768) autostereoscopic color display. For interaction, mouse and keyboard were used. Subjects did not wear any special glasses.

Tasks

In the first experiment, the task that participants should accomplish was to select the correct thumbnail from a large set of 2D and 3D thumbnails for a given content name in a reasonable time. This user study had 60 steps. For the first 30 step, 3D thumbnails were randomly located on the 3D environment and displayed in a 3D grid layout. In addition to this, 2D thumbnails were randomly positioned on the 3D grid layout for the rest. For each step, a target thumbnail with a given content name was asked to be browsed by subjects. The second experiment was similar to the first, but in this case, subjects accomplished the selection of the target thumbnail from a set of 3D mesh or

Figure 4. The target 2D, 3D and parallax-mapped thumbnails used in experiments

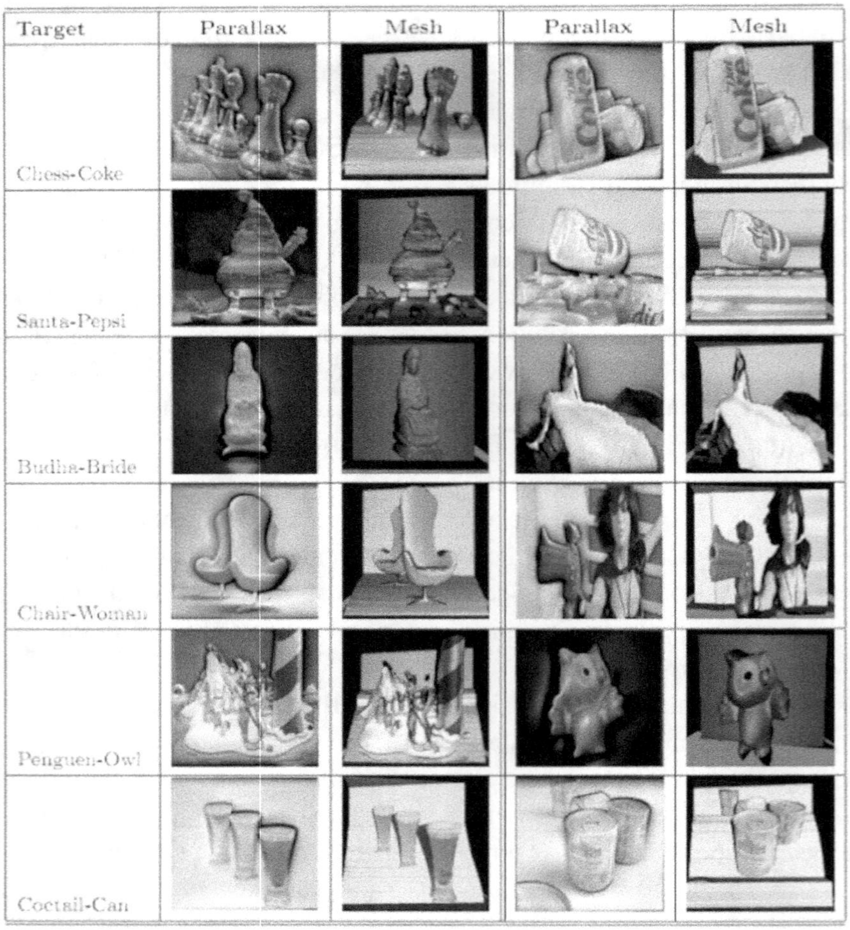

parallax-mapped based thumbnails. Other than this, the structure of the experiment was similar with the first one.

For all tests, 12 different target content names as illustrated in Figure 4, were asked to be browsed and no text labels were satisfied for thumbnails.

Results and Discussion

For the first experiment, our hypothesis *was 3D thumbnails are illustrative than 2D thumbnails.* In order to prove this, we recorded the search time and the number of clicks performed to reach the target thumbnail. The comparison results are illustrated in Figure 5. By using 3D thumbnails, subjects accomplished 30 experiment steps in 142.138 seconds with 78.92 clicks, while they performed 87.304 clicks in 159.184 seconds with 2D thumbnails. From the figure and total results, it is clearly occurred that recognition time for 3D is shorter than 2D. With the aim of better indication of the statistically significant difference of 3D thumbnails, we have also performed a *paired samples t-test* on the experimental data. The mean error of each test case of 2D thumbnails was compared to the mean error of 3D thumbnails, and it showed that the difference between 3D thumbnails and 2D thumbnails is statistically significant with $p < 0.05$.

Figure 5. 2D thumbnails vs. 3D thumbnails (based on search time)

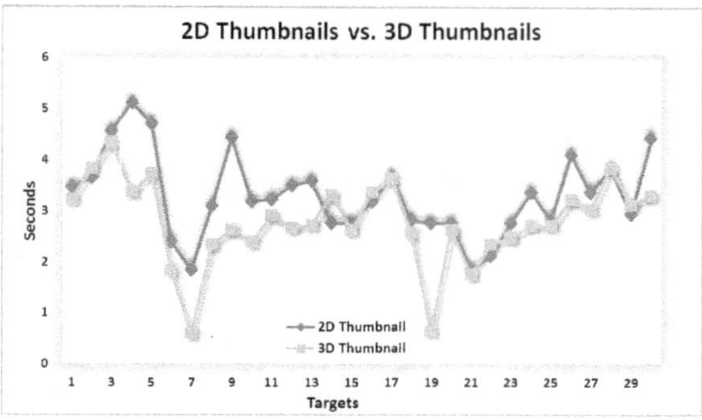

Figure 6. 3D thumbnails based on 3D mesh generation vs. parallax-mapped technique (based on search time)

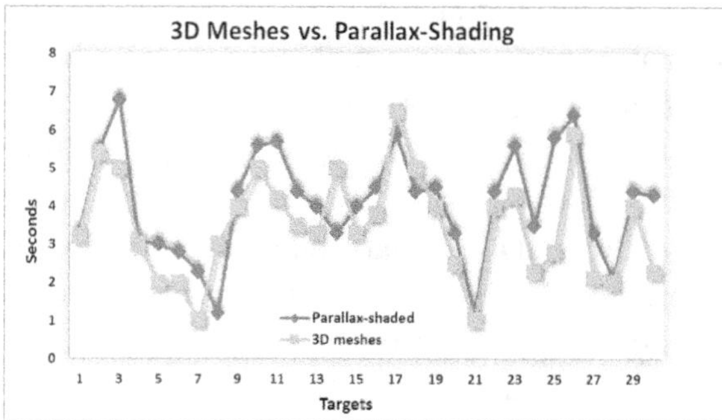

The second experimental study was based on *3D meshes are illustrative than parallax-mapped images* hypothesis. By using 3D mesh-based thumbnails, subjects completed 30 experiment steps in 139.476 seconds while they performed the test with parallax-mapped thumbnails in 142.965 seconds. Moreover, 96.17 clicks were acquired to complete tasks for finding targets with 3D mesh-based thumbnails and 101 clicks were obtained with parallax-mapped thumbnails. Figure 6 and total result show that thumbnails that are based on 3D mesh generation technique are more recognizable than parallax-mapped thumbnails. However, from the *paired samples t-test* results, this difference is not significant and our hypothesis was rejected because the statistically significant difference between the two methods was not acceptable ($p > 0.05$).

FUTURE RESEARCH DIRECTIONS

For the future work, 3D thumbnails should be generated for CSV and MDV formats and a comparison between 4 methodologies should be accomplished. Thence, the suitable format that provides a fast 3D thumbnail creation approach can be determined. Moreover, as our video frame selection is based on a saliency-based approach, a stronger and efficient video summarization technique should be applied to our 3D thumbnail generation system in order to get the most meaningful frame that represents the entire video. As a result, our thumbnail generation methodology can be improved and more illustrative thumbnails for all kinds of 3D videos can be generated. Finally, additional user studies should be performed with the aim of the proof for the efficiency of 3D thumbnails.

CONCLUSION

A framework that generates 3D thumbnails for 3D videos with depth is proposed in this chapter. The goal of the framework is to create meaningful, illustrative and efficient thumbnails by preserving the important parts of the selected frame of the input video. The inputs of the system are two different video formats: V+D and LDV, and the generation approaches are different for each format. For V+D, the important objects are extracted by a saliency-depth based method. In other words, the visually salient and closer objects are important and necessary to be preserved. However, the foreground objects that are on the background layer are assumed to be important for LDV formats. This method is called layer-based. After determining the important objects, remaining steps are same for all formats. Important objects are extracted from the color map and resulting gaps are filled. After that, newly created background image is resized to a standard thumbnail size and important objects are pasted on it by a special algorithm that has 4 constraints: aspects ratios of the important objects are maintained, the background color and the positions of the important objects should be same as the original color map and objects should not be overlapping if they are not overlapping in the original image. In the final stage, two techniques are used to generate the resultant 3D thumbnail: 3D mesh and parallax mapping methods.

Finally, we have performed two user experiments in order to test the efficiency of 3D thumbnails. In the first experiment, we compared 2D thumbnails and 3D thumbnails. The experiment results show that 3D thumbnails are statistically ($p < 0.05$) illustrative than 2D thumbnails. Moreover, the second experiment's aim is to test the efficiencies of the two proposed methods for generating 3D thumbnails: 3D mesh and parallax mapping techniques. From the experiment results, it is indicated that there is no significantly difference between two techniques since $p > 0.05$. Thus, either the 3D thumbnail which is generated by 3D mesh or parallax-shading technique can give the 3D visual effect and enhance the depth perception.

ACKNOWLEGMENT

We gratefully acknowledge support from the "All 3D Imaging Phone" project under contract FP7-213349.

REFERENCES

Adobe. (2010). *Website.* Retrieved August 12, 2010, from http://www.adobe.com

Bregler, C. (1998). Tracking people with twists and exponential maps. *IEEE Computer Society Conference on Computer Vision and Pattern Recognition*, (pp. 8-15).

Comaniciu, D., & Meer, P. (2002). Mean shift: A robust approach toward feature space analysis. *IEEE Transactions on Pattern Analysis and Machine Intelligence*, *24*(5), 603–619. doi:10.1109/34.1000236

Farin, D., Effelsberg, W., & De, H. N. P. (2002). Robust clustering-based video-summarization with integration of domain-knowledge. In *International Conference on Multimedia and Expo (ICME): Vol. 1,* (pp. 89-92). Lausanne, Switzerland.

Felzenszwalb, P., & Huttenlocher, D. (2004). Efficient graph-based image segmentation. *International Journal of Computer Vision*, *6*(2), 167–181. doi:10.1023/B:VISI.0000022288.19776.77

Garland, M., & Heckbert, P. (1998). Simplifying surfaces with color and texture using quadric error metrics. In *IEEE Conference on Visualization*, (pp. 263-269).

Gimp. (2010). *Website.* Retrieved August 12, 2010, from http://www.gimp.com

Gong, Y., & Liu, X. (2000). Generating optimal video summaries. In *International Conference on Multimedia and Expo (ICME): Vol. 3* (pp. 1559-1562).

Gundogdu, R. B., Yigit, Y., & Capin, T. (2010). 3D thumbnails for mobile media browser interface with autostereoscopic displays. *Springer Lecture Notes in Computer Science, Special Issue on IEEE Multimedia Modeling 2010.* Chongqing, China.

Harel, J., Koch, C., & Perona, P. (2007). Graph-based visual saliency. *Advances in Neural Information Processing Systems*, *19*, 545–552.

Harrison, P. (2001). A non-hierarchical procedure for re-synthesis of complex textures. In *Proc. WSCG*, 190-97.

F Institute. (2008). *D5.1 – Requirements and specifications for 3D video*. All 3D Imaging Phone.

Itti, L., Koch, C., & Niebur, E. (1998). A model of saliency-based visual attention for rapid scene analysis. *IEEE Transactions on Pattern Analysis and Machine Intelligence*, *20*(11), 1254–1259. doi:10.1109/34.730558

Jojic, N., Frey, B., & Kannan, A. (2003). Epitomic analysis of appearance and shape. *IEEE International Conference on Computer Vision*, (pp. 34-41).

Kaneko, T., Takahei, T., Inami, M., Kawakami, Y. Y. N., Maeda, T., & Tachi, S. (2001). *Detailed shape representation with parallax mapping* (pp. 205–208). ICAT.

Lienhart, R., Pfeiffer, S., & Effelsberg, W. (1997). Video abstracting. *Communications of the ACM*, *40*(12), 55–62. doi:10.1145/265563.265572

Longhurst, P. (2006). A GPU based saliency map for high-fidelity selective rendering. In *International Conference on Computer Graphics, Virtual Reality, Visualization and Interaction*, (pp. 21-29). Africa.

Müller, K., Smolic, A., Dix, K., Kauff, P., & Wiegand, T. (2008). Reliability-based generation and view synthesis in layered depth video. In *Proc. IEEE International Workshop on Multimedia Signal Processing (MMSP2008)*, (pp. 34-39). Cairns, Australia.

Mundur, P., Rao, Y., & Yesha, Y. (2006). Keyframe-based video summarization using Delaunay clustering. *International Journal on Digital Libraries*, *6*, 219–232. doi:10.1007/s00799-005-0129-9

Pantofaru, C. (2005). *A comparison of image segmentation algorithms*. Robotics Inst., Carnegie Mellon University.

Setlur, V., Takagi, S., Raskar, R., Gleicher, M., & Gooch, B. (2005) Automatic Image retargeting. In *International Conference on Mobile and Ubiquitous Multimedia*, (pp. 59-68). New Zealand.

Suh, B., Ling, L., Bederson, B., & Jacobs, D. (2003). Automatic thumbnail cropping and its effectiveness. *ACM Symposium on User interface Software and Technology*, (pp. 95-104). Vancouver, Canada.

Yow, K. (1998). *Automatic human face detection and localization.* Unpublished doctoral dissertation, University of Cambridge, Cambridge.

ADDITIONAL READING

Comaniciu, D., & Meer, P. (2002). Mean shift: A robust approach toward feature space analysis. *IEEE Transactions on Pattern Analysis and Machine Intelligence*, *24*(5), 603–619. doi:10.1109/34.1000236

Gundogdu, R. B., Yigit, Y., & Capin, T. (2010). 3D thumbnails for mobile media browser interface with autostereoscopic displays. *Springer Lecture Notes in Computer Science, Special Issue on IEEE Multimedia Modeling 2010*. Chongqing, China.

Harel, J., Koch, C., & Perona, P. (2007). Graph-based visual saliency. *Advances in Neural Information Processing Systems*, *19*, 545–552.

Harel, J., Koch, C., & Perona, P. (2007). Graph-based visual saliency. *Advances in Neural Information Processing Systems*, *19*, 545–552.

Itti, L., Koch, C., & Niebur, E. (1998). A model of saliency-based visual attention for rapid scene analysis. *IEEE Transactions on Pattern Analysis and Machine Intelligence*, *20*(11), 1254–1259. doi:10.1109/34.730558

Setlur, V. (2005). *Optimizing computer Imagery for more effective visual communication.* Unpublished doctoral dissertation, Northwestern University, Illinois.

Talton, J. O. (2004). *A short survey of mesh simplification algorithms. University of Illinois*. Urban-Campaign.

KEY TERMS AND DEFINITIONS

3D Video: Kind of a visual media that satisfies 3D depth perception that can be provided by a 3D display.

Auto-Stereoscopic Display: A 3D display that helps user to see the content without help of any 3D glasses on the flat screen.

Depth Map: A grey-scale map that includes depth information of the content for every pixel. In this map, the object that is nearest is the bright one.

Grid Layout: A type of a layout that divides the container into equal-sized rectangles and each rectangle holds one item.

Mean-Shift Segmentation: A powerful image segmentation technique that is based on non-parametric iterative algorithm and can be used for clustering, finding modes etc.

Parallax Shading: A shading technique that displaces the individual pixel height of a surface, so that when the resulting image is seen as three-dimensional.

Saliency: Refers to visual saliency which is the distinct subjective perceptual quality that grabs attention.

Thumbnail: A small image that represents the content and used to help in recognizing and organizing several numbers of contents.

About the Contributors

Aamir Saeed Malik has a BS in Electrical Engineering from University of Engineering & Technology, Lahore, Pakistan, MS in Nuclear Engineering from Quaid-i-Azam University, Islamabad, Pakistan, another MS in Information & Communication and PhD in Information & Mechatronics from Gwangju Institute of Science & Technology, Gwangju, Republic of Korea. He has more than 10 years of research experience and has worked for GoP, IBM, and Hamdard University in Pakistan, and Yeungnam and Hanyang Universities in South Korea. He is currently working at Universiti Teknologi PETRONAS in Malaysia. He is Senior Member of IEEE. His research interests include image processing, 3D shape recovery, medical imaging, EEG signal processing and content based image retrieval (CBIR).

Tae-Sun Choi received the BS degree in Electrical Engineering from the Seoul Nation University, Seoul, Korea, in 1976, the MS degree in Electrical Engineering from the Korea Advanced Institute of Science and Technology, Seoul, Korea, in 1979, and the PhD degree in Electrical Engineering from the State University of New York at Stony Brook, 1993. He is currently a Professor in the School of Information and Mechatronics at Gwangju Institute of Science and Technology, Gwangju, Korea. His research interests include image processing, machine/robot vision, and visual communications.

Humaira Nisar received the BE (Honors) in Electrical Engineering from University of Engineering and Technology, Lahore, Pakistan in 1993. She received MS degree in Nuclear Engineering from Quaid-e-Azam University, Islamabad, Pakistan in 1995. She received MS degree in Mechatronics and PhD in Information and Mechatronics from Gwangju Institute of Science and Technology, Republic of Korea in 2000 and 2008 respectively. Currently, she is Assistant Professor at Department of Electronics Engineering, Universiti Tunku Abdul Rahman, Perak, Malaysia. Her research interests include image processing, motion estimation, video compression, and signal processing.

* * *

Fakhreddine Ababsa received the PhD degree in Robotics from the University of Evry Val d'Essonne (France) in 2002. He joined the Complex Systems Laboratory in December 1999. Since 2004, he is Assistant Professor in Computer Science at the University of Evry Val d'Essonne France. His research interests include augmented reality, pattern recognition, motion tracking, sensor fusion, and image processing.

Mongi A. Abidi (S'83–M'85) received his MS and PhD degrees in electrical engineering from the University of Tennessee, Knoxville, in 1985 and 1987, respectively. He is currently a Professor with the Department of Electrical Engineering and Computer Science, University of Tennessee, directing research activities at the Imaging, Robotics, and Intelligent Systems Laboratory. He has published more than 300 papers and edited/written four books in the area of imaging and robotics. He received The Most Cited Paper Award in Computer Vision and Image Understanding for 2006 and 2007.

Li-Minn Ang completed his Bachelor of Engineering and PhD at Edith Cowan University in Perth, Australia in 1996 and 2001, respectively. He then taught at Monash University before joining The University of Nottingham Malaysia Campus in 2004. His research interests are in the fields of signal, image, vision processing, and reconfigurable computing.

Francesco Bellocchio received the Computer Science Degree from the Università degli Studi di Milano, Milano, Italy, in 2007. He is currently working towards the PhD degree at the Department of Information Technology, Università degli Studi di Milano, Crema, Italy. His research interests are related mainly to neural networks and soft-computing paradigms and their application to the real-time 3-D surface reconstruction and signal and image processing.

Zarrad Boubaker is an Associate Professor in Biology and Medical Engineering at Higher School of Health Sciences and Techniques, Monastir, Tunisia. He has PhD in Biology and Medical Engineering from Claude Bernard University, Lyon I, France (1995). His research interests include medical image processing, filtering, tomographic reconstruction, quantification, computation of x-ray beam quality parameters, and Radioprotection.

Tolga Capin is an Assistant Professor at the Department of Computer Engineering in Bilkent University. He has received his B.S. Computer Engineering and Information Sciences in 1991 and MS Computer Engineering and Information Sciences in 1993 at Bilkent University, Turkey. He has received his PhD Computer Science at EPFL (Ecole Polytechnique Federale de Lausanne), Switzerland in 1998. Dr. Capin worked as a research manager and principal scientist in Nokia Research Center (Irving, Texas) between 2000 and 2006. He has more than 25 journal papers and book chapters, 40 conference papers, and a book. He has 3 patents and 10 pending patent applications. His research interests include networked virtual environments, mobile graphics, computer animation, and human-computer interaction.

Tian-Sheuan Chang received the BS, MS, and PhD degrees in Electronics Engineering from National Chiao-Tung University, Hsinchu, Taiwan, in 1993, 1995, and 1999, respectively. He is currently with Department of Electronics Engineering, National Chiao-Tung University, as an Associate Professor. During 2000 to 2004, he worked at Global Unichip Corp. Hsinchu, Taiwan. His research interests include IP and SOC design, VLSI signal processing, and computer architecture.

Wei Jen Chew received her Bachelor of Engineering degree (with honours) in the field of Electrical and Computer Systems and Masters of Electrical and Computer Systems Engineering from Monash University in 2005 and 2007, respectively. She is currently a Research Assistant and PhD student at The University of Nottingham, Malaysia Campus. Her research interest is in the field of image processing.

Dimitrios Chrysostomou received the Diploma degree from the Democritus University of Thrace, Xanthi, Greece, in 2006. He is currently working toward the PhD degree with the Laboratory of Robotics and Automation, Department of Production and Management Engineering, Democritus University of Thrace. His areas of interest include 3D reconstruction, multi view algorithms, multi camera systems and robotics. He is involved in several national (Greek) and international (European) research projects in the field of machine vision systems. Mr. Chrysostomou is a member of the IEEE, ACM, SPIE, the Technical Chamber of Greece (TEE), and the National Union of Production and Management Engineers.

Dinu Coltuc received the Diploma of Engineer (1982) and the PhD degree (1997) in Electronics and Telecommunications, both from the Politechnica University of Bucharest, Romania. He is currently a Professor in the Electrical Engineering School, Valahia University of Targoviste, Romania. Prof. Coltuc is the Head of the Electronics Department and the Director of the Research Center for Electrical Engineering, Electronics and Information Technology of Valahia University. He served also as invited Professor in France at University of Savoie, INP Grenoble, Jean Monnet University. His research interests lie in the areas of image and signal processing and include watermarking, image enhancement, fast algorithms.

Peter H.N. de With, an IEEE Fellow, obtained his MSc in Electrical Engineering from the Eindhoven University of Technology, and his PhD from Delft University of Technology, The Netherlands. He joined Philips Research Labs Eindhoven in 1984, where he worked on video coding for digital recording. From 1985 to 1993, he was involved in several European projects on SDTV and HDTV recording. In this period, he contributed as a principal coding expert to the DV standardization for digital camcording. Between 1994 and 1997, he was leading the design of advanced programmable video architectures at the same lab. In 1996, he became senior TV systems architect and in 1997, he was appointed as Full Professor at the University of Mannheim, Germany, at the faculty Computer Engineering. In Mannheim, he was heading the chair on Digital Circuitry and Simulation with the emphasis on video systems. Between 2000 and 2007, he was with LogicaCMG in Eindhoven as a principal consultant and also Professor at the Eindhoven University of Technology, at the faculty of Electrical Engineering. He is now with CycloMedia Technology, The Netherlands. He has written and coauthored over 200 papers on video coding, architectures and their realization. Regularly, he is a teacher of the Philips Technical Training and for other post-academic courses. In 1995 and 2000, he coauthored papers that received the IEEE CES Transactions Paper Award, and in 2004, the VCIP Best Paper Award. In 1996, he obtained a company Invention Award. Professor de With is an IEEE Fellow, advisor to Philips, scientific advisor of the Dutch Imaging School ASCII, IEEE ISCE, and board member of various working groups.

Jean-Yves Didier received his M.S. degree in Virtual Reality and Computer Science from University of Evry Val d'Essonne, France, and his Ph.D. degree in robotics from the same university in 2002 and 2005 respectively. Since 2005, he is a temporary teacher in a computer engineering school named Institut d'Informatique d'Entreprise (France), and a researcher at the IBISC Laboratory. His research works are focused on software architecture for rapid prototyping of augmented reality applications.

Luat Do obtained his MSc degree in Electrical Engineering in 2009, at the Eindhoven University of Technology (TU/e), Eindhoven, The Netherlands. In September 2009, he joined the Video Coding and Architectures group at the TU/e as a PhD student and is currently working on free-viewpoint interpolation algorithms which are a part of the European iGlance project. Luat's research interests include different aspects of 3D video: view synthesis, compression, efficient implementation and quality assessment of free-viewpoint interpolation, and stereo image generation for 3DTV.

Mercedes Farjas graduated at the Universidad Politécnica de Madrid (UPM) in 1982 as an Engineering Surveyor. After a few years with Universidad Politécnica de Las Palmas de Gran Canarias, she returned to Madrid as Surveyor to the City Council. In 1989, she took a post as Assistant Professor at UPM, and during these years she completed first BSc and the PhD in Educational Science and then BSc in Geodesy and Cartographical Engineering. In 1996, she was made Professor in Cartography and Surveying at UPM, where she was instrumental in setting up the research group Cartography applied in Archaeology and Heritage. She is leading the research topic New Technologies in Cultural Heritage, in the UPM Laboratory Cultural Heritage Management and New Technologies.

Stefano Ferrari received the MSc degree in Computer Science from the Università degli Studi di Milano, Milano, Italy, in 1995 and the Ph.D. in Computer and Automation Engineering from the Politecnico di Milano, Milano, Italy, in 2001. Since 2002, he has been an Assistant Professor at the Department of Information Technology, Università degli Studi di Milano. His research interests are related mainly to neural networks and soft-computing paradigms and their application to the computer graphics, signal processing, and measurement systems.

Velappa Ganapathy was born on 1st May 1941 at Singalandapuram, Salem, Tamil Nadu, India. He had obtained his Bachelor of Engineering in Electrical & Electronics Engineering and Master of Science in Electrical Engineering both from the University of Madras, India. He obtained his PhD in Electrical Engineering (Digital Signal Processing) from the Indian Institute of Technology, Madras, India. He had worked in various capacities as Associate Lecturer, Lecturer, Assistant Professor, Associate Professor, and Professor in institutions like Government College of Technology (12 years), Coimbatore, Anna University Chennai (18 years), Multimedia University (3 years), Malaysia and Monash University (9 ½ years) Malaysia. His research interests are digital signal processing, robotics, power systems, artificial intelligence, and image processing. Currently, he is with the University of Malaya, Kuala Lumpur, Malaysia as Professor of Electrical Engineering.

Antonios Gasteratos is an Assistant Professor of "Mechatronics and Artificial Vision" at the DPME. He teaches the courses of Robotics, Automatic Control Systems, Measurements Technology and Electronics. He completed Diploma and Ph.D. from the Department of Electrical and Computer Engineering, DUTH, Greece, in 1994 and 1999, respectively. During 1999-2000, he was a Post-Doc Fellow at the Laboratory of Integrated Advanced Robotics (LIRA-Lab), DIST, University of Genoa, Italy. He has served as a reviewer to numerous of scientific journals and international conferences. He is the Greek Associate High Level Group (HLG) Delegate at EUREKA initiative. His research interests are mainly in mechatronics and in robot vision. He has published one textbook, 3 book chapters, and more than 90 scientific papers. He is a member of the IEEE, IAPR, ECCAI, EURASIP, and the Technical Chamber of Greece (TEE). Dr. Gasteratos is a member of EURON, euCognition and I*PROMS European networks. He had organized the International Conference on Computer Vision Systems (ICVS 2008).

Yuanzheng Gong came to ISU in Spring 2010 for his PhD in Mechanical Engineering. He received his BS in Mechanical Engineering from University of Science and Technology of China in 2009. Currently, he is working with Dr. Zhang on crime tool mark characterization, and improvement of high-speed 3D measurement system. His research interests include computer vision, pattern recognition, and optical metrology.

Armin Grasnick was born in Berlin in 1965. After an apprenticeship in precision optics at Carl Zeiss, he graduated in technical optics and precision engineering. With his enormous interest in 3D, he founded his first company in 1998 where he invented, developed, and produced various 3D displays. Further company foundations followed. Today, he has released more than 60 papers and patent publications that are based on 3D and characterize current developments. Numerous awards acknowledged his merits. In 2000, he gained the "Gruenderpreis Thueringen," in 2001 the Mario Technology Award NAB in Las Vegas (together with the company DDD) and in 2002 the Order of Merit of the Federal Republic of Germany.

Raja Guedouar is a Research Assistant in the Department of Medical Imaging at Higher School of Health Sciences and Techniques of Monastir in Tunisia. She has a MS in Medical Application of Nuclear Physics (2001) and another MS in Electronic devices (2002). Her research interests are medical image processing, filtering, tomographic reconstruction, quantification, dosimetry, and radioprotection.

José María Hierro, born in Madrid in 1971, Surveyor Engineer from the Polytechnic University of Madrid, is currently carrying out a postgraduate degree in Management and Project Management at the National University of Distance Education. In 1996, he joined the engineering firm Lonjas Tecnología, Energía y Medioambiente, S.A., where he has developed his career in the field of renewable energy. Since 1999, he has been responsible for the management of the Technical Office´s Department and Area of mechanical plant installation. He has participated in the Engineering, Erection and Commissioning of numerous industrial plants producing electrical power, mainly cogeneration, photovoltaic, and biomass, both in Spain and Brazil.

Chao-Ching (Burt) Ho received his B.S. and M.S. in mechanical engineering from National Taiwan University, Taipei, Taiwan, R.O.C., in 1995 and 1997, respectively, and his Ph.D. in electrical engineering from the National Taiwan University of Science and Technology, Taipei, Taiwan, 2008. He served as a naval engineer officer from 1997 to 1999. From 1999 to 2003, he was RD manager of the 3DFamily Inc. From 2006 to 2009, he was a senior designer of home entertainment department at STMicroelectronics. Since 2009, he has been with the Department of Mechanical Engineering, National Yunlin University of Science and Technology, where he is currently an Assistant Professor. His research interests are visual servo control, implementation of manipulators, embedded system and SoC integration, as well as stereo vision.

S. Fatih Isler is a Research and Teaching Assistant at the Department of Computer Engineering in Bilkent University, Turkey since August 2009. He has received his BS Computer Engineering and Information Sciences on July 2009 at Department of Computer Engineering in Bilkent University, Turkey. S. Fatih Isler worked as a part-time Software Engineer at Protel GSM in 2009, and as an intern at Aselsan Inc. and Central Bank of Turkish Republic in 2008. He is still working as a software engineer in the 3DPhone project which is funded by the European Union 7th RTD Framework Programme since August 2009. His research interests include human-computer interaction and computer graphics.

Ray Jarvis, an IEEE Fellow, completed a BE (Elec.) and PhD (Elec.) at the University of Western Australia in 1962 and 1968, respectively. After two years at Purdue University, he returned to Australia and took up a Senior Lectureship at the Australian National University where he was instrumental in establishing the Department of Computer Science. In 1985, he took up a Chair in the Department of Electrical and Computer Systems Engineering at Monash University and established the Intelligent Robotics Research Centre in 1987 and continues to be its Director. He is a Fellow of the IEEE (from 1992). His research interests include Artificial Intelligence, Computer Vision, Pattern Recognition and Intelligent Robotics. Between 2003 and 2007, he was the Director of the Australian Research Council Centre for Perceptive and Intelligent Machines in Complex Environments.

Licheng Jiao received the BS degree from Shanghai Jiao Tong University, China, in 1982, and the MS and PhD degrees from Xian Jiao Tong University, China, in 1984 and 1990, respectively. From 1990 to 1991, he was a Postdoctoral Fellow in the National Key Lab for Radar Signal Processing at Xidian University, China. Since 1992, he has been with the School of Electronic Engineering at Xidian University, China, where he is currently a distinguished professor. He is also the Dean of the School of Electronic Engineering and the Institute of Intelligent Information Processing at Xidian University, China. His current research interests include signal and image processing, nonlinear circuit and systems theory, learning theory and algorithms, computational vision, computational neuroscience, optimization problems, wavelet theory, and data mining.

Cheolkon Jung received the BS, MS, and PhD degrees in Electronic Engineering from Sungkyunkwan University, Republic of Korea, in 1995, 1997, and 2002, respectively. He was with the Samsung Advanced Institute of Technology (Samsung Electronics), Republic of Korea, as a research staff member from 2002 to 2007. He was a research professor in the School of Information and Communication Engineering at Sungkyunkwan University, Republic of Korea, from 2007 to 2009. Since 2009, he has worked for the Institute of Intelligent Information Processing at Xidian University, China, as a Professor. His main research interests include computer vision, pattern recognition, image and video processing, multimedia content analysis, and management.

Nikolaus Karpinsky was, at the time of writing, a second year Master's student in Human Computer Interaction and Computer Engineering at Iowa State University. He has published two journal articles and two peer reviewed conference publications. Currently, Nik is working on integrated techniques of different disciplines such as optical engineering, computer engineering, and software engineering, to develop ways of making cost effective 3D scanning, compression, streaming, and display technologies. Nik hopes to increase exposure to 3D scanning and 3D technology, pushing it into mainstream markets effectively replacing 2D. His research interests include augmented reality, computer vision, virtual reality, parallel computing, and human computer interaction.

Tae-Seong Kim received the BS in Biomedical Engineering (BME) in 1991, MS in BME and EE in 1993 and 1998, and PhD in BME in 1999 from the University of Southern California (USC). Dr. Kim worked as a Research Assistant Professor at the Alfred Mann Institute for Biomedical Engineering and Department of Biomedical Engineering at USC. Currently, he is an Associate Professor in the Department of BME at Kyung Hee University in Korea. His research interests have spanned various areas of biomedical imaging including MRI, functional MRI, E/MEG imaging, and ultrasound. Lately, he has started research work in proactive computing at the u-Lifecare Research Center. Dr. Kim has published more than 60 peer reviewed papers and 150 proceedings, and holds 3 international patents. He is a member of IEEE and Tau Beta Pi, and listed in Who's Who in the World ('09,'10,'11) and Who's Who in Science and Engineering ('11-'12).

Andreas Koschan (M'90) received the Diploma (MS) degree in computer science and the Dr.-Ing (PhD) degree in computer engineering from the Technical University Berlin, Berlin, Germany, in 1985 and 1991, respectively. He is currently a Research Associate Professor with the Department of Electrical Engineering and Computer Science, University of Tennessee, Knoxville. His research work has primarily focused on color image processing and 3-D computer vision, including stereo vision and laser range finding techniques. He is a coauthor of two textbooks on 3-D image processing and one textbook on color image processing. Dr. Koschan is a member of the Society for Imaging Science and Technology.

Gauthier Lafruit received the Engineering and PhD degree in Electrical Engineering from the Free University of Brussels, Brussels, Belgium, in 1989 and 1995, respectively. Between 1989 and 1994, he was with the Belgian National Foundation for Scientific Research as a Research Scientist, active in the area of wavelet image compression. In 1996, he joined the Multimedia Group with IMEC, Leuven, Belgium, as a Senior Scientist, and received the Barco Scientific Award on the initiative of Barco NV and the Belgian National Foundation for Scientific Research, for his contribution to efficient hardware implementations in wavelet image processing applications. Currently, he is Principal Scientist Vision Systems with IMEC, in the Smart Systems and Energy Technology (SSET) department, Leuven, Belgium. He has acquired image processing expertise in various applications: video coding, multicamera acquisition, multiview rendering, image analysis and video stereoscopy, where he has applied the "Triple-A" philosophy in Application-Algorithm-Architecture trade-off studies. Dr. Lafruit is Associate Editor for *IEEE Transactions on Circuits and Systems for Video Technology*.

Sungyoung Lee received his BS from Korea University, South Korea. He got his MS and PhD in Computer Science from Illinois Institute of Technology (IIT), Chicago, Illinois, USA in 1987 and 1991, respectively. He has been a Professor in the Department of Computer Engineering, Kyung Hee University, Korea since 1993. He is a founding Director of the Ubiquitous Computing Laboratory, and has been affiliated with a director of Neo Medicinal u-Lifecare Research Center, Kyung Hee University since 2006. Before joining Kyung Hee University, he was an Assistant Professor in the Department of Computer Science, Governors State University, Illinois, USA from 1992 to 1993. His current research focuses on Ubiquitous Computing, Cloud Computing, Intelligent Computing, Context-Aware Computing, WSN, Embedded Realtime and Cyber-Physical Systems, and eHealth. He has authored/coauthored more than 315 technical articles (101 of which are published in archival journals). He is a member of the ACM and IEEE.

Young-Koo Lee received his BS, MS, and PhD degrees in Computer Science from the Korea Advanced Institute of Science and Technology, South Korea. He is a Professor in the Department of Computer Engineering at Kyung Hee University, South Korea. His research interests include ubiquitous data management, data mining, and databases. He is a member of IEEE, the IEEE Computer Society, and the ACM.

Jiangbo Lu (M'09) received the B.S. and M.S. degrees in electrical engineering from Zhejiang University, Hangzhou, China, in 2000 and 2003, respectively, and the Ph.D. degree in electrical engineering, Katholieke Universiteit Leuven, Leuven, Belgium, in 2009. From April 2003 to August 2004, he was with VIA-S3 Graphics, Shanghai, China, as a Graphics Processing Unit (GPU) Architecture Design Engineer. In 2002 and 2005, he conducted visiting research at Microsoft Research Asia, Beijing, China. Since October 2004, he has been with the Multimedia Group, IMEC, Leuven, Belgium, as a Ph.D. Researcher, where he pioneered and led research on real-time stereo matching and view synthesis. Since September 2009, he has been working with the Advanced Digital Sciences Center, Singapore, which is a joint research center between the University of Illinois at Urbana-Champaign, Urbana, and the Agency for Science, Technology and Research (ASTAR), Singapore, where he is currently a Research Scientist. His research interests include multimedia signal processing, computer vision, visual computing, video coding, interactive multimedia applications and systems, GPU-based computing, and embedded computer vision.

Muhammad Tariq Mahmood received the MCS degree in Computer Science from AJK University of Muzaffarabad Pakistan in 2004 and the MS degree in intelligent software systems from Blekinge Institute of Technology Sweden in 2006. He is currently pursuing his PhD degree at Signal and Image Processing Lab, School of Information and Mechatronics, Gwangju Institute of Science and Technology, Gwangju, Korea. His research interests include image processing, 3D shape recovery, object detection, machine learning, and stochastic methods.

Malik Mallem received a PhD degree in robotics and computer sciences from Paris XII University, in 1990. His research deals with Augmented Reality (AR) applied to robotics and telerobotics at IBISC-Complex System Laboratory CNRS FRE 2873, Evry, France. Since 1999, he is Professor at Evry University and a head of AR team which is implied in several projects in cooperation with French industries (CEN, ALSTOM, RENAULT) and research laboratories (http://lsc.univ-evry.fr/techno/EVRA/index.html).

Jesús Martínez-Frías is Senior Research Scientist at the "Centro de Astrobiologia" (CSIC/INTA), associated to the NASA Astrobiology Institute (NAI), where he was former Director of the Planetary Geology Laboratory. In addition, he is Professor "Ad-Honorem" at the Polytechnic University of Madrid, and Founder and Director (CSIC) of the Unit on Spectroscopy, Astrobiology, and Cosmogeochemistry. He has been visiting Scientist in the Universities of Leeds, Heidelberg, Toronto, California and Autónoma de México. He has published 6 books and more than 200 scientific articles and book chapters, and has given more than 100 invited talks in numerous countries. He has been member of high level commissions (e.g UNCSTD (Vice-Chair), ESF-IMPACT, NAI's Mars Focus Group (Co-Chair)). At present, he is Spain's coordinator of AGID (Geoethics) and The Planetary Geology, Chair of the IUGS COGE and official Co-I of the NASA-MSL-REMS and ESA-ExoMars-Raman. He has received various awards (e.g. NASA Group Achievement Award, 2006).

Yasuyuki Matsuura is currently a Postdoctoral Fellow at the Graduate School of Information Science, Nagoya University since 2010. He is a guest researcher at the RIKEN (The Institute of Physics and Chemical Research). He received PhD in science from Nagoya City University in 2009. Broadly, his research lies in Medical Engineering (especially, Biological Signal Processing). Applying this theoretical background, he enjoys doing research in Environmental Physiology for preventive medicine. His broad research interests are in the development of signal processing algorithms for analysis of biological systems. He is currently focusing his studies on computational modeling of electrogastrogram, and on application of non-linear analysis methods to characterize biological signals (Electrocardiogram, Electrogastrogram, etc.). He is a Member of IEEE and the Engineering in Medicine and Biology Society.

Hao Men is a PhD student in the Department of Mechanical Engineering at Stevens Institute of Technology, Hoboken, NJ, USA. He received his B.S. degree in Mechanical Engineering from Xi'an Jiaotong University and M.S. degree from Beijing University of Technology in China, in 2003 and 2006 respectively. His current research interests include robotics for remote and autonomous mapping, algorithms for processing point clouds data for map reconstruction, sensor fusion for accurate dimensional, color, temperature, and other parameters. His interests include design of hardware and firmware for embedded systems and related application software development.

Fabrice Meriaudeau was born in Villeurbanne, France, on March 18, 1971. He received both the master degree in physics at Dijon University, France as well as an Engineering Degree (FIRST) in material sciences in 1994. He also obtained a PhD in image processing at the same university in 1997. He was a postdoc for one year at The Oak Ridge National Laboratory. He is currently Professeur des Universités at the Le2i (www.le2i.com), head of the University Center Condorcet and deputy Director of the Le2i (UMR CNRS). His research interests are focused on image processing for artificial vision inspection and particularly on non conventional imaging systems (UV, IR, polarization). He has coordinated an Erasmus Mundus Master in the field of Computer Vision and Robotics from 2006 to 2010 and he is now the Vice President for International Affairs for the University of Burgundy. He has authored and co-authored more than 150 international publications and holds three patents. He was the Chairman of SPIE's conference on Machine Vision Application in Industrial Inspection and member of numerous technical committees of international conferences in the area of computer vision.

Masaru Miyao is currently a Professor at the Graduate School of Information Science, Nagoya University since 2002. He is also a Professor for the Information Engineering Department at School of Engineering, Nagoya University. He received his MD from Nagoya University in 1977 and his PhD in medicine from Nagoya University in 1982. Broadly, his research lies in Human-Computer Interaction (HCI). More specifically, he enjoys doing research in ergonomics for 3-D display technology and mobile interaction including Head Mounted Displays (HMDs). His research is focused on building and evaluating systems designed to human vision, especially accommodation and convergence for stereoscopic displays, and presently, studying on how to make comfortable 3-D displays and 3-D movie contents. He is a director and an editor of Japanese society for social medicine and councilor of Japanese society for occupational health and Japanese society for hygiene.

Yannick Morvan received his MS in Electrical Engineering from the Institut Supérieur d'Electronique et du Numérique (ISEN), France in 2003. During his undergraduate studies, he worked, in 2002, at Philips Research on embedded image processing software and, in 2003, at Philips Medical Systems on X-ray image quality enhancement algorithms. In 2004, he joined, as a Ph.D. candidate, the Video Coding and Architectures research group at the Eindhoven University of Technology, The Netherlands. During his Ph.D. project, he was involved in a joint project of Philips Research and the Eindhoven University of Technology about the development of a multi-camera video acquisition and compression system for 3-D television. In 2006, he co-organized with Philips Research the "IEEE workshop on Content Generation and Coding for 3D-television". His research interests include multi-view coding, 3D reconstruction and image rendering. One of his papers on multi-view coding was a Best Paper Finalist at the 2007 Picture Coding Symposium in Lisbon, Portugal. In 2008, Yannick Morvan became 3D Imaging Scientist at Philips Healthcare, Best, The Netherlands.

Lazaros Nalpantidis holds a PhD (2010) from the Department of Production and Management Engineering, Democritus University of Thrace, Greece in the field of robotic vision. He holds a BSc degree (2003) in physics and the MSc degree (2005) (with Honors) in electronics engineering from the Aristotle University of Thessaloniki, Greece. He has participated in various European, as well as in national research projects. He has served as a reviewer and committee for various international conferences and journals and is co-author of 18 scientific papers in various conferences and 6 in international journals. His current research interests include vision systems for robotic applications such as depth perception, obstacle avoidance and SLAM. He is a member of IEEE and IEEE Robotics and Automation Society.

Oon-Ee Ng is currently completing his PhD candidature at Monash University Sunway Campus in Malaysia. He graduated from Monash University Malaysia with a Bachelor of Engineering (Mechatronics) with First Class Honours in 2006. He has been a student member of IEEE for four years. His research interests center primarily around stereo vision, with particular emphasis on algorithm development. He is also interested in computer algorithms in general and how they are applied to theoretical and practical problems.

Maria Petrou studied Physics at the Aristotle University of Thessaloniki, Greece, Applied Mathematics in Cambridge, UK, and obtained her PhD and DSc degrees both from Cambridge University in Astronomy and Engineering, respectively. She is the Director of the Informatics and Telematics Institute of CERTH, Thessaloniki, Greece, and the Chair of Signal Processing at Imperial College London, UK. She has co-authored two books, "Image Processing, the fundamentals" and "Image Processing dealing with texture", in 1999 (second edition 2010) and 2006, respectively, and co-edited the book "Next generation artificial vision systems, reverse engineering the human visual system." She has published more than 350 scientific articles on astronomy, computer vision, image processing and pattern recognition. She is a Fellow of the Royal Academy of Engineering.

Kishore Pochiraju is an Associate Professor in the Department of Mechanical Engineering at Stevens Institute of Technology, Hoboken, NJ, USA. He is also the Director of Design and Manufacturing Institute, a research center focusing on design methodologies, real time mechatronic systems and advanced materials. He received his PhD in 1993 from Drexel University and joined Stevens after working as a postdoctoral fellow at the University of Delaware. His research focuses on computational methods for advanced materials and systems design. He is currently working on predicting long-term durability of lightweight composite structures with multi-scale computational methods and on design of real-time electro-mechanical systems. He is an author of 3 book chapters and nearly 125 journal and conference proceedings papers. He is a member of ASME and IEEE.

S. G. Ponnambalam is an Associate Professor in the School of Engineering at Monash University, Sunway Campus, Malaysia. He is heading the Mechatronics Engineering Discipline at Sunway Campus. He is an Associate Editor of IEEE-Transaction on Automation Science and Engineering, International Journal of Robotics and Automation, International Journal of Computers and Applications, and Journal of Mechatronics and Applications. He is also serving as editorial board member for many international journals. He is holding a Senior Member status of IEEE, Fellow of IMechE(UK), and CEng(UK). He has over 200 articles published in various referred journals, refereed conferences and chapters in edited books. His articles are published in different peer-reviewed journals, including *International Journal of Production Research, International Journal of Advanced Manufacturing Technology, Production Planning and Control, Robotics and Computer-Integrated Manufacturing, Computers & Industrial engineering, Journal of Material Processing Technology,* and *International Journal of Intelligent Systems, Technology and Applications.*

Surendra Ranganath received the B. Tech. degree in Electrical Engineering from the Indian Institute of Technology (Kanpur), the ME degree in Electrical Communication Engineering from the Indian Institute of Science (Bangalore) and the PhD degree in Electrical Engineering from the University of California (Davis). From 1982 to 1985, he was with the Applied Research Group at Tektronix, Inc., Beaverton, OR, where he was working in the area of digital video processing for enhanced and high definition TV. From 1986 to 1991, he was with the medical imaging group at Philips Laboratories, Briarcliff Manor, NY. From 1991- 2009, he was with the Department of Electrical and Computer Engineering at the National University of Singapore. He is currently a Professor at the Indian Institute of Technology – Gandhinagar. His research interests are in digital signal and image processing, computer vision, and machine learning with focus on human-computer interaction and video understanding applications.

Francisco Rovira-Más received a degree in Agricultural Engineering in 1996 from the Polytechnic University of Valencia, Spain, where he was an Assistant Professor from 1997 to 2000. He obtained a Ph.D. in Agricultural Engineering in 2003 from the University of Illinois at Urbana-Champaign in the United States of America. Between 2003 and 2005, Francisco was a member of the Intelligent Vehicles System group at the John Deere Technology Center in Moline (Illinois) and at the John Deere Intelligent Vehicle Systems unit in Urbandale (Iowa), both in the USA. In 2006, he returned to the Polytechnic University of Valencia where he currently is an Associate Professor. His research interests include autonomous vehicles, machine vision, controls, stereoscopic vision, off-road equipment automation, robotics, and artificial intelligence. Many of his ideas and previous projects are described in the monograph Mechatronics and Intelligent Systems for Off-road Vehicles (Springer, 2010).

Daniel (Danny) Ruijters is employed by Philips Healthcare since 2001. Currently he is working as Sr. Scientist 3D Imaging at the iXR innovation department in Best, the Netherlands. He received his engineering degree at the University of Technology Aachen (RWTH), and performed his master thesis at ENST in Paris. Next to his work for Philips, he has recently finished a joint PhD thesis at the Katholieke Universiteit Leuven and the University of Technology Eindhoven (TU/e). His primary research interest areas are medical image processing, 3D visualization, image registration, fast algorithms, and hardware acceleration. Daniel has acted as session chair during the 2006 WSCG conference and the 2008 IASTED Conference on Computer Graphics and Imaging (CGIM), and was invited for the panel discussion of the 2008 MICCAI workshop on Augmented Environments for Medical Imaging and Computer-Aided Surgery (AMI-ARCS). He served as reviewer for the 2008 MICCAI High Performance Computing workshop, Computer Methods and Programs in Biomedicine, IEEE Transactions on Visualization and Computer Graphics, IEEE Transactions on Medical Imaging, and European Radiology.

Kah Phooi Seng received her PhD and Bachelor degree (first class honours) from University of Tasmania, Australia in 2001 and 1997 respectively. Currently, she is a member of the School of Electrical & Electronic Engineering at The University of Nottingham Malaysia Campus. Her research interests are in the fields of intelligent visual processing, biometrics and multi-biometrics, artificial intelligence, and signal processing.

Yan Shuicheng (SM'09) is currently an Assistant Professor in the Department of Electrical and Computer Engineering at National University of Singapore, and the founding lead of the Learning and Vision Research Group (http://www.lv-nus.org). Dr. Yan's research areas include computer vision, multimedia, and machine learning, and he has authored or co-authored about 190 technical papers over a wide range of research topics. He is an Associate Editor of IEEE Transactions on Circuits and Systems for Video Technology, and has been serving as the Guest Editor of the special issues for TMM and CVIU. He received the Best Paper Awards from ACM MM'10, ICME'10, ICIMCS'09, and PREMIA 2008 Best Student Paper Award, the winner prize of the classification task in PASCAL VOC2010, and the honorable mention prize of the detection task in PASCAL VOC2010.

Terence Sim received his Bachelor's degree from Massachusetts Institute of Technology, in 1990, Masters from Stanford University in 1991 and PhD from Carnegie Mellon University in 2002. He is currently an Asst. Prof. at the School of Computing, National University of Singapore. He serves as Vice-Chairman of the Biometrics Technical Committee (BTC), Singapore, and Chairman of the Cross-Jurisdictional and Societal Aspects Working Group (WG6) within the BTC. He is also the Vice-President of the Pattern Recognition and Machine Intelligence Association (PREMIA), a national professional body for pattern recognition. He won the 4th Temasek Young Investigator's Award in 2005. His primary research areas are face recognition, biometrics, and computational photography. He is also interested in computer vision problems in general, such as shape-from-shading, photometric stereo, object recognition and also in some aspects of music processing, such as polyphonic music transcription.

Peng Song received his B.S. degree in Automation from Harbin Institute of Technology (2007), M.S. degree in Control Science and Engineering from Harbin Institute of Technology Shenzhen Graduate School (2010). He is pursuing his Ph.D. degree in the School of Computer Engineering, Nanyang Technological University since 2010. His research interests include computer vision and graphics, image-based modeling, and human computer interaction.

Rajeev Srivastava was born in 1974 in Jaunpur, Uttar Pradesh, India. He received his B.E. in Computer Engineering from Gorakhpur University, India; M.E. in Computer Technology and Applications from University of Delhi, India and completed his Ph.D. in Computer Engineering from the same University. He has about 13 years of teaching and research experience. Currently he is working as an Associate Professor in the Dept. of Computer Engineering, ITBHU, Varanasi, India since November 2007. His research interests include image processing and computer algorithms. He has published around 30 research papers in international journals and conferences. He is listed in Marquis Who's Who in Science and Engineering-2011; and "2000 Outstanding Intellectuals of the world-2010" by IBC, Cambridge. He has been selected for the award of "Top 100 Educators of the World-2010," and "Man of the year award-2010" both by IBC, Cambridge. He is reviewer and member of editorial board of 2 international journals, member of technical program committee of 7 international conferences, and on the reviewer panel of Tata McGraw Hill and Oxford University Press, India. He also received research grant from Ministry of HRD, New Delhi, India for his project on "E-content development for the subject Digital Image Processing and Machine Vision."

Ruchir Srivastava has been pursuing his PhD degree in Department of Electrical and Computer Engineering, National University of Singapore, since 2007. He obtained his Bachelor of Technology (B. Tech) in Electrical Engineering from Indian Institute of Technology, Roorkee, India in 2007. His research interests include image processing, facial expression recognition from 3D models, and emotion recognition using multimodal approaches.

Sreenivas Rangan Sukumar is an exploratory researcher with wide interests in science and engineering. After graduating with a Doctor of Philosophy degree in Electrical Engineering from the University of Tennessee, Knoxville in 2008, he is currently a Research Scientist at the Oak Ridge National Laboratory. He has over 20 publications spanning areas of system design for 3D/4D sensing systems, semi-supervised and autonomous 3D reconstruction of scenes, uncertainty minimization in mobile imaging systems and spatio-temporal optimization applied to defense and security applications. His recent research interests are in deriving and implementing search and analysis methods for time-varying multi-variate sensor data streams generated by networked interconnected systems.

Hiroki Takada is currently an Associate Professor at the Graduate School of Engineering, University of Fukui since 2010. He is also an Associate Professor for Department of Human & Artificial Intelligent systems, University of Fukui and a Guest Researcher for Aichi Medical University School of Medicine. He received many awards including an award for encouragement from Society for Science on Form in 2002 and PhD in science in 2004. Broadly, his research lies in Mathematical Physics (especially, Stochastic Process Theory). Applying this theoretical background, he enjoys doing research in Environmental Physiology for preventive medicine. His research is focused on aging, fainting, and motion sickness which is also induced by stereoscopic images. There have been eye strain issues in stereoscopic movies. He is an editor of "FORMA" and Members of IEEE and International Society for Gerontechnology.

Nguyen Duc Thang received his B.E. degree in Computer Engineering from Posts and Telecommunications Institute of Technology, Vietnam. He is currently working toward his M.S. leading to Ph.D. degree in the Department of Computer Engineering at Kyung Hee University, South Korea. His research interests include artificial intelligence, computer vision, and machine learning.

Xiaojun Wu is an Associate Professor at the Division of Control and Mechatronics, Harbin Institute of Technology Shenzhen Graduate School. He received his B.S. and M.S. degree from Jilin University in 1998 and 2001 respectively, and PhD in Mechatronics from Shenyang Institute of Automation, Chinese Academy of Sciences in 2004. He received the Best Paper Award (with M.Y. Wang) of 2007 International CAD Conference & Exhibition and the Best Paper Award in Information (with P. Song and M.Y. Wang) of IEEE ICIA (2009). He is engaged in research projects concerned with 3D scanning, surface reconstruction, image-based modeling, and heterogeneous object modeling and visualization.

Yeliz Yigit is a Graduate Student and Teaching Assistant at the Department of Computer Engineering in Bilkent University, Turkey. She has received her B.S. Computer Engineering and Information Sciences in 2007 and M.S. Computer Engineering and Information Sciences in 2010 at Department of Computer Engineering in Bilkent University, Turkey. She is still working as a Researcher in the 3DPhone project which is funded by the European Union 7th RTD Framework Programme since 2008.

Her research interests include computer and mobile graphics, virtual reality and environments, 3D media, and human-computer interaction.

Iman Maissa Zendjebil received his PhD degree in computer engineering from the University of Evry Val d'Essonne (France) in 2010. His research works are focused on 3D localization for outdoor augmented reality applications.

Ke Zhang received the B.S. and M.S. degrees from Zhejiang University, Hangzhou, China, in 2005 and 2007, respectively, both in electrical engineering. He is currently working toward the Ph.D. degree in the Department of Electrical Engineering at Katholieke Universiteit Leuven, Leuven, Belgium. Since 2007, he has been a PhD researcher at IMEC. His research interests include multimedia processing systems, video coding, and computer vision.

Lu Zhang received his B.E. degree in electrical engineering from Shandong University, Ji'nan, China, in 2007 and the M.S. degree, with distinction honor, in embedded systems from Delft University of Technology, Delft, The Netherlands, in 2010. In his M.S. thesis he studied state-of-the-art stereo matching algorithms and investigated the VLSI architecture for parallel and pipelined processing, and achieved both high accuracy and frame rates with a single Stratix-III FPGA implementation. His current research and career interests include high performance and parallel computing, application specific hardware and software design, real-time systems and computer architecture. He is currently working at Intel in Eindhoven, the Netherlands.

Song Zhang is an Assistant Professor of Mechanical Engineering at Iowa State University (ISU). He is also affiliated with the Human Computer Interaction (HCI) graduate program at ISU. His research interests include real-time 3D machine/computer vision, 3D video processing, human computer interaction, and virtual reality. He has published more than 60 papers including 27 journal articles and 3 book chapters in optics, computer science, medical science, et cetera. Among the journal papers he published, five of them were featured on their covers. One of his papers was awarded the best of SIGGRAPH by the Walt Disney Co., and reported by the media including The-Scientist Magazine, Photonics, Physorg, First Science, and Futurity: Discover and the Future. He currently serves as a reviewer for over twenty journals, and is a member of IEEE, SPIE, OSA, and ASME.

Mohammad Zia Uddin received his BS degree in Computer Science and Engineering from International Islamic University Chittagong, Bangladesh. He is currently working toward his MS leading to Ph.D. degree in the Department of Biomedical Engineering at Kyung Hee University, Republic of Korea. His research interest includes pattern recognition, image processing, computer vision, and machine learning.

Svitlana Zinger received the MSc degree in computer science in 2000 from the Radiophysics faculty of the Dnepropetrovsk State University, Ukraine. She received the Ph.D. degree in 2004 from the Ecole Nationale Superieure des Telecommunications, France. Her Ph.D. thesis was on interpolation and resampling of 3D data. In 2005 she was a postdoctoral fellow in the Multimedia and Multilingual

Knowledge Engineering Laboratory of the French Atomic Agency, France, where she worked on creation of a large-scale image ontology for content based image retrieval. In 2006-2008, she was a postdoctoral researcher at the Center for Language and Cognition Groningen and an associated researcher at the Artificial Intelligence department in the University of Groningen, the Netherlands, working on information retrieval from handwritten documents. She is currently a postdoc at the Video Coding and Architectures Research group in the Eindhoven University of Technology.

Index

C

D

E

F

G

H

I

J

K

L

M

N

O

P

R

S

T

U

V

W

Z

www.ingramcontent.com/pod-product-compliance
Lightning Source LLC
LaVergne TN
LVHW080924210926
848485LV00004B/722